ENVIRONMENTAL INSTRUMENTATION AND ANALYSIS HANDBOOK

ENVIRONMENTAL INSTRUMENTATION AND ANALYSIS HANDBOOK

RANDY D. DOWN
Forensic Analysis & Engineering Corp.

JAY H. LEHR
The Heartland Institute
Bennett and Williams, Inc.

WILEY-INTERSCIENCE
A JOHN WILEY & SONS, INC., PUBLICATION

Published by John Wiley & Sons, Inc., Hoboken, New Jersey.
Published simultaneously in Canada.

For general information on our other products and services please contact our Customer Care Department within the U.S. at 877-762-2974, outside the U.S. at 317-572-3993 or fax 317-572-4002.

Wiley also publishes its books in a variety of electronic formats. Some content that appears in print, however, may not be available in electronic format.

Library of Congress Cataloging-in-Publication Data:

Down, Randy D.
   Environmental instrumentation and analysis handbook / Randy D. Down, Jay H. Lehr.
      p.   cm.
   Includes index.
   ISBN 978-0-471-46354-2
 1. Environmental monitoring–Instruments–Handbooks, manuals, etc.
      2. Pollution–Measurement–Handbooks, manuals, etc.   I. Lehr, Jay H., 1936-   II. Title.
   TD193.D685   2004
   628–dc22   2004005656

CONTENTS

PART IV   WASTEWATER MONITORING

PART V   AIR MONITORING

PART VI   FLOW MONITORING

PREFACE

It has been two decades since environmental science, environmental engineering, and environmental consulting took root as major disciplines and professions throughout the developed world. The learning curve has been steep as it relates to the previously unrecognized physics of contaminant transport. Today those principles are usually well understood by a mature army of environmental professions.

An area that has lagged in full comprehension among the practitioners in these fields is an understanding and awareness of the hardware for measuring the physical and chemical characteristics of contaminated sites. The application of these instruments and methodologies to characterize the solid, liquid, and gaseous chemical content within a transport media are not well understood.

Professionals have long relied on personal experience, diverse journal articles, and manufacturer's advertisements and catalogs to choose efficient and accurate means of obtaining the necessary field data to characterize a site. This has resulted in too narrow a focus in the development of appropriate remediation programs and monitoring protocols.

More than three dozen talented environmental professionals who are experienced and adept at extracting the most telling and accurate data from the "field" have come together in this book to catalog nearly all the equipment and techniques that are available to modern scientists, engineers, and technicians.

This has been a fulfilling and rewarding effort: the gathering of the best and brightest professionals across many continents to share their expertise. We have asked them to describe the basic science, be it physics, chemistry, biology, hydrology, or computer data logging, that supports their field analysis followed by detailed explanations of the various hardware in use today. In most cases the authors offer descriptions of typical case studies in which the equipment was successfully utilized.

Of significant value are the pitfalls and foibles of the procedures and equipment that may not always measure up under less than ideal conditions. What may go wrong is often as valuable as what should correctly occur.

For ease of access, we have divided the description of field instruments and procedures into six basic categories: instrumentation and methodologies, water quality parameters, groundwater monitoring, wastewater monitoring, air monitoring, and flow monitoring. Some sections could have fit neatly in more than one category, but we trust the reader will have no trouble identifying the information being sought regardless of the category in which it is placed.

It is rare to have an opportunity to add a truly innovative package of information to the literature that has not been previously attempted. We are confident that this has been achieved through the cooperation and dedication of the many contributors to this book to whom we are eternally grateful.

We made no strong effort to confine the authors to a special format of presentation in length, depth, or breadth of their subject matter. We only asked that they enable the reader to fully understand the conditions under which field instruments and procedures were applicable and how to implement their use.

Some authors felt their specialty was in need of a comprehensive detailed expose not readily found in existing literature. They saw this book as an opportunity to supply just such a treatise. Other authors felt only a brief "how-to" manual approach was sufficient in their area of expertise. The reader will recognize these differences and likely benefit equally from both approaches.

Our profession has needed this handbook for the past decade. We hope it will fill the need for the next decade.

JAY H. LEHR

I

INSTRUMENTATION METHODOLOGIES

1

INFLUENCE OF REGULATORY REQUIREMENTS ON INSTRUMENTATION DESIGN

Randy D. Down, P.E.

Forensic Analysis & Engineering Corp.
Raleigh, North Carolina

Environmental Instrumentation and Analysis Handbook, by Randy D. Down and Jay H. Lehr
ISBN 0-471-46354-X   Copyright © 2005 John Wiley & Sons, Inc.

1.1 INTRODUCTION

Federal, state, and local regulatory requirements have long played an important role in driving the advancement of new technologies for the measurement and control of environmental pollution. They will continue to do so. The same can be said for competitive advancements in measurement and control technology—that they drive the regulatory requirements. As this chapter will illustrate, regulations and competitive, technological development ultimately work hand in hand to influence the future of environmental instrumentation—thus the rapidly changing nature of environmental instrumentation and controls.

This handbook will serve as a valuable guide in the application of new and emerging environmental instrumentation and control technologies needed to meet current and future regulatory requirements.

1.2 ENVIRONMENTAL REGULATORY REQUIREMENTS

It was not intended that this handbook serve as a reference for environmental regulatory requirements. Regulatory requirements vary from state to state throughout the United States and abroad and are periodically updated and revised. Any information regarding regulatory requirements that pertains to your geographical location should be obtained directly from the appropriate local governing agencies. It is advisable to work with a local or regional environmental consultant or directly with the regulatory agency to determine which regulations apply to your specific application. Doing so will greatly reduce your risk of misapplying expensive instruments and potential incurring fines that may be imposed for failing to meet all regulatory requirements. Such fines can be very costly and embarrassing.

When involved in the development, specification, or selection and application of instrumentation, as it relates to environmental applications, this book will serve as a very useful technical resource. It will aid you in asking the right questions and avoiding some of the many potential pitfalls that can occur when trying to select and specify appropriate instrumentation for a measurement or control application.

1.3 KEY FACTORS INFLUENCING DEVELOPMENT

Two key factors drive the development of new technology as it applies to environmental measurement and control:

- Steps required to cost effectively meet compliance requirements dictated by federal and state regulatory agencies
- An opportunity to be highly profitable by being the first firm to develop and market a new, more cost-effective and reliable technology (sensor, transmitter, analyzer, telemetry device, and/or controller). Statistically, those companies that are first to market with a new technology tend to capture and retain 70%

or more of the total market share. Therefore, great emphasis is placed on being the first firm to market an innovative or more cost-effective, new product or technology.

New environmental measurement and control technology comes from many areas of science and industry. Government and private investments made in the development of new alloys and synthetic materials as well as smaller and lighter electromechanical components are one example. Sensor technology for the aerospace and auto industries is a good example of a major source of new technology and products. Spin-off applications, if applicable to industrial and commercial applications (and relatively cost effective), can have dramatic results in advancing control technology. Major aerospace and automobile manufacturers as well as government agencies often have greater resources with which to fund in-depth research and development.

1.4 EMERGING SENSOR TECHNOLOGY

Advancement of new sensor technology is by far the most influential factor in the evolution of regulatory requirements as well as instrumentation and control technologies. Keeping up with this technology is a major challenge for regulatory experts, scientists, and engineers who are tasked with providing clients and the general public with the optimal means of pollution measurement and abatement.

When establishing the minimum human exposure limit for known and suspected carcinogens, the regulatory minimum exposure level is often established by the minimum *measurable* concentration. The minimum measurement level established by the government must be achievable in terms of measurment accuracy and repeatability. Unattainable regulatory limits would be meaningless.

The ability of a measurement system to accurately monitor an environmental variable (such as humidity, temperature, pressure, flow and level) or to detect and analyze a specific chemical substance and its concentration over time is crucial if we are to successfully measure and control pollutants and preserve the health and safety of our environment.

Measurement, as discussed in greater depth later in this book, is a function of accuracy, precision, reliability, repeatability, sensitivity, and response time. As new sensor technology evolves, its value to the industry will be judged by its ability to meet these criteria and by its relative cost in relation to currently used technology.

1.5 OTHER ADVANCING TECHNOLOGIES

Closely following the rapidly advancing sensor technology and further influencing sensor development is the continuing development of solid-state electronics and large-scale integration of electronic circuitry into microcircuitry. Development of microminiature electronic components (such as resistors, diodes, capacitors, transistors, and integrated circuits) and "nanotechnology" (the development of microminiature mechanical/electrical devices) has positively influenced the measurement and controls industry in multiple ways:

- Electronic and mechanical components are now physically much smaller.
- Being smaller, these devices require less electrical energy to function.
- Using less energy, they also produce less heat, allowing them to be housed in more compact, better sealed, and in some cases nonventilated enclosures.
- Allowing them to be tightly enclosed makes them better suited for use in harsh environments and means they are less likely to be influenced by variations in temperature, vibration, and humidity.
- manufacturing and assembly costs are significantly reduced.
- consumer prices are reduced.

If we look deeper, we find that other technological advances have allowed and supported the continued development of these microcircuits and components. A prime example is the advancement of clean-room technology. A dust particle, spore, strand of human hair, or chafe (particles of dry skin) will appear quite large under a microscope when examined alongside some modern-day miniaturized components and circuitry. Such environmental "contamination" can damage or impair the reliability and performance of these microminiature components.

Advancements in clean-room design and packaging technology have significantly reduced the risk of such contamination. This has largely been accomplished through the development of high-efficiency air filtration systems and better guidelines for proper "housekeeping," such as

- wearing low-particulate-producing disposable suits, booties, and hair nets;
- providing pressurized gown-up areas, airlocks, and positively pressurized clean-room spaces (to prevent contaminated air from migrating into the cleaner space); and
- providing "sticky" mats at entering doorways to pick up any particulate that might otherwise be "tracked in" on the bottom of footwear.

Conversely, advancements in clean-room technology have largely occurred through improved accuracy in measuring and quantifying the presence of airborne contaminants.

Improved accuracy of particulate monitoring instrumentation is a good example of advancing sensor technology that is aiding the advancement in measuring and certifying clean-room quality, which in turn has aided advancements in sensor technology. This is a good example of different technologies that are ultimately working hand in hand to accelerate the advancement of environmental instrumentation technology.

Nanotechnology (the creation of functional materials, devices, and systems through control of matter on a length scale of 1–100 nm) may very well have the greatest impact of any technological advancement in measurement and control technology over the next 10–15 years. The manipulation of physical properties (physical, chemical, biological, mechanical, electrical) occurs at a microminiature scale. To put it in perspective, 10 nm is approximately a thousand times smaller than the diameter of a human hair. A scientific and technical revolution is beginning based on the newfound ability to systematically organize and manipulate matter at the nanoscale.

1.6 REGULATORY TRENDS

Regulatory agencies tend to avoid direct specification of a technology to meet a regulatory requirement. They wisely prefer to define performance criteria (accuracy and reliability) that must be achieved in order to be in regulatory compliance. In so doing, regulatory agencies can avoid specifying a level of system performance that exceeds readily available technology. It also reduces the risk of specifying technologies that are available but are so cost prohibitive that they would create undue financial hardship for those companies found to be out of compliance.

Regulatory agencies must weigh the potentially high cost of available technology against the value derived by enforcing a cleaner environment and ultimately determine what is in the public's best interest. These decisions are often controversial and may be challenged in the courts. At risk are thousands of jobs, as companies are required to spend millions of dollars to significantly reduce their air emissions (or pretreat wastewater) and remain competitive with overseas companies. This burden on manufacturers must be weighed against the potential long-term (and perhaps yet-unknown) impact of the exposure of people and the environment to human-generated contaminants.

An effective approach to working with industry to continuously improve our nation's air and water quality while not financially crippling U.S. companies (which in some cases compete with overseas firms facing fewer environmental restrictions) is to employ a MACT (maximum achievable control technology) or BACT (best available control technology) analysis and gradually increase restrictions on certain pollutants over a period of several years.

Graduated environmental restrictions allow several things to occur that aid industry: They allow industrial firms time to determine and budget for the cost of compliance, schedule downtime (if necessary), and investigate methods of changing their internal production processes to lower the level of emitted pollutants requiring control. They also allow system developers and pollution control system manufacturers additional time to develop methods to meet compliance requirements that are more cost effective than current technology may allow.

1.7 MACT/BACT ANALYSIS

Addressing the issue of abatement costs versus the benefits to the environment requires a methodology that will establish the best approach based on present-day technology. Typically, the approach that has been adopted by many state regulatory agencies is either a BACT or MACT study and report.

In a MACT or BACT analysis, the feasible alternatives for pollution control are examined and compared in a matrix, weighing factors such as pollutant removal efficiency, capital costs, operating costs, life expectancy, reliability, and complexity. Ultimately, a *cost per unit volume of pollutant removed*, ususally expressed in *dollars per ton*, is established for each viable option. The option having the best projected cost per volume of removed pollutant is usually selected unless there are extenuating reasons not to (such as a lack of available fuel, insufficient space for the equipment, or a lack of trained or skilled support staff needed to operate or maintain the system).

As an example, a BACT or MACT analysis for abatement of pollutant air emissions will often include evaluation of such technologies as carbon recovery systems, thermal oxidizers, scrubbers, dust collectors, and flares.

Evaluating these various technology options requires a detailed determination of their cost of construction, operation, maintenance, waste disposal, and salvage. Typically, an environmental consultant is contracted to perform an independent BACT or MACT analysis. This helps avoid potential public concerns over a perceived conflict of interest if the analysis were performed in-house.

Pollution abatement system costs often range well into the thousands, in some cases millions, of dollars. Cost of abatement systems is largely dependent upon:

- Type and controllability of substances to be abated
- Supplemental fuel costs
- Disposal of the removed pollutant (if any)
- Quantity (volumetric flow rate) of the pollutant

Some pollutants, such as mercury, are much more difficult to remove than others, such as volatile organic compounds (VOCs). They may also need to be handled and disposed of differently (further driving up the total cost of abatement).

As a general rule, the larger the volume of pollutants generated, the physically larger the equipment needed to handle it, and perhaps the more equipment is needed to control it. All of these characteristics serve to drive up the cost of abatement.

Various pollution abatement systems are described in greater depth later in this book.

1.8 PRODUCT DEVELOPMENT

As mentioned earlier, many factors influence the development of pollution control technology and environmental instrumentation. Among the major factors influencing instrumentation development are:

- Increasingly stringent regulatory requirements
- Continuing advancements in microelectronics

- Advancing clean-room technology
- Sensor development for the aerospace and automotive industries ("spin-offs")
- Emerging networking technologies (Internet, Ethernet, Fieldbus, wireless telemetry)
- A globally competitive market for instrumentation and controls

Manufacturers of sensors used in environmental control applications and pollution abatement technologies will continue to develop and market new sensor technology, with no apparent end in sight. As the technology advances, allowing more cost-effective measurement of pollutants, more stringent environmental limits will be imposed until the general public is satisfied that sufficient pollution control strategies have been established to reverse the concerns over a clean environment.

1.9   NETWORKED SYSTEMS

The current trend in the instrumentation industry as a whole is to network what are referred to as "intelligent" transmitters and analyzers that contain their own microprocessors. Typically, they have a unique device address on a daisy-chained, twisted/shielded pair of wires. The transmitter or analyzer onboard microprocessor has the ability to run self-diagnostics and identify hardware problems (e.g., sensor failure). This ability to diagnose a transmitter or sensor problem from a remote location (using the workstation PC, a laptop, or a hand-held diagnostic tool) has particular value when the transmitters are mounted in such difficult-to-access locations as the top of a high exhaust stack without the use of special equipment.

Smart transmitters are also capable of transmitting output data at a much faster "digital" speed than an analog signal that requires a longer scanning period for a computer or processor to determine and update its output value. This gives them a faster system reaction time (or scan time). Typically, such networked "intelligent" transmitters are connected to a DCS (distributed control system) by a single network cable (typically a single twisted and shielded pair of wires). In such a system, the central processor at the front end of the system (also referred to as operator workstations) communicates over a daisy-chained serial connection (twisted pair of wires) directly with each field device. Large systems can have hundreds of such field devices. As mentioned earlier, each smart device is assigned a unique address for identification.

Portable instruments have also advanced a great deal in the last 20 years. This is largely due to the advancements in microelectronics and batteries. Battery technology has allowed battery size, weight, and cost to be reduced. This further enhances the performance of portable instruments, which will be discussed in much greater depth later in this handbook. Portable instruments offer many advantages as well as some disadvantages when compared to stationary instruments.

1.10   FUTURE CONSIDERATIONS

Looking into our future (albeit without the aid of a crystal ball), it would seem that the greatest advances in environmental instrumentation development over the next 10 years will likely center around advanced technology and cost reductions in the areas of:

- Faster data processing
- More intelligent field devices
- Lower cost analyzers
- Microminiature electromechanical applications using nanotechnology
- Extensive use of wireless telemetry
- Faster sensor response times (greater sensitivity, quicker response to changes in the measured variable)

An improved ability to detect and measure concentrations of airborne pollutants and reduced costs for these instruments will likely lead to more stringent clean-air requirements. An analogy is the introduction and development of electric and combination gas-and-electric cars (cleaner fuel-burning transportation), which will continue to drive clean-air technology as well as slow down the depletion of our finite supply of fossil fuels.

Another potential change in our near future will be greater efforts by pollution control system manufacturers to market packaged abatement systems, with instrumentation and controls a pretested part of the overall system package.

1.11   INTERNATIONAL ORGANIZATION FOR STANDARDIZATION

With significant advances in science and technology and a greater sharing of technologies in large part due to the end of the Cold War and creation of a worldwide web, the engineering community worldwide has grown much closer. It has thus become apparent that international quality standards are needed in order to provide a consistent level of quality engineering and management standards worldwide. The International Organization for Standardization (ISO) develops standards for this purpose. Its standards are being adopted by many international firms and professional organizations. These quality standards may someday be adopted by regulatory agencies as a minimum level of quality that must be attained. This will influence the development and quality of instrumentation worldwide in a very positive way.

An example of these standards, ISO 9001, specifies requirements for a quality management system for any organization that needs to demonstrate its ability to consistently provide product that meets customer and applicable regulatory requirements and aims to enhance customer satisfaction. Standard ISO 9001:2000 has been organized in a user-friendly format with terms that are easily recognized by all business sectors. The standard is used for certification/registration and contractual purposes by organizations seeking recognition of their quality management system.

The greatest value is obtained when the entire family of standards is used in an integrated manner. This enables companies to relate them to other management systems (e.g., environmental) and many sector-specific requirements (such as ISO/TS/ 16949 in the automotive industry) and will assist in gaining recognition through national award programs.

1.12 CONCLUSION

Keeping up with technological advancements in environmental instrumentation and control systems is a daunting task for scientists, engineers, and technicians. There is no end to this process. Aside from the laws of physics, we can anticipate a continuation of major technological advances in the years ahead.

The good news is that the cost of the technology will likely continue to decline while sensitivity and capability of the instruments will continue to rise. Much in the way that PC technology has evolved in recent years, advancing technologies will drive capability up and at the same time, through healthy competition, drive costs down.

Environmental instrumentation will continue to play a vital role in monitoring and protecting our health and in our very existence. The better we understand how to correctly apply it, the better our opportunity to understand and manage the impact on our environment.

1.13 ADDITIONAL SOURCES OF INFORMATION

Information contained in this handbook will aid you in understanding the various applications and solutions most often encountered in this field. It would not be possible to provide all of the answers in a single volume.

Supplemental information can be found by reading technical articles in trade journals and through lectures and conferences conducted by organizations such as the Instrumentation, Systems and Automation Society (ISA), formerly known as the Instrument Society of America.

Equipment suppliers and system houses, although somewhat biased toward their own supplier's line of instruments and components (because that is how they generate the most revenue), are often a good source of information, particularly regarding availability and cost. They tend to be more knowledgable than consultants about the technical aspects of their specific product lines. They often are involved with the installation, programming, and servicing of the equipment, not just the performance and specification aspects, as are many consultants.

Independent engineering consultants offering expertise in environmental *and* instrumentation-and-control applications are another good resource. Good consultants offer an advantage (over system houses) of not being biased toward a particular manufacturer's equipment. They tend to act more on the client's behalf because they do not stand to profit by convincing the client to use a particular manufacturer's product line. They can also help establish budgetary costs and scheduling requirements for the design, purchase, and installation of the instruments and controls.

Another source of useful information is the technical library. Most public libraries are virtually devoid of any helpful or current literature on environmental instrumentation and controls. University libraries are much more likely to retain the technical of books and other literature that can be useful.

Helpful guidance in selecting instrumentation for an envrionmental application can be found on the Internet using one of many very good search engines. The Internet can be used to find useful information on:

- New technological inovations
- New product releases
- Equipment pricing and specifications
- Available literature and catalogs
- Consultants
- Manufacturers

In conclusion, seek out the information that you will need in order to obtain the proper instrumentation for your application *before* making a substantial investment. In the long run, it will save a great deal of time and help avoid costly and sometimes embarassing rework.

2

IN SITU VERSUS EXTRACTIVE MEASUREMENT TECHNIQUES

Gerald McGowan

Teledyne Monitor Labs, Inc.
Englewood, Colorado

Environmental Instrumentation and Analysis Handbook, by Randy D. Down and Jay H. Lehr
ISBN 0-471-46354-X   Copyright © 2005 John Wiley & Sons, Inc.

Both extractive and in situ gas analysis systems have been used successfully in a wide variety of applications. Similarly, they have both been misapplied, which often results in poor performance coupled with maintenance and reliability problems. It is the intent of this chapter to provide sufficient information that potential users can better understand the system capabilities and limitations and avoid the problem installations. Extractive systems are characterized by a sample extraction and transport system in addition to the gas conditioning systems and analyzers required for the actual measurement. In situ systems are characterized by their ability to measure the gas of interest in place, or where it normally exists, without any sample extraction or transport systems. To be useful in today's environmental and process monitoring environment, such measurement systems must be augmented with automated calibration and diagnostic systems that enhance the accuracy and reliability of such systems. Calibration systems must provide a complete check of the normal measurement system to ensure the integrity and accuracy of resulting measurements. Specifications for such systems often require the ability to calibrate with known (certified) calibration gases or, alternately, a gas cell or similar device of repeatable concentration indication. Gas samples of particular interest for this application typically originate in a combustion process that dictates much of the sample extraction, transport, and conditioning systems. The typical applications of interest are those associated with U.S. Environmental Protection Agency (EPA) continuous emission monitoring system (CEMS) requirements, and such regulatory compliance is assumed as a basis of comparison for these systems of interest. The EPA requires that the performance of each CEMS be individually certified in the actual installation. The performance certification requirements are described in US 40 CFR 60, Appendix B, and 40 CFR 75, Appendix A. In Europe, a type approval for a given CEMS and application are often required for use in government-regulated applications.

(*Note*: Some of the suppliers of the various components discussed herein have been identified in those sections. For a complete list of suppliers, please search the Internet using one of the search engines or consult web pages with industry information, such as www.awma.org, www.isadirectory.org, www.thomasregister.com, www.manufacturing.net, www.industry.net, and www.pollutiononline.com. EPA emission measurement and regulatory information is available at www.epa.gov/ttn/emc, www.epa.gov/acidrain, www.epa.gov/acidrain/otc/otcmain.html, www.access.gpo.gov, and http://ttnwww.rtpnc.epa.gov/html/emticwww/index.htm.)

2.1 EXTRACTIVE MEASUREMENT TECHNIQUES

Extractive measurement systems have been widely used for many years to allow gas analyzers to be located remote from the sampling point. Sampling points are often in hostile environments such as duct, pipes, and smoke stacks, which are exposed to the weather and where it is difficult to maintain good analyzer performance. By extracting a sample from the containment structure of interest, cleaning it up to the degree necessary to transport it, transporting it for tens or hundreds of meters to an analyzer cabinet, further conditioning the sample as needed for the analyzers, and feeding it to an analyzer in an analyzer-friendly environment, it is much easier to achieve the desired degree of measurement accuracy and reliability. This section will address the critical issues involved in extractive systems. Conventional extractive systems are herein defined as those using a gas cooler/drier to reduce acid gas and water dewpoints in the sample of interest, thereby providing dry-basis concentration measurements. They do not necessarily remove all water content, but they reduce it to a level of about 1% v/v or less. Other extractive systems include hot, wet extractive systems where the gas sample is maintained in a hot, wet state from the point of extraction through the point where it is discharged from the analyzer and dilution extractive systems where the extracted gas sample is diluted with a clean air supply prior to measurement. Both of the latter systems provide wet-basis concentration measurements.

GASES OF INTEREST. Air pollutants of most common regulatory interest include SO_2 (sulfur dioxide), NO_X (oxides of nitrogen to include NO and NO_2), and CO (carbon monoxide). The EPA has established primary ambient air quality standards for these gases as well as for particulate, ozone, and lead. Other pollutant, hazardous, or toxic gases of interest include TRS (total reduced sulfur), HCl (hydrogen chloride), NH_3 (ammonia), and THCs (total hydrocarbons). In addition to the above pollutants, a diluent measurement is usually required that may be either O_2 (oxygen) or CO_2 (carbon dioxide). The diluent measurement allows the pollutant measurement to be compensated or corrected for the dilution or reduction of pollutant concentration that occurs when excess air is applied to the burners or otherwise added to the exhaust stream of a monitored process. Of this group of gases, it is important to recognize that HCl and NH_3 cannot be accurately measured in a system that cools/dries the sample gas in the presence of water condensate since these gases are water soluble and are removed with the water. The THC measurements are also affected by moisture removal as only the light hydrocarbons may pass through the system without substantial degradation. It should be further noted that O_2 cannot be measured in a dilution system if the dilution air supply contains O_2, as with instrument air. In such cases O_2 is measured by a separate sensor or probe installed at the stack where the O_2 is measured in the undiluted gas stream. Stack gas velocity or flow measurements are often used in conjunction with SO_2, NO_X, and CO_2 in order to determine emissions in pounds per hour or tons per year; however, such gas flow measurement devices will not be described herein.

ANALYZERS/SENSORS. Sulfur dioxide is typically measured with UV (ultraviolet) absorption, UV fluorescence, or IR (infrared) absorption. Infrared absorption analyzers, often called NDIR (nondispersive infrared) analyzers (meaning without a spectrally dispersive element such as a grating), are often enhanced with GFC (gas filter correlation), which reduces their sensitivity to interference from H_2O and other gases. Gas filter correlation is most effective in reducing interferences when the gas of interest exhibits sufficient fine structure in the spectral region of interest. In general, GFC provides high specificity for diatomic molecules but may not be as effective in measuring triatomic molecules, which have more broadband absorption characteristics and less fine structure. Because of the very broad absorption characteristics of H_2O and CO_2 in much of the IR region, they are common interferents for many NDIR-measuring instruments. Special care is often required to ensure that either virtually all the H_2O is removed from the sample or a very specific detection technique is used, such as GFC, in order to provide the required measurement accuracy. The UV-measuring instruments are largely immune from this problem since H_2O does not absorb in the UV spectral region. Ultraviolet fluorescence–based SO_2 instruments have been developed for ambient air monitoring at low levels (<20 ppm) and with minimum detectable levels of less than 1 ppb. Such analyzers are based on illuminating the sample gas with a UV light source, which excites the SO_2 molecules, and monitoring the resulting fluorescence (emission of UV energy at longer wavelengths than the excitation wavelength), which is produced when the molecule relaxes to its normal state. Proper selection of excitation and detection wavelengths is required to minimize interference due to quenching agents such as CO_2, O_2, and H_2O and other coexisting gases. For most applications, a hydrocarbon removal device (permeable membrane) is used to minimize the potential interference from such gases.

Oxides of nitrogen are defined as the sum of NO and NO_2 as measured by a reference method (EPA 40 CFR 53) that incorporates an O_3-based chemiluminescent method and an NO_2-to-NO converter. With this method, O_3 and the unknown sample gas are mixed in a reaction cell attached to the front of a photomultiplier tube where the $NO + O_3$ reaction produces an excited NO_2 molecule. When this excited molecule relaxes, it emits light that is detected by the photomultiplier tube and provides a signal from which the concentration of NO is calculated. Thus the instrument basically measures NO, but with the NO_2-to-NO converter (molybdenum catalyst heated to about 600°F), it also measures NO_X. The NO_2 is calculated as the difference between the raw sample measurement of NO and the converter-processed sample, which provides a measure of NO_X. For most combustion sources NO_X consists of 95–99% NO with the remainder being NO_2. Thus, in many applications, it is possible to measure NO and report it as NO_X. The EPA requires lb/mmBtu and lb/h of NO_X to be reported as if it were in the form of NO_2 since that is the final oxidized state of NO after it is discharged into the atmosphere and subjected to atmospheric chemical reactions. The predominance of NO in many applications is significant since NO is best measured in the UV or with chemiluminescence and NO_2 is a weak absorber in both the UV and in the IR. The NO_2 also absorbs in the visible light spectra and is one of the few gases visible to the human observer, but the measurement of NO_2 using visible light has proven difficult. Chemiluminescent analyzers that depend on the reaction of ozone and NO have been developed for ambient air quality measurements and have excellent low-level (<0.1–20-ppm) measurement capability. Minimum detectable levels for such analyzers are less than 1 ppb. They can be ranged for higher concentrations by reducing the sample flow rate into the reaction cell. In the application of chemiluminescent analyzers to CEMS applications, care must be exercised to ensure that the analyzer interference rejection characteristics are suitable for the major stack gas constituents of interest. The O_2 and CO_2 in the sample stream can cause quenching of the chemiluminescent reaction, the effect of which is dependent on specific analyzer design details. In combustion applications the concentrations of O_2 are lower, and CO_2 is much higher, than in the typical ambient air monitoring applications. Special CEMS configurations of NO_X analyzers have been developed to minimize the CO_2 quenching effect by increasing the sample dilution with the ozone carrier gas. Calibration gases may need to contain the typical level of CO_2 observed in the sample in order to maximize the accuracy of the system. Further, if substantial NO_2 is present in NO_X, one must ensure that the converter included in the NO_X analyzer can handle such high NO_2 concentrations. The IR measurement of NO and/or NO_2 is particularly difficult because their primary spectral absorption signatures are overlaid by H_2O absorption. Thus GFC or dispersive techniques are usually required to obtain reasonable H_2O interference rejection. The UV absorption measurement of NO is much more specific with respect to water interference and provides good sensitivity, but not down to ambient levels without special techniques.

Carbon monoxide and CO_2 are usually measured in the IR using a combination of NDIR and GFC measurement techniques. This technique has been perfected for use in CO ambient air quality analyzers and has been extended to higher ranges as well as to CO_2 measurement. The optical measurement path length is the primary variable to scale analyzer measurement ranges, with near 5 m path length used for ambient CO measurements down to 0.1 ppm or less and progressively shorter path lengths associated with higher concentration measurements. Multipass cells are used to reduce the physical size and volume associated with such long path lengths. With proper design, such instruments can be made relatively insensitive to typical levels of H_2O in dry-sample systems. However, CO is affected by CO_2 since they spectrally overlap each other, but proper selection of the measurement band and the use of GFC reduces this effect to near negligible proportions.

Note: If other gases cause substantial interference with the measurement of the gas of interest, the interference can generally be reduced by either of three techniques: (1) physically removing it from the sample by using a specific filter targeted at the interfering species; (2) measuring the interfering gas and mathematically subtracting its contribution from the measurement of the gas of interest; and (3) calibrating it out by using a calibration gas(es) which includes the known interferent so that its effect is taken into account during the calibration of the instrument. The latter only works well if the concentration of the interfering gas is reasonably stable in time.

Oxygen is measured either by a solid-state electrolyte (zirconium oxide) sensor technique or with paramagnetic techniques in most CEMS applications. The solid-state electrolyte technique, also known as the fuel cell technique, has been perfected over many years as both an in situ analyzer and a conventional bench/rack extractive-type analyzer. It is extremely rugged and reliable and easily meets CEMS performance requirements. It has no known interferences, although care must be exercised in some applications since it provides a measurement of net

O_2, that is, the O_2 left over after the coexisting combustibles have been combusted. The zirconium cell is operated at near $1500°F$, which causes the combustibles in the gas stream to combust on the surface of the cell and leaves the residual O_2 to be measured by the cell. This is not normally of any significance since combustibles are on the order of tens or hundreds of ppm, and O_2 is on the order of several percent (tens of thousands of ppm) of the gas. Paramagnetic techniques are also very well developed and do not have the complication of a hot measurement cell associated with the zirconium oxide approach. They can exhibit some NO interference and must be applied properly for reliable operation. Electrochemical cells are also used for O_2 measurements. Their lifetime is limited, necessitating periodic replacement.

Electrochemical cells can also be used to measure SO_2 and NO_X as well as other gases of interest. Current technology has produced cells that reduce the potential poisoning and interference effects of older designs and are quite robust. When properly used, they can provide accurate and reliable measurements over short-(hours) to medium-term (days) time periods. They do require periodic ambient air refresh cycles so that measurements must be interrupted from time to time, or dual cells can be used that alternate between the refresh and measurement cycles. They have been used very successfully in hand-held and portable test instruments but have never gained significant market acceptance in U.S. CEMS applications.

The catalytic sensor is another gas detection technique that has been used extensively for combustion control applications and, with further refinements, has been recently applied to continuous emission monitoring. Historically, catalytic sensors were used for CO and combustible monitoring where it was relatively easy to provide a heated catalyst that when exposed to combustible gases would facilitate combustion and a rise in temperature of the catalyst. This temperature rise was then correlated with the concentration of the combustible. Such sensors were also subject to poisoning of the catalyst and were not highly specific. A prepackaged sensor-based system has been marketed that uses advanced catalytic techniques to measure CO and NO_X (both NO and NO_2) with a zirconium oxide sensor for O_2. Such catalytic sensors are inherently simple to operate and eliminate much of the complexity associated with traditional optical-based analyzers, but they must be carefully configured for the specific application.

KEY CONCERNS. In any extractive gas measurement system, there are two key concerns: sample integrity and minimizing maintenance of the sample train. Sample integrity is typically maintained by ensuring that the sample is kept in a hot, wet condition up until the point that it is either diluted or where the water is deliberately removed. If water is removed, one must consider the potential degradation of some of the acid gases and hydrocarbons. As the temperature of the gas stream is reduced, condensables are removed along with the liquid water. Further, the water-soluble gases in the sample will be reduced by virtue of their interaction with water on the walls of the cooler. Verifying that the sample train is leak free is, of course, basic to the operation of an extractive system. For parts of the sample train under vacuum, this is of particular concern. Sample train temperature is also key in maintaining the sample integrity since a drop in temperature below either the acid or water dewpoint will quickly cause near-immediate restrictions in the transport system and may cause corrosion that can destroy sample train components. Heated sample lines are typically provided for conventional extractive systems, but dilution extractive systems may be able to use only freeze-protected lines. Further, the sample dewpoint temperature of the gas entering the analyzers after being cooled/dried must be below the temperature of the analyzer measurement cell or condensation can destroy the integrity of the analyzer. The difficult part of this situation is estimating the acid dewpoints associated with SO_2 and SO_3, NO_X, HCl, and NH_3 given their variability. In general, a sample temperature of near 375°F may be needed for relatively high acid gas concentrations while a sample temperature of near 300°F may be adequate for other combustion applications. The sample temperature is, of course, key to minimizing maintenance but maintenance is also affected by how the particulate is handled. Periodic blowback of the primary filter is one approach for minimizing fouling. Several stages of particulate filtering are often necessary to ensure that fine particulate is removed prior to entering the analyzers or dust contamination may shorten the maintenance-free life of the analyzer. Maintaining adequate flow rates (velocity) so that particulate does not settle out in the sample lines is another key factor. Sloping sample lines in a downward direction with no sharp bends is also helpful. In dirty-gas applications, frequent maintenance of the sample system may be necessary.

It should also be noted that heated Teflon sample lines may outgas sufficient CO to be troublesome in low-CO-level applications. In such cases, heated stainless steel lines are required.

ENCLOSURE/INSTALLATION ISSUES. Extractive systems are typically mounted in 19-in. enclosed cabinets. A typical system is shown in Figure 2.1. Consideration must be given to the environmental conditions surrounding the intended location of the analysis rack/cabinet. All such systems must be installed such that they will operate properly and can be maintained under year-around weather conditions. A compromise is often required between long sample lines, which are required to allow the analysis rack/cabinet to be located in a convenient indoor location such as the control room, and shorter sample lines, which require either a separate HVAC (heated, ventilated, and air-conditioned) temperature-controlled house near the stack or duct of interest or a ruggedized HVAC-controlled cabinet that can be located near the stack or duct. Wall-mount enclosures that can be mounted outdoors may allow the shortest sample lines, as they can be mounted very near the point of sample extraction. Some type of protection from the weather may be needed to facilitate maintenance under adverse weather conditions. Sample line lengths should be minimized to reduce the power consumption and cost associated with heated lines, simplify installation, shorten sample transport lag time, reduce sample pump vacuum/ flow capacity requirements, and generally minimize maintenance. The HVAC requirements generally are more demanding for analyzer systems associated with very low level concentration measurements as opposed to higher level gas concentration measurements.

Figure 2.1   Extractive system. (Courtesy of Monitor Labs., Inc.)

2.1.1   Conventional Extractive Systems

Conventional extractive systems are characterized by the preparation of a cool, dry sample that results in a dry-basis measurement of the gas of interest. The designation of such systems as providing dry-basis measurements is not totally accurate since most water removal systems only reduce the moisture level to near 1% H_2O absolute humidity. Not only is the sample gas measured at full strength, but the concentration of the gases of interest is actually elevated by removing the water from the sample. Under this heading the sampling probes, sample lines, gas coolers/driers, flow controls, controller, and analyzers will be discussed.

2.1.1.1   Sampling Probe.   Most gas samples must be kept hot until the moisture is deliberately removed. If cold spots develop in the sample train, water and acids will condense out of the sample, ultimately causing the line to plug. Thus, the sampling probe is typically heated so that the sample temperature is maintained above the water and acid dewpoints at the point where it is passed on to a heated sample line. Sampling probes are characterized by a pipe or tube sometimes designated as the "straw" that extends into the stack, duct, or pipe of interest and an attached enclosure that includes a heated filter on the back end of the probe. If gases of interest are relatively clean, it may be possible to just use a filter attached to the input end of the straw. However, if gases are fairly dirty (substantial particulate), the in-stack filter is often supplanted with an out-of-stack filter located in the heated enclosure on the back end of the probe. In this way, the filter is easier to maintain and does not require probe removal from the stack, as do the in-stack filters. Further, the calibration gas can be injected into the sample transport system before or after the filter depending on the specific probe design. The most accurate check of the sample system is obtained by injecting it before the filter. A port is often provided for periodic blowback, which minimizes maintenance requirements. More sophisticated systems may be interlocked so that no sample is drawn from the source until all heated parts of the system are at normal operating temperature. With a heated probe, some entrained water in the condensed phase can be accommodated in the sample stream so long as probe operating temperatures are sufficient to vaporize it and keep it in the vapor phase.

2.1.1.2 Sample Transport Lines. Sample transport lines can be provided in three different types. First, an unheated line is the simplest method of transport; however, it requires that the sample be dried prior to being transported. Second, a heated transport line can be used that will maintain the sample at a selectable temperature between 300 and 400°F. This is the most generally used configuration as it is applicable to raw gas samples for which relatively high temperatures must be maintained to preclude acid gas condensation. Third, freeze-protected lines are also available that are used with gas samples that have been dried to an intermediate dewpoint and only need to be kept above freezing to prevent condensation and line pluggage. Heating of sample lines is achieved with self-limiting built-in heating elements that are constructed to maintain a given temperature or with resistive in-line heating elements under the control of a temperature controller. Some heated lines can be trimmed to length in the field without compromising the heating system, and others must be ordered to length and not altered. Steam-heated lines are available for special situations. To enhance reliability and simplify installation, sample lines are often supplied as part of a combined umbilical cord that contains primary and backup sample lines, calibration gas lines, blowback line, and electrical connections for the probe. Tubing sizes are selected according to length and flow rate to keep the pressure drop to a value that is compatible with the pump capabilities and to maintain and provide adequate velocity in the line to shorten transport delay times and prevent settling out of the entrained particulate.

2.1.1.3 Gas Coolers/Driers. Gas coolers/driers are often installed in the analyzer cabinet, which is remotely located from the sample point. In such applications, a heated sample line is required between the probe and analyzer cabinet. In special situations, it may be advantageous to install the gas cooler/drier immediately after the sample probe and remove sufficient water that an unheated sample line, or a merely freeze-protected line, can be used to transport the sample. In either case, the gas cooler/drier generally consists of one or more of the following cooling techniques: passive air-to-air heat exchangers, active refrigerant, or thermoelectric coolers. In addition, sample gases may be dried by use of membrane-type (Nafion) driers, molecular sieves, and desiccant materials. Molecular sieves and dessicants require frequent replacement or drying/regeneration.

Passive cooling may be used as the first stage of cooling to reduce the 300–400°F sample temperature to somewhere near room temperature. This can then be followed by an active cooling stage that will reduce the gas temperature and dew point to about 35–40°F. The sample gas temperature cannot be reduced much below this temperature without causing the condensed water to freeze and ultimately plug the system unless special configurations are incorporated. Condensed water is usually removed by a peristaltic pump that maintains the pressure seal but still allows the liquid water to be removed in a quasi-continuous manner. Periodically opening a drain valve also can be used to discharge water accumulations. A typical thermoelectric cooler is shown in Figure 2.2. At 35°F the gas sample still contains about 0.7% H_2O, which will vary with operating conditions. Special cooler configurations have been supplied that provide dew points well below the freezing point; however, these systems require two separate thermoelectric coolers. They are cycled so that when ice builds up on one, the sample is switched to the other cooler, which has completed a defrost cycle and is cooled down and ready for sample again. Thermoelectric coolers can be obtained with a combination of passive and active cooling stages. A typical configuration may incorporate one stage of passive cooling and two stages of active cooling. The active cooling stages are often plumbed so that one stage is on the input side of

Figure 2.2 Thermoelectric gas cooler/drier. (Courtesy of Baldwin Environmental, Inc.)

the sample pump and the other on the outlet side of the sample pump to maximize overall effectiveness. Diagnostic temperature and water slip sensors may be included to ensure recognition of abnormal operating conditions. The NO_2 and SO_2 are somewhat water soluble and coolers need to be designed to minimize the water–gas interface to minimize absorption into the water discharge stream. Tests have shown that NO_2 absorption can be less than 1 ppm in well-designed gas cooler/driers. Sulfur dioxide absorption is normally of no concern but may require consideration for very low emission applications. The IR-based analyzers may be adversely affected by the substantial water content and variability associated with active coolers used with dew points above the freezing level and may force installation of a membrane drier, which can reduce the dew point to near $-40°F$. Such devices are typically available under the trade name Perma Pure (Perma Pure, Inc.) and are very effective in drying the sample, albeit at the expense of having to drive the membrane drier with a very dry, clean air supply.

2.1.1.4 Sample Pump and Flow Controls. Typically, the sample pump is located in the analyzer cabinet and is plumbed to pull the sample from the probe and discharge the sample into a sample manifold where it can be distributed to various analyzers. As a result, analyzers can vent to the atmosphere and basically operate at atmospheric pressure. Typical flow rates of 2–10 L/min are established by the pump in association with pressure drop in the transport lines, filters, and manually set flow controls. Flow rates are based on several trade-offs. Higher flow rates generally load up particulate filters at a faster rate but provide faster response times. Higher nominal flow rates also necessitate higher calibration gas flow rates, which consume calibration gas bottles at a faster rate than at low flow rates. Higher flow rates also provide more pressure drop for a given length and diameter of sample line, which may necessitate a larger sample pump. Pumps are usually of the diaphragm type and are unheated unless the pump is used with a hot, wet sample. Since most pumps are used in series with the analyzers, it is important that the gas wetted parts of the pump are inert with respect to the gases of interest. Teflon and stainless steel are common materials of construction.

2.1.1.5 Controller. The utility of such extractive systems is greatly enhanced by the addition of a controller. The controller often takes the form of a programmable logic controller (PLC), datalogger, special-purpose sequencing device, or PC with sufficient input/output (I/O) to provide equivalent functionality. The controller basically provides automated programmed activation of zero/span calibration check cycles (solenoid valves for each calibration gas bottle), probe blowback or backpurge cycles, and sequential control of multiple-sample probes in a multipoint measurement system. It may optionally provide for data acquisition and buffering, correction for zero/span drift, correction for temperature/pressure effects, conversion of measurement to units of the standard/limit, monitoring and indicating the status of system internal components, and monitoring and/or implementing certain diagnostic and/or maintenance procedures. In high-reliability applications, the controller can be used to buffer measurement data for a week or more, thus allowing substantial DAHS (data acquisition and handling systems, basically a special-purpose PC-based data acquisition and reporting system) downtime without loss of data.

Systems can be developed for either analog or digital systems. In older systems, analyzers provided analog outputs that were read and processed by the controller and output as 4–20-mA signals. In newer systems, analyzers may feed a serial RS232, RS422, or RS485 output to the controller, which processes the information and outputs resultant data either in similar serial form to a PC or in one of the typical data bus formats for use in larger network applications. The analyzer span for an analog-based system defines the level at which the maximum analog output (20 mA for a 4–20-mA system) is achieved. Measurements much above the span level will peg the output (or put the output in the rails or stops, as some describe it) making it impossible to determine the actual values. For serial digital interface systems, the meaning of a span value is not significant since the analyzer outputs measurement values from zero to its maximum range capability in serial digital format without regard to an intermediate span value. Thus, as long as the required span is less than the maximum-range capabilities of the instrument, the serial output is valid. In EPA-regulated applications, the declared span value defines the allowable zero and span calibration gas values as well as calibration drift/error. Similarly, status information such as in-calibration, fault, maintenance request, failed calibration, or down for service can be indicated with either relay contact closures or appropriate serial digital data transmittals.

Some applications require the use of local and/or remote control and indicator panels. In such cases, panels can be provided that allow for manual activation/deactivation of valves, pumps, calibration cycles, and so on, and provide indicators for the mode of operation, appropriate system level flow, and pressure measurements as well as analyzer concentration measurements and diagnostics.

2.1.1.6 Analyzer Considerations. Since the gas samples in conventional extractive systems are at full strength, that is, not diluted, the analyzers generally do not have to be as sophisticated as the low-level analyzers used in ambient air quality or dilution extractive measurement systems. In some cases this allows less expensive analyzers to be used. Care must be exercised to ensure that the H_2O level and variability in the dried sample of such systems are compatible with the interference rejection characteristics of the gas analyzers. If ambient air analyzers are used because of extremely low pollutant levels, consideration must be given to the fact that dried stack gas samples contain over 100 times more CO_2 than ambient air, which may require special consideration when selecting calibration gases for the chemiluminescent NO_X analyzers.

2.1.1.7 Advantages and Disadvantages

2.1.1.7.1 Advantages. Conventional extractive systems have been well accepted in many process and environmental compliance applications. Maintenance of such systems is very straightforward without the need for highly specialized training. Pumps, valves, and sample lines can be diagnosed and corrective action taken without much sophistication. The systems are very flexible in that it is relatively easy to change or add other gas measurements since separate analyzers are typically provided for each gas of interest and concentrations are full strength. This type of system is also amenable to multipoint applications since it is relatively easy to time share the analyzers with multiple sampling points. The EPA regulations generally require only one measurement at each point every 15 min, allowing time to sequence between several measurement points. Sample lines that are not being sampled should generally not be left dormant or stagnant but should be kept in an active state with dedicated sample pumps or eductors that keep fresh sample flowing through the lines. Further, this type of system is most adaptable to high-temperature sample gas streams since the temperature can be reduced in passive heat exchangers to temperatures that can be handled via heated sample lines. Entrained water in the gas stream can also be accommodated with the use of heated sample probes. For those applications that require reporting of dry-basis pollutant concentrations, the conventional extractive system is particularly applicable.

2.1.1.7.2 Disadvantages. Chronic fouling problems may occur with moderate to high concentration levels of the acid gases, such as SO_2, SO_3, NO, NO_2, HCl, and NH_3. The solution to such problems is often not obvious and may be application and site specific. Maintenance of the sample train in dirty-gas applications will require some dedicated service time. These systems are not appropriate for measuring highly water soluble gases. Also, they produce dry-basis measurements, which are difficult to use in applications requiring a measurement of stack gas flow rate in order to establish the mass flow of emissions in pounds per hour. In situ gas flow measurement devices provide wet-basis measurements and require a measurement of moisture in order to be used with dry-basis pollutant/diluent measurements. Heated sample lines are reasonably reliable but ultimately have heater and/or tubing failures, thereby requiring replacement. Further, long sample line runs can be expensive to purchase and install. Sample pumps are mechanical devices that are continuously operated and also limit system reliability unless replaced or refurbished on a periodic schedule.

2.1.2 Hot, Wet Extractive Systems

2.1.2.1 Basic Characteristics. Hot, wet extractive systems are used when it is necessary to keep the sample hot, without any moisture removal, in order to preserve the gases of interest. If measurements of HCl, HF, NH_3, H_2O, or heavy hydrocarbons are required, among other gases, it is expected that a hot, wet extractive system would be required. Since the sample train, including probe, filter, and sample lines, is typically heated in both conventional and hot, wet extractive systems, there is no significant difference until the sample is ready to be delivered to the analyzer cabinet. All of the sample train up to and through the analyzer must be kept hot in order to maintain the integrity of the gases of interest in a hot, wet system. This includes valves, particulate filters, cabinet internal tubing, pumps, manifolds, as well as the analyzer sample inlet and measurement cell or cavity. At hot temperatures of near 375°F, this can present a variety of problems, which is the reason such systems are generally avoided unless there is no other realistic choice of measurement techniques. The choice of analyzers and ranges that are compatible with such temperatures is greatly reduced from those that can be used in a conventional dry-sample extractive system. After the sample has been measured, the sample temperature is not so critical and condensation may be allowed if care is exercised in the selection of materials used in the wet phase of the sample train. Liquid moisture can also cause pluggage if the sample train is not appropriately designed.

2.1.2.2 Analyzers. The analyzers that can be used in a hot, wet extractive system must be specifically designed for such hot conditions. Further, since the original moisture content of the sample gas is preserved, the analyzers must be capable of coping with large quantities of H_2O. For IR-based spectroscopic instruments this means that the detection technique must provide a high degree of H_2O interference rejection. Instruments that have been used with hot, wet systems include NDIR/GFC, UV, and FTIR (Fourier transform infrared). The UV region is preferred since it has no H_2O interference problems.

2.1.2.3 Advantages and Disadvantages. The advantage of such systems is that, by virtue of having a hot, wet sample, the concentration and integrity of the acid gases are preserved. The disadvantage of such systems is the high cost of purchasing and maintaining a hot, wet sample train that extends all the way through the analyzer measurement cavity/cell. You should note that in such systems maintenance is much more time consuming. First, it is necessary for the sample train to cool off before one can easily work with such components. Second, if you adjust some parameters when cool, the condition of the analyzer and sample train may change as it gets hot. Third, when completing a maintenance operation, it is necessary to wait for some time until the sample train components have all reached their normal operating temperature. They do provide wet-basis measurements that are needed in some applications.

2.1.3 Dilution Extractive Systems

2.1.3.1 Basic Characteristics. Dilution extractive systems were developed to avoid the sample transport and handling problems associated with high acid gas concentrations and high acid and water dewpoint temperatures. The promise of dilution systems was that you could make a stack gas as easy to handle and measure as an ambient air sample and that ambient air analyzers were so well developed, tested, and available that such systems could provide better performance at similar cost than available from other techniques. The reality is that they are complex systems that require careful attention to detail but can provide excellent accuracy when well maintained.

Dilution of a stack sample is typically done using an orifice under critical (sonic flow) conditions to fix the sample flow at a specific flow rate and mixing the same with a controlled flow of dilution air. The most common systems are based on the

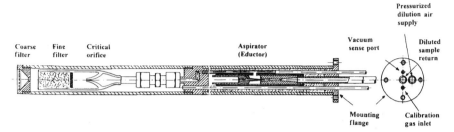

Figure 2.3   In-stack dilution probe diagram. (Courtesy of EPM Environmental.)

EPM environmental dilution probe technique, which uses an eductor (or aspirator) generated vacuum to create a critical pressure ratio that ensures the constant (sonic) flow of sample gas through the orifice. Basically, once the absolute pressure ratio (downstream/upstream) across the orifice is less than 0.47 for air, the volumetric flow through the orifice will be independent of whether the downstream/upstream pressure ratio changes provided it remains below the critical pressure ratio of 0.47. A dilution probe is illustrated in Figure 2.3. Here, positive pressure of about 30–60 psig is applied to the input port on an eductor. This pressure creates sufficient flow through the eductor venturi that the vacuum in the throat of the venturi exceeds the critical pressure ratio needed to operate the sample orifice in the critical flow regime. This assumes that the sample is obtained from a near-atmospheric pressure source such as a stack. The critical orifice and drive pressure to the eductor are selected to provide the desired dilution ratio. Typical dilution ratios, which are defined by the ratio of the sample flow plus dilution flow (flow of pressurized air supply into the eductor) divided by the sample flow, range from 50 to 250. Further, the dilution ratio multiplied by the measured diluted concentration yields the original concentration. Ultimately, the dilution ratio is obtained by dividing the known concentration of a calibration gas by the resultant measured diluted value.

With this probe design, the port labeled calibration gas inlet (Fig. 2.3) feeds gas into the stack side of the fine filter allowing calibration gas to be measured in the same way as the sample. This port can also be used to blow back the coarse filter. Each probe is typically associated with a dilution flow control module that allows the user to adjust the pressure to the eductor, monitor the vacuum applied to the back side of the critical orifice, and set calibration and/or blowback gas flow rates. The control module is usually installed in the analyzer rack.

Particulate filtration of the sample gas is critical to maintaining the orifice in a clean and unobstructed condition. Several stages of filtering are usually incorporated, as shown in Figure 2.3. Periodic blowback through the coarse filter may extend the service interval. In addition, inertial filtering has also been used to further reduce the amount of particulate in the gas stream flowing through the orifice.

Systems can be designed such that the dilution, which occurs when the small sample flow is mixed with a much greater flow of dilution air, takes place either in the stack, called in-stack dilution, or on the outside end of the sampling probe, in which case it is designated as out-of-stack dilution, or after the sample has been transported to the analyzer cabinet, in which case it is designated as at-the-cabinet

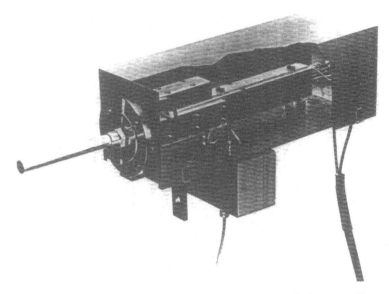

Figure 2.4   Out-of-stack dilution probe. (Courtesy of EPM Environmental.)

dilution. The dilution on the stack approaches (either in-stack or out-of-stack dilution) have historically been the preferred technique, most likely due to the simpler sample transport requirements. A typical out-of-stack dilution probe is shown in Figure 2.4. It uses the same dilution mechanism as described above.

All three approaches yield the desired dilution as far as the analyzers are concerned; however, there are other concerns. If dilution is done on the stack, either in stack or out of stack, the resultant sample is diluted prior to transport. Thus the water and acid dew points may be sufficiently reduced that the sample transport lines can be either unheated, particularly for southern climes, or only freeze protected. This greatly reduces the complexity of sample transport. The range of H_2O dew points available with different stack H_2O concentrations and dilution ratios is shown in Table 2.1.

Table 2.1 H_2O Diluted Sample Concentration and Dewpoint

		By Dilution Ratio				
H_2O	Original Sample	30	50	100	175	250
Stack Concentration, %	30	1.00	0.60	0.30	0.17	0.12
Dewpoint, °F	157	46	32	16	3	−4
Stack Concentration, %	20	0.67	0.40	0.20	0.11	0.08
Dewpoint, °F	141	35	22	6	−6	−13
Stack Concentration, %	10	0.33	0.20	0.10	0.06	0.04
Dewpoint, °F	115	18	6	−8	−18	−26

Note: Percent moisture (v/v) based on sea level atmospheric pressure.

Obviously, the dewpoint of the diluted sample must be well below the minimum ambient temperature to which the sample line is exposed if unheated sample lines are to be used. If dilution is done at the cabinet, the sample must be transported hot and wet up until the point of dilution. Note that all dilution systems provide a wet-basis measurement since the moisture is not removed from the sample. Obviously, with dilution at the cabinet, the transported sample must be of much greater volume than that which is required for the diluted sample; that is, some of the sample stream must be vented to ensure adequate velocity in the sample line and response time. Also, it should be noted that the sample line is operated at positive pressure for dilution on the stack whereas the sample line will operate at negative pressure when dilution takes place at the cabinet. The tolerance to small leaks in the sample train from the stack to the analyzer cabinet is therefore quite different for the two approaches. When comparing in-stack versus out-of-stack dilution, it should be observed that for in-stack dilution systems the sample probe must be removed from the stack for maintenance, whereas in an out-of-stack dilution system the dilution components are located on the back end of the sampling probe, where they may be more easily accessed and maintained without probe removal. Further, in-stack dilution results in the entire dilution module temperature floating up and down with the stack temperature, potentially causing other problems. Heated probes have been supplied to minimize such temperature variations. Out-of-stack dilution systems generally provide a controlled (heated) constant-temperature environment for the dilution system components, allowing potentially greater accuracy.

All dilution extractive systems depend on the purity of the dilution air supply to maintain reasonable accuracy. To make sure that the dilution air supply does not contribute to an error in the measurement, it is reasonable to demand that the dilution air supply contain no more than 1% of the typical concentration, after dilution, of each gas to be measured. Thus, if stack NO_X is to be measured at a level of 100 ppm, with a dilution ratio of 100, the dilution air supply must contain no more than 0.01 ppm or 10 ppb of NO_X. This often requires complex dilution air supply cleanup packages to remove all the pollutants of interest to an acceptable level.

Vacuum-driven dilution systems are also available that utilize a vacuum pump instead of a pressurized air supply and eductor to create the vacuum necessary to provide critical flow in the sample and dilution orifices. If a vacuum pump is placed at the cabinet end of the sample transport line and reduces the inlet pressure to well below half an atmosphere, critical orifices can be placed in the sample probe or near the pump inlet to provide controlled flow of dilution air and sample into the vacuum pump. The outlet of the vacuum pump then supplies a diluted sample to the manifold that supplies the analyzers. Thus the entire sample line is operated under vacuum and with sufficient vacuum that the dilution ratio does not have to be as high to accomplish the equivalent reduction in acid and water dew points in the sample line. This approach has been used in applications where a venturi/eductor-based system would provide too much dilution.

Since both vacuum-driven and pressurized air/eductor-driven dilution systems depend on the constancy of the sonic flow through a critical orifice, let us examine the dependencies of the sonic velocity associated with an orifice operated at or below the critical pressure ratio.

2.1.3.2   Dilution Ratio Considerations. With a typical dilution probe–based extractive CEMS, the sample of the stack gas is extracted through a probe with an orifice that is operated below the critical pressure ratio, a ratio of downstream absolute pressure to upstream pressure of approximately 0.47 or less, so that the flow through the orifice is critical, or at the speed of sound. The vacuum drive for the orifice is obtained from an eductor, or ejector, and should be monitored to ensure satisfactory operation. The sample is pulled through the eductor and mixed with the dilution air used to supply pressure to the eductor. Thus the dilution ratio (DR) can be defined as

$$DR = \frac{Q_s + Q_d}{Q_s} \tag{2.1}$$

where Q_s is the sample flow and Q_d the dilution flow. Further, the measured value of the pollutant in a diluted sample is related to the original sample concentration by the DR as

$$DR = \frac{concentration(original)}{concentration(measured)} \tag{2.2}$$

In typical CEMS applications the DR is set to values in the range of 50–250. A given dilution ratio is obtained by properly sizing the critical orifice and matching the eductor to the required total flow rate and sample vacuum. The pressure of the dilution air supply to the eductor is adjusted to fine tune the dilution ratio over a small adjustable range. The high dilution ratios make the sample easier to handle and compatible with the measurement ranges of conventional ambient air analyzers. Total diluted sample flow rates $(Q_s + Q_d)$ are often in the range of 5–10 L/min, resulting in sample flows in the range of 0.2–0.02 L/min, which correspond to orifice sizes of a few thousandths of an inch. The sonic flow rate associated with a given gas is well defined as shown by the equation

$$V_c = \left(\frac{kP}{D}\right)^{0.5} \tag{2.3}$$

where   k = isentropic exponent with a nominal value of 1.4
      P = absolute pressure, Pa
      D = gas density, kg/m^3
      V_c = sonic velocity, m/s

As confirmation of the above theory, the 71st edition of the *Handbook of Chemistry and Physics* (CRC, Boca Raton, Florida) lists sonic velocity of pure SO_2 [molecular weight (MW) of 64] at 699 ft/s, oxygen (MW of 32) at 1037 ft/s, and methane (MW of 16) at 1411 ft/s, with the velocity measured at 0°C. The sonic velocity of dry air (MW of 28.75) is 1087 ft/s, and the addition of moisture increases that value since H_2O has a molecular weight of 18.

Thus, the sonic velocity of a gas through a thin-plate orifice is pretty well defined, but the mass flow rate through a specific orifice is dependent upon the exact orifice shape/size and thermal conditions, which make it much more difficult to calculate. This is further complicated since neither air nor stack gases are "ideal gases" and both require considerable sophistication to characterize under conditions involved in sonic flow through a nozzle. A calculation of the mass flow rate through a specific orifice under critical pressure conditions is complicated and well beyond the scope of this discussion. Further, the mass flow rate through an eductor is dependent upon a large number of variables, making it very difficult, if not impossible, to accurately calculate. Thus, the dilution ratio for a given dilution apparatus operating under some specific conditions cannot be readily calculated. Consequently, it is much easier and more accurate to measure the actual Q_s for a given orifice under controlled conditions than to calculate it Similarly, an eductor can be selected based upon measured characteristics. However, after establishing the dilution ratio corresponding to a critical orifice operating under some known reference conditions, the dilution ratio at actual operating conditions must be determined. Jahnke and Marshall[1] have shown that the dilution ratio under a new set of conditions can be calculated from the dilution ratio established under some previous conditions (subscript 0) by use of the equation

$$DR = 1 + \left[(DR_0 - 1) \left(\frac{P_0}{P} \frac{\sqrt{T}}{\sqrt{T_0}} \frac{\sqrt{MW}}{\sqrt{MW_0}} \right) \right] \qquad (2.4)$$

where temperatures and pressures are absolute with respect to the critical orifice upstream gas conditions.

Note that the T_{act} term is theoretically the temperature of the gas stream entering the orifice but that temperature may or may not be the measured stack temperature depending on the specific thermal design of the system. Further, the temperature of the dilution air supply applied to the eductor will influence the overall dilution ratio. It has become accepted practice to make the orifice and eductor as well as the associated gas temperatures as close to the same temperature as possible and as constant as possible if accuracy is to be maximized. Establishing empirical data relating dilution ratio to stack temperature is recommended for a specific design to ensure the accuracy of the dilution ratio prediction model. The pressure and molecular weight dependence as shown above appear to be consistent for many dilution systems.

In actual operating conditions, the stack gas pressure and temperature as well as typical molecular weight are easily determined and form the basis for establishing the sample flow rate at any given conditions. The actual value of the dilution ratio in a CEMS is typically obtained from the periodic gas calibration sequence. By feeding a known upscale calibration gas into the front end of the dilution probe, it is diluted and measured as would be the actual stack sample. This assumes, of course, that sufficient calibration gas is supplied that there is a slight backflow of calibration gas out of the stack inlet port on the dilution probe and that the calibration gas is conditioned so that it is at very nearly the same temperature as the stack gas. Under these conditions, we obtain the calibrated dilution ratio as follows:

$$DR(\text{calibration}) = \frac{\text{known gas concentration}}{\text{actual measured concentration}} \qquad (2.5)$$

Care must be exercised, however, since from the previous equations it is known that the composite molecular weight of the gas determines the sonic velocity through the orifice. Thus, either the composite weight of the calibration gas must be the same as the stack gas or compensation should be made for the differences. It should also be noted that after a given calibration cycle, if the DR is different than the previous value, it is not easily determined whether that change is caused by a change in the flow rate through the eductor or a change in the calibration of the analyzer. If the system incorporates automatic span compensation, the source of the change is of no concern since both effects are taken into account with the calculation of the new dilution ratio based on the analyzer measurement of the diluted gas stream.

In some dilution systems, a mass flow meter has been used to provide a continuous measure of the total diluted sample flow rate. This provides a good indication of the stability of the dilution system and the maintenance of proper settings for the given dilution ratio. Further, if we can accurately model the flow through the sample orifice, we can assume that the remaining changes in the mass flow indication are the result of changes in the eductor throughput and make a corresponding change in the applied dilution ratio. Regardless of the model selected for DR compensation, empirical data should be obtained to verify the model under different ambient and stack gas temperatures and pressures.

2.1.3.3   *Analyzers.*

Analyzers used with dilution extractive systems are typically the classical ambient air analyzers for SO_2, NO_X, and CO or a derivative of the same. Most ambient analyzers have a wide dynamic range; that is, the minimum range that will meet EPA CEMS performance criteria may be as low as 2.5% of the maximum range of the analyzer. At lower ranges the drift of the instrument, in ppm, may exceed the regulatory requirements specified in terms of percentage of span. In other words, the drift of an analyzer in ppm becomes a greater percent of span as the span is reduced. But, there is a wide range of stack concentrations that may be measured with a given dilution ratio because of their low-level measurement capabilities. Such analyzers typically provide minimum detectable limits of less than 1 ppb for SO_2 and NO_X. This is shown in Table 2.2, which illustrates the capabilities of typical ambient air analyzers used with different dilution ratios. Of course, CO_2 is not an ambient air pollutant, but CO_2 analyzers have been developed specifically for these dilution systems.

From this table you will note that the measurement of low levels of CO with high dilution ratios may be a problem. If other pollutants are to be measured in the

Table 2.2   Maximum/Minimum Measurement Ranges for Given Dilution Ratio

Gas Specie	Analyzer Ranges	Typical Full Scale	By Dilution Ratio				
			30	50	100	175	250
SO_2, ppm	Max	20	600	1,000	2,000	3,500	5,000
	Min	0.5	15	25	50	88	125
No_X, ppm	Max	20	600	1,000	2,000	3,500	5,000
	Min	0.5	15	25	50	88	125
CO, ppm	Max	200	6,000	10,000	20,000	35,000	50,000
	Min	5	150	250	500	857	1,250
CO_2, %	Max-3	0.3	9	15	30	53	75
	Max-2	0.15	5	8	15	26	38
	Max-1	0.075	2	4	8	13	19
	Min-1	0.0075	0.2	0.4	0.8	1.3	1.9

Note: (1) Analyzer ranges are for typical ambient analyzer, other ranges available. (2) Gas concentrations can be measured accurately down to 20% of typical minimum range.

diluted state, they will require analyzers with comparable sophistication to the ambient analyzers. When applying ambient analyzers to a dilution measurement system, care should be exercised to ensure that the specified performance, including interference rejection, is compatible with expected gas measurement conditions. It will be noted that the diluent measurement must be CO_2 if clean, dry instrument air is used as the dilution air supply since it contains a large percentage of O_2.

Ambient analyzers must be type approved by the EPA (40 CFR 53) to be used in regulatory ambient monitoring applications in the United States. Similar regulations apply to the TUV/UBA/UMEG Eignunsgepruft type approval for Germany, Mcert for the United Kingdom, and others. This approval process guarantees a specified level of performance in ambient monitoring applications, and such analyzers, when properly adapted and applied to stack dilution systems, can provide similarly excellent accuracy. In the United States, the EPA approval for ambient monitoring applications is not required and has no applicability to stack monitoring applications, which are individually certified as installed.

2.1.3.4   *Advantages and Disadvantages.*   The advantages of dilution extractive systems primarily relate to the effect of relatively high dilution ratios (>100), which greatly reduce sample-handling difficulties. In the best case, they eliminate the need for continuously heated sample lines that substantially enhance the reliability of the system. In the other cases, freeze-protected or heated sample lines may still be required. Dilution systems can be time shared between multiple points; however, the replication of a dilution probe and dilution flow controller for each sampling location is not a trivial cost. For systems with dilution on the stack, the sample line is under positive pressure and exhibits a degree of insensitivity to small leaks. Such systems provide wet-basis measurements and are easily adapted to systems

requiring flow monitoring to determine mass emission rates in pounds per hour. Relatively simple dilution systems work well in applications requiring emissions in lb/mmBtu because of the tolerance of these measurement units to variations in the dilution ratio. Ambient air analyzers are generally well tested and proven designs offering comparatively higher reliability and performance for their cost.

The disadvantages of dilution systems also relate to the inherent characteristics of the dilution mechanism. First, there is the very small orifice that is exposed to the stack gas with only limited particulate filtration protection. Sample orifices are typically in the range of 0.003–0.006 in. in diameter. Dirt or condensed hydrocarbons or other compounds can build up in the orifice over time, altering the sample flow, if not blocking the sample flow. Entrained water droplets or condensation is obviously not consistent with reliable performance without special heated configurations. Second, because of the dependency of sonic velocity on input gas parameters, including temperature, pressure, and molecular weight, one must measure and/or compensate for the same if system accuracy is to be maximized. Any variation of the flow rate of the pressurized dilution air supply can provide significant errors if left uncorrected. Mass flow measurement of the diluted sample can reduce such errors at the expense of another measurement device. Further, the gas cleanup system for the dilution air supply must purify the instrument air to a greater degree than necessary for other systems and it must be well maintained. These systems are complex, requiring substantial maintenance and attention, especially in dirty-gas applications. Well-controlled HVAC enclosures/houses are typically required to ensure good analyzer performance. Dry nitrogen must be used as the supply for the dilution air if O_2 is to be measured as the diluent for pollutant lb/mmBtu and similar excess-air corrected calculations.

2.1.4 Special Systems

Extractive systems have been modified for special applications such as for measuring TRS, NH_3, and THC and for NO_X, CO, and O_2, which are associated with the NO_X Budget Program. These special systems will be described below:

1. *Total reduced sulfur* (TRS) measurements are required for certain pulp and paper facilities and are a measure of the odorous emissions often experienced around such facilities. The TRS is defined by the EPA to include hydrogen sulfide, methyl mercaptan, dimethyl sulfide, and dimethyl disulfide. A typical approach is to subject the sample gas to a high-temperature oxidizing environment (oven), which results in the conversion of the various compounds to SO_2. However, since SO_2 is not part of TRS, one must establish the difference between the original SO_2 level and the elevated SO_2 level in the oxidized gas stream. This is further complicated by the fact that background SO_2 levels may be several hundred ppm and TRS levels are limited to less than 30 ppm. Thus a differential SO_2 measurement is required to obtain the desired accuracy. An addition to this technique is to use a column of SO_2-absorbing material to remove SO_2 from the sample, leaving only the TRS compounds to be oxidized and measured. This

eliminates the need for a differential SO_2 measurement. Often two columns are used so that one can be regenerating while the other is in use. With appropriate cycling, one can obtain a reasonable measure of only the TRS compounds. Dilution techniques have also been used to reduce some of the sample-handling problems and to reduce probe and sample-line-fouling problems.

2. *Ammonia* (NH_3) is a very difficult gas to measure using extractive techniques. It has a high propensity to stick to almost any surface at reasonable temperatures and is highly reactive and soluble. It requires very high temperatures (possibly 500–600°F) to ensure the integrity of the transported sample. As a result, other approaches have been developed to make the measurement amenable to conventional measurement techniques. Both of the following techniques use a differential NO_X measurement. Such measurements can be achieved with the difference of two NO_X analyzer readings or with a single differential NO_X measurement, which is readily available from many chemiluminescent analyzers. Most such analyzers have a measurement and zero reference cycle. By inputting one of the gas streams as the zero reference, you can obtain a differential measurement with respect to the other input. In one approach a high-temperature converter (1450°F) is used on the stack to convert NH_3 to NO. Then by transporting the converted sample and raw sample to the analyzer cabinet and using a differential NO_X measurement, one can establish the original NH_3 concentration without concern for the transport of NH_3 through the sample lines. A conventional gas cooler/drier is used prior to the analyzers to minimize NH_3 and H_2O content in the gas entering the analyzer(s). The transported raw sample stream can also be scrubbed of NH_3, which eliminates the potential contamination of the sample line with NH_3 and the associated precipitates and condensates. The differential analyzer measures the difference between the converted sample, which contains NO resulting from both the basic NO_X in the sample stream and the NH_3, and the raw sample stream, which only contains the basic NO_X.

The other NH_3 measurement approach is to use the background NO_X, or some added NO_X, to convert the NH_3 in the sample to N_2 and H_2O using a heated catalyst (SCR) on the stack. The selective catalytic reduction (SCR) of NO_X involves the reduction of NO_X to N_2 and H_2O when mixed with NH_3 in the presence of a heated catalyst. It is one of the most often used techniques for the reduction of NO_X in combustion applications. But, since NO_X is used up in the conversion of NH_3 to N_2 and H_2O and the resulting change in NO_X can be correlated with the NH_3 in the sample stream, a measurement of the reduction of NO_X is indicative of the NH_3 present in the input gas stream. Note that there must be an excess of NO_X over that required for the stoichiometric conversion of NH_3 for the process to work. Prior to the SCR reactor a portion of the sample stream is diverted to the same high-temperature NH_3-to-NO converter as used previously, which converts the NH_3 and NO_X to NO. Both this sample (with the full NO_X content) and the SCR reacted sample (with the NO_X reduced by the amount of NH_3) are transported to a differential NO_X analyzer. The resulting difference is a direct indication (with proper scaling) of the NH_3 in the original sample, again without concern for the transport of NH_3 through the sample lines. Both systems can be implemented using conventional heated sample lines and a gas cooler/drier prior to measurement. Such systems are complex and the potential sample transport problems provide justification for considering in situ techniques that avoid the sample-handling and conversion problems. In situ analyzers using TDLs (tunable diode lasers) in the near IR with derivative measurement techniques have recently been used for these applications.

3. *Total hydrocarbons* (THCs) are sometimes monitored in association with some of the other more common pollutants. In some applications it is important to differentiate between THCs and TNMHCs (total nonmethane hydrocarbons), since methane results from many nonindustrial (agricultural) processes. Flame ionization detectors (FIDs) are often used for the detection and measurement of hydrocarbons, with scrubbers or gas chromatograph (GC) columns used to separate the methane from the remaining hydrocarbons. The FID typically has a wide measurement range from fractions of a ppm to hundreds or thousands of ppm. It requires a hydrogen-based fuel to sustain the flame. The FID does not have uniform sensitivity for all hydrocarbon compounds, in terms of the number of molecules of carbon, but can be individually calibrated for a given compound of interest. The FIDs are often calibrated in ppm of methane or propane or equivalent. Hot, wet extractive systems and heated FIDs are required if the full range of hydrocarbons are to be measured, for example, C1–C12 or higher. Cool, dry samples and an unheated FID may be sufficient if only the low-molecular-weight (C1–C6 or C8) compounds are desired. The measurement of BTX (benzene, toluene, and xylene) typically requires a hot-sample train and FID. Photo ionization detectors (PIDs) are also used for monitoring VOCs (volatile organic compounds) and THCs.

2.2 IN SITU MEASUREMENT TECHNIQUES

In situ measurement systems are available in three fundamentally different configurations. Across-the-stack systems project an interrogating light beam completely across the stack and measure the spectroscopic absorption of specific gases of interest in the line of sight. Probe-type systems use a stack-mounted probe with diffusion cell. The diffusion cell may either contain a gas sensor or, for optical instruments, an optical cavity where the stack gases absorb light at the wavelengths characteristic of the gases involved. Open-path or fenceline monitors are long-path optical devices that monitor the presence of gases of interest along a fenceline or other strategic boundary. Their use has been very limited, and they will not be covered further here. In situ optical devices depend on the fundamental principle involved in Beer's law (also called Beer–Lambert law) to establish concentration measurements as shown below:

$$\frac{I}{I_0} = 10^{-acl}$$

where I is the intensity of the light detected at the output of a measurement cell (or path) of length l, with a concentration of the gas of interest of c and with that gas exhibiting an absorption coefficient a. Here, I_0 is the intensity of light detected at the output of the same measurement cell with none of the absorbing gas(es) present in the measurement cell.

It should be noted that Beer's law is based on the use of a monochromatic light (single-wavelength) source that is never practically realized unless a laser is used as the light source. Further, the absorption coefficient is a unique characteristic of each specific gas as well as its temperature and pressure and is further dependent on the specific wavelength(s) of light used for the measurement. To keep the exponential terms as independent as possible, the concentration c should be expressed in units of density, which are representative of the number of molecules of the gas of interest per unit volume. For most spectroscopic measurement systems, the absorption coefficient is empirically established from experimental data. If other gases absorb at the same wavelength, the product of the absorption coefficient and concentration (ac) shown above becomes the sum of all the individual ac products of the gases that absorb at that wavelength.

2.2.1 Across-the-Stack-Systems

Across-the-stack systems use optical measurement techniques to sense the spectroscopic absorption of each gas of interest that is converted to gas concentration values. They are supplied in two different versions—single pass and double pass. Single-pass systems have a transmitter on one side of the stack and a receiver/detector on the opposite side. A typical single-pass system is shown in Figure 2.5.

Double-pass systems have a transceiver (transmitter and receiver) on one side of the stack and a reflector on the opposite side. Thus the optical measurement path

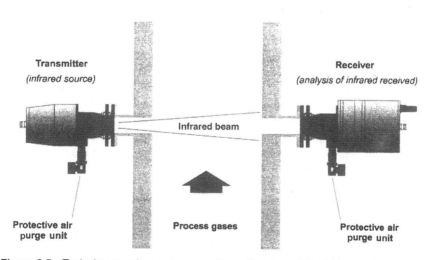

Figure 2.5 Typical across-the-stack gas analyzer. (Courtesy of Codel International, Ltd.)

length is twice the physical path length. On stacks with small diameters, the double-pass systems may be preferable because of the longer optical measurement path length and greater sensitivity. For large-diameter stacks the double-pass systems may be at a disadvantage because of the very long optical path length, which may make it difficult to obtain enough light, or optical energy, at the detector to make accurate measurements. Path lengths can be reduced by installing mounting flanges with long barrel lengths that project into the stack. Across-the-stack systems must measure the concentration of the desired gases in the presence of substantial and variable quantities of dust and H_2O.

The most significant problems associated with across-the-stack systems are three-fold: (1) maintaining a degree of insensitivity to the misalignment that may occur between the two halves of the systems and (2) providing an accurate means of calibrating the gas analyzer, and (3) accommodating the widely different ranges of stack diameters, stack gas temperatures, ambient temperatures, stack gas particulate loading, concentrations of the gases of interest and coexisting background gases. By designing the optical system to provide a relatively uniform beam that overfills the receiver or reflector and ensuring that the duct or stack attachment is stable, an instrument can be made to withstand the small changes in alignment that inevitably occur without substantial degradation in accuracy. Other instruments have included automatic self-aligning optical systems to resolve this potential problem. Further, by using a dual-channel (dual-wavelength) ratiometric measurement technique (measurement/reference), the effects of misalignment can be minimized. The calibration problem is the most difficult to solve. To calibrate the instrument under ideal conditions, one would fill the stack or duct with zero gas and several upscale gases in sequence so that they could be measured at the same path length and under the same temperature and pressure conditions as the normal sample. Even with these conditions, the sample would not be the same as stack gas since the dust and H_2O content would not be the same. But, since this calibration situation is not possible, various compromises have been incorporated. In some cases, a separate zero pipe is installed across the stack that allows the optical system to maintain a zero reference for the analyzer. Either a dual-beam or beam-switching mechanism must be used so that the system can interrogate either the normal stack optical path or the zero path. This pipe across the stack is cumbersome to install but does provide a zero reference and a clear path upon which gas cells can be superimposed for upscale checks. Gas cells inserted into the normal measurement path are difficult to quantify and/or calibrate since the stack gas background is nonzero and highly variable. Note that a completely valid check of the instrument calibration can only be done when the calibration gases are measured at the same temperature and pressure as the stack gas. In another case a slotted pipe has been used within a rotating sleeve that can be positioned so as to block the entrance of stack gases and allow the pipe volume to be periodically purged with zero air. The attachment of the two portions of the analyzer to opposite ends of the pipe also reduces alignment problems. However, the mechanical complexity of such systems results in other reliability problems. Others have tried a variety of indirect techniques for establishing simulated zero and span calibration values. These include using additional gas cells and/or filters superimposed on the stack background and using the resulting measurement information to calculate calibration check values. Light pipes (fiber optics) have also been used to bring a reference light beam around the stack to obtain a zero reference, and various attributes of the measurement system have been used to estimate zero and span values. All such techniques, however, fall short of providing a complete check of the normal measurement system under the same measurement conditions as the normal measurement, and accuracy is subject to question. It should be noted that it is quite possible to design a calibration system that yields acceptable calibration data, but the question must be asked as to whether the results of the calibration check indicate the performance of the real stack gas measurement system. The mechanical, optical, electronic, and/or software components involved in the calibration system may be somewhat different than those involved in the normal measurement system.

2.2.1.1 *Analyzers.*

Analyzers for such applications must be specifically designed to accommodate the associated measurement conditions. This results from the fact that across-the-stack systems must tolerate not only a substantial loss of light but also a variable light intensity that can result from variations in particulate density or alignment. If dust loading is high, sufficient light may be scattered, absorbed, and/or reflected that the remaining detected light is marginal for an accurate measurement of a gas of interest. Also, it is necessary to distinguish between the absorption resulting from particulate versus the gas of interest. Because of this, some systems use dual-wavelength differential measurements. In such cases a light intensity measurement is made at a wavelength where there is minimal absorption by the gases of interest and at another nearby wavelength where the absorption is near maximum. By ratioing the two measurements, the resulting concentration calculation can be made nearly independent of the particulate density, light level, and/or alignment. Since relatively large amounts of vapor-phase moisture are present in most applications of interest, gas filter correlation or similar detection techniques are often used in IR-based systems to reduce moisture interference problems to acceptable values. The UV systems avoid such problems. Liquid-phase moisture is very efficient at scattering light and may preclude across-the-stack measurements.

2.2.1.2 *Advantages and Disadvantages.*

One of the advantages of across-the-stack systems is that they provide an integrated measurement over the path of interest. In fact, they do provide a line-integrated average measurement that tends to reduce the problems associated with stratification But, to be most accurate, one would prefer an area-weighted average that would be representative of the total emissions. Further, when properly installed, both point and path measurements can provide accurate results in most applications. The EPA 40 CFR 60 and 75 regulations contain installation recommendations designed to ensure accurate measurements of total emissions. Reasonable qualitative (relative) accuracy can be achieved with well-designed instruments in selected applications, but quantitative (absolute) accuracy is difficult to prove. Such systems are typically air purged on both ends of the system to protect the exposed optical surfaces, and as a result there is no material in direct contact with the stack gases. With no sample train to maintain, analyzer maintenance is often minimal.

Disadvantages include the problems of alignment and calibration noted above. Without a zero pipe or similar capability, calibration integrity is always suspect. With a zero pipe, installation of the system becomes cumbersome. Additionally, these systems are subject to the problems of light scattering and absorption by coexisting particulate or entrained water droplets, which may preclude accurate and reliable measurements under such conditions. Further, these systems require air purge systems to protect exposed optical surfaces, all of which must be periodically maintained.

2.2.2 Probe-Type Systems

The first probe-type in situ system to achieve a high degree of market acceptance was the Westinghouse/Hagan zirconium oxide/fuel cell–based oxygen analyzer. This analyzer utilized a ceramic thimble approximately 6 in. long attached to the end of a typical 6-ft stainless steel probe that was inserted into the stack. Inside the ceramic thimble was located the zirconium oxide oxygen sensor in a heated environment. Stack gases diffused through the ceramic thimble and came into contact with the measurement side of the zirconium oxide cell and were measured in accordance with the Nernst equation, which describes the zirconium oxide cell response. Because of the diffusion mechanism involved in the transport of gases into the sensor area, these types of devices were called diffusion cell–based analyzers. It is important to note that there is no net flow in or out of the cell since diffusion brings the gas inside the cell in equilibrium with gases outside of the cell. As a result, the ceramic thimble does not become plugged in the course of time, like a similar filter in an extractive measurement system. The only exception to the no-net-flow rule is during calibration when calibration gas is flowing into the diffusion cell from a calibration tube within the probe. In this case the cell is slightly overfilled and there is a net flow of gas out of the cell. Such calibration has been designated as dynamic calibration since the calibration gas is measured at actual stack temperature and stack pressure, which may be highly variable. This provides the ultimate in calibration accuracy and integrity since the calibration gas is measured exactly the same as the stack gas. The biggest advantage of the diffusion cell is that it allows the near-free diffusion of stack gases into the measurement cell but effectively eliminates the transport of particulate into the measurement cell. This particulate filtering can be enhanced by proper choice of the pore size of the ceramic thimble material.

Typical pore sizes are from 2 to 5–10 μm depending on application. Later developments have used sintered stainless steel, which performs similarly, for the diffusion cell construction. A feature of the original in situ probes was also that they included a metal v-bar flow deflector located adjacent to the upstream-facing surface of the diffusion cell. This flow deflector separates the flow around the diffusion cell and creates wind shear on the sides of the diffusion cell, keeping it clear of

Figure 2.6   In situ oxygen analyzer. (Courtesy of Monitor Labs., Inc.)

particulate even in extremely dirty applications. These devices have been used on the upstream side of baghouses and electrostatic precipitators used with coal-fired boilers without any significant deterioration of performance. In some cases, the ceramic material must be protected from erosion by extending the flow deflector so that it prevents the direct impact of particulate on the sides of the diffusion cell. A variety of zirconium oxide–based in situ analyzers have been marketed using the same basic principles described above. They have become a de facto standard for combustion control applications and many EPA-required CEMS diluent-monitoring applications. Most have achieved the excellent reliability and low maintenance for which in situ analyzers have become known. A typical in situ probe-type analyzer is shown in Figure 2.6.

2.2.2.1 Basic Characteristics. The same basic techniques employed above have been incorporated into other gas analyzers. The most classical instruments of this type are the SM81XX and EX4700 analyzers developed originally by Lear Siegler, now Teledyne Monitor Labs. The SM81XX uses a second-derivative UV measurement technique to measure the spectroscopic absorption of SO_2 and NO. This instrument is pictured in Figure 2.7, which shows the transceiver and short probe, a J-box used for external gas and electrical connections, and the instrument controller. The second-derivative measurement technique provides a narrow-band

Figure 2.7 In situ So$_2$ and NO probe-type analyzer. (Courtesy of Monitor Labs., Inc.)

ratiometric (measurement/reference) measurement that is insensitive to light level variations and broadband (nonstructural) interfering gases. The EX4700 was similar in approach but used NDIR (nondispersive infrared) multiwavelength techniques to measure CO_2, CO, and H_2O. Instead of placing the sensor in the diffusion cell, these instruments use the diffusion cell to contain the optical measurement cavity. In such optical instruments, a transceiver is attached to the out-of-stack end of the stainless steel probe. The transceiver contains the light source and detector and a means for projecting the interrogating light beam down the probe through the measurement cavity and back. Weather covers are used to protect the transceiver from the elements. The measurement cavity is typically defined by an optical window on the transceiver side and a mirror or retroreflector on the stack side. As stack gases permeate the volume contained in between these two ends of the cavity, they absorb optical energy characteristic of those gases. That absorption is sensed in the transceiver and ultimately converted to a measurement of the concentration of the gases of interest. Since these measurements are made in place with respect to the total volume of stack gas, they are wet-basis volumetric measurements.

Since most optical/spectroscopic instruments are essentially molecule counters (i.e., they sense the number of molecules of the gas of interest contained within the measurement cavity), the measurements must be corrected for the effects of gas density as expressed by the ideal gas law. In addition, they are spectroscopic devices that conform to a limited degree to Beer's law and are therefore dependent on the temperature and pressure as they affect the absorption coefficient for the gas and wavelength of interest. As a result, gas temperature and pressure must be measured and used in the measurement compensation scheme if accurate measurements are to be obtained over the wide range of temperatures encountered in stack gas monitoring applications.

To get higher sensitivity, longer measurement cavities have been incorporated into many instruments. Measurement cavities up to a meter in length have been provided in some instruments. If such long cavities are completely enclosed with filter material, some portions of the diffusion cell filter area may have to be blinded to reduce the calibration gas requirements to a reasonable value. Further, they may become so long in comparison to the stack diameter that the measurement becomes a path measurement instead of a point measurement as defined by EPA specifications. The installation requirement for a point measurement system, the point being less than 10% of the stack diameter, is only that the point must be at least 1 m from the wall, which eliminates wall effects. The installation requirement for path measurement systems, the path being greater than 10% of the stack diameter, is that 70% of the path must be contained within the inner 50% of the cross-sectional area of the stack. See EPA 40 CFR 60, Appendix B, Performance Specification 2, for more details regarding installation criteria.

2.2.2.2 Analyzers. Analyzers for probe-type systems have traditionally been either the zirconium oxide–type of sensor or the spectroscopic-based devices, which may use either UV or IR spectral regions. Because of the diffusion cell construction, the optical systems do not have to accommodate the wide variations of particulate density and light level that occur in across-the-stack systems. However, to maximize the time between maintenance intervals, most analyzers for probe-type systems are designed to accommodate a slow deterioration of light level associated with the buildup of a slight film of particulate that eventually deposits on the exposed optical surfaces. When the light level decreases to a given level, some analyzers provide a warning that maintenance is required. The maintenance-free interval may be as long as 6–12 months.

Controllers for such instruments are typically microprocessor driven and provide the advantages of easily programmed calibration cycles; measurement averaging, correction, and processing; and automatic diagnostics. In some cases a single controller can be used with multiple instruments, thereby reducing cost and panel space requirements.

The classical problems of potential H_2O and CO_2 interference are particularly applicable to IR in situ systems since the measurements are wet basis. The UV-based measurements avoid such potential problems.

2.2.2.3 Advantages and Disadvantages. The classical advantages of in situ probe-type gas analyzers include the ability to handle very dirty gas applications; minimal maintenance due to the diffusion cell technique, which uses, for example, no sample probes, sample lines, pumps, or gas conditioning; and unquestioned sample integrity. Further they have the ability to scale the measurement cavity size for required sensitivity (for spectroscopic instruments) and dynamic gas calibration, which checks the complete measurement system and measures calibration gases at the same temperature and pressure as the stack gas, thereby assuring unqualified calibration accuracy. Reliability has been excellent for most types of analyzers/sensors. Some analyzers can share a remote control unit among multiple analyzers, thereby minimizing requirements for panel space in the control room. Measurements are typically corrected for gas temperature and pressure, which is an inherent requirement with variable stack operating conditions. Being wet-basis measurements, they are easy to combine with wet-basis flow measurements to provide the mass flow rate of emissions in pounds per hour. Single-port on-stack attachment/insertion simplifies installation.

The disadvantages of such measurement techniques include the fact that they must be installed in a reasonably accessible location since the transceiver is located on the stack with the probe. If multiple gases are measured with a single measurement cavity, the cavity length must be a compromise to obtain required sensitivity for all gases to be measured. The diffusion cell may blind if used with gases containing substantial entrained water. Further, it may be difficult to handle high stack temperatures (above 500°F) due to gasket material limitations in glass-to-metal probe seals. Measurement cavity lengths are limited given a typical 6-ft probe, making very low measurement ranges (single-digit ppm levels) difficult for some spectroscopic analyzers. Systems require experienced and/or trained instrument technicians for proper service.

2.3 KEY APPLICATION DIFFERENCES

First, it should be noted that no single measurement technology is "best" for all applications. Each measurement approach is better suited to some applications than others. The selection of the best system must be determined for each application based on site-specific conditions and monitoring requirements. The following generalities must be considered as such, since there are always exceptions to the general rule.

2.3.1 Conventional Extractive Systems

Such systems can accommodate high-gas-temperature and wet-gas applications and are best for very low concentration measurements. They are heavily favored if dry-basis measurements are required. They have the least sophisticated maintenance requirements, but significant routine maintenance may be required in dirty-gas or high-acid gas concentration applications. They are the best choice if analyzers can be shared with multiple sampling points and often require a corresponding gas moisture measurement if combined with flow measurements. They are the easiest systems to reconfigure for additional gases or different concentrations after field installation.

2.3.2 Hot, Wet Extractive Systems

Such systems are a virtual necessity if water-soluble acid gases are to be measured. Wet-basis measurements can easily be combined with gas flow measurements. Maintenance is more difficult due to use of hot components throughout the sample train and analyzer measurement cavity/cell. They are complex systems that generally are not justified for use with gases that can be measured using other techniques.

Selection of analyzers that can accommodate the hot sample train within the analyzer is limited. The complexity of such systems results in relatively higher purchase prices.

2.3.3 Dilution Extractive Systems

Such systems are great for high-pollutant concentration measurement applications. They significantly reduce sampling handling and transport problems. They can be used with ambient air or similar derivative analyzers that offer excellent performance and reliability. The systems are complex, requiring significant periodic maintenance and trained personnel. They are inherently accurate when measuring emission rates in lb/μBtu, which are independent of the variations in dilution ratio. They do require considerable sophistication to obtain highly accurate individual gas concentration measurements. Very low concentrations may not be measurable when diluted using a high dilution ratio. Air cleanup packages for dilution air supply must provide very high purity gas and are complex and require periodic maintenance. Wet-basis measurements can easily be combined with gas flow measurements. The potential use of unheated or freeze-protected sample lines is a major benefit compared to conventional extractive systems. Care must be exercised when condensed water may be present in the sample gas.

2.3.4 Across-the-Stack Systems

Such systems are good for trend monitoring applications and those installations where a high degree of stability and accuracy is not required to be demonstrable on stack. With the potential limitations of simulated calibration check systems (which do not utilize known stack level calibration gases measured at stack temperature and pressure), their use in high-accuracy compliance applications has been limited. An across-the-stack zero pipe may be required to maximize accuracy. They may be advantageous in highly stratified gas stream applications. Large stack diameters, high particulate loading, high water content, and high stack temperatures may be troublesome. Sensitivity is determined by stack diameter unless extended mounting flanges are used to reduce path length. Entrained water droplets in the gas stream are particularly difficult to accommodate. The systems are often capable of multigas measurements with a single analyzer. Protective air purge systems require periodic maintenance. Equipment is sophisticated, requiring trained personnel. Wet-basis measurements can be easily combined with gas flow measurements.

2.3.5 Probe-Type In Situ Systems

Such systems excel in high-particulate-concentration applications. They require minimal routine maintenance and have long maintenance-free intervals, but maintenance must be performed on the stack where the analyzer/probe is located. Most changes in measurement range are accommodated in optical instruments by straightforward measurement cavity and/or probe replacement. Unquestionable sample integrity and an excellent calibration technique, which measure known calibration gases at stack temperature and pressure, assure highly accurate measurements. They have single-mounting ports for simple installation. Most optical analyzers have multigas capability. Very low pollutant concentrations, entrained condensed water in the sample gas, or very high stack temperatures may preclude use of such systems. Ultraviolet-based systems easily accommodate high H_2O vapor-phase concentrations. Wet-basis measurements can be easily combined with gas flow measurements.

2.4 GENERAL PRECAUTIONS

One of the most common causes of unreliable gas measurement systems and associated high maintenance costs is the lack of a reliable high-quality supply of instrument air. Instrument air is used as a zero calibration gas source in some applications, as a source of diluent air for dilution systems, as a backpurge supply for extractive probes, and as a purge air supply for many analyzers. Analyzers are often purged when a large part of the optical path in spectroscopic instruments is outside the actual measurement path and is not sealed. In such cases, the exposed part of the optical path must be purged to ensure accurate measurements. This is of particular importance if ambient air is likely to introduce gases that are being measured, interfere with gases being measured, have constituents that may condense out of the gas stream under conditions seen by the sample stream, or have constituents that may chemically react with sample train components or gases. Instrument air must generally be dry to dew point of near $-40°C$, be free of particulate, with no entrained water, oil, or hydrocarbons, and be free of gases being exhausted or vented from the process or smoke stack stream. It is not sufficient that this is true most of the time—*these conditions must be met all the time.* Occasional lapses in the instrument air supply purity can contaminate sample lines, valves, pumps, filters, and most importantly analyzers. Repair of such equipment under these conditions can be very expensive. When there is any doubt about the consistency of the instrument air supply, an air cleanup package should be installed as a preventative measure.

It is also very important that personnel involved in the maintenance of such gas-measuring equipment be trained so that their maintenance and diagnostic procedures are consistent with factory recommendations. When people are not trained on the particular equipment for which they are responsible, there is a significant risk that improper adjustments and component repair and/or replacement will impair the reliability and accuracy of the system.

Spare parts are, of course, a must for those components that are most likely to fail or need replacement. Downtime of the system is often very inexpensive, and adequate spare parts are required if high uptime is of concern.

Then, there are those concerns about the location of the sample extraction point(s) or sample measurement with in situ equipment. Stratification of gases can cause system measurements to differ from stack-sampling measurements.

This occurs when the measurement system samples from a point or short path that is not representative of the complete cross section of the duct or stack. It is important to maximize the upstream and downstream distances of the measurement point/path from flow disturbances and minimize air inleakage into the gas stream of interest. Further, it is always helpful to locate the analyzer portions of the system in areas where ambient temperatures are not at the extremes of their specifications and where the temperatures are relatively consistent. Additionally, installation locations should be easily accessible for maintenance personnel, relatively free of shock and vibration, and not located next to high-power RF transmitters or other sources of major electromagnetic interference. Also, special precautions are necessary in the CEMS design if stack gases are likely to include entrained water, high particulate loading, gas components that may condense out of the stream under conditions encountered in the sample train, or large variations in temperature or pollutant levels. Consult with the manufacturer when in doubt.

REFERENCES

1. J. A. Jahnke, R. P. Marshall, and G. B. Maybach, *Pressure and Temperature Effects in Dilution Extractive Continuous Emission Monitoring Systems—Final Report*, Electric Power Research Institute, Palo Alto, CA, 1994.

2. G. F. McGowan, A Review of CEM Measurement Techniques, paper presented at Northern Rocky Mountain ISA Conference, ISA, May 1994.

3. J. A. Jahnke, *Continujous Emission Monitoring*, 2nd ed., Wiley, New York, 2000.

4. R. L. Myers, Field Experiences Using Dilution Probe Technniques for Continuous Source Emission Monitoring. Transactions—Continuous Emission Monitoring—Advances and Issues, *Air Pollut. Control Assoc.*, 431–439 (1986).

5. J. A. Jahnke, Eliminating Bias in CEM Systems, in *Acid Rain & Electric Utilities–Permits, Allowances, Monitoring & Meteorology*, Publication VIP-46, Air & Waste Management Association, Pittsburgh, PA, 1995.

3

VALIDATION OF CONTINUOUS EMISSION MONITOR (CEM) SYSTEM ACCURACY AND RELIABILITY

Todd B. Colin, Ph.D.

Eastman Kodak Company
Rochester, New York

Environmental Instrumentation and Analysis Handbook, by Randy D. Down and Jay H. Lehr
ISBN 0-471-46354-X   Copyright © 2005 John Wiley & Sons, Inc.

3.1 INTRODUCTION

Online measurements are becoming a routine part of industrial culture. They are used for process control, process verification, health and safety systems, and emissions monitoring. As people become increasingly dependent upon those measurements to run and manage their processes, it becomes critical that the quality of those measurements be assessed and documented.

3.1.1 What Is System Validation?

System validation is the process of ensuring that the analyzer system does what it was intended to do. This includes validation of every component from sample to data. The two parameters most important to validate are accuracy and reliability. Accuracy usually requires two parts: accuracy and precision. Consider a dart board. Accuracy is a measure of how close to the bullseye you get, and precision is a measure of how close together the darts are. It is possible to have high precision and low accuracy and vice versa. Both must be characterized to determine the amount of confidence in a particular data point. Reliability is a measure of uptime of the system. All systems require maintenance and calibration, during which time the system cannot be making measurements. A good analyzer system should have uptime more than 99% and minimal unscheduled maintenance events.

3.1.2 Why Validate an Analytical System?

There are a number of reasons why an analyzer system should be validated. First and most important is that the user of the data needs to have confidence in the data generated by the equipment. Whether the application is driven by process control, health and safety, or environmental issues, reliable data are critical to make good decisions and take action. The second reason to validate an analytical system is to document the system capabilities for compliance with International Organization for Standardization (ISO), quality system, or environmental regulations.

3.1.3 Types of Validations

Three main types of installations require validation: new installations, replacement of existing installations, and installations that fall under government regulation. All three are similar, but each has its own unique aspects that must be addressed.

3.1.3.1 New Installation. New installations start with a blank canvas and allow the system designer the most flexibility in system design. This can be advantageous as state-of-the-art technologies can be employed to make the validation process easier and more reliable. Disadvantages can include a relative lack of historical data on the process being measured and good reference methods in place to use as a comparison.

3.1.3.2 Replacement of Existing System. Existing systems are often more challenging from a design aspect since it is necessary to work around an existing system. The advantage is that there is a measurement in place that can be used to cross over the new system, making the validation process easier.

3.1.3.3 Government Agency–Regulated Installations. Government-regulated installations are either new or existing installations but have the added burden of fulfilling the documentation requirements for a government agency. The U.S. Environmental Protection Agency (EPA) and many state environmental agencies will accept good laboratory practice and quality assurance programs as the foundation of the required documentation. The EPA publishes methods that are approved for use in compliance situations.* Method 301 covers the validation of emissions testing.

3.2 VALIDATION OF NEW CEM INSTALLATIONS

The best approach to validation of a new analytical system is to break it down into its component parts and then begin validating the function of each component. Consider the example analyzer system shown in Figure 3.1. A process is being monitored at three independent points (1–3). A sample is drawn from the process, conditioned by filtration, and then provided to a multiplexer. The multiplexer selects the stream to be measured and provides it to the analyzer. The analyzer can subsequently measure the sample and report its findings through the data system to the process control computer, which can then use the available data to control the process. In this example, system validation can be divided into three parts, beginning with the analyzer, followed by the data system, and ending with the sampling system. In some cases, the analyzer technology may require the combination of the analyzer and data system validations into a single step.

3.2.1 Analyzer Calibration and Verification

The first component that the analyzer system must validate is the analyzer itself. There are a wide variety of analytical technologies used for online measurements, each having their calibration requirements. The most common types of analyzers

---

*EPA methods are available from the Emission Measurement Technical Information Center (EMTIC), Mail Drop 19, U.S. Environmental Protection Agency, Research Triangle Park, NC 27711.

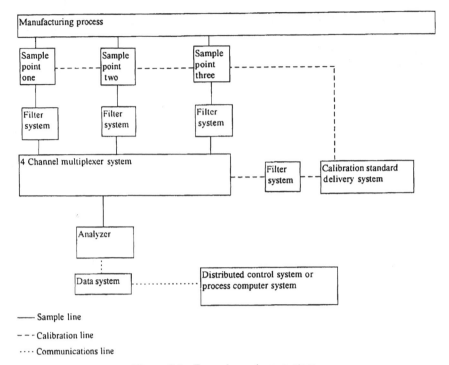

Figure 3.1   Example analyzer system.

can be divided into two subclassifications: simple and advanced. A simple analyzer consists of a sensor that responds to a chemical or group of chemicals and is calibrated by setting a zero and a span value. Total hydrocarbon analyzers, toxic gas monitors (phosphine, hydrogen cyanide, etc.), and chlorinated solvent detectors are examples of simple analyzers. Advanced analyzers are multicomponent analyzers capable of distinguishing a number of analytical components from more complex matrices. These analyzers require more complex calibrations. Gas chromatography (GC), infrared (IR) spectrometry, near-infrared (NIR) spectrometry, and mass spectrometry (MS) are examples that use advanced analyzers. The complexity of the calibration will depend upon the complexity of the matrix of components to be resolved. Gas chromatography systems typically require one or two mixtures that contain known concentrations of the analytes in a matrix similar to the process. A linear regression model is usually generated to predict concentration from the area count for each analyte peak. A good resource for process GC is the text by Annino and Villalobos.[1] Infrared and NIR calibrations can be similar to the GC case if the spectra of the analytes and sample matrix do not overlap. If overlap is present, a multivariate approach is required. The two most common multivariate calibrations are partial least-squares regression (PLS) and multiple linear regression (MLR). In both cases, a series of calibration mixtures is required that brackets the concentrations of all the analytes and captures the total variability of

Actual versus predicted plot

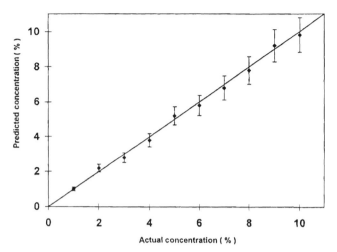

Figure 3.2   Example actual versus predicted plot.

the sample matrix. Martens provides a good treatise on multivariate calibration.[2] Mass spectrometry calibrations use a matrix inversion (MI) technique that is multivariate but only requires that the analyzer be calibrated on a binary mixture for each component and one mixture similar to the process matrix. The binary mixture is defined as the analyte in the sample matrix, typically air or nitrogen. Each type of calibration is prone to its own pitfalls; however, the multivariate PLS and MLR techniques can be the most difficult to validate since the total process variability must be captured in the calibration set.

In general, the technique to evaluate the validity of a calibration is to compare the measured concentrations of a set of standards to their actual concentrations. Figure 3.2 shows an example of an actual-versus-predicted plot. The predicted values are plotted on the y axis, and the actual values are plotted on the x axis. If the calibration is exact, all points will fall on a line with a slope of 1.0. The amount of error from this line is a measure of the accuracy of the measurement. The two forms of error that cause deviation from the line are bias or systematic error and random error. There are two types of bias: positive and negative. Positive bias occurs when all the points are below the line. In some cases it is possible to generate a correction factor to eliminate a bias in the measurement. Random error is the precision of the measurement and is evenly distributed above and below the line. The analytical precision of the measurement should be determined by analyzing the same standard over a period of time. The period should reflect the magnitude of time the analyzer is required to be stable. The samples should be run at times when the full process variability can be assessed. For example, the time of day should be varied to account for daily temperature variation. Figure 3.3 shows an example of a control chart. These plots are useful to assess the repeatability of an analyzer and indicate that the instrument is about to fail.

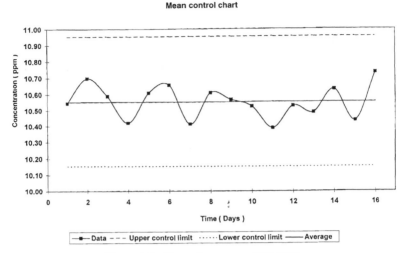

Figure 3.3   Example mean control chart.

The plots contain several elements. First, there is a trend of the measured values. Second, a line is drawn representing the mean of those values. Third, two lines representing the upper and lower control limits are plotted from the calculated data. The rules of statistical process control can be used to evaluate any point or series of points to determine if the die analyzer is still in control or, in analytical terms, stable. A full treatise of process control statistics is beyond the scope of this text. However, a number of books are available on the subject, including the text by Kaoru Ishikawa.[3]

3.2.2   Sampling System Verification

Once the analyzer has been validated, the next task is to validate the sample system. An analyzer is only as good as the sample it receives from the process. Several methods are available to validate the sample system integrity. If the analyzer is a mass spectrometer, isotopic spiking can be employed. This involves introducing a known quantity of isotopically tagged analyte at the sample point that is subsequently measured on the analyzer. The amount of recovery of the spiked sample is then calculated. All analytical techniques can employ a comparison to either a validated test method or a validated standard. Comparison to a validated test method is the best method of validation. Samples can be extracted from the same sample point at the same time and are run on the analyzer system, and the valided test method results can be compared statistically. A statistical F test is used to determine if the precision is acceptable. The bias between the analyzer system and the validated test can be evaluated by calculating the t statistic. This will identify a bias for any given level of confidence. The confidence level is typically 80%.

When a validated test method is not available, the next best option is to introduce a calibration standard at the sample point in the process and then verify that the analyzer returns the correct value for the standard. Once the analyzer has been validated, any discrepancies must be in the sample handling system. If the nature of the process is such that it is not possible to introduce a standard at the sample point in the process, then a point as close to the sample point as possible should be chosen. The remainder of the sample system should then be validated by verification of mechanical integrity. This would include a pressure test of sample lines and any sample probes used in the process.

Once the sample system is validated, it is a good idea to validate the location of the sample point. For example, in a large exhaust duct, measurements should be made across the diameter in both the x and y directions to identify any stratification that might exist. If the process stream is stratified, another more suitable location should be used. In some cases, alternate sample locations are not possible. This situation requires a selection of the part of the stream that is deemed to be most representative.

3.2.3 Data System Verification

Classical data system validation involves introducing a standard electronic reference signal to the electronics that produces known results. This type of validation works well on simple analyzers and chromatographs, where detector signals can be simulated easily. Near-IR, mid-IR, or mass spectrometer signals are not easily simulated. Fourier transform instruments are nearly impossible to simulate. In these cases, the data system should be validated as a part of the analyzer. Quality system documentation should reflect the integration of validation.

3.2.4 Validating the Entire System

Once the components have been validated, it is useful to validate the system end to end. A standard should be introduced at the sample point and the end data checked to make sure that the system provides the correct result. This also allows an opportunity to verify the system lag time, which is particularly useful in process control applications. The lag time between the time the sample passes the sample point and when the analyzer actually reports the result can severely impact closed-loop control algorithms such as proportional-integral-differential (PID) routines. If control parameters such as lag time are disturbed, the control system must be retuned to the process, which can be a difficult task.

3.3 VALIDATION OF CEM REPLACEMENT INSTALLATIONS

3.3.1 Analyzer Calibration and Verification

When a functional analyzer is to be replaced, it is necessary to install the new system in parallel with the existing system. This serves two purposes. First, it limits the analyzer downtime since the new analyzer is installed and validated prior to the removal of the old system. Second, it offers an opportunity to compare the two online systems to make sure that the results are consistent. The new analyzer is validated the same way as a new analyzer installation prior to beginning the crossover validation. Figure 3.4 illustrates how an analyzer system can be installed to allow for crossover testing.

3.3.2 Sampling System Verification

The sampling system is the first component that may be fundamentally altered by the co-installation of a new and existing analyzer system. If an extractive sampling system is employed by the existing system, it can be used to support both analyzers provided the following conditions are met. First, the system must be in good condition. Second, the system must have sufficient sample capacity to support both analyzers during the crossover. Third, addition of the new analyzer must not disrupt the existing system longer than the allowable downtime for the existing system. If these conditions are able to be met, this is an excellent way to cross over two analyzers. Since the sample system is common, the sample provided to both analyzer systems is always the same.

In cases where the sample system cannot be used in common, the new sample point should be chosen close to the existing sample point. The proximity of the sample points will provide a similar sample to both instruments simultaneously.

3.3.3 Data System Verification

The data system for a replacement analyzer system is done the same way as for a new analyzer system. The only difference is that the data from the new system is compared to the existing system. The task is easy if the results are the same; however, if discrepancies are discovered, a great deal of good detective work is usually required to determine which system, if any, is correct.

3.3.4 Validating the Entire System

As with a new installation, standards need to be introduced at the sample point to validate the entire analytical system. The system lag time from both the existing system and the new system should be measured, as it is necessary to correct for lag time when comparing online results.

3.3.5 Crossover Validations to Existing Analyzer

Crossover is used to verify that the new system responds to the process the same way as the existing system. Calibration standards typically do not reflect the full range of variability seen in the process. Crossover is required to make sure that

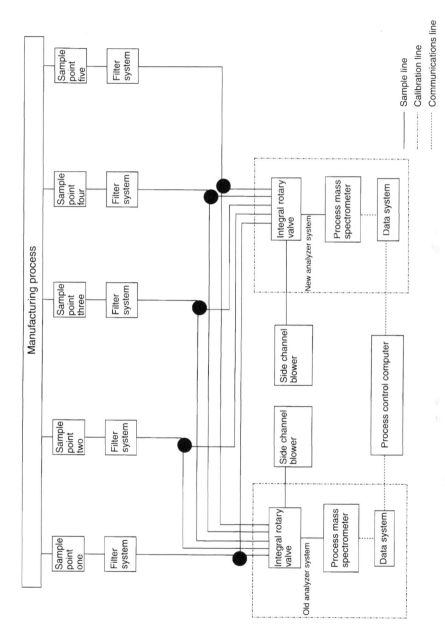

Figure 3.4 Example analyzer system installed for cross over testing.

there are no unusual conditions that result in analytical differences between the analyzers. Typically it is good to run analyzers side by side two to three times longer than they are expected to be stable so that the effects of recalibration can be included in the crossover test. The same techniques used to compare the analyzer system to a validated test method should be used to compare the new and old analyzer systems. The old analyzer system serves as the validated test method in this case. The F and t statistics should be monitored to identify differences. When differences are found, they should be documented and, if an assignable cause exists, corrected. If no assignable cause exists, then the new system's performance should be documented so that the change will not affect any process decisions that are made based on those data. In some cases it may be possible to calculate a correction factor to compensate for a system bias.

3.3.6 Practical Illustration of Analyzer Validation

The best way to illustrate the principle of validation is to consider the example analyzer system in Figure 3.4. This is a replacement of an existing analyzer system that was previously validated using the EPA method 301.[4] Therefore, the old analyzer system can be considered to be a validated test method. First, the new analyzer is installed in parallel with the old analyzer. The sample points and filtration systems have adequate capacity to supply sample to both analyzers at the same time, which eliminates the need to a colocated sample point. Using the same sample point for both analyzers ensures the sample supplied to each will be identical. Each analyzer is equipped with its own rotary sampling valve and high-volume sample pump. Once the two systems are installed in parallel, the new analyzer is calibrated and validated independently using the gas standards used for calibration of the old analyzer system. The analyzer employed here is a mass spectrometer, which required a combination of the analyzer and data system validations. Once the new system is calibrated and meets the specifications of 5% relative standard deviation for precision and 5% relative bias, the crossover testing can begin. The two analyzers are placed on line simultaneously, and the standard gas introduced at the sample point is analyzed. The analyzers are allowed to collect a data point every hour over a 9-h period. The F test is then conducted per the calculations in EPA method 301. In this example the F value was 0.8. Since the critical value for F is 1.0, the precision of the new analyzer system is not statistically different from the old analyzer system.

The bias is then addressed by calculating the t statistic using the calculations outlined in method 301. In this example the critical value for t with eight degrees of freedom is 1.397. The t value calculated for the data was 1.102. Since the t value is less than the critical value, the bias is not significant. Therefore, the new analyzer system meets or exceeds the capabilities of the old analyzer system.

The calibration frequency of the old analyzer system was every 24 h. Therefore, the analyzers were run side by side on the process for four days to assess the long-term stability of the systems. Figure 3.5 shows the trend chart of the data. Note that

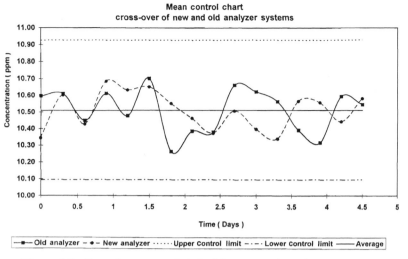

Figure 3.5 Example mean control chart from example analyzer cross-over.

drift does occur but that the two analyzers remain within the 5% relative error specification.

3.4 VALIDATIONS INVOLVING GOVERNMENT AGENCIES

3.4.1 Types of Government Regulation

Government regulation can come from a number of different sources. The Food and Drug Administration (FDA), the EPA, and the Occupational Safety and Health Organization (OSHA) are a few examples of organizations that might require documentation validating the use of process monitoring equipment. These organizations and their state counterparts have requirements that must be met. The best way to achieve compliance is to work directly with the organization in question. The EPA method 301 is a detailed method for validation of analytical methods. It covers the validation of bias, precision, sampling system, reference materials, and performance auditing as well as provides a detailed listing of the appropriate methods for calculating statistics during the validation process.

3.4.2 Documentation Required

A good foundation for documenting the validation of an analyzer system is to follow good-quality practices. The ISO guide 25 contains an excellent platform from which to build a quality system.[5] Records should include all calibration and maintenance records, validation test records, authorized users of the system, and standard operating procedures.

REFERENCES

1. R. Annino and R. Villalobos, *Process Gas Chromatography: Fundamentals and Applications*, Instrument Society of America, Research Triangle Park, NC, 1992.

2. H. Martens, *Multivariate Calibration*, Wiley, New York, 1989.

3. K. Ishikawa, *Guide to Quality Control*, Quality Resources, White Plains, NY, 1991.

4. Environmental Protection Agency (EPA), *Method 301—Field Validation of Pollutant Measurement Methods from Various Waste Media*, Emission Measurement Technical Information Center (EMTIC), EPA, Research Triangle Park, NC.

5. International Organization for Standardization (ISO), *ISO Guide 25*, ISO, Geneva, Switzerland, 1990.

4

INTEGRATION OF CEM INTO DISTRIBUTED CONTROL SYSTEMS

Joseph A. Ice

Control Systems Engineer
Cypress, Texas

Environmental Instrumentation and Analysis Handbook, by Randy D. Down and Jay H. Lehr
ISBN 0-471-46354-X   Copyright © 2005 John Wiley & Sons, Inc.

4.1 INTRODUCTION

Continuous-emission monitoring (CEM) is being used and installed in order to monitor atmospheric emissions from chemical processing plants, emissions from combustion units, and many other types of process plants. Industry, in the United States and many other countries, has increased the monitoring and reporting requirements because of governmental regulations. The U.S. Environmental Protection Agency (EPA), regulations on compliance assurance monitoring (CAM), requires or encourages the use of CEM systems. While the CAM rule, at this time, does not require existing plants to install CEM systems, it does require monitoring of many operating parameters to ensure the plant is "compliant." A new plant or equipment can expect that some type of CEM system will be required. To be "in compliance," the operating parameters and their relationship to the actual emission rates must be proven and periodically reproven through a series of point source sample testing.

The first step toward successfully utilizing a CEM is to determine which CFR (U.S. Code of Federal Regulations), state, or national regulation(s) apply. The ultimate objective is to utilize the CEM information in a format that is usable by plant operations and the plant environmental staffs. Also the information should be formatted so that its submittal to the appropriate regulatory agencies will be accepted. In many applications this will be the most difficult part of the job. The two major federal regulations are 40 CFR 60 for the process industry and 40 CFR 75 for the power industry.

The next step to a successfull project is to review the federal agency requirements, state or province agency requirements, and the operating company objectives, users, and capabilities. The basic assumption in our discussion is that the facility and CEM system is in North America. However, the same rationale applies to projects for the rest of the world. Up to now there has been no mention of analyzer types and distributed control systems (DCS). The omission has been deliberate. The initial steps are to know the plant permit requirements from the local regulatory agency. Then determine the EPA or equivalent agency requirements.

4.2 AGENCY PERMIT AND REPORT REQUIREMENTS

Now the task is to read and interpret the permit and understand what is required. Expect to find many new monitoring requirements from the existing plant or equipment. If one is lucky, the new operating parameters to be monitored will already be included in the existing DCS. This then establishes the first technical requirement for the CEM. The communications between the DCS and the CEM will need to be bi-directional. The communications requirements will be discussed later. At this time the task is to determine the operating parameters required by the permit. As an example, one plant added 4 furnaces to the existing 12 furnaces. The permit required a CEM for the 4 new furnaces. However, for the agency report, all of the furnaces needed to be monitored and included in the report. A 1-h average furnace-firing rate (flow × heating value) was required for all 16 furnaces. Then a rolling 12-month average firing rate was required for all 16 furnaces. For the first 12 furnaces, the firing rate exceeding the 1-h maximum and the firing rate exceeding the annual average limits were also required for the report. All of the existing parameters now need to be retained as data points for 2 years. This topic will need to be revisited later when the data acquisition system (DAS) is discussed.

At this point let us determine the permit submittal requirements to the regulatory agency and how the report will be submitted. Will the report be submitted daily, weekly, monthly, by exception, or what? Can the report be mailed, faxed, e-mailed, or sent by modem? Several states require a daily modem dial-in, usually just after midnight. This means the state gets to see the information from the raw data. If there is a malfunction in the system, the report should flag the problem. Also a primary concern is CEM availability time. Remember that the agency can visit the facility at any time and look at any data at any time. Who has the master time, is it the DCS or the CEM? Do you want to explain to a state inspector why the data saved in a DCS history module is 15 s or 3 min different from the time stamp on the CEM? The personal computer's (PC) clock is inherently more unstable than the time function in a DCS. This is an introduction to a later topic of time synchronization between the CEM and DCS and which one should be the master clock.

Let us also determine the various alarms required for the report. CEM hardware, calibration gas, process, PC communications, and DCS communication are some of the alarm points that may be necessary. Do not forget to include a data point to show that any of the new emission sources are on. Now all of the operating parameters can be added to the report format. Is the fuel's Btu content needed? Will the hydrogen content be necessary? The old operating parameters in the DCS for flow, firing rates, dry fuel factor, and fuel high heating factor could be some of the data that need to added to the report. If the requirement is to analyze CO, NO_X, and O_2, then the following analyzer points are needed: CO in ppm, corrected CO in ppm, NO_X in ppm, corrected NO_X in ppm, O_2 in percent, and exhaust stack flow rate. Now the report is ready for the difficult calculation task of going from the analyzer data in ppm and converting to the units the regulatory agency wants on the report. For example, CO in pounds per hour and also pounds per MMBtu and NO_X in pounds per hour and pounds per MMBtu are the required units. Calibration information needs to be included such as the daily zero and span, the quarterly cylinder gas audit (CGA), and any corrective action taken with the CGA if there are exceedances of ±15%.

4.3 CEM TECHNICAL REQUIREMENTS

4.3.1 Analyzer Specifications

We can start to discuss the requirements for the new analyzer equipment that the CEM will need. The most common types of CEM analyzers are NO_X, CO, CO_2, O_2,

SO_2, HCL, TRS, and opacity. The permit will define the exact types of analyzers required. All of the analyzers will be required to meet the EPA reference method testing and performance standard tests for the gas being measured. There are other CRF volumes that the permits may require. The gas analyzers can be selected for technical determinations of the sample detection method in accordance to the sample gas. The permit will usually specify a sample cycle time. This information will be important for the analyzer specifications. Also know the sample line lengths. The line length will be important in answering several questions.

However, the following list shows the most used analyzer types for the process industry.

Volume	Method	Gas	Type of Measurement
CFR 40 Part 60	10	CO	Infrared
CFR 40 Part 60	3A	O_2	Zirconium oxide or paramagnetic
CFR 40 Part 60	7E	NO_X	Chemiluminescent
CFR 40 Part 60	6C	SO_2	Ultraviolet
CFR 40 Part 60	3A	CO_2	Infrared
CFR 40 Part 60	23	HCL	Infrared
CFR 40 Part 60	15A/16A	TRS	Ultraviolet

One question that should be considered is if the analyzer can be shared between two or more samples. First, find out what is the permit sample cycle time requirement. Then make a quick check of the sample line length and the associated sample lag time. If the analyzer equipment cycle time and sample lag time are sufficiently shorter than permit cycle time, then the possibility of using one set of analyzers for two or more sample points should be considered. This will save equipment costs and calibration times. All of the analyzer hardware time-share capability is dependent upon the permit.

The next items to that should be technically defined are the sample system and sample probe. Here it is better to take a very conservative approach. As an example, if there are four furnaces to be fitted with a CEM systems, then have one sample probe and one sample system for each furnace stack. The rationale for this configuration is that if the permit agency does not approve the time sharing of the analyzers, then the only parts that will need to be added will be the analyzers.

4.3.2 Sample System

The sample system should be designed to deliver a filtered but otherwise unmodified sample to the individual analyzers within the CEM system. It should be designed for long-term, low-maintenance operation and contain no moving parts. The sample acquisition and condition can be achieved by proper operation of the following subsystems: (1) probe, (2) sample transport, and (3) sample system. The probe has to be designed to allow for extraction and filtering of the gas sample. The probe needs to have as a minimum the following features: a 0.5-μm filter element, a heater jacket for the filter element, provision for cleaning or replacing the filter element without having to remove the probe, provision for back-purging the filter element, stainless steel or similar material for the enclosure construction, and a NEMA 4X weather enclosure.

The sample transport system should be designed to transport the unconditioned sample from the probe location to the CEM analysis enclosure. The sample transport system should be a two-tube umbilical (sample and calibration) with the following design features: All sample line core tubing should be Teflon FEP tubing with a $\frac{3}{8}$-in. o.d. for the sample, a $\frac{1}{4}$-in. o.d. for the test gas, and a 0.40-in. wall thickness. The sample transport should be heat traced to prevent condensation of the unconditioned sample. The sample transport umbilical temperature must be monitored and a fault signal provided for abnormal temperatures. The sample transport umbilical should be polyvinyl chloride (PVC) or polypropylene jacket. All compression fittings and interconnects should be Swagelok (316SS or Teflon) or equivalent. Most analyzer manufacturers use only one vendor for these types of fittings. If the manufacturer is ISO 9000 certified, the paper work required to track two different fitting vendors is so monumental that the manufacturer would not be able to provide a product at a competitive price. The sample system needs to extract the representative sample in an unmodified state so that all measurements can be performed on a dry basis. This is done by drawing the unmodified sample through the CEM system via a diaphragm vacuum pump. The sample flow must be sensed with an in-line flow sensor. A thermoelectric sample cooler with an integral moisture monitor and condensate removal system should be used to condition the sample to a dry basis by condensing the entrained moisture vapor. The cooler should provide a constant dew point of $40°F \pm 0.5°F$. The sample system components need to be of a modular design in order to facilitate field replacement. An indicator panel needs to be provided for local indication of sample flows and vacuum pressure. Flow control valves need to be accessible and are used to set and control the sample flow rate to the analyzers. Individual in-line flow sensors provide real-time digital indication of flow through each analyzer. The system should be designed so system alarm and a loss of power will automatically shut down and vent the system to prevent uncontrolled sample condensation.

4.3.3 Basic Analyzer Requirements

The CEM system analyzers need to be designed to accurately determine the concentrations of the required components on a dry-basis measurement. The following characteristics are most of the main items needed by the analyzers: The analyzers need to be of a modular design to permit the interchangeability of parts. This implies that only one manufacturer is preferred for all of the different analyzers. A domestic manufacturer can reduce training time and spare-parts requirements. A full-time national service group is a requirement by many users in determining the bidders list. Many users also require a 24-h toll-free telephone service. These topics will be discussed later. Utilizing a modular design allows for easier maintenance and a quicker change out. This is important since some state agencies do not allow for long time periods of no data. Analyzers and sample handling systems should have bi-directional digital communications to increase reliability and for ease of maintenance. Each analyzer should have the capability to transmit all concentration, calibration, and diagnostic information via the bi-directional digital communications network. If the analyzers are time-shared, then there is a requirement for a programmable logic controller (PLC) to do the time-sharing switching functions, communications, and the automatic calibrations. Some analyzer systems can do the PLC functions internally within the analyzer system. Other analyzers will need a separate PLC. The stand-alone PLC will need to be powered with a small uninterruptible power supply (UPS). This will eliminate any inadvertent power-up problems when the power is restored after a power outage. Manual and automatic zero and span calibration adjustments are required. This means that a microprocessor-based analyzer must be specified that also includes autocalibration features. The analyzers have to have a very high degree of stability during normal operations. Span and zero drift rates of less that 1.0% full-scale per day are required in order to meet the permit accuracy requirements. Using less accurate equipment will increase the plant operating costs in several ways. First, the plant will not be close to its optimum operating point. Thus, less product is being produced resulting in lower revenue. Also if the analyzers are not accurate enough when compared to the stack test, then this happens. If the analyzers test within 7.5% of the stack test or RATA (relative accuracy test audit), then the next RATA is conducted annually. However, if the analyzer accuracy is between 7.5 and 20%, the RATA is conducted semiannually. To conduct a RATA, the cost can go up to $30,000 per test session. This is just the cost of the test and does not include any costs associated with product loss and lost revenues.

In existing plants, when CEM systems are installed, the plant soon can determine how good its operating procedures are. Sometimes the plant is not being operated as is thought. If the temperatures, pressures, or flows seem off or different, it is possible. The process engineers need to do their mass balances. This may show that the added analyzer information can assist in developing more through-put because the plant is able to operate closer to the optimum.

If an oxygen analyzer is used, it may need to measure O_2 in a range of 0–25%. A built-in flow sensor is required. The analyzer may have modular architecture and be distributed from the power supply and the local interface. The type of O_2 analyzer may use paramagnetic measurement technology. The analyzer should provide sample pressure, sample flow rate, case temperature and control, and power supply voltages as a minimum for diagnostic information.

For nitrogen oxide (NO_X) analyzer applications the measuring range of 0–100 ppm is a good general range. It is best to also include the dry-basis measurement requirement. In this case the dry-basis measurement range could be 0–0.15 lb/MMBtu. This number is required for the data acquisition system and reports section and will be discussed later. A built-in flow sensor is required. The analyzer may have modular architecture and be distributed from the power supply and the local interface. The type of NO_X analyzer may use chemiluminescence measurement principles. The analyzer should provide sample pressure, sample flow rate, ozonator pressure, detector signal and temperature, and power supply voltages as a minimum for diagnostic information.

For carbon monoxide (CO) analyzer applications, the measuring range of 0–100 ppm is a good general range. A dry-basis measurement equivalent range is in the area of $0 - 0.075$ lb/MMBtu. This number is required for the data acquisition system and reports section and will be discussed later. A built-in flow sensor is required. The analyzer may have modular architecture and be distributed from the power supply and the local interface. The type of CO analyzer may use nondispersive infrared (NDIR) measurement technology. The analyzer should provide sample pressure, sample flow rate, IR source current, detector temperature and control, detector signal, case temperature and control, and power supply voltages as a minimum for diagnostic information.

The above analyzers are examples of three different types of analyzers and their basic requirements. The different types are paramagnetic, chemiluminescence, and NDIR measurement technologies. An alternate for an O_2 measurement could be a zirconium oxide analyzer. If a sulfur dioxide measurement is required, then it could be done with a nondispersive ultraviolet (NDUV) analyzer. If an opacity analyzer is needed, it could be a dual-pass in situ transmissometer utilizing techniques similar to NDIR and NDUV analyzers. This generally covers most of the analyzers used in CEM systems.

4.3.4 Plant Environmental Requirements

The plant environmental engineer has a very large input into the CEM system at this point. After all, the plant report needs to contain *all* of the parameters as defined in the permit. The environmental engineer has a vested interest because he or she is the first person the permitting agency will attempt to find if there is a major problem. Remember, if convicted of providing fraudulent data, there is a good possibility of spending time in prison. The permit may require that all data measured or calculated be printed in the reports and permanently recorded in electronic files on the server. Which server or server(s) are affected will be discussed later. For one case, the addition of 4 new furnaces resulted in the CEM system being placed on the 4 new furnaces. No retrofitting of the 12 existing furnaces was required. However, data from the existing areas was required by the permit to be in the CEM report.

4.3.5 CEM System Report Specifications

First will be all of the new furnace report requirements. The four new furnaces needed the following basic data printed in the report: the exhaust stack flow rate for each stack, the exhaust stack percent O_2 for each stack, the furnace firing rate in MMBtu/h for the annual average and 1-h maximum, the CO concentration in ppm on a dry basis of at least four equally spaced data points per hour, and the NO_x concentration in ppm on a dry basis of at least four equally spaced data points per hour. The CO and NO_X cycle times are important. Specifically, according to the permit, the data cycle time is 15 min. If the analyzer data cycle time can be shortened to 5 or 6 min for both CO and NO_X, then the possibility of time-sharing analyzers may exist. In this example only two sets of analyzers may be needed instead of four sets. The ultimate cost savings would be to have only two sample systems instead of four sample systems. The cautious approach would have four sample systems in case the permit agency does not sample sharing. Next is the H_2S concentration in ppm on a dry volume basis. This must be representative for the fuel burned at all burners.

Next is the necessary data reduction for each new furnace on each daily printout. The concentration data is reduced to hourly averages once per day. The pounds per hour of CO is updated daily. The pounds per MMBtu of CO is updated daily and a 30-day rolling average is maintained. The pounds per hour of NO_X is updated daily. The pounds per MMBtu of NO_X is updated daily, and a 30-day rolling average is maintained. Each new furnace annual average firing rate is required for a rolling 12-month average that is updated daily. Also a 1-h maximum firing rate is required on a rolling 24-h period that is updated hourly.

All the alarms and notifications need to be recorded on the report. CEM system downtime is required for all new furnaces. The furnace firing rate that exceeds the 1-h maximum limit is an alarm in MMBtu per hour units. The furnace firing rate that exceeds the annual average limit is an alarm in MMBtu per hour. The heat-specific CO alarm limit is in lb/MMBtu. The heat-specific NO_X alarm limit is in lb/MMBtu. The maximum CO alarm limit is in pounds per hour. The maximum NO_X alarm limit is in pounds per hour. The actual specific data is a function of the type of application and the permit specifications. This example is only for a specific furnace application for a specific state permit.

The final requirement for this specific situation is the reporting of calibration data: Update daily the zero and span of each analyzer, reports the corrective action when within a 24-h period the span drift exceeds twice the specification per 40 CFR 60 App. B, report the quarterly cylinder gas audit (CGA) per 40 CFR 60 App. F, Procedure 1, Section 4.1.2, and report the corrective action taken with the CGA when exceedances are in excess of $\pm15\%$. This completes the requirements for the new furnaces.

The existing furnaces need to have the following data for the daily report: Each of the old furnaces needs the furnace firing rate updated daily in MMBtu/h for the annual average and 1-h maximum. Next is the H_2S concentration in ppm on a dry volume basis. This must be representative for the fuel burned at all burners. Each old furnace annual average firing rate is required for a rolling 12-month average that is updated daily. Also a 1-h maximum firing rate is required on a rolling 24-h period that is updated hourly. CEM system downtime is required for all old furnaces. The furnace firing rate that exceeds the 1-h maximum limit is an alarm in MMBtu/h units. The furnace firing rate that exceeds the annual average limit is an alarm in MMBtu/h. The H_2S concentration that exceeds the maximum limit is an alarm in either mg/dry scm or grains dry scf. The final requirement for this specific situation is the reporting of calibration data: Update the zero and span of each analyzer, report the corrective action when within a 24-h period the span drift exceeds twice the specification per 40 CFR 60 App. B, report the quarterly CGA per 40 CFR 60 App. F, Procedure 1, Section 4.1.2, and report the corrective action taken with the CGA when exceedances are in excess of ±15%. This completes the requirements for the old furnaces.

4.3.6   Data Acquisition System Requirements

The data acquisition system (DAS) provides the centralized control monitoring and reporting of all CEM functions. This will include the sample/blowback cycle and system calibrations. Also the DAS will acquire all data and status information from the measurement system (analyzers) (including calibration adjustments) digitally. As discussed earlier, this information will be processed along with the system status outputs in preparation of the report. The DAS will be capable of handling and processing data from all vendor-supplied analyzers. The system will edit the data and generate process emissions reports in conformance with the EPA and the permitting state. To do this task the DAS should consist of a data-handling PC that is interfaced to the analyzer digital communications network. The DAS software should be a Windows® application, utilizing dynamic color graphics displays and a relational database for data storage and report generation.

For communications, the DAS should have a serial interface for direct digital communications between the data-handling PC and the analyzer communications network. The data-handling PC could be remotely located up to 4000 ft from the CEM. The communications must be able to operate over a single pair of standard metallic wires. The serial interface shall be capable of bi-directional communications using standard Microsoft Windows® DDE protocol. The DAS must have the capability to monitor all pertinent analyzer diagnostic variables on the analyzer network. The DAS must monitor all necessary sample controller variables available on the analyzer network.

The DAS software should be a true Microsoft Windows NT application. The software needs to be capable of real-time, multiuser, and multitasking. Several other operating systems are able to provide these functions. The next requirement is to provide the data in either Excel® spreadsheet or Access® database file formats. After all, the permit report is just one of the tasks that the environmental engineer does. In making an educated guess, most major companies have selected Microsoft Office® as their standard PC software package. Acceptable file formats will save a lot of keystrokes. Being able to link and invoke macros is also a major time saver for the environmental engineer. There are several other reasons for specifying Microsoft® software. Most programmers and knowledgeable personnel computer users can program in C Language or Visual Basic software or with the macro in the applications programs. Programmers with this background are available in most companies. On the opposite end of the spectrum are those capable of programming for specific DCS applications. DCS programmers are not readily available and in many cases their time is usually spent on high-priority tasks only associated with the DCS and specific control application assignments. Additionally most analyzer technicians are PC literate and can function very well in a Windows environment. Thus the amount of extra training is usually less that for any other operating system and associated software. If Microsoft Windows NT® is used, then the operating system can be made to always boot directly into the CEM data acquisition application program. The data acquisition software should be a true Microsoft Windows® application. The dynamic color graphics should be developed in Microsoft Visual Basic® for Windows. The relational database should be Microsoft Access. The operator interface must use pull-down menus. The operator interface must have a Windows-based help facility. The data-handling microprocessor should be an Intel Pentium chip and an IBM-compatible hardware system. For the data-handling PC, the pointing device should be Microsoft Windows® compatible. This specifies the basic operating system and applications software. At the same time the basic system hardware is also determined.

4.4 DAS FEATURES

The DAS has many functions that it needs to perform. The DAS needs to display all of the pollutant concentrations, calculated emissions values, required emissions averages, alarm conditions, calibration results, and the CEM status. The analyzer diagnostic variables need to be displayed as part of the information display format and allow for alarming on any of the variables and also shown in an alarm display. The DAS should permit the user the ability to modify any of the sample system control parameters (such as start, duration, and frequency times for probe blowback and calibration sequences). Sample system control parameters should be displayed on a sample system information screen. The DAS should allow for the addition of future analyzers or process instrumentation variables without software program code modifications. The data-handling PC's primary storage device should be capable of storing a minimum of 2 years of data files and reports. The DAS needs to provide for the automatic backup (archival) of the database and report files to any type of compatible storage device. The retrieval and display of historical data from the files must be both in table and strip chart format. The files may be retrieved from the PC hard drive, floppy disk, or other media. A set of standard reports that will meet all of the permit demands and be available either automatically at preset time intervals or on user demand. These reports will be daily, monthly, and quarterly summary reports, exceedence reports, and calibration results reports. User-defined custom reports need to be available through the use of the Access® database software. The database files should be available on a read-only basis for viewing or by copying by other Windows® programs. The DAS needs to be able to be supported on a telephone basis for factory support and troubleshooting purposes.

4.5 COMMUNICATIONS INTERFACES

Although not explicitly stated earlier, some of the communications between the CEM system and the DCS needs to be bi-directional. Why? In the previous example, all of the data for the existing furnaces will be sent from the DCS to the CEM system. In some cases, the DCS will calculate constants for the CEM system. The major data point item that has to be synchronized is the time. Since PC clocks are known not to keep accurate time, the DCS clock is the primary time source. If any governmental agency wants to check CEM data with the plant data, it is best if all of the time stamps are in agreement. That requires a DCS watchdog timer on the DCS to CEM system communications line and a CEM system watchdog timer on the CEM system to DCS communications line. This gets the main technical consideration out of the way. In some cases, the DCS will also keep a complete file in the history module of the total environmental data.

Because of the bi-directional communications requirements between the CEM and the DCS, the communications work out better if this was a Modbus® link. The trend for many of the DCS vendors is to use Modbus® communications between the DCS and many other third-party vendors. This is a very established communications protocol. Some examples of additional third-party Modbus® devices are analyzer systems, emergency shutdown systems, compressor speed controllers, vibration and temperature monitors, antisurge controllers, programmable logic controllers, specialized rotating equipment controllers, and other miscellaneous equipment devices. Many of these interfaces are critical to the plant operation.

There are some other data communications requirements. The internal communications from the CEM analyzers to the digital network is one. Another communications line is for the standard RS-232 modem communications. This line allows factory troubleshooting into the CEM system. Now the plant has the option of having its technicians do either troubleshooting or maintenance remotely. Another option is to have the analyzer/CEM vendor provide remote troubleshooting or maintenance. The communications options are starting to increase. Just by using the standards of the data communications industry, the connectivity now is opened to include modems, telephones, cellular phones, and satellites. Now internal analyzer performance information is externally available to remote sites. This could allow an analyzer technician in Texas the ability to monitor, download internal information, and reset and adjust many analyzers. From one location a technician could work on analyzers from New York to Texas to California to Alaska. The line can send the daily report to a specific environmental agency, if this is a permit requirement. Several states require that the daily report be automatically transmitted at a certain time and to a specific number. Some countries are requiring the same automatic data transmission of the plant environmental reports on a daily basis. This requirement is being placed in an increasing number of the environmental permits.

An expanding requirement is placement of the CEM system data in the information system (IS) network. The CEM data information is just another management tool. Interfacing with the IS department needs to be documented so that the exact type of communications hardware is used. One of the biggest problems today is the lack of standardized communication protocols and data highways and plug-and-play connectivity. The Modbus interface between the CEM system and the DCS is good but also needs to be customized for each application. Many applications in addition to the CEM system to DCS connectivity now are requiring the connectivity to the plant or enterprise network(s). A standard interface protocol such as

Ethernet can be used. Here the application is a little different. This is the most convenient way for the plant environmental engineer to obtain unit data and then add all of the individual units data together to compile the plant data. This is why file formats become important. If there is an enterprise network, the individual data and the compiled plant data are network accessible to the corporate environmental staff. Now a variety of reports can be developed. Why is Ethernet of interest? The cost benefit of standardized information is now possible. Market competition is adding this type of function to all types of equipment. Being Ethernet compatible and transmission control protocol/Internet protocol (TCP/IP) addressable eliminates the need for custom device drivers. Using standard protocols instead of custom device drivers decreases the software costs. Open-architecture-based systems are now providing control, measurement, maintenance functions, communications support, and remote access capability to corporate databases. These are just some of the reasons to start building CEM systems with several types of communications capabilities. The future outlook is increased communications and networks. More information is both an internal and external requirement and more systems are starting to deliver that capability. In order not to impede the workflow, standard protocols, communications, and data highway networks are just some of the productivity enhancements.

Why is CEM system and analyzer data in general becoming a high demand commodity? As data or information becomes available to more groups in usable formats, it can help explain situations or events. In the process industry, data can be used to optimize the process, to use less incoming material, and manufacture more outgoing material. Also the process can be set up to run as close to optimum as possible. If the optimal point is approached, then the efficiency is close to being maximized with respect to input and output. This should be close to the providing the lowest operating and material costs for the most profit. Process control parameters can be set to attempt to maximize throughput or whatever are the company objectives. The networking of CEM systems and analyzers in general are in the early stages of product development. In another 5–10 years, the specific products will probably be network compatible. Whether the network will be Ethernet, Fieldbus, Profibus, or a combination of any and all is subject to debate at this time. In 5–10 years the debate should be resolved. Then much of the technical specifications will be network oriented and provide the next advancement in system capabilities.

4.6   TOTAL GROUP SOLUTIONS

Once the decision has been made that a CEM system is required, the work to acquire a consensus starts. This may be the most difficult part of the project. As discussed in the earlier paragraphs, it now is obvious that a CEM system is a project-specific system. Regulatory-related requirements vary for each state or country. Internal company requirements usually are different. Since the formulation of plant products can be different, specifications usually cannot be to a rigid one specification fits all conditions. Even if vendor R has supplied over 100 units to company C in California for product G, it does not mean the total system for the same company in Texas for product E will be identical. The hardware may be very similar. The software may be similar. But nothing will be totally identical between the two CEM systems. The analyzer hardware can be standard products for most components such as CO, O_2, NO_X, CO_2, and SO_2. Also the sample systems usually are made from standard parts, although the configuration may vary.

The software is the most custom part of the system. As we have already discussed, variations in communications requirements is a major reason for the custom software. Varying permit requirements also account for custom software. In many CEM products, the software is produced by a third-party vendor. After all, it is convenient to have access to highly qualified professional programmers. For very specialized applications, it generally is better to have the software code written by experts that write code 8 h or more a day.

The plant analyzer engineer and technician are the first of the group that need to buy into the type of CEM system. For instance, the hardware will be extractive for the probe, if there will be sample switching, and if the DAS provides system control, daily calibration, data gathering, and reporting functions. The sample probe is placed directly in contact with the process environment and is used to extract a representative sample from the process stream. A filter in the probe prevents particulates from entering the sample line. The sample line is heated at a temperature above the sample gas dew point for transporting the sample from the sample probe to the sample conditioning system. The sample conditioning system draws a continuous sample from the sample probe through the heated sample line. The sample conditioning system cools the sample gas causing a condensate to form. This "dry" sample is then sent to the analyzer units. Then the analyzers provide a concentration measurement for the desired components. For the sample conditioning system, it should have the capability to route the test gas either remotely (to the sample probe and back through the sample line and sample system). Daily calibrations must be done remotely because of regulatory requirements. For diagnostic reasons, the testing of just the analyzer can be done by injecting the test gas directly into the analyzer. Calibration of the analyzers and the system can be done either manually or automatically. The system can calibrate the analyzers automatically once each day. The DAS controls the autocalibration start time. The DAS monitors the response of each analyzer module to the test gas and computes the correction factors that are used to correct the analyzer readings. The sample system has the following alarms that are monitored by the DAS:

Sample probe temperature low
Heated sample line temperature low
Sample gas dryer temperature high
Condensate in sample
Sample pressure low

The alarms for the sample probe temperature low, heated sample line temperature low, sample gas dryer temperature high, and condensate in sample are interlocked with the sample pump to shut down the pump and prevent moisture from getting into the analyzers. Any of these alarms will energize a "common system alarm" output contact. Another output contact is provided for the communications failure between the DAS and the DCS. This information provides some insight into the requirements of the analyzers, sample systems, DAS, and internal and common alarms for a CEM system for furnaces. The stack gases measured for a specific project were O_2, NO/NO_X, and CO.

The DAS software should provide a minimum of four functions. The first function is the operation screens that are the main operator interface screens in the system. The second function is the reports generated by the DAS automatically and manually. The third function is the procedures that outline the daily, weekly, and monthly tasks required to maintain the DAS. The software should be designed to accept keyboard and mouse operations. The operations screens should contain a minimum of six folders to present vital information for the system operation. These folders show a system overview, trending, CEM system setup, analyzer information, data acquisition system setup, and reports.

A system overview display should provide a quick status check of the current mode of operation. Areas on the screen should include the sample line, status area, current alarms, and thumb tabs for selecting the stack of interest (if a furnace application). The status area should contain the following information: *emission source operating/not operating; data invalid* message for a system fault on invalid reportable data; *calibration failed* when the calibration captures values that are out of specified limits: *out of calibration* displayed when the calibration has drifted out of specified limits; *pump fail* occurs on a low pump discharge pressure; *moisture in sample* occurs when moisture in the sample system increases to a point where the analyzer may become damaged; *high dryer temperature* alarms to prevent damage to the sample system dryer; *low line temperature* alarms when the sample line temperature is not high enough to maintain all moisture in the line as a vapor; *low probe temperature* alarms when the sample probe temperature is lower than the specified value; *maintenance mode; remote/local sample mode* displays the current sample path; *calibration in progress* is displayed in either auto or manual calibration mode; *blowback in progress* is displayed as either auto or manual blowback.

The system overview for the sample system is a graphical representation of the sample gas traveling from the stack and calibration gases through the sample pump to the analyzers (for NO_X, CO, and O_2 as some examples). The system overview for calibration and blowback status displays the status, and manually controls the calibration and blowback procedures. The display shows the current time, date, and schedule for the next autocalibration. Also current analyzer drift information can be displayed. This value is the difference of the set calibration gas value from the value of the analyzer recorded during the last calibration. The alarms section provides the analyzer technician with a quick overview of the alarms currently active on the CEM system. Alarms are displayed for *CEM out of control* analyzers that are out of calibration for 5 consecutive days. *System alarm* is displayed for the following problems: pump stopped, moisture in sample, high dryer temperature, low line temperature, and low probe temperature.

The system overview for trending provides a graphical access to the historical data collected by the DAS. The stacks trend data is visible at all times. There is a 1-min average trending for CO, diluent corrected CO, NO_X, and diluent corrected NO_X and O_2. These trends are shown simultaneously on the same graph. A historical display of any combination of consecutive days may be selected.

Some analyzer systems are capable of providing similar functions as a programmable logic controller (PLC). Other analyzer systems do not have this capability. If the analyzer systems require an external PLC, then the DAS needs to provide an operator interface to allow and view the adjustment for certain functions such as:

Drift correction

Stack control

PLC battery condition

PLC clock time and date

The CEM system allows the operator to define the calibration procedure. Some configurable parameters include:

Calibration time

Zero gas range duration

Span gas range duration

Stabilize time

Zero gas reference values

Span gas reference values

Error (% range)

Out of control (% range)

The CEM system allows the operator to define the blowback procedure. Configurable parameters may include:

Blowback start time

Blowback duration

Stabilize duration time

With the CEM system the operator needs to view the fuel firing sources for each stack. Some operation and indication functions include:

Source firing rate value—the current rate of fuel usage for the stack from the DCS

Stack operate function—indicated if the source is firing or not from the DCS

Fuel heating value—the value of heat output in MMbtu per standard cubic foot of fuel used from the DCS

H_2S concentration—the value for the hydrogen sulfide emission from the DCS

The CEM system provides an analyzer screen that gives the analyzer technician information on the analyzers and their current operation. Information can include voltages, pressures, and temperatures as a troubleshooting tool. If any values for the analyzers are out of specification, a status indication will show in the alarm area.

The CEM system should provide a DAS overview. A general DAS setup screen should provide the analyzer technician the capability to select and set various parameters and data entries within the system. Generally, the topics may include maintenance, general information, library, stack information, and diagnostics. Most of the topics are self-explanatory and very specific for only the analyzer technicians. One of the screens of interest to the environmental engineer is a Constants screen. This screen allows the operator to view the constants used in the various calculations for the emissions volumes by the stack. These values are entered during the system software design and normally can only be viewed by this screen. Other values are shown next to the various equations shown on the Constants screen.

A Stack Setup screen allows the technician to define the DAS alarm set points. Configurable parameters include:

CO and NO$_X$ lb/h warning: Any emissions higher than this value will cause a CO or NO$_X$ approaching limit alarm.

CO and NO$_X$ lb/h limit: Any emissions higher than this value will cause a CO or NO$_X$ exceeding limit alarm.

CO and NO$_X$ lb/MMBtu warning: Any emissions higher than this value will cause a CO or NO$_X$ approaching limit alarm.

CO and NO$_X$ lb/MMBtu limit: Any emissions higher than this value will cause a CO or NO$_X$ exceeding limit alarm.

The Reports screen is one of the most important screens. This is important to the analyzer technician, the environmental engineer, the plant operations, and many others. This option allows the operator to select the type of report to be displayed concerning stack emissions, calibration information, and reasons for any abnormal conditions that need to be addressed. Some typical selection options are System Configuration Reports, Calibration Data Reports, Operations Reports, and Assignment Reports. Also the Reports screen allows the operator to print copies of all reports generated by the DAS. The system configuration menu provides the technician the ability to select an analyzer and then print the analyzer values. Emission summarys per stack can be selected and printed with either hourly or daily averages. Alarm conditions are explained as well as any loss of data. This is a very brief summary of what the DAS can provide. The main requirements are to be able to have the software control all of the analyzer functions and still gather data for reports in a timely manner. Reports or data requests should take only minutes to process and not hours.

The Alarm screen allows the operator to view and acknowledge all alarms that occur in the CEM system. All the alarms are displayed until the operator acknowledges the alarm and the alarm conditions clear. The Alarm screen should show all alarms that are related to the data collection and reporting of the DAS. Several of the alarms displayed are:

Analyzer did not pass calibration
Blowback on
CEM data invalid
CO exceeded emission level
Emission source not operating
Analyzers out of control
Blowback inhibited at local panel
CEM in maintenance mode
Autocalibration

The DAS can generate daily emissions by the hour, alarm, and calibration reports for as many days as requested on demand. A 24-h report is automatically generated daily and contains diluent-corrected CO and NO_X averages and hourly O_2 and hourly CO and NO_X values for the previous 24 h. A calibration report is generated for the first calibration procedure that is run that day. Subsequent calibrations done the same day are appended to the file. This shows how many of the hardware functions, reports, calibrations, and operations gets tied together by the software. This is why many of the plant personnel are required to have input into the CEM system.

The DCS hardware and software team will now be discussed. This interface between the DCS and the CEM system is not very complicated. Once the specific type of communications is determined, then it is easily accomplished. With most DCS vendors, the preferred third-party interface at this time is with a Modbus serial link. The specific software requirements are defined for each vendor. A mapping of the Modbus register with a definition of the data points to be exchanged between the DAS and the DCS is an absolute requirement. The sooner the better this information is exchanged, both the DCS and CEM system vendors can confirm the requirements are met. This requires the tag name, description Modbus address, rang, and process units be exchanged between the vendors.

In a past project the only stumbling block between the two software groups involved synchronization of the time stamp. The time stamp based on the DCS time was a major requirement. None of the operations, environmental, or analyzer engineers wanted to explain to the state why each piece of hardware had a different time. After all, which one was right? What happened if one clock was either faster or slower than the other? This is why the DCS and CEM system each had a watchdog timer checking on the other system. However, passing the current DCS time to the CEM was a software problem. The initial time format was in hours, seconds, and minutes (HH:MM:SS) based on a 24-h format (0–23 : 59 : 59). This information could not be produced using one Modbus[R] register. The easy solution was to break up the time into three Modbus® registers. One register representing hours. The next register representing minutes. The last register seconds. The change worked. Due to the custom software required by both vendors, the CEM system was given a factory acceptance test (FAT). Then during the DCS FAT, the interface was completely checked out between both vendors with no problems. The actual CEM hardware and software was used at the DCS FAT.

Another situation can exist with the brand of analyzer used throughout the plant. In the furnace stack application, we have talked about an H_2S analyzer and measurement. How is this data to be acquired? The easy way is with only one communications link. Normally all of the plant analyzers are directly linked to the DCS through an analyzer communications network. Since the DCS has the H_2S data, this is just one of the points sent from the DCS to the CEM system. Make use of all the existing data links as much as possible.

Three interested groups have been discussed in the last several pages. What is their role in the CEM system? Plant operations have ultimate authority over the project, since they operate the plant. That is the main reason for their involvement in a project. Naturally, the project team is concerned with the design, schedule, construction, and costs. The analyzer specialists are involved to provide a system that will meet the technical requirements. After all they are the ones that will be required to provide maintenance and fix the system if it goes down in the middle of the night. The analyzer specialists will be the ones required to provide any explanations if the system is not operating correctly. They could become involved in any meetings with the regulatory agency. While the analyzer specialists are not the final authority on the report requirements, they do have a significant input as to the report data and its accuracy. In some situations, the CEM system will be providing additional information to operations. Sometimes the data may not be believed. The CEM may provide data that is contrary to how operations thinks the furnaces are being operated. To be able to verify the data, the analyzer specialists will need to go back to the stack test data and start from the data provided by the independent third-party stack tester.

Corporate engineering will be involved as they will want to ensure that system standardization occurs between the various plants in the corporation. In some cases they may have developed a long-term working relationship with a vendor. If this is the case, then a sole source agreement will provide similar equipment and construction quality standards. Even if the applications may be different, the technical approach will be consistent. Spare parts may be more readily available at a nearby plant than from the vendor. Also important is the vendor field support staff. With resources stretched very thin, the fastest help available may not be the closest. Also, pricing agreements associated with alliance agreements could prove to be lower than those achieved through competitive bidding. Corporate engineering staff can be either a big help or a major hindrance. The corporate types attend many meetings, technical shows, and national standards associations and publish technical articles. To support the corporate engineer, any will assist in giving presentations, writing articles, and writing national standards.

The role of the plant environmental engineers has been discussed in providing the information of the exact requirements of the daily report and other data they need for the plant. The environmental engineer's involvement is an absolute must if the project is to succeed. This is an early and continued involvement in providing the permit information, report requirements, and any other information required.

The plant information systems (IS) specialist is needed to provide the technical requirements for connecting to the IS network. The idea is to provide the daily CEM data to the IS network for the plant environmental engineer and other designated parties on the local network. The local network needs to be defined and the type of hardware connection specified. For this specific requirement the file format is of importance. This is the philosophy for requiring data in Excel file formats. Keystrokes saved will save time and may provide reports in a more timely fashion.

This discussion shows how many different groups have an involvement in developing a new CEM system. This is why many meetings are required in order to establish all of the technical requirements for the CEM system. Many different department's inputs are necessary. Also two or more vendor companies may have major interfaces requirements with each other. For a successful project, many totally separate functions need to be integrated together.

4.7 DATA EXCHANGES

We have had some discussion of the different data that needs to be exchanged between the DCS and the CEM system. Since each state permit is different, applications may be similar but are seldom identical. Table 4.1 shows a typical data exchange between the CEM system and a DCS.

The pollutant concentration is generated by the various analyzers and then goes to the DAS. A separate analyzer usually located at the fuel gas knockout drum generates the fuel gas sulfur concentration. The time of day is usually synchronized by the DCS to WWV for time correlation. This signal has a watchdog timer in both the DCS and CEM system. This will provide an alarm if either system times out.

Now that the data have been exchanged between the CEM system and the DCS, what happens to the data? Table 4.2 presents a description of the various conversion calculations, where they are done, and the input source and output destination.

Table 4.2 highlights the plant environmental specialist's involvement. First, the calculation conversions for the analyzers are very important for the units required by the permit. For this example, another plant analyzer input is required for H_2S, and the DCS needs to be programmed to provide this input to the CEM system. Probably the most difficult data manipulation is the F factor. The factor can be either determined by an equation or by a constant number (as determined by the permit). Due to the number of inputs required for the equation, it is much easier to use the constant if permitted.

Table 4.1   Inputs

Description of Data	Units	Data Generator	Data Receiver	Purpose of Data
Pollutant concentration	ppmv	CEM	CEM	"Raw" emissions data
Furnace firing rate	Btu/h	DCS	CEM	Input to mass emissions rate calculation
Fuel gas sulfur concentration	ppmv	DCS	CEM	Calculation of SO_2 emissions
Heating value of fuel gas	Btu/scf	DCS	CEM	input to mass emissions rate calculation
Fuel gas combustion gas rate	scf/Btu	DCS	DCS	Calculation of stack gas volumetric rate
Stack gas density	lb/scf	DCS	DCS	Calculation of stack gas mass flow rate
Stack gas volumetric flow rate	scf/h	DCS	CEM	Input to mass emissions rate calculation
Stack gas mass flow rate	lb/h	DCS	CEM	Input to mass emissions rate calculation
Time of day	WWV	DCS	CEM	Correlation of compiled data
Hardware malfunction	Alarm	CEM	DCS	CEMS not performing per specifications
Software malfunction	Alarm	CEM	DCS	CEMS not performing per specifications
Calibration drift	Alarm	CEM	DCS	CEMS not performing per specifications

Some states require the daily report be automatically sent at a predetermined time each day. This is when the environmental specialists has to trust the total CEM system. The CEM analyzers have to be working correctly. The equations, constants, and report have to be accepted by the state. After all, there is no external intervention allowed. With the situation as outlined, a modem (with a dedicated

Table 4.2   Data Manipulation

Description of Calculation	Location	Source of Input	Output Destination
Convert from ppmv to lb/h	CEM	DCS and CEM	CEM (report, storage)
Convert from ppmv to lb/MMBtu	CEM	DCS and CEM	CEM (report, storage)
Heating value of fuel gas	DCS	Fuel gas analyzer	DCS
Combustion gas density	DCS	DCS	DCS
Combustion gas volume	DCS	"F" factor	DCS (calculation of stack gas volume)
Stack gas volumetric rate	DCS	DCS	CEM

voice phone line) is required for both data transmissions to the state agency and for troubleshooting. With some applications, four RS-232 ports could be used. For PCs, this is usually the maximum number of usable communications ports. The environmental specialist usually is only concerned with the daily modem communications with the permit agency. The analyzer technician is concerned with the Modbus RS-232 communications between the DAS and the DCS. If there is a PLC used for a time-sharing application, the analyzer tech is also concerned with the analyzer 4–20 mA communications from the analyzers to the PLC and then the RS-232 communications from the PLC to the PC (DAS). Analyzer diagnostics could be another RS-232 signal. To get the data from the DAS, a 10 base T Ethernet card could be added to the PC.

The analyzer maintenance tech now needs to maintain at least three different types of communications software and hardware. To help, a second telephone voice line is required. Why? If factory assistance is required, a direct voice link is established with the second phone line. Then the factory can use the modem link to interrogate the PC and associated analyzer hardware and software. The analyzer technician could use this feature to solve problems from other locations than onsite. If a software patch is required, it could be downloaded directly from the factory to the CEM system. This is an example of several of the communications applications being used today. Networking will be discussed later.

4.7.1 Data Retention

What data need to be retained, how long, where? With the state permit agencies having the capabilities of examining anything, it is best to retain more data than required. Any data to help answer a state question will always be useful. Save all the raw data (15-min, hourly, daily, monthly and yearly averages) in the DCS History module. Save all the raw data (15-min, hourly, daily, monthly, and yearly averages) and agency reports on the PC hard drive. The environmental specialist may even save the daily permit report and raw data to the IS network. What does the permit require? It may be prudent to save 2 years or more of raw data. No matter how long the mean time between failures (MTBF) is, electronic equipment will fail at the most inopportune time. Make use of all the various interfaces the CEM system has for saving any and all data. The DCS time stamp will be on all of the data that is to be saved. The only question will be: Which set of data is the official version? This is a company decision. Most companies will prefer to stay with the DCS History module as the official data storage location. If environmental officials actually go and look at the plant data, the most complete data will come from the DCS. The CEM will store the daily report because of the software file format.

4.7.2 Factory Acceptance Testing

The factory acceptance test (FAT) is the culmination of all of the work to make the CEM system usable by all the various groups that have been discussed. The vendor should provide a detailed testing procedure for the FAT. Assuming the testing procedure has been approved prior to the start of the FAT, the first day of testing should start with a meeting to make sure everyone's objectives are in agreement and consistent with the purchase order. Before any testing starts, point-to-point checks of the wiring should be done to assure the wiring drawings are accurate. This includes continuity testing of all wires, correct terminal block and wiring terminations, and contact configuration of all switches and relays. Usually, the first devices tested are the individual analyzers. The analyzers should be checked with the appropriate calibration gases and input and output emulation devices for all field devices. The functional operation of all of the analyzers and directly associated hardware is a must. The factory certification tests should be a requirement in the specification document. Any changes and all exceptions have to be part of the official signed documentation. One of the most important CEM analyzer test is the drift test. All analyzers must be tested for the customary 24-h a day for 7 days. The drift measurement will verify the equipment's capability to be used in determining the emission's specific concentrations. This is the easy part of the testing. The next part requires simulating the various emissions conditions so that the DAS reports can be verified. It is recognized that some tests and equivalent report functions cannot be simulated and can only be done with the plant running.

If a vendor has a Modbus communication link as a standard product, it may not be necessary to test the interface with the DCS. However, if the Modbus software is either a new product or custom made for each application, it is best to test the interface with the DCS. The interface testing should be done at the DCS vendor's FAT. If any software code needs to be changed, it is better to do it where both systems can be tested with each other for as long as necessary. All of the Modbus addresses from both systems need to be individually checked. Both the CEM system and the DCS need to be tested together for a sufficient amount of time. To be able to determine that the DCS inputs to the CEM system work in the CEM and CEM outputs to the DCS work in the DCS need to be tested during the DCS FAT. How else will the DCS History module be tested to determine that it is saving the required data in the correct format or units. Three or 4 years after the DCS FAT is done and the state permit agency is in the plant reviewing data is not the time to find out that the DCS has not been saving the correct data or all of the required data.

Once the CEM system has been installed in the field, the vendor should provide on-site technical assistance during the installation, checkout, field calibration, and startup of the CEM equipment. After the equipment is running, a test needs to be developed to allow an evaluation of the CEM system performance in the field. These performance verification procedures will provide an evaluation of the analyzer and system relative accuracy, low- and high-level drift, and calibration drift. Now the CEM system is working as the plant starts up.

Once the plant is running at its design capacity, the next item to work on is the site certification testing. This is the ultimate test for any CEM system. In many cases the CEM certification can be done in conjunction with other compliance testing. The CEM vendor needs to submit a CEM System Certification Test Protocol to meet all the requirements of 40 CFR 60 Appendix B, Performance Specification, for the process industry or 40 CFR 75 for the power industry. The protocol is subject to the review and approval of the state agency. Part of the field certification procedure requires the CEM vendor to attend a precertification meeting. This meeting will help at the plant site in advance of the testing dates in order to comply with the state's notification requirements. The meeting will cover in detail the protocols to be used by the stack sample tester in conducting the certification testing. This allows the state to question and/or modify the protocols. As a minimum, the vendor representatives, plant representatives, state officials, and the testing company officials should be in attendance. When the CEM System Certification Test Protocol is approved, then the stack tester can conduct the tests using the approved procedures. The tests need to be witnessed by the various responsible representatives. At the test completion, the CEM vendor has 30 days to submit the Certification Test Program Report results for approval by the state. Results of a 7-day drift test will be compiled by the plant and submitted to the CEM vendor for inclusion in the certificate report. Now it is obvious why the CEM vendor is required to provide a certified system. If the CEM vendor's equipment certification testing results are not accepted by the state, the vendor is responsible to remedy the situation. If the certification testing cannot be either completed or is not accepted by the state for any reason that is attributable to the facility, the plant is responsible to incur any additional costs. Once the certificate is accepted by the state, the plant can now start thinking about the next RATA tests, cylinder gas audits (CGA), and daily calibration testing.

What else should the CEM vendor provide? One phone line is directly connected to a modem in the PC. This modem can automatically send data to a state agency or receive an incoming call from the CEM vendor's modem for diagnostic testing. The CEM vendor should have a 24-h, toll-free telephone service to provide prompt response to minor issues and follow-up on service and startup situations. Also the vendor should have a full-time direct employed national service group. To support the service group should be a centralized stocking warehouse with the 24-h 7-days a week service with the ability to provide same-day parts service. If the vendor has either regional or local service facilities, the normal commodity item parts should be stocked there. For plants requiring a maintenance contract with the CEM vendor, the local service person would be a logical source of the normal commodity parts.

What are possible CEM system upgrades? The one hardware upgrade that is immediately available is with the personnel computer. Instead of having a regular consumer PC, use a server PC. The cost is indeed higher. But perhaps the benefits are worth the cost. If a hard drive fails, PC servers can swap out the parts on a hot swap-out basis. This helps decrease CEM system downtime. State agencies do not like any large amounts of downtime on their environmental reports. Also on the hardware, purchase as much random-access memory (RAM) as possible. A minimum amount of RAM is 128 Megs. An unmentioned side benefit of using a PC server is the computer maintenance. Usually the IS group will lay claim to any PC server. That is a good deal. If the PC server does have a malfunction, the IS group now will be the one called out to do the troubleshooting. Other potential areas for improvement would be the use of an SCSI hard drive instead of an IDE hard drive. Software program tuning is one area that needs a critical review. The normal analyzer application program needs to be running all the time. Compiling a special report from the database needs to make very good use of the multitasking aspects of the operating system. If little attention is given to the secondary application tasks, the report could take literally up to 30 min or longer to generate. This is an unacceptable length of time for a report to be generated. Working with the software operating system and applications programs as one total system is the most logical way to optimize the multitasking features.

At the start of this chapter, the Windows NT® operating software was discussed as the operating system of choice. There are several reasons for this statement. First is the cost of the package. Second is that many very good programmers are proficient in using Windows NT® and the various applications software such as Access and Excel. Availability is another consideration since most software resellers have these packages on the shelf. Compatible file formats are a must. Since the data is exchanged between departments, agencies, and other individuals, it is much easier to have the data in either an Access or Excel file format for reports or other special projects. This is the reasoning up to now for not mentioning Unix or even Linux operating systems. It can be debated forever which is the best of the operating systems. Technical superiority was not one of the selection criteria. Data file manipulation was as major point. Most computer users are reasonably knowledgeable about wordprocessing, spreadsheet, and database applications software. Which company has the majority of the market share of these application programs? Microsoft is the company. There are similar applications programs for Unix and Linux. Their file formats are not as widely known as are the various Microsoft file formats. While the Unix and Linux file formats may be compatible with Excel and Access, the problem will be that no IS help desk will provide help for non-Microsoft application programs. In the next year or two the trend may change, and major corporations will be supporting the Linux operating system and applications programs.

Today there are more options available such as Fieldbus®, Profibus®, Control-Net®, DeviceNet®, LonWorks®, Smart Distributed System (SDS)®, and World-FIP® to name several types of networks. This shows how dynamic the network technologies are progressing in control systems. This does not mean that the new technologies are being installed in large quantities in the plants as yet. There is interest in the open "device" networks for control system applications. A single international standard is not going to be available in the near future. However, the benefits of selecting some type of networking technology should be considered. There are direct benefits that can be established now.

The major gas chromatograph manufacturers also have dedicated analyzer networks. Until recently most networks were proprietary data highways. While diagnostic information is available through the analyzer network, it is not remotely accessible without a lot of difficulty. Analyzer networks are changing because of the user requirements. At this time there is no universal communications protocol for remote communications offered by the analyzer vendors. The requirement to do diagnostics and maintenance remotely is becoming a major technical requirement. Major analyzer users are starting to require remote communications capabilities for analyzers. The analyzer technicians can dial-in and troubleshoot the problems from a PC at home. This can decrease the downtime of the analyzer just by not requiring any travel time to the plant. In other cases, the plant may contract to the vendors to do the remote dial-in and troubleshooting online. The plant users are now demanding one standard communications protocol for remote communications. This standard remote communications protocol is being requested for CEM systems, gas chromatographs, moisture, oxygen and combustibles, pH, conductivity, and H_2S analyzers, just to name some analyzers. The plant user wants to know the last bit of information from all of the analyzers. The plants want to connect every type of analyzer to a highway with the same communications software.

4.8 CONCLUSION

Now analyzer vendors are developing new products. The new analyzer products have Fieldbus[R] communications capabilities as a standard feature. Analyzers with Profibus will be available. With the new bus features built into the analyzers, the integration into a DCS will become a standard. More options can be selected. The DCS will be able to acquire and store more field data. The connection of most third-party devices will become more of a network question. The best part is that the software interface will become more fill-in-the-blanks software instead of writing new code for each application. Common software applications will help reduce troubleshooting time. Soon the plug-in and use features will become part of the standard features for the vendors.

Several DCS vendors are now going to open systems. This will encourage the analyzer vendors to provide products with features that use TCP/IP or Ethernet protocols. If the use of Microsoft operating systems for multiuser and multitasking applications continue, then the transfer of data for various reports will be in either spreadsheet or database application software formats across many different platforms. This would allow data in the report formats to go from the analyzers to the DCS, to the plant IS network. At the same time a 24-h report could be generated and sent to environmental agencies at a specified time. The corporate levels could then have an instantaneous check of each plant's output, efficiencies, and product types and be able to have a total global report of sales, costs, production, and potential profit/loss positions on a daily basis. With the open system using a spreadsheet or database application software corporate-wide data can be acquired and reviewed on a continuing bases. Then if any adjustments are required, the corporation can do it with almost instant results.

The future data requirements will establish the type of networks directly involved with the end instruments and the DCS. In some situations, the corporate requirements will be added to the plant needs. This will dictate the types of control system networks and other hierarchical data networks. The ability to interface to a DCS and other data networks are just starting to develop with both standard hardware and software. Having international standards for both the software and hardware is speeding the product development by both the DCS and analyzer vendors.

5

INFRARED ABSORPTION SPECTROSCOPY

Tye Ed Barber and Norma L. Ayala

Tennessee Technological University
Cookeville, Tennessee

John M.E. Storey

Oak Ridge National Laboratory
Oak Ridge, Tennessee

G. Louis Powell, William D. Brosey, and Norman R. Smyrl

Oak Ridge Y-12 Plant
Oak Ridge, Tennessee

Environmental Instrumentation and Analysis Handbook, by Randy D. Down and Jay H. Lehr
ISBN 0-471-46354-X   Copyright © 2005 John Wiley & Sons, Inc.

5.1 INFRARED ABSORPTION SPECTROSCOPY

Infrared (IR) spectroscopy is based on the measurement of infrared radiation absorbed, emitted, or reflected by the sample of interest. The first IR experiments were performed in 1800 when Herschel detected the heat changes due to IR radiation with a thermometer.[1] In the 1940s, commercial IR instruments became available, and IR spectroscopy rapidly became an important method for elucidation of molecular structure.[2] IR methods for air emission monitoring and analysis have been in use for over 50 years, but the usage has dramatically increased in the past 2 decades because of the availability of better instrumentation.[3] The goal of this chapter is to introduce the fundamental concepts of IR instruments so that readers can make informed decisions when comparing IR techniques and purchasing IR spectrometers for gas analysis. Since the vast majority of IR techniques are based on absorption of IR radiation, only absorption methods will be discussed in this chapter.

5.1.1 Absorption of IR Radiation

When electromagnetic radiation is absorbed, it increases the energy of the molecule. In general, IR radiation is only energetic enough to cause changes in the vibrational and rotational energy of molecules.[4] To absorb IR radiation, a change in the net dipole moment of the molecule must occur, and the energy of the radiation must be equal to the difference between allowed energy transitions of the molecule.[5] The only common molecules that do not absorb IR radiation are homonuclear diatomics such as oxygen, O_2, nitrogen, N_2, and hydrogen, H_2.[6] Because these molecules do not absorb IR radiation, their concentrations cannot be measured directly using IR spectroscopy. Theoretically, IR spectroscopy could be used to measure the concentrations of all compounds that absorb IR radiation provided no spectral interference occurs. This makes IR absorption techniques some of the most powerful and versatile methods for gas analysis.

    The frequency at which a molecule absorbs IR radiation is determined by the types of atoms composing the molecule, the types of bonds forming the molecule, the location of the bonds and atoms in the molecule, and their interaction within the molecule.[5,7] When IR radiation is absorbed, it causes a change in the vibration or rotational motion of a molecule.[7] Figure 5.1 shows the IR vibrational modes for molecules containing three or more atoms. In this figure, the circles designate atoms and the connecting lines chemical bonds. The stretching vibrations involve a change in the length of the bond. Stretching can be symmetric or asymmetric as indicated in the figure. Bending vibrations involve a change in angle between two bonds.[7] Depending on the symmetry or relative location of atoms and bonds within a molecule, not all vibrational modes will be observed. In some molecules, the various types of vibrations may occur at the same frequency and are indistinguishable.[5]

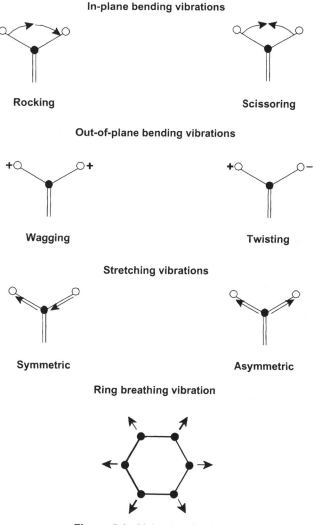

Figure 5.1   Molecular vibrations.

Infrared absorption is observed not only at the fundamental frequency of a vibration but also at overtones and at combination frequencies. An overtone band will occur at approximately integral multiples of fundamental frequencies.[1,5] When a photon of IR radiation excites two vibrational modes simultaneously, a combination band may be observed.[6] The frequency of the combination band will be at the sum of the fundamental frequencies of the two excited vibrational modes. Difference bands can occur when an excited molecule absorbs IR radiation.[6] The intensities of the overtone, combination bands, and difference bands are usually significantly less intense than the fundamental bands.[6]

The exact position of an absorption band cannot be determined by the type of vibration alone due to vibrational coupling.[6] Energy of a vibration is determined by the interaction of all the vibrations of the molecule. While most of the interactions may be extremely weak, certain vibrational modes can be strongly coupled.[7] Vibrational coupling can cause the frequency of an absorption band from the same type of chemical bond and atoms in one molecule to be shifted to a higher or lower frequency than the same type of chemical bond and atoms in a different molecule.[6] The combination of the different types of possible vibrational modes, overtones, difference bands, combination bands, and coupling results in the uniqueness of the IR spectrum for differing compounds.[4-7]

5.1.2 Infrared Spectra

The IR spectral region covers the spectral range from 0.78 to 1000 μm. This spectral range is frequently divided into three ranges: near infrared from 0.78 to 2.5 μm, mid-infrared from 2.5 to 50 μm, and far infrared from 50 to 1000 μm.[7] Spectral features in an absorption spectrum are commonly referred to as absorption bands or peaks. When working with IR data, spectra and spectral features may be described in several different systems of units related to the abscissa (x axis) used to describe the data. Table 5.1 shows the equations used to convert between

Table 5.1 Equations Used for Converting Between Most Commonly Encountered Scales in IR Spectroscopy

	To Convert to Wavelength in nm (λ_{nm})	To Convert to Wavelength in μm ($\lambda_{\mu m}$)	To Convert to Wavenumber, $\bar{v}$, cm^{-1}	To Convert to Frequency v, s^{-1}
Wavelength, nm (λ_{nm})	—	$\lambda_{\mu m} = \lambda_{nm}/10^3$	$\bar{v} = 10^7/\lambda_{nm}$	$v = 3 \times 10^{17}/\lambda_{nm}$
Wavelength, μm ($\lambda_{\mu m}$)	$\lambda_{nm} = \lambda_{\mu m} \times 10^3$	—	$\bar{v} = 10^4/\lambda_{\mu m}$	$v = 3 \times 10^{14}/\lambda_{\mu m}$
Wavenumber, cm^{-1} ($\bar{v}$)	$\lambda_{nm} = 10^7/\bar{v}$	$\lambda_{\mu m} = 10^4/\bar{v}$	—	$v = \bar{v} \times 3 \times 10^{10}$
Frequency, s^{-1} (v)	$\lambda_{nm} = 3 \times 10^{17}/v$	$\lambda_{\mu m} = 3 \times 10^{14}/v$	$\bar{v} = v/(3 \times 10^{10})$	—

the different units often encountered in infrared spectroscopy. The wave-number scale is the most commonly used scale. It is sometimes referred to as frequency because wave numbers are proportional to frequency.[7] While this is not absolutely correct, it is widely used in the literature and in speech. The ordinate (y axis) of an IR spectrum can be expressed in arbitrary units, transmittance, or absorbance. When expressed in an arbitrary scale such as volts, counts, or amps, the axis is referring to instrumental readout units. This scale is usually arbitrary since a scaling factor is often employed, and the signal is dependent on instrumental factors.

Most IR gas analysis is based on the use of the mid-IR region. This is the region where the fundamental vibrational energy transitions occur resulting in absorption bands of the highest intensity. The mid-IR region is also selected for gas analysis because of the availability of instrumental components.[3,5–7] The mid-IR is often divided into two regions when discussing organic spectra: the group frequency region (1200–$3600 \, \text{cm}^{-1}$) and the fingerprint region (600–$1200 \, \text{cm}^{-1}$).[7] In the group frequency region, absorption bands are observed that are not greatly influenced by the overall molecule. The absorption bands in the group frequency region can often be used to classify the molecule by attributing bands to certain types of bonds. In the fingerprint region, the absorption bands observed are that characteristic of the entire molecule. Since the bands in the fingerprint region are dramatically influenced by the overall molecule, absorption bands in this region can be used for qualitative interpretation of IR spectra of pure compounds or very simple mixtures.

5.1.3 Beer's Law

In order to quantitatively measure the absorption of IR radiation, two measurements are required. It is necessary to determine the incident power P_0 and the power transmitted P by the sample.[4,6] The transmittance T is defined as the fraction of radiation passing through a sample as shown in Figure 5.2. Attenuation of the beam can be expressed as absorbance A, given by the equation

$$A = -\log_{10} T$$

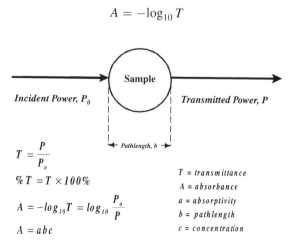

$$T = \frac{P}{P_0}$$

$$\%T = T \times 100\%$$

$$A = -\log_{10} T = \log_{10} \frac{P_0}{P}$$

$$A = abc$$

$T = transmittance$
$A = absorbance$
$a = absorptivity$
$b = pathlength$
$c = concentration$

Figure 5.2 Measurement of absorption.

The absorbance can be related to concentration by Beer's law, sometimes called the Beer–Lambert law:

$$A = abc$$

where a is the absorptivity constant, b is path length, and c is concentration. The absorptivity will have dimensions expressing unit length per unit concentration of the analyte.

Under ideal conditions, Beer's law could be used to determine the concentration of any absorbing species until total absorption of the radiation occurs. However, Beer's law is based on assumptions that are impossible in practice to achieve. It is assumed that the incident radiation is monochromatic, the absorbing molecules are independent of each other, the path length of the radiation is uniform for the entire beam consisting of parallel rays, the sample is homogenous, the sample does not scatter the radiation, and the flux of photons of IR radiation is not sufficient to cause saturation of the absorption transition.[4–7] Although it is not possible to construct an absorption spectrometer that will meet all these assumptions, it is possible to design analytical systems that will allow the application of Beer's law.

In real systems, the attenuation of the transmitted beam is due not only to the absorption of radiation by the sample but also to the scattering of radiation by the sample, and losses resulting from optical components necessary to contain the sample such as reflections from windows of a gas cell.[4,6,7] The drawing in Figure 5.3 shows some of the types of losses occurring in a gas cell. When measuring the absorption, it is necessary to minimize or correct for all losses in the intensity of the radiation beam passing through the sample except for losses due to absorption by the species of interest.

In practice, the beam intensity is seldom measured before and after a sample as is implied by Figure 5.2. Usually, the intensity of the beam passing through the sample is compared to the intensity of a beam passing through a reference. The manner in which the reference is collected is dependent on the instrument used to measure

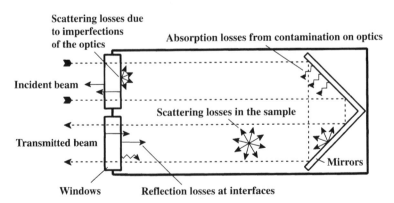

Figure 5.3 Losses in transmitted intensity of radiation in gas cell due to unwanted scattering and absorption.

absorption. Some instruments collect the reference and sample sequentially, whereas other instruments collect the reference and sample simultaneously. In both cases, it is necessary for the reference to be as similar to the sample as possible. By using the same cell, or two identical gas cells, for the reference and sample, the decrease in transmitted intensity caused by unwanted reflection and scattering from optics will be canceled or greatly reduced when the ratio of the sample and reference is taken.[4,6,7]

Beer's law assumes that the radiation beam passing through the sample is monochromatic. This cannot be achieved with any spectrometer because this would require infinitely high resolution.[8] This results in apparent deviations from Beer's law due to the use of polychromatic radiation.[4,6,7] The deviation will increase as the resolution of the spectrometer decreases. In general, the deviation is not significant as long as the resolution of the spectrometer is sufficient to fully resolve the absorption band.

Even when the resolution of the spectrometer is sufficient to fully resolve the absorption band, deviation from Beer's law will be observed due to interactions between molecules. These interactions may be chemical or physical.[4,7] The apparent deviations are a result of changes in the analyte's absorption spectrum.[7] Deviations due to chemical and physical interactions can be minimized by adjusting the pressure of a system or working at low concentrations.

More difficult to deal with is the absorption by species in the sample other than the analyte. At a given wavelength, the total absorbance based on Beer's law is equal to the summation of the absorbance of all the species absorbing in the system, assuming there is no interaction among the various species.[4] This can be expressed as

$$A_t = A_1 + A_2 + A_3 + \cdots + A_n = a_1 b c_1 + a_2 b c_2 + a_3 b c_3 + \cdots + a_n b c_n$$

where the subscripts refer to absorbing species $1, 2, 3, \ldots, n$. Since instruments measure the total absorbance of the sample, it is apparent from the above equation that the concentration of the analyte cannot be determined by measuring absorbance at a single frequency when multiple absorbing species are present. Methods for dealing with absorption by interferences are discussed in a later section.

5.1.4 Differential Absorption

The application of Beer's law requires the measurement of the transmitted intensity through a reference and through the sample to determine the transmittance. In many applications, it is not possible to measure a reference. For example, in remote-sensing applications measurements are made in the atmosphere. In these cases, it is possible to determine concentration from the ratio of the transmitted intensity measured at the wavelength where the analyte absorbs and the transmitted intensity at a nearby wavelength(s) where the analyte does not.[6,9-11]

For most cases, the same measurement consideration of absorbance applies to the measure of the differential absorption. Measurement of differential absorption has several advantages. The effects of optical scattering by the sample are reduced since the scattering at the analytical and reference wavelengths will be very similar.[9] Also, there can be little or no time delay between the measurement of the reference and analytical wavelengths. This can reduce the impact of instrumental and sample drift.[9]

5.2   COMPONENTS COMMON TO ALL INFRARED INSTRUMENTS

All instruments used to measure IR absorption have several components in common.[4–7] Figure 5.4 illustrates generic IR absorption instruments. Every device for measuring IR absorption will have a source of IR radiation, a device for selection of the frequency of radiation to be measured, a sample or reference, a transducer for converting IR radiation into an electrical signal, and a readout device. In some instruments, two or more of the components shown in Figure 5.4 may be combined into a single element that performs multiple functions. The sample may also be located between the source and the frequency selection device.

An absorption spectrometer can use either a single-beam configuration or a double-beam configuration. In the single-beam configuration, a single compartment or gas volume is used for collection of sample and reference data. Most methods using a single-beam instrument use stored reference data for calculation of absorbance. A few single-beam methods estimate absorbance by comparing the transmitted intensity at the frequency that the analyte absorbs to the transmitted intensity at a frequency that the analyte does not absorb.[6,9–11] In the double-beam configuration, separate optical paths are used to measure the reference and sample at the same time. The advantage of the single-beam configuration is that the instrumental design is simpler, and the same optical path is used for both the sample and reference.[6] The advantage of the double-beam configuration is that variations in the instrument's performance can potentially be minimized.[4,6,7]

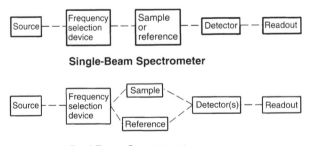

Figure 5.4 Components of single- and dual-beam spectrometer configurations used for measuring absorbance.

5.2.1 Optical Materials

A variety of optical materials are available for working in the IR region. For many applications, the primary considerations are the spectral range and durability of the material. In the majority of IR systems, mirrors are used to direct, focus, and collimate the IR radiation. Great care must be exercised when handling these mirrors to prevent the mirror surface from being damaged. Various components of IR instruments require the use of materials that transmit IR radiation. The primary selection most users have to make is selecting IR windows. Since windows are generally exposed to the harshest conditions, they have to be either durable or inexpensive enough that they can be frequently replaced. A very complete coverage of IR reflecting and transmitting materials can be found in *The Infrared Handbook*.[12]

5.2.2 Sources

Several considerations must be made in the selection of an infrared source. In all cases, the source used in an IR instrument must produce adequate radiant power at the frequency or frequencies of interest so that it is possible to detect the radiation above a given noise level.[7] The output of the source must also have sufficient stability over the measurement time frame of the instrument. Other important considerations include the cost and lifetime of the source.

5.2.2.1 Continuum Sources. Except for laser-based instruments, IR instruments use continuum sources. The most commonly used continuum IR sources are electrically heated resistive materials. Heated materials can be used as IR sources because everything with a temperature above absolute zero radiates IR radiation.[13] The emitted frequencies and intensities are determined by the temperature of the material and the surface conditions of the object.[13] Infrared sources are often described in terms of a blackbody. A blackbody absorbs all radiant energy regardless of wavelength and emits as much energy as it absorbs to maintain thermal equilibrium.[6] This makes a blackbody the most efficient thermal radiator possible.[6] Figure 5.5 shows the calculated radiant intensity for a blackbody emitter at several temperatures. It can be seen from the curves in Figure 5.5 that the higher the temperature of the source, the higher the radiant intensity. The information in Figure 5.5 can be used to obtain a crude estimate of a source behavior based on changes in temperature. When comparing two IR instruments with almost identical components, the instrument with the hotter source will generally have better signal-to-noise performance.

The radiant intensity of real sources will differ from the ideal blackbody. Real sources are not perfect emitters and are sometimes enclosed in an external housing. The window in the enclosure will attenuate some frequencies of radiation by absorption or by reflection. To distinguish their behavior from blackbodies, real sources are often called gray bodies.[6]

5.2.2.2 Light-Emitting Diodes. Infrared light-emitting diodes (LEDs) emit a continuum of IR radiation over a limited spectral range. LEDs are made from

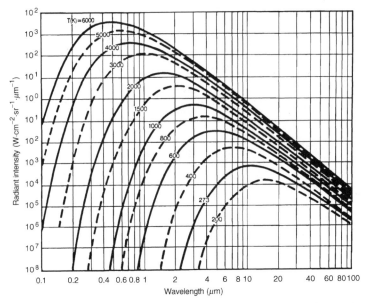

Figure 5.5   Blackbody radiant intensity. (Courtesy of Hamamatsu Corporation.)

pn semiconductor materials. When the appropriate voltage is applied across the junction, the holes and electrons combine to emit light.[6,14] By using different semiconductor materials, LEDs can be fabricated that emit at various wavelengths over most of the near-IR and mid-IR regions. LEDs offer advantages over electrically heated sources. They are more efficient in converting electricity into light and can have higher radiant power at the wavelengths of interest. LEDs are sometimes used in small dedicated single-component analyzers.[15]

5.2.2.3 Lasers. Unlike continuum sources, lasers emit over very narrow frequency ranges. Lasers can either have fixed wavelengths or can be tuneable over a wavelength range. Fixed wavelength lasers are used to measure absorption at a fixed wavelength, whereas tuneable lasers can be used to obtain an absorption spectrum over the tuning range of the laser. Because lasers emit over very narrow frequency bands, a wavelength selection device is not needed as is shown in Figure 5.4.[6,10] The resolutions of laser systems are several orders of magnitude better than is obtainable with continuum sources and frequency selection devices.[6,10,16] High power and coherence of lasers also make it possible to do detection schemes that would be impossible using continuum sources. A variety of IR lasers are commercially available. Carbon dioxide and diode lasers are frequently used for IR spectroscopy in research applications. A detailed discussion of lasers and laser spectroscopy can be found in *Laser Spectroscopy* by Demtröder.[10]

Table 5.2   Characteristics of Infrared Detectors (Courtesy of Hamamatsu Corporation)

Type Operating Temp. (K)			Detector D* (cm·Hz$^{1/2}$/W)			Wavelength (μm) Rise time (μs)	
Thermal-type detector		Thermal		No Wave-	300	$D^*(\lambda, 10, 1) = 6 \times 10^8$	10000
		Thermopile		length	300	$D^*(\lambda, 10, 1) = 1 \times 10^8$	5000
		Bolometer		dependency	300	$D^*(\lambda, 10, 1) = 1 \times 10^9$	500
		Pneumatic cell	Golay cell, Condenser-		300	$D^*(\lambda, 10, 1) = 2 \times 10^8$	40000
		Pyroelectric detector	microphone TGS,LiTaO$_3$				
Quantum-detector type detector	Intrinsic-type	Photo-conductive type	Pbs	1–3	300	$D^*(500, 600, 1) = 1 \times 10^9$	200
			PbSe	1–4.5	300	$D^*(500, 600, 1) = 1 \times 10^8$	5
			HgCdTe	2–12	77	$D^*(500, 1000, 1) = 1 \times 10^{10}$	1
		Photo-voltaic type	Ge	0.6–1.9	300	$D^*(\lambda_p, 1000, 1) = 1 \times 10^{11}$	2
			InAs	1–3	77	$D^*(500, 1200, 1) = 1 \times 10^{10}$	1
			InSb	2–5.5	77	$D^*(500, 1200, 1) = 2 \times 10^{10}$	1
			HgCdTe	2–12	77	$D^*(500, 1000, 1) = 1 \times 10^{10}$	0.005
	Extrinsic-type detector		Ge : Au	1–10	77	$D^*(500, 900, 1) = 1 \times 10^9$	0.05
			Ge : Hg	2–14	4.2	$D^*(500, 900, 1) = 6 \times 10^9$	0.05
			Ge : Cu	2–30	4.2	$D^*(500, 900, 1) = 8 \times 10^9$	0.05
			Ge : Zn	2–40	4.2	$D^*(500, 900, 1) = 5 \times 10^9$	0.05
			Si : Ga	1–17	4.2	$D^*(500, 900, 1) = 5 \times 10^9$	0.05
			Si : As	1–23	4.2	$D^*(500, 900, 1) = 5 \times 10^9$	0.05

5.2.3   Detectors

Infrared detectors can be classified as either thermal detectors or quantum detectors. Quantum detectors are also called photon detectors. Thermal IR detectors respond to the heating effect of the IR radiation.[7] The response of a thermal detector is determined by the change in some physical property due to changes in the detector's temperature.[5] Quantum detectors for IR are semiconductor materials that change electrical properties upon the absorption of IR photons.[5] Table 5.2 lists common types of IR detectors.

5.2.3.1   Thermal Detectors.   To maximize the temperature change of a thermal detector, the active element is kept as small as possible and is thermally insulated from the substrate material so that the IR heat will cause the greatest change in the active elements temperature.[5] When a thermal detector absorbs radiation, its temperature will increase and the physical change can be measured. After the radiation ceases, the element will return to the temperature of the substrate at a rate determined by the thermal conductance of the insulation.[5] Since the response of a thermal detector is determined by changes due to the heat of the radiation, the response of

thermal detectors is independent of wavelength, assuming the absorption of the detector approximates that of a blackbody.[13] One advantage of thermal detectors is that they can be operated at room temperature. The disadvantages of thermal detectors are their slow response time and poor response relative to quantum detectors.[5]

Several different physical properties can be used to measure IR radiation. In pneumatic detectors, the expansion of a material is used to detect IR radiation. Most pneumatic detectors are based on a thin black membrane placed in an air-tight gas-filled chamber. When radiation is absorbed by the membrane, the gas expands. By measuring the change in the pressure of the gas cell, the intensity of the IR radiation can be determined. A common pneumatic detector is the Golay cell.[5,6] Pneumatic detectors are also often used in nondispersive instruments.[5-7] Thermocouples are based on the small voltage induced at the junction of two dissimilar materials. The voltage is proportional to the temperature at the junction.[5] Radiation absorbed by a thin coating on the thermocouple will cause a change in the junction temperature and the voltage produced. Thermopiles are multiple thermocouples connected in series to produce a higher output voltage.[5] Materials that have a relatively large change in resistance as a function of temperature can be coated with a thin absorbing layer to construct IR detectors.[4] These detectors are called bolometers or thermistors.

For applications requiring a relatively fast response time from a thermal detector, pyroelectric detectors are used. Pyroelectric crystals are dielectric materials that have a permanent electrical polarization because of the alignment of the molecules in the crystal.[17] By placing the pyroelectric crystals between an IR transparent electrode and another electrode, the pyroelectric crystal can be incorporated into a capacitor circuit that is dependent on the polarization of the pyroelectric crystal. The absorption of IR radiation causes thermal changes in the crystal lattice spacing affecting the electrical polarization.[5] The change in polarization results in change in the surface charge of the capacitor generating a current in the circuit.[5] The magnitude of the generated current is determined by the surface area of the pyroelectric crystal and the rate of change of polarization with temperature.[7] Because pyroelectric detectors are based on changes in the polarization, they do not respond to constant IR radiation levels and can only be used to detect modulated signals.[6] One of the most common type of pyroelectric detectors is the DTGS (deuterated triglycine sulfate) detector.[17]

5.2.3.2 Quantum Detectors.

Except at the shorter wavelengths of the near-IR region (<950 nm) where photomultier tubes can be used, IR quantum detectors are based on semiconductor materials.[13] Semiconductor materials respond to IR radiation when the absorption of IR radiation causes the promotion of an electron from a nonconducting state to a higher energy conducting state.[7] To raise the electron to the conducting state, the photon must have sufficient energy determined by the type of semiconductor materials used to fabricate the sensor. Because of the energy required to promote the electron and the absorption properties of the material, the response of quantum detectors is dependent on the wavelength of the radiation.

The primary advantages of quantum detectors over thermal detectors are fast response time and better sensitivity.

Quantum detectors can be configured to operate in photoconductive or photovoltaic modes.[5] In a photoconductive detector, the absorption radiation promotes electrons to the conducting state, increasing the conductivity of the semiconductor.[5] Photovoltaic detectors are based on *pn* semiconductor junctions. Absorption of radiation causes the formation of electron–hole pairs that produces a current and changes the voltage across the junction.[17] To minimize the background thermal noise of IR quantum detectors, it is sometimes necessary to cool the detector.

5.2.3.3 *Selection Factors for Infrared Detectors.*

One way to compare detectors is to examine the D-star values or specific detectivity D^*.[18] D-star is a normalized measure of the detectable power taking into account the bandwidth of the detector and the area of the active element of the detector.[12] Figure 5.6 shows the D-star for common infrared detectors. This information can be used to estimate the signal-to-noise performance of a detector.[18] The ratio of two D-star values can also be used to compare relative performance of two different types of detectors with the same area operated at the same wavelength, bandwidth, and radiant power. For example, the ratio of the D-star for a pyroelectric detector and a mercury cadmium telluride (MCT) detector reveals that the performance of a pyroelectric detector is between 10 and 150 times poorer than an MCT detector, depending on the type of MCT detector and wavelength. The signal-to-noise performance can be improved by averaging, but the reduction in noise level will improve by the square root of the number of averages.[3,6,7]

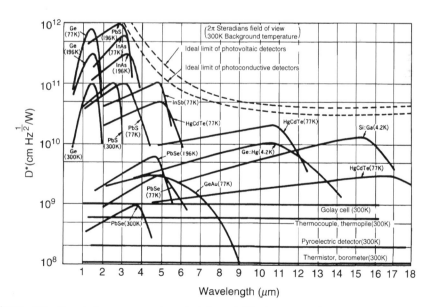

Figure 5.6 Spectral response characteristics of IR detectors. (Courtesy of Hamamatsu Corporation.)

However, the selection of an IR detector cannot be based on D-star alone. It is also necessary to consider other factors before selecting a detector. In general, the time constant of the detector and associated electronics should be around one tenth the time of the transient signal event being measured.[6] Besides the response time of a detector, it is also necessary to consider the linearity of variation in the signal from the detector with variation in radiation levels.[18] All detectors will exhibit nonlinear behavior at some incident power level. For IR systems using quantum detectors, it is often necessary to attenuate the radiant power reaching the detector to be in the linear range of the detector. To test for linearity, a reduction in radiation should result in an equal reduction in detector output. For wavelengths where neutral density filters are available, this is easily tested by placing the filter in the optical path of the instrument. A neutral density filter is a filter that uniformly attenuates the transmitted beam regardless of the wavelength. The other major consideration in selecting a detector for use in an IR instrument is the operating cost over the lifetime of the detector. Detectors requiring cooling for operation can be very expensive to maintain. This is especially true for detectors that require cryogenic liquids.

5.2.4 Gas Cells and Remote Sensing

While there are numerous techniques used in IR spectroscopy, the two basic sampling approaches are the measurement of absorption with the gas in an enclosed cell and the measurement of absorption of the gas directly without sample collection.[3,6,9,11,16,19] Which approach is used is determined by the type of information required from the analysis. In applications using gas sampling, the sample is transferred into a gas cell. This can involve pumps, cryogens, vacuum systems, and so on, depending on the sample type and the analyte. In applications in which sample collection is not practical or possible, the absorption can be measured directly by passing the infrared radiation beam through the region of interest.

5.2.4.1 *Gas Cells.* Gas cells used in IR analysis can range from simple tubes with windows on the end to a complex optical system.[6] In its simplest form, a gas cell is constructed by placing IR transmitting windows at the end a tube with the necessary connections to allow the introduction of the gas sample (Fig 5.7a).[6] These cells can be made very inexpensively and can be extremely rugged. This type of gas cell is used primarily for the detection of gas at high concentrations. For the detection of gas at lower concentrations, it is necessary to increase the path length of the cell to increase the measured absorbance.[3,6] With single-pass gas cells, the required increase in length rapidly makes single-pass cells impractical for the detection of species at low concentrations.

In order to detect species in the part-per-million range and lower, it is necessary to use multipass gas cells. These cells use optics to pass the beam of radiation through the sample volume multiple times (Fig. 5.7b).[3,6,16,20] Multipass cells often have mirrors inside the cell to reflect the beam. Multipass cells using internal mirrors have been used to pass the beam over 100 times through the cell volume. These

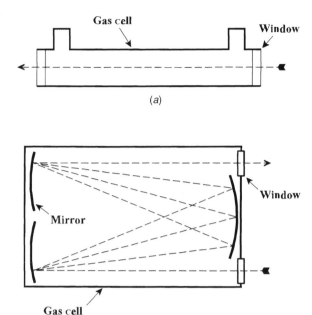

Figure 5.7 Gas cells used for measurement of absorption: (*a*) single-pass gas cell; (*b*) multipass White cell.

cells allow several meters of path length to be obtained using a very short cell with a small volume. Commercially available multipass gas cells can be obtained with path lengths up to 1 km.[21]

Care must be taken with use of gas cells. If the gas temperature is greater than ambient, losses due to condensation on the cell walls and tubing can occur. This can result in a loss of analyte and cause fogging of the windows. Furthermore, analytes can react with cell materials, tubing, and fittings. The presence of another substance in the sample stream can also cause chemical reactions during the transfer process to the gas cell that may influence the measurement of the analyte concentration.[3,21,22]

5.2.4.2 *Remote Sensing.* Remote-sensing techniques allow the collection of IR spectra without sample collection. This can be advantageous when it is desired to monitor analytes in systems where sample collection is impractical or impossible. It is especially useful for detection of atmospheric pollutants that are at low concentrations spread over large areas.[11,19,22] Figure 5.8 shows two different approaches used to remotely measure absorption. In bistatic systems, the source and detector are located on opposite sides of the region of interest. This configuration can be difficult to properly align.[19] In the unistatic system, a retroreflector is used to send the beam back toward the source. Because a retroreflector does require precise alignment and the source/detector can be packaged together, unistatic systems are often easier to implement.[19]

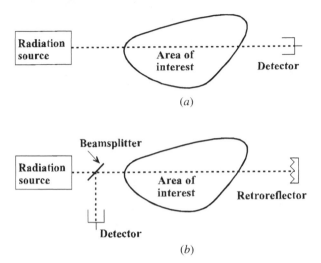

(a)

(b)

Figure 5.8 Different approaches used for remote sensing of absorption. The radiation source indicates all the components necessary to generate radiation of a known intensity at the desired wavelength. (*a*) Bistatic—single-pass measurement with the detector and source mounted on opposite sides of the area of interest. (*b*) Unistatic—double-pass measurement with the radiation reflected back through the area of interest.

5.3 INFRARED SPECTROMETERS

The primary difference between types of IR spectrometers is the frequency selection device. The purpose of the frequency selection device is to control the wavelengths of radiation reaching the detector.[4,6,7] Because spectroscopy is based on the measurement of the interactions of specific wavelengths of radiation with matter, the frequency selection device is the most important part of a spectrometer. It will ultimately determine the signal-to-noise performance of the instrument and resolution of the spectrometer. When comparing spectrometer types based on the frequency selection device, the parameters to consider are spectral range, radiation throughput, scanning speed, and resolution. Radiation throughput is how much radiation from the source at a given wavelength passes through the frequency selection device at a given resolution.[6] For the best signal-to-noise performance, the highest radiation throughput is desired. Scanning speed is how much time is required to cover a certain spectral range.[6] Resolution is a measure of the ability of an instrument to distinguish between two closely spaced spectral features. A high-resolution instrument can distinguish between closer spaced spectral features better than low-resolution instruments, often giving the high-resolution instrument superior selectivity.[6]

5.3.1 Nondispersive Infrared Spectrometers

Nondispersive infrared (NDIR) spectrometers are used to obtain quantitative information for one or a few constituents in a mixture.[5–7] Several of the most common emission gases are monitored with NDIR spectrometers including CO, CO_2, SO_2, C_2H_2, NH_3, and CH_4.[23] The U.S. Environmental Protection Agency (EPA) requires many sources to continuously monitor their emissions and the NDIR spectrometer is an accepted method for doing so.[23] There are two basic types of NDIR instruments. They are nonfilter photometers and filter photometers or spectrometers.[5-7] Laser and interferometer systems are also sometimes called NDIR instruments, but they are discussed separately in this chapter.

5.3.1.1 *Nonfilter NDIRs Instruments.* The basic design of nonfilter NDIR instruments has not changed in five decades.[3] Figure 5.9 shows a schematic of a typical design of a nonfilter NDIR instrument. The sample is typically pumped through the sample cell continously, and the reference cell is either filled with a nonabsorbing gas such as N_2 or contains a flowing reference gas. The detector converts the difference in absorbance between the sample and reference cells to an electrical signal that is then amplified. From the amplified signal, the concentration of the analyte can be determined.[5-7]

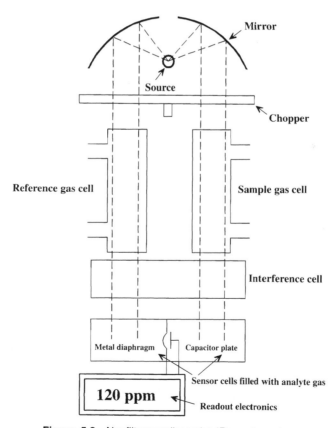

Figure 5.9 Nonfilter nondispersive IR spectrometer.

In its simplest form, the detector is filled with the gas of interest and contains a diaphragm (a two-plate capacitor) separating the two sides of the detector.[5–7] If light is absorbed on the sample side, the diaphragm in the detector distends toward the sample side of the detector, changing the capacitance. The chopper modulates the radiation from the IR source to provide a modulated signal, or waveform, which is easier to process electronically.[6]

Nonfilter NDIR instruments are used over a broad range of concentrations from part-per-million to percentage levels of components depending on the path length of the sample cell.[3] To improve the selectivity of nonfilter NDIR instruments, a specific filter may be used to limit the influence of components that absorb in the same region as the analyte. This can be accomplished by completely absorbing the radiation at wavelengths where the interfering compound absorbs.[5–7] For example, a gas cell filled with carbon dioxide may be used with a low carbon monoxide instrument in order to prevent carbon dioxide from being measured as carbon monoxide (Fig. 5.9). Since no radiation reaches the detector at wavelengths where the carbon dioxide absorbs, it is not detected in the gas sample.[5–7]

5.3.1.2 *Filter NDIRs Instruments.*

Filter instruments can be more selective than nonfilter instruments and more flexible. Filter instruments use bandpass filters to transmit a narrow portion of the IR spectrum to the detector.[5–7] They typically have a high radiation throughput, but they have low resolution. The filter wavelength is usually selected for a portion of the IR spectrum where only the analyte absorbs for a given sample matrix. Because of their low resolution, filter NDIR instruments are subject to interferences.

Filter-based instruments may have a single-filter, multiple filters mounted on a filter wheel, or a circular variable filter.[6,7] Single-filter instruments are used to

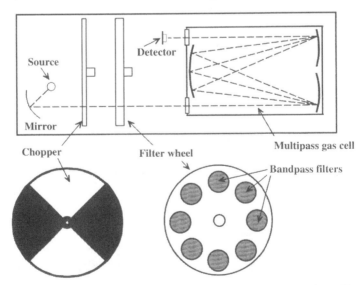

Figure 5.10 Filter wheel nondispersive IR spectrometer. Wavelengths are selected by rotating the filter wheel to the desired bandpass filter.

measure a single wavelength range and can only be used to detect one species. Filter wheel and circular variable filter instruments can be used to measure multiple wavelengths. Thus, a single instrument can be used to monitor several different gases. Figure 5.10 shows a filter wheel instrument. The wavelength and analyte is selected by rotating the filter wheel that contains multiple bandpass filters.

5.3.2 Dispersive Spectrometers

Dispersive spectrometers were once the primary IR instruments, but they have been replaced in almost all applications by Fourier transform instruments.[7] However, they are still used for some dedicated applications. Dispersive instruments are based on the use of a dispersive element to select the wavelengths of interest. Figure 5.11 shows a monochromator. Infrared radiation is focused onto the entrance slit of the monochromator and a mirror is used to collimate the radiation beam. The grating causes the radiation to constructively and destructively interfere at different angles depending on the wavelength of the radiation. By adjusting the angle of the grating relative to the exit slit, it is possible to control what wavelengths of radiation pass through the exit slit.

The resolution of dispersive systems is determined by the width of the slits and the dispersion of the dispersive element.[6,7] Wider slits allow more wavelengths of radiation to pass through the exit slit for a given dispersion. Decreasing the dispersion will decrease the resolution of the spectrometer. When using a dispersive system, a trade-off must be made between the spectral resolution and radiation throughput. Narrow slit widths will give higher resolution, but the signal-to-noise performance of the system will decrease due to the lower transmission of radiation to the detector.

Dispersive instruments are still manufactured for some gas monitoring applications, and there has been renewed interest in dispersive instruments for specialized miniature instruments for gas detection. Typically, these instruments operate at fixed wavelengths depending on the analyte. Some newer instruments use an array

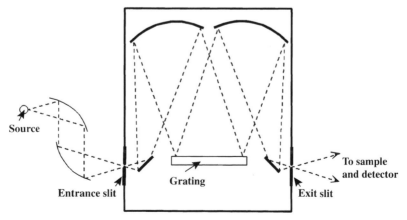

Figure 5.11 Monochromator. By rotating the grating, the wavelengths passing through the exit slit can be varied.

of detectors to replace the exit slit allowing a spectrum to be collected without moving the grating.[6,7] Recent advances in dispersive technology have allowed dispersive instruments to be made commercially available for the detection of specific gases in a package that weighs approximately 50 g.[24]

5.3.3 Fourier Transform Infrared Spectrometers

For wide spectral coverage, high resolution, and excellent signal-to-noise performance, interferometer-based instruments are unsurpassed.[6,7,25,26] Fourier transform infrared (FTIR) spectrometers detect all wavelengths over their spectral range simultaneously and have large apertures allowing high light throughput. This gives them excellent signal-to-noise performance for most applications. With the rapid developments in optics and computers, FTIR spectrometer technology has changed dramatically during the past several years. Modern inexpensive FTIR instruments can outperform expensive research systems that are just a few years old.[27] FTIR spectrometers can be used in almost all air emission and monitoring applications.[3]

There are many designs of interferometers that can be used for gas analysis. The majority of commercial instruments are based on a moving mirror design using some variation of the interferometer first developed by Michelson in 1891.[26] In this type of interferometer, the collimated radiation beam from the source is divided into two beams using a beamsplitter. One beam is directed to a fixed mirror, and the other beam is directed to a moving mirror. The reflected beams are then recombined at the beamsplitter. When the recombined beam is focused onto a detector, the interference resulting from the radiation following different optical paths will be observed.

Figure 5.12 shows a drawing of a Michelson-type interferometer as well as the interference pattern resulting from a monochromatic perfectly collimated beam of radiation and an ideal interferometer. The difference in the optical path followed by the two beams is called the retardation.[26] At zero retardation, the two beams travel the same distance. In this position, the beams will interfere constructively. If the mirror is moved out one quarter of the wavelength of the radiation, the reflected beam from the moving mirror will travel one-half of a wavelength longer. At one-half wavelength retardation, the combination of the two beam will give total destructive interference. Further movement of the mirror until the retardation is one wavelength will again result in total constructive interference.

For monochromatic radiation, the beam observed at the detector will oscillate from total constructive to total destructive interference as the mirror position moves. Constructive interference will be observed at zero retardation and retardations equal to integral multiples of the wavelength.[26] Destructive interference will occur at retardations equal to 0.5, 1.5, 2.5, and so on multiples of wavelength. When polychromatic radiation from a source is measured using an interferometer, the polychromatic interferogram is the summation of all the interferograms of each wavelength.[6,26] With a broadband continuum source, the interferogram will appear as a decaying ringing pattern as shown in Figure 5.13.

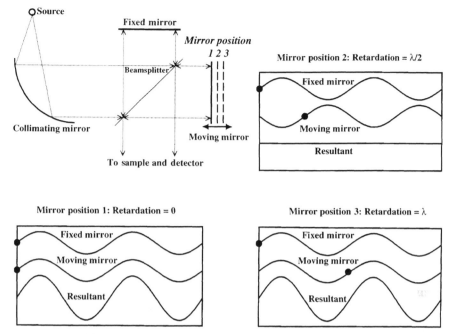

Figure 5.12   Michelson interferometer with interference resultant for monochromatic radiation shown for different mirror positions. The circle indicates the same point in time of radiation emitted by the source. In mirror position 1, the distance to the moving mirror and fixed mirror is the same. When the beams divided by the beamsplitter are reflected back by the mirrors and recombined, they will both have had to travel the same distance and will interfere constructively. In mirror position 2, the distance traveled by the beam to and from the moving mirror is a half of a wavelength longer than the fixed mirror. The beams at the beamsplitter are now half of a wavelength different, and the beams will be recombined destructively.

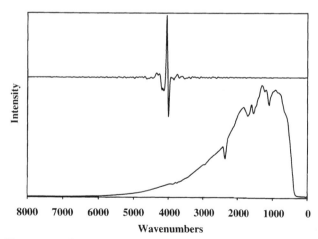

Figure 5.13   Upper trace shows an interferogram from a continuum source, and lower trace is the resulting spectrum after performing the Fourier transform.

Interferograms encode the spectral information in a domain relative to the moving mirror position.[6,7,26] Since most interferometers scan the mirror at a constant velocity, the interferogram is often treated as a time domain function. To convert the interferogram into a domain relative to the frequency of the radiation, the cosine Fourier transform is used.[5,6,7,26] Figure 5.13 shows the resulting spectra from the fast Fourier transform on the interferogram. The resolution of a spectrum obtained with a Michelson interferometer is determined by the distance traveled by the moving mirror and the quality of the collimated beam. The longer the distance traveled the higher the resolution of the resulting spectrum.[6,26]

The quality of a moving mirror interferometer is partly determined by how well it can maintain the spacial relationship between the beam of the moving mirror and the beam of the fixed mirror over the required scan distance of the moving mirror.[26] The quality is also determined by how accurately the mirror can be moved. If the moving mirror tilts during a scan, the distance traveled by the infrared radiation will not be uniform across the collimated beam. This will result in loss of resolution. To overcome problems related to mirror movement, manufacturers use both optical and mechanical techniques. Since the interference pattern is determined by the wavelength of the radiation, better scanning mechanisms are required to achieve acceptable performance in the near-IR.

The spectral coverage of an FTIR instrument is determined by the materials used to construct the optics, the source, and the scan mechanism.[26] The selection of the source is determined by the desired spectral region. For the mid- and far-IR regions, sources closely approximating a blackbody emitter are usually selected. In the near-IR, tungsten filament lamps are often used. The majority of FTIR methods are based on the use of rapid scan interferometers operating in the mid-IR.[3] A rapid scan interferometer moves the mirror at sufficiently high velocities that each wavelength interferes at modulation rates in the audio-frequency range.[6,26] For specialized applications, step-scanning interferometers that move the mirror in steps rather than at constant velocity are sometimes used.[26] The scan speed of an FTIR instrument may be quoted in time required for a single scan at a given resolution or at the frequency of data acquisition rate determined by the position of the mirror relative to a frequency reference often expressed in kilohertz.

The spectral range coverage of FTIR instruments are often limited by the beamsplitter.[25,26] When selecting a beamsplitter, it is important to consider not only spectral coverage and efficiency but also to consider the hygroscopic nature of the beamsplitter. Some beamsplitters will rapidly deteriorate when exposed even to low levels of humidity. Some of most common type of mid-IR beamsplitter are made from halide salts and should not be exposed to higher than 30% humidity.[25] Replacement of a beamsplitter can cost almost half the price of the entire FTIR system. For that reason, FTIR interferometer compartments must be kept sealed and desiccated or purged with dry air or nitrogen at all times. Several manufacturers are now offering zinc selenide beamsplitters that are not hygroscopic.[25] These beamsplitters cut off at 500 wave numbers, but this does not affect the detection of most gas compounds.

The selection of the detector for an FTIR instrument requires consideration of operation cost, data aquisition rate, and spectral range coverage. A comparison of infrared detectors was discussed in a previous section. For gas analysis, the two detectors most often used are the DTGS and MCT detectors. MCT detectors will allow rapid data acquisition rates at high signal-to-noise. Unfortunately to achieve this performance, MCT detectors have to be cooled.[13] For best performance, MCT detectors are cooled with liquid nitrogen.[3,13] When operating and purchasing cost is a primary limitation, DTGS detectors can be used. However, DTGS detectors limit the scan speed, and they also have very poor signal-to-noise performance compared to an MCT cooled to 77 K.[3,13] Sometimes it is suggested that scan averaging can be used to improve the performance of a room temperature DTGS detector to that of a MCT at 77 K. For many systems, this is incorrect. The number of scans required is often prohibitive. It has been reported that more than 10,000 scans with a DTGS detector would be required to obtain the same signal-to-noise as 1 scan with an MCT, assuming the system is detector noise limited.[3]

5.3.4 Laser Spectrometers

For many research and specialized applications, laser-based spectrometers are used.[6,10,11,16,22] In a laser spectrometer, the source and frequency selection device is the laser. This greatly simplifies the design of the spectrometer, and the resolution of the spectrometer is determined by the spectral linewidth of the laser.[10] Some laser spectrometers use tuneable lasers that can be tuned over a spectral range to obtain a spectrum.[16] Others have several lasers that emit at different wavelengths or single laser that emits at several wavelengths.[6,10,22] The spectrum can be obtained by measuring the signal from the different wavelengths from the laser or lasers, but the spacing between spectral data points is determined by the wavelengths of the lasers.

The primary advantages of laser spectrometers are their high resolution and their ability to emit high radiant intensity at the wavelength of interest.[6,10,16] This allows laser spectrometers to be used in applications where other types of spectrometers cannot be used or are not practical. The major disadvantages of laser-based systems are their limited spectral range and difficulty of use depending on the laser.[22] Laser technology is rapidly improving, and it is anticipated that within a few years laser-based spectrometers will be almost as common as other IR spectrometers.

5.4 GAS CALIBRATION STANDARDS

Because calibrations are so essential to obtaining accurate measurements, the selection of calibration gases is just as critical to the success of an infrared measurement as the selection of the instrument. The cost of the calibration gases over the lifetime of the instrument will often surpass the cost of the instrument. Many of the industrial gas suppliers have departments devoted to the production and analysis of calibration gases for a host of compounds. The customer should expect to talk with a knowledgeable technical person about their particular application, the reliability of the standard, and any special equipment requirements.

5.4.1 Generation of Gases

For compounds not stored in cylinder wells, or that prefer to adsorb to surfaces, it is critical to generate reproducible concentrations of the compound over the range of concentrations that is expected. Water is one example of a compound that often needs to be measured with IR instrumentation, if only to remove its interference from the spectrum in order to see other compounds of interest. Common techniques for generating reproducible water vapor concentrations include using a bubbler at a given temperature, or passing the carrier gas over a fabric wick under fixed conditions of temperature and pressure.[28] Compounds that are solid at room temperature are often dissolved in a carrier solvent with a high volatility and then atomized into a flowing stream of carrier gas.[28,29] The flow of the gas and the liquid can both be carefully controlled, resulting in the desired ranges of concentrations.

5.4.2 Mixing Gases

It is often necessary to mix several gases or gas mixtures to generate a gas mixture standard that is similar to the sample matrix or has the desired concentration. The mixing of gases is often based either on the measurement of pressure (static) or based on the control of flow (dynamic).[28]

5.4.2.1 Pressure Mixing. In pressure-based mixing, a closed system is used as shown in Figure 5.14. The quantity of the desired gases or gas mixtures introduced into the closed system is controlled by monitoring the pressure of the system. From

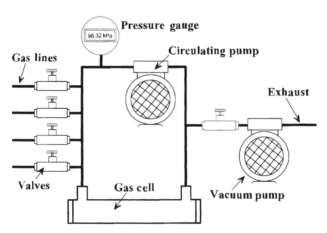

Figure 5.14 Apparatus for mixing gases based on pressure. The system is first evacuated, then filled with the desired pressures of the gases. From the partial pressures, the concentrations of the gas can be determined. The circulating pump ensures complete mixing of the gases.

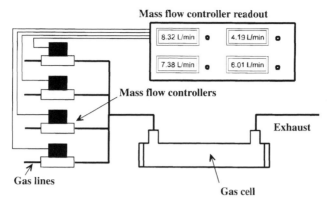

Figure 5.15 Apparatus for mixing gases based on flow. From the ratio of the flow of a gas to the total flow, the concentration of the gas can be determined.

the partial pressure of the gases, their concentrations can be calculated. Mixing gases using pressure is very inexpensive, and the mixing apparatus can be easily fabricated.[28] The accuracy of mixing is determined by the accuracy of the pressure measurement and temperature control. Although care must be taken to ensure that the gases are throughly mixed. Depending on the design of the mixing apparatus, plugs or regions of nonuniform concentrations may be created in the container and tubing used to mix the gases. To avoid this problem, circulating pumps can be used to mix the gases. The disadvantages of using pressure mixing are the time required to generate gas mixtures, only limited quantities of gases can be generated at a time, and gases that are not easily stored in containers are generally difficult to mix based on pressure.[28]

5.4.2.2   Flow Mixing. In flow mixing, the concentration of gases is controlled by adjusting the relative flows of multiple gases or gas mixtures (Fig. 5.15).[28] By increasing the flow of a given gas relative to the total flow of the system, the concentration of that gas can be increased. Flow mixing systems can use a variety of techniques to control the flows. Needle valves, gas dividers, mass flow controllers, and so forth have all been used to mix gases. The accuracy of the flow controls will determine the accuracy of the generated gas mixture. Flow mixing is often superior to pressure mixing because gas mixtures can be generated faster, mixed more rapidly, and concentrations varied more easily.[28] The primary disadvantage of flow mixing is the cost of the flow controllers.[28]

5.5   CALIBRATION CONSIDERATIONS

Before a technique for calibration can be selected, it is first necessary to consider the absorption characteristics of the sample matrix relative to the absorption characteristics of the analyte.[4–6,7,17] There are two approaches that can be taken for the determination of analyte concentration. The first involves the selection of a wavelength where the analyte absorbs, but no interfering absorption band occurs. The second is to use some mathematical method to determine the analyte concentration from absorption data even though the absorption of the analyte is overlapped with the absorption of interfering species.

The apparent overlap of the absorption bands is determined by both the real absorption characteristics of the sample and the resolution of the instrument. As previously discussed, the absorption characteristics of a species is determined by the molecule's vibrations and the different types of band broadening. In simple gas mixtures, it is often possible to find wavelengths where the analyte absorbs without interference.

Frequently, the ability to measure the absorption band without interference is limited by the resolution of the instrument. The resolution of an absorption spectrometer can be defined as its ability to distinguish two closely spaced absorption bands.[6] When the resolution greatly exceeds the bandwidth of the absorption peaks, it is possible to fully resolve the absorption spectrum. As the resolution of the instrument decreases, the absorption bands will appear to increase in width. Figure 5.16 shows the changes in the absorption spectrum of air as resolution of the instrument decreases. As a result of decreasing resolution, the bands appear to be wider and the measured absorption intensity decreases. With decreasing resolution, the overlapping of the absorption peaks increases.

To avoid the problems of peak overlap, it is necessary to work at a higher resolution, but as the resolution increases the signal-to-noise ratio decreases.[4,6,7] To maintain the signal-to-noise performance, longer data acquisition times to allow for signal averaging is required. Since the signal-to-noise ratio only increases by the square root of the averages, doubling the resolution will increase the acquisition time by at least a factor of 4 to obtain the same signal-to-noise levels.[6,7] For this reason, it is necessary to make trade-offs between resolution and signal-to-noise ratio.[5-7] The optimal resolution for an analysis will be determined by the instrument type and sample matrix, and not every IR spectrometer can be used for analysis of every gas mixture. Typically, it is desired to work at the lowest possible resolution for a given analysis.

Simple filter-based instruments and dispersive instruments have relatively poor resolution. They are best suited for the analysis of simple gas mixtures or the detection of analytes with strong absorption bands that are significantly greater than any interferences in the sample matrix.[6,7] These instruments are well suited for industrial hygiene applications such as the detection of gas leaks or the detection of carbon dioxide levels. Several companies manufacture these instruments for monitoring hydrocarbons, carbon dioxide, and carbon monoxide in stack gases. Their calibration is often based on measurement of absorption at a single wavelength or the ratio of two wavelengths.[5-7,23]

For the greatest flexibility in analysis, interferometer-based instruments should be used.[3] Moderately priced FTIR instruments are now available that have resolution of $0.125 \, cm^{-1}$. This resolution is sufficient to resolve the spectral lines of most gas phase absorption bands that have a bandwidth of around $0.2 \, cm^{-1}$.[3] Many

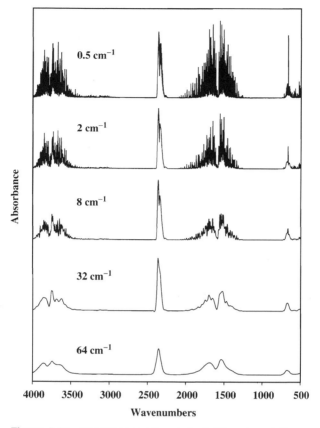

Figure 5.16 Absorption spectra of air at different resolutions.

inexpensive FTIR instruments have a resolution limited to 0.5, 1.0, or 2.0 cm^{-1}. For signal-to-noise ratio, resolution, time, and cost trade-offs, 0.5 cm^{-1} has been reported to be the best selection for trace gas measurements.[3]

5.5.1 Single-Wavelength Calibration

A single wavelength calibration requires the selection of the analytical wavelength to minimize any background absorption due to interferences.[4,6,7,17,26] By detecting a single wavelength, it is not possible to distinguish between variation in absorption because of changes in analyte concentration and variation in absorption of interferents. This limits single-wavelength techniques to applications where a wavelength can be selected where only the analyte absorbs or where the concentration of the background absorbing species does not change. Single-wavelength calibrations are always subject to error as a result of their inability to distinguish the analyte from interferents.

5.5.2 Spectral Subtraction

Subtraction has been widely used in gas analysis when the absorption of the analyte is hidden by absorption of the sample matrix. A complete discussion of subtraction is given by Hanst.[3,26] In subtraction, reference spectra are used to remove the contribution of interfering absorptions to allow the identification and quantification of analyte(s) of interest.[3,26] When using subtraction, it is assumed that at a given wavelength the absorption contributions are additive, so that

$$A_{T(\lambda)} = A_{A(\lambda)} + A_{I1(\lambda)} + A_{I2(\lambda)} + \cdots + A_{IN(\lambda)}$$

where $A_{T(\lambda)}$ is the total absorption, $A_{A(\lambda)}$ is the absorption of the analyte, and $A_{IN(\lambda)}$ is the absorbance of the interfering species. If reference spectra R are available for each interfering compound, then the absorption of the analyte can be calculated by

$$A_{A(\lambda)} = A_{T(\lambda)} - f_1 R_{I1(\lambda)} - f_2 R_{I2(\lambda)} - \cdots - f_n R_{IN(\lambda)}$$

where f_n is the factor used to adjust the reference spectra to match the contribution of the interferent. Subtraction can be performed manually by the analyst adjusting the subtraction factor f or automatically using a software routine to determine the best f factor.[3,26] Once the interferents absorption contribution has been removed, it is possible to determine the analyte concentration using Beer's law.

   In order to obtain good subtractions, it is necessary for the reference spectra to closely match the conditions of the sample matrix. The reference spectra must have the same deviations from ideal Beer's law behavior as the sample. It is also necessary for the reference spectra to be collected using the same instrumental parameters as the sample spectra. In addition, care must be taken that the wavelengths of the reference spectra and the sample are not shifted from each other. For best performance, the reference spectra should have very low noise levels since the noise in the reference spectra will be propagated in the resulting subtracted spectra.[3]

5.5.3 Multivariate Calibration

Another approach to address overlapping peaks is to use multivariate calibration techniques.[26,30–33] Multivariate calibration methods are used primarily for quantitative analysis and not for qualitative determination. In these methods, absorption measurements are made at multiple wavelengths and multiple equations are solved to determine the contribution of the analyte(s) to absorption spectra. From the solutions, the concentration of the analyte is determined. There are many software packages available that perform multivariate calibrations using several different multivariate techniques. The most common calibration procedures are classical least squares (CLS) sometimes called the K-matrix method, inverse least squares (ILS) sometimes called the P-matrix or multiple linear regression, principal component regression (PCR), and partial least squares (PLS).[32]

   The application of these methods usually requires the generation of calibration gas mixtures that closely resemble the sample matrix or the independent analysis of samples to determine the concentration.[33] Absorption spectra from the known calibration gases or samples are then used to determine the relationship between the spectra and concentrations. The absorption spectra used for the calibration are called the training sets.[33] How carefully the training set is developed will determine how useful the multivariate calibration will be. The number of spectra required for the training set will be determined by the complexity of the sample matrix.[33] For very simple mixtures analyzed using a spectrometer with good signal-to-noise performance, only a few calibration spectra may be required. For complex mixtures, hundreds of calibration spectra may be required. Once the calibration variables have been determined from the training set, the concentration of the analyte in the sample can be determined.

Which calibration technique is best for a given sample matrix is determined by the absorption characteristics of the sample and the spectrometer.[32,33] The difference between the multivariate techniques is how the relationship between absorbance and concentration is determined and how much spectral data is used. CLS, PCR, and PLS allow the entire absorption spectra to be used for the calibration.[32] By using all the data in a spectrum that contains concentration information, a dramatic improvement in the precision of the calibration can be obtained. ILS is limited in the number of wavelengths that can be used for the calibration to the number of spectra in the training set.[33] For best performance, CLS requires that all the components comprising the overlapping spectra be known.[32] ILS, PCR, and PLS can be used to analyze samples where only the concentration of analyte(s) are known as long as the other components are taken into account during the development of the training sets.[32,33] For this reason, ILS, PCR, and PLS often prove to be the most useful for complex sample matrixes. Since ILS can use a limited number of wavelengths in a spectrum, PCR and PLS almost always give better results.[32,33] It is normally found that PLS is superior to PCR for most samples because of the way the calibration calculations are performed.[32,33]

It has become a popular trend for manufacturers to sell instruments precalibrated using some multivariate method. The user should be cautious and make sure the calibration gas matrix matches the sample matrix or the calibration may not be relevant. In addition, great care must be taken when calibration parameters are transferred from one instrument to another.

5.6   SELECTION OF AN INFRARED TECHNIQUE

The quality of the data obtained from any analytical instrument is ultimately determined by how carefully the instrument's calibration is performed. While improvements in instrumentation have dramatically improved the reliability and performance of IR spectrometers, the user should always be cautious in selecting an instrument.

It must always be remembered that ultimately all IR absorption instruments do the same thing. They all provide an output that is related to measurement of the intensity of radiation passing through a sample and/or reference. The quantity expressed by the instrument's readout may be expressed as transmittance, absorbance, an arbitrary scale, or concentration, but the fundamental physical parameters are always the same. Regardless of how elaborate an instrument's user interface appears, the system can never provide more information than is contained in the raw absorption data. The amount of information that can be extracted from absorption data will be dependent on the instrument's spectral resolution, signal-to-noise performance, spectral range, and reproducibility.

Estimates of cost need to include not only expenses related directly to the instrument's operation, but also include an estimate of expenses related to how the instrument's data will be utilized. If data from the instrument is used to document emissions, protect workers from exposure, or to control a product process, it is also necessary to consider how an instrument's failure, incorrect calibration, and the like could impact safety and liability. It does not make sense to try to use a cheaper instrument or method that cannot provide the desired accuracy, precision, reliability to save a few thousand dollars when data from the instrument could have an impact of several hundred thousand dollars.

REFERENCES

1. C. E. Meloan, *Elementary Infrared Spectroscopy*, Macmillian, New York, 1963.

2. R. C. Lord, Infrared Spectroscopy, in S. P. Parker (Ed.), *Spectroscopy Source Book*, McGraw-Hill, New York, 1987, pp. 137–145.

3. P. L. Hanst and S. T. Hanst, Gas Measurements in the Fundamental Infrared Region, in M. W. Sigrist (Ed.), *Air Monitoring by Spectroscopic Techniques*, Wiley, New York, 1994, pp. 335–470.

4. D. A. Skoog, D. M. West, and F. J. Holler, *Fundamentals of Analytical Chemistry*, 7th ed., Saunders College Publishing, Philadelphia, PA, 1996.

5. H. H. Willard, L. L. Merritt, Jr., J. A. Dean, and F. A. Seattle, Jr., *Instrumental Methods of Analysis*, 6th ed., Wadsworth Publishing, Belmont, CA, 1981.

6. J. D. Ingel, Jr. and S. R. Crouch, *Spectrochemical Analysis*, Prentice-Hall, Englewood Cliffs, NJ, 1988.

7. D. A. Skoog, F. J. Holler, and T. A. Nieman, *Principles of Instrumental Analysis*, 5th ed., Saunders College Publishing, Philadelphia, PA, 1998.

8. C. Zhu and P. R. Griffiths, *Appl. Spectrosc.* **52**(11), 1403–1408 (1998).

9. U. Platt, Differential Optical Absorption Spectroscopy (DOAS), in M. W. Sigrist (Ed.), *Air Monitoring by Spectroscopic Techniques*, Wiley, New York, 1994, pp. 27–84.

10. W. Demtröder, *Laser Spectroscopy Basic Concepts and Instrumentation*, Springer-Verlag, New York, 1988.

11. S. Svanberg, Differential Absorption LIDAR (DIAL), in M. W. Sigrist (Ed.), *Air Monitoring by Spectroscopic Techniques*, Wiley, New York, 1994, pp. 85–162.

12. W. L. Wolfe and G. J. Zissis (Eds.), *The Infrared Handbook, rev. ed.*, Environmental Research Institute of Michigan, Ann Arbor, MI, 1993.

13. *Characteristics and Use of Infrared Detectors—Technical Information SD-12*, Hamamatsu Corporation, Bridgewater, NJ, 1989.

14. J. Hecht, *Understanding Lasers*, Howard W. Sams & Company, Carmel, IN, 1988.

15. *IR Detectors–IR Emitters–Optical Gas Cells*, Product Literature, Laser Monitoring System, Devon, United Kingdom.

16. H. I. Schiff, G. I. Mackay, and J. Bechara, The Use of Tuneable Diode laser Absorption Spectroscopy for Atmospheric Measurements, in M. W. Sigrist (Ed.), *Air Monitoring by Spectroscopic Techniques*, Wiley, New York, 1994, pp. 239–334.

17. D. C. Harris, *Quantitative Chemical Analysis*, 5th ed., W. H. Freeman and Company, New York, 1999.

18. J. D. Vincent, *Fundamentals of Infrared Detectors—Operation and Testing*, Wiley, New York, 1990.

19. E. R. Altwicker, L. W. Canter, S. S. Cha, K. T. Chuang, D. H. F. Liu, G. Ramachandran, R. K. Raufer, P. C. Reist, A. R. Sanger, A. Turk, and C. P. Wagner, Air Pollution, in D. H. F. Liu and B. G. Lipták (Eds.), *Environmental Engineers' Handbook*, 2nd ed., Lewis Publishing, Boca Raton, FL, 1997, pp. 227–448.

20. J. U. White, *J. Opt. Soc. Am.* **32**, 285–288 (1942).

21. Product Literature, Infrared Analysis, Anaheim, CA.

22. P. L. Kebabian and C. E. Kolb, The Neutral Gas Laser: A Tool for Remote Sensing of Chemical Species by Infrared Absorption, in E. D. Winegar and L. H. Keith (Eds.), *Sampling and Analysis of Airborne Pollutants*, Lewis Publishers, Boca Raton, FL, 1993, Chapter 16.

23. J. A. Jahnke, *Continuous Emission Monitoring*, Van Nostrand Reinhold, New York, 1993.

24. tapIR Data Sheet, Product Information, American Laubscher, Farmingdale, NY.

25. Bomen MB155, Product Information, Bomen, Québec, Canada.

26. P. R. Griffiths and J. A. de Haseth, *Fourier Transform Infrared Spectrometry*, Wiley, New York, 1986.

27. D. Noble, *Anal. Chem.*, **67**(11), 381A–385A (1995).

28. W. J. Woodfin, *Gas Vapor Generating Systems for Laboratories*, DHHS (NIOSH) Publication 84–113, Department of Health and Human Services, Washington, DC, 1984.

29. J. F. Pankow, W. Luo, L. M. Isabelle, D. A. Bender, and R. J. Baker, *Anal. Chem.*, **70**(24), 5213–5221 (1998).

30. D. M. Haaland and E. V. Thomas, *Anal. Chem.*, **60**(11), 1193–1202 (1988).

31. D. M. Haaland and E. V. Thomas, *Anal. Chem.*, **60**(11), 1202–1208 (1988).

32. E. V. Thomas and D. M. Haaland, *Anal. Chem.*, **62**(10), 1091–1099 (1990).

33. *PLSplus*, Galactic Industries Corporation, Salem, NH, 1991.

6

ULTRAVIOLET ANALYZERS

Jeffrey E. Johnston
Zetetic Institute
Tucson, Arizona

Marc M. Baum
Oak Crest Institute of Science
Pasadena, California

Environmental Instrumentation and Analysis Handbook, by Randy D. Down and Jay H. Lehr
ISBN 0-471-46354-X   Copyright © 2005 John Wiley & Sons, Inc.

Air is a complex matrix of major, minor, and trace chemical species. Usually only a few of the trace constituents are of interest to the observer, the remaining substances being potential interferents. Optical analyzers have the capability of accurately quantifying some species in volumetric concentrations as low as the part-per-billion (ppb_v) level with minimal effects from other components.

The key to recognizing the advantages and limitations of optical techniques is to understand the spectroscopy of the mixture's components. Matter can absorb, emit, or scatter light. Each species interacts with light differently as a function of wavelength (color), presenting a unique spectral "fingerprint" that allows it to be distinguished from other components of the mixture. The objective of any optical analysis method is to use this unique signature to quantify the species of interest.

In comparison with the infrared (IR) region of the electromagnetic spectrum, the ultraviolet (UV) region offers the advantages of sensitivity and practicality for pollutant measurement. Ultraviolet spectral signatures are usually more intense, resulting in a lower concentration of sample yielding a similar signal. Practically, the large variety of commercially available measurement methods and optical components in the UV facilitates greater flexibility in instrument design.

In absorption spectroscopy—the most widely used spectroscopic method for pollutant observation—both the advantage and limitation of UV observations become apparent. Figure 6.1 shows the transmission spectrum of a typical sample of air at ground level.[1] The 200–400 nm region of the UV experiences little attenuation by both the major (N_2 and O_2) and minor (H_2O, CO_2, etc.) components of the ambient atmosphere, providing a relatively clear spectral region for measuring trace species. Although the 800–20,000-nm region of the IR is free from absorption by the major species, minor species provide significant absorption, which often

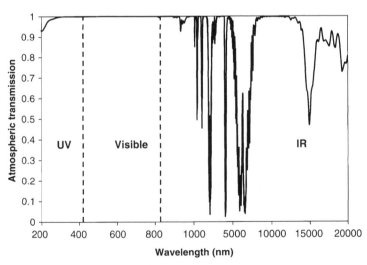

Figure 6.1   Transmission function of an unpolluted 10-m optical path in the atmosphere. Below 200 nm the transmission quickly drops to zero.

interferes with trace species measurements. In fact, the near-ubiquitous nature of the H_2O absorption structure makes low-level IR measurements of other compounds difficult. The advantages of the UV observations are often accentuated when a continuous emission monitor (CEM) is used to measure the exhaust of combustion processes. Typically, in this application, high concentrations (percent level) of both CO_2 and H_2O are common.

Ultraviolet methods are also limited by atmospheric transmission. Similar to the omnipresent absorptions in the IR, both the O_2 and H_2O features cause nearly complete absorption in the UV below 200 nm. Most species have both UV and IR absorption spectra, but only a portion of them are present in the usable region of the UV. For this reason, UV and IR measurements are both necessary and complementary.

As a general philosophy, one measurement method is not "better" than another. Variable requirements of sensitivity, selectivity, response time, convenience, size, and price make the choice of analyzer application specific.

This chapter describes the capabilities of UV analyzers, allowing potential users to determine what type of UV instrument will provide the optimal solution to the problem at hand.

6.1 ULTRAVIOLET SPECTROSCOPY

6.1.1 Physical Principles

There are three observable ways by which light can interact with matter: absorption, emission, and scattering. Optical gas analyzers employ these effects to observe pollutants directly or, following a chemical transformation, indirectly.

These three mechanisms are best illustrated with energy level diagrams shown in Figure 6.2. Energy levels represent the total energy possessed by an atom or molecule and are well defined, discrete, and species specific as a result of quantum mechanical principles. An energy level diagram schematically illustrates the location of these allowed energy levels and how light can interact with a given species by promoting transitions from one quantum state to another. The vertical axis of the diagram is the energy of the species and the horizontal bars are the energy levels. Arrows between the energy levels indicate the excitations or de-excitations; squiggly arrows indicate interactions with light.

Energy is conserved in radiative processes. Emitted or absorbed light possesses energy equal to the difference between the upper and lower excitation states. The energy of a photon is conventionally described in terms of wave number (v, cm^{-1}) or wavelength (λ, nm). The conversion between energy (E, ergs), wave number, and wavelength is

$$E = hcv = \frac{hc10^7}{\lambda}$$

where h = Planck's constant (erg·s)

c = speed of light (cm/s)

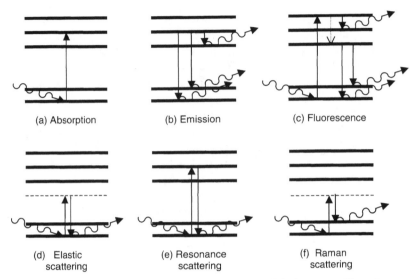

Figure 6.2 Energy level diagrams illustrating how light interacts with matter.

A spectrum is a superposition of spectral lines from specific transitions. The transition energy between levels determines the wavelength of the line; the number of atoms or molecules (population) in the initial state and the transition probability between states determine the intensity of the line.

6.1.1.1 *Absorption.*

In the absorption process (Fig. 6.2a), energy from a photon excites a species to a higher energy state. At room temperature, most analyte molecules will be in their ground (lowest energy) state, and excitation from that state via absorption is expressed in terms of an absorption cross section (probability), $\sigma(\lambda)$. This absorption coefficient is the sum of all absorption transition probabilities between the ground and excited states at every wavelength.

Quantitatively, the magnitude of the absorption features is used to determine species concentration by the Beer–Lambert law:

$$I_t(\lambda) = I_0(\lambda)e^{-\sigma(\lambda)[c_m]\ell} \tag{6.1}$$

where $I_0(\lambda)$ = light intensity (photons/s) before passing through sample

$I_t(\lambda)$ = light intensity (photons/s) transmitted through sample

$\sigma(\lambda)$ = absorption cross section (cm^2)

$[c_m]$ = concentration of lower state (cm^{-3})

ℓ = path length of light through sample (cm)

The Beer–Lambert law can be applied as long as the resolution of the spectrum is fine enough so that the narrowest features are adequately resolved.

6.1.1.2 Emission. Emission, the converse of absorption, is the radiational decay of an excited state (Fig. 6.2*b*). An external excitation mechanism, such as a chemical reaction or electrical discharge, is required to produce emission spectra. Fluorescence, a special case of emission, occurs when the excitation is caused by radiative absorption (Fig. 6.2*c*).

The emission intensity is expressed by the formula

$$I_e(\lambda) = N_n A_{n,m}$$

where $I_e(\lambda) =$ photon flux at transition wavelength (photons/s)
$N_n =$ total population of excited state
$A_{n,m} =$ emission transition probability between states n and m (s^{-1})

Once a species is in an excited state, it will usually be depopulated by one of a number of processes, such as radiative emission, collisional deactivation, or chemical transformation. These competing mechanisms must be taken into account when accurate measurements of population, hence species concentration, are made. For many radiative transitions at atmospheric pressure, collisional deactivation (quenching) is the dominant decay mechanism.

6.1.1.3 Scattering. Scattering is a process where light is absorbed and instantaneously reemitted. This process is most efficient in the UV because the scattering cross section is proportional to the fourth power of the incident photon energy.

There are four types of scattering processes: elastic, resonance, ordinary Raman, and resonance Raman. In elastic scattering, the target atom or molecule absorbs light and is temporarily excited into a "virtual state" (shown as a dashed line in Fig. 6.2*d*). Resonance scattering occurs when this virtual state is coincident with an excited-state energy level (Fig. 6.2*e*). For resonance processes the scattering cross section is greatly enhanced. In coherent processes (non-Raman), the energy of the scattered photon is the same as the incident photon. In Raman processes (Fig. 6.2*f*), it is shifted by a discrete amount of energy dependent on the chemical makeup of the scatterer.

Elastic scattering is used to measure particles and aerosols. Gases can be speciated in two different ways. Resonant processes tune the wavelength of the incident beam to the resonance features of the target species. Raman processes observe the wavelength shift of the returned signal.

6.1.2 Spectral Structure

Atoms generally possess a relatively small number of energy levels and, consequently, emit or absorb light as a modest number of strong, sharp lines.[2] This makes it fairly easy to identify and quantify their presence in gaseous form (e.g., elemental mercury). For air quality monitoring, elemental measurements play a relatively minor role in comparison with molecular compounds.

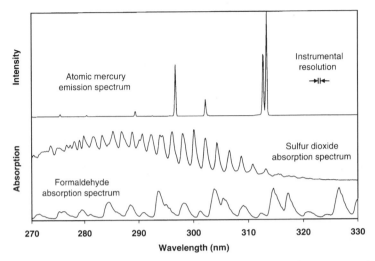

Figure 6.3 Comparison of atomic and molecular spectra. Atomic features are sharp and localized; molecular features are diffuse and widespread.

Molecules have a more complex spectral structure (Fig. 6.3) than their atomic counterparts because of their ability to vibrate and rotate.[3] Thus, each electronic energy level is split into numerous vibrational and rotational energy "sublevels." Transitions among this multitude of states result in molecular spectra being very broad and diffuse compared to atomic spectra. The breadth of these bands can compromise selectivity because features from several species may overlap in the measurement region. The diffuseness of the bands lowers sensitivity because a weaker signal is generated, as compared with a single absorption line.

Gas analysis poses the challenge of providing consistent, accurate, and automatic quantification of pollutants from complex, faint, and overlapping spectra. How this is achieved is discussed in Section 6.2.

6.2 MEASUREMENT METHODS

Optical instrumentation can be classified into two categories differentiated by the directness of the pollutant observation: spectroscopic and nonspectroscopic. Spectroscopic instruments measure the actual pollutant via optical methods. Nonspectroscopic optical techniques quantify the pollutant after it has been chemically transformed. Although nonspectroscopic instrumentation functions in the visible as well as the UV, we restrict the scope of this work to the UV because the measurement principles are the same in both spectral regions. The emphasis of this discussion is on the spectroscopic measurement of atmospheric pollutants.

6.2.1   Absorptive Techniques

Absorption is the most common optical method for the measurement of atmospheric pollutants. Nearly any species that absorbs UV light can be measured by this method. These techniques are categorized by the type of instrumentation used to perform the observation.

6.2.1.1 Nondispersive. Nondispersive instruments provide a simple and relatively inexpensive way of quantifying pollutants. In this instrumentation, a photometer and a light source are used to measure the light absorbed by a sample. A photometer is a device that uses an optical filter to quantify light intensity in a narrow spectral window. The light source can have a broad emission, such as a deuterium lamp, or can be a line source, such as an atomic discharge lamp. Lasers provide another alternative source, but at the present time they are rarely used because of cost and flexibility constraints. Atomic and laser sources provide narrow lines of a constant wavelength. The precision of these features provides good instrument linearity and measurement stability.

Photometric analyzers may be either spectroscopic or nonspectroscopic. Photometric methods are popular with indicator-based nonspectroscopic instruments because they provide a relatively inexpensive and accurate way to "read" a chemical indicator. These instruments usually measure the absorbance of a liquid indicating reagent or the surface reflectance of a chemical reagent. Treatment of this nonspectroscopic instrumentation is discussed elsewhere.[4]

Although individual designs of spectroscopic nondispersive analyzers may vary in layout and composition, a description of the function and merit of several common designs[5] follows.

6.2.1.1.1 Single-Wavelength Filter Photometers. The simplest of all UV analyzers, the single-beam filter photometric analyzer, is shown in Figure 6.4. Light from an appropriate source is collimated (made parallel), filtered, directed through a sample, and subsequently focused on the surface of a detector. The filter transmits at a wavelength where a strong absorption from the sample exists. In essence, only $I_t(\lambda)$ from Equation (6.1) is measured with a sample in the cell; $I_0(\lambda)$ is updated by periodically purging the cell with "zero" gas. The Beer–Lambert law is applied to these measurements to determine gas concentration.

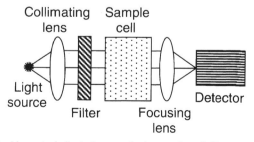

Figure 6.4   Optical layout of single-beam, single-wavelength filter photometric analyzer.

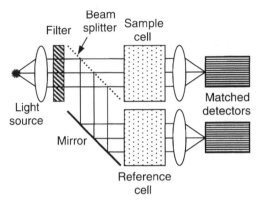

Figure 6.5 Optical layout of beam-split, dual-beam, single-wavelength photometric analyzer.

The dual-beam filter photometric analyzer shown in Figure 6.5 provides a continuous reading of both $I_t(\lambda)$ and $I_0(\lambda)$. The beam is filtered, split, and projected through two different media of identical optical path lengths: a reference cell and a sample cell. Light is captured by two matched detectors to generate electronic signals that are logarithmically amplified and subtracted to yield a quantity proportional to the gas concentration. Periodic purging of the sample cell provides a "zero" reading to correct for instrumental drift.

Figure 6.6 shows a second embodiment of this instrument where light is sequentially gated through each path to facilitate measurements with one detector and one amplifier. Beam splitting offers better time resolution, simplified electronics, and no moving parts. On the other hand, sequential gating may be cost advantageous and minimizes drift by using the same electronics for both measurements.

6.2.1.1.2 Dual-Wavelength Filter Photometers. The dual-wavelength filter photometric analyzer measures the intensity of light at a reference wavelength, $I_t(\lambda_{ref})$,

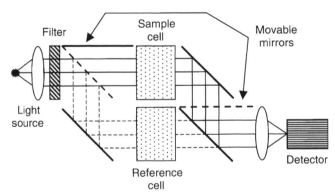

Figure 6.6 Optical layout of sequentially gated, dual-beam, single-wavelength photometric analyzer.

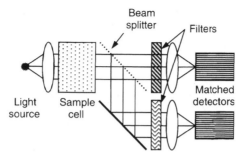

Figure 6.7 Optical layout of split-beam, dual-wavelength photometric analyzer.

and at a measurement wavelength, $I_t(\lambda_{meas})$. The filters are chosen so that λ_{meas} is a region of strong absorption and λ_{ref} is in a region of weak or no absorption. Assuming $I_0(\lambda_{ref})$ is proportional to $I_0(\lambda_{meas})$, $\ln[I_t(\lambda_{ref})/I_t(\lambda_{meas})]$ is linearly related to the concentration of target species in the sample cell. There are two different types of dual-wavelength instruments: split-beam and filter wheel photometric analyzers.

In the split-beam dual-wavelength design shown in Figure 6.7, the light traverses the sample and is divided into two beams that are optically filtered at two different wavelengths before striking two separate matched detectors. The configuration is similar to the split-beam single-wavelength analyzer, except that $I_t(\lambda_{ref})$ is measured instead of $I_0(\lambda_{meas})$. However, the dual-wavelength approach has the benefit of being impervious to variations in sample opacity, as both measurements are obtained from the same sample cell.

Filter wheel photometric analyzers (Fig. 6.8) have certain similarities with the sequentially gated systems discussed above. Measurements at different wavelengths are time multiplexed and performed by the same detector. Variations in sample concentration and opacity must be slower than the filter wheel cycle time for the readings to be valid. The filter wheel can hold a number of filters, giving this technique the ability to measure multiple analytes or collect more than one reference to increase selectivity.

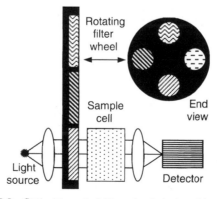

Figure 6.8 Optical layout of filter wheel photometric analyzer.

*6.2.1.2 **Dispersive.*** Unlike systems discussed in the preceding section, most dispersive instruments have the ability to observe complete absorption spectra over a wide wavelength range and at an optimal spectral resolution. Consequently, very low detection limits are achievable with good selectivity.

An absorption-measuring dispersive gas analyzer consists of a broadband light source, a sample, a dispersive wavelength selection device, and a detector. Wavelength separation is achieved in a spectrometer or in a spectrograph. Both devices use an optical element such as a diffraction grating or prism to separate the light into its component colors.

6.2.1.2.1 Spectrometers. A spectrometer determines light intensity as a function of wavelength. The information contained in the spectrum allows multiple species to be identified and measured through the mathematical removal of interfering features.

An analyzer employing a Czerny–Turner spectrometer to measure gas absorption is shown in Figure 6.9. Light from a continuum source is directed sequentially through the spectrometer and sample to the detector. Placing the spectrometer before the sample minimizes photodissociation. Light entering the spectrometer through the entrance slit is collimated by the first mirror. The dispersive optic preserves the beam's parallelism and separates it into its component wavelengths. The focusing mirror images the light from the dispersive optic onto the exit slit plane. The narrow-bandwidth light passing through the exit slit is directed through the sample and then to the detector. The spectrometer is scanned in wavelength by either rotating the dispersive element or changing the lateral position of the entrance or exit slit. The slit widths control the instrument's resolution (wavelength bandwidth), which is selected to balance a linear response to analyte concentration with light throughput.

Once the spectrum is obtained, chemometric data reduction techniques are applied to yield species concentrations. Chemometrics is the process of spectral analysis that goes beyond the measurement of peak height. Spectral data can be untangled by a host of signal-processing tools, including second-order derivatives,

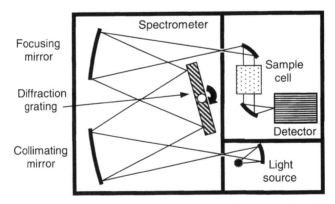

Figure 6.9 Optical layout of gas analyzer employing spectrometer.

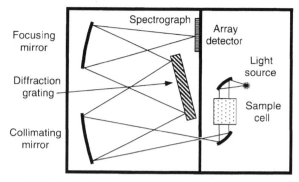

Figure 6.10   Optical layout of gas analyzer employing spectrograph.

fast Fourier transforms, classical and partial least-squares fits, neural networks, and pattern recognition algorithms. A detailed discussion of chemometrics is beyond the scope of this text. A good review[6] is available to the interested reader.

6.2.1.2.2   Spectrographs. Figure 6.10 shows an absorptive gas analyzer employing a Czerny–Turner spectrograph. A spectrograph is similar to a scanning spectrometer in function and form except that it employs an array detector in the focal plane. This design utilizes light as efficiently as possible by allowing observation of the entire spectrum at once. In a scanning spectrometer the exit slit rejects most of the light, decreasing the signal-to-noise ratio of the system. The simultaneity of the measurement makes an analyzer utilizing a spectrograph insensitive to short-term variations in analyte concentration and changes in sample opacity. The lack of reliance on moving parts makes this system more rugged for field applications.

Although a spectrograph is at times advantageous, it is not ideal for all applications. A multiple-pass spectrometer is the best solution when large absorptions need to be measured. Also, the expense of the array detector can be prohibitive, although small, moderately priced systems are now available.

6.2.1.2.3   Correlation Spectrometers. Correlation spectrometry is a measurement technique that provides fast response time without any chemometric data analysis. A correlation spectrometer[7] is identical to the scanning spectrometer design discussed previously, except that multiple slits are used to allow light at several points in the spectrum through to a single channel detector. The shape and placement of the slits are chosen to allow transmission of light only at the local extrema of the target species' absorption spectrum (Fig. 6.11). The transmitted wavelengths are modulated, by either oscillating the slit mask or rocking the dispersive optic, to produce a signal that is successively correlated with the local maxima and minima. In the case where the absorption is small $[\sigma(\lambda)[c_m]\ell \ll 1]$, the absorption intensity is

$$I_a(\lambda) = I_0\left[1 - e^{-\sigma(\lambda)[c_m]\ell}\right] \approx I_0\sigma(\lambda)[c_m]\ell$$

This implies the modulation amplitude is proportional to the amount of gas present.

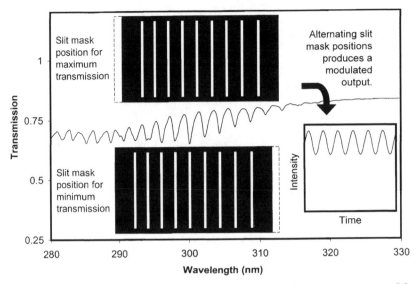

Figure 6.11   Slit mask position for correlation spectrometer designed to measure SO_2.

The technique of correlation spectroscopy provides a very fast response time with a sensitivity on par with methods employing chemometric data analysis. Selectivity is better than nondispersive methods but not as good as methods that use chemometrics. The gas-specific slit mask usually limits the capability of this type of analyzer to the measurement of a single gas species with a roughly periodic absorption structure.

6.2.1.3 *Interferometric.* Similar to spectroscopy, interferometry characterizes absorption signatures by analyzing patterns of light that have passed through the sample. The difference is that the patterns are arranged in the spatial as opposed to the frequency domain. For practical purposes, the information content of interferometric data can be considered equivalent to spectral data.

The two applicable UV measurement methods are correlation interferometry and Fourier transform spectroscopy. Both techniques involve the use of the "polarization interference filter"[8] (PIF) as an interferometric element, shown in Figure 6.12. In the PIF, light passing through a small hole is collimated and directed through a polarizer, birefringent crystal, second polarizer, and detector. The polarizers set up a pattern of interference in the crystal that gives the PIF a transmission function that is periodic with respect to the energy of the incident photons. The length of the crystal determines the length of the period. Although other instruments can be used, the robustness of the PIF makes it the only commercially practical interferometer at this time for UV measurements outside of the laboratory.

6.2.1.3.1 Correlation Interferometers. The correlation interferometer[9] provides a very fast, dedicated reading of one pollutant species with a roughly periodic

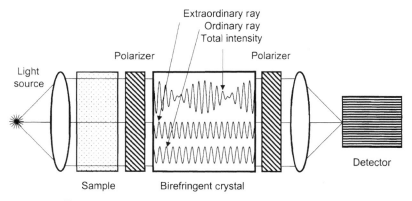

Figure 6.12   Optical layout of polarization interference filter.

absorption structure. It is the interferometric analog to the correlation spectrometer and has a similar sensitivity and selectivity.

The correlation interferometer is essentially a PIF with two additional optical components: a modulator and an optical filter. It has a transmission function (Fig. 6.13) with a bandpass component determined by the optical filter bandpass and a periodic component determined by the PIF tuning. The filter is selected with a transmission function that overlaps the target species absorption, and the PIF is tuned to match the periodicity of the absorption feature. The modulator adjusts the phase of the PIF transmission function so it is alternately in sync and

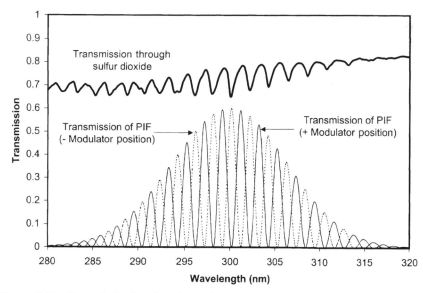

Figure 6.13   Transmission function of correlation interferometer designed to measure SO_2.

out of sync with the absorption features of the gas. Thus, the correlation interferometer "gates" the light much like the correlation spectrometer. This will produce a modulation amplitude proportional to the gas present in the sample when the absorption is small.

6.2.1.3.2 Fourier Transform Spectrometers. Fourier transform spectrometers are interferometers that use Fourier analysis to obtain spectra from interferograms. As with other spectral methods, the spectrum is analyzed by chemometric techniques to quantify the pollutants.

In the UV, Fourier transform spectroscopy is more or less equivalent to dispersion spectroscopy. Fourier transform spectroscopy usually covers a larger spectral range, offers a slight light-gathering advantage (Jacquinot's advantage), and does not suffer from light contamination. Dispersive methods enable measurements to be made in a smaller spectral region and do not require as large a dynamic range as Fourier transform methods.

6.2.2 Emissive Techniques

Analyzers relying on light emission as their measurement principle generally are dedicated to a single species to optimize the excitation mechanism and the observation wavelength.

All emissive instruments are similar in optical design. Light from the excited species is quantified by a photometer. Unlike with absorption spectroscopy, the signal increases with pollutant concentration, so the measurement is essentially observed against a "black" background. This avoids the large dynamic range requirements of noncorrelative, absorptive detectors and electronics. Efficient high-gain detectors, such as photomultiplier tubes, are used to measure the low light levels typical of emissive techniques; these configurations can even detect single photon events. Thus, with similar sample path lengths, emissive instruments can have sensitivities many times higher than their absorptive counterparts.

Emissive techniques are differentiated on the basis of their excitation mechanism, as detailed in the following sections.

6.2.2.1 Fluorescence. Fluorescence[10] uses a radiative source to produce the excited state. Fluorescence is the only truly spectroscopic emissive technique because it does not chemically transform the measured species. Emission may be observed from the original excited state or from a lower state populated through collisional or radiative cascade. Regardless of the process, the fluorescent intensity will be proportional to the amount of light absorbed if the absorption is small.

A typical fluorescence gas analyzer is shown in Figure 6.14. Light from a UV light source is collimated, filtered, and directed through a sample and then blocked. The source can be either a broadband emitter or an atomic discharge lamp. Light from the fluorescence process is collected at right angles to the excitation beam by a photometer.

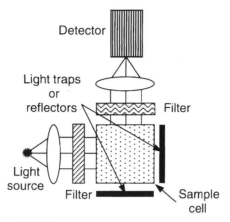

Figure 6.14 Optical layout of fluorescence gas analyzer.

The source filter selects wavelengths of light that will excite the target molecule, whereas the photometer filter passes light from the emission band of interest to the detector and rejects light from the excitation source. The most popular application of fluorescent analyzers of this type is for the measurement of sulfur dioxide.

Laser-induced fluorescence (LIF) relies on a coherent source, in lieu of the traditional UV light source and filter, to target the specific absorption feature of interest. The practical limitations of UV lasers, as discussed previously, limit the usefulness of LIF in commercial analyzers at the present time.

6.2.2.2 *Chemiluminescence.*

6.2.2.2 Chemiluminescence. Chemiluminescent[11] gas analyzers rely on a chemical reaction to produce the electronically excited state. As the chemical transformation is the key to producing the excited state, these nonspectroscopic instruments are usually designed specifically for one compound.

Pollutant measurement by a chemiluminescence technique is a two-step process. In the first step, the sample is mixed with a known amount of reactant to generate the excited species. In the second step, light produced by the decay of the excited state is detected. The chemiluminescent intensity is quantitatively expressed from stoichiometry and reaction kinetics. In some cases, varying the pressure in the reaction cell can increase the efficiency of the chemiluminescent process.

Figure 6.15 illustrates a chemiluminescent analyzer for the detection of NO_X. In this common application, NO is measured through a reaction with O_3. Through different sample pretreatments, the NO_2 concentration can also be inferred. The analyzer alternately measures the NO content of samples directly and after they have been cycled through a converter. The converter reduces NO_2 to NO, so that the processed samples have an NO concentration equal to the sum of the NO and NO_2 concentrations in the original sample. Subtraction of these two measurements yields the NO_2 concentration.

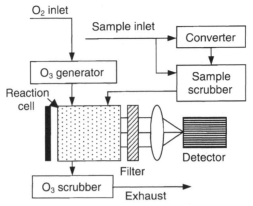

Figure 6.15 Optical and flow diagram of chemiluminescent analyzer for measurement of NO_X.

Chemiluminescence techniques[11] are also used to measure O_3 (reacted with ethylene), sulfur compounds (reacted with H_2), or phosphorous compounds (reacted with H_2).

6.2.3 Scattering Techniques

Scattering methods play a unique role in the measurement of atmospheric pollutants because they are the only techniques that yield range-resolved information. These methods can quantify the distribution of either particulates or chemical species in the outdoor environment. The following section will focus on the use of UV scattering methods for trace-gas measurement.

6.2.3.1 Lidar. Lidar (light detection and ranging) quantifies both the location and concentration of the measured species. Analogous to radar, a lidar (Fig. 6.16) sends out a pulse of light and times its return from targets. The targets considered here consist of pollutant atoms or molecules. Although various forms of lidar exist, differential absorption lidar (DIAL) is the most applicable technique for trace-pollutant analysis in the UV.

In DIAL,[12,13] a laser source is tuned to an absorption feature of the target species. A second laser is tuned to an adjacent wavelength where little or no absorption occurs. Although the scattering cross sections (probabilities) of these two beams are quite different, the wavelength separation is sufficiently small that, in every other way, the beams are identical. Pulses from the two sources are transmitted through the atmosphere separated by a very short time interval, guaranteeing the same atmospheric conditions. The backscattered signal is received by a telescope, passed through a narrow-bandpass optical filter, and detected by a photomultiplier. The lidar receiver is sensitive to extremely small signals because the detector is looking at an inherently "black" background, as with emission spectroscopy. Numerically differentiated data are used to determine concentration as a function of distance.

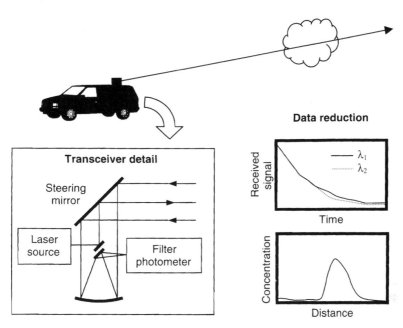

Figure 6.16 Principle of DIAL operation.

DIAL can also be used to determine path-integrated concentrations between the transceiver and a physical target. This technique, termed *topographic target lidar*, is similar in concept to monostatic open-path instrumentation discussed later.

In order to ensure accurate measurements, the time scale of constituent variability must be smaller than the volumetric scan time of the instrument, and the atmosphere must be free from significant amounts of interferents that absorb at the DIAL wavelengths.

6.3 STANDARD APPLICATIONS

The stringent needs of environmental measurements drive the design of highly specific instrumentation. The requirements of specific applications determine the optimal measurement principle, design, and configuration of a UV gas analyzer. The following discussion addresses three common types of applications and the way UV analyzers are designed to fit these very different requirements.

6.3.1 Continuous Emission Monitors

A continuous emission monitor[5] (CEM) measures the chemical composition of a fluid on an uninterrupted, automated basis. In the field of gas analysis, CEMs are mainly used to measure emissions from industrial processes, such as stack gas. Data from these instruments are used to assure environmental compliance and to achieve efficient process control. The demanding measurement requirements and environmental conditions typical of most CEM applications dictate the need for highly specialized equipment.

6.3.1.1 Challenges. The conditions surrounding the process exhaust stream can present numerous design challenges for the CEM. Gas temperatures can be in excess of 1000°C, with dew points above 250°C. Highly corrosive compounds are often present. As it is often located at the stack, the analyzer must be designed to withstand the industrial environment where vibration, power fluctuations, high dust levels, and wide temperature variations are encountered.

The above requirements are addressed through a series of design approaches. Extractive probes are made of materials that are impervious to the high temperatures and chemical composition of the stream; optical probes are designed to avoid the stack environment through the circulation of clean air flows. Analyzer functionality is maintained through sealed, temperature-controlled enclosures and power-regulated electronics.

6.3.1.2 Approaches. There are two main measurement philosophies for a CEM: extractive and *in situ*. An intermediate approach known as *ex situ* is also employed, although it is far less common than the other two. Most of the technologies discussed in the preceding sections can be applied to either of these configurations, each with its own merits and drawbacks.

6.3.1.2.1 Extractive. Extractive systems pipe effluent samples to the analyzer, typically located a distance from the stack. This technique is best suited to analytes that are easily transported, especially if they are difficult to measure without pretreatment.

Transporting the sample to the analyzer has several advantages. Practically any UV measurement technique can be used. Extractive CEMs are easily zeroed in the field by flowing N_2 through the sample line. Sample temperature and pressure are accurately controlled during the measurement, reducing errors arising from environmental changes. Even the sample cell is designed with a path length optimized for the required measurements.

The main challenge associated with the extractive methodology lies with preserving sample integrity as it is transported over long distances (typically 20 m or more). This is especially difficult when corrosive gases are measured. NH_3, for instance, adsorbs to the walls of the sampling system, even if the extraction lines are heated. The probe, filters, lines, dilution/drying systems, pumps, valves, and other components of the sampling equipment also make extractive systems more maintenance intensive and prone to failure than their noninvasive counterparts.

Typical gases measured by extractive UV CEMs are NO, NO_2, SO_2, and H_2S. NO_X measurements are performed using chemiluminescence (often the preferred method) or absorption methods. The chemiluminescent NO_X measurement technique was described previously in Section 6.2.2.

SO_2 measurements are made by UV fluorescence or absorption techniques. Fluorescence requires the sample to be diluted because of the quenching efficiency of other flue gas constituents.

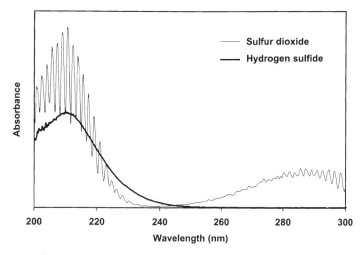

Figure 6.17 H₂S and SO₂ spectra. Although H₂S has a strong absorption feature at 210 nm, the possible presence of SO₂ limits its measurement to a region of relatively weak absorption near 240 nm.

H₂S measurements can be made either directly by absorption or via chemical conversion to SO₂. The conversion process is preferred because the SO₂ absorption spectrum is both stronger and better defined than the H₂S spectrum (Fig. 6.17), resulting in a much lower cross-sensitivity to other stream constituents. Analogous to chemiluminescent NO_X determinations, SO₂ is measured before and after conversion to infer the H₂S concentration.

6.3.1.2.2   In Situ. Analyzers mounted to the stack or duct that monitor pollutants directly in the process exhaust without conditioning are commonly referred to as *in situ* systems. They are best suited for constituents that are difficult to transport and can be measured by absorption spectroscopy. *In situ* CEMs are also popular for process control applications where fast response times are needed.

The noninvasive approach offers several important advantages over the extractive method. Errors associated with transporting and conditioning the sample are eliminated. The UV beam samples a line integral across the stack and, thus, is not prone to errors arising from gas stratification in the stack. *In situ* systems have fast response times because they observe the stream directly, as opposed to transporting the sample and refreshing a measurement cell. Finally, because there are fewer parts, *in situ* CEMs have lower maintenance requirements.

However, these benefits come at a price. *In situ* analyzers can be difficult to zero as the stack cannot be purged with N₂. Scintillation due to temperature variations and scattering of the light by dust, water droplets, and aerosol are other common problems. Scattering effects are especially troublesome for UV systems because they are more pronounced for short-wavelength radiation.

In situ systems primarily rely on absorptive techniques to measure analyte concentration. Fluorescence, although feasible in principle, is not normally used. Nonspectroscopic methods cannot be used because chemical transformations cannot be performed on the sample without removing it from the stack. Thus, for gases where chemiluminescence or chemical conversion provides a better measurement (e.g., H_2S), extractive systems may be more appropriate.

In situ UV CEMs are used to measure NO, NO_2, SO_2, NH_3, and BTEX (benzene, toluene, ethylbenzene, and xylene) isomers. Multiple gas analyzers are usually preferred when more than one species needs to be analyzed, for they only require one set of mounting holes and flanges at the stack.

6.3.1.2.3   Ex Situ. A rare, yet attractive, alternative to the extractive and *in situ* approaches is the *ex situ*, or bypass, system. This method (Fig. 6.18) is a compromise that overcomes many of the drawbacks typically associated with more traditional methods.

The analyzer is located adjacent to the stack where a bypass from the process diverts the sample through the instrument's in-line measurement cell. By allowing stack conditions to be reproduced, sample conditioning is avoided because the gases can be measured at very high temperatures (up to 1000°C). Due to the low volume of the sampling system, the instrument response time is nearly as fast as that of an *in situ* system without the zeroing and alignment problems. Gases measured by *in situ* systems can also be measured *ex situ* because the measurement principle is identical.

The *ex situ* technique is unique to UV analyzers. Although the path length is small (typically less than 30 cm long), the strong absorption of many gases in the UV often gives a system detection limit below 1 part per million (ppm_v).

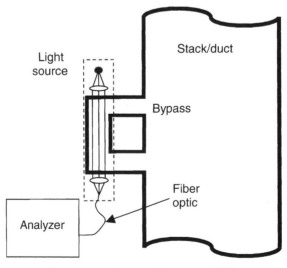

Figure 6.18   Layout of an *ex situ* CEM.

Such short path lengths do not necessarily provide a reasonable absorption in the IR.

6.3.2 Ambient Point Source Monitors

Ambient measurements deal predominantly with the effect of pollutants on human and ecosystem health. There are two major applications for these analyzers: environmental air quality, and health and safety. Environmental air quality monitoring usually determines pollutant levels in the ppb_v range; health and safety monitoring typically requires ppm_v sensitivity.

6.3.2.1 Challenges. Although many ambient point source analyzers have similar design characteristics to extractive CEMs, the detection limits of ambient analyzers are roughly three orders of magnitude lower. The large variety and variability of other gases in the atmosphere can complicate ambient monitoring.

The stringent stability and sensitivity requirements are met by increasing optical path length, controlling sample temperature and pressure, and controlling instrument temperature. Selectivity is achieved by using scrubbing and sophisticated chemometrics. The scrubbing processes can remove either interferents or, in the case where measurements are taken both prescrubbing and post-scrubbing, analytes.

In addition to these techniques, species-specific chemical techniques using liquid reagents, chemically treated paper tape, or chemiluminescence can provide the required sensitivity and selectivity for a particular application. These nonspectroscopic instruments are predominantly used in the health and safety field.

6.3.2.2 Approaches. Specific applications of health and safety UV analyzers are too numerous to mention in this text. Both absorptive and emissive techniques are employed to measure a number of gases, including aromatics, chlorine, ozone, as well as nitrogen, phosphorous, and sulfur compounds.

Environmental air quality UV monitors[4] focus on four specific compounds: SO_2, NO_2, NO, and O_3. SO_2 is monitored with the fluorescence technique, much like extractive CEMs (except without dilution), as well as through absorption methods. NO and NO_2 are monitored by the chemiluminescent technique discussed in Section 6.2.2. O_3 is measured either by absorption techniques or by chemiluminescence.

6.3.3 Ambient Open-Path Monitors

Open-path monitors[13,14] probe a cross section of the atmosphere hundreds of meters deep. As with ambient point source analyzers, the open-path approach can be used for the purposes of health and safety and environmental air quality. It is particularly useful in averaging the concentration of species that have spatial variability (such as fugitive emissions) or simply measuring compounds at low levels.

There are two subcategories of open-path instruments: range resolved and path-integrating. Range-resolved instruments were discussed in Section 6.2.3. Although range-resolved systems show great potential, the associated costs limit their utilization at the present time. Path-integrating instruments are essentially identical to point source absorptive gas analyzers, except that the sample cell is replaced by a large open section of the atmosphere. The ensuing discussion is restricted to path-integrating instrumentation.

6.3.3.1 *Challenges.*

Open-path monitors have similar measurement sensitivity requirements to ambient point source monitors. They need to measure samples that can be diffuse or have large spatial variability, as found in a plume. The instruments need to be waterproof, weatherproof, and insensitive to large temperature variations. The open measurement path necessitates that the monitor be insensitive to extraneous light sources and atmospheric scintillation. The system should be able to compensate for adverse optical conditions such as fog, snow, or rain.

Both the sensitivity and sample variability measurement requirements are addressed by the geometry of the system. The long optical path gives high sensitivities. The integrative technique is insensitive to spatial variability. All weather operation is achieved by using temperature-controlled, weather-tight enclosures or indoor analyzers linked by fiber-optic cables to sealed outdoor optics. Narrow field-of-view telescopes are used to curtail the effect of unwanted light. Atmospheric scintillation effects are overcome by acquiring data quickly or simultaneously. The effects of aerosol and particulate influences are minimized by using insensitive measurement/data analysis techniques.

6.3.3.2 *Approaches.*

Open-path instruments use two different measurement geometries, bistatic and monostatic, as shown in Figure 6.19. Open-path instruments can also be used as passive sensors for SO_2 and NO_2 when the sky is used as a light source.

Ultraviolet open-path measurements are performed by a number of instruments, including spectrometers, spectrographs, correlation interferometers, and correlation spectrometers.

Spectrometer-based systems are specially designed to analyze thousands of summed spectra, each scanned within a fraction of a second to avoid the variability induced by atmospheric scintillation. Spectrograph-based systems avoid scintillation effects by acquiring data at all wavelengths simultaneously. Spectra acquired by these systems are reduced by a chemometric technique called differential optical absorption spectroscopy (DOAS), which accounts for beam attenuation effects. These instruments have a large measurement range with good selectivity and response times on the order of 1 min.

As with spectrometers, correlation techniques acquire their data quickly to avoid atmospheric scintillation effects. Because sophisticated data analysis is not used, correlation techniques can have shorter response times (on the order of seconds) than chemometric methods at the cost of lower selectivity.

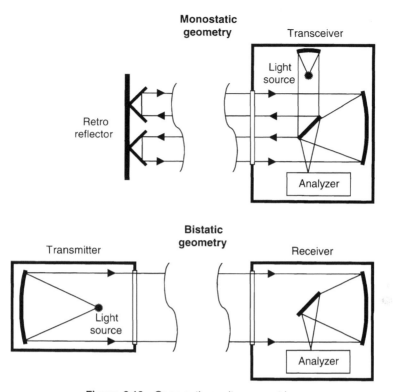

Figure 6.19 Open-path monitor geometries.

Both DOAS and correlation techniques require a high-frequency structure in the target species absorption spectrum, leading to a similar list of potential analytes for all measurement methods. Some of these gases are SO_2, NO_2, NO, O_3, NH_3, CH_2O, and aromatic compounds. The sensitivity of all active open-path UV instruments is similar, with low $ppm_v \cdot m$ detection limits for most measurable species.

6.4 PRACTICAL CONSIDERATIONS

In most applications, the instrument's end user requires continuous characterization of a gas sample at minimal cost. In this section, the requirements that determine instrument cost are examined:

- Measurement requirements
- Physical requirements
- Operational requirements

Measurement requirements determine the applicable measurement technique and instrument configuration, hence the up-front cost of the system. Physical requirements concern the needs related to instrument siting and installation, hence the installation cost of the system. Operational requirements concern the need for consumables and staffing, hence the yearly cost of ownership of the system.

Often the physical and operational requirements represent hidden expenses that can be quite variable from instrument to instrument and yet dominate the total cost of ownership of the system. The key to maximizing the utility of the analyzer at minimum cost is in precisely matching the specifications of the analyzer to the requirements of the application.

6.4.1   Measurement Requirements

6.4.1.1   Nature of the Emission. The most accurate and least expensive measurement method is the most desirable. The choice of measurement method is dependent on the nature of the emission. A portable point source monitor can be a good choice when searching for leaks or localized emissions. Open-path monitoring is appropriate for measuring average trends of an area source, such as fugitive emissions. Lidar is suitable when a map of the concentration at various points in space is needed.

6.4.1.2   Number of Gases and Expandability. The choice of system also depends strongly on the number of gases to be monitored and the relative merits of the various optical UV techniques for the application. For instance, a host of instrument permutations can be proposed for monitoring NO_X and SO_2 in stacks. In some applications, a dispersive analyzer capable of measuring multiple gases may be used instead of a number of single-species analyzers.

Future requirements of the system should also be taken into account. A system that has capacity to measure at multiple locations or can easily be adapted to add more species may be cost advantageous in the long term.

6.4.1.3   Sample Composition. Instrumental cross-sensitivity to other components in the sample may be the practical measurement limitation, invalidating specifications such as lower detection limits. The expected concentrations and variability of the sample constituents must be known in order to predict how the monitor will respond under real measurement conditions.

In many cases, this is the determining factor of whether a specific analyzer will work in a certain application. For this reason, a performance track record is a valuable tool for determining the suitability of a monitor in a given situation.

6.4.2   Physical Requirements

Installation costs are related to how well an analyzer is adapted to its operational environment. Construction of instrument shelters/shacks on-site can be more expensive than the capital cost of the analyzer. Outdoor or industrial environments may require heated, insulated, weatherproof, explosion-proof, or chemically resistant enclosures. Environments with excessive vibration, radiation, or strong electric/magnetic fields may affect instrument performance and require isolation or shielding.

6.4.3 Operational Requirements

The operational costs of an instrument are closely related to the amount of attention it requires from staff members.

6.4.3.1 *Operator Skill Level.* An ideal analyzer should be easily operated with no specialized knowledge or skills. Usually, a few days of training from the manufacturer should give users sufficient background to run the instrument. The analyzer can have high operational costs if specially trained personnel need to be on staff or contracted.

6.4.3.2 *Maintenance.* Instruments should have a maintenance schedule, with spare parts and consumables on hand to minimize downtime. Preventative maintenance and good system diagnostics are also helpful. As with operation, maintenance should require minimal specialized knowledge and skills.

6.4.3.3 *Calibration.* Calibration procedures are important in maintaining the quality of the instrument's data. Periodic calibration should be used to correct the analyzer and validate the accuracy recorded between calibration cycles. Diagnostics can be coupled with these routine checks to determine if the instrument requires servicing.

Recalibration should be convenient and consume minimal effort and material resources. Automated calibration systems are often used in conjunction with certified cylinder gases. This ensures that the instrument is performing correctly with minimal labor.

6.4.3.4 *Alternatives.* An increasingly popular solution to the problem of operation expenses are to contract measurement service agencies. In this case, data, as opposed to instrumentation, is purchased from the subcontractor. This is very popular with instruments that have a large capital cost and/or need specialized staff for operation, such as DIAL.

6.4.4 Example—Reduction of NO_X

The discussed requirements are best demonstrated in the example of NO_X reduction in an industrial gas stream. Physical and operational requirements of this system are not discussed because they are standard for a CEM application. Instead, the focus will be on the measurement requirements. A nonstandard UV analyzer is proposed, so an accurate determination of measurement requirements is critical to the proper function of the instrument.

6.4.4.1 *Application Overview.* The reduction of NO_X in the exhaust streams of combustion processes is an application that has received considerable attention in recent years. Combustion of fuels at high temperatures leads to the formation of NO and NO_2. NO_X levels are reduced by injecting NH_3 (either in the form of the free gas or as urea) into the stack gas treatment system,[14] leading to a reaction producing H_2O and N_2. The process requires control. Insufficient reagent results in partial NO_X reduction leading to high emission levels. Excess NH_3 can damage downstream gas treatment systems and will lead to unacceptable NH_3 emissions to the environment.

6.4.4.2 *Measurement Requirements.* The objective of the analysis is to provide feedback to the de-NO_X system and measure NO_X and NH_3 emissions to the environment. Additionally, SO_2 measurements may also be desirable for certain fuel types.

For the instrument to meet the application's goals, it must have a sub-ppm_v practical lower measurement limit (including drift and cross-sensitivity effects) for NO_X and NH_3. To provide proper feedback, the response time must be on the order of a few seconds. Accuracy and repeatability requirements are typical of a CEM.

6.4.4.3 *Sample Composition.* Typical measurement conditions will vary dramatically depending on fuel type, regulatory requirements, and the gas treatment system. The NO levels will usually be one order of magnitude greater than NO_2 concentrations, and NH_3 emissions should be below 10 ppm_v. The SO_2 levels can be anywhere between 0 and 1000 ppm_v, sometimes more as the SO_2 scrubber is often downstream of the NH_3 injection system. In addition, large and variable quantities of CO, CO_2, and H_2O will be present. The temperature of the stream will be consistent with most CEM applications.

6.4.4.4 *Proposed Solution.* SO_2, NO, NO_2, and NH_3 all possess characteristic absorption signatures in the UV. The other principal stream components (CO_2, CO, and H_2O) do not have interfering bands, making UV absorption the technique of choice.

The required fast response times as well as the need to measure the highly adsorptive NH_3 favor an *in situ* or *ex situ* configuration over an extractive CEM.

The absorption spectra of these gases are shown in Figure 6.20. The rapid response and strong spectral overlap suggest that a spectrograph with an array detector would be the best choice. Chemometric analysis is required to minimize cross-sensitivities and optimize precision. However, even a sophisticated signal-processing approach cannot compensate for a dramatic imbalance in analyte concentrations; 1%$_v$ SO_2 will completely occlude the absorption structure from 3 ppm_v NH_3. For this reason, the utility of this analyzer is application specific.

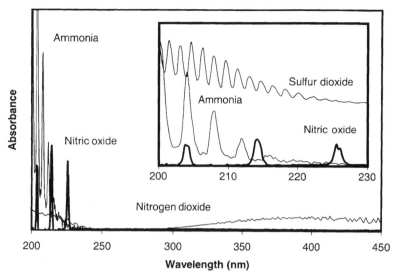

Figure 6.20   Spectra of NH_3, NO, and NO_2. Inset shows detail in the region where both NH_3 and NO absorb. An overlay of SO_2 is also shown in the inset to demonstrate its role as a potential interferent.

6.5   SUMMARY

Absorptive, emissive, and scattering techniques are all used to measure pollutants in the UV. These techniques are used in a wide range of instruments, with varying degrees of sophistication ranging from single-wavelength observations to high-resolution spectroscopic measurements that quantify multiple species simultaneously. The characterized region can be as small as a few cubic centimeters to sections of the atmosphere hundreds of meters in size. Analyte concentration can be as high as percent levels to below the ppb_v threshold.

The key to choosing the appropriate analyzer lies with a precise match of instrument specifications and application requirements. If these conditions are met, the unit will be optimal, both functionally and economically.

LIST OF SYMBOLS

ppb_v	Volumetric part per billion
ppm_v	Volumetric part per million
E	Energy of a photon
c	Speed of light
ν	Frequency of light
h	Planck's constant
λ	Wavelength of light

$I_0(\lambda)$	Initial intensity of light at a given wavelength
$I_t(\lambda)$	Transmitted intensity of light at a given wavelength
$I_a(\lambda)$	Absorbed intensity of light at a given wavelength
$I_e(\lambda)$	Emitted intensity of light at a given wavelength
$\sigma(\lambda)$	Absorption cross section at a given wavelength
$[c_m]$	Concentration of a species in energy state m
ℓ	Path length of light through a sample
N_n	Total amount of a species in energy state n
$A_{n,m}$	Radiative transition probability between states n and m

REFERENCES

1. A. Berk, L. S. Bernstein, and D. C. Robertson, *MODTRAN: A Moderate Resolution Model for LOWTRAN 7*, GL-TR-89-0122, 1989.

2. G. Herzberg, *Atomic Spectra and Atomic Structure*, 2nd ed., Dover, New York, 1944.

3. G. Herzberg, *Molecular Spectra and Molecular Structure*, 2nd ed., Krieger, Malabar, 1989.

4. M. L. Woebkenberg and C. S. McCammon, Direct Reading Gas and Vapor Instruments, in B. S. Cohen and S. V. Hering (Eds.), *Air Sampling Instruments for Evaluation of Atmospheric Contaminants*, 8th ed., American Conference of Governmental Industrial Hygienists, Cincinnati, OH, 1995, Section 19.

5. J. A. Jahnke, *Continuous Emission Monitoring*, Van Nostrand Reinhold, New York, 1993, pp. 92–94.

6. P. Pelikén, M. Ceppan, and M. Liska, *Applications of Numerical Methods in Molecular Spectroscopy*, CRC Press, Boca Raton, FL, 1994.

7. J. H. Davies, A. R. Barringer, and R. Dick, Gaseous Correlation Spectrometric Measurements, in D. A. Killinger and A. Mooradian (Eds.), *Optical and Laser Remote Sensing*, Springer Series in Optical Science, Vol. 39, Springer-Verlag, Berlin, 1983, Section 2.5.

8. A. Yariv and P. Yeh, *Optical Waves in Crystals*, Wiley, New York, 1984, pp. 131–143.

9. J. E. Johnston and E. I. Ivanov, Ultraviolet Interferometry Applied to Open Path Gas Monitoring, in *Proceedings of the 42nd Annual ISA Analysis Division Symposium*, Vol. 30, Instrument Society of America, Research Triangle Park, NC, 1997, pp. 17–24.

10. J. Greyson, *Carbon, Nitrogen, and Sulfur Pollutants and Their Determination in Air and Water*, Dekker, New York, 1990, pp. 138–180.

11. D. A. Skoog and J. J. Leary, *Principles of Instrumental Analysis*, 4th ed., Saunders, 1992, pp. 191–195.

12. M. W. Sigrist, *Air Monitoring by Spectroscopic Techniques*, Wiley, New York, 1994.

13. R. J. H. Clark and R. E. Hester (Eds.), *Spectroscopy in Environmental Science*, Advances in Spectroscopy, Vol. 24, Wiley, Chichester, 1995.

14. W. T. Davis, A. Pakrasi, and A. J. Buonicore, Combustion Sources, in A. J. Buonicore and W. T. Davis (Eds.), *Air Pollution Engineering Manual*, Van Nostrand Reinhold, New York, 1992, pp. 243–244.

7

TOTAL HYDROCARBON ANALYSIS USING FLAME IONIZATION DETECTOR

John Kosch

AMKO Systems, Inc.
Richmond Hill, Ontario, Canada

Environmental Instrumentation and Analysis Handbook, by Randy D. Down and Jay H. Lehr
ISBN 0-471-46354-X   Copyright © 2005 John Wiley & Sons, Inc.

7.1   APPLICATIONS

The flame ionization detector (FID) is a device which is used to continuously monitor total hydrocarbon concentrations in ambient air samples and industrial emissions.

Continuous emission monitoring (CEM) applications measure and report the total hydrocarbons (THCs) released to the atmosphere. Hydrocarbons react with NO_X and sunlight to form ozone, which is a major component of smog. Ozone is beneficial in the upper atmosphere, where it protects the earth by filtering out ultraviolet radiation, but at ground level it is a noxious pollutant. Ozone irritates the lung tissue and causes coughing, choking, and stinging eyes. Continuous monitoring ensures that corrective actions can be taken when THC concentrations do not comply with environmental regulations.

Threshold limit value (TLV) monitoring is performed in industries where humans may be exposed to volatile organic compounds (VOCs). Many VOCs, especially those used as solvents, have toxic effects. Some hydrocarbons, such as benzene, are known carcinogens, and others, such as 1,3-butadiene, are suspected carcinogens. A FID can be used to constantly measure and signal an alarm when the VOC concentration becomes unsafe.

In coating or drying processes, when solvents are present, VOC levels in the air could be high enough to present an explosion hazard. The VOC concentration in such processes must be controlled to keep it below the lower explosion limit (LEL). With no VOC monitoring system in place, the VOC concentration is kept below 25% of the LEL by ventilating with excess air (10,000 cfm for every gallon evaporated as a rule of thumb). A significant cost saving can be realized when a monitoring system is in place. Utilizing a FID allows the process to be run safely and economically by maintaining the VOC level at 40–45% of the LEL, signaling an alarm or triggering an emergency shutdown if necessary.

7.2   OPERATING PRINCIPLE

A FID incorporates regulated fuel, air, and sample delivery systems, an internal burner, and associated electronics for measuring the ion current produced by species introduced into the flame. A schematic of a typical burner is shown in Figure 7.1. Hydrogen is mixed with the gas stream at the bottom of the jet and air or oxygen is supplied axially around the jet. The hydrogen flame burns at the tip, perhaps a 20-gauge capillary, which also functions as the cathode and is insulated from the body by a ceramic seal. The collector electrode, located about 6 mm above the burner tip, consists of a loop of platinum. The charged particles generated in the flame are collected at these two terminals, resulting in a small flow of current, which is amplified for output.

In a FID the major ionization mechanism is *chemi-ionization*. The diffusion flame is illustrated in Figure 7.2 and an explanation of the salient regions is provided below.

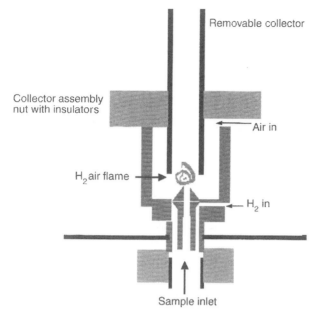

Figure 7.1 Hydrogen mixed with gas stream at the bottom of jet.

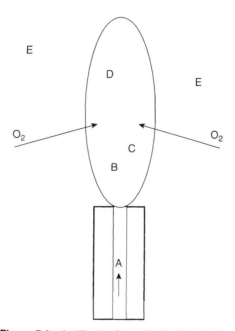

Figure 7.2 A diffusion flame of chemi-ionization.

7.2.1   Burner Jet

The flame is fed by a hydrogen–sample mixture coming through the burner jet and combustion air flows outside the flame.

7.2.2   Inner Flame Zone (Pyrolysis Zone)

After being heated, the hydrocarbons in the hydrogen–sample mixture are cracked. For example,

$$C_2H_6 + H^* \rightarrow C_2H_5^* + H_2$$
$$C_2H_5^* + H^* \rightarrow 2CH_5^*$$

Thus radicals (e.g., CH_2^*, CH^*, C^*) are created.

7.2.3   Outer Flame Zone (Oxidation Zone)

The major reactions for ion creation occur in this area, which is only approximately 1/100 mm thick. Oxygen is coming from the outside combustion air by diffusion and hydrogen and hydrocarbon-cracking fragments are flowing in from the pyrolysis zone. The major reaction here is

$$CH^* + O^* \rightarrow CHO{\sim}(\text{excited !}) \rightarrow CHO^+ + e^-$$

This reaction is called *chemi-ionization*. The CHO^+ ion is primarily responsible for the FID signal.

7.2.4   Surrounding Oxidizing Atmosphere

Combustion products and ions pass into this zone. Because of the electrostatic field, the ions have to flow in the direction of the electrodes. On their way to the electrodes, recombination processes occur; for example,

$$CHO^+ + OH^- \rightarrow CHO + OH$$
$$H_3O^+ + e^- \rightarrow H_2O + H$$

Table 7.1   FID Nonresponsive Compounds

Ar	He	Ne	$SiCl_4$
HCHO	H_2O	NH_3	SiF_4
CO	H_2S	NO	$SiHCl_3$
CO_2	Kr	N_2O	SO_2
COS	N_2	NO_2	Xe
CS_2	HCOOH	O_2	

Table 7.2   Effective Carbon Numbers

Type of Atom	Occurrence	Effective Carbon Number
Carbon	In aliphatic compound	+1.0
	In aromatic compound	+1.0
	In olefinic compound	+0.95
	In acetylenic compound	+1.30
	In carbonyl radical	0.0
	In nitrile	+0.3
Oxygen	In ether	−1.0
	In primary alcohol	−0.6
	In secondary alcohol	−0.75
	In tertiary alcohol, ester	−0.25
Chlorine	As two or more chlorine atoms on single aliphatic carbon atom	−0.12 each
	In olefinic carbon atom	+0.05
Nitrogen	In amine	Value similar to oxygen atom in corresponding alcohol

The detector responds to compounds that produce ions when burned in a hydrogen–air flame. These include all organic compounds, although a few exhibit poor sensitivity (e.g., formic acid, acetaldehyde). Response is greatest with hydrocarbons and diminishes with increasing substitution of other elements. The system is linear for most organic compounds, from the minimum detectable limit through concentrations greater than 10,000,000 times the minimum detectable limit. The linear range depends on each specific compound. It is directly proportional to the sensitivity of the FID toward the given compound. Apart from vapors of elements in groups I and II of the periodic table, these being elements ionized in flames, it does not respond to inorganic compounds. Table 7.1 lists the compounds that give little or no response.

In a rough approximation, the FID signal is proportional to the number of carbon atoms in the sample gas. The response factor to a particular compound may be estimated by summing the effective carbon numbers for every atom in the molecule, as listed in Table 7.2.

In practice, the response factors predicted by the sums of the effective carbon numbers for constituent atoms in most cases deviate significantly from the actual, empirical values observed by the instrument. As response factors are dependent on gas flow and detector geometry, they have to be ascertained for each FID individually. The following are examples of response factors for various hydrocarbons relative to methane:

Substance	Number of carbon Atoms	Relative Output	Substance	Number of carbon Atoms	Relative Output
Normal paraffins			Ethylcyclopentane	7	6.5
Methane	1	1.0	*cis*-1,2-Dimethylcyclopentane	7	6.5
Ethane	2	2.0	*trans*-1,3-Dimethylcyclopentane	7	6.5
Propane	3	3.0	*cis*-1,3-Dimethylcyclopentane	7	6.5
n-Butane	4	3.8	Methylcyclohexane	7	6.6
Isobutane	4	3.8	*trans*-1,2-Dimethylcyclopentane	7	6.6
Pentane	5	5.0	1,1-Dimethylcyclopentane	7	6.7
Hexane	6	6.1	1-Methyl-*trans*-3-ethylcyclopentane	8	7.3
Heptane	7	7.0	*n*-Propylcyclopentane	8	7.3
Octane	8	7.8	*trans*-1,2-*cis*-4-Trimethylcyclopentane	8	7.3
Nonane	9	8.8	*cis*-1,2-*trans*-3-Trimethylcyclopentane	8	7.3
			Isopropylcyclopentane	8	7.3
			cis-1,2-Dimethylcyclohexane	8	7.4
Halogen-substituted paraffins			*trans*-1,4-Dimethylcyclohexane	8	7.4
Dichlorodifluoromethane	1	0.38	*cis*-1,2-*trans*-4-Trimethylcyclopentane	8	7.4
Dichloromethane	1	0.87	1-Methyl-*cis*-2-ethylcyclopentane	8	7.5
Chloroform	1	0.68	1-Methyl-*cis*-3-ethylcyclopentane	8	7.5
Carbon tetrachloride	1	0.48	Ethylcyclohexane	8	7.6
Trichloroethylene	2	2.1	1,1-Dimethylcyclohexane	8	7.7
Tetrachloroethylene	2	2.2	1,1,2-Trimethylcyclopentane	8	7.7
1,3-Dichloropropane	3	2.4	1,1,3-Trimethylcyclopentane	8	7.8
			1-Methyl-*cis*-4-ethylcyclohexane	9	8.1
Isoparaffins			1-Methyl-*trans*-4-ethylcyclohexane	9	8.3
2-Methylpropane	4	3.8	Isopropylcyclohexane	9	8.3
2,2-Dimethylpropane	5	4.7	1,1,2-Trimethylcyclohexane	9	8.6

(*Continued*)

Substance	Number of carbon Atoms	Relative Output	Substance	Number of carbon Atoms	Relative Output
2-Methylbutane	5	5.3			
2,2-Dimethylbutane	6	6.1	*Olefins*		
3-Methylpentane	6	6.3	Acetylene	2	2.6
2-Methylpentane	6	6.3	Ethene	2	1.9
2,3-Dimethylpentane	7	6.9	Propene	3	2.9
2-Methylhexane	7	7.1	1-Butene	4	3.8
3-Methylhexane	7	7.1	2-Butene	4	3.7
2,2-Dimethylpentane	7	7.1	1,3-Butadiene	4	3.8
2,4-Dimethylpentane	7	7.1	Dimethylbutadiene	6	5.8
3-Ethylpentane	7	7.1			
2,2,3-Trimethylbutane	7	7.1	*Halogen-substituted olefins*		
3,3-Dimethylpentane	7	7.2	Vinylchloride	2	1.7
2-Methylheptane	8	7.7	3-Chloropropane	3	2.9
2-Methyl-3-ethylpentane	8	7.8			
2,3-Dimethylhexane	8	7.9	*Aromatic compounds*		
2,4-Dimethylhexane	8	7.9	Benzene	6	5.8
3,4-Dimethylhexane	8	7.9	Toluene	7	7.0
3-Ethylhexane	8	8.0	*p*-Xylene	8	7.4
2,2,4-Trimethylpentane	8	8.0	*o*-Xylene	8	7.6
3-Methylheptane	8	8.1	Ethylbenzene	8	7.6
2,2-Dimethylhexane	8	8.1	*m*-Xylene	8	7.7
2,5-Dimethylhexane	8	8.1	1,2,4-Trimethylbenzene	9	8.1
4-Methylheptane	8	8.1	Isopropylbenzene	9	8.1
2,2,3-Trimethylpentane	8	8.1	1,2,3-Trimethylbenzene	9	8.2

(*Continued*)

Substance	Number of carbon Atoms	Relative Output	Substance	Number of carbon Atoms	Relative Output
2,3,5-Trimethylhexane	9	8.6	1,3,5-Trimethylbenzene	9	8.2
2,2-Dimethylheptane	9	8.7	1-Methyl-4-ethylbenzene	9	8.4
2,2,4-Trimethylhexane	9	8.9	1-Methyl-3-ethylbenzene	9	8.5
2,2,5-Trimethylhexane	9	8.9	*n*-Propylbenzene	9	8.5
2,3,3,4-Tetramethylpentane	9	8.9	1-Methyl-2-ethylbenzene	9	8.6
2,3,3-Trimethylhexane	9	9.0	*n*-Butylbenzene	10	9.2
3,3-Diethylpentane	9	9.0	1-Methyl-2-isopropylbenzene	10	9.3
2,2,3,3-Tetramethylpentane	9	9.0	1-Methyl-4-isopropylbenzene	10	9.3
2,2,3-Trimethylhexane	9	9.0	*sec*-Butylbenzene	10	9.4
2,2,4-Trimethylhexane	9	9.0	1-Methyl-3-isopropylbenzene	10	9.5
3,3,5-Trimethylheptane	10	9.8	*tert*-Butylbenzene	10	9.6
2,2,4,5-Tetramethylhexane	10	9.9			
2,2,3,3-Tetramethylhexane	10	10.1	*Alcohols*		
			Methanol	1	0.77
Cyclic compounds			Ethanol	2	1.8
Cyclopropane	3	2.9	*n*-Propanol	3	2.6
Cyclopentane	5	4.7	*i*-Propanol	3	2.2
Methylcyclopentane	6	5.7	*n*-Butanol	4	3.5
Cyclohexane	6	5.8			

7.3 MEASURING RANGE

The wide linear range (approximately six to seven orders of magnitude) is an important advantage of the FID. Using a gas chromatograph, the detection limit is near 1 ppb. The upper range of the FID-based continuous THC monitor is typically 10%.

7.4 POTENTIAL SHORTFALLS

Any compound capable of ionizing in the burner flame is a potential interference. When analyzing for hydrocarbons, it should be taken into account that other organic types such as alcohols, aldehydes, and ketones will give a response.

The FID cannot be used to measure individual hydrocarbon components without prior separation steps, such as a GC column. The measurement of multiple components in gas chromatography analysis usually is based on each compound having a unique retention time. The possibility exists, however, of more than one compound having the same retention time for any given set of operating parameters.

7.5 TIPS

Most continuous hydrocarbon analyzers utilize a pressure drop across a capillary to regulate the flow. The use of a fine filter is recommended to protect the fine capillaries and needle valves of the instrument against particles that may change flow characteristics.

Recombination of the electrons and positive ions occurs after the ionization process to a degree determined by the ion concentrations and electrode voltage. At low voltages, the ion current is proportional to the applied voltage, while at high voltages the maximum number of ions are accelerated to the electrodes. The best detector performance is obtained with the electrode over the flame as the electrode collector. In general, electrode spacing and size are not critical (spacing determines only the minimum voltage required to reach a saturation current). A fixed-position, cylindrical collector electrode is recommended to assure a uniform spacing between the electrodes and provide long-term reproducibility.

Fuels can range from 40% hydrogen in nitrogen to pure hydrogen depending on FID design. Bottled hydrogen should be prepared from water-pumped sources. An electrolytic generator, which produces ultrapure hydrogen under high pressure, is recommended as it provides a degree of safety compared with the explosive hazard of bottled hydrogen. All connections should be made with stainless steel tubing and thoroughly tested for leaks.

Compressed air in cylinders, low in hydrocarbon content, should be used to supply the burner. Because hydrocarbon-free air is difficult to obtain, if this air is also used to calibrate the zero point, a zero adjustment should be performed for each new cylinder. Compressed commercial-grade oxygen in cylinders is suitable for detectors requiring oxygen.

Hydrocarbon analyzers should have the zero and span calibrations set on a regular basis, at least once daily. At optimum operating conditions, only a one-component, one-point calibration is required since response is linear to the carbon content of the sample.

After continued operation, the detector may become contaminated, resulting in increased noise and decreased sensitivity. The FID should be carefully disassembled and cleaned with a soft brush and suitable solvent such as methylene chloride. Contamination from fingers should be avoided when replacing the jets. Care must be taken in accurately repositioning the electrodes as the FID is reassembled. It is recommended that the FID be cleaned every 3–6 months to assure proper maintenance and optimum performance.

7.6 CONCLUSION

Clearly the FID is a powerful tool for the analysis of total hydrocarbons. It is applicable to a wide range of compounds and has been demonstrated to be highly sensitive, stable, relatively insensitive to flow and temperature changes, rugged, and reliable.

8

GAS CHROMATOGRAPHY IN ENVIRONMENTAL ANALYSIS

John N. Driscoll
PID Analyzers, LLC
Walpole, Massachusetts

Environmental Instrumentation and Analysis Handbook, by Randy D. Down and Jay H. Lehr
ISBN 0-471-46354-X   Copyright © 2005 John Wiley & Sons, Inc.

8.1   INTRODUCTION

This chapter is written from the perspective of an environmental professional. It focuses on those aspects of gas chromatography (GC) that aid in the selection of instrumentation and columns for both field and laboratory methods. In addition, it should be a useful resource for anyone who is responsible for interpreting data collected in the field.

We start by describing the theory of GC; then we discuss the selection of columns, gas–solid and gas–liquid chromatography, phases, and packed and capillary columns. This material is intended to provide the reader with sufficient information to select a proper column for analysis of a particular site or a difficult sample. In the next section, we describe the hardware required for GC. The fourth section describes the need for good temperature control even for field gas chromatographs.

The GC detectors that are described include the photoionization detector (PID), the flame ionization detector (FID), the thermal conductivity detector (TCD), the electron capture detector (ECD), the far-UV absorbance detector (FUV), and the flame photometric detector (FPD). For each detector, the theory of operation, range, detection limits, and characteristics are described. Individual species can be measured from ppt to percentage levels with either a specific- or universal-type detector.

Finally, we discuss the analysis of volatile organic (VOC) and semivolatile organic (SVOC) compounds, dual detectors (PID/FID), headspace, and concentrators for gas chromatographs that can be used for monitoring low- or sub-ppb levels of toxic species at the fenceline.

8.2   GAS CHROMATOGRAPHY THEORY

Gas chromatography is a method of continuous chemical separation of one or more individual compounds between two phases. One phase remains fixed (stationary phase); the other, the mobile phase (carrier gas), flows over the stationary phase. The components enter the stationary phase simultaneously at the injector but move along at different rates. The lower the vapor pressure of the compound (higher boiling point), the longer the compound will remain in the stationary phase. The time that each compound remains in the stationary phase depends on two factors: the vapor pressure of the compound and its solubility in the stationary phase. These compounds are then detected at the end of the column. A plot of the output of the detector response versus time is termed a chromatogram.

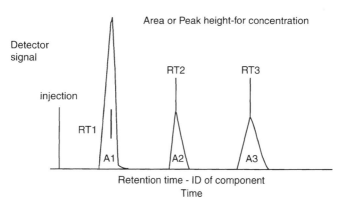

Figure 8.1   Use of a chromatogram for identification of components and concentration.

Elution times may be reduced by increasing the temperature of the GC oven. Gas chromatographs can be run isothermally (constant temperature) to separate a narrow boiling range of solutes. If the separation of low- and high-boiling compounds is necessary, temperature programming (linear increase of column temperature) is used.

The *retention time* is defined as the time measured from the start of injection to the peak maximum and can be used to identify *resolved* components in mixtures. The times measured as RT1, RT2, and RT3 shown in Figure 8.1 would be the retention times for components A1–A3. The retention time is characteristic for a compound and the stationary phase at a given temperature and is used for identification when the mixture of compounds is completely resolved. To confirm that a particular component is present requires the identification on two columns with different polarities of stationary phases. Some environmental methods allow confirmation of compound identity by comparing both retention times and detector response factors with known standards. Instruments that are configured for either dual columns with a single detector or a single column with dual detectors (PID/FID) can combine analysis *and* confirmation in a single run.

8.2.1   Column Selection

There are a large number of GC packings available. Each exhibits specific retention characteristics for specific compounds. Many times, a better separation is obtained more easily by changing the liquid phase than by increasing the length of the column. A properly made capillary column of 5 m in length will have about 12,000–15,000 effective plates, more than 100 times the resolving power of a short packed column. Interestingly enough, with all of the developments in capillary column technology, at a recent symposium,[1] one researcher was still talking about the utility of short, packed columns at ambient temperature. With the minimum of

Table 8.1   Comparison of Packed and Capillary Columns

Parameter	Packed Column	Capillary Column
Length (m)	1.5–6	5–100
i.d. (mm)	2–4	0.15–0.53
Flow rate (mL/min)	10–60	0.5–30
Total plates (length, m)	5000 (2 m)	75,000 (25 m)
Film thickness (µm)	1–10	0.1–5

separating power (efficiency), *many peaks could still be unresolved under a single peak*. A comparison of packed and capillary columns is shown in Table 8.1.

Methyl silicone stationary phase is considered nonpolar, generally eluting compounds in boiling point order. When polar functional groups are added in the stationary phase. The *number of theoretical plates* is a term taken from chemical engineering originally used to describe the efficiency of a distillation apparatus. This theory was applied to columns in GC to describe the efficiency (separation ability) of a column. Separation occurs as a result of continuous movement between the stationary phase and the mobile phase. Clearly, the larger the number of plates, the greater the resolving power of the column.

The number of theoretical plates is given by

$$n = 5.545(t/w)^2$$

where t = retention time

   w = peak half height

The number of effective theoretical plates is given by

$$N_{\text{eff}} = 5.545(t'/w)^2$$

where t' = adjusted retention time, $= t - t_m$

   t_m = retention time for inert peak, e.g., methane

The height equivalent to a theoretical plate is given by

$$h = \frac{L}{n}$$

where L is the length of the column.

8.2.1.1   *Gas–Solid Chromatography.*

Solid packings are generally used to separate gases and compounds with boiling points below propane. Polymers which are derivatives of styrene divinyl benzene, cross-linked acrylic ester, cross-linked polystyrene, etc., are small particles with pores and variable surface areas. These porous polymers are available in a variety of polarities for specific separations of low-molecular-weight compounds (methane, ethane, ethylene, H_2S). It would be difficult to analyze ethylene (gas) and benzene (very long retention time) on these porous polymers; similarly, it is difficult to analyze ethylene on a short (5 m) capillary column since it would be an unretained (would elute very quickly) compound.

Zeolytes or molecular sieves employ size exclusion for separation. Certain molecules that are small enough to enter the pores exit the stationary phase later than larger molecules that cannot enter the pores readily. These phases are commonly used for the separation of permanent gases (including O_2 and N_2).

8.2.1.2 Gas–Liquid Chromatography. Columns with a liquid stationary phase are generally used to separate compounds with boiling points above propane (C_3H_8). More that 70% of all separations in GC can be accomplished with a methyl silicone liquid phase (OV1, OV101, SE30). However, there are more than 1000 packed-column liquid phases available attesting to their versatility for specific separations.

8.2.1.3 Types of Column Phases. The stationary phase is the most influential column parameter since it determines the final resolution and has an influence on the sample capacity and the analysis time. The most important thing to remember is that "like dissolve like." Nonpolar compounds should be separated on a nonpolar column and polar compounds on a polar column. In Figure 8.2, the range of polarity of a group of organic compounds is compared with the polarity of different phases.

In other words, if one has nonpolar hydrocarbons to separate, a nonpolar phase such as SE3000, NBW30; with more polar compounds such as alcohols and esters, a polar phase such as carbowax should be used. The data in Table 8.2 list the optimum liquid phases on a packed or capillary column for a variety of analytes. The terminology in Table 8.2 is that of Ohio Valley Specialities (OV). In order of

Polarity of Compounds and Phases

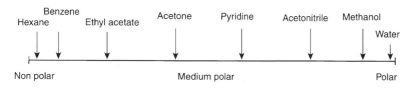

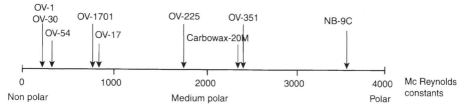

Figure 8.2 Comparison of the polarity of polar and nonpolar compounds and phases.

Table 8.2   GC Applications and Column Phases

Applications	Column Phases
Alcohols	Carbowax 20M, OV1701
Aldehydes	Carbowax 20M, OV1, SE30
Amines	OV54
Aromatic hydrocarbons	Carbowax 20M
Dioxins	OV54
Glycols	Carbowax 20M, OV1701
Halogenated hydrocarbons	OV54, OV1701
Ketones	OV1, OV54
PAHs	OV54, OV1701
PCBs	OV54, OV1701
Pesticides	
Triazine herbicides	OV351, OV225
EPA 608	OV54, OV1701
Phenols	
Free	OV1, OV225
Acetylated	OV54, OV1701
Solvents	OV54, OV1701

polarity, these silicone phases are least polar (OV1, OV101), medium polar (OV1701), and most polar (OV275). Their composition is as follows:

Many of the phases used in packed columns are also used in capillary columns with much greater effect on the latter. Figure 8.3 is a schematic of packed and capillary columns. A comparison of the separation of packed and capillary columns is given in Figure 8.4. Note that a significantly larger number of peaks is detected with the capillary column.

8.2.2   Capillary Columns

Capillary columns were first used in GC during the 1960s. The early columns were long (50 m), narrow bore, and stainless steel or soft (soda lime) glass tubing. With the latter, breakage was a problem, but these columns were more inert than stainless steel. Fused silica was also used as column material, but it was more difficult to work with a flame than with glass and it was easily broken. The coatings (stationary phase) on the columns were adsorbed but not bonded.

In 1979, Dandeneau and Zerenner[2] described a new type of fused silica capillary column that had a coating of polyimide on the outside, which made the column relatively flexible. At the same time, the use of bonded (to the fused silica) stationary phases increased dramatically because of the longer lifetime, inertness, and reduced bleed. This created a surge in the use of capillary columns, particularly in the United States. Some years ago, glass-lined stainless steel columns were introduced. These again improved the utility of capillary columns, particularly in the field.

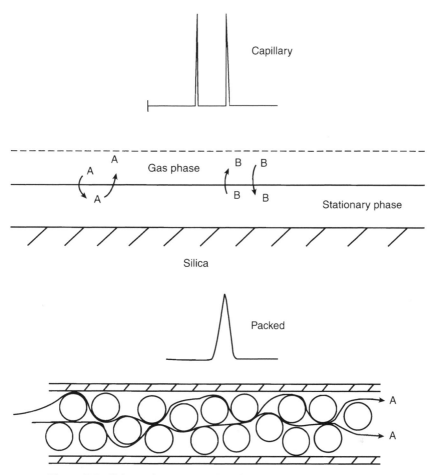

Figure 8.3 Schematic representation of packed and capillary columns.

Capillary columns have a high resolution (3000 plates/m) and vary from 5 to 100 m in length. The liquid phases (polar or nonpolar) are bonded to the fused silica. The columns can be made of fused silica (coated with polyimide so that they are flexible or stainless steel (lined with silicon oxide). Column diameters can be 0.53, 0.32, 0.20, or 0.15 mm. Capillary columns can also be packed with porous polymers (bonded to the fused silica) to form highly efficient PLOT columns for separation of low-molecular-weight compounds or fixed gases. A comparison of columns and their characteristics is given in Figure 8.5.

Packed columns have a relatively high sample capacity (difficult to overload column) because the liquid (stationary) phase coating is quite high compared to capillary columns. With bonded capillary columns, the film thickness of the station-ary phase can be controlled. A thin layer of stationary phase will provide a faster column that is better for high-molecular-weight compounds. Here, one has to

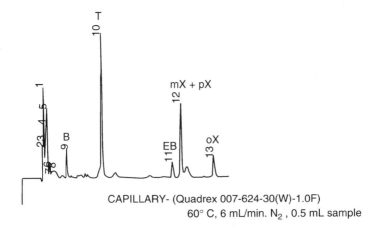

CAPILLARY- (Quadrex 007-624-30(W)-1.0F)
60° C, 6 mL/min. N₂ , 0.5 mL sample

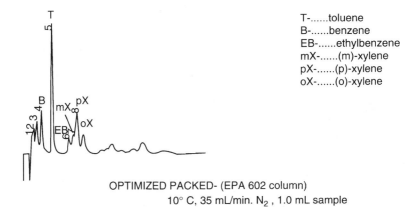

T-......toluene
B-......benzene
EB-......ethylbenzene
mX-......(m)-xylene
pX-......(p)-xylene
oX-......(o)-xylene

OPTIMIZED PACKED- (EPA 602 column)
10° C, 35 mL/min. N₂ , 1.0 mL sample

Figure 8.4   Chromatogram of packed and capillary columns.

Comparison of Different Column Types

Column type	Length (M)	I. D. (MM)	Resolving power	Sample capacity	Inertness
WCOT Narrow bore	5 - 100	0.20 - 0.32	Excellent	Low	Excellent
WCOT Wide bore	5 - 100	0.50 - 0.75	Good	Low	Good
Micro packed	0.5 - 10	0.5 - 1.0	Acceptable	Medium	Acceptable
Packed	0.5 - 5	2.0 - 6.0	Bad	High	Bad

Figure 8.5   Comparison of types and performance of columns.

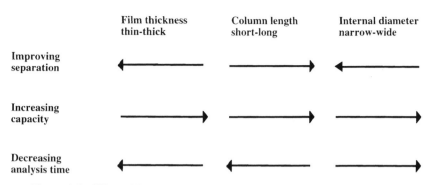

Figure 8.6 Effect of film thickness and other parameters on column performance.

consider the capacity factor k which is the ratio of the time the solute spends in the stationary and mobile phases:

$$k = \left(t - \frac{t_m}{t_m} \right) = \frac{t'}{t_m}$$

A number of factors affect column performance:

Inner Diameter (i.d.). The smaller the i.d., the higher the efficiency and the shorter the analysis time

Film Thickness. The higher the film thickness, the greater the capacity; the higher the film thickness, the longer the analysis time; thickness ranges from 0.1 to 10 μm.

Length. Increasing the length will increase resolution, the analysis time, and the capacity

The effect of these parameters is shown in Figure 8.6.

In summary, the selection of a column involves a number of trade-offs and specific knowledge of the compounds to be analyzed,

8.3 GC HARDWARE

A schematic of the typical GC hardware is shown in Figure 8.7.

1. *Injector.* A sample is introduced into the heated injector, where it is vaporized and carried to the column via, for example, a liquid or gas syringe, liquid or gas valve, concentrator, or purge and trap.

2. *Packed Columns.* These are $\frac{1}{4}$, $\frac{1}{8}$, or $\frac{1}{16}$ in. (micropack), 2–3 m in length, 300–500 plates/m packing material: porous polymer, liquid phase (1–3%) on diatomite.

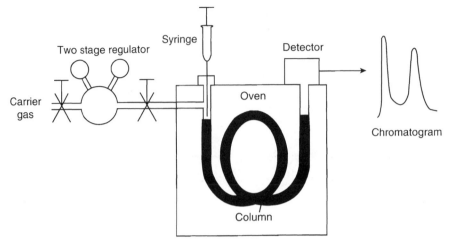

Figure 8.7 Schematic of a gas chromatograph.

3. *Capillary Columns.* These are 0.53, 0.32, 0.20, or 0.15 mm with liquid phase bonded to the fused silica and are available in fused silica lined stainless steel with the liquid phase bonded to the silica; efficiency is 1000–3000 plates/m with typical lengths of 15–30 m.

4. *Carrier Gas.* This is mobile phase used to move the components through the column to the detector; note that the high-sensitivity detectors (PID, FID, ECD, FUV, FPD) require high-purity carrier; the ECD requires that oxygen and water be eliminated (trap is usually required) from the carrier since these species can absorb electrons and affect sensitivity

5. *Oven.* This is an isothermal or temperature-programmed heated device; the higher the temperature, the shorter the retention time; good temperature stability ($\pm 0.1 - 0.2°C$) is required (see Section 8.4).

6. *Detector.* This produces a response proportional to the component that is separated by the column. Detectors may include a PID, FID, TCD, ECD, FPD, or FUV detector.

7. *Amplifier.* This receives an output from a detector (typically picoamperes for an ionization detector) and amplifies it so that the signal can be detected by a recorder or integrator

8. *Integrator.* This takes the signal from the amplifier and produces an output (chromatogram) and peak height or area (used for quantitation). If we note Figure 8.1, The height of the peak measured from the baseline to the peak maximum and the area which is determined by integrating the area underneath the peak are proportional to the concentration. Generally integrators will provide both area and height values. At low concentrations with packed columns, peak height may provide a better value.

8.4 TEMPERATURE CONTROL

Many portable gas chromatographs of the 1980s were typically ambient temperature instruments *with no temperature control* and short chromatographic columns. Even today, some of these portable instruments do not have very good temperature control. The problem with these instruments was that a change of just a few degrees celsius in the temperature of the column could result in a significant change in the retention time of the species of interest. The chromatographic separation depends upon solute (material being analyzed) partitioning between the stationary and mobile phases. This controls the efficiency or separating power of the column. Temperature control is very important; therefore we have added this section.

Giddings[3] has developed an expression for the efficiency or plate height (*H*) as follows:

$$H = \frac{2D}{v} + d_{\text{p}} + 2R(1 - R)vt_{\text{t}}$$

where D = ordinary diffusion coefficient

R = ratio of zone velocity to mobile phase velocity

t_{d} = lifetime in the stationary phase

d_{p} = particle diameter

v = velocity

The partition or distribution constant (*K*) has a temperature coefficient (related to *R*) given by

$$K = k(e^{-dH/RT})$$

where   k = const

dH = enthalpy of sorption

R = ideal gas constant

T = absolute temperature

In addition to the temperature dependence of K, the ordinary diffusion coefficient (D) has a temperature dependence, as does the term t_{d}, the lifetime in the stationary phase. The latter can be approximated through the Arrhenius equation.[4]

Retention time is defined as the time from injection to the peak maximum and can be used to identify resolved components in mixtures. Since the retention times are used to identify the species of interest, a shift in temperature could lead to the wrong species being identified, particularly in a complex mixture. Ambient temperatures, as anyone knows, are anything but constant.

If the separation of low- and high-boiling compounds is required, temperature programming (linear increase of column temperature) is needed.

The difficulty with temperature control is that it takes power to maintain the temperature, and the higher the temperature, the greater the power consumption. Thus, in the design of field portable gas chromatographs, much flexibility is lost if battery operation is the most important criterion. Alternatives to internal batteries are generators and batteries in vehicles. The GC311 has been designed to operate from generators or vehicle batteries. Using these alternative methods, for the HNU Systems GC311, one has to make few concessions in the performance of the instrument.

In the 1980s, a number of portable gas chromatographs were introduced that employed temperature control and capillary columns. Since resolution is proportional to column length, considerably better performance can be obtained with a 5-m column than with a 0.3-m column. The longer the column, the better the separation. Some of the portable gas chromatographs maintain temperatures of only 50°C (to minimize power consumption) and are limited in the variety of species that can be analyzed.

8.5 GC DETECTORS

The detectors selected for this section include the most popular detectors for field work. We have not included the mass-selective detector (MSD) in this section.

8.5.1 Photoionization Detector

In 1976, the first commercial PID was described by HNU Systems.[5] The process of ionization which occurs when a photon of sufficient energy is absorbed by a molecule results in the formation of an ion plus and an electron:

$$R + h\nu \Rightarrow R^+ + e^-$$

where   R = an ionizable species

$h\nu$ = a photon with sufficient energy to ionize species R

In the ion chamber, the ions (R^+) formed by absorption of the UV photons are collected by applying a positive potential to the accelerating electrode, and measuring the current at the collection electrode. A PID consists of an ion chamber, a UV lamp with sufficient energy to ionize organic and inorganic compounds, a voltage source for the accelerating electrode, and an amplifier capable of measuring down to 1 pA full scale. A schematic of a PID is shown in Figure 8.8. A list of ionization potentials is given in Chapter 10 on photoionization.

The PID is a concentration-sensitive detector (sample is not destroyed) where the sensitivity is increased as the flow rate is reduced. Thus, the sensitivity can

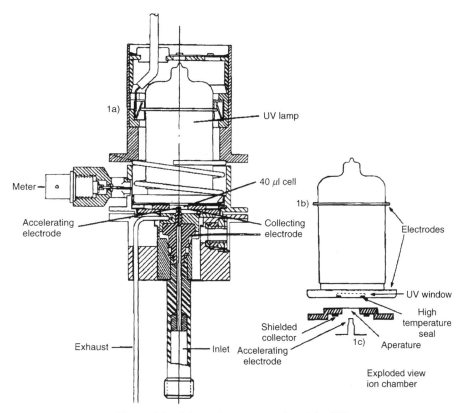

Figure 8.8 Schematic representation of the PID.

be improved by operating the PID at lower flow rates; however, one must have sufficient flow to sweep the sample through the PID:

$$C_{PID} \propto \frac{1}{F}$$

where C = concentration
   F = flow rate of carrier gas

In terms of sensitivity, the PID is from 10 to 100 times more sensitive than the FID, making this detector ideal for environmental applications. This results from the higher ionization efficiency of the PID. The apparent ionization efficiency of the PID is approximately 10^{-4} while that of the FID is 10^{-5}.

Some characteristics of the PID are given in Table 8.3. The sensitivity of the PID response with the structure of organic compounds[6] is given in Table 8.4.

Table 8.3   PID Characteristics

Sensitivity increases as the carbon number increases (carbon counter).

For 10.2-eV lamp, responds to carbon aliphatic compounds $> C_4$, all olefins, and all aromatics.

The PID also responds to inorganic compounds such as H_2S, NH_3, Br_2, I_2, PH_3, AsH_3, e.g., any compound with an ionization potential of <10.6 eV.

The PID is more sensitive than the FID—40 times more sensitive for aromatics, 20 times for olefins, 10 times for alkanes $> C_6$.

Nondestructive detector; other detectors can be run downstream.

Only carrier gas (prepurified nitrogen or helium) is required for operation.

Concentration-sensitive detector.

Table 8.4   PID Sensitivity for Organic Compounds

Sensitivity increases as carbon number increases.

For n-alkanes, $SM = 0.715n - 0.457$, where $SM =$ molar sensitivity relative to benzene (benzene $= 1.0$) and $n =$ carbon number.

Sensitivity for alkanes $<$ alkenes $<$ aromatics.

Sensitivity for alkanes $<$ alcohols esters $\leq$ aldehydes $<$ ketones.

Sensitivity for cyclic compounds $>$ noncyclic compounds.

Sensitivity for branched compounds $>$ nonbranched compounds.

Sensitivity for fluorine substituted $<$ chlorine substituted $<$ bromine substituted $<$ iodine substituted.

For substituted benzenes, ring activators (electron-releasing groups) increase sensitivity.

For substituted benzenes, ring deactivators (electron-withdrawing groups) decrease sensitivity (exception: halogenated benzenes).

For soil and water samples that involve solvent extraction, a number of solvents can be used to produce a small or negative response with the PID. These are shown in Table 8.5. The advantage of these solvents is that many of the volatile hydrocarbons can still be detected since the solvent peak is similar to an unretained compound and elutes very quickly. The FID, for example, does not respond to carbon disulfide. This can be used for a similar purpose, but a hood will be needed to minimize odor problems.

Table 8.5   PID Response with Various Solvents

Solvent	Ionization Potential (eV)	Response
Water	12.35	Negative peak
Methanol	10.85	Negative peak
Chloroform	11.42	Negative peak
Carbon tetrachloride	11.47	Negative peak
Acetonitrile	12.22	Negative peak
Pentane	10.35	Small positive peak

Table 8.6   FID Characteristics

Sensitivity increases as the carbon number increases (carbon counter).

Sensitivity to substituted species depends on the mass of carbon present and the ability to break the carbon bonds.

The FID is most sensitive to hydrocarbons.

Detector is destructive since sample is burned.

Requires the use of zero-grade (high-purity) hydrogen and air to produce the flame.

8.5.2   Flame Ionization Detector

The process of ionization occurs in organic compounds when the carbon–carbon bond is broken via a thermal process in the flame that results in the formation of carbon ions. These ions are collected in the flame by applying a positive potential to the FID jet and the ions are pushed to the collection electrode where the current is measured. The response (current) is proportional to the concentration and is measured with an electrometer/amplifier. An FID consists of a combustion/ion chamber, a flame, a voltage source for the accelerating electrode (usually applied to the jet), and an amplifier capable of measuring down to 1–5 pA full scale.

The FID is a mass-sensitive detector, the output of which is directly proportional to the ratio of the compound's carbon mass to the total compound mass. Thus, the sample is destroyed in the flame. Some characteristics of the FID are listed in Table 8.6.

8.5.3   Electron Capture Detector

The ECD consists of an accelerating and collection electrode as well as a radioactive source. The source, ^{63}Ni, is a beta (electron) emitter and produces a high background level of free electrons, in the carrier gas. Any compounds that enter the detector which are electron absorbing reduce the background level of free electrons, and there is a resultant drop in the current which is measured by an electrometer. The newer type of electronics (pulsed constant current) have improved the performance dramatically, increasing the linear response from 10^2 to $>10^5$. With no sample, the pulse frequency is low. When an electron-absorbing compound passes through the detector, the frequency increases to compensate for the current loss to the sample. The concentration is then proportional to the pulse frequency.

Earlier ECDs [with direct-current (DC) electronics] had problems with saturation of the current and subsequent reduction of the linear range of the detector to just over 100. The most sensitive compounds for this are chlorinated hydrocarbons, which have sensitivities as low as 0.1 ppb of lindane.

8.5.4   Thermal Conductivity Detector

The TCD measures the difference between the thermal transfer characteristics of the gas and a reference gas, generally helium, but hydrogen, argon, or nitrogen

Table 8.7   Thermal Conductivities for Selected Compounds

Component	Thermal Conductivity[a]
Acetylene	0.78
Ammonia	0.90
Butane	0.68
Carbon dioxide	0.55
Chlorine	0.32
Ethane	0.75
Helium	5.97
Hydrogen	7.15
Methane	1.25
Sulfur dioxide	0.35
Xenon	0.21

[a]Relative to air, 0°C.

can be used depending on the application. The sample and reference filaments are two legs of a Wheatstone bridge. A constant current is applied with a resultant rise in filament temperature. As the sample passes through the detector, the resistance changes as the reference gas is replaced by the sample, which has a lower thermal conductivity. The thermal conductivities for a number of compounds are given in Table 8.7. This difference in resistance is proportional to the concentration. The response is universal since the detector responds to any compound that conducts heat. The minimum detection limit is in the range 100–200 ppm or $\frac{1}{20}$ that value for H_2. The maximum concentration is 100%.

8.5.5   Flame Photometric Detector

The sample is burned in a hydrogen-rich flame which excites sulfur or phosphorus to a low-lying electronic level. This is followed by a resultant relaxation to the ground state with a corresponding emission of a blue (S) or green (P) photon. This type of emission is termed *chemiluminescence*. The emission is at 394 nm for sulfur and 525 nm for phosphorus. The S–C selectivity ratio is >10,000 : 1. This detector can use rare earth filters instead of interference filters for sulfur and phosphorus to improve detection limits and eliminate some of the deficiencies of interference filters.[7] Detection limits are in the 5- and 20-pg range for phosphorus and sulfur, respectively.

8.5.6   Far-UV Absorbance Detector

Most organic and inorganic species absorb strongly in the far UV. Notable exceptions are the inert gases helium and nitrogen, which absorb very weakly in this region. Certain diatomic species such as O_2, which have low absorption in

Table 8.8   Detection Limits for the FUV Detector

Compound	Detection Limit (ng)
Sulfur dioxide	0.7
Methane	0.3
Oxygen	14
Water	3
Propane	1
Chloroform	5
Ethylene	1
Hydrogen sulfide	3

the region of the lamp energy (124 nm), will have a poor response, but low-ppm levels can still be detected.

The FUV detector is relatively new to GC (compared to other GC detectors) since it was introduced by HNU Systems in 1984. It is frequently compared with the TCD since it will respond to any compound that absorbs in the far or vacuum UV. The latter name is a misnomer since with a carrier gas flowing through the cell a vacuum is not needed. Thus, the detector has a response that is nearly universal, a low dead volume (40 μL), and a fast electrometer time constant. The primary emission from this lamp is the 124-mn line. Although there are visible lines from this lamp, the photodiode is unresponsive to any long-wavelength UV or visible emissions, and only the absorption at 124 nm needs to be considered[8] for the absorption process.

The minimum detection limits for organic compounds, oxygen, water, and inorganic compounds are in the range 50 ppb–10 ppm. A summary of the detection limits for organic and inorganic compounds is given in Table 8.8.

A summary of the response and range of the various detectors is shown in Table 8.9. Note that the response ranges from universal (TCD) to selective (FPD for sulfur and phosphorus) while the detectors span a concentration range of about a billion to 1.

Table 8.9   Summary of Detector Characteristics

Type	Response	Carrier Gas	Range[†]
PID	Organic, inorganic	Nitrogen,[a] helium,[a] hydrogen[a]	0.5 ppb–low %
FID	Organic	Nitrogen,[a] helium,[a] hydrogen[a]	50 ppb–%
FUV	Organic, inorganic, fixed	Nitrogen,[a] helium,[a] hydrogen	50 ppb–low %
ECD	Halogenated, nitrogen compounds	Argon–methane,[a] helium, nitrogen[a]	0.1 ppb–1 ppm
TCD	Organic, inorganic, fixed	Helium, hydrogen	200 ppm–100%
FPD	Sulfur, phosphorus	Nitrogen,[a] helium,[a] hydrogen	25 ppb–100 ppm

[a]High purity.
[†]Recent improvements.

8.6   DISCUSSION

The framework of the Environmental Protection Agency (EPA) methodology involves five levels of investigative screening or analyses. Level I involves field screening of VOCs with hand-held analyzers (EPA protocol specifies a photoionization detector like the HNU model PI or DL101) and other site characterization equipment such as an oxygen meter, explosimeter, radiation survey equipment, and chemical testing tubes.[9] This type of measurement is described in Chapter 11 on photoionization and is not treated here.

Level II screening can establish the identity of the compound(s) and relative concentrations. In the early to mid-1980s, this was done predominantly by sending samples to a laboratory for detailed analysis. It is interesting to note that >50% of the samples returned to the laboratory during the 1980s for the EPA CLP program were identified as *"no detects."* This demonstrates how important field methods are. The intermediate level II analysis was introduced by the EPA to reduce both the time required to start remedial actions and the high costs associated with laboratory analysis. An additional factor was the cost of keeping trained personnel in the field waiting for results.[9] Level II measurements involve field analysis with more sophisticated instrumentation (i.e., portable GC or GC in a laboratory in a trailer) to provide identification (as far as possible) of specific components present.

The final three levels (levels III–V) use laboratories located "off-site" and frequently involve CLP analysis.[9] We will not be concerned with these latter techniques. Of course, a certain percentage of field samples should be returned and analyzed in a laboratory with standard EPA methods. Semivolatile hydrocarbons

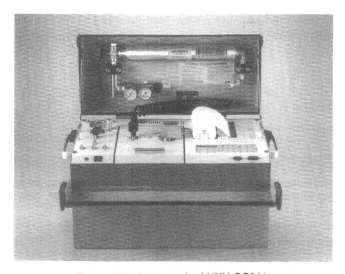

Figure 8.9   Photograph of HNU GC311.

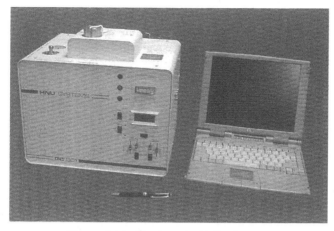

Figure 8.10 Photograph of HNU GC321.

do not migrate but may have to be removed as a result of their proximity to a source of drinking water. The two most serious threats from the volatiles involve evaporation into the air and migration away from the original source of contamination through the soil and into a source of groundwater. Remediation of the groundwater to EPA levels generally takes years.

During the 1970s and 1980s, the passage of the Resource, Conservation and Recovery Act (RCRA) and Comprehensive Environmental Response, Compensation and Liability Act (CERCLA, or Superfund) expanded the list of chemicals under EPA regulation. This led to the development of field screening methods for volatile organics to augment the CLP program.[10] The portable gas chromatograph was one of the stars of the EPA's field screening programs for the analysis of volatiles.[11] In 1988, the EPA published *Field Methods Catalogue*,[10] which described simplified methods for volatile and semivolatile hydrocarbons which had been used for field screening. It is clear that a portable or compact gas chromatograph for an on-site trailer best meet the needs for field measurement. A portable gas chromatograph, the HNU model GC311, is shown in Figure 8.9 and a compact gas chromatograph, the PID model 322 is shown in Figure 8.10. Both are capable of analyzing volatile and semivolatile hydrocarbons.

8.6.1 Sampling

8.6.1.1 Air, Water, Soil. Volatile hydrocarbons can be present in a variety of matrices in field samples, including air, water, soil, soil gas, and sludge. Of course, the air samples can be analyzed directly by manually injecting 1 or 5 mL of air into the gas chromatograph. Many portable gas chromatograph have an automatic mode where the air is injected into the gas chromatograph at a fixed interval. Water or soil samples cannot be measured by directly injecting into the gas chromatograph since the former would quickly overload the column and the possible the detector. Instead, methods such as headspace, purge and trap (volatiles), or solvent extraction (for VOCs or SVOCs) are used to change the environment of the sample for analysis by GC.

8.6.1.2   *Headspace.* To measure VOCs with good precision and accuracy, the sample has to be in a dilute (ppm-level or below) solution. Henry's law applies as long as solute molecules never interact significantly, because then the escaping tendency is strictly proportional to the number of solute molecules in the fixed amount of solvent. The measurement of low concentrations of organics in water can be accomplished through the application of Henry's law, which states that, at equilibrium, the solubility of a gas in a liquid is proportional to the partial pressure of a gas in contact with a liquid as given below:

$$\text{VOC(aq)} \Leftrightarrow K\,P_{\text{VOC}}$$

where VOC(aq) is the concentration of benzene in the liquid phase, K is the Henry's law constant which governs the solubility of gases in water, and P_{VOC} is the partial pressure of benzene in the gas phase.

As a result of the above equation, it can be seen that if the concentration of benzene in the *gas phase* and *at equilibrium* is measured, this is related to the concentration of benzene in the dilute aqueous solution by a proportionality constant (K) that can be determined by calibration.

Simple headspace measurements can be made by equilibrating the liquid or soil sample in a sealed container (jar, VOA vial, or plastic bag) with a small headspace. Stewart and colleagues have developed the static headspace method, which provides a useful and reproducible methodology for field measurements. This is described in Section 8.6.2.3.

8.6.1.3   *Soil Gas.* Although headspace analyses[13] are common for volatile hydrocarbons, one of the most commonly used field analysis techniques for site characterization is soil gas analysis,[13] where the sample is collected by in situ pumping of a well. These wells are relatively inexpensive to drill and can be surveyed rapidly (as many as 35–50 per day). This is a useful procedure to quickly evaluate the extent and source (since specific pollutants can be identified with GC) of contamination for a site. With an HNU GC311, which has a built-in sampling system, the sample stream can be sampled and analyzed directly.

8.6.1.4   *Carbon Bed.* The gas chromatograph is an ideal device for monitoring the output of a carbon bed. These devices are used to remove the residual hydrocarbons from the air pumped into the soil and pulled our (pump-and-treat method). The EPA requires a monitor on the output of these devices. To obtain a faster response, the GC column can be replaced with a piece of $\frac{1}{8}$- or $\frac{1}{16}$- in. tubing. Then the instrument will be a monitor for total VOCs.

8.6.2 Extraction Methods

8.6.2.1 Purge and Trap. This method is for VOCs which are not very soluble in water. This technique was adopted by the EPA[11] for water analysis and is the basis for most of the water methodology. A 15-mL sample is purged (10–15 min) with clean nitrogen or helium to sweep the VOCs out of the water sample. The VOCs in the nitrogen are collected on a tenax trap, which absorbs the hydrocarbons. Once the purging is complete, the tenax trap is rapidly heated and the sample is injected onto the GC column for analysis. This method can be used for water and soil samples and will detect low-ppb concentrations of many hydrocarbons.

8.6.2.2 Solvent Extraction. In Europe and other parts of the world, the purge-and-trap method has not been accepted for the analysis of volatiles. Instead, solvent extraction is the method of choice. The sample can be water or soil, and an organic solvent is used the extract the trace organic compounds from the sample. Then the solvent can be injected into the heated injection port of the gas chromatograph.

Field methods for the extraction and analysis of volatiles and semivolatiles [pesticides, polychlorinated biphenyls (PCBs), polyaromatic hydrocarbons (PAH)] have been described in detail previously.[10] Provided that the gas chromatograph has sufficient versatility, all of these samples can be analyzed with the same instrument. The GC oven temperature control at 150–200°C and a heated injection port are required for analysis of the semivolatiles.

In Table 8.10, we compare the detection limits for soil and water samples for the extraction and headspace methods.[14]

8.6.2.3 Static Headspace. This field method (static headspace) will allow the analyst in the field to rapidly screen the soil or ground water samples and if a no

Table 8.10 Detection Limits for Soil and Water Samples by GC[a]

Method	HC Concentration in Soil–Water Extract	HC Concentration Injected into GC	GC Detection Limit FID	GC Detection Limit PID
Headspace[b]	10 ppm	10 ng[c]	0.1 ng	0.005 ng
Solvent extraction	1 μg/mL	1 ng	0.1 ng	0.005 ng
Static headspace, soil[d]			<0.5 ppm	<10 ppb
Static headspace, water			<50 ppb	<1 ppb

[a]HC, hydrocarbons.
[b]1 g of soil or water in 100-mL container, heated mildly and cooled; assume aromatic HC.
[c]Assume 1-mL gas sample injected.
[d]4 g in 25 mL DI water, 50 μL headspace injected.

Table 8.11 Comparison of Static Headspace and Purge-and-Trap Techniques

Method[a]	Benzene	Toluene
Static HS	17.3	61.8
Purge and trap	12.7	40.6
Static HS	1144	4320
Purge and trap	709	2170
Static HS	496	3180
Purge and trap	330	2800
Static HS	21.2	ND
Purge and trap	18.0	ND
Static HS	10.5	ND
Purge and trap	8.1	ND

[a]HS, headspace.

detect is found, another sample can be taken and analyzed. One obvious advantage of this technique is that the equipment needed is minimal compared to the purge-and-trap technique, yet Robbins and Stuart[14] have shown that comparable results can be obtained with detection limits of the order of 1 ppb.

This method was developed by Robbins and Stuart[15] at the University of Connecticut for the extraction of low levels of volatile organics from water. A 4-g soil sample is added to 25 mL of water in a 40-ml VOA vial and 100 μL of mercuric chloride (2.4 g/L) was added as a preservative. Each vial was shaken for 10 s, inverted, and placed in a water bath for 30 min at $25 \pm 0.3°C$ to reach thermal equilbrium. A 50-μL gas sample is injected into the gas chromatograph. An example of the comparison between static headspace and purge and trap for benzene and toluene is shown in Table 8.11. The correlation coefficients (r^2) for the static headspace and purge-and-trap data in Table 8.11 was 0.999 for benzene and 0.89 for toluene. A chromatogram of VOCs in water is shown in Figure 8.11.

8.6.3 Volatile Organic Compounds

Typically, at a site, GC can initially be used for industrial hygine surveys to evaluate the level of toxic VOCs and implement a plan to protect the workers. Then it can be used for soil gas surveys and checking contaminated soil and water. An example of a sample containing BTX and other VOC's by photoionization detection is shown in Figure 8.12.

Several years ago, some underground gasoline tanks ruptured in Falmouth, Massachusetts. Since the soil is sandy, the contents of the tanks spread quickly over a considerable area. Initially, the site was investigated using a portable PID (HNU model 101) by measurements in a number of soil gas wells to determine the extent of the plume. The plume had migrated more than 300 yards from the original source. This type of level I screening could be used to determine the extent of contamination of the soil ("total" but not individual hydrocarbons) and

Detector: PID

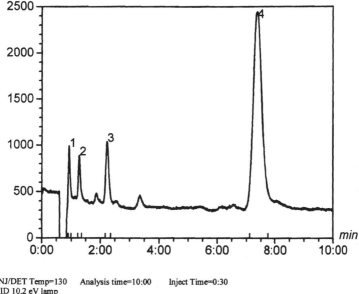

INJ/DET Temp=130   Analysis time=10:00   Inject Time=0:30
PID 10.2 eV lamp
Oven Temp=70   Sample Time=0:30
Syringe Inj=Off   Alarms=Off
Cont. Pump=Off
311 Plotter=Off
311 TWA Cycle=1   Calibration Cycle=2   Repeat Cycle=1
Notes: MXT-1 30m x 0.53 mm, 5.0 u film, 28 cc/min N2 70 C

	Name	Conc. ppb	Height	Area	Time
1	Isopropanol	1998.260	467	32990	0:56
2	Benzene	0.288	374	31170	1:17
3	TCE	0.296	523	62890	2:14
4	Ethyl Benzene	7.500	1936	736840	7:23

Figure 8.11 Chromatogram of VOCs in wastewater with HNU GC311.

groundwater which occurred. Following this, a portable gas chromatograph (GC311) was used to characterize the composition of the fuel detected and the ultimate source of the contamination. When multiple sources are present, this fingerprinting data can be used to identify the source of a leak.

8.6.4 Semivolatile Organic Compounds

Semivolatiles, including pesticides, herbicides, PCBs, and PAHs, account for nearly 30% of the field samples. Field methods for sample preparation and analysis have been described.[10] Some chromatograms of PCBs and PAHs are given in Figures 8.13 and 8.14.

One field method[10] involves taking an 800-mg soil sample or 10-mL water sample, adding 1 mL of a 1:4 water–methanol mixture, adding 1 mL of hexane, shaking

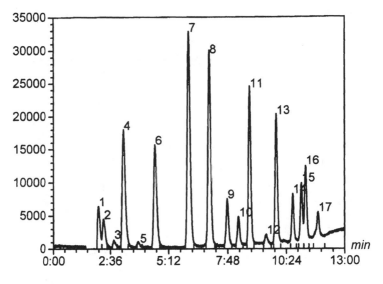

INJ/DET Temp=140   Analysis time=13:00   Inject Time=0:25
Oven Temp=60I, TP to 180C @ 9C/min.   Sample Time=0:30
Cont. Pump=Off
311 Plotter=Off
Calibration Cycle=0   Repeat Cycle=1
Notes: TO14 precursors at 1 ppm

	Name	Conc	Height	Area	Time
1	vinyl chloride		5735	928250	2:05
2	methylene chloride		3771	682260	2:19
3	unknown		779	113150	2:46
4	3 chloro propene		17215	3351610	3:12
5	unknown		167	86320	3:56
6	cis 1,2 dichloro ethene		14945	2766070	4:37
7	benzene		31992	5727230	6:05
8	trichloroethylene		28740	4957430	7:00
9	cis 1,3 dichloropropene		6826	1107910	7:47
10	trans 1,3 dichloropropene		4109	680870	8:17
11	toluene		23548	3601010	8:46
12	1,2 dibromo ethane		1458	365550	9:31
13	perchloroethylene		19492	3058870	9:57
14	chlorobenzene		7594	1207250	10:42
15	unknown		9304	1388770	11:05
16	o-xylene & styrene		11906	1865190	11:16

Figure 8.12   Sample of ozone precursors (1 pm) via HNU GC 311 with PID.

for 30 s, letting stand for 30 s (if the mixture emulsifies, centrifuging the sample), and injecting the top layer (hexane) into the gas chromatograph. This technique is useful for extraction of PAHs, PCBs, and other nonvolatile hydrocarbons.

This method was modified[16] and used for the determination of DDT in soil. A GC311 with a PID (10.2 eV) was employed for the field analysis. This site was one of the best examples of the need for a field analysis. The site had been visited two times previously, and samples had been sent back to a laboratory for analysis. Although this was the third time in the field, new areas of contamination were discovered that had not been encountered previously.[16] Forty-four samples were

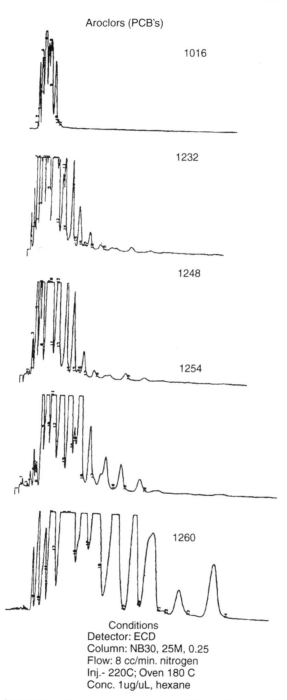

Figure 8.13 Chromatograms of various arochlors (PCBs) with HNU GC311 with ECD.

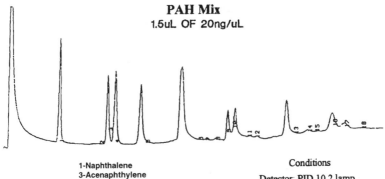

PAH Mix
1.5uL OF 20ng/uL

1-Naphthalene
3-Acenaphthylene
4-Acenaphthene
5-Fluorene
6-Phenathrene/anthracene
9-Fluoranthene
10-Pyrene
13-Chrysene/benzo(a)anthracene
16-Benzo(b,k)fluoranthene
17-Benzo(a)pyrene

Conditions

Detector: PID 10.2 lamp
Column: NB30, 5M, 0.32 mm
Flow: 15 cc/min helium
Inj. temp.:250C
Oven: 1 min. @80C, 16C/min. to
140C, hold 1min., 14C/min to 200 C, hold
8 min.

Figure 8.14 Chromatogram of PAHs.

collected and analyzed in a three-day time period and enough information was gathered to finally clean up the site. Nearly a year had passed since the first visit to the site, which demonstrates the need for good field methodology for semivolatiles. Excellent agreement was observed between the field (PID) and laboratory (ECD) methods, even though the methodology and detectors were different.

A useful detector combination for sample confirmation is the PID and ECD. Detector response ratios are used to provide additional confirmation of the presence or the structure of a particular compound in a peak. For example, trichlorobenzene would be expected to have a strong response on both the PID and ECD while another compound with the same retention time would produce a very different response ratio. As observed above, the need for field screening is obvious even for semivolatiles. Driscoll and Atwood[17] evaluated the 8000 series of EPA methods and found that essentially all of the samples, including phenols, pesticides, herbicides, nitrosamines, PAHs, nitroaromatics, PCBs, and phthalate esters, can be analyzed by GC using a PID and ECD individually or in common. A typical chromatogram for some of the semivolatiles is shown in Figure 8.11.

8.6.5 Dual Detectors

Dual detectors are an important consideration for field analyses because they are the minimum needed for confirmation of a particular compound. With GC, one has to run a sample on two columns of different polarity (e.g., polar and nonpolar) to confirm the identity of a particular compound. This is not necessarily something that should be done in the field. Instead, it is more useful to identify compounds by comparing both retention times and detector response factors with known standards. This is the basis of a number of environmental methods.

Gasoline hydrocarbons are one of the most common contaminants found in the field. The PID and FID response ratios[18] can be used to identify alkanes (PID/FID ratios of 8–10), alkenes (PID/FID ratios of 18–24), and aromatics (PID/FID ratios of 40–50) in complex mixtures.[18]

For groundwater applications in the vicinity of gasoline stations, it is necessary to measure levels of aromatic hydrocarbons in the presence of gasoline- or fuel oil–contaminated samples. The EPA method 602 or 8020 does not have adequate selectivity for this particular analysis since high-molecular-weight alkanes can coelute with the aromatic hydrocarbons, resulting in an interference. The approach we took[18] following purge and trap involved the use of a highly polar capillary column (carbowax initially but then DB5 because of the improved long-term stability) which would elute the nonpolar alkanes quickly (and in one broad peak) while providing adequate resolution for the aromatic hydrocarbons, particularly the xylenes. An added feature of this method is that alkanes and aromatics can be quantitated, if desired. One advantage of this technique is that we can identify interferences from aliphatic hydrocarbons in the determination of aromatic hydrocarbons from the differences in their relative responses on the PID and FID.[19] For example, it is possible for C14 or C15 (from fuel oil) hydrocarbons to coelute with the aromatic hydrocarbons. The PID/FID response ratio will be an order of magnitude lower if an aliphatic hydrocarbon is present in place of the aromatic hydrocarbon, making the identification process relatively easy.

Dual detectors have been used in the laboratory for many years to analyze difficult "unknown" environmental samples. The PID has interchangeable lamps and the 11.7-eV lamp can detect the low-molecular-weight chlorinated hydrocarbons, which are so prevalent in wells and groundwater. The PID with a 10.2-eV lamp is used for hydrocarbons aromatic, olefinic, and alkanes > butane.

The FUV detector has a more general response and is very useful for landfills since it responds to CH_4, CO, and CO_2. None of these compounds respond with the PID. The detector is also useful for the detection of low-molecular-weight chloroalkanes, which are not detected by the PID (10.2 eV). These latter species are quite common on hazardous waste sites.

8.6.6 Site or Fenceline Monitoring

When working with air samples at the ppb levels, severe errors can be introduced by carryover from the Teflon in the syringe. With an unskilled analyst precision as poor as 20–30% would not be unusual. The latter technique does not depend on the operator since it is automatic. This sample introduction mode can be used for air, headspace (soil, water, sludge), and soil gas. The precision at low-ppm levels is ±1–2%; at ppb levels ±5–10%. The instrument can be run in a continuous mode or one sample at a time. Automatic calibration at a specified time interval can be programmed if an area is to be monitored over a period of time.

During the remediation process, pockets of pollution can be stirred up and VOCs and semivolatile hydrocarbons can be released to the atmosphere. Since many of

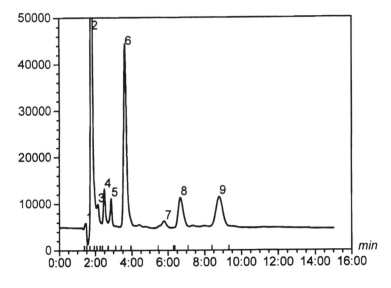

INJ/DET Temp=140   Analysis time=15:00   Inject Time=0:45
Detector: PID 10.2 eV lamp
Oven Temp=100   Sample Time=1:00
Syringe Inj=Off   Alarms=Off   Paper Save=Off
Cont. Pump=Off
311 Plotter=Off
Calibration Cycle=0   Repeat Cycle=8
Notes: desorber, mxt-624 column, 60 m x 0.53 mm id 3.0 u film 15 cc/min N2 @ 32 psi @ 100 C

	Name	Conc	Height	Area	Time
1	unknown		96	10980	1:28
2	unknown		20732	2285410	1:47
3	unknown		508	85990	2:09
4	benzene	0.404 ppb	821	142900	2:31
5	unknown		631	108760	2:53
6	toluene	2.466 ppb	3879	897870	3:38
7	unknown		133	57690	5:49
8	o-xylene	0.568 ppb	650	252340	6:41
9	unknown		668	360400	8:48

Figure 8.15 Ambient air analysis with HNU GC 311 with concentrator and PID.

these sites are in urban areas, it is important to continuously monitor the fenceline to minimize the exposure of surrounding neighbors to these pollutants.

A concentrator described previously[20] is available as an option for the GC311. This system allowed the detection of ppt levels of aromatic hydrocarbons in the atmosphere on an automatic basis. A typical chromatogram of an ambient air sample is shown in Figure 8.15. This system improves chromatography by eliminating any air or water peaks which would interfere with the early eluting peaks at low-ppb levels. The concentrator also improves the performance of the FUV detector, as shown in Figure 8.6. This detector is useful down to 0.1 ppm without preconcentration, but the interference from both water and oxygen is very significant since both these species absorb UV and thus produce a detector response. The material in the concentrator is hydrophilic and the water can be swept through without any loss of volatiles. Applications for this accessory include fenceline monitoring, background soil checks, following the emissions from or to a particular source, checking carrier gases for contamination, and any applications where additional sensitivity is required.

In this chapter we have described a number of basic aspects of chromatography in order to provide the reader with a reasonable understanding of both field and laboratory methods. Should the reader be interested in learning more about chromatography or detectors, Ref. 3 and 19 are recommended.

REFERENCES

1. Symposium on Air Quality Measurement, MIT, October 1993.

2. R. Dandeneau and E. H. Zerenner, *J. HRC & CC* **2**, 351 (1979).

3. J. C. Giddings, Theory of Chromatography, in Heftmann (Ed.) *Chromatography: A Laboratory Handbook*, Van Nostrand Reinhold, New York, 1975.

4. F. Daniels and R. A. Alberty, *Physical Chemistry*, Wiley, New York, 1965.

5. J. N. Driscoll and F. F. Spaziani, Development of a New Photoionization Detector for Gas Chromatography, *Res. Devel.* **27**, 50 (1976).

6. M. L. Langhorst, Photoionization Detector Sensitivity for Organic Compounds, *J. Chrom. Sci.* **19**, 98 (1980).

7. J. N. Driscoll and A. W. Berger, A New FPD with Rare Earth Filters, *J. Chromatogr.* (1989).

8. J. N. Driscoll, M. Duffy, S. Pappas, and M. Sanford, Analysis of ppb Levels of Organics in Water by Means of Purge and Trap Capillary GC and Selective Detectors, *J. Chromatogr.* **441**, 73 (May 1988).

9. Anonymous, *Data Quality Objectives for Remedial Response Activities*, EPA, Development Process, PB88-131370, EPA, Washington, DC, March 1987.

10. Anonymous, *Field Screening Methods Catalogue, Users Guide*, EPA/540/2/88/005), Hazardous site Evaluation Div., EPA Office of Emergency and Remedial Response, Washington, DC, 1988.

11. H. Fribush and J. Fisk, *Survey of U.S. EPA Regional Field Analytical Needs*, American Laboratory, October 1990, p. 29.

12. J. N. Driscoll, J. Hanby, and J. Pennaro, Review of Field Screening Methodology for Analysis of Soil, in P. Kostecki, E. Calabrese, and M. Bonazountas (Eds.), *Hydrocarbon Contaminated Soils*, Lewis Publishing, Ann Arbor, MI, 1992.

13. Anonymous, *Field Measurements, Dependable Data When You Need It*, EPA/530/UST-90-003, EPA, Washington, DC, September 1990.

14. J. N. Driscoll, Review of Field Screening Methods for the Analysis of Hydrocarbons in Soils and Groundwater, *Int. Labmate.*, October 1992, pp. 27–32.

15. V. Roe, M. J. Lucy, and J. D. Stuart, Manual Headspace Method to Analyze for the Volatile Aromatics in Gasoline in Groundwater and Soil Samples, *Anal. Chem.* **61**, 2584 (1989).

16. J. N. Driscoll, M. Whelan, C. Woods, M. Duffy, and M. Cihak, Looking for Pesticides, *Soils*, January/February 1992, pp. 12–15.

17. J. N. Driscoll and E. S. Atwood, Use of Selective Detectors for Field Analysis of Semivolatiles in Soil and Water, *J. Chromatogr.* (July 1993).

18. J. N. Driscoll, J. Ford, and E. T. Gruber, Gas Chromatographic Detection and Identification of Aromatic and Aliphatic Hydrocarbons in Complex Mixtures, *J. Chromatogr.* **158**, 171 (1978).

19. J. N. Driscoll, Far UV Inoization (Photoionization) and Absorbance Detectors, in H. Hill and D. G. McMinn (Eds.) *Detectors for Capillary Chromatography*, Wiley, New York, 1992.

20. M. Whalen, J. N. Driscoll, and C. Wood, Detection of Aromatic Hydrocarbons in the Atmosphere at PPT Levels, *Atmos. Environ.* **28**, 567 (1993).

9

ONLINE ANALYSIS OF ENVIRONMENTAL SAMPLES BY MASS SPECTROMETRY

Raimo A. Ketola, Ph.D.

VTT Processes
Espoo, Finland

Environmental Instrumentation and Analysis Handbook, by Randy D. Down and Jay H. Lehr
ISBN 0-471-46354-X   Copyright © 2005 John Wiley & Sons, Inc.

9.1   INTRODUCTION

9.1.1   Background

The driving force for the use and development of analytical measurement methods for online/on-site environmental analysis includes the demand of waste reduction, treatment, and remediation of contaminated sites and a greater need for pure drinking water.[1–17] These demands are also helping to obtain optimized and cost-effective production, since environmental pollution, energy consumption, and raw material usage are minimized with optimized production. That is why many offline and online analytical methods have been developed for the analysis of hazardous components in water, air, and soil. Mass spectrometry (MS) is one of the most used methods for analyzing organic compounds in a variety of matrices. In this technique analyte molecules are vaporized and ionized and the ions are separated according to their mass-to-charge ratio (m/z) and finally detected. Mass spectrometric methods provide numerous advantages over other commonly used methods, such as gas chromatography (GC) or infrared spectroscopy. The most important advantages of MS are excellent sensitivity, high speed of analysis, and real-time measurement capability. Other good analytical characteristics of MS are good precision, wide linear range, ease of automation, and capability for real-time analysis of several compounds simultaneously.

The most significant advantages obtained by online/on-site environmental MS are (1) reduction of costs of contaminated site assessment and remediation monitoring, (2) increased amount of information about a contaminated site, (3) decreased possibilities for sample contamination and for alteration of composition of samples during storage, and (4) possibility for interactive sampling.

9.1.2   Definition and Scope

Many of the properties required from a mass spectrometer capable for online/on-site analysis are the same as required from a mass spectrometer placed in a stationary laboratory. However, many additional properties are required from on-site/in situ mass spectrometers. The special desirable properties for online/on-site mass spectrometers can be summarized as follows: (1) capability for long-term automatic monitoring, (2) minimum maintenance requirement, (3) fast automatic identification and quantitation and clear presentation of the obtained data, (4) operation by nonexpert personnel, (5) suitability for field installation in harsh and hazardous environments, (6) minimum weight, power consumption, and size, (7) quick and easy transport, and (8) rapid system setup.

The scope of this chapter is to give an overview of the most suitable and often used mass spectrometric methods and instruments in online environmental organic analysis. It includes the instruments which can be used in real time or almost real time and continuous analysis of air, water, soil, or particles. In this context it excludes all hyphenated techniques such as chromatographic devices connected to MS. A few examples using online MS for the analysis of environmental samples are also given.

9.2 MASS ANALYZERS USED IN ONLINE MASS SPECTROMETRY

9.2.1 Quadrupole

The quadrupole is a device which uses the stability of the trajectories to separate ions according to their m/z. Quadrupole analyzers are made up of four rods with circular or, ideally, hyperbolic section (Fig. 9.1). The quadrupole principle was described by Paul and Steinwedel in 1953.[18] Ions travelling along the z axis are subjected to the influence of a total electric field made up of a quadrupolar alternative field superposed on a constant field resulting from the application of the potentials upon the rods:

$$\Phi_0 = +(U - V \cos \omega t) \qquad \text{and} \qquad -\Phi_0 = -(U - V \cos \omega t)$$

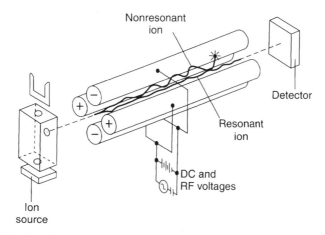

Figure 9.1 Quadrupole instrument made up of the source, quadrupole cylindrical rods, and detector.

In this equation, Φ_0 represents the potential applied to the rods, ω the angular frequency [in $\text{rad/s} = 2\pi f m$, where f is the frequency of the radio-frequency (RF) field], U is the direct potential, and V is the zero-to-peak amplitude of the RF voltage. In practice, the highest detectable mass is 4000 amu, and the resolution is around 3000. Usually, a quadrupole mass spectrometer is operated at unit resolution, that is, a resolution that is sufficient to separate two peaks one mass unit apart. A quadrupole analyzer is the most common analyzer used in online/on-site environmental measurements because it is small, easy to operate, and commercially available as a portable instrument and needs only a little maintenance. The mass range of portable instruments is usually from 1 to 500 amu; thus it is best suited for the analysis of atmospheric gases and low-molecular-weight compounds.

9.2.2   Ion Trap

An ion trap was first introduced in 1960 by Paul and Steinwedel.[19] It is made up of a circular electrode with two spherical caps on the top and the bottom (Fig. 9.2). Conceptually, the ion trap can be imagined as a quadrupole bent on itself in order to form a closed loop. The inner rod is reduced to a point, the outer rod is the circular electrode, and the top and bottom rods make up the caps. The overlapping of a direct potential with an alternative one gives a kind of three-dimensional quadrupole in which the ions of all masses are trapped on a three-dimensional eight-shape trajectory. Ions with different masses are present together inside the trap and are expelled according to their masses so as to obtain the spectrum. By selecting appropriate values of voltages allows trapping of only the ions with a given m/z. These ions fragment in time within the trap, which then contains all of the product ions. Time-dependent tandem mass spectrometry (MS/MS) is thus carried out. The process can be repeated several times, selecting successive fragment ions. This capability of the ion trap to perform tandem mass spectrometry makes it suitable for analysis, in which selectivity is needed.

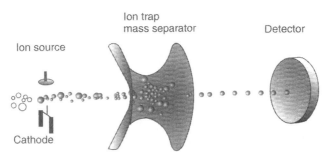

Figure 9.2   Ion trap mass spectrometer. Reprinted with permission from G. Matz, W. Schröder, and T. Kotiaho, Mobile Mass Spectrometry Used for On-Site/In Situ Environmental Measurements, in R. A. Meyers (Ed.), *Encyclopedia of Analytical Chemistry*, Wiley, Chichester, 2000. pp. 3783–3804, Copyright 2000 by John Wiley & Sons.

The performance of the ion trap MS in the environmental analysis has been evaluated by the U.S. Environmental Protection Agency (EPA).[20] With this system, mounted in a van, volatile organic compounds (VOCs) may be measured online in the MS/MS mode. In this mode precursor ions are further fragmented by collision with a reaction gas. From these fragments product ions are recorded, displaying structure details of the precursor ion from the other spectrum. The MS/MS procedures enable compound identification even in complex mixtures without chromatographic separation. Measuring with MS/MS requires a rather complicated measuring strategy and is used when target compounds are known before the measurements start.

9.2.3   Time-of-Flight Analyzers

Time-of-flight (TOF) analyzers were introduced by Wiley and McLaren[21] in 1955, and several reviews of development of TOF instruments and their performance have been published.[22,23] A schematic picture of TOF with a reflectron (an electrostatic mirror) is presented in Figure 9.3.[24] In a TOF instrument ions are expelled from the ion source in bundles which are either produced by intermittent process such as plasma desorption or expelled by a transient application of the required potentials on the source-focusing lenses. They are accelerated by a potential V_s and fly a distance d before reaching the detector. With this technique, all the ions are produced in a short time span and all are in principle analyzed; normally the efficiency is 5–30%. The reflectron is used to compensate both the energy and spatial deviation of similar ions, resulting in a high-resolution instrument. The mass range is much larger than with quadrupoles or ion traps, being thousands or ten thousands even with a short flight path. The TOF instrument is not a scanning device, but the ions are detected in parallel or semisimultaneously. That is why TOF is suitable for applications in which short transients have to be measured, such as pulsed laser desorptions. Also atmospheric pressure ionization devices [atmospheric pressure chemical ionization (APCI) or electrospray ionization (ESI)] are combined with TOF, like a hybrid quadrupole–TOF, in which the ion source is orthogonal, that

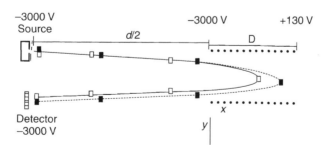

Figure 9.3 Schematic of TOF mass spectrometer equipped with reflectron. Reprinted with permission from E. De Hoffmann, J. Charette, and V. Stroobant, *Mass Spectrometry: Principles and Applications*, Wiley, Chichester 1996. Copyright 1996 by John Wiley & Sons.

is, the axis of the ion beam from the source is perpendicular to the axis of the flight path in the TOF. In this way a pulsed sampling of ions is possible at several kilohertz, resulting in good transmission. For this reason TOF instruments are very popular in the aerosol analysis, in which the particles must be analyzed rapidly.

9.2.4 Sector Instruments

Double-focusing sector instruments are made of two parts, magnetic and electric sectors. In a magnetic field the ion follows a circular trajectory with a radius r so that the centrifugal force equilibrates the magnetic force:

$$qvB = \frac{mv^2}{r} \qquad \text{with} \qquad mv = qBr$$

where B is the magnetic field. For every value of B, the ions with the same charge and the same momentum (mv) have a circular trajectory with a characteristic r value. Thus, the magnetic analyzer selects the ions according to their momentum. Taking into account the kinetic energy of the ions at the source outlet leads to

$$\frac{m}{q} = \frac{r^2 B^2}{2V_s}$$

where V_s in the potential difference in the source. When the radius r is fixed for a certain instrument, changing B as a function of time allows successive observations of ions with various values of m/q provided that they all have the same kinetic energy. On the other hand, it is possible to use the characteristic of the magnetic sector that ions with the same kinetic energy but different ratios m/q have trajectories with different r values. Such ions emerge from the magnetic field at different positions. These instruments are said to be dispersive. The result is that ions with identical charge and mass are dispersed by a magnetic field according to their kinetic energy. To avoid this dispersion, which alters the mass resolution, the kinetic energy dispersion must be controlled. This is achieved with an electrostatic analyzer. In a cylindrical electrostatic field the trajectory of the ion is circular and the velocity is constantly perpendicular to the field. Thus the centrifugal force equilibrates the electrostatic force according to the following equations, where E is the intensity of the electrostatic field:

$$qE = \frac{mv^2}{r}$$

Introducing the entrance kinetic energy, we obtain

$$r = \frac{2E_c}{qE}$$

Figure 9.4 Combination of two sectors, electrical and magnetic, in which magnetic sector is turned over so as to obtain double focusing. Reprinted with permission from E. De Hoffmann, J. Charette, and V. Stroobant, *Mass Spectrometry: Principles and Applications*, Wiley, Chichester 1996. Copyright 1996 by John Wiley & Sons.

The trajectory being independent of the mass, the electric field is not a mass analyzer but rather a kinetic energy analyzer, just as the magnetic field is a momentum analyzer. With suitably chosen geometry of both magnetic and electrostatic fields the incoming beams are focused in direction but energy is dispersed. If two sectors with the same energy dispersion are oriented as is shown in Figure 9.4, the first sector dispersion energy is corrected by the second convergence, which results in a double-focusing instrument.[24] A mass resolution of 10,000 is easily achievable, thus allowing a very specific analysis, which is demanded, for example, in the analysis of polychlorinated dibenzodioxins. However, the sector instruments are rather large and complicated, and so far their use as online/on-site instruments has been very limited.

9.2.5 Miniature and Portable Mass Analyzers

All the major mass analyzers have been miniaturized.[25] These miniaturized analyzers are especially suitable for online and on-site applications due to their small size and reduced power consumption. Although the performance of hand-portable analyzers is almost as good as that of normal-sized analyzers, the usage of these hand-portable mass spectrometers is restricted because the electronics and the pump systems have not been miniaturized to the same level as the analyzer itself. This means that, for example, the overall size and the power consumption are nearly the same with normal-sized and portable instruments, and thus portability is the only advantage of the hand-held instruments. The miniature analyzers, in which the size is in millimeters, have a good tolerance for high pressures due to a short mean free path of ions, but usually the sensitivity and the resolution are worse than with normal-sized instruments. However, Figure 9.5 shows a mass spectrum of *o*-dichlorobenzene measured with a cylindrical ion trap with 2.5 mm inner radius.[26] The resolution of 100 (50% valley) was measured for m/z 146 and the detection limit of 500 ppb was measured for toluene. These values are normally enough for online analysis of contaminated environmental samples.

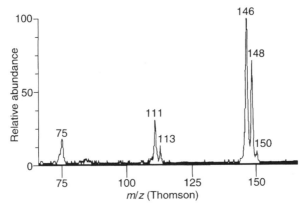

Figure 9.5 Mass spectrum of *o*-dichlorobenzene measured with cylindrical ion trap with 2.5 mm inner radius. Reprinted with permission from E. R. Badman, R. C. Johnson, W. R. Plass, and R. G. Cooks, A Miniature Cylindrical Quadrupole Ion Trap: Simulation and Experiment, *Anal. Chem.* **70**, 4896 (1998). Copyright 1998 by the American Chemical Society.

A thorough comparison of different miniature mass analyzers for on-site analysis has been made.[27] This comparison (Table 9.1) shows that the development of miniature mass analyzers is still at the early stages; thus huge improvement is expected in the future. The parameters which demand most improvement include the response time and the system weight. In most cases the trend has been that when the overall size of the system is minimized the performance of the analyzer (resolution, limit of detection, etc.) is getting worse. However, the need for miniature mass analyzers is growing, and viable mass spectrometers for online analysis of environmental samples will be developed in the near future. Another comparison of four mass analyzers for on-site applications has been made by Kotiaho.[15,28] This comparison (Table 9.2) reveals that none of these mass analyzers is superior to others; rather it reveals that the choice of mass analyzer to a specific measurement depends not only on the matrix and the compound to be analyzed but also on other aspects such as size, price, power consumption, and speed of measurement.

9.3 SAMPLE IONIZATION AND INTRODUCTION

9.3.1 Electron and Chemical Ionization

In the ion source of a mass spectrometer, the products are ionized prior to analysis by the mass analyzer. The most common ion source is an electron impact source which was already invented in 1929.[29] This source consists of a heated filament giving off electrons. The electrons are accelerated, normally with 70 eV, toward an anode and collide with gaseous molecules injected into the source. The molecules are ionized, normally by expelling an electron. On average, one ion is produced for every 1000 molecules entering the source under the usual spectrometer

Table 9.1 Comparison of Various Miniaturized Mass Analyzers

Analyzer[a]	Accuracy	Precision	Limit of Detection	Response	Recovery	Scan Rate	System Volume	System Weight	Score
SRS	2	2	1	2	6	8	6	7	4.3
XPR-2	3	8	2	7	5	5	2	6	4.8
Ferran	10	10	7	10	10	9	3	6	8.1
Polaris-Q	7	3	2	8	8	2	10	10	6.3
UF-IT	3	3	6	3	3	2	7	8	4.4
TOF	9	6	7	9	7	2	9	8	7.1
MG-2100	9	7	10	10	5	10	6	7	8.0
CDFMS	4	5	9	8	6	2	2	4	5.0
Average	5.9	5.5	5.5	7.1	6.3	5.0	5.6	7.0	—

Source: Reprinted with permission from C. R. Arkin, T. P. Griffin, A. K. Ottens, J. A. Diaz, D. W. Follistein, F. W. Adamsand, and W. R. Helms, Evaluation of Small Mass Spectrometer Systems for Permanent Gas Analysis, *J. Am. Soc. Mass Spectrom.* 13, 1004 (2002). Copyright 2002 by Elsevier Science.

[a]SRS = quadrupole RGA-100 from Stanford Research Systems, XPR-2 = quadrupole from Inficon, Ferran = quadrupole array from Ferran, Polaris-Q = quadrupole ion trap from ThermoFinnigan, UF-IT = quadrupole ion trap from University of Florida/KSC, TOF = o-TOF from IonWerks, MG-2100 = cycloidal focusing sector from Monitor Instruments, CDFMS = double focusing sector from University of Minnesota/KSC.

Table 9.2   Comparison of Properties of Quadrupole, Ion Trap, Magnetic Sector, and Time-of-Flight Analyzer for Use as Online/On-Site Mass Spectrometers

Property	Quadrupole	Ion Trap	Magnetic Sector	TOF
Size	+++	++	++	++
Simplicity, durability	+++	+++	++	+++
Weight	+++	++	+	++
Price	+++	++	+	++
Mass Range	+++	++	+++	+++
Pressure requirement	++	+++	+	+
Sensitivity	++	+++	++	+++
Scanning modes	++	+++	++	+
Commercial availability	+++	++	+	++
Precision	++	+	+++	++
Power requirement	++	++	+++	+
Speed of measurement	++	++	++	+++
Resolution	++	++	+++	++

Note: +++, very good; ++, good; +, fair.

Source: Reprinted with permission from T. Kotiaho, On-line/On-Site Environmental Analysis by Mass Spectrometry, paper presented at Seventh International Symposium, Chemistry Forum 2001, 14.-16.5.2001, Warsaw, Poland.

conditions at 70 eV. Also the sample pressure is directly correlated with the resulting ion current, and therefore the electron impact ion source can be used in quantitative measurements. The electron ionization (EI) is called a hard ionization technique, as in EI the molecule receives a lot of internal energy, depending on the accelerating potential of ionizing electrons, thus leading to fragmentation of the molecule. The fragmentation of organic molecules is rather unique, and therefore the EI mass spectra are commonly used for identification purposes. Most commercial mass spectral libraries are made of EI mass spectra of compounds at 70 eV.

A softer ionization technique called chemical ionization (CI) is sometimes used.[30] In this technique primary ions are produced by electron impact in high pressure, in millitorr, by introducing a suitable reagent gas into the source. These primary ions then collide with the molecules to be analyzed. Various chemical reactions can occur in the plasma, such as proton transfer, hydride abstraction, additions, and charge transfers, yielding both positive and negative ions. Almost all neutral substances are able to yield positive ions, whereas negative ions with an even number of electrons require the presence of acidic groups or electronegative elements to produce them. The most common reagent gases used in CI are methane, ammonia, and isobutane. In CI with molecules these reagent gases produce mainly the protonated molecular ion or some adducts of the molecular ion, thus allowing determination of the molecular mass of the molecules in the sample. The choice between different reagent gases depends on the nature of the molecules to be analyzed. Methane can ionize nearly all organic compounds, isobutane is much more inefficient in ionizing hydrocarbons, and ammonia is the most selective, being able to ionize polar compounds.

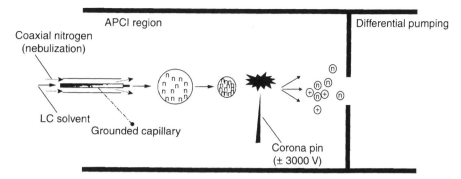

Figure 9.6   A typical atmospheric pressure chemical ionization source.

9.3.2   Atmospheric Pressure Chemical Ionization

Normally the ionization of molecules occurs in a reduced pressure in the ion source, the pressure being 10^{-7}–10^{-5} mbar in the EI ion source. When the ionization can be performed in atmospheric pressure, the ionization efficiency increases 10^3–10^4 times compared to that in an EI ion source,[31,49] and this leads to much more sensitive instruments. The principle of the APCI[32,50] instrument is presented in Figure 9.6. The APCI source consists of a capillary interface for liquid introduction, a heated nebulization system, and a high-voltage corona discharge needle. The liquid sample is nebulized with a heated gas flow, generating gas-phase molecules of the analytes and the solvent. With the help of the corona discharge needle, solvent molecules are ionized, which then ionize the analyte molecules. Usually APCI is used in connection with liquid chromatography–mass spectrometry, but it also can be used for online analysis of environmental liquid and gaseous streams. To reduce the atmospheric pressure to a reduced pressure in the analyzer section, a high-capacity pumping system is needed. Therefore these instruments are limited to operation in large containers or mobile laboratories. On the other hand, polar and aggressive substances may be detected, such as reactive trace compounds in the atmosphere[33–35,51–53] and polar indoor air pollutants.[36,54] The main advantages of the APCI method in on-site/in situ applications are (1) good sensitivity, in favorable cases of the order of a few ppt; (2) fast response times and small memory effects; (3) real-time monitoring capability; and (4) relatively light maintenance. The main limitations of the method include factors such as (1) ionization is dependent on the sample content and (2) sensitivity is good only for compounds with high proton affinity, high electron affinity, high gas-phase acidity, and/or low ionization potential.

9.3.3   Direct Capillary Inlets

For introduction of gaseous samples, a capillary inlet with a two-stage pressure reduction or an adjustable leak valve is very often used. Advantages of capillary inlets and other simple direct methods for sample introduction are (1) very fast response times (on the order of seconds) for gaseous samples, (2) simple and rugged techniques, and (3) suitability for long-term monitoring experiments. In addition, capillary inlets can be easily interfaced to any kind of mass spectrometer, the sampling can be easily automated using commercial sampling valves, and samples can be pumped from relatively long distances. Limitations of the methods are that they are applicable only to gaseous and volatile compounds, short-lived reactive spices may react in the capillary tubing and thus the real composition of the gas mixture/chemical reaction cannot be measured, and relatively high detection limits are achievable for organic compounds in the analysis of environmental samples.

9.3.4   Membrane Introduction Mass Spectrometry

Membrane introduction (inlet) mass spectrometry (MIMS) is a technique often used in environmental analysis.[35,37–41] It is a relatively old technique as it was introduced by Hoch and Kok[36,42] in 1963 as a method for the continuous monitoring of gases dissolved in water. The basis of this technique is the selective transport of analyte molecules across a semipermeable membrane into the ion source of a mass spectrometer, where they are ionized and subsequently analyzed. Because the flow of the analyte matrix, usually water or air, through the membrane is proportionally smaller than the flow of the organic analytes, a certain degree of analyte enrichment is obtained. Due to this enrichment of analyte molecules entering the ion source of a mass spectrometer (on the order of 10–100 times), MIMS allows lower levels of detection than other liquid or air introduction methods.

Volatile organic compounds can be sampled and analyzed directly from soils, liquids, and gases by MIMS. This unusual flexibility in matrices that can be analyzed is achieved through an inert polymer membrane that is used as the only interface between the sample and a mass spectrometer. In this fashion most VOCs can be analyzed directly, and in most cases a changing chemical environment in the sample can be monitored online. Normally the membranes used in MIMS are very hydrophobic, and hydrophobic compounds dissolve very well in the membrane. They have low detection limits (nanograms per liter in water and nano- or micrograms per cubic meter in air), whereas hydrophilic compounds have detection limits in the high-microgram-per-liter range in water and milligrams per cubic meter in air. Currently the detection of VOCs [boiling point (bp) < 200°C] has become a routine matter. However, effort is put into the development of methods that will allow the direct detection of less and nonvolatile organic compounds.[37–40,43–46]

Various membrane inlets have been designed to fit a specific application. However, most of the many designs belong to one of four categories (Fig. 9.7): (a) a direct-insertion membrane probe, where the membrane is mounted inside the mass spectrometer and the sample (gas or liquid) has to be transported to it; (b) a flow-over design, where the sample is flushed over the membrane, which is directly attached to the vacuum chamber of the mass spectrometer; (c) a helium-purged inlet, where permeate molecules at the inside of the membrane are purged

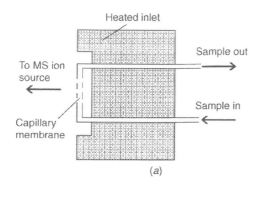

(a)

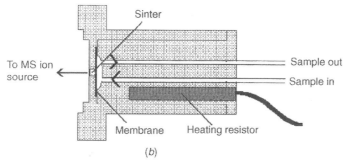

(b)

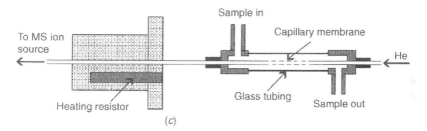

(c)

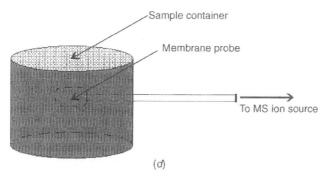

(d)

Figure 9.7 Most commonly used designs of membrane inlets used in online analysis: (a) flow-through or direct insertion probe, (b) flow-over inlet, (c) helium-purged inlet, and (d) membrane probe.

by helium to the mass spectrometric ion source; and (d) a membrane probe, where the membrane is mounted at the end of a probe to be inserted directly into the sample matrix. The membrane probe is probably the simplest membrane inlet to use. It consists of a capillary (steel) perforated and covered by a polymer membrane in one end, whereas the other end is connected via an evacuated tube to the mass spectrometer. Probes can be inserted directly into almost any sample. For example, they have been inserted into plants[41,47] and sediments.[42,48] The drawback of the membrane probe is that it can only be used to analyze gases and highly volatile organic compounds (bp < 100°C). Compounds of a lower volatility interact with the surfaces in the evacuated tube that connects the inlet to the mass spectrometer, and very long response times are often obtained. To circumvent the problem with condensation of sample molecules in the tube that connects a membrane probe to the mass spectrometer, the direct-insertion membrane probe was invented.[43,49] In this inlet the membrane is mounted at the end of a probe that is inserted directly into the mass spectrometric ion source. The liquid or gas sample is then flushed through the probe. The advantage of the inlet is the elimination of the connection tube between the membrane and the ionizing region, and VOCs with a boiling point up to around 200°C can be measured. Furthermore, the inlet has the advantages that the sample can be chemically modified online prior to the detection and a calibration using an external standard is straightforward. An alternative to the close coupling of the membrane inlet to the ionizing region of the mass spectrometer is found with the helium-purged membrane inlet.[44,50] In this inlet a liquid or gas sample is flushed across one side of the membrane, whereas the other side is continuously purged by a helium stream that carries the permeated molecules to the ion source of the mass spectrometer. The purging of molecules from the membrane to the ion source reduces the problem with condensation effects in connection tubes, and the inlet has almost the same applications as the direct-insertion membrane probe.

To improve either selectivity or sensitivity, many devices have been developed that combine the membrane inlet with another analytical technique for separation or concentration. The two-stage inlets can be split into systems using two steps of membrane separation, a preseparation or preconcentration device before the membrane inlet, and a postconcentration device after the membrane inlet. One combination that is frequently used is the combination of a helium-purged membrane inlet and a jet separator.[45,51] In this combination the jet separator is used for selective removal of helium from the purge stream before it reaches the mass spectrometer.

The advantages of membrane inlets in online/on-site analysis can be summarized as follows: (1) a simple and rugged technique; (2) capability for long-time operation; (3) detection limits at or below ppb, directly from water or air samples without any sample pretreatment; (4) good automation capability using flow injection analysis (FIA) methods; and (5) compatibility with any kind of mass spectrometer.[45,46,52,53] Limitations of the method are that high molecular weight and more polar compounds are difficult to analyze with the standard MIMS methods and analysis of complicated mixtures is difficult. However, relatively intense development work to reduce the limitation effect of these problems is ongoing in the MIMS field. For example, mixtures can be analyzed using tandem mass spectrometry or mathematical methods to deconvolute the identity and the concentration of the individual compounds from a multicomponent mass spectrum.[39,40,45–48,52–54]

9.3.5   Flow Tube Techniques

9.3.5.1   Selected Ion Flow Tube Mass Spectrometry.   Selected ion flow tube mass spectrometry (SIFT-MS) relies on chemical ionization of trace gases in air samples using precursor ions that can be rapidly changed to allow the analysis of transient or limited-volume samples.[55] The precursor ion species of choice are H_3O^+, NO^+, and O_2^+ because they do not react with the major constituents of air. Figure 9.8 shows a schematic diagram of a typical SIFT-MS instrument. In this technique the precursor ions are produced in an ion source, such as a gas discharge ion source, and certain precursor ions can be selected individually from this continuous ion source by a quadrupole mass filter. The precursor ions are then mixed with a fast-flowing buffer gas, usually helium. The sample is introduced into this buffer gas, where trace gases of the sample are chemically ionized by the precursor ions species during the flow through the ion flow tube to the mass analyzer.

The analysis of trace gases in the air samples requires the knowledge of several prerequisites: (1) the carrier gas flow dynamics, including the flow speed and diffusion effects; (2) the sample flow rate into the carrier gas; (3) the discrimination in the mass analyzer against product ions of different masses; and (4) the rate coefficients and ionic products of the ion–molecule reactions involved in the analysis. The first three parameters are usually fixed with a given instrument, so accurate quantitation depends on proper measurement of the rate coefficients. With dry air samples understanding ion–molecule reactions is simpler and more straightforward, but with samples with varying humidity the chemistry in the flow tube is more

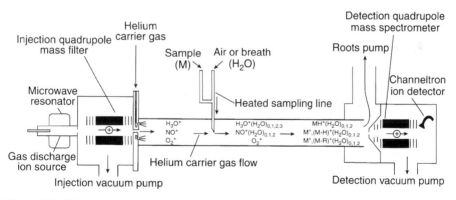

Figure 9.8   Schematic of SIFT-MS instrument. The precursor ions are injected into the helium carrier gas and, with the introduction of water vapor, the H_3O^+ and NO^+ ions are partially converted to their hydrates. These various precursor ions then react with the trace-gas molecules, μ, producing the types of product ions shown. Reprinted with permission from D. Smith, A. M. Diskin, Y. Ji, and P. Španěl, Concurrent Use of H_3O^+, NO^+ and O_2^+ Precursor Ions for the Detection and Quantification of Diverse Trace Gases in the Presence of Air and Breath by Selected Ion-Flow Tube Mass Spectrometry, *Int. J. Mass Spectrom.* **209**, 81 (2001). Copyright 2001 by Elsevier Science.

complicated due to water adducts of the analyte molecules. The H_3O^+ ions are the most useful precursor for organic compounds, but their concentration changes a lot along with the water adducts when humid air is analyzed. On the other hand, the use of H_3O^+ ions can give the water content of the sample. The NO^+ ions are also very valuable since they do not form hydrates and thus react with most organic compounds to form a single product ion. However, they sometimes react via three-body association, a reaction which is both pressure and temperature dependent. The O_2^+ ions are valuable as precursors because they do not form hydrates to a significant extent and are especially useful for the detection and quantitation of small inorganic molecules such as NO, NO_2, and NH_3, which do not react at significant rates with either H_3O^+ or O_2^+. For online analysis, the instrument is rather simple but the ion chemistry involved might sometimes be complicated and straightforward identification of the constituents in the sample is not always possible.

9.3.5.2 Proton Transfer Reaction Mass Spectrometry.
Proton transfer reaction mass spectrometry (PTR-MS) is a modification of SIFT-MS.[56] A schematic view of a PTR-MS instrument is shown in Figure 9.9. The major differences between these two techniques are that the sample is used as the buffer gas in PTR-MS and the only precursor ion used in PTR-MS is H_3O^+, making proton transfer the major ionization reaction in this technique. The density of product ions of reactant R [RH^+] is given by

$$[RH^+] = [H_3O^+]_0(1 - e^{-k[R]t}) \approx [H_3O^+]_0[R]kt$$

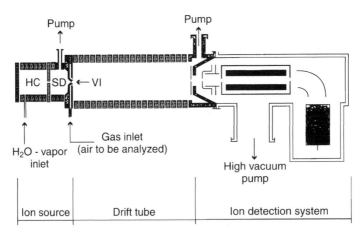

Figure 9.9 Schematic of PTR-MS instrument: HC, hollow cathode; SD, source drift region; VI, Venturi-type inlet. Reprinted with permission from W. Lindinger, A. Hansel, and A. Jordan, and On-Line Monitoring of Volatile Organic Compounds at pptv Levels by Means of Proton-Transfer-Reaction Mass Spectrometry (PTR-MS), Medical Applications, Food Control and Environmental Research, *Int. J. Mass Spectrom. Ion Proc.* **173**, 191 (1998) Copyright 1998 by Elsevier Science.

where $[H_3O^+]_0$ is the density of H_3O^+ ions in the absence of reactant neutrals in the buffer gas, k is the reaction rate constant for the proton transfer reaction, and t is the average time the ions spend in the reaction region. To reach a high density of $[RH^+]$ requires that the density of $[H_3O^+]$ be kept high. The high density of primary ions H_3O^+ of purity of about 99.5% is produced by a hollow cathode in the source. This has two advantages: high concentration and, therefore, high count rates of primary ions are obtained, which in turn means high count rates for protonated reactants $[RH^+]$ and better sensitivity, and no quadrupole system needs to be installed to pre-select the reactant ions H_3O^+ before entering a short drift region filled with water vapor, unlike in the SIFT-MS system. From this region, the ions (H_3O^+) reach a large reaction region, which is in the form of a drift section, filled with the air (pressure a few 10^{-1} torr) containing trace constituents to be analyzed. No further buffer gas is needed, and therefore the original mole fraction of R in air is retained in the reaction region. On the way, H_3O^+ ions undergo nonreactive collisions with any of the common components in air, but a small fraction (typically on the order of a percent) react with trace constituents.

For quantitative purposes, a reaction rate constant for the proton transfer reaction from H_3O^+ to a reactant must be measured. When the reaction rate constant of a reactant is known, the concentration of the reactant can be calculated without further calibration of the instrument. The PTR-MS technique can also be used for identification of different isobaric molecules (molecules having the same molecular mass but different molecular formula) based on their different ion mobilities in the drift tube and different behaviors in collision-induced dissociation (CID).

9.3.6 Laser Techniques

9.3.6.1 *Resonance-Enhanced Multiphoton Ionization Mass Spectrometry.* Laser ionization techniques have widely been used nowadays, especially in the analysis of biomolecules (e.g., proteins, peptides), in matrix-assisted laser desorption ionization mass spectrometry (MALDI-MS).[57] The advantages of laser ionization are fast ionization and thus good time resolution, soft ionization with little or no fragmentation, good selectivity in some cases, and good spatial resolution. Laser ionization combined with a TOF instrument results in a fast analyzer suitable for online measurements.

In resonance-enhanced multiphoton ionization (REMPI) mass spectrometry two or three photons are usually used. The first photon is absorbed by the molecule, causing an excitation of the molecule. The other photons are used for ionization of the excited molecule. A schematic of a REMPI-TOFMS instrument is shown in Figure 9.10.[58] In this experiment the REMPI-TOFMS instrument was used for the real-time characterization of combustion-generated polycyclic aromatic hydrocarbons (PAHs). For example, anthracene was selectively detected using a two-color REMPI scheme; that is, anthracene was excited at 361.1 nm and ionized at 308 nm. Phenanthrene, an isomer of anthracene, also is excited at 361.1 nm, but it absorbs weakly at 308 nm; thus the signal detected at m/z 178 is attributed entirely to anthracene.

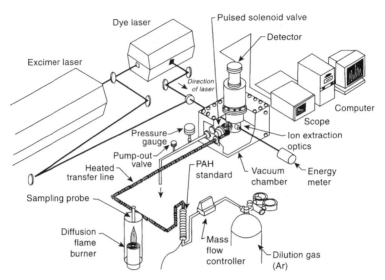

Figure 9.10 Schematic of REMPI-TOFMS flame sampling system configured for two-color REMPI detection. Reprinted with permission from C. M. Gittins, M. J. Castaldi, S. M. Senkan, and E. A. Rohlfing, Real-Time Quantitative Analysis of Combustion-Generated Polycyclic Aromatic Hydrocarbons by Resonance-Enhanced Multiphoton Ionization Time-of-Flight Mass Spectrometry, *Anal. Chem.* **69**, 286 (1997). Copyright 1997 by the American Chemical Society.

9.3.6.2 Aerosol Laser Mass Spectrometry.

A thorough review of aerosol MS was presented by Suess and Prather in 1999.[59] Several different analytical techniques have been used for offline measurement of aerosol particles, but many of them have limitations and disadvantages due to loss of time resolution and filter sampling in which the sample might be changed during filtration and storage. This has led to development of online aerosol MS, which has the advantage of measurement of single particles on a real-time basis. The first online aerosol mass spectrometric device was invented by Davis in 1977.[60] Since then other surface/thermal ionization techniques have also been used for online analysis of aerosol particles.[61,62] With these techniques single aerosol particles were introduced into a mass spectrometer using nozzles and skimmers, enabling a real-time analysis. However, they suffered from some drawbacks: For example, the aerosol particle was assumed to comprise from a single compound, which is not true for ambient particles; the ionization might change the composition of the particle; and the scanning mass spectrometers used were not able to acquire the whole mass spectrum for a single particle. For these reasons the emphasis of online aerosol MS has lately been laid to TOF instruments equipped with laser ionization. The apparent benefit of these instruments is the capability of acquiring a complete mass spectrum for single particles. Figure 9.11 shows a schematic diagram a dual-polarity, online, and field transportable TOFMS instrument equipped with a particle beam interface.[53] With this instrument it is possible to obtain both positive and negative mass spectra for

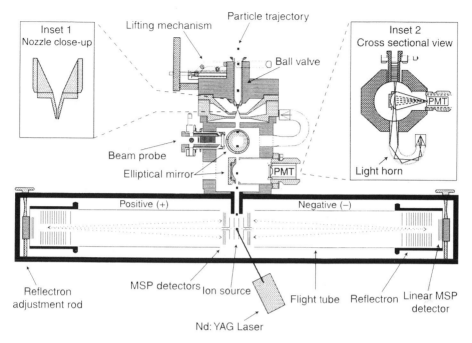

Figure 9.11 Schematic showing particle beam interface and particle-sizing region joined to mass spectrometer region of field-portable instruments. Reprinted with permission from E. Gard, J. E. Mayer, B. D. Morrical, T. Dienes, D. P. Fergenson, and K. A. Prather, Real-Time Analysis of Individual Atmospheric Aerosol Particles: Design and Performance of a Portable ATOFMS, *Anal. Chem.* **69**, 4083 (1997). Copyright 1997 by the American Chemical Society.

each aerosol particle. Two extra lasers can be used for aerodynamic particle size determination.

9.4 DETECTORS

After the mass analyzer has separated the ions, the ions have to be detected in order to receive the signal abundance of each ion. Two types of detectors are used, those which directly measure the charges that reach the detector and those which can multiply the intensity of the original signal. The first ones, such as a Faraday cup detector, can be used for the measurement of high concentrations, as is the case in the analysis of the composition of atmospheric gases. The latter ones, such as electron and photomultipliers or array detectors, can be used for the measurement of low ion currents (low concentrations) because the signal can be amplified by 10^4–10^8 times. Thus, they are suitable for most environmental applications, in which very low detection limits are required. The dynamic linear ranges of both types of detectors are four to five orders of magnitude, and therefore it is possible to cover a variety of compounds with varying concentrations with a single detector without changing amplification settings.

9.5   ANALYSIS OF ENVIRONMENTAL SAMPLES

9.5.1   Water

Quality control of both drinking water and wastewater has become more and more important lately. Due to lack of pure groundwater in many places, natural water must be treated somehow to get edible water. Purification of drinking water is sometimes performed using chlorine, producing various toxic chlorinated compounds such as cyanogen chloride. The analytical method recommended by the EPA for the quantitation of cyanogen chloride (CNCl) (method 542.2) employs purge-and-trap gas chromatography/mass spectrometry (P&T-GC/MS). For the comparison, the MIMS method was applied in the analysis of aqueous CNCl solutions, as shown in Figure 9.12.[63] With this online method it was possible to reach a detection limit at low ppb levels, using selected ion monitoring of m/z 61. Another study with MIMS was performed to evaluate the capabilities of MIMS to measure VOCs listed in EPA method 524.2 for drinking water.[64] Table 9.3 shows the detection limits obtained with MIMS in direct analysis of water with a sampling rate of 6 min or less. For most compounds the detection limit was well below 1 ppb, except for the most volatile compounds such as chloromethane. In addition to monitoring drinking water, the quality control of wasterwater is equally important in order to minimize pollution of the environment. Figure 9.13 presents results of continuous online monitoring of a wastewater stream of an oil refinery,[65] clearly showing the advantage of MIMS in online measurements when the compounds are well known beforehand. The concentrations of various organic compounds could be monitored for 18 days without any interruption in the measurement. The temporal changes in the concentrations were detectable on a real-time basis, making it possible to change the purification procedure online.

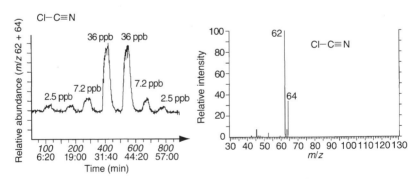

Figure 9.12 Analysis of cyanogen chloride (CNCl) at ppb level using jet separator–silicone membrane–ion trap mass spectrometer system. Jet separator and GC transfer line were kept at 110°C while silicone membrane was kept at 90°C. Reprinted with permission from N. Srinivasan, R. C. Johnson, N. Kasthurikrishnan, P. Wong, and R. G. Cooks, Membrane Introduction Mass Spectrometry, *Anal. Chim. Acta* **350**, 257 (1997). Copyright 1997 by Elsevier Science.

Table 9.3 Detection Limits of Volatile Organic Compounds Listed in EPA Method 524.2 Measured by MIMS

Compound Name	(m/z) Quantitation Ion	EPA Quantitation Ion	Detection Limit (ppb)	Molecular Weight
Toluene	91	91	0.01	92
Chloroform	83	83	0.025	118
o-Xylene	91	106	0.025	106
p-Xylene	91	106	0.025	106
m-Xylene	106	106	0.05	106
o-Chlorotoluene	91	91	0.05	126
p-Chlorotoluene	91	91	0.05	128
1,2-Dichloroethane	62	62	0.1	98
1,3-Dichloropropene	75	75	0.1	110
Benzene	78	78	0.1	78
Carbon tetrachloride	117	117	0.1	152
Ethylbenzene	91	91	0.1	106
Styrene	104	104	0.1	104
n-Butylbenzene	91, 92	91	0.1	134
n-Propylbenzene	91	91	0.1	120
1,1-Dichloroethane	63, 83	63	0.2	98
1,1-Dichloroethene	96, 60	96	0.2	96
1,1-Dichloropropene	75	n/a[a]	0.2	110
1,2,3-Trichloropropane	75	75	0.2	146
1,2,4-Trimethylbenzene	105	105	0.2	120
1,2-Dibromo-3-chloropropane	75	75	0.2	234
1,2-Dibromoethane	107	107	0.2	186
1,3-Dichloropropane	76	76	0.2	112
Bromodichloromethane	83	83	0.2	162
Bromoform	173	173	0.2	25
Chlorobenzene	112	112	0.2	112
Dibromochloromethane	129	129	0.2	206
Isopropylbenzene	105	105	0.2	120
Tetrachloroethene	166	166	0.2	164
Trichlorofluoromethane	101, 103	101	0.2	136
Vinyl chloride	62	62	0.2	62
cis-1,2-Dichloroethene	96	96	0.2	96
sec-Butylbenzene	105	105	0.2	134
trans-1,2-Dichloroethene	96	96	0.2	96
1,2-Dichlorobenzene	146	146	0.3	146
1,4-Dichlorobenzene	146	146	0.3	146
2,2-Dichloropropane	77	77	0.3	112
Bromobenzene	156, 158	156	0.3	156
1,3,5-Trimethylbenzene	105	15	0.4	120
1,3-Dichlorobenzene	146	146	0.4	146
Naphthalene	128	128	0.4	128
Trichloroethene	130	96	0.4	130

Table 9.3 *(Continued)*

Compound Name	(m/z) Quantitation Ion	EPA Quantitation Ion	Detection Limit (ppb)	Molecular Weight
p-Isopropyltoluene	119	119	0.4	134
1,1,1-Trichloroethane	96	97	0.5	132
1,2,4-Trichlorobenzene	180, 182	180	0.5	180
Dibromomethane	174, 176	93	0.5	172
Methylene chloride	49	84	0.5	84
tert-Butylbenzene	119	119	0.5	134
1,1,2-Trichloroethane	96	83	0.6	132
1,2,3-Trichlorobenzene	180, 182	180	0.7	180
1,1,1,2-Tetrachloroethane	131	131	0.8	166
1,1,2,2-Tetrachloroethane	83	83	0.8	166
1,2-Dichloropropane	76	63	0.8	112
Bromochloromethane	49	128	1	128
Bromomethane	94, 96	94	1	94
Dichlorodifluoromethane	85, 87	85	1	120
Hexachlorobutadiene	224	225	1	258
Chloroethane	49	64	5	64
Chloromethane	49	50	10	50

[a]Not applicable.

Source: Reprinted with permission from S. Bauer and D. Solyom, Determination of Volatile Organic compounds at The parts Per Trillion level in Complex Aqueous Matrices Using Membrane Introduction Mass Spectrometry, *Anal. Chem.* **66**, 4422 (1994). Copyright 1994 by the American Chemical Society.

9.5.2 Air

Monitoring air is simpler than monitoring water because the sampling is usually more straightforward as there is less interference from particles, high viscosity, and so on. On the other hand, very low detection limits should be reached to be able to detect hazardous, toxic chemicals at low enough levels. A combination of a membrane inlet and a jet separator has been used for direct analysis of organic compounds in air.[66] This combination yielded an efficient concentration of the analytes from an air sample to the ion trap mass spectrometer, providing low detection limits for organic compounds, (e.g., 20 ppt for toluene). Furthermore, the sampling probe used was versatile, as shown in Figure 9.14, in which the same probe was applied for the analysis of toluene in three different matrices—air, water, and soil—in a successive way in less than 50 min.

The applicability of PTR-MS for long-term online monitoring of VOCs has been demonstrated by measurements of the diurnal variations of isoprene and pinene in ambient air near Innsbruck, Austria, as shown in Figure 9.15.[67] The figure clearly shows that the concentrations of both compounds varied during the measurement period, the maximum for isoprene being during the day and the maximum for

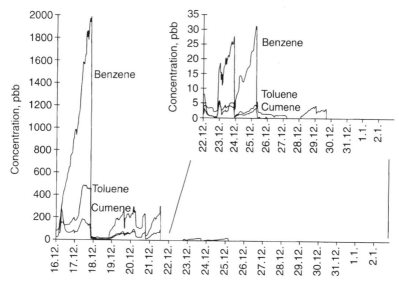

Figure 9.13 Measurements made over 18-day period of wastewater stream of oil refinery. Ions measured were m/z 78 for benzene, m/z 92 for toluene, and m/z 120 for cumene (isopropylbenzene). Results show daily variations in concentrations of these compounds, which correlated well with off-line gas chromatographic measurements, which were performed once a day (data not shown). Reprinted with permission from T. Kotiaho, R. Kostiainen, R. A. Ketola, T. Mansikka, I. Mattila, V. Komppa, T. Honkanen, K. Wickström, J. Waldvogel, and O. Pilviö, Development of a Fully Automatic Membrane Inlet Mass Spectrometric Measurement System for On-Line Industrial Waste Water Monitoring, *Process Control and Quality*, **11**, 71 (1998). Copyright 1998 by VSP International Science Publishers.

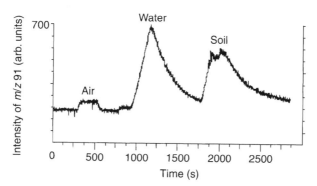

Figure 9.14 Ion chromatogram showing signal for m/z 91 from 1 ppb toluene in air, water, and soil obtained by simply moving inlet to membrane separator from one sample to another. Reprinted with permission from M. E. Cisper, C. G. Gill, L. E. Townsend, and P. H. Hemberger, On-Line Detection of Voltaile Organic Compounds in Air at Parts-per-Trillion Levels by Membrane Introduction Mass Spectrometry, *Anal. Chem.* **67**, 1413 (1995). Copyright 1995 by the American Chemical Society.

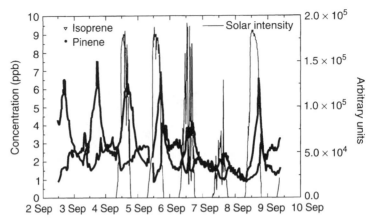

Figure 9.15 Concentrations of isoprene and pinene as measured from September 2 to September 9, 1997 in ambient air on the Western outskirts of Innsbruck. Also shown is relative solar intensity. Reprinted with permission from W. Lindinger, A. Hansel, and A. Jordan, On-Line Monitoring of Volatile Organic Compounds at pptv Levels by Means of Proton-Transfer-Reaction Mass Spectrometry (PTR-MS), Medical Applications, Food Control and Environmental Research, *Int. J. Mass Spectrom. Ion Proc.* **173**, 191 (1998). Copyright 1998 by Elsevier Science.

pinene being during the night. Also the solar intensity had an effect on the concentration of isoprene—the brighter the day, the higher the concentration of isoprene observed. Isoprene was also analyzed and monitored using a membrane inlet, chemical ionization, and an ion trap.[20] To obtain further selectivity against other hydrocarbons present in ambient samples, chemical ionization with vinyl methyl ether to produce a cyclic adduct via [2 + 4] Diels–Adler cycloaddition and tandem mass spectrometry were used. The instrumentation applied for the analysis is shown in Figure 9.16. The sampling period lasted for 5 min, after which the analytes were desorbed from the membrane into a cryo trap, from which the analytes were transported via a heated transfer line into the ion trap. Total analysis time was 7–8 min. This system was applied to the measurement of isoprene in a greenhouse room filled with velvet bean plants (Fig. 9.17). The figure shows how the temporal changes affected by switching off the sun lamp could be monitored, which would have been impossible or difficult with traditional offline analyses.

Most direct mass spectrometric techniques can be used for the measurement of VOCs but only a few techniques can be applied to direct measurement of semivolatile organic compounds in air. One of those techniques is APCI-MS.[68] Normally APCI-MS is used for liquid samples, such as an effluent from liquid chromatography, but it is also possible to use the sheath gas line as the gaseous sample interface, allowing a direct analysis of air samples. Due to the chemical ionization, the method is suitable for semivolatile compounds and detection limits at ppt levels can be reached. An example of monitoring two semivolatile compounds, methyl salicylate and dimethyl methylphosphonate, is shown in Figure 9.18. In addition, the selectivity of the method can be adjusted to suit the targeted analyte by an

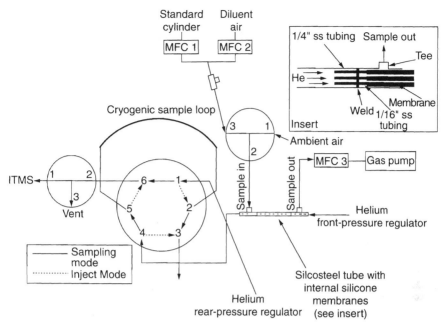

Figure 9.16 Schematic diagram of membrane preconcentration sample inlet used for analysis of isoprene in ambient air. Reprinted with permission from A. Colorado, D. J. Barket Jr., J. M. Hurst, and P. B. Shepson, A Fast-Response Method for Determination of Atmospheric Isoprene Using Quadrupole Ion Trap Mass Spectrometry, *Anal. Chem.* **70**, 5129 (1998). Copyright 1998 by the American Chemical Society.

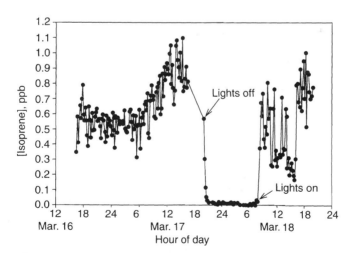

Figure 9.17 Greenhouse measurements of isoprene from velvet beans by vinyl methyl ether chemical ionization ion trap MS/MS. Reprinted with permission from A. Colorado, D. J. Barket Jr., J. M. Hurst, and P. B. Shepson, A Fast-Response Method for Determination of Atmospheric Isoprene Using Quadrupole Ion Trap Mass Spectrometry, *Anal. Chem.* **70**, 5129 (1998). Copyright 1998 by the American Chemical Society.

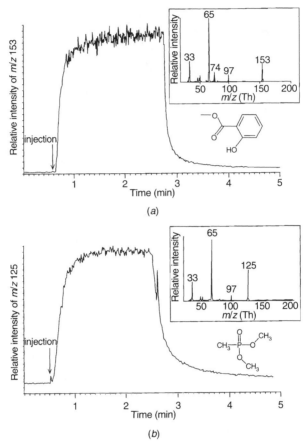

Figure 9.18 Single ion monitoring traces of (a) 20 ppb methyl salicylate (m/z 153) and (b) 90 ppb dimethyl methylphosphonate (m/z 125) in air. Insets: mass spectra extracted from total ion current trace. $[M+H]^+$ ions of each analyte (m/z 153 for methyl salicylate and m/z 125 for dimethyl methylphosphonate) are observed in the corresponding mass spectra, as are protonated clusters of methanol $[CH_3OH]_nH^+$, $n = 1$–3, formed in APCI source at m/z 33 ($n = 1$), 65 ($n = 2$), and 97 ($n = 3$). Reprinted with permission from L. Charles, L. S. Riter, and R. G. Cooks, Direct Analysis of Semivolatile Organic Compounds in Air by Atmospheric Pressure Chemical Ionization Mass Spectrometry, *Anal. Chem.* **73**, 5061 (2001). Copyright 2001 by the American Chemical Society.

appropriate choice of CI reagent and its flow rate in relation to a sheath gas (sample gas) flow rate.

Semivolatile organic compounds in a gaseous sample can also be measured with REMPI-TOF. For instance, due to its selective ionization, REMPI-TOF is applicable to a specific detection of PAH compounds in gas samples with varying concentrations of other, interfering compounds.[69] Figure 9.19 shows concentration–time profiles of selected PAH compounds obtained by REMPI-TOF at a wavelength of

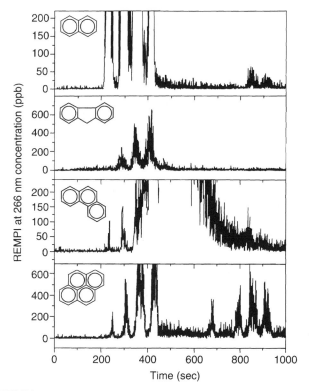

Figure 9.19 REMPI at 266 nm concentration–time profiles of selected PAHs during changing conditions of combustion process. Profiles generated by iteration of single-shot mass spectra within window of 1 amu width for selected masses, here 128, 166, 178, and 202, for different compounds. Reprinted with permission from H. J. Heger, R. Zimmermann, R. Dorfner, M. Beckmann, H. Griebel, A. Kettrup, and U. Boesl, On-Line Emission Analysis of Polycyclic Aromatic Hydrocarbons Down to pptv Concentration Levels in the Flue Gas of an Incineration Pilot Plant with a Mobile Resonance-Enhanced Multiphoton Ionization Time-of-Flight Mass Spectrometer, *Anal. Chem.* **71**, 46 (1999). Copyright 1999 by the American Chemical Society.

266 nm. It can be seen from the figure that the matrix did not interfere with the analysis, and the individual PAH compounds could be selectively detected by their masses.

9.5.3 Other Matrices

Direct online methods or instruments for the analysis of soil samples have not been invented yet, but a few nearly real-time and online instruments and methods have been developed. One of these methods employs a membrane interface connected to a purge-and-trap technique called purge-and-membrane (PAM) technique (Fig. 9.20).[70] The soil sample is flushed with inert gas, and the flushed eluent is

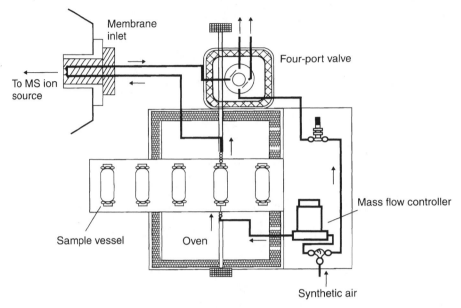

Figure 9.20 Schematic of PAM sampler showing sampling mode of operation. Arrows indicate direction of gas flow. Reprinted with permission from M. Ojala, I. Mattila, V. Tarkiainen, T. Särme, R. A. Ketola, A. Määttänen, R. Kostiainen, and T. Kotiaho, Purge-and-Membrane Mass Spectrometry, a Screening Method for Analysis of VOCs from Soil Samples, *Anal. Chem.* **73**, 3624 (2001). Copyright 2001 by the American Chemical Society.

analyzed with the membrane interface to obtain sensitivity in the direct measurement. This way it is possible to reach detection limits of 2–250 µg/kg for many organic compounds, which are low enough when analyzing contaminated landsites. Another technique applied to soil analysis is two-step laser mass spectrometry.[71] The soil sample is pressed to a small rod which is placed into the ion source where the direct laser desorption is carried out. Or the desorption is carried out in front of a pulsed nozzle with subsequent acceleration of the desorbed material toward the ionization region by means of a supersonic jet. The ionization is then performed with another laser. Figure 9.21 shows mass spectra of pyrene desorbed from three different types of soil. The ionization wavelength was 266 nm, which is suitable for PAH compounds, and this produces very selective ionization since no interfering peaks were detected from any of the soil matrices.

Laser mass spectrometric techniques are also suitable for online characterization of organic aerosols. Figure 9.22 presents the detection of a single wood smoke particle in both positive and negative modes simultaneously using an aerosol-TOFMS system.[53] The positive ions present are assigned as carbon/hydrogen envelopes ranging from C_2^+ to C_5^+, a very characteristic potassium signal, and a prominent CH_3CO^+ peak, and the negative ion signal is comprised of oxygen-containing hydrocarbon peaks as well as small $C_2H_4^-$ and HSO_4^- signals.

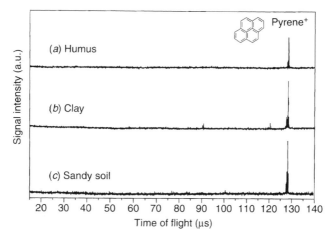

Figure 9.21 Laser mass spectra of pyrene (concentration 10 μg/kg) detected in different types of soil by laser desorption laser mass spectrometry at wavelength of 266 nm. Note absence of possibly disturbing signals. Reprinted with permission from C. Weickhardt, K. Tönnies, and D. Globig, Detection and Quantification of Aromatic Contaminants in Water and Soil Samples by Means of Laser Desorption Laser Mass Spectrometry, *Anal. Chem.* **74**, 4861 (2002). Copyright 2002 by the American Chemical Society.

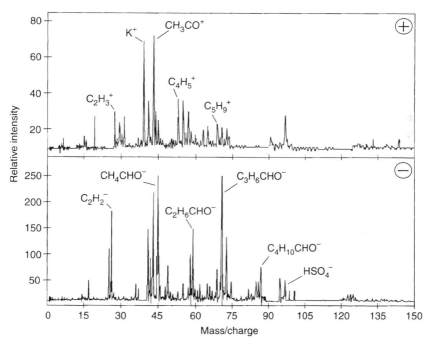

Figure 9.22 Simultaneous dual-ion mass spectra from single wood smoke particle having aerodynamic diameter of 1.2 μm. Reprinted with permission from E. Gard, J. E. Mayer, B. D. Morrical, T. Dienes, D. P. Fergenson, and K. A. Prather, Real-Time Analysis of Individual Atmospheric Aerosol Particles: Design and Performance of a Portable ATOFMS, *Anal. Chem.* **69**, 4083 (1997). Copyright 1997 by the American Chemical Society.

9.6 FUTURE PERSPECTIVES

The future development of online/on-site environmental analysis by mass spectrometry is happening in two interconnecting fields; the instrument development field and the application field. One of the most important instrument development fields is to obtain lower and lower detection limits. This is largely driven by the continuously lowering regulatory levels set by authorities in environmental or occupational health legislation. Other important demands for the instruments are that they should be more versatile, cheaper, smaller, and faster. To achieve these goals, very intensive development of miniaturized mass spectrometers which maintain the performance characteristics of traditional laboratory mass spectrometers is currently going on.[25] It is also expected that a large increase both in the volume and scope of applications will take place. The use of mass spectrometry in applications such as rapid analysis in emergency situations such as spills, fires, natural disasters, industrial accidents, detection of explosives and drugs, and monitoring of chemical weapons will certainly increase. A great challenge in the development of online mass spectrometry is to manufacture very cheap, small, and hand-held mass spectrometers for daily use at work or at home, just like mobile phones are used today—or the mass spectrometer can even be a part of a mobile phone or other similar device.

REFERENCES

1. K. R. Beebe, W. W. Blaser, R. A. Bredeweg, J. P. Chauvel, Jr., R. S. Harner, M. LaPack, A. Leugers, D. P. Martin, L. G. Wright, and E. D. Yalvac, Process Analytical Chemistry, *Anal. Chem.* **65**, 199R (1993).

2. W. W. Blaser, R. A. Bredeweg, R. S. Harner, M. A. LaPack, A. Leugers, D. P. Martin, R. J. Pell, J. Workman, Jr., and L. G. Wright, Process Analytical Chemistry, *Anal. Chem.* **67**, 47R (1995).

3. J. Workman, Jr., D. J. Veltkamp, S. Doherty, B. B. Anderson, K. E. Creasy, M. Koch, J. F. Tatera, A. L. Robinson, L. Bond, L. W. Burgess, G. N. Bokerman, A. H. Ullman, G. P. Darsey, F. Mozayeni, J. A. Bamberger, and M. Stautberg Greewood, Process Analytical Chemistry, *Anal. Chem.* **71**, 121R (1999).

4. V. Lopez-Avila and H. H. Hill, Field Analytical Chemistry, *Anal. Chem.* **69**, 289R (1997).

5. R. E. Clement, P. W. Yang, and C. J. Koester, Environmental Analysis, *Anal. Chem.* **69**, 251R (1997).

6. S. D. Richardson, Environmental Mass Spectrometry, *Anal. Chem.* **72**, 4477 (2000).

7. P. Nicholas, Process and Environmental Monitoring Using Mass Spectrometry, *Spectroscopy* **6**, 36 (1991).

8. G. Baykut and J. Franzen, Mobile Mass Spectrometry; a Decade of Field Applications, *Trends Anal. Chem.* **13**, 267 (1994).

9. W. C. McDonald, M. D. Erickson, B. M. Abraham, and A. Robbat Jr., Developments and Applications of Field Mass Spectrometers, *Environ. Sci. Technol.* **28**, 336A (1994).

10. M. R. Walsh and M. A. LaPack, On-Line Measurements Using Mass Spectrometry, *ISA Trans.* **34**, 67 (1995).

11. European Council, directive 98/83/EC of November 3, 1998 on the quality of water intended for human consumption.

12. Safe Drinking Water Act, http://www.epa.gov/OGWDW/sdwa/sdwa.html.

13. M. B. Wise and M. R. Guerin, Direct Sampling MS for Environmental Screening, *Anal. Chem.* **69**, 26A (1997).

14. M. B. Wise, C. V. Thompson, R. Merriweather, and M. R. Guerin, Review of Direct MS Analysis of Environmental Samples, *Field Anal. Chem. Technol.* **1**, 251 (1997).

15. T. Kotiaho, On-Site Environmental and In-Situ Process Analysis by Mass Spectrometry, *J. Mass Spectrom.* **31**, 1 (1996).

16. G. Matz and W. Schröder Fast GC/MS Field Screening for Excavation and Bioremediation of Contamined Soil, *Field Anal. Chem. Technol.* **1**, 77 (1996).

17. G. Matz, W. Schröder, and T. Kotiaho, Mobile Mass Spectrometry Used for On-Site/In-Situ Environmental Measurements, in R. A. Meyers (Ed.), *Encyclopedia of Analytical Chemistry*, Wiley, New York, 2000.

18. W. Paul and H. S. Steinwedel, A new Mass Spectrometer Without Magnetic Field, *Z. Naturforsch.* **8a**, 448 (1953).

19. W. Paul and H. S. Steinwedel, Separation of Ions of Different Specific Charge, DE Patent No. 1071376, 1959.

20. A. Colorado, D. J. Barket, Jr., J. M. Hurst, and P. B. Shepson, A Fast-Response Method for Determination of Atmospheric Isoprene Using Quadrupole Ion Trap Mass Spectrometry, *Anal. Chem.* **70**, 5129 (1998).

21. W. C. Wiley and I. H. McLaren, Time-of-Flight Mass Spectrometer with Improved Resolution, *Rev. Sci. Instrum.* **26**, 1150 (1955).

22. R. J. Cotter, Time-of-Flight Mass Spectrometry for the Structural Analysis of Biological Molecules, *Anal. Chem.* **64**, 1027A (1992).

23. I. V. Chernushevich, A. V. Loboda, and B. A. Thomson, An Introduction to Quadrupole-Time-of-Flight Mass Spectrometry, *J. Mass Spectrom.* **36**, 849 (2001).

24. E. De Hoffmann, J. Charette, and V. Stroobant, *Mass Spectrometry: Principles and Applications*, Wiley, Chichester, 1996.

25. E. R. Badman and R. G. Cooks, Miniature Mass Analyzers, *J. Mass Spectrom.* **35**, 659 (2000).

26. E. R. Badman, R. C. Johnson, W. R. Plass, and R. G. Cooks, A Miniature Cylindrical Quadrupole Ion Trap: Simulation and Experiment, *Anal. Chem.* **70**, 4896 (1998).

27. C. R. Arkin, T. P. Griffin, A. K. Ottens, J. A. Diaz, D. W. Follistein, F. W. Adamsand, and W. R. Helms, Evaluation of Small Mass Spectrometer Systems for Permanent Gas Analysis, *J. Am. Soc. Mass Spectrom.* **13**, 1004 (2002).

28. T. Kotiaho, On-Line/On-Site Environmental Analysis by Mass Spectrometry, paper presented at the Seventh International Symposium, Chemistry Forum 2001, Warsaw, Poland.

29. W. Bleakney, A New Method of Positive-Ray Analysis and Its Application to the Measurement of Ionization Potentials in Mercury Vapour, *Phys. Rev.* **34**, 157 (1929).

30. A. G. Harrison, *Chemical Ionization Mass Spectrometry*, CRC, Boca Raton, FL, 992.

31. E. C. Huang, T. Wachs, J. J. Conboy, and J. D. Henion, Atmospheric Pressure Ionization Mass Spectrometry. Detection for the Separation Sciences, *Anal. Chem.* **62**, 713A (1990).

32. F. L. Eisele and H. Berresheim, High-Pressure Chemical Ionization Flow Reactor for Real-Time Mass Spectrometric Detection of Sulfur Gases and Unsaturated Hydrocarbons in Air, *Anal. Chem.* **64**, 263 (1992).

33. O. Möhler, Th. Reiner, and F. Arnold, A Novel Aircraft-Based Tandem Mass Spectrometer for Atmospheric Ion and Trace Gas Measurements, *Rev. Sci. Instrum.* **64**, 1199 (1993).

34. A. A. Viggiano, In Situ Mass Spectrometry and Ion Chemistry in the Stratosphere and Troposphere, *Mass Spectrom. Rev.* **12**, 115 (1993).

35. E. Gard, J. E. Mayer, B. D. Morrical, T. Dienes, D. P. Fergenson, and K. A. Prather, Real-Time Analysis of Individual Atmospheric Aerosol Particles: Design and Performance of a Portable ATOFMS, *Anal. Chem.* **69**, 4083 (1997).

36. N. S. Karellas, G. B. De Brou, and A. C. Ng, Real-Time Detection of Airborne PCP by Mobile APCI MS/MS, paper presented at the Fortieth ASMS Conference on Mass Spectrometry, Washington DC, 1992.

37. T. Kotiaho, F. R. Lauritsen, T. K. Choudhury, R. G. Cooks, and G. T. Tsao, Membrane Introduction Mass Spectrometry, *Anal. Chem.* **63**, 875A (1991).

38. F. R. Lauritsen and T. Kotiaho, Advances in Membrane Inlet Mass Spectrometry (MIMS), *Rev. Anal. Chem.* **15**, 237 (1996).

39. T. Kotiaho, R. Kostiainen, R. A. Ketola, M. Ojala, I. Mattila, and T. Mansikka, Membrane Inlet Mass Spectrometry, in the Past and in the Future, *Adv. Mass Spectrom.* **14**, 501–527 (1998).

40. R. C. Johnson, R. G. Cooks, T. M. Allen, M. E. Cisper, and P. H. Hemberger, Membrane Introduction Mass Spectrometry: Trends and Applications, *Mass Spectrom. Rev.* **19**, 1 (2000).

41. R. A. Ketola, T. Kotiaho, M. E. Cisper, and T. M. Allen, Environmental Applications of Membrane Introduction Mass Spectrometry, *J. Mass Spectrom.* **37**, 457 (2002).

42. G. Hoch and B. Kok, A mass Spectrometric Inlet System for Sampling Gases Dissolved in Liquid Phase, *Arch. Biochem. Biophys.* **101**, 160 (1963).

43. M. Leth and F. R. Lauritsen, A Fully Integrated Trap-Membrane Inlet Mass Spectrometry System for the Measurement of Semivolatile Organic Compounds in Aqueous Solution, *Rapid Commun. Mass Spectrom.* **9**, 591 (1995).

44. G. Matz and P. Kesners, Thermal Membrane Desorption Application (TMDA) Method for On-Line Analysis of Organics in Water by GC-MS, *Anal. Mag.* **23**, M12 (1995).

45. M. E. Cisper and P. H. Hemberger, The Direct Analysis of Semi-Volatile Organic Compounds by Membrane Introduction Mass Spectrometry, *Rapid Commun. Mass Spectrom.* **11**, 1449 (1997).

46. M. H. Soni, J. H. Callahan, and S. W. McElvany, Laser Desorption-Membrane Introduction Mass Spectrometry, *Anal. Chem.* **70**, 3103 (1998).

47. S. Bohátka, Process Monitoring in Fermentors and Living Plants by Membrane Inlet Mass Spectrometry, *Rapid Commun. Mass Spectrom.* **11**, 656 (1997).

48. J. Benstead and D. Lloyd, Direct Mass Spectrometric Measurement of Gases in Peat Cores, *FEMS Microbiol. Ecol.* **13**, 233 (1994).

49. M. Bier and R. G. Cooks, Membrane Interface for Selective Introduction of Volatile Compounds Directly into the Ionization Chamber of a Mass Spectrometer, *Anal. Chem.* **59**, 597 (1987).

50. L. E. Slivon, M. R. Bauer, J. S. Ho, and W. L. Budde, Helium-Purged Hollow Fiber Membrane Mass Spectrometer Interface for Continuous Measurement of Organic Compounds in Water, *Anal. Chem.* **63**, 1335 (1991).

51. L. E. Dejarme, S. J. Bauer, R. G. Cooks, F. R. Lauritsen, T. Kotiaho, and T. Graf, Jet Separator/Membrane Introduction Mass Spectrometry for On-Line Quantitation of Volatile Organic Compounds in Aqueous Solutions, *Rapid Commun. Mass Spectrom.* **7**, 935 (1993).

52. R. A. Ketola and F. R. Lauritsen, Detection of Dicarboxylic Acids In Aqueous Samples Using Membrane Inlet mass Spectrometry with Desorption Chemical Ionization, *Rapid Commun. Mass Spectrom.* **13**, 749 (1999).

53. F. R. Lauritsen and J. Rose, Determination of Steroid Hormones by Membrane Inlet Mass Spectrometry and Desorption Chemical Ionization, *Analyst* **125**, 1577 (2000).

54. R. A. Ketola, M. Ojala, V. Komppa, T. Kotiaho, J. Juujärvi, and J. Heikkonen, A Non-Linear Asymmetric Error Function-Based Least Mean Square Approach for the Analysis of Multicomponent Mass Spectra Measured by Membrane Inlet Mass Spectrometry, *Rapid Commun. Mass Spectrom.* **13**, 654 (1999).

55. D. Smith, A. M. Diskin, Y. Ji, and P. Španěl, Concurrent Use of H_3O^+, NO^+ and O_2^+ Precursor Ions for the Detection and Quantification of Diverse Trace Gases in the Presence of Air and Breath by Selected Ion-Flow Tube Mass Spectrometry, *Int. J. Mass Spectrom.* **209**, 81 (2001).

56. A. Hansel, A. Jordan, R. Holzinger, P. Prazeller, W. Vogel, and W. Lindinger, Proton Transfer Reaction Mass Spectrometry: Online Trace Gas Analysis at the ppb Level, *Int. J. Mass Spectrom. Ion Proc.* **149/150**, 609 (1995).

57. R. J. Cotter (Ed.) *Time-of-Flight Mass Spectrometry: Instrumentation and Applications in Biological Research*, ACS Professional Reference Books, Washington DC, 1997.

58. C. M. Gittins, M. J. Castaldi, S. M. Senkan, and E. A. Rohlfing, Real-Time Quantitative Analysis of Combustion-Generated Polycyclic Aromatic Hydrocarbons by Resonance-Enhanced Multiphoton Ionization Time-of-Flight Mass Spectrometry, *Anal. Chem.* **69**, 286 (1997).

59. D. T. Suess and K. A. Prather, Mass Spectrometry of Aerosols, *Chem. Rev.* **99**, 3007 (1999).

60. W. D. Davis, Continuous Mass Spectrometric Analysis of Particulates by Use of Surface Ionization, *Environ. Sci. Technol.* **11**, 587 (1977).

61. J. J. Stoffels and C. R. Lagergren, On the Real-Time Measurement of Particles in Air by Direct-Inlet Surface-Ionization Mass Spectrometry, *Int. J. Mass Spectrom. Ion. Phys.* **40**, 243 (1981).

62. M. P. Sinha, C. E. Griffin, D. D. Norris, T. J. Estes, V. L. Vilker, and S. K. Friedlander, Particle Analysis by Mass Spectrometry, *J. Colloid Interface Sci.* **87**, 140 (1982).

63. N. Srinivasan, R. C. Johnson, N. Kasthurikrishnan, P. Wong, and R. G. Cooks, Membrane Introduction Mass Spectrometry, *Anal. Chim. Acta* **350**, 257 (1997).

64. S. Bauer and D. Solyom, Determination of Volatile Organic Compounds at the Parts per Trillion Level in Complex Aqueous Matrices Using Membrane Introduction Mass Spectrometry, *Anal. Chem.* **66**, 4422 (1994).

65. T. Kotiaho, R. Kostiainen, R. A. Ketola, T. Mansikka, I. Mattila, V. Komppa, T. Honkanen, K. Wickström, J. Waldvogel, and O. Pilviö, Development of a Fully Automatic Membrane Inlet Mass Spectrometric Measurement System for On-Line Industrial Waste Water Monitoring. *Process Control Qual.* **11**, 71 (1998).

66. M. E. Cisper, C. G. Gill, L. E. Townsend, and P. H. Hemberger, On-Line Detection of Volatile Organic Compounds in Air at Parts-per-Trillion Levels by Membrane Introduction Mass Spectrometry, *Anal. Chem.* **67**, 1413 (1995).

67. W. Lindinger, A. Hansel, and A. Jordan, On-Line Monitoring of Volatile Organic Compounds at pptv Levels by Means of Proton-Transfer-Reaction Mass Spectrometry (PTR-MS), Medical Applications, Food Control and Environmental Research, *Int. J. Mass Spectrom. Ion Proc.* **173**, 191 (1998).

68. L. Charles, L. S. Riter, and R. G. Cooks, Direct Analysis of Semivolatile Organic Compounds in Air by Atmospheric Pressure Chemical Ionization Mass Spectrometry, *Anal. Chem.* **73**, 5061 (2001).

69. H. J. Heger, R. Zimmermann, R. Dorfner, M. Beckmann, H. Griebel, A. Kettrup, and U. Boesl, On-Line Emission Analysis of Polycyclic Aromatic Hydrocarbons Down to pptv Concentration Levels in the Flue Gas of an Incineration Pilot Plant with a Mobile Resonance-Enhanced Multiphoton Ionization Time-of-Flight Mass Spectrometer, *Anal. Chem.* **71**, 46 (1999).

70. M. Ojala, I. Mattila, V. Tarkiainen, T. Särme, R. A. Ketola, A. Määttänen, R. Kostiainen, and T. Kotiaho, Purge-and-Membrane Mass Spectrometry, a Screening Method for Analysis of VOCs from Soil Samples, *Anal. Chem.* **73**, 3624 (2001).

71. C. Weickhardt, K. Tönnies, and D. Globig, Detection and Quantification of Aromatic Contaminants in Water and Soil Samples by Means of Laser Desorption Laser Mass Spectrometry, *Anal. Chem.* **74**, 4861 (2002).

10

PHOTOIONIZATION

John N. Driscoll
PID Analyzers, LLC
Walpole, Massachusetts

Environmental Instrumentation and Analysis Handbook, by Randy D. Down and Jay H. Lehr
ISBN 0-471-46354-X   Copyright © 2005 John Wiley & Sons, Inc.

10.1 INTRODUCTION

This chapter is written from the perspective of an environmental professional. It focuses on those aspects of photoionization that aid in the understanding and selection of portable instrumentation and their relationship to laboratory methods. In addition, it should be a useful resource for anyone who is responsible for interpreting data collected in the field.

By the late 1970s, Love Canal was one of the world's first recognized hazardous waste sites. As a result of the attendant publicity and the discovery of a multitude of additional sites throughout the United States, this led to the Comprehensive Environmental Response, Compensation and Liability Act (CERCLA), now known as Superfund, which has been used to clean up hazardous waste sites. The most serious problems with hazardous waste sites involve contamination of the soil and migration of toxic organics into the surrounding groundwater. The HNU model PI101 and HNU gas chromatography (GC) detector model PI52 (laboratory GC detector using photoionization) were used extensively at this site by the State of New York to determine the extent and nature of the contamination. This led to the choice of the HNU photoionization detector (PID) for the first rapid-response team assembled by the Environmental Protection Agency (EPA) in the late 1970s.

The passage of the Resource, Conservation and Recovery Act (RCRA) in 1976 and the CERCLA, or Superfund, in 1980 also led to an expansion of the list of hazardous chemical compounds under EPA regulation. There are in excess of 250,000 gasoline stations in the United States. It is estimated that 30–40% of these stations have leaking fuel storage tanks contributing significantly to the problem of groundwater contamination. The leaking underground storage tank (UST) program was initiated in the 1980s to deal with this problem. Many of the stations have had tanks replaced along with a cleanup, but there are still a number of problem "tanks" and areas that have to be cleaned up.

For the Superfund program, the large number of contaminants or "priority pollutants" resulted in the emphasis of analytical methods with multianalyte capabilities such as gas chromatography/mass spectrometry (GC/MS) for organic analysis and subsequent laboratory analysis. The EPA and its regional Superfund contractors, for example, field investigation teams (FITs) and technical assistance teams (TATs),[1] saw a need for and developed faster, lower cost field screening methods to supplement and in some cases replace the laboratory analytical methods, depending on how the data were to be used.[2] Answers were needed quickly and 15–30-day turnarounds were just not acceptable for the many critical and timely decisions that had to be made in the field. This led to the evolution of field screening and analysis methods in the 1980s by the EPA and the FITs and TATs. In order to support these new methods, portable versions of analytical laboratory instruments were needed to provide similar, if not identical, analytical capabilities in the field environment.

The star performers of these Superfund methods for hydrocarbons were portable field instruments such as the total volatile organics analyzer and gas chromatographs. Some rapid and simple methods for screening organics in water and soil using photoionization analyzers will be described in this chapter.

Table 10.1   Most Frequently Found Organic Compounds in Air at Superfund Sites

Compound	Found at Percentage of Sites
Trichloroethylene	33
Toluene	28
Benzene	26
Polychlorinated biphenyl (PCB)	22
Chloroform	20
Tetrachloroethylene	16
Phenol	15
1,1,1-Trichloroethane	14

10.2   STRUCTURES OF SELECTED ORGANIC COMPOUNDS AT HAZARDOUS WASTE SITES

The most common organic compounds found at hazardous waste sites are gasoline hydrocarbons, chloroalkanes such as trichloroethane and tetrachloroethylene, and polychlorinated polychlor biphenyls (PCBs). These compounds are from leaking underground storage tanks (gasoline stations), solvent usage, (chlorinated solvents) and transformer insulation (PCBs). The chlorinated compounds are from dry cleaning and other industrial operations. In Table 10.1, the most common organic compounds are given along with their frequency at 546 Superfund sites. The structures for some hydrocarbons and common toxic organic compounds are shown in Figures 10.1 and 10.2.

Gasoline, for example, is a complex mixture of aromatic, aliphatic (alkanes), and olefinic (alkenes) compounds that may contain 200 or more compounds produced by distillation of crude oil. It is not surprising, then, that benzene and toluene,

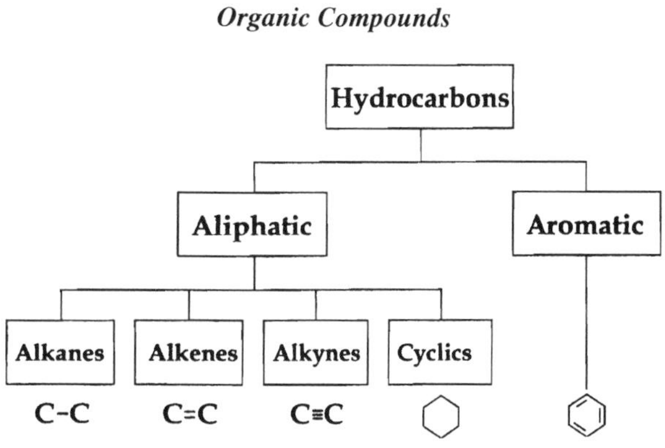

Organic Compounds

Figure 10.1   Structures of hydrocarbons.

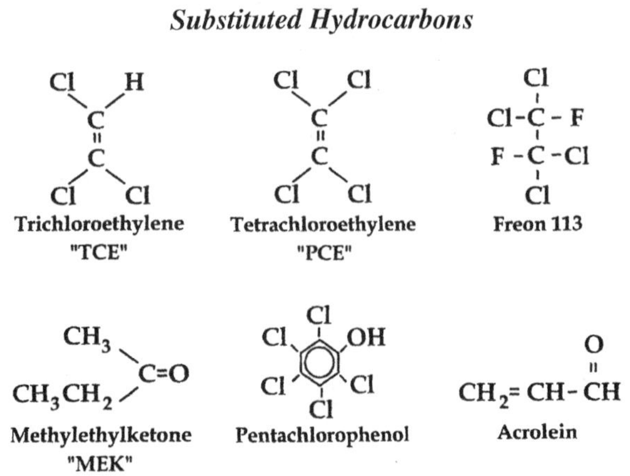

Substituted Hydrocarbons

Trichloroethylene "TCE"

Tetrachloroethylene "PCE"

Freon 113

Methylethylketone "MEK"

Pentachlorophenol

Acrolein

Figure 10.2 Structures of toxic organic compounds.

which were added to gasoline to replace lead compounds, are among the most common organic compounds found in Table 10.1. Other compounds such as oxygenated species may be blended into gasoline to produce more efficient combustion. Recent studies have shown that methyl *tert*-butyl ether (MTBE) is being detected in water supplies. This compound will migrate faster than benzene or toluene and can be an early warning indicator of a leaking gasoline tank. Diesel fuel, a higher boiling mixture than gasoline, contains alkanes and many cyclic compounds. It is necessary to use a gas chromatograph to differentiate these two fuel types (see, e.g., Chapter 8 on gas chromatography).

10.3 PHOTOIONIZATION

In 1974, the first commercial photoionization instrument (PID) was described by HNU Systems.[3] The process of ionization which occurs when a photon of sufficient energy is absorbed by a molecule results in the formation of an ion plus an electron:

$$R + h\nu \Rightarrow R^+ + e^-$$

where   R = an ionizable species

$h\nu$ = a photon with sufficient energy to ionize species R

In the ion chamber, the ions (R^+) formed by absorption of the UV photons are collected by applying a positive potential to the accelerating electrode and measuring the current at the collection electrode. A PID consists of an ion chamber, a UV lamp with sufficient energy to ionize organic and inorganic compounds, a voltage source for the accelerating electrode, an amplifier, and a readout. A schematic of a PID is

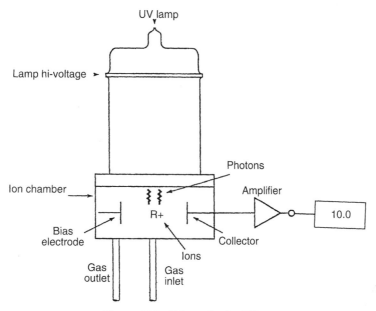

Figure 10.3 Schematic of a PID.

shown in Figure 10.3. Three PID analyzers are shown in Figures 10.4a–c. The latter instrument, the model 102, is the only one still in production.

The PI101 (Fig. 10.4a) was used at the Love Canal site in 1975 and thousands of hazardous waste sites worldwide since that time. It was produced until 2002 when a

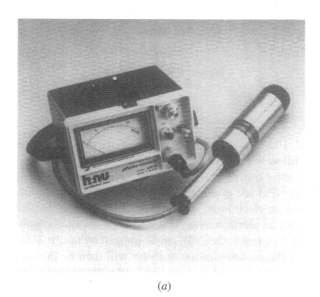

(a)

Figure 10.4 Photographs of PIDs.

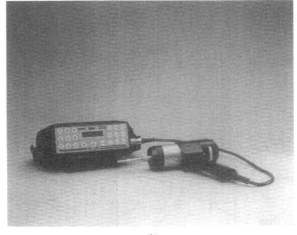

(b)

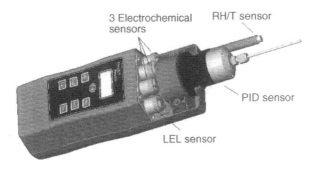

Model 102+

(c)

Figure 10.4 (Continued)

new model, the 102, was introduced. Many of the instruments produced 20–25 years ago are still in use today.

Some of the major constituents in the air at a hazardous waste site and their ionization potentials (IPs) are given in Table 10.2. The IP is the amount of energy required to move an electron an infinite distance from the nucleus, thus creating a positive ion plus an electron. Note that all of the major components of air have IPs above 12 eV. As a result, they will not be ionized by the 10.2- or 11.7-eV lamps. The response for the photoionization analyzer will then be the sum of the organic and inorganic compounds in air that are ionized by the appropriate lamp (9.5, 10.2, or 11.7 eV). Ionization potentials and response factors for 9.5-, 10.2-, and 11.7-eV lamps are given in Table 10.3.

Table 10.2 Ionization Potentials of Major and Minor Components in Air

Compound	Ionization Potential (eV)	Response
Benzene	9.25	High
Carbon dioxide	13.79	None
Carbon monoxide	14.01	None
Methane	12.98	None
Nitrogen	14.54	None
Oxygen	13.61	None
Trichloroethylene		High
Water	12.35	None

To what compounds will the 10.2-eV lamp respond? Generally, the response will extend to compounds with IPs up to 0.3 eV higher than 10.2 or 10.5 eV. The 10.2-eV lamp will respond to H_2S, which has an IP of 10.5 eV. While most of the measurement with a PID can be done with a 10.2-eV lamp, the 9.5-eV lamp does provide more selectivity (responds to fewer compounds) and the 11.7-eV lamp will respond to more compounds, as shown in Figure 10.5. The chloroalkanes, in particular, are of interest to measure at a hazardous waste site since these compounds do not have a very high response on the FID.

Table 10.3 List of Response Factors for Various Lamps

| Compound | Lamp | | |
	9.5 eV	10.2 eV	11.7 eV
Acetone		4.20	3.80
Benzene	10.00	10.00	10.00
1, 3-butadiene		6.99	7.60
Carbon disulfide	3.37	4.90	27.70
Ethane			0.69
Ethylene		0.89	
Ethylene		0.60	
Ethylene dichloride			10.50
Ethylene oxide		0.30	13.20
Ethylbenzene		12.0	
Formaldehyde			1.07
Hydrogen sulfide		1.20	10.10
Isobutylene		5.50	
Methanol			2.34
Methylene chloride		0.14	6.83
Phenol	7.70	1.05	
Styrene	15.34	14.87	10.70
1,1,2-tetrachloroethane			8.60
Tetrachloroethylene		8.60	
Vinyl chloride		3.20	7.80
p-xylene		11.40	

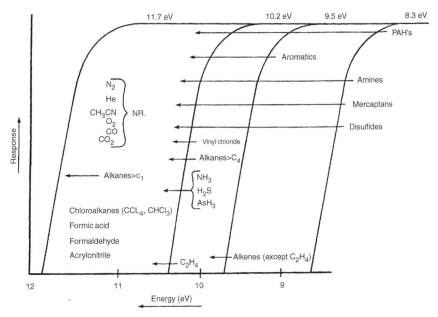

Figure 10.5 Effect of lamp energy on response.

Some characteristics of the PID are given in Table 10.4. Note that the PID is a carbon counter (molar response increases with carbon number), like the flame ionization detector (FID). There are several differences in that the PID is nondestructive (the sample is burned in the FID) and the PID has no response to ambient methane. The methane background measured on the FID is 3–5 ppm. Thus, the PID is an order of magnitude more sensitive than the FID since it can measure down to 0.1 ppm, which is the background in clean air.

A schematic of the PID probe is shown in Figure 10.6. As a result of the level of contamination around a hazardous waste site, it is not unusual for the ion chamber to require cleaning on a regular basis. The instrument was designed so that the ionization chamber is removable and easy to clean. The lamp is also removable

Table 10.4 PID Characteristics

Sensitivity increases as the carbon number increases (carbon counter).

For 10.2-eV lamp, responds to carbon aliphatic compounds $>C_4$, all olefins, and all aromatics; the PID also responds to inorganic compounds such as H_2S, NH_3, Br_2, I_2, PH_3, AsH_3, e.g., any compound with an ionization potential of <10.6 eV.

For 9.5-eV lamp, the response is more selective; higher response to aromatic hydrocarbons and less response to aliphatic hydrocarbons.

For 11.7-eV lamp, responds to more compounds, e.g., carbon tetrachloride, chloroform, methanol, formaldehyde, which have ionization potentials >10.6 and <11.7 eV.

Non destructive—sample (tube or bag) can be collected downstream of the detector.

Concentration sensitive—relatively insensitive to flow rare over a wide range.

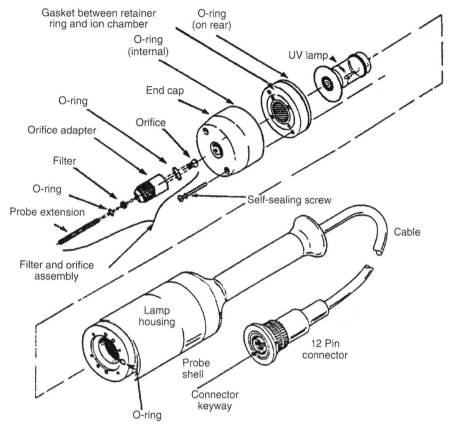

Figure 10.6   Probe assembly for PID.

(no wires to attach/detach) and the window is easy to clean. If the sensitivity of the PID is low (determined by the isobutylene reading), one of the first things to do is to clean the lamp. This process takes only a few minutes to complete.

The PI101 is extremely rugged and has only three controls: zero, span, and range selection switch. The DL101 is under microprocessor control but has a survey mode that is similar in performance and ease of use to the PI101. These instruments are designed for a long lifetime and ability to be used by personnel with a minimum of training. Model 102 is the newest version. It is a single-piece construction and weighs only 1.9 lb. This PID has a library of newly 300 compounds that can be accessed via a few key strokes. It has a removable ion chamber.

10.4   SCREENING AND ANALYSIS LEVELS

The framework of the EPA methodology involves five levels of investigative screening or analyses (Fig. 10.7). Level I involves field screening with hand-held analyzers

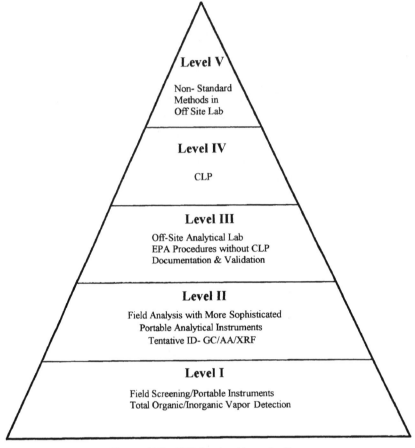

Figure 10.7 EPA levels of investigation.

(EPA protocol specifies a PID like the HNU model PI or DL101) shown in Figure 10.4 and other site characterization equipment such as an oxygen meter, explosimeter, radiation survey equipment, and chemical testing tubes. The new model 102+ has both LEL and O_2 incorporated in the PID. Level I effort is designed to determine the real-time total level of contaminants present (i.e., total volatile organics), which allows determination of the appropriate level of on-site respiratory protection and evaluation of air quality for existing or potential threats to surrounding populations.[4] Additional benefits include more efficient and cost-effective sampling and analysis and assistance in guiding quick and suitable cleanup efforts. A summary of the precision and accuracy expected for the various levels is given in Table 10.5.

Examples of tools for level II (and sometimes level III) screening are the HNU GC311 and X-ray fluorescence (XRF) instruments.

Level II involves field analysis with more sophisticated instrumentation to provide identification (as far as possible) of specific components present. The final

Table 10.5   Precision and Accuracy Requirements for Various EPA Methods

Level	Precision	Accuracy	Requirement
I	±10%	±50%	Screening (1000 ppm detection limit)
II	±10%	±15%	Evaluation of human exposure (detection limit, low ppm)
III	±5%	±10%	Litigation and regulatory enforcement (detection limit, ppb)

level discussed, level III, uses laboratories located "off-site" and frequently involves CLP analysis.[2] Levels IV and V will not be discussed in this chapter.

10.4.1   Level I Screening

The purpose of level I field screening is to find total contaminant levels (i.e., total volatile organics). Using a screening/survey instrument, such as the HNU model PI or DL101 (see Fig. 10.4), for example, it is possible to accomplish the following during level I screening:

- Identification of contamination sources
- Monitoring soil vapor wells to determine the extent of the pollutant plume (headspace)
- Measuring the total concentration profile in a borehole to determine contaminant migration on the groundwater
- Protecting the health of workers involved in the investigation and remedial work

10.4.2   Level II Screening

Once a level I screening identifies a contaminated area and delineates its extent, a level II screening can establish the identity of the compound(s) and relative concentrations. In the 1980s, this was done predominantly by sending samples to a laboratory for detailed analysis. In the 1990s, many of these measurements were done in mobile laboratories. The intermediate level II analysis was introduced by the EPA in order to reduce both the time required to start remedial actions and the high costs associated with laboratory analysis and keeping trained personnel in the field waiting for results.

Field analytical equipment is currently used for on-site detection and identification of organic and inorganic contaminants in air, water, and soil. Portable gas chromatographs (i.e., HNU model 311; see Fig. 10.1) are used for characterization of volatile organic compounds, semivolatile organics, polyaromatic hydrocarbons (PAHs), and PCBs. Methods for level II screening are described in more detail in Chapter 9 on gas chromatography.

10.5 TOTAL HYDROCARBON MEASUREMENT TECHNIQUES

10.5.1 Calibration

10.5.1.1 Air Samples. One of the first things to do after calibration with a span gas and a preliminary survey of the "hot spots" is to develop a set of calibration curves for the PID if some new or toxic compounds are found, and a direct-reading value for these compounds in ambient air is required. A very useful tool for this application is a 1- or 2-L Hamilton syringe. This device has a number of uses, from calibration to sample collection. The syringe is made of plexiglass, and some caution should be observed with sample preparation or collection at high percentage levels. In general, ppm levels can be prepared or sampled with no difficulty.

A method for sample preparation was described by Becker et al.[5] To prepare a gas sample with a liquid, one needs a liquid syringe (10 μL) and the 1- or 2-L syringe. The formula for calculation of the values produced is

$$\frac{[\text{volume (mL)}][\text{density (g/mL)}][\text{molar volume (mol}^{-1})]}{[\text{molecular weight (g/mol)}][1 \text{ g syringe volume(L)}]} = \text{ppm v/v}$$

A calibration curve can be prepared by using five to six injections of various volumes to cover the range desired and recording the values produced by the PID. This is a fairly easy method to prepare a calibration curve. The measurements should be within 5–10% provided that the user can accurately deliver the known values of the liquid. This will be the largest error in the method. This procedure works best with volatile liquids, but the syringe can be mildly heated with a hair dryer to help a less volatile liquid evaporate.

One of the most frequent problems with collecting vapors out of a stack or out of the ground involves the condition of the sample, for example, saturation with water at ambient (soil gas) or elevated temperatures (stack). Many volatile organic compounds are soluble in water to a greater (acetone, miscible) or lesser extent (benzene). Condensation in the sampling line can lead to losses from a few percent to as great as 100% for an organic compound that is miscible with water. Any losses greater than a 5% are clearly unacceptable. If the syringe is half filled with clean dry air prior to sampling, the water content will be reduced by half. If this is insufficient, then a larger quantity of dry air should be used to prevent condensation. The model 101's can measure the level in the 1- or 2-L syringe. The final value should be adjusted for the volume of dilution air.

10.5.1.2 Headspace Samples. Soil or water samples can be prepared on either a volume or weight basis. Liquid samples are generally prepared on a volume basis (ppm v/v) and soil samples are prepared on a weight basis (mg/kg). To prepare a 1-ppm sample by volume in water, measure 1 μL of liquid (using a 10-μL syringe) and dissolve in 1 L of water (in a volumetric flask). To prepare a soil sample, use a 10-μL syringe (multiply by the density of the liquid to get the weight) and weigh 1000 g of soil on a balance with a readability of 0.1 g. Mix thoroughly to ensure that the volatile hydrocarbon standard is evenly dispersed in the soil.

10.5.2   Industrial Hygiene and Survey Measurements

One of the primary uses of the PID at hazardous waste sites is for worker protection. The air at the site is initially screened, and if values above a specified level are observed, the teams are required to wear self-contained breathing apparatus. Then areas are screened for high levels in the soil, surface, or groundwater by measuring the concentrations a short distance above the source. At this point, a sample could be taken for analysis by a level II method or further measurements can be made with the PID. Clearly, the latter measurements can be made more quickly.

10.5.3   Soil Gas Measurements

Although headspace analyses[6] are common for volatile hydrocarbons, one of the most commonly used field analysis technique for site characterization is soil gas analysis[7] where the sample is collected by in situ pumping of a well using a probe such as that shown in Figure 10.6. In many cases, such as with an PI101 or DL101. This type of level I screening could be used to determine the extent of contamination of the soil ("total" but not individual hydrocarbons) and groundwater which occurred. Site characterization is done rapidly and in "real time."

10.5.4   Headspace

The headspace method is easy to perform, requires a minimum of equipment, and requires only that the sample and standards be at equilibrium to obtain accurate results. A typical procedure involves weighing (or measuring) 1 g of soil into a weighing boat which is placed in a container of about 100 mL volume which can be sealed (a series of half-pint clean paint cans could suffice), placed in an oven at 60°C for 15 min, and cooled to room temperature.[6] The sample can be left at ambient (e.g., in a trailer on-site) for a specified period of time (approximately 1 h). Standards bracketing the samples should be run at the same time and under exactly the same conditions. For a total hydrocarbon measurement, the vessel top is removed and a reading (maximum) is taken and recorded. The concentration can then be determined by comparison to a calibration curve generated from standards. Liquid samples can be prepared by making solutions as described in Section 10.5.1, placed in a sealed bottle (250 mL), and allowed to equilibrate. Then, the bottle is uncapped and the maximum reading is observed (and recorded) on the PID. A typical calibration curve for toluene in water is shown in Figure 10.8. Note the linear response and sensitivity of this method. Robbins[8] described a headspace method using a polyethylene bag where samples are collected in a modified 1-quart bag. Twenty-five grams of soil are added to the bag and the bag is inflated until taut. After 3 or 4 min of agitation, the bag is ready to be sampled. The method will also work with water samples.

If the determination of individual hydrocarbons is required, then a portable gas chromatograph is the instrument of choice. Here, the vessels should be modified to include a Swagelock connector that incorporates a septum. A gas syringe is used to

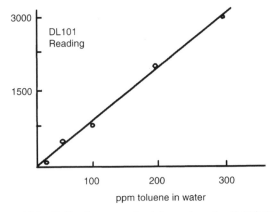

Figure 10.8 Calibration curve for toluene in water (headspace).

remove a sample (1 mL) and inject it into the gas chromatograph (see also Chapter 8 on gas chromatography).

10.6 REPORTING DATA

After developing and demonstrating the ability of field screening and analytical methods to supplement or replace laboratory analytical methods, the EPA and its contractors had the task of developing "field analyses." These results were to be reported as a "tentative identification" rather than a definitive identification as field analysis was not meant to replace laboratory analysis using greater sophistication and analytical controls (i.e., CLP analysis). Field analytical data seemed to best supplement laboratory methods when laboratory analysis had already identified the contaminants present at a site. Field analytical methods could be used to set worker safety levels, determine the extent of contamination and the placement of monitoring wells, optimize sampling grids, establish cleanup levels during treatability studies, and identify critical samples for CLP confirmatory analysis.[2] Generally, at least 10% of the samples analyzed in the field are returned for laboratory analysis unless one uses a gas chromatograph in a mobile laboratory where level II or III methods are used.

10.7 CONCLUSIONS

The EPA has shown that field and laboratory methods can yield similar results for volatile hydrocarbons. It was determined that when very little is known about a site or its contaminants, it is generally more cost effective to use the CLP (or an equivalent, like a state-certified laboratory) as a screen than to conduct extensive field analyses designed for analyzing a limited number of target compounds. Field analytical methods were determined to be most useful when the contaminants of concern have already been identified, so that, for example, appropriate methods, dilutions, and calibration ranges can be employed.[2] It was also found that credibility could be lent to field data by using quality control techniques similar to laboratory methods (e.g., duplicates, standards run at regular intervals, percentage of samples sent back to the lab).

The EPA has published a field screening methods catalog as a reference (see Ref. 7). Field screening and analytical methods have been extensively developed through the 1980s and into the 1990s due in large part to the Superfund program. These methods have been used to obtain real-time data for health and safety considerations during site investigations and determine the presence or absence of contamination. They are also used to obtain qualitative data relative to a primary calibration standard if the contaminants being measured are unknown or obtain quantitative data if a contaminant is known and the instrument is calibrated to that contaminant. A final use is to identify soil, water, air, and waste locations that have a high likelihood of showing contamination through subsequent CLP confirmatory analysis.[2]

The PID has been shown to be a useful tool for industrial hygiene, soil gas monitoring, finding contaminated areas, and screening soil and water by headspace. The PID is easy to use and, by changing the lamp, can be a selective or more general detector (11.7-eV lamp) for chloroalkanes (e.g., dichloroethane, chloroform, carbon tetrachloride).

In this chapter we have described a number of basic aspects of photoionization and portable instruments in order to provide the reader with a basic understanding of field methodology.

REFERENCES

1. L. R. Williams, Guest Editorial, American Environmental Laboratory, October 1990, p. 6.

2. *Data Quality Objectives for Remedial Response Activities*, Development Processes, PB 88-131370, EPA, Washington, DC, March 1987.

3. J. N. Driscoll and F. F. Spaziani, *Anal. Inst.*, October 1974.

4. J. O. Morin, Development and Application of an Analytical Screening Program to Superfund Activities, In *Proceedings of the National Conference on Management of Uncontrolled Hazardous Waste Sites*. Hazardous Materials Control Research Institute. Silver Springs, MD, 1985, p. 97

5. J. H. Becker, J. N. Driscoll, D. Renaud, and P. Tiffany, Instrument Calibration with Toxic and Hazardous Materials, *Ind. Hyg. News*, July 1983.

6. J. Driscoll and J. H. Becker, Rapid Screening Techniques for Determination of Residual Organics in Food, Polymers and Soil, Pitt. Conf. Paper # 603, 1981.

7. *Field Screening Methods Catalogue, Users Guide*, EPA/540/2)88/005, Hazardous Site Evaluation Division, EPA, Office of Emergency and Remedial Response, Washington, DC, 1988.

8. G. Robbins, *Groundwater Monitoring Rev.* (1989).

11

PORTABLE VERSUS STATIONARY ANALYTICAL INSTRUMENTS

Randy D. Down, P.E.
Forensic Analysis & Engineering Corp.
Raleigh, North Carolina

Environmental Instrumentation and Analysis Handbook, by Randy D. Down and Jay H. Lehr
ISBN 0-471-46354-X   Copyright © 2005 John Wiley & Sons, Inc.

11.1 INTRODUCTION

When dealing with certain types of environmental applications, such as groundwater monitoring, surface water sampling, continuous emissions monitoring, and leak detection (to name a few), we need to use the best means at our disposal to obtain accurate and reliable measurement of pollutant constituents and concentrations.

Locations where the measurement must take place may not be readily accessible, and the cost of installing and maintaining multiple stationary instruments can be high. In a case such as this, it may make more sense to use a portable instrument rather than a permanently installed instrument at each location.

The purpose of this chapter is to identify some typical environmental applications that require a decision regarding the use of portable versus stationary instruments, provide the pros and cons of each approach as well as selection guidelines, and identify the potential pitfalls of both portable and stationary instruments.

11.2 PORTABLE INSTRUMENTS

Analytical instruments classified as "portable" are those capable of being easily moved by an individual from one measurement location to another without requiring a substantial amount of dismantling or posing a high risk of damage during transport (Figs. 11.1 and 11.2). Often, such devices are designed and constructed to be compact and light enough for a technician to safely carry. If delicate (sensitive to shock or vibration), the instrument is fitted into some type of protective carrier, such as a padded equipment case. Many portable instruments are designed for rugged use.

Portable analyzers are not new, although they are more common now than 20 years ago. Advances in material science and microelectronics have led to smaller and lighter instruments that have superior performance over their predecessors. Materials of construction (metal alloys, ceramics, plastics) are now stronger and lighter than those available 20 years ago. Due to large-scale integration (LSI), electronic semiconductor devices are much smaller, requiring less heat dissipation to function. This allows them to be enclosed in smaller enclosures without as great a concern about overheating as in the past. Sensor technology has also continued to develop at a very rapid pace. More precise measurement, often at a lower cost, is now possible through the development of high-technology sensors, many of which were developed for the aerospace and auto industries.

Figure 11.1 Old-style portable instrument (inclined manometer).

Figure 11.2 New-style portable instrument (digital manometer).

Examples of portable instruments (including a few that are small enough to fit into a shirt pocket) include:

- Portable gas analyzers
- Clamp-on (nonintrusive) liquid flowmeters
- Data loggers
- Turbidity meters
- TDS (total dissolved solids) meters
- Conductivity meters
- pH meters
- DO (dissolved oxygen) meters
- ORP (oxygen reduction potential) meters
- CO (carbon monoxide) monitors
- Thermometers
- Hygrometers
- Pressure gauges
- Velometers
- Anemometers
- Manometers
- Barometers

- Altimeters
- Psychrometers
- Luminescence (light) meters
- Range (distance) meters
- Colorimeters
- Salinity (salt content) meters
- Tensiometers (soil moisture)
- Metal detectors

Combination meters, often referred to as multimeters, can be a cost-saving approach to acquiring the portable instruments you will need to perform certain types of measurement. Good examples of combination meters are the pH–ORP–°C meter and a combination thermometer–hygrometer–velometer–light meter. The pH–ORP–°C meter combines three instruments that are used in the analysis of water quality. The combination thermometer–hygrometer–velometer–light meter combines four instruments that are used to analyze the ambient environment. The advantages of these multifunctional instruments include lower cost (because they share the same electronics and power source), the convenience of a single instrument to transport and protect, and the compact size and versatility of these devices. A disadvantage, and an important one when considering this type of instrument, is the risk of having none of these measurement capabilities available to you should the shared electronics fail or become damaged. It is an "all-of-the-eggs-in-one-basket" approach. Generally speaking, the reliability of these multi-purpose instruments is very good, and therefore the risk is not particularly high that they will fail if the instrument is probably stored, protected, and calibrated. Some of these instruments are reliable enough to allow the manufacturer to offer workmanship warranties ranging up to 5 years.

11.3 POWER SOURCES

Another distinguishing trait of portable instruments is their self-contained power source. Many portable instruments are equipped with replaceable batteries or rechargeable battery packs. Battery technology has advanced in recent years due to very aggressive competition in the battery market. Each battery manufacturer is competing to produce a battery with a longer life, less potential for leakage, and lower cost. As a result, batteries now have a longer life and greater reliability. They are also better sealed in order to protect the instrument they are installed in from the effects of corrosion due to battery acid.

Solar energy is another viable source of power for portable instruments. Solar cells can be placed in any location where there is sufficient radiation from sunlight to directly power the instrument. They can also be used to charge a battery for operation during the hours in which there is no available sunlight. Solar power is particularly useful when the instrument is to be used in a remote location where there is no available electrical power and it is undesirable to have to frequently return to the monitoring site to recharge or replace the batteries. A solar-powered supply can also be used to power telemetry devices that will allow you to remotely monitor transmitted data from a portable instrument or to power a local data logger to digitally record measured data for future use.

A consideration when selecting an instrument that will require the use of disposable batteries is the availability of replacement batteries. Ideally, the instrument uses batteries that are commercially available (such as sizes A, AA, AAA, C, or D) in the event that they are needed right away. Commercially available batteries are almost always less expensive to procure than custom manufactured batteries.

Battery selection is important as it plays a part in the reliability of the measurement instrument that it is powering. The most common commercially available batteries are lithium ion, nickel–metal hydride (NiMH), and nickel–cadmium (NiCd) batteries.

1. *Lithium Ion Batteries.* Lithium ion batteries are at least 30% lighter in weight and carry at least 30% more capacity than NiMH or NiCd batteries. However, lithium batteries require a special type of charger because of their unique charging requirements. While lithium batteries do not build a "memory effect," they need to be charged at a rate that is somewhere between that of a conventional trickle charger and a rapid charge. The battery will accept a rapid charge but must be slow charged for the last 15% of the charging cycle. Overheating (due to overcharging the battery) will damage the battery and can potentially result in a fire.

2. *Nickel–Metal Hydride Batteries.* Nickel–metal hydride batteries provide the same voltage as NiCd batteries and have at least 30% more capacity than NiCd batteries. They also take approximately 20% longer to charge. The first few times that you charge a NiMH battery, use a trickle charge (slow charge), as this will condition the battery. Unlike a NiCd battery, NiMH batteries can withstand random charging. Nickel–metal hydride batteries tend to have a shorter life cycle than NiCd batteries, which are the most common type of battery. Nickel–metal hydride batteries can be twice as expensive as the NiCd version. Avoid overheating, as heat quickly degrades a NiMH battery.

3. *Nickel–Cadmium Batteries.* Nickel–cadmium batteries are the oldest and most rugged commercially available batteries on the market. They also exhibit a long service life. Always fully discharge NiCd batteries before recharging them. This will "exercise" the cells in the battery so that they are less likely to build dendrides, which are the cause of "memory effect." Memory effect is a condition in which a battery loses capacity and begins to use only the cells that are fully charged and discharged regularly, resulting in poor battery performance and eventual failure.

Before taking any portable instrument into the field, it is highly recommended that you first remove the instrument from its protective carrying case and thoroughly examine it to ensure that all of the necessary components and accessories are with it, *including fresh batteries.* Test the functionality and calibration of the unit before taking it into the field. It takes a few extra minutes to do this, but it will help ensure that you do not wind up at a remote measurement location only to find that your instrument does not work.

11.4 CALIBRATION

Electronic measurement instruments often require periodic calibration to ensure that the displayed values of the measurements are valid and accurate to the manufacturer's rating of the instrument. This requires the use of a certified calibration laboratory that is NIST (National Institute of Standards and Technology) traceable.

As posted on its website, the mission of NIST is to develop and promote measurement, standards, and technology to enhance productivity, facilitate trade, and improve the quality of life. To help meet the measurement and standards needs of U.S. industry and the nation, NIST provides calibrations, standard reference materials, standard reference data, test methods, proficiency evaluation materials, measurement quality assurance programs, and laboratory accreditation services that assist a customer in establishing traceability of results of measurements or values of standards.

Traceability requires the establishment of an unbroken chain of comparisons to stated references. NIST assures the traceability of results of measurements or values of standards that it provides either directly or through an official NIST program or collaboration. Other organizations are responsible for establishing the traceability of their own results or values to those of NIST or other stated references. NIST has adopted this policy statement to document its role with respect to traceability.

When used in critical environmental measurement or control applications (as needed to meet state or federal requirements) or in an applications such as a forensic analysis (where the measurement may be challenged in a court case), credibility of the measurement may be contingent upon the traceability of that instrument's calibration to a documented standard. NIST provides the necessary paper trail.

11.5 DATA LOGGERS

Although covered in greater detail in another chapter of this book, it is worthwhile to mention data loggers when we discuss portable instruments because they are frequently used with portable instruments to record the measured data if they do not have an integral recording capability (Fig. 11.3). Data loggers will typically accept an industry standard 4–20-mA DC signal or 1–10-V DC signal.

Figure 11.3 Portable data logger.

The resolution of the data logger's recording is often programmable, allowing you to record a measurement every second up to recording a sample measurement once every hour or longer between samples. The longer the period between samples, the longer the battery charge will last and sampling will continue.

The recorded data are stored in solid-state memory and can be uploaded at a later time to a personal computer, often in the form of a software spreadsheet. This typically requires the purchase of software needed to perform the data transfer from memory to the spreadsheet through the computer's USB port.

Most data loggers are very compact and housed in weather-resistant enclosures and have a relatively low power requirement. Some are designed to be intrinsically safe for use in hazardous locations.

11.6 PROS AND CONS

Portable instruments have some inherent advantages and disadvantages when compared to stationary analyzers. It is important to consider the pros and cons as they may influence your decision.

Advantages of Portable Instruments Over Stationary (Fixed-Place) Instruments

- Easier relocation from one measurement location to another
- Easier servicing and calibration
- Potentially avoiding a need to purchase multiple instruments
- Serving as a backup unit in the event that a permanent instrument fails

Disadvantages of Portable Instruments Over Stationary Units

- Higher risk of damage when transporting the unit (such as dropping it)
- Higher risk of theft (it is easy to walk away with)
- Higher risk of tampering (because of its accessibility)
- Higher risk of being out of calibration (due to exposure to vibration and handling)
- Higher risk of unavailability (when the instrument is being used by others)

11.7 STATIONARY INSTRUMENTS

Permanently mounted and semipermanently mounted instruments installed in a fixed location are engineered and designed specifically for the type of harsh environment they will be exposed to for their unique application. They are referred to as "stationary" instruments (Fig. 11.4).

1. *Advantages of Stationary Instruments Versus Portable Instruments.* Since they may not need to be as ruggedly built as a portable instrument (depending on

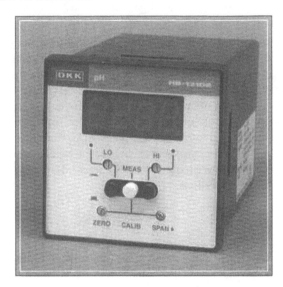

Figure 11.4   Stationary (panel-mounted) instrument.

their operating environment), manufacturers can reduce the net cost of the instrument by using less expensive materials that still afford sufficient protection to perform reliably in that application. For example, a commercial-grade thermostat will work very well in a relatively clean, dry room. The same device would likely fail after a short time if placed in a harsh environment where moisture condensed, corrosive vapors such as chlorinated compounds were present, or there was a high incidence of vibration.

Portable instruments, by contrast, are generally of a more rugged design, so they may be applied in a variety of areas and under a range of conditions and may include areas rated as "hazardous." This "generic" design of portable instruments contributes to their often higher unit cost than that of a stationary equivalent. They must be designed to be durable under a wide range of conditions.

2. *Disadvantages of Stationary Instruments Versus Portable Instruments.* Unlike portable devices, stationary instruments are usually hard wired or hard piped in place. The wiring or piping (depending on whether the device is electronic or pneumatic) provides two things: a source of power to the instrument from an electrical power supply and a connection to a remote display device (meter or gage) and/or control device. A transducer may need to be installed between the instrument and the display device or controller depending upon the signal compatibility of the measurement instrument.

If the device is electronic, the connection to remote devices will be wiring in the form of a twisted and/or shielded, multiconductor cable or individual insulated wires. Depending on the location, the cable or wires may be installed in a protective metal or plastic conduit. Permanently installed electric and electronic devices must be installed in accordance with NEC (National Electric Code) standards. These are the *minimum* standards for installation that have been established to protect the safety of human life and to safeguard against fires.

If pneumatic, the devices are provided with a compressed (instrument) air supply and produce a pneumatic output signal (typically within a 0–20-psig range) to a remote measurement or control device.

11.8 HARSH ENVIRONMENTS

Harsh environments can be the result of any combination of the following conditions:

- *High humidity* (moisture condensing on the surface of the instrument) is present.
- *Wet areas* (where instruments can be splashed with water or another liquid) are present.
- *High ambient air temperature* (affecting the electronics or mechanical seals) is present.
- *Airborne dust* (fine particulate) is present (which can plug orifices).
- *Corrosive vapors or gases* are present (causing surface corrosion).
- *Vibration* occurs (resulting in metal fatigue or abrasion).
- *Flammable or explosive vapors* are present.

11.9 HAZARDOUS AREAS

When selecting a stationary instrument for a specific location, you must determine that location's hazard classification. Hazardous area classifications are defined by NFPA (National Fire Prevention Association) regulations and should be clearly identified on building plans and instrument loop diagrams (control drawings).

"Explosion-proof" instruments are designed to be used in Class I, Division I or Class I, Division II areas. Devices that are designed to be explosion proof are tightly sealed in an enclosure to prevent a volatile liquid or vapor from entering as well as contain an explosion should one occur within the enclosure (the instrument were to act as a source of ignition). This, in theory, would only occur if the seals on the enclosure failed or the enclosure's design was flawed.

A Class I, Division II area has a high potential for a hazard (fire or explosion) should a system fail (a leaking pipe fitting or tank allows a volatile gas, liquid, or dust into the area) and a source of ignition (a spark, flame, or hot surface) is present. This is considered an "abnormal" condition.

In a Class I, Division I area, volatile gases, vapors, or dust is present under "normal" working conditions, such as an open (vented) tank or bath containing a flammable solvent. This area classification is considered more hazardous than Division II because of the greater potential for a fire or explosion if a source of ignition is present.

Fixed instruments installed in a hazardous area are typically rated for use in that area by the Industrial Research Institute (IRI) or Factory Mutual (FM). These insurance underwriters have established safety criteria for facilities where they provide insurance coverage. They have also established safety design requirements for instruments and other devices that will be installed in a hazardous area. Most other insurance underwriters have adopted the FM or IRI standards.

When selecting and installing any devices that can generate a source of ignition (not intrinsically safe), it is very important to:

1. Determine the hazard rating of the area (or that it is not rated a hazardous area)
2. Select an appropriate device (by conferring with the device's manufacturer) that is designed to be installed in a hazardous area or that can be installed in an explosion-proof enclosure

11.10   SAMPLING PUMPS

Accurate measurement of the concentrations of airborne particulate and vapors requires that a measured amount of sampled air (milliliters per second) pass over a sensor. This is done using a sampling pump. A peristaltic pump draws air through a sampling tube and passes it across the analyzer's sensor at a precisely controlled rate of flow. The resulting measurement is then displayed in engineering units of parts per milliliter of air. This type of measurement is useful in determining the concentration of airborne contaminants and level of human exposure to the measured contaminants if the space is occupied.

11.11   MOUNTING REQUIREMENTS

Mounting hardware is often included with a fixed instrument so that it can be securely clamped or bolted onto a section of pipe, flat wall surface, or foundation. The instrument has mounting holes in its enclosure (housing) that accept the mounting hardware.

Some instruments have standard mounting hardware that can be used for various types of mounting surfaces and positions. This is referred to by instrument manufactures as "universal" mounting hardware.

Mounting locations are dictated by sensing requirements and the environment. For example, if a flow sensor probe must be positioned in the center of a measured flow pattern, then the device must be mounted in such a way as to obtain a proper reading. In some cases, the installer may be required to custom fabricate a mounting bracket or pipe stand to ensure that the instrument is securely mounted in a proper position. Some field-mounted devices are also position sensitive and must be mounted horizontally or upright to function properly (Fig. 11.5). Position-sensitive instruments may need to be rezeroed (adjusted) to correct for the position.

If the instrument is to be mounted where the electronics can be damaged by exposure to high ambient temperature (typically the upper limit for electronic instruments is 87°C (140°F), the electronics portion of the device must be located in a remote location from the sensor. As an example, a flow transmitter may have its probe in a high-pressure steam pipe. The electronic "head" (electronics enclosure) is mounted on a wall or pipe stand far enough from the steam pipe to prevent the

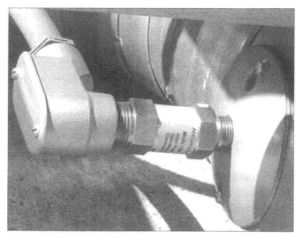

Figure 11.5 Field-mounted stationary transmitter.

electronics from overheating from the heat given off by the steam pipe. A shielded cable interconnects the electronics and probe. In some cases, the transmitter head (electronics) is attached to the probe but has an extension section with a thermal insulation between the two to prevent heat from being conducted from the probe to the head.

11.12 PROTECTION

It may also be necessary to protect a field-mounted device from damage caused by the impact of moving objects and equipment, such as forklift trucks and ladders. Protective bollards (posts consisting of stand pipes anchored in a concrete foundation and filled with concrete) are positioned between the instrument and the area where the damage would originate (such as an aisle way, service corridor, or truck-unloading area).

11.13 ACCESSIBILITY

Instruments, whenever feasible, should be mounted where they are readily accessible for inspection, servicing, and field calibration. Often this is not possible, as the sensor or instrument must be located where it will reliably measure the desired variable.

Problems arising from a lack of instrument accessibility have led to the recent development of "smart" transmitters. These instruments allow an operator or programmer to remotely perform test and calibration functions using remote digital communications to a smart transmitter's onboard microprocessor. This is accomplished over the same 4–20-mA DC twisted/shielded cable used to transmit an output signal from the instrument and provide power to it (referred to as the "current loop"). A hand-held device serves this purpose (by getting in series with the loop) or the data can be downloaded from a distributed control system's main processor. With this type of system, a service technician can be hundreds of feet from the instrument and still configure and calibrate it. This avoids technicians having to periodically get to a difficult location, such as a confined space or high location, where they may be exposed to possible safety hazards.

11.14 CALIBRATION

Depending upon the nature of the instrument, it may require periodic calibration to provide accurate information. The frequency of calibration required is often a function of how harsh an environment the instrument is exposed to and the design of the device. For example, a gas detector may use a chemical sensor that must be replaced after a certain period of time because the chemical reaction that occurs at the sensor as part of the detection process shortens its life expectancy. The need to periodically replace the sensor should be considered when selecting a mounting location for the analyzer. To preserve the warranty, follow the instrument manufacturer's guidelines when storing or using the instrument.

Calibration techniques differ with the instrument's design. Some devices can be successfully calibrated in the field by qualified instrumentation technicians. Others must be removed from service in order to calibrate them properly. This may require shipping them to a certified calibration laboratory with NIST-traceable calibration standards. These standards ensure that all instruments are calibrated in the same fashion and to the same degree of accuracy.

Recent design enhancements now allow some instruments (such as combustible gas monitors) to be field programmed and calibrated without opening their explosion-proof enclosures. A simple permanent magnet, like those sometimes found on a pocket screwdriver, is used to calibrate the device by waving it in front of the glass face of the instrument enclosure. A magnetically sensitive (*Hall effect*) electronic switching device within the instrument allows the calibration to occur by stepping through a selection menu without risk of compromising the sealed enclosure and creating a potential hazard or having to interrupt power to the device.

11.15 CONCLUSIONS

The decision to use a stationary or a portable instrument will be a function of all of the things that we have considered in this chapter. The decision is not as simple as it might first appear. When considering which option to choose, consider the cost, the physical requirements, and logistics involved.

It is highly recommended that you work with a qualified consultant who has extensive experience in the instrumentation and control field when making a selection. Independent consultants are a good resource because of their generally unbiased position. Manufacturers and manufacturers' representatives (the regional distributors) have more intimate knowledge of their products, but their primary objective is to sell you an instrument from their inventory. That is how they make a living. A reputable and knowledgeable instrument supplier can also still help you by advising you of the best device they can provide for your application. But be aware that they may not always carry the best instrument for your application.

12

APPLICATION OF XRF TO THE ANALYSIS OF ENVIRONMENTAL SAMPLES

John N. Driscoll

PID Analyzers, LLC
Walpole, Massachusetts

Environmental Instrumentation and Analysis Handbook, by Randy D. Down and Jay H. Lehr
ISBN 0-471-46354-X   Copyright © 2005 John Wiley & Sons, Inc.

12.1 INTRODUCTION

This chapter is written from the perspective of an environmental professional. It focuses on those aspects of X-ray fluorescence (XRF) that aid in the understanding and selection of instrumentation for both field and laboratory methods. In addition, it should be a useful resource for anyone who is responsible for interpreting data collected in the field.

Energy-dispersive X-ray fluorescence (EDXRF) is a relatively new analytical technique that began with the development of the lithium-drifted silicon detector. Thirty years ago, it could be used to identify elements and estimate the quantity of the element from the intensity of the peak (intensity vs. energy) in the spectrum. It was not until the early 1980s that software was developed that could deconvolute the peaks and correct for absorption and emission from neighboring elements. This was the beginning of truly quantitative (corrected) results for XRF from complex sample.

Energy-dispersive X-ray fluorescence has a number of advantages for environmental analysis when compared to other spectroscopic techniques such as inductively coupled plasma (ICP) or atomic absorption (AA):

1. Spectral or chemical matrix interferences can be eliminated through calibration and spectral deconvolution.

2. Samples can be analyzed for metals *nondestructively* and utilized for additional measurements of organic compounds or inorganic ions.

3. Elements can be analyzed over a wide range from low-ppm to percentage levels.

4. Qualitative and/or semiquantitative analysis requires little sample preparation and reduces the cost per test.

5. Rapid analysis of a wide variety of environmental samples can be performed.

A comparison of these techniques is given in Table 12.1.

X-ray fluorescence has been used in the field or laboratory for a variety of environmental applications, including metals in soil,[1–3] water,[3] and air.[4] The advantages of XRF include ease of use, minimum sample preparation, and rapid on-site analysis. The latter feature cannot be underestimated when a number of trucks loaded with excavated soil are waiting for a proper destination.

Table 12.1   Comparison of EDXRF with Other Spectroscopic Methods for Metals

Technique	EDXRF	AA	ICP
Sample preparation, water	None	Acid dissolution	Acid dissolution
Sample preparation, soil/sludge	Grind (if necessary), sieve to 100 mesh	Acid dissolution	Acid dissolution
Range	ppm to %	ppb	ppb
Multielement	Yes	No, generally single	Yes
Ease of use	Yes	No skilled person	No skilled person

Another advantage of XRF is that this technique is ideally suited for field screening work[4] as a result of the minimization of sample preparation and the portability. With an X-ray tube and filters, XRF can be used for a variety of environmental analyses, including particulate matter in air, as well as metals in soil, and water. We will be discussing some environmental applications in the following sections.

12.2   BASICS OF XRF SPECTROSCOPY

X-ray fluorescence is the characteristic emission resulting from the absorption of high-energy radiation by atoms in the X-ray region. This technique is used for the analysis of materials (heavy metals, sulfur, chlorine) in the environment (soil, water, air particulate) and is based on fundamental principles of atomic spectroscopy.

What is the origin of the spectra? The data in Table 12.2 provide the electronic configuration of some of the lighter elements. Note that the K shells are filled first, then the L, and finally the M. Each of these elements has a different electronic configuration. Therefore each will have different energy levels and will produce a different X-ray emission spectrum.

Inner electrons (K, L, or M shell) can be knocked out of their orbital by the high-energy radiation from an X-ray tube or a radioactive source (^{109}Cd, ^{55}Fe, ^{241}Am, etc.). These energies have to be higher than the K, L, or M shell binding energies. Vacancies in the K shell are replaced by electrons from the L or M shell with the

Table 12.2   Electronic Configuration of Some Light Elements

Element	Z	Ionization Potential (eV)	K	L	M
Na	11	5.14	2	8	1
Mg	12	7.64	2	8	2
Al	13	5.98	2	8	3
Si	14	8.15	2	8	4
P	15	10.6	2	8	5

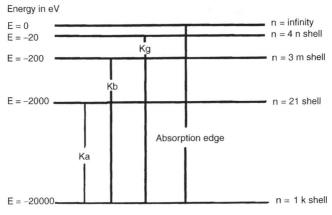

Figure 12.1   Energy level diagram.

resulting characteristic X-ray emission. Since we are dealing with atoms, absorption and emission lines are at the same wavelength or frequency. A typical energy level diagram for an atomic species is shown in Figure 12.1 and a typical X-ray spectrum is shown in Figure 12.2.

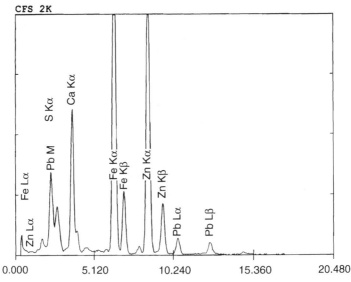

Figure 12.2   Characteristic X-ray spectrum of soil.

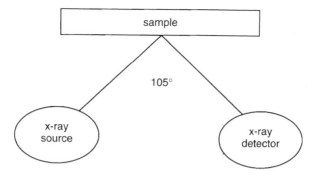

Figure 12.3 Typical XRF geometry.

12.2.1 Detection Geometry

The maximum intensity of fluorescence occurs at 90° from the incident radiation source. Therefore, for XRF instrumentation, the source and the X-ray detector are at angles of 80°–110° from each other. A typical geometry is shown in Figure 12.3.

12.2.2 X-Ray Sources

The types of sources used for EDXRF instrumentation include both radioactive sources (^{55}Fe, ^{109}Cd, ^{241}Am, ^{57}Co, etc.) and X-ray tubes. The former sources are all beta emitters. A high-energy particle is emitted with an energy of 5.9 keV (Mn corresponding to ^{55}Fe) and is used for excitation. Typical source levels are in the range 10–100 mCi. Instruments using these sources do require a license, and licensing requirements vary from state to state. Most states require notification prior to bringing an instrument into the state. The main advantages of a source-excited instrument is that these sources are monochromatic; for example, the ^{55}Fe source produces a spectrum of the Mn K_α and K_β lines similar to those in Figure 12.4. Another advantage is that the sources do not require any power but the X-ray tube requires 10–20 kV and 10–20 W. This makes battery operation more difficult because of the high-power utilization. In order to analyze heavy metals in soil, at least two sources (^{109}Cd and ^{241}Am) are needed. The ^{55}Fe is needed for the lighter elements in the soil. The disadvantage is that the sources, particularly ^{109}Cd and ^{55}Fe, do decay and have to be replaced every few years. An EDXRF instrument based on radioactive sources was described by Driscoll et al.[5]

The emission spectrum for an X-ray tube is polychromatic or broadband rather than monochromatic like the radioactive sources. A typical emission spectrum for an X-ray tube with a Rh anode is shown in Figure 12.5. The count rate (counts per second) or intensity and therefore the sensitivity of the tube-based XRF are considerably higher than for the source base instrument. With the tube, the anode can be changed (different tube), the excitation conditions can be optimized, and filters can

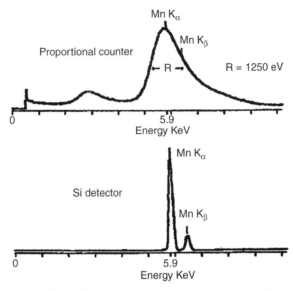

Figure 12.4 Effect of detector type on spectral resolution.

be used to compensate for the X-ray continuum and improve detection limits. Tube lifetimes are generally in excess of 5000 h and low-power tubes (10 W or less) can be quite stable. The clear disadvantage of the tube is its power consumption, which makes the development of a battery-operated instrument more challenging. It is possible to operate an instrument like the XR1000 off an inverter (converts

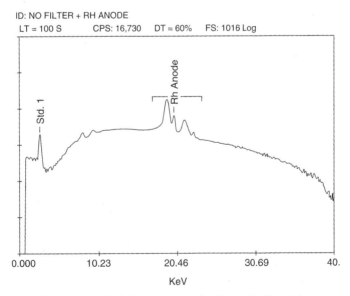

Figure 12.5 Emission spectrum of a Rh anode X-ray tube.

12 V DC to 120 V AC). Thus a 12-V battery can be the source of power for this instrument. A second choice is a gasoline-powered generator.

12.2.3 Detectors

Two types of detectors have been used in XRF instrumentation: proportional counters, which are the more traditional, and cooled [by liquid nitrogen or electronically (Peltier)] *high*-resolution semiconductor detectors, which have only been used in the field very recently. The results obtained with the two types of detectors can be quite different, as will be shown below.

The relative standard deviation (σ_Q) of the pulse amplitude distribution for a detector is given by

$$\frac{\sigma_Q}{Q} = \frac{C^{1/2}}{E}$$

where Q = distribution in pulse amplitude

$\quad C$ = constant for particular fill gas, = $W(F + b)$

$\quad$ F, b = constants that depend on ion chamber design

$\quad W$ = energy required to form one ion pair

$\quad E$ = energy deposited by incident radiation

If the standard deviation is multiplied by 2.35, the energy resolution (R) of the proportional counter can be calculated. At 10 keV, the energy resolution of a proportional counter is 12.5% or 1250 eV.[6] The resolution is given by

$$\text{Energy resolution } (R) = \frac{\text{FWHM}}{H_0}$$

where FWHM = frequency width at half maximum

$\quad H_0$ = peak centroid

The Si(Li) detectors have been used since the 1960s for detecting X-rays and low-energy gammas ($<$100 keV). These semiconductor detectors are useful in that dimensions can be kept smaller than proportional counters because solid densities are 1000 times greater than for a gas. As ionizing radiation passes through the semiconductor, many electron–hole pairs are produced along the track. *The low value of the ionizing energy required (3 eV) for a semiconductor detector (about $\frac{1}{10}$ that of a proportional counter) creates 10 times the number of charge carriers than a proportional counter and is one of the factors that leads to increased resolution for the semiconductor detectors.* The resolution for a semiconductor detector is about 1.5% at 10 keV or 150 eV. The effect of detector resolution is shown in Figure 12.4, where the energy spectra of an ^{55}Fe source are compared for a Si(Li) detector (resolution 185 eV) and a proportional counter (resolution 1200 eV).

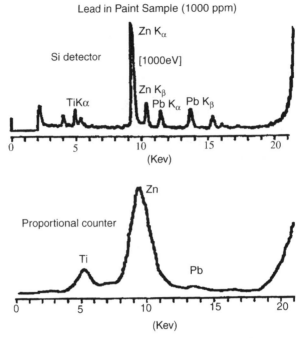

Figure 12.6 Comparison of a silicon Pin detector and a proportional counter spectrum for lead paint.

For these tests, only the detector was changed and all the electronics, including the MCA, were the same. The actual resolution of the proportional counter was 1250 eV, in agreement with the value calculated above. The Mn K_α and K_β peaks at 5.9 and 6.4 keV are well resolved with the Si(Li) detector, which had a resolution of 180 eV, also in close agreement with the predicted value. *It is clear from the spectra in Figure 12.6 that with the poor resolution (>1000 eV) of a proportional counter, adjacent element peaks are difficult to detect. Typically adsorption edge filters are used to compensate for this problem. However, these filters reduce the count rate and greatly affect the precision and accuracy of the results.* A comparison of the detector characteristics is given in Table 12.3.

Table 12.3   Comparison of X-Ray Detector Performance

Detector	Si(Li)	Si Pin	Gas Proportional
Resolution, eV	165	220–240	700–1200
Range	Na–U	Na–U	Element range limited
Efficiency	High	Intermediate	Low
Count rate	High	Intermediate	Low
Cooling	Liquid N_2	Electronic—Peltier	None

Clearly, the accuracy depends on the resolution of the detector used and the presence or absence of neighboring elements which may interfere. For the lead paint sample, the peaks of lead and a neighboring element such as zinc are sharply resolved for the solid-state detector. Note the significant quantity of zinc in this paint sample. The spectrum for the XRF instrument with the proportional counter is shown in Figure 12.6. The high concentration of zinc and its overlap with the lead L_α peak make it difficult to quantify the lead peak.

The Si(Li) detector will provide the highest resolution, count rate, and efficiency of the three detectors. The disadvantage of this detector is that it requires liquid nitrogen to operate and 30–40 min to cool down. The Si Pin detector operates over the whole range of elements Na–U, has an acceptable resolution of 230 eV, and only requires about 5–10 min to cool down. The count rate and efficiency are lower than the Si(Li) detector. The proportional counter is limited to only the simplest applications where elements are not close together; otherwise significant corrections and/or filters are required. If the user wants to cover a significant range, more than one (two or three) proportional detectors (filters) would be required. The count rate is considerably lower than the solid-state detectors, as explained earlier in this chapter.

12.3 SAMPLE PREPARATION

12.3.1 Sampling

One of the first considerations in the field is to ensure that the samples taken are representative of the site. This is particularly true for soil samples that are typically nonhomogeneous.

Sometimes multiple levels of contaminants are present at the site, and it is not unusual to find data such as that shown in Figure 12.7, which profiles the arsenic

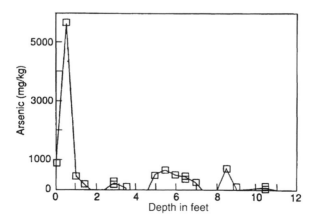

Figure 12.7 Arsenic concentration (*f*) depth in the soil.

concentration as a function of depth in soil at a hazardous waste site. There is no set pattern to disposing of hazardous waste.

Even in a shovel full of dirt there can be a significant variation in the concentration over a range of a few inches. *You have to ensure that you have a representative sample (well mixed). The next consideration is that the sample must be homogeneous or all of the analyses will be incorrect whether in the field or in the laboratory!* One should use a consistent technique in gathering and preparing the sample. An example is the arsenic in soil shown in Figure 12.7. The standards and unknowns should be of similar composition to minimize or eliminate matrix effects from the samples. Bad standards can also affect the accuracy of the result. In the software, it is important to be able to view the calibration curve and add or remove standards. The accuracy of the results will depend upon the accuracy of the standards.

12.3.2 Soils

This is a key element for accurate XRF analysis. Samples to be analyzed must be similar to standards (e.g., particle size, composition). In general, samples may have to be sieved to a uniform particle size (sandy soils), ground with a mortar and pestle (soils with a high clay content, sludge) to a uniform particle size, placed in a ball mill or shatter box (for geological sample such as rocks), and so on. Soil samples are analyzed directly after preparation by placing in a special low-volume polyethylene sample cup fitted with a 0.15 film Mylar window and counting for 300 s. It is essential that the standard(s) and the sample have the same surface characteristics and similar composition. Particle size, compactness, and state (liquid, solid, etc.) can affect the intensity of XRF.

Soils contain trace heavy metals, with the most common elements being Pb, As, Hg, Fe, Cu, Zn, V, Ti, Cr, Mn, Cd, and Ni. The reasons for sieving the sandy soil samples noted above are twofold: The majority of the metals are concentrated in the fine particles and large particles can cause scattering.

12.3.3 Liquid Samples

Aqueous samples at ppm levels can be analyzed directly by placing in a special low-volume polyethylene sample cup fitted with a 0.15 film Mylar window and counting for 300 s. These samples can be analyzed directly without any pretreatment. Heavy metals can be found in surface water, groundwater, rivers, lakes, or ponds. Other environmental applications involve measurement of sulfur in oil, diesel fuel, or gasoline; chlorine in process or liquid waste [polychlorinated biphenyls (PCBs) or chlorinated hydrocarbons], and heavy metals in waste oil.

The detection limits are in the low- or sub- (heavy metals) ppm range. One problem with liquids is that X-rays are not completely absorbed so that these samples may yield a large continuum background. Some software contains a background subtraction algorithm than can eliminate the continuum and produce peaks for low-ppm levels of metals in oil.

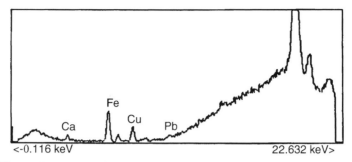

<-0.116 keV   22.632 keV>

Figure 12.8 Low-ppb levels of metals in tap water after preconcentration.

Metals in groundwater, lakes, and ponds should only contain background levels of heavy metals that are generally low- or sub-ppb levels. These are usually analyzed by ICP or AA down to < 1 ppb, but XRF can generally go down to < 1 ppm. What can be done about this? If we take 1 L of water and pour it through 1 g of ion exchange resin, then dry and analyze that resin, we can detect 1 ppb of metals in water. A spectrum at low-ppb levels is shown in Figure 12.8. This process takes only about 5 min and can be easily done in the field.

12.3.4 Filter Paper

Particulate matter in the air is collected on filter paper (FP) and can be analyzed by XRF. One problem frequently encountered is that glass fiber contains trace metals that vary from batch to batch and could interfere with the analysis of samples. Some XRF software contains features such as a strip function (HNU Systems, XR1000 software) that will automatically subtract a blank value from each of the filter analyzers. The net difference will then be reported. Note that the blank FP should be from the same batch as the FPs being analyzed since there can be considerable batch-to-batch variations. The blank can also be useful for oil samples. A typical XRF spectrum is shown in Figure 12.9.

A summary of the sample preparation for soil, water, and air samples is given in Table 12.4. The sample concentration method is described elsewhere[7] and is extremely easy to execute in the field. It involves pouring the water sample (1 L if low-ppb levels are to be detected) through 1 g of cation exchange material (in a funnel with a plug of glass wool to keep the resin in the filter), drying, and measuring by XRF. The standards are treated in a similar manner. This improves the detection limit by about three decades. With a soil sample, one has to use acid dissolution to get the metals into a liquid phase, but unless one takes an extraordinarily large soil sample, there will be only about one order-of-magnitude improvement in the detection limit.

```
Sample ID: 6733633A
Energy Range: 0 - 40 keV 10 eV/ch 6 usec
Preset: Livetime      0   Seconds
Realtime:   965.74 Seconds
Livetime:   600.00 Seconds
 37% Deadtime          9325 Counts/Second
Acquisition Date: 09-Mar-98
Acquisition Time: 23:11:47
```

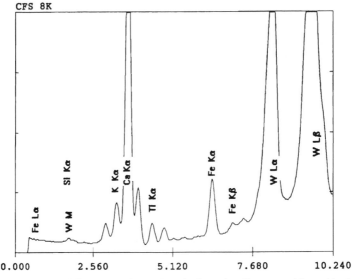

Figure 12.9   XRF Spectrum of metal particulates (on filter paper) in ambient air.

Table 12.4   Sample Preparation for Environmental Samples

	Water	Oil	Soil	Air
Preparation	None	None	Grind, sieve to 100 mesh	Collect on filter paper
Analysis	Direct	Direct	Direct	Direct
Detection limit	Low to sub-ppm	Low to sub-ppm	Low ppm	$\mu g/m^3$
Methods to improve detection limits	Concentrate on ion exchange resin	Acid dissolution, then concentrate on ion exchange resin	Acid dissolution, then concentrate on ion exchange resin	Acid dissolution, then concentrate on ion exchange resin
Improvement	1000 : 1; low-ppb levels	10–20 : 1; mid-ppb	10–20 : 1; high ppb	10–20:1

12.4   XRF EQUIPMENT AND OPERATIONAL CONSIDERATIONS

12.4.1   XRF Equipment

The major components of an XRF system are the source (X-ray tube or radioactive source), the X-ray and high-voltage power supply, the sample compartment, the detector and preamplifier, the amplifier, the multichannel analyzer (MCA), and the microprocessor or computer.

12.4.2 Operational Considerations for Tube-Excited XRF

- The accuracy *and* precision of XRF measurements *depend upon the count rate;* the higher the count rate, the better the precision and accuracy; for a Si Pin detector, a rate of >5,000 counts provides optimum results.
- Choose XRF conditions so that the voltage is 1.5–2 × the energy of the heaviest element to be detected; that is, Fe K_β is 7.06 keV HV = 10.5–14 kV.
- Use the tube current to bring the value of the dead time to approximately 40–60%.
- The anode should be higher in the periodic table than the K lines to be excited; otherwise, the L line will have to be used for analysis.
- Most common anode is Rh; it cannot be used to excite, for example, the K lines of Sn (see the periodic table) and the L line interferes with Cl. The next most common anode is W; the L line interferes with K lines of Cu and Zn.
- Use vacuum or He if analyzing elements below Z = 20 (Ca).
- Vacuum, helium, or prepurified nitrogen must be used to eliminate argon in the air, which absorbs X-rays.
- Samples may have to be *compressed* into pellets if the major components to be determined are light elements (e.g., limestone and cement).
- XRF is very dependent upon the preparation of the surface of the sample being analyzed.

12.4.3 Minimum Detectable Concentration

The minimum detectable concentration (MDC) is the concentration which produces a sufficient number of peak counts that can be recognized, with a 95% confidence, as a positive deviation above the normal variations in the background interference. The minimum number of detectable counts that will yield a 5% probability of a false-positive result is 1.64 $2\sigma_B$, where σ_B is the error in the background. The MDC is given by

$$MDC = 2.33 \frac{\sqrt{B}}{\varepsilon}$$

where ε is the counting efficiency (i.e., counts per second per unit concentration), as determined from a single measurement or the slope of the calibration curve, and B is an estimate of the background interference error based on the background counts or the blank counts obtained from the zero intercept of the calibration curve.

12.5 ERRORS IN EDXRF

The most common errors encountered can have two very different meanings which are frequently confused with each other:

1. *Accuracy*—the difference between the *measured* value and the *true* value.
2. *Precision*—the degree of agreement among replicate determinations made under identical conditions.

12.5.1 Total Error

The evaluation of an analytical instrument or method often requires the determination of the instrumental precision or reproducibility. Such determinations, however, are often confusing because of terminology. This is particularly true for EDXRF analyzers because a major source of random error is associated with counting statistics which may be unfamiliar to many analytical chemists.

The total error in a measurement is the result of systematic and random errors. Since XRF is a comparative technique, it is assumed the systematic errors have been eliminated in the instrument calibration procedure. In this case, the total error (σ_T) is given by the relation

$$\sigma_T = \sqrt{\sigma_c^2 + \sigma_I^2 + \sigma_p^2 + \sigma_s^2}$$

where σ_c is the counting error, σ_I represents the instrument-related errors, and σ_p is the error related to procedure, such as sample preparation, nonhomogeneous, particle size differences, and other miscellaneous errors which would be present regardless of what instrument is used for the analysis. The error in the concentration supplied by the alternate reference method is σ_s.

Thus, the total error includes the X-ray analysis error and the error of the alternate technique against which the X-ray method is being calibrated or compared.

12.5.2 Counting Errors

Counting errors are a fundamental phenomenon of the random nature of the radiation process and are *not* a property of the instrument. They obey the normal laws of random events and constitute the *best* possible precision attainable in an X-ray experiment. The counting error, unlike other errors, can be predicted from a single measurement. It depends only on the total number of accumulated counts and is equal to the square root of the total counts. For example, if 10,000 counts were accumulated in a 100-s counting interval, the statistical counting error would be 100 counts ($\sqrt{10,000}$). This represents the best precision that can be obtained in a hypothetical 100-s counting interval and implies that the counts accumulated in 68% of a large number of 100-s measurements would fall within about ±100 counts of the true mean ($10,000$).

The statistical counting error can be minimized by counting for the longest interval consistent with the requirements of the total analysis problem. However, the error cannot be reduced by dividing this interval into segments and taking the mean of the individual measurements. For example, the standard deviation of the *mean* of eight 100-s measurements of the above sample would be ± 35 counts (100/8). If the same sample had been counted continuously for 800 s, 8000 counts would have been accumulated with a standard deviation of ± 285 counts (80,000). The percent error, however, is identical in both cases (0.35%). The standard deviation of the mean value can be improved by making more measurements, and the standard deviation of a measurement can be improved by counting for a longer time.

The magnitude of the other types of random errors (instrumental and procedural) must be evaluated experimentally by making a sufficient number of replicate measurements to define their frequency distribution. One of the best methods of evaluating the total random error is to prepare and analyze a large number of replicate samples of the same material. This will provide an accurate estimate of the precision of the *total procedure* but will not necessarily be an accurate estimate of the instrumental precision because it will include sampling and sample preparation errors, which may be significant. The instrumental precision or reproducibility may be determined by calculating the standard deviation of a series of measurements of the *same* sample as a function of time or sample position using long counting intervals to minimize the counting error contribution. The procedural errors can be estimated by subtracting the counting and instrumental errors from the total errors.

12.5.3 Systematic Errors

Systematic errors yield values which deviate from the true value by a constant repeatable amount and include errors such as calibration, environmental, technique, and personal habit errors. These can be either avoided or corrected for and will primarily affect the accuracy.

12.5.4 Random Errors

Random errors cause repeated Individual measured values to disagree and may be caused by errors in judgment, fluctuating conditions, small disturbances, and counting errors. These errors will determine the precision, and although they cannot be avoided, they can be determined by statistical methods.

12.5.5 Average Absolute Error

The average absolute error is the sum of the absolute values of the deviations from the mean divided by the number of measurements. The standard deviation approaches 1.25 times the average absolute error as the number of measurements approaches infinity.

12.6 RESULTS AND DISCUSSION

12.6.1 EPA Results with SEFA-P

The HNU source-excited fluorescence analyzer (SEFA-P) was used by an EPA contractor at a number of Superfund sites in the United States.[7] The SEFA-P is a radioactive source-excited XRF instrument with a Si(Li) liquid nitrogen–cooled detector. A SITE demonstration of the SEFA-P was conducted, and the instrument was used to determine the concentrations of ppm to percentage levels of metals in soils. The results compared favorably with an EPA-approved reference laboratory method. It was found possible to analyze 30–50 samples in an 8–10-h day. With the XR1000 and its automated 10-position sample tray, it should be possible to analyze 50–75 samples in the same time period.

12.6.2 Results with Tube-Excited XRF[8]

Clearly, there is a significant difference in performance between an instrument with a solid-state detector and a proportional counter. In addition, many of the XRF instruments *with* proportional counters utilize radioactive sources that produce one-third to one-half the count rate produced by an X-ray tube, further reducing the count rate, detection limits, and accuracy. The effect of count rate on the minimum detectable concentration (MDC) was mentioned previously and is given as

$$MDC = 2.33 - \sqrt{B}/\text{count rate}$$

where B is the background interference level.

Thus, the higher the count rate, the lower the detection limit. Of course, resolution will also have an effect on detection limits since the higher resolution detector will produce sharper peaks, which will be more easily detected. This can be seen in the XRF spectrum of lead paint shown in Figure 12.6. A table-top XRF analyzer, the XR1000 (HNU Systems) is shown in Figure 12.10. This instrument has a small footprint, Windows XRF software, and a Peltier (electronically cooled) Si Pin detector and provides high performance in a compact package.

Resolution also becomes important for the detection of lighter elements, as shown in Table 12.5. The energy spacing for adjacent lighter elements is about one-half that for transition elements, as shown in Table 12.5.

Another area where detector resolution becomes important is in the use of software for qualitative (auto identification) and quantitative results such as fundamental parameters and techniques (FPTs) and other software where corrections are made for absorption and enhancement effects for neighboring elements. With broad peaks, it becomes exceedingly difficult to apply these corrections. The calibration curve for Mg in cement in Figure 12.11 was obtained with the HNU XRF analyzer. Figure 12.11 has no interelement corrections. In Figure 12.12, the absorption/enhancement correction for silicon was included. The difference in the calibration curves shown in Figures 12.11 and 12.12 demonstrated the improvement in the results when the corrections are used.

Figure 12.10   HNU model XR1000 table-top XRF.

Table 12.5   Comparison of Energy Level Spacing for Transition Metals and Light Elements

Atomic Number	Element	Energy (KeV)
26	Fe	K_α—6.4 KeV
27	Co	K_α—6.93 KeV
		Difference = 0.53 KeV
13	Al	K_α—1.487 KeV
14	Si	K_α—1.740 KeV
		Difference = 0.253

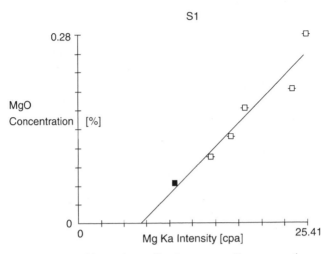

Figure 12.11   Magnesium calibration curve with no corrections.

There are two different software programs that can be used: empirical and FPT. The empirical method requires the use of multiple standards (minimum of three plus one for each element used for correction) to bracket the unknowns. This software uses a linear regression fit with a Simplex variation to minimize the error in x and y coordinates. Standards (use the AA or ICP data as the actual concentration) can be prepared by using the AA or ICP data from the same site. *Note*: If this procedure is used, a representative aliquot of the soil samples should be available. Now the soil data by XRF should correlate with the AA or ICP data for the site. Some

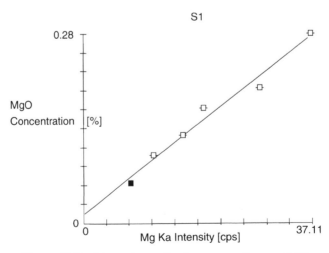

Figure 12.12   Magnesium calibration curve with corrections.

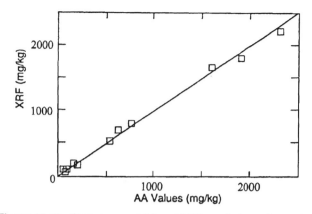

Figure 12.13 Comparison of AA and XRF results for soil samples.

site data in Figure 12.13 compares AA with XRF data. The correlation coefficient is 0.995. The empirical method is useful from ppm to percentage levels, but multiple calibration curves would be required to cover this range. An accuracy of 98+% may be obtained, but this depends upon the accuracy of the standards, the count rate, matrix effects, and other factors.

The importance of interelement corrections in environmental analysis can be seen with the analysis of lead and arsenic. The K_α line of arsenic and the L_α line of lead are both at 10.5 keV. If both are present, there are several options: (a) use interelement corrections (will need standards with known arsenic and lead concentrations) or (b) use the K_β line of arsenic and the L_β line of lead. The lower limit of detection will not be as low with the β lines, but the interference will be eliminated.

The FPT standardized program can be used for high ppm to 100%. All of the elements in the sample must be known and must add up to 100%. In this program, corrections are made for detector efficiency, tube output, other elements (absorption and enhancement), and so on. At least one standard is needed. Typical accuracy ranges from 90 to 95%.

Driscoll and Johnson[9] have shown the neural networks can be successfully applied to qualitative and quantitative identification of elements such as lead (Pb) and arsenic (As). the Pb L_α and As k_α X-ray spectra overlap at 10.6 keV. Using a single neuron and 15 inputs per element, they were able to predict the shape of the As and Pb spectra (qualitative). For quantitative analysis 45 inputs were required with three processing neurons to determine As and Pb concentrations in complex soil mixtures. This technique has significant promise for improving algorithms used in spectroscopic software.

In this chapter we have described a number of basic aspects of XRF in order to provide the reader with a basic understanding of both XRF theory and sample preparation. Should the reader be interested in learning more about XRF applications or detectors and electronics, Ref. 1 and 6 are recommended.

REFERENCES

1. V. Valkovic, *X-Ray Spectroscopy in the Environmental Sciences*, CRC Press, Boca Raton, FL, 1989.

2. Anonymous, *Data Quality Objectives for Remedial Response Activities, RI/FS Activities at a Site with Contaminated Soils & Ground Water*, PB 88-131388, EPA, Washington, DC, March 1987.

3. Anonymous, *Data Quality Objectives for Remedial Response Activities, Development Processes*, PB 88-131370, EPA, Washington, DC, March 1987.

4. R. K. Stevens and W. F. Herget, *Analytical Methods Applied to Air Pollution*, Ann Arbor Science Publishers, Ann Arbor, MI, 1974.

5. J. N. Driscoll, J. K. Marshall, E. S. Atwood, C. Wood, and T. Spittler, A New Instrument for the Determination of Heavy Metals & Radioactivity at Mixed Waste Sites, *Am. Lab.*, July 1991, p. 131.

6. G. F. Knoll, *Radiation Detection and Measurement*, Wiley, New York, 1989.

7. EPA, *Superfund Innovative Technology Evaluation: Characterization and Monitoring Program*, http://www.clu-in.com/site/charmon.htm.

8. J. N. Driscoll, Field Analysis of Metals in the Environment by XRF, *Int. Env. Tech.*, September/October 1998.

9. J. N. Driscoll and W. Johnson, Use of Neural Networks in XRF, Pittsburgh on Anal. Chem. & Applied Spectroscopy, Paper 1968P New Orleans, LA, March 2002.

13

LABORATORY ANALYSIS

Paul J. Giammatteo, Ph.D. and John C. Edwards, Ph.D.

Process NMR Associates, LLC
Danbury, Connecticut

13.1 NUCLEAR MAGNETIC RESONANCE

"Nuclear Magnetic Resonance (NMR) spectroscopy is one of the most powerful experimental methods available for atomic and molecular level structure elucidation. It is a powerful technique in that it is ... noninvasive [and is] used to identify individual compounds [analyze mixtures], aid in determining the structures of large molecules ... and examine the kinetics of certain reactions" (Ref. 1, p. 3). "Of all

Environmental Instrumentation and Analysis Handbook, by Randy D. Down and Jay H. Lehr
ISBN 0-471-46354-X   Copyright © 2005 John Wiley & Sons, Inc.

techniques available for determining structures, NMR spectroscopy is the most valuable. It's the method that organic chemists turn to first for information" (Ref. 2, p. 381). Nuclear magnetic resonance has now expanded well beyond being the principal tool of the organic chemist. Its ability to analyze a variety of sample types, from the simplest molecules to the intact human body, makes NMR a powerful tool for the environmental analyst.

Nuclear magnetic resonance can analyze any type of sample regardless of the physical (liquid, solid, gas) and environmental (temperature, pressure) conditions of that sample. Diverse applications of NMR analyses include:

- Solid-state analyses of coals, soils, catalytic materials, polymers, and composite materials
- Monitoring of molecular interactions, including drug–deoxyribonucleic acid (DNA) interactions, protein–substrate binding, protein folding, catalytic site activity monitoring, reaction mechanism studies, and graphite-to-diamond transitions
- Imaging applications, including magnetic resonance imaging (MRI) for medical diagnosis, intact cellular chemistry monitoring, material imaging for composite material manufacturing and performance evaluation, and flow imaging

For those unfamiliar with NMR, we refer the reader to any number of college level organic chemistry texts as these offer good introductions to the fundamentals of NMR spectroscopy. As our intent is to introduce areas of environmental analysis for NMR, we will selectively highlight the key underlying principles of the technique as these principles are required for any application of NMR.

13.1.1 Basic Concepts

The property of nuclear spin, discovered by Rabi and co-workers in 1939, is the fundamental property for observing NMR. Similar to a child's spinning top, a spinning, positively charged nucleus acts like a tiny magnet with poles (i.e., magnetic dipole) perpendicular to the spin of the nucleus. Since a dipole is present, a magnetic moment exists for that nucleus. Magnetic moments are defined as the torque felt by the nucleus when in the presence of an external magnetic field. The "intensity" of the magnetic moment is dependent on the strength of the applied magnetic field. In the absence of an applied field, the nuclear spins and their magnetic moments are randomly oriented. When a sample is placed in a strong magnetic field, a torque is applied to the nuclear spins until each magnetic moment either aligns with (lower energy state) or against (higher energy state) the applied field. This torque exists until the spins are perturbed from alignment with the external magnetic field or the external field is removed. As there are always slightly more spins aligned with the external magnetic field (lower energy) than aligned against the external magnetic field (higher energy), there is an effective net magnetic moment aligned with the external magnetic field (Fig. 13.1).[1]

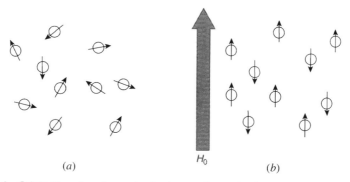

Figure 13.1 Orientation of nuclear spins in absence (*a*, random) and presence (*b*, oriented) of applied external magnetic field (Ref. 2, p. 382).

Once aligned with the external magnetic field, the oriented nuclei can be irradiated with electromagnetic radiation of the proper frequency. For NMR, this is radio-frequency (RF) irradiation in the megahertz (MHz) frequency range. When irradiated, energy absorption occurs, "flipping" the spins from either a low to a high or a high to a low energy state. The net effect is to move the net magnetic moment off the axis of alignment. The off-axis distance moved, or tip angle, is dependent on the strength and duration of the applied RF pulse. For a conventional liquid-state NMR experiment, the application of a short (5–15-μs), low-energy (~ 1-W) RF pulse is all that is required to move the net magnetic moment off the equilibrium axis 90°. Once the RF pulse is removed, the net magnetic moment returns (relaxes) back to equilibrium and alignment with the external magnetic field. This is shown schematically in Figure 13.2. The resonance condition, that

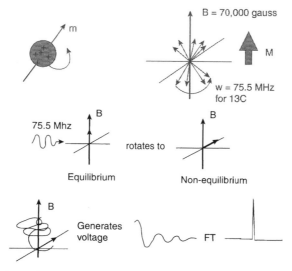

Figure 13.2 Schematic representation of NMR experiment showing effect on magnetic moments (*m*) under influence of applied external magnetic field (*B*) and RF pulses.

Table 13.1 Comparison of Resonance Frequencies of Select NMR Active Nuclei at Varying Magnetic Field Strengths

NMR Active Nucleus	Resonance Frequency at Various Magnetic Field Strengths (MHz)					
	1.4 T	4.7 T	7.05 T	9.4 T	11.75 T	14.1 T
^1H	60 MHz	200.0	300.0	400.0	500.0	600.0
^{13}C	15.1	50.3	75.4	100.5	125.7	150.8
^{31}P	24.3	80.9	121.4	161.9	202.4	242.9
^{15}N	6.8	20.3	30.4	40.5	50.7	60.8
^{29}Si	11.9	39.7	59.6	79.5	99.3	119.2

is, the RF frequency in which the nuclei absorb energy and are perturbed from equilibrium, is dependent on the strength of the applied magnetic field. For a 1.4-T magnet, hydrogen resonates at 60 MHz, ^{13}C resonates at 15 MHz, and ^{31}P resonates at 24 MHz. For a 7.0-T magnet, hydrogen resonates at 300 MHz, ^{13}c resonates at 75 MHz, and ^{31}P resonates at 121 MHz. Table 13.1 lists several NMR active nuclei and their resonance frequencies at different magnetic field strengths.

As one can see, the adsorption frequency is not the same for all nuclei at a given magnetic field strength. The strength of NMR as a molecular structure elucidation tool comes from its ability to resolve differences in the magnetic environment of the nucleus in the molecule. All nuclei in molecules are surrounded by electron clouds, effectively giving each nucleus in the molecule its own small effective magnetic field. These small local fields act in opposition to the applied magnetic field so that the effective field felt by each nucleus is slightly weaker than the applied magnetic field (Ref. 2, p. 383):

$$H_{\text{effective}} = H_{\text{applied}} - H_{\text{local}}$$

The electron clouds surrounding the nuclei shield the nuclei from the applied magnetic field. All nuclei in the same molecular and/or chemical environment are equivalent. Since they are effectively in a similar electronic environment, they will experience the same level of shielding from the applied magnetic field. For example, water has two hydrogens bound to a single oxygen. As these hydrogens are electronically equivalent, there will be one resonance line for the hydrogen atoms in water. Nuclei that exist in different molecular/structural environments, both within a molecule as well as between molecules, are nonequivalent and will give rise to different NMR absorption. For example, ethanol (CH_3CH_2OH) has three nonequivalent proton types, three protons on the methyl group, two protons on the CH_2 group, and one proton on the hydroxyl group. Therefore, ethanol gives rise to three resonance lines. As an example of nonequivalence between molecules, benzene has six equivalent hydrogens and water has two equivalent hydrogens. However, the electronic environment of the hydrogens attached to the aromatic carbons of benzene is different than the electronic environment of the hydrogens in water. Benzene and water are nonequivalent; therefore a mixture of benzene and water gives rise to two NMR absorptions. Finally, *n*-heptane and isooctane both contain CH_3 and CH_2 groups, while isooctane also contains a CH group. All CH_3 groups are equivalent whether they are in *n*-heptane or isooctane. The same holds for the CH_2 groups. However, the CH_3 hydrogens are not equivalent to the CH_2 hydrogens, which are also not equivalent to the CH of isooctane. Therefore, there are three NMR absorptions for a mixture of *n*-heptane and isooctane that are the same three NMR absorptions for isooctane alone. However, the intensity of each absorption would be different if one were analyzing isooctane alone or if one were analyzing a mixture of isooctane and *n*-heptane

The differences in NMR absorption frequencies based on the above description define chemical shift. The integrated intensity of each chemical shift position is proportional to the number of nuclei at that chemical shift position. As chemical shift is defined as ppm (parts per million) in NMR, this does not represent a concentration. In NMR ppm is defined as the difference in hertz from a reference frequency over the resonance frequency (megahertz) of the observed nuclei. For example, CH_3 groups of heptane have a proton chemical shift of 0.8 ppm while benzene protons have a chemical shift of 7.1 ppm. The chemical shift remains the same for each independent of the field strength of the magnet in which the NMR experiment was performed. That is, for the CH_3 groups of heptane in NMR spectrometers at 1.4 and 7.0 T, respectively,

$$\frac{48 \text{ Hz from TMS reference}}{60 \text{ MHz}} = 0.8 \text{ ppm} \qquad \frac{240 \text{ Hz from TMS reference}}{300 \text{ MHz}} = 0.8 \text{ ppm}$$

Increasing field strength does not effect chemical shift but does improve the resolution of NMR absorptions where the differences in chemical shifts are small. Figure 13.3 lists proton and ^{13}C chemical shifts of various molecular functional groups. By using chemical shift information as well as integrated intensity, coupling interactions, and other quantum mechanical transitions inherent in the NMR experiment, numerous quantitative applications of NMR for liquid-sample analyses are possible (see Figs. 13.3 and 13.4). The reader is referred to more detailed descriptions of environmentally important NMR applications in the references listed at the end of this chapter.

13.2 LIQUID-STATE NMR

Most NMR experiments are performed on samples that are in the liquid state. Typically, a liquid or solid sample is dissolved in an appropriate perdeuterated NMR solvent [i.e., deuterated chloroform ($CDCl_3$), deuterated water (D_2O), deuterated acetone (C_2D_6CO)] to form a homogeneous solution. This solution is placed in a glass NMR tube [typically 5 or 10 mm inside diameter (i.d.)] that is then placed in the NMR probe:

Comparison of ^1H and ^{13}C chemical shifts.

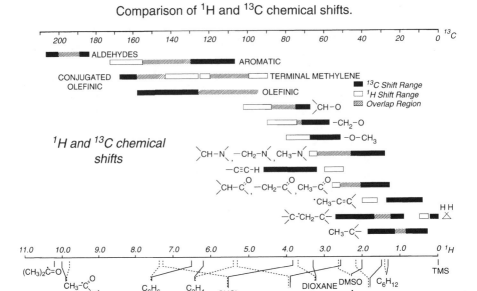

Figure 13.3 Comparison of ^1H and ^{13}C chemical shifts for various chemical functionalities (Ref. 3, p. 288).

1. Deuterated solvents are used for assuring that a proton signal from the solvent, especially when analyzing small sample quantities, does not interfere.

2. The deuterium resonance frequency is used to frequency lock the spectrometer, eliminating frequency drift when signal averaging is used.

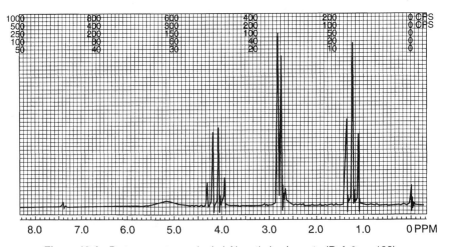

Figure 13.4 Proton spectrum of ethyl *N*-methylcarbamate (Ref. 3, p. 198).

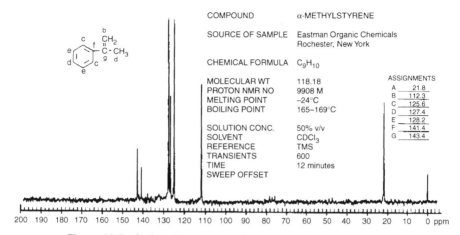

	ASSIGNMENTS	
COMPOUND		α-METHYLSTYRENE
SOURCE OF SAMPLE		Eastman Organic Chemicals Rochester, New York
CHEMICAL FORMULA		C_9H_{10}
MOLECULAR WT		118.18
PROTON NMR NO		9908 M
MELTING POINT		−24°C
BOILING POINT		165–169°C
SOLUTION CONC.		50% v/v
SOLVENT		$CDCl_3$
REFERENCE		TMS
TRANSIENTS		600
TIME		12 minutes
SWEEP OFFSET		

ASSIGNMENTS

A	21.8
B	112.3
C	125.6
D	127.4
E	128.2
F	141.4
G	143.4

Figure 13.5 Carbon-13 spectrum of α-methylstyrene (Ref. 3, p. 299).

Signal averaging and proper solvent selection enable NMR to routinely analyze samples from microgram to gram quantities without difficulty.

Figures 13.4 and 13.5 show the typical organic chemistry applications of NMR. Inorganic applications are highlighted in Figures 13.6 and 13.7, which resolve the interactions of aluminum chloride with oxalic and lactic acids. Figure 13.8 shows typical proton and ^{13}C spectra of a natural polysaccharide, and Figure 13.9 shows proton spectra of an aqueous amino acid complex.

"The major advantage of NMR is its value as a tool for elucidating the structures of complex or unusual . . . compounds. Furthermore, the structural information is

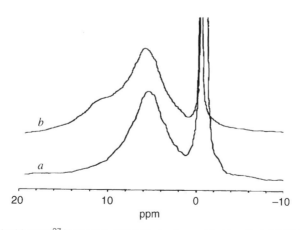

Figure 13.6 Liquid-state ^{27}Al spectra of 0.1 M aluminum chloride with (a) 0.05 M and (b) 0.1 M oxalic acid (Ref. 1, p. 167).

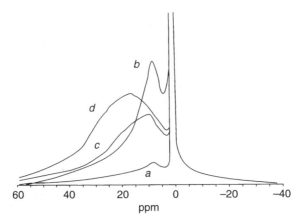

Figure 13.7   Liquid-state ^{27}Al NMR spectra of 0.1 M aluminum chloride with (*a*) 0.01 M, (*b*) 0.1 M, (*c*) 0.2 M, and (*d*) 0.3 M lactic acid (Ref. 1, p. 168).

acquired under ... non destructive conditions and in solutions where the compounds are more apt to appear in the form in which they exist in the environ-ment ... For compounds of environmental significance ... NMR is more likely than GC/MS to reveal their structure as they appear in a natural aqueous environment and without the risk of thermolytic degradation" (Ref. 2, p. 138).

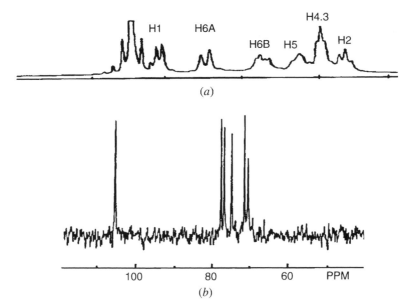

Figure 13.8   (*a*) Proton and (*b*) ^{13}C liquid-state NMR spectra of naturally occurring polysaccharide Pustulan (Ref. 4, p. 27).

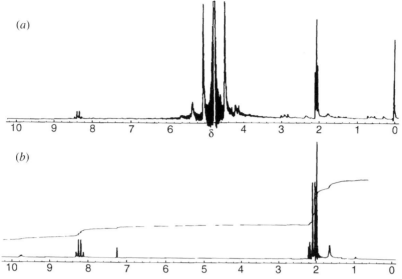

Figure 13.9 Proton NMR spectra of alanine in (*a*) buffered D_2O and (*b*) $CDCl_3$ solutions (Ref. 1, p. 133).

13.3   SOLID-STATE NMR

"There are a number of contributions to the NMR spectra of solids . . . the critical feature of these interactions is that they are anisotropic (i.e., orientation dependent) in the solid due to the relatively fixed orientations of the molecules, whereas in solution average values are obtained due to the fast isotropic motion of the molecules. In the case of chemical shifts and scalar couplings, discrete average values are found in solution and can be used for structure elucidation. In the case of dipolar and quadrupolar couplings, the average values are exactly zero and these effects are not observed at all in solution . . . [In a solid sample, these] solid state interactions are affected by the degree of ordering and motion present in the solid" (Ref. 5, p. 25).

The ability of NMR to analyze structures of compounds as they exist in the environment can also be extended to analyzing solid samples directly, including solid samples themselves and/or as samples bound, adsorbed, intercalated, and so on within a solid matrix. Analyzing solid samples directly by NMR involves more extensive experimental techniques than analyzing liquids due to the interactions of the magnetic moments with the applied magnetic field in a solid matrix. These interactions tend to reduce signal intensities and broaden the NMR lines on the order of the entire spectral width. Experimental techniques designed to reduce solid NMR line widths to those approaching liquid NMR line widths as well as increasing sensitivity include:

- Magic angle spinning
- Cross-polarization
- High-power decoupling

Typically, solid-state NMR is used to obtain spectra of NMR nuclei other than proton (i.e., ^{13}C, ^{31}P, ^{15}N). While the following sections will introduce several techniques used in obtaining NMR spectra on solid samples, the reader is again referred to the references for more detailed descriptions.[5,6]

13.3.1 Magic Angle Spinning

The orientation of magnetic moments of nuclei in the presence of an external field is uniform in solution samples as the motion of the molecules in solution is rapid (compared to the NMR timescale), allowing reorientation and alignment of the magnetic moments with the applied magnetic field. In contrast, molecular motion and therefore magnetic moment realignment are restricted in solid samples. For a pure solid sample, such as frozen chloroform, off-axis alignment of the individual magnetic moments will exhibit a random, angular dependence with respect to the applied magnetic field—a dependence that is related to the random orientation of each individual molecule to that field as a result of packing in the solid matrix. That is, for a frozen, amorphous solid sample of chloroform, one molecule may be aligned at 10° off the applied external field axis, another molecule parallel to the axis, another perpendicular, and so on (Fig. 13.10). As a result, while one would obtain a single-line ^{13}C NMR spectrum in solution, in a solid, one obtains a powder pattern spectrum that represents the distributions of individual magnetic moment orientations with respect to the applied magnetic field.[7] This phenomenon, called *chemical shift anisotropy*, represents the change in energy of the system that arises from shielding the magnetic moment from the applied magnetic field by the electrons. This is represented by the chemical shift tensor as

$$\sigma_{obs} = 3/2\sigma_{iso}\sin^2\theta + 1/2(3\cos^2\theta - 1)$$

If one sets $\sigma_{obs} = \sigma_{iso}$ (the isotropic chemical shift tensor as observed in liquids), then $\theta = 54.7°$. When a solid sample is rapidly spun at this "magic angle," 54.7°

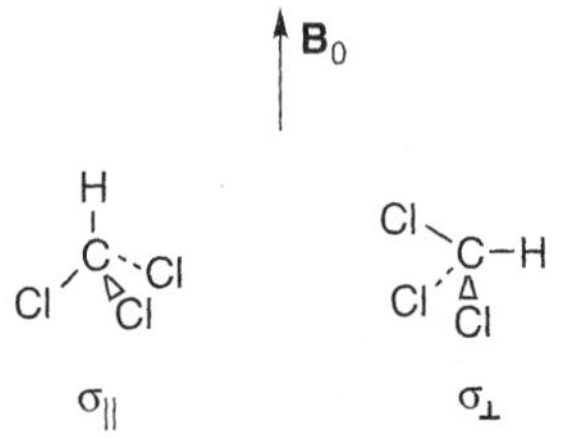

Figure 13.10 Orientations of individual chloroform molecules in frozen, amorphous solid with respect to applied external magnetic field (Ref. 7, p. 153).

from the applied magnetic field axis, the powder pattern collapses and the solid spectrum approaches the isotropic chemical shift.

13.3.2 Cross-Polarization

Another detrimental consequence in observing solid samples by NMR is the effect on both signal intensity and resolution due to dipole–dipole interactions. That is, the spin flip of a magnetic moment from one nucleus affects all nuclei in proximity to it. For abundant protons, the dipole–dipole interactions are very strong and non-uniform due to the multiple orientations to the applied magnetic field. This causes varying rapid relaxation rates that increase line broadening and loss of signal intensity for protons in the solid state. Conversely, ^{13}C nuclei are very dilute and therefore isolated from ^{13}C–^{13}C dipole interactions. Dipole–dipole relaxation pathways are minimal, and ^{13}C relaxation rates are exceedingly long, adversely affecting sensitivity. Under the right conditions, one can "spin lock" the proton and ^{13}C magnetization, transferring the dipole interactions from the abundant spins (protons) to the dilute spins (^{13}C). That is, under the proper pulse conditions, one can place the net polarized magnetization of the abundant proton spins into "contact" with the net polarized magnetization of the dilute ^{13}C spins (Fig. 13.11). During the contact period, proton magnetization is transferred to the carbon magnetization, resulting in a large increase in carbon intensity with a minimum loss of proton intensity. Further, "in contrast to the simple pulse-FT [Fourier transform] experiment, the carbon magnetization for the FID [free induction decay] which yields the spectrum does not depend at all on the regrowth of the carbon magnetization in the delay period but arises entirely from the contact with the proton spins. This means that the intensity of the carbon spectrum effectively depends on the relaxation of the proton spin system" (Ref. 5, p. 186).

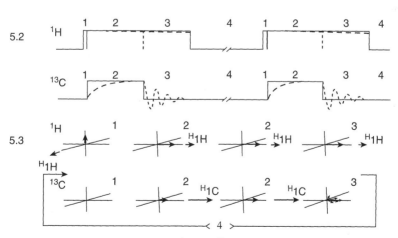

Figure 13.11 Pulse sequence (5.2) and ^{1}H and ^{13}C magnetization behavior (5.3) for cross-polarization (Ref. 5, p. 185).

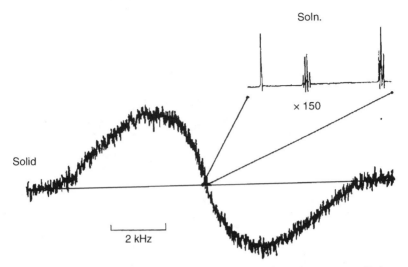

Figure 13.12 Comparison of liquid and nonspinning solid NMR spectra of ethanol (Ref. 5, p. 31).

13.3.3 High-Power Decoupling

Figure 13.12 "shows the spectrum of solid ethanol . . . contrasted with the high-resolution spectrum of liquid ethanol. The spectra indicate clearly that there are interactions present in the solid state which are orders of magnitude larger than those in solution. Thus the normal solution spectrum of ethanol exhibiting spin-spin couplings and chemical shifts occupies less than 1 KHz (at 90 MHz) of the approximately 100 KHz range which . . . [is needed] to properly define the solid-state spectrum. The interaction which gives rise to these very broad, usually feature-less adsorptions . . . are the dipole–dipole interactions between nuclei" (Ref. 5, p. 30). To remove these dipolar interactions in the liquid state, that is, to remove ^{1}H–^{13}C couplings in the carbon spectrum, typical proton decoupling power levels of 1–10 W are all that is required. To remove similar interactions in the solid state requires kilowatt power levels and potentially one can "fry" one's sample under such extreme conditions. Fortunately, dipole–dipole interactions (defined by the Hamiltonian H) have a similar angular dependence as the chemical shift anisotropy such that

$$H_{\text{dipole}} \propto (3\cos^2 = -1)/r_{ij}^3$$

where r_{ij} is the distance between two dipolar coupled nuclei (Ref. 5, p. 32). Therefore, magic angle spinning ($\theta = 54.7°$) also facilitates the reduction of large dipole–dipole interactions reducing decoupling power requirements to approximately 0.1–1-kW. Even at these levels, decoupling using continuous excitation presents two difficulties: (1) sample damage due to excessive decoupling power is possible and (2) variations in decoupling intensity over the spectral range are possible, causing variations in the effective decoupling in the spectrum. To overcome these limitations, a number of pulsed decoupling sequences have been developed.

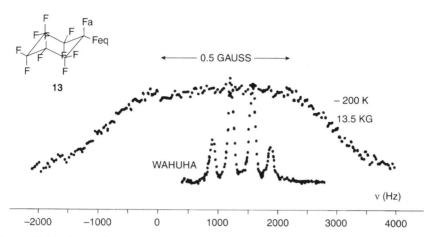

Figure 13.13 Comparison of NMR spectra of perfluorocyclohexane with (lower) and without (upper) decoupling (Ref. 5, p. 154).

The net effect of these sequences is to effectively provide high-power decoupling uniformly over the entire spectral range at much lower power levels, typically 100–500 W (see Fig. 13.13) (Ref. 5, Chapter 4).

Figure 13.14 shows that combining cross-polarization, magic angle spinning, and high-power decoupling enables solid-state NMR analyses to approximate the high-resolution features of liquid-state NMR. The combination of these solid-state techniques opened the door for NMR to probe the chemical nature of many systems in the solid state directly, including:

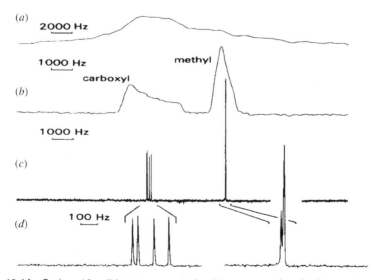

Figure 13.14 Carbon-13 solid-state spectra of calcium acetate hemihydrate under (a) static conditions, no decoupling; (b) static conditions with decoupling; (c) magic angle spinning with decoupling; (d) expansion of (c) to show spectral fine structure (Ref. 7, p. 156).

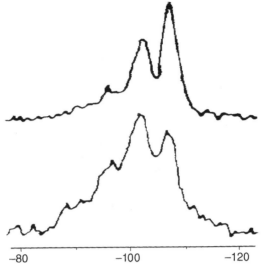

Figure 13.15 Silicon-29 NMR of dealuminated zeolite Y samples with (lower) and without (upper) cross-polization (Ref. 5, p. 350).

- Active site–substrate interactions in catalysts, proteins, and drug binding sites
- Three-dimensional structure elucidation of amorphous materials not amenable to X-ray crystallographic analyses
- Interactions of compounds in in situ environments without the need for extraction (i.e., pollutants in soils, oil content of shales, organic compounds in coal)

These techniques are shown in Figures 13.15–13.19

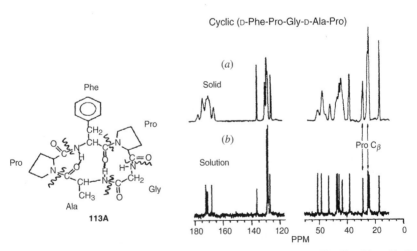

Figure 13.16 Carbon-13 NMR spectra of a cyclic peptide (cyclo-D-Phe-Pro-Gly-D-Ala-Pro) comparing liquid (lower) and solid (upper) NMR analyses (Ref. 5, p. 422).

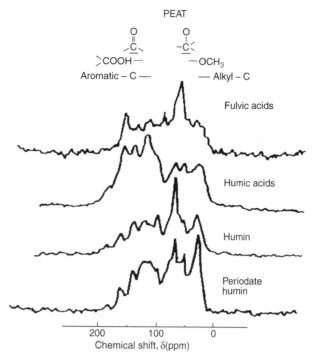

Figure 13.17 Carbon-13 solid NMR spectra of fulvic acids, humic acids, and humin from Okefenokee peat (Ref. 6, p. 243).

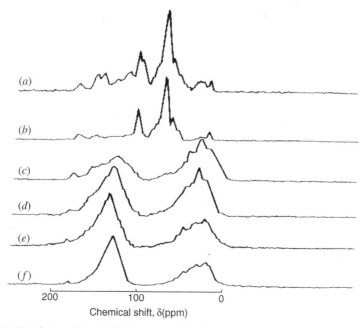

Figure 13.18 Carbon-13 solid NMR spectra of (a, b) peat, $(c, d, e,)$ lignite, and (f) bituminous coal (Ref. 6, p. 242).

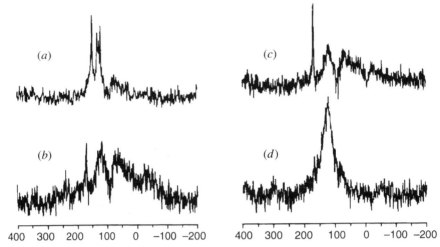

Figure 13.19 Carbon-13 solid NMR spectra of pentachlorophenol (*a*) technical grade, (*b*) adsorbed on TVA coal fly ash, (*c*) after leaching fly ash with water, and (*d*) same as (*c*) but collected using different NMR experimental conditions (Ref. 1, p. 115).

13.4 ENVIRONMENTAL APPLICATIONS OF NMR

"Over the past 20 years, gas chromatography/mass spectroscopy (GC/MS) has been widely used to identify trace organic environmental contaminants and to study the mechanisms of the formation or transformation of organic compounds either by natural or man-made processes. In the area of water and wastewater disinfection, GC/MS has been highly successful in identifying numerous volatile organic chlorination by-products ... However, ... much of the chemistry continues to be poorly understood. Analysis of trace organics by GC/MS relies on the assumption that the compounds analyzed are (1) volatile and (2) thermally stable to GC temperatures ... Because nuclear magnetic resonance spectroscopy (NMR) is a mild and nondestructive method of analysis it can reveal reactions that occur in water that cannot be observed by GC/MS" (Ref. 1, p. 130).

"Many amino acids are very polar or thermally labile, and, when chlorinated, are not amenable to GC/MS analysis" (Ref. 1, p. 134). However, reactions of amino acids with aqueous chlorine are readily extracted into deuterated chloroform, making the reaction mixture readily observable by NMR (Fig. 13.20). By studying various amino acid aqueous chlorine reactions this way, a mechanism for chlorination and decomposition has been proposed (Fig. 13.21).

Albaret et al.[8] used proton, ^{31}P, and combined ^{1}H–^{31}P NMR techniques to analyze extracted soil samples contaminated with organophosphorus from chemical weapons. While the proton NMR was extremely sensitive, overlapping resonances made compound identification difficult, even after spiking samples with known materials. The ^{31}P spectra could resolve many of the components, but low spectral sensitivity made characterization of the trace levels of the organophosphorus

Figure 13.20 Proton NMR spectrum of threonine in CDCl₃ after extraction from aqueous solution containing excess aqueous chlorine (Ref. 1, p. 136).

compounds inadequate. Using an NMR experimental protocol called *two-dimensional NMR*, this study combined the ¹H and ³¹P analyses into a series of two-dimensional analyses. In doing so, the combined sensitivity of the ¹H analyses and the resolution of the ³¹P facilitated the identification of the organophosphorus contaminants.

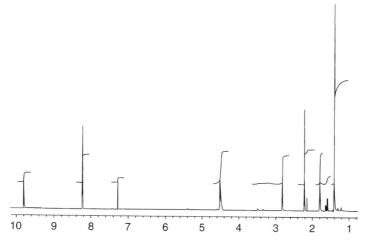

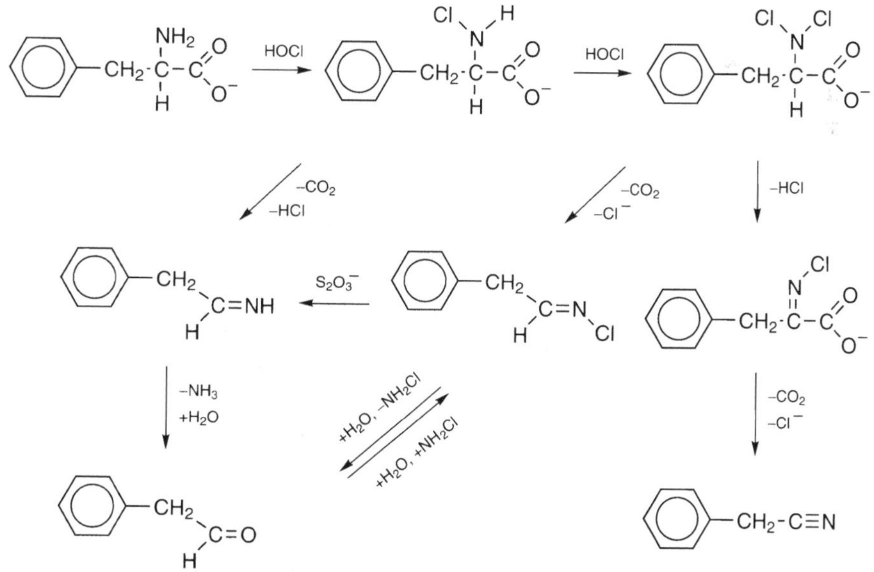

Figure 13.21 Reaction pathways for aqueous chlorination and decomposition of phenylalanine (Ref. 1, p. 135).

Phosphorus is a key element in several classes of compounds of environmental significance:

1. Detergents
2. Fertilizers
3. Pesticides
4. Chemical and biological weapons

Phosphorus-31, the NMR active form of phosphorus, readily lends itself to NMR analyses to a wide range of samples, including wastewater, groundwater, soils, clays, and other subsurface materials. The ability to selectively resonate ^{31}P coupled with excellent chemical shift resolution makes compound/contaminant identification by ^{31}P NMR highly effective.

Phosphorus-31 NMR was used to study the interactions of orthophosphates in various aqueous environments. In a study designed to monitor the effects of those compounds on lake and reservoir eutrophication, seasonal variations of total soluble phosphorus were observed (Fig. 13.22). In this study, the author observed that "in

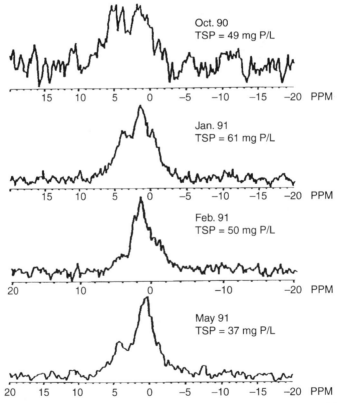

Figure 13.22 Phosphorus-31 NMR spectra of total soluble phosphorus in lake samples showing seasonal variation in phosphorus demand (Ref. 1, p. 230).

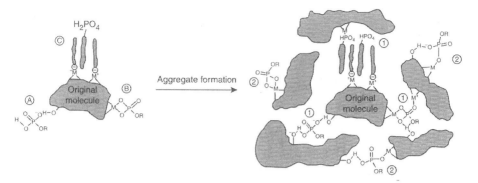

Figure 13.23 Proposed aggregate growth model for incorporation of organic phosphorus into humic and fulvic molecules by (*a*) hydrogen bonding, (*b*) metal ion bridging, and (*c*) hydrophilic bonding. Interior (1) and surface (2) organic phosphorus molecules are observed (Ref. 1, p. 235).

the late winter and early spring when algal blooms occur, there is a high demand for phosphorus. Orthophosphate will be rapidly consumed . . . Dissolved organic phosphorus that becomes incorporated with micelles or aggregates could be more difficult to degrade" (Ref. 1, p. 231). Changes in the spectra shown in Figure 13.22 result from the following seasonal variations:

(a) Algal growth phases (late winter to early spring)

(b) Algal and plant cell lysis (summer)

(c) Fallen leaves (fall)

(d) Particle settling (early to midwinter)

The incorporation of organic phosphorus from the aqueous phase into the solid/soil phase can also be probed with NMR. Figure 13.23 shows a proposed mechanism for this transition. Figure 13.24 shows the ^{31}P NMR of a soil sample with little to no pretreatment prior to analysis.

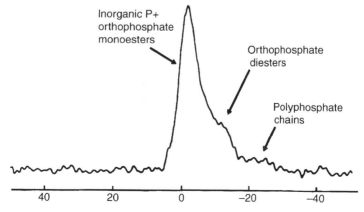

Figure 13.24 Phosphorus-31 spectrum of a Pomare soil sample (Ref. 1, p. 254).

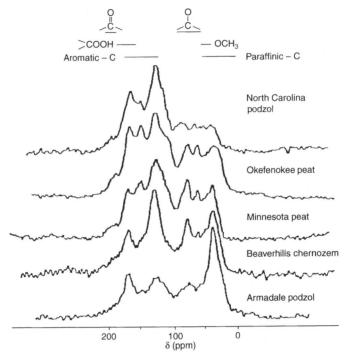

Figure 13.25 Carbon-13 solid-state spectra of variety of humic materials (Ref. 6, p. 204).

Whole-soil analysis by NMR is uniquely suited to observing:

(a) The natural organic matter in the soil (Fig. 13.25),

(b) The natural inorganic soil matrices (Fig. 13.26),

(c) Contaminant material, both organic and inorganic (Fig. 13.27),

(d) The adsorption/desorption processes and mechanisms

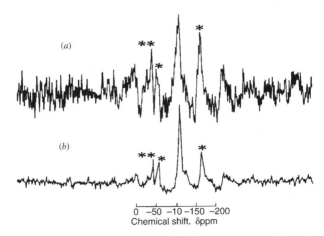

Figure 13.26 Silicon-29 solid-state NMR spectra of Urrbrae pasture soil spectra (a) before and (b) after iron removal by dithionite reduction (Ref. 6, p. 179).

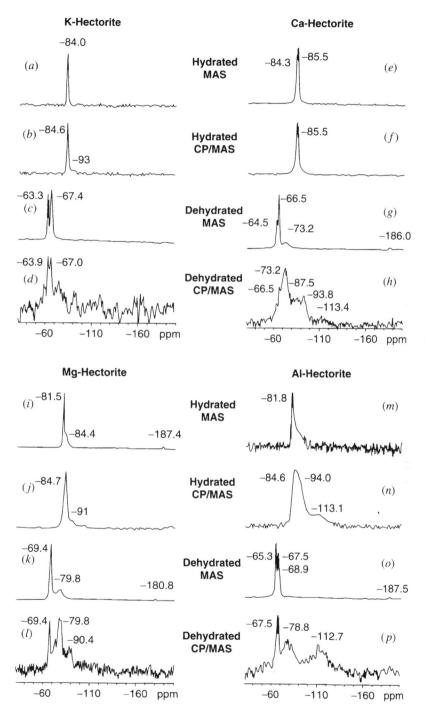

Figure 13.27 Nitrogen-15 NMR spectra of various cation bindings in pyridine–hectorite complexes (Ref. 8, p. 3170).

13.5 TECHNIQUES AND APPLICATIONS ON THE HORIZON

Recent advances in NMR probe technology, computer-assisted analyses and structure determination, and coupling of NMR to other analytical techniques are expanding NMR's utility for environmental analysis. Godejohan et al.[9] describe the use of high-performance liquid chromatography (HPLC) proton NMR in the analysis of contaminated groundwater samples. In combining HPLC with NMR, the study was able to detect and rapidly identify nitroaromatic compounds and other explosives in the microgram-per-liter range in groundwater samples obtained near a former munitions plant: "Mixture analysis by NMR (without chromatographic separation) may be used for a first overview of the contaminants present in samples from hazardous waste sites. However, if these mixtures are very complex, continuous flow HPLC-NMR should be used in a second step. This method allows the identification and quantitation of contaminants down to the microgram-per-liter range in a single run, ... even the determination of as-yet unknown compounds could be achieved without a second injection of the sample by structure elucidation of the desired pollutant" (Ref. 9, pp. 3836–3837).

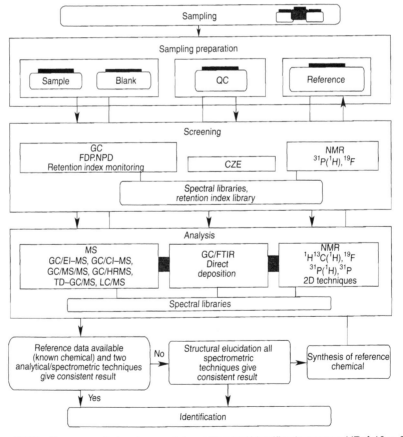

Figure 13.28 Recommended environmental sampling and identification protocol (Ref. 10, p. 953).

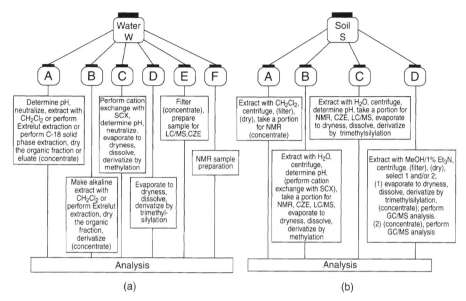

Figure 13.29 Recommended operating procedures for preparation of (*a*) aqueous samples and (*b*) soil samples (Ref. 10, p. 954).

Further combinations of NMR with GC, MS, and infrared (IR) spectrometry are also possible, arming the environmental analyst with an effective analytical arsenal. Used in combination with well-defined analytical protocols (Figs. 13.28 and 13.29), sample/contaminant identification, even on the smallest quantities, becomes achievable, accurate, and quantitative.

Combinatorial chemistry, a technology used in pharmaceutical development, is a technology waiting to be exploited in environmental chemistry. Combinatorial chemistry was designed to create new compounds en masse as well as test them for desirable properties.[11] Typically, syntheses occur on solid-phase substrates. Multitudes of minireactors are loaded, reacted under controlled variable conditions, and analyzed automatically using a variety of robotics, auto samplers, and computer-aided analyses. By adapting such technologies such that solid-phase substrates can be actual soils, reactions can be replaced by automated extractions, and temperatures, pressures, pH, and so on, accurately controlled, one can rapidly profile the environmental impact of any contaminant. Combining combinatorial techniques with automated analyses such as HPLC–NMR and/or HPLC—NMR–MS would enable environmental impact studies to be carried out in a laboratory setting, under multiple conditions, such that mechanisms of degradation, adsorption, contamination, and so on could be well understood.

13.6   SUMMARY

In environmental analysis one advantage in using NMR "is the examination of chemical and physical interactions between contaminants and the environmental matrix, especially for heterogeneous and complex matrices ... because NMR can be used as an in-situ and non-invasive probe" (Ref. 1, p. 313). Another advantage is that NMR "can specifically follow the chemistry occurring in complex environments and matrices" (Ref. 1, p. 313). Whether the sample is liquid or solid, organic or inorganic, free or adsorbed, extractable or bound, an NMR technique can usually be found to probe the physical/chemical nature of the sample. As analytical and computational technologies continue to develop, NMR's utility as a key environmental analytical technology will also continue to expand.

REFERENCES

1. M. A. Nanny, R. A. Minear, and J. A. Leenheer, *Nuclear Magnetic Resonance Spectroscopy in Environmental Chemistry*, Oxford University Press, New York, 1997.

2. J. McMurry, *Fundamentals of Organic Chemistry*, Wadsworth, Belmont, 1990.

3. R. M. Silverstein, G. C. Bassler, and T. C. Morrill, *Spectrometric Identification of Organic Compounds*, Wiley, New York, 1981.

4. P. J. Giammatteo, *Determination of the Primary, Secondary, and Tertiary Structures of Polysaccharides: An NMR Study*, Ph.D. Thesis, Wesleyan University, Middletown, CT, 1989.

5. C. A. Fyfe, *Solid State NMR for Chemists*. CFC Press, Ontario, 1983.

6. M. A. Wilson, *NMR Techniques and Applications in Geochemistry and Soil Chemistry*, Pergamon, Oxford, 1987.

7. R. K. Harris, *Nuclear Magnetic Resonance (NMR) Spectroscopy: A Physicochemical View*, Longman Scientific & Technical, Essex, 1986.

8. C. Albaret, D. Loeillet, P. Auge, and P. L. Fortier, Application of Two-Dimensional ^{1}H–^{31}P Inverse NMR Spectroscopy to the Detection of Trace Amounts of Organophosphorus Compounds Related to the Chemical Weapons Convention, *Anal. Chem.* **69**, 2694–2700 (1997).

9. M. Godejohann, A. Preiss, C. Mugge, and G. Wunsch, Application of On-Line HPLC-^{1}H NMR to Environmental Samples: Analysis of Groundwater Near Former Ammunition Plants. *Anal. Chem.* **69**, 3832–3837 (1997).

10. M. Mesilaakso, Chemical Weapons Convention, NMR Analysis, in R. A. Meyers (Ed.), *Environmental Analysis and Remediation: Wiley Encyclopedia Series in Environmental Science*, Wiley, New York, 1998, pp. 950–971.

11. S. Borman, Combinatorial Chemistry, *Chem. Eng. News*, April 6, 1998, pp. 47–67.

12. L. Ukrainczyk and K. A. Smith, Solid State ^{15}N NMR study of Pyridine Adsorption on Clay Minerals, *Environ. Sci. Technol.* **30**, 3167–3176 (1996).

14

SOLID-PHASE MICROEXTRACTION

Yong Chen and Janusz Pawliszyn

University of Waterloo
Waterloo, Ontario, Canada

Environmental Instrumentation and Analysis Handbook, by Randy D. Down and Jay H. Lehr
ISBN 0-471-46354-X   Copyright © 2005 John Wiley & Sons, Inc.

14.1   INTRODUCTION

The analytical procedure for complex samples consists of several steps, typically including sampling, sample preparation, separation, quantification, statistical evaluation, and decision making. Each step is critical for obtaining correct and informative results. The sampling step includes deciding where to get samples that properly define the object or problem being characterized and choosing a method to obtain samples in the right amounts. Sample preparation involves extraction procedures and can also include "cleanup" procedures for very complex "dirty" samples. This step must also bring the analytes to a suitable concentration level for detection, and therefore sample preparation methods typically include enrichment. The current state of the art in sample preparation techniques employs multistep procedures involving organic solvents. These characteristics make it difficult to develop a method that integrates sampling and sample preparation with separation methods for the purpose of automation. The result is that over 80% of analysis time is currently spent on sampling and sample preparation steps for complex samples.

   Solid-phase microextraction (SPME) was developed to address the need to for rapid sample preparation both in the laboratory and on-site where the investigated system is located. It presents many advantages over conventional analytical methods by combining sampling, sample preparation, and direct transfer of the analytes into a standard gas chromatograph. Since its commercial introduction in the early 1990s, SPME has been successfully applied to the sampling and analysis of environmental samples. This chapter presents the basic principle of SPME and its application for environmental studies.

14.2   EVOLUTION OF SOLID-PHASE MICROEXTRACTION TECHNOLOGY

The early work on laser desorption/fast gas chromatography (GC) resulted in rapid separation times, even for very high molecular mass species.[1] However, the preparation of samples for this experiment took hours, which was over an order of magnitude longer than the separation times. In this experiment, optical fibers were used to transmit laser light energy to the GC instrument. The sample preparation process was analogous to standard solvent extraction procedures. The fiber tip was coated with the sample by dipping one end of the optical fiber in the solvent extract, coating the fiber, and then removing volatile solvents through evaporation. The fiber tip, prepared in such a way, was inserted into the injector of a gas chromatograph and analytes were volatilized onto the front of the GC column by means of a laser pulse. During that work, a need was recognized for rapid sample preparation techniques to retain the time efficiency advantages made possible by the use of the laser pulse and a high-speed separation instrument. The challenge was addressed using fibers, since optical fibers could be purchased coated with several types of polymeric films. The original purpose of these coatings was simply to protect the fibers from breakage. Because of the thin films used (10–100 μm), the expected extraction times for these systems were very short. In addition, novel films could be prepared, since chromatographers have a good knowledge based on fused silica coating methods gained from capillary column manufacturing experience.

In the initial work on SPME, sections of fused silica optical fibers, both uncoated and coated with liquid and solid polymeric phases, were dipped into an aqueous sample containing test analytes and then placed in a GC injector.[2] The process of introducing and removing the fibers required the opening of the injector, which resulted in loss of head pressure at the column. Despite their basic nature, those early experiments provided very important preliminary data that confirmed the usefulness of this simple approach, since both polar and nonpolar chemical species were extracted rapidly and reproducibly from aqueous samples.

The development of the technique accelerated rapidly with the implementation of coated fibers incorporated into a microsyringe, resulting in the first SPME device.[3] Figure 14.1 shows an example of an SPME device based on the Hamilton 7000 series microsyringe. The metal rod, which serves as the piston in a microsyringe, is replaced with stainless steel microtubing having an inside diameter (i.d.) slightly larger than the outside diameter (o.d.) of the fused silica rod. Typically, the first 5 mm of the coating is removed from a 1.5-cm-long fiber, which is then inserted into the microtubing. High-temperature epoxy glue is used to permanently mount the fiber. Sample injection is then very much like standard syringe injection. Movement of the plunger allowed exposure of the fiber during extraction and desorption and its protection in the needle during storage and penetration of the septum. SPME devices do not need expensive syringes like the Hamilton syringes. As Figure 14.2 illustrates, a useful device can be built from a short piece of stainless steel microtubing (to hold the fiber), another piece of larger tubing (to work as a "needle"), and a septum (to seal the connection between the microtubing and the needle). The design from Figure 14.2 is the basic building block of a commercial SPME device described later and illustrated in Figure 14.3.

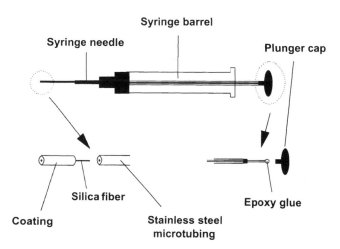

Figure 14.1   Custom-made SPME device based on Hamilton 7000 series syringe.

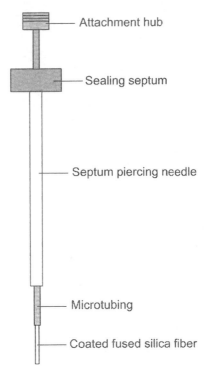

Figure 14.2 Simple versions of SPME device using coated fiber.

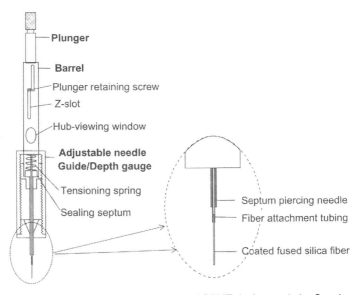

Figure 14.3 Design of first commercial SPME device made by Supelco.

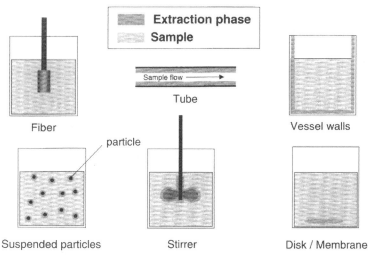

Figure 14.4   Configurations of solid-phase microextraction.

The extraction coating has also been coated on other elements of the analytical system. Recently there have been reports of coating the interior of vessels,[4] the exterior of magnetic stirring bars,[5] or even pieces of poly(dimethylsiloxane) (PDMS) tubes and thin membranes.[6] Several of these implementations are shown in Figure 14.4. The main reason for developing these alternative approaches is to enhance sensitivity by using a larger volume of the extraction phase (PDMS) and improving kinetics of the mass transfer between sample and PDMS by increasing the surface-to-volume ratio of the extraction phase. The main disadvantage of these approaches, however, is loss of convenience associated with syringe configuration, in particular in introduction to the analytical instrument. This step necessitates use of high-volume desorption devices and creates difficulties in automation of the extraction process as well as handling volatile compounds which are lost during transfer of the extraction phase from the sample to the injection system. Since high sensitivies are obtained for hydrophobic high-molecular-weight compounds with fiber SPME, the advantages of using larger volume phases are limited, especially for small sample volumes.[7]

Although to date SPME devices have been used principally in laboratory applications, more current research has been directed toward remote monitoring, particularly for clinical, field environmental, and industrial hygiene applications. In their operating principles, such devices are analogous to the devices described above, but modifications are made for greater convenience in given applications.

An important feature of a field device is the ability to preserve extracted analytes in the coating. The simplest practical way to accomplish this goal is to seal the end of the needle with a piece of septum. Additionally, cooling extends the storage time. Polymeric septum material, however, may cause losses of analytes from the fiber. Therefore, a more appropriate approach is to use metal-to-metal (or solid polymer)

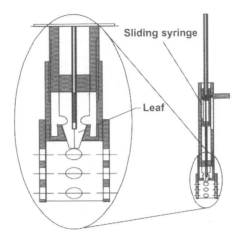

Figure 14.5 SPME device modified for field application.

seals. Figure 14.5 illustrates an example of a device construction based on a "leaf" closure. It is anticipated that future devices designed for field applications will be more rugged than the current laboratory versions and will look more like "sticks" or "pens" than syringes.

Recently, different configurations of field samplers were evaluated for the analysis of volatile compounds.[8] The PDMS fiber demonstrated the lowest ability to retain these compounds. Carboxen–PDMS had the highest and PDMS–DVB (divinylbenzene) was intermediate. The devices employed various sealing methods to preserve the samples on the fiber and to protect from external contamination. The Supelco field sampler seals by retracting the outer needle behind a silicone septum. Two prototype devices constructed were sealed either by a leaf system (Fig. 14.5), which opens automatically when the outer needle is exposed, or by capping the outer needle after sampling with a Teflon cap.

Another promising technique uses a valve syringe, where the fiber is withdrawn into the barrel of the syringe, along with a sample of the air being analyzed, by means of retracting the syringe plunger. The valve is then sealed until analysis. Initial data describing the new approaches clearly demonstrate advantages of the new designs to seal the fiber in the needle, compared to the commercial devices, by eliminating the losses and contamination to the septa material of volatile components.

14.3 PRINCIPLES OF SOLID-PHASE MICROEXTRACTION

14.3.1 Equilibrium Extraction

In SPME a small amount of extracting phase associated with a solid support is placed in contact with the sample matrix for a predetermined amount of time. If the time is long enough, a concentration equilibrium is established between the

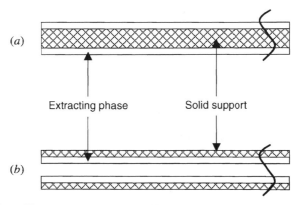

Figure 14.6 Two different implementations of SPME technique: (*a*) polymer coated on outer surface of fiber; (*b*) polymer coated on internal surface of capillary tube.

sample matrix and the extraction phase. When equilibrium conditions are reached, exposing the fiber for a longer time does not accumulate more analytes. There are two different implementations of the SPME technique explored extensively to date, which are shown in Figure 14.6. One is associated with a tube design and the other with fiber design. The tube design can use very similar arrangements as solid-phase extraction (SPE), however, the primary difference, in addition to volume of the extracting phase, is that the objective of SPME is never an exhaustive extraction but is to produce full breakthrough as soon as possible, since this indicates equilibrium extraction has been reached.

A more traditional approach to SPME involves coated fibers. The transport of analytes from the matrix into the coating begins as soon as the coated fiber has been placed in contact with the sample (Fig. 14.7). Typically, SPME extraction is considered to be complete when the analyte concentration has reached distribution

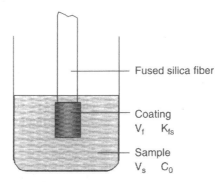

Figure 14.7 Microextraction with SPME: V_f, volume of fiber coating; K_{fs}, fiber/sample partition coefficient; V_s, volume of sample; C_0, initial concentration of analyte in sample.

equilibrium between the sample matrix and the fiber coating. The equilibrium conditions can be described as[9]

$$n = \frac{K_{fs} V_f V_s C_0}{K_{fs} V_f + V_s}$$

(14.1)

where n is the number of moles extracted by the coating, K_{fs} is a fiber-coating/sample matrix distribution constant, V_f is the fiber-coating volume, V_s is the sample volume, and C_0 is the initial concentration of a given analyte in the sample.

Strictly speaking, this discussion is limited to partitioning equilibrium involving liquid polymeric phases such as PDMS. The method of analysis for solid sorbent coatings is analogous for low analyte concentrations, since the total surface area available for adsorption is proportional to the coating volume if we assume constant porosity of the sorbent. For high analyte concentrations, saturation of the surface can occur, resulting in nonlinear isotherms. Similarly high concentration of a competitive interference compound can displace the target analyte from the surface of the sorbent.

Equation (14.1), which assumes that the sample matrix can be represented as a single homogeneous phase and that no headspace is present in the system, can be modified to account for the existence of other components in the matrix by considering the volumes of the individual phases and the appropriate distribution constants. The extraction can be interrupted and the fiber analyzed prior to equilibrium. To obtain reproducible data, however, constant convection conditions and careful timing of the extraction are necessary.

Simplicity and convenience of operation make SPME a superior alternative to more established techniques in a number of applications. In some cases, the technique facilitates unique investigations. Equation (14.1) indicates that, after equilibrium has been reached, there is a direct proportional relationship between sample concentration and the amount of analyte extracted. This is the basis for analyte quantification. The most visible advantages of SPME exist at the extremes of sample volumes. Because the setup is small and convenient, coated fibers can be used to extract analytes from very small samples. For example, SPME devices are used to probe for substances emitted by a single flower bulb during its life span; the use of submicrometer-diameter fibers permits the investigation of single cells. Since SPME does not extract target analytes exhaustively, its presence in a living system should not result in significant disturbance. In addition, the technique facilitates speciation in natural systems, since the presence of a minute fiber, which removes small amounts of analyte, is not likely to disturb chemical equilibrium in the system. It should be noted, however, that the fraction of analyte extracted increases as the ratio of coating to sample volume increases. Complete extraction can be achieved for small sample volumes when distribution constants are reasonably high. This observation can be used to advantage if exhaustive extraction is required. It is very difficult to work with small sample volumes using conventional sample preparation techniques. Also, SPME allows rapid extraction and transfer to an analytical instrument. These features result in an additional advantage when investigating intermediates in the system. For example, SPME was used to study biodegradation pathways of industrial contaminants.[10] The other advantage is that this technique can be used for studies of the distribution of analytes in a complex multiphase system[11] and speciates different forms of analytes in a sample.[12]

In addition, when sample volume is very large, Equation (14.1) can be simplified to

$$n = K_{fs}V_fC_0 \qquad (14.2)$$

which points to the usefulness of the technique for field applications. In this equation, the amount of extracted analyte is independent of the volume of the sample. In practice, there is no need to collect a defined sample prior to analysis as the fiber can be exposed directly to the ambient air, water, production stream, and so on. The amount of extracted analyte will correspond directly to its concentration in the matrix without being dependent on the sample volume. When the sampling step is eliminated, the whole analytical process can be accelerated, and errors associated with analyte losses through decomposition or adsorption on the sampling container walls will be prevented.

Three basic types of extractions can be performed using SPME: direct extraction, headspace configuration, and a membrane protection approach. Figure 14.8 illustrates the differences among these modes. In the direct-extraction mode (Fig. 14.8a), the coated fiber is inserted into the sample and the analytes are transported directly from the sample matrix to the extracting phase. To facilitate rapid extraction, some level of agitation is required to transport analytes from the bulk of the solution to the vicinity of the fiber. For gaseous samples, natural convection of air is sufficient to facilitate rapid equilibration. For aqueous matrices, more efficient agitation techniques such as fast sample flow, rapid fiber or vial movement, stirring, or sonication are required.[6,13] These conditions are necessary to reduce the effect caused by the "depletion zone" produced close to the fiber as a result of fluid shielding and small diffusion coefficients of analytes in liquid matrices.

In the headspace mode (Fig. 14.8b), the analytes need to be transported through the barrier of air before they can reach the coating. This modification serves

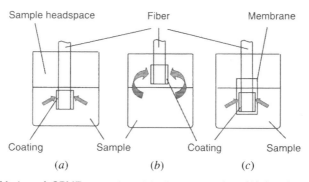

Figure 14.8 Modes of SPME operation: (a) direct extraction; (b) headspace SPME; (c) membrane-protected SPME.

primarily to protect the fiber coating from damage by high-molecular-mass and other nonvolatile interferences present in the sample matrix, such as humic materials or proteins. This headspace mode also allows modification of the matrix, such as a change of the pH, without damaging the fiber. Amounts of analyte extracted into the coating from the same vial at equilibrium using direct and headspace sampling are identical as long as sample and gaseous headspace volumes are the same. This is caused by the fact that the equilibrium concentration is independent of fiber location in the sample/headspace system. If the above condition is not satisfied, a significant sensitivity difference between the direct and headspace approaches exists only for very volatile analytes.

Figure 14.8c shows the principle of indirect SPME extraction through a membrane. The initial purpose of the membrane barrier was to protect the fiber against damage, similar to the use of headspace SPME when very dirty samples are analyzed. Membrane protection is advantageous for determination of analytes having volatilities too low for the headspace approach. In addition, a membrane made from appropriate material can add a certain degree of selectivity to the extraction process. However, the kinetics of membrane extraction is substantially slower than for direct extraction, because the analytes must diffuse through the membrane before they can reach the coating. The use of thin membranes and increased extraction temperatures will result in faster extraction times.[14] The thicker membranes can be used to slow down the mass transfer through the membrane, resulting in the time-weighted average (TWA) measurement for aqueous and gaseous samples.

The choice of sampling mode has a very significant impact on extraction kinetics, however. When the fiber coating is in the headspace, the analytes are removed from the headspace first, followed by indirect extraction from the matrix, as shown in Figure 14.8b. Overall mass transfer to the fiber is typically limited by mass transfer rates from the sample to the headspace. Therefore, volatile analytes are extracted faster than semivolatiles since they are at a higher concentration in the headspace, which contributes to faster mass transport rates through the headspace. Temperature has a significant effect on the kinetics of the process by determining the vapor pressure of analytes. In fact, the equilibration times for volatiles are shorter in the headspace SPME mode than for direct extraction under similar agitation conditions. This outcome is produced by two factors: A substantial portion of the analyte is in the headspace prior to extraction and diffusion coefficients in the gaseous phase are typically four orders of magnitude larger than in liquid media. Since concentrations of semivolatiles in the gaseous phase at room temperature are typically small, however, overall mass transfer rates are substantially lower and result in longer extraction times. They can be improved by using very efficient agitation or by increasing the extraction temperature.[15]

14.3.2 Diffusion-Based Calibration

There is a substantial difference between the performance of liquid and solid coatings. With liquid coatings, the analytes partition into the extraction phase, in which the molecules are solvated by the coating molecules. The diffusion coefficient in the

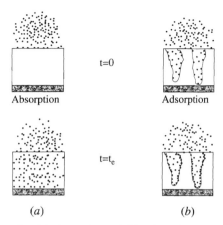

Figure 14.9 Extraction using (a) absorptive and (b) adsorptive extraction phases immediately after exposure of phase to sample ($t = 0$) and after completion of extraction ($t = t_e$).

liquid coating enables the molecules to penetrate the whole volume of the coating within a reasonable extraction time if the coating is thin (see Fig. 14.9a). With solid sorbents (Fig. 14.9b), the coating has a glassy or a well-defined crystalline structure which, if dense, substantially reduces diffusion coefficients within the structure. Within the time of the experiment, therefore, sorption occurs only on the porous surface of the coating. During exaction by use of a solid phase and high analyte/ interference concentration, after long extraction times, compounds with poor affinity toward the phase are frequently displaced by analytes characterized by stronger binding or those present in the sample at high concentrations. This is because only a limited surface area is available for adsorption. If this area is substantially occupied, competition occurs and the equilibrium amount extracted can vary with the concentrations of both the target and other analytes.[16] In extraction with liquid phases, on the other hand, partitioning between the sample matrix and extraction phase occurs. Under these conditions, equilibrium extraction amounts vary only if the bulk coating properties are modified by the extracted components; this occurs only when the amount extracted is a substantial portion (a few percent) of the extraction phase, resulting in a possible source of nonlinearity. This is rarely observed, because extraction/enrichment techniques are typically used for analysis of trace contaminants.

One way to overcome this fundamental limitation of porous coatings in a microextraction application is use of an extraction time much less than the equilibration time, so that the total amount of analytes accumulated by the porous coating is substantially below the saturation value. When such experiments are performed, it is critical to control not only extraction times precisely but also convection conditions, because they determine the thickness of the diffusion layer. One way of eliminating the need to compensate for differences in convection is to normalize (i.e., use consistent) agitation conditions. The short-term exposure measurement has an advantage in that the rate of extraction is defined by the diffusivity of analytes through the boundary layer of the sample matrix and, thus, the corresponding

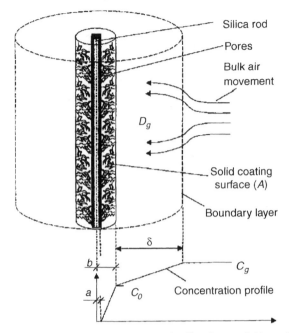

Figure 14.10 Schematic diagram of diffusion-based calibration model for cylindrical geometry. Terms are defined in text.

diffusion coefficients rather than by distribution constants. This situation is illustrated in Figure 14.10 for the cylindrical geometry of the extraction phase dispersed on the supporting rod.

The analyte concentration in the bulk of the matrix can be regarded as constant when a short sampling time is used and there is a constant supply of analyte as a result of convection. The volume of sample is much greater than the volume of the interface, and the extraction process does not affect the bulk sample concentration. In addition, adsorption binding is frequently instantaneous and essentially irreversible. The solid coating can be treated as a "perfect sink" for analytes. The analyte concentration on the coating surface is far from saturation and can be assumed to be negligible for short sampling times and the relatively low analyte concentrations in a typical sample. The analyte concentration profile can be assumed to be linear from C_g to C_0.

The function describing the mass of extracted analyte with sampling time can be derived[17] by the use of the equation

$$n(t) = \frac{B_3 A D_g}{\delta} \int_0^t C_g(t)\, dt \tag{14.3}$$

where n is the mass of analyte extracted (ng) in a sampling time t, D_g is the gas-phase molecular diffusion coefficient, A is the outer surface area of the sorbent, δ is the thickness of the boundary surrounding the extraction phase, B_3 is a geometric factor, and C_g is the analyte concentration in the bulk of the sample. It can be assumed that the analyte concentration is constant for very short sampling times and, therefore, Equation (14.3) can be further reduced to

$$n(t) = \frac{B_3 D_g A}{\delta} C_g t \qquad (14.4)$$

It can be seen from Equation (14.4) that the mass extracted is proportional to the sampling time, D_g for each analyte, and the bulk sample concentration and inversely proportional to δ. This is consistent with the fact that an analyte with a greater D_g will cross the interface and reach the surface of the coating more quickly. Values of D_g for each analyte can be found in the literature or estimated from physicochemical properties. This relationship enables quantitative analysis. As mentioned above, the discussion assumes nonreversible adsorption. Equation (14.4) can be modified to enable estimation of the concentration of analyte in the sample for rapid sampling with solid sorbents:

$$C_g = \frac{n\delta}{B_3 D_g A t} \qquad (14.5)$$

the amount of extracted analyte (n) can be estimated from the detector response.

The thickness of the boundary layer (δ) is a function of sampling conditions. The most important factors affecting δ are the geometric configuration of the extraction phase, sample velocity, temperature, and D_g for each analyte. The effective thickness of the boundary layer can be estimated for the coated fiber geometry by the use of Equation (14.6), an empirical equation adapted from heat transfer theory:

$$\delta = 9.52 \times \frac{b}{Re^{0.62} Sc^{0.38}} \qquad (14.6)$$

where the Reynolds number $Re = 2u_s b / v$, u_s is the linear sample velocity, v is the kinematic viscosity of the matrix, Schmidt number $Sc = v / D_s$, and b is the outside radius of the fiber coating. The effective thickness of the boundary layer in Equation (14.6) is a surrogate (or average) estimate and does not take into account changes of the thickness that can occur when the flow separates, a wave is formed, or both. Equation (14.6) indicates that the thickness of the boundary layer will decrease with increasing linear sample velocity. Similarly, when sampling temperature (T_s) increases, the kinematic viscosity decreases. Because the kinematic viscosity term is present in the numerator of Re and in the denominator of Sc, the overall effect on δ is small. Reduction of the boundary layer and an increased rate of mass transfer for analyte can be achieved in two ways—by increasing the sample velocity and by increasing the sample temperature. Increasing the temperature will, however, reduce the efficiency of the solid sorbent (reduce K). As a result, the sorbent coating might not be able to adsorb all molecules reaching its surface and it might, therefore, stop behaving as a "perfect sink" for all the analytes.

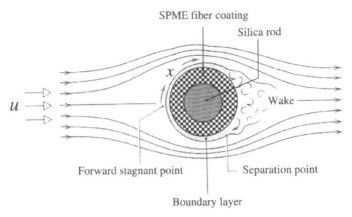

Figure 14.11   Schematic diagram of rapid extraction with SPME fiber in cross flow.

This on-site calibration method was validated for air sampling.[18,19] However, it was found that this method tended to underestimate the mass uptake of extracted analyte (*n*) for water sampling.[20] To address this problem, a new mass transfer model was proposed recently[21] (illustrated in Fig. 14.11). When a SPME fiber is exposed to a fluid sample whose motion is normal to the axis of the fiber, the fluid is brought to rest at the forward stagnation point from which the boundary layer develops with increasing *x* under the influence of a favorable pressure gradient. At the separation point, downstream movement is checked because fluid near the fiber surface lacks sufficient momentum to overcome the pressure gradient. In the meantime, the oncoming fluid also precludes flow back upstream. Boundary layer separation thus occurs, and a wake is formed in the downstream, where flow is highly irregular and can be characterized by vortex formation. Correspondingly, the thickness of the boundary layer (δ) is minimum at the forward stagnation point. It increases with the increase of *x* and reaches its maximum value right after the separation point. In the rear of the fiber where a wake is formed, δ decreases again.

Instead of calculating δ, the average mass transfer coefficient was used to correlate the mass transfer process. According to Hilpert,[22,23] the average Nusselt number $\overline{\text{Nu}}$ can be calculated by the equation

$$\overline{\text{Nu}} \equiv \frac{\bar{h}d}{D} = E\,\text{Re}^m\,\text{Sc}^{1/3} \tag{14.7}$$

where $\bar{h}$ is average mass transfer coefficient, *d* is the outside diameter of the fiber, *D* is the diffusion coefficient, Re is the Reynolds number, and Sc is the Schmidt number. Constants *E* and *m* are dependent on the Reynolds number and are available in the literature.[22,23]

Once $\bar{h}$ is known, the amount of extracted analytes *dn* during sampling period *dt* can be calculated by the equation

$$dn = \bar{h}A \int_0^t \left(C_{\text{bulk}} - C_{\text{sorbent}} \right) dt \tag{14.8}$$

where A is the surface area of the fiber, C_{bulk} is the bulk analyte concentration, and $C_{sorbent}$ is the analyte concentration at the interface of the fiber surface and samples of interest. If the sorbent is highly efficient toward target analytes and also is far away from equilibrium, $C_{sorbent}$ is assumed to be zero. Under constant bulk analyte concentration, integration of Equation (14.8) results in

$$n = \bar{h}AC_{bulk}t \qquad (14.9)$$

Inspection of Equation (14.9) shows that the product of $\bar{h}A$ has the units of cubic centimeters per second, which corresponds to the sampling rate used in active and passive sampling, while the product of $\bar{h}C_{bulk}$ is the mass flux ($ng/cm^2/s$) toward the fiber.

Rearrangement of Equation (14.9) results in

$$C_{bulk} = \frac{n}{\bar{h}At} \qquad (14.10)$$

Equation (14.10) indicates that the concentration of samples can be determined by the mass uptake n onto a SPME fiber during sampling period t. The new method was proven to be very accurate for both rapid air and water sampling.

14.3.3 Passive Time-Weighted Average Sampling

Consideration of different arrangements of the extraction phase is beneficial. For example, extension of the boundary layer by a protective shield that restricts convection would result in a TWA measurement of analyte concentration. A variety of diffusive samplers have been developed based on this principle. One system consists of an externally coated fiber with the extraction phase withdrawn into the needle (Fig. 14.12). When the extraction phase in an SPME device is not exposed directly to the sample but is contained within protective tubing (a needle) without any flow of sample through it, diffusive transfer of analytes occurs via the static sample (gas phase or other matrix) trapped in the needle. This geometric arrangement is a very simple method capable of generating a response proportional to the integral of the analyte concentration over time and space (when the needle is moved through space).[24] Under these conditions, the only mechanism of analyte transport to the extracting phase is diffusion through the matrix contained in the needle. During this process, a linear concentration profile is established in the tubing characterized by a surface area A and the distance Z between the needle opening and the position of the extracting phase. The amount of analyte extracted, dn, during time interval dt can be calculated by considering Fick's first law of diffusion:

$$dn = AD_m \frac{dc}{dZ} dt = AD_m \frac{\Delta C(t)}{Z} dt \qquad (14.11)$$

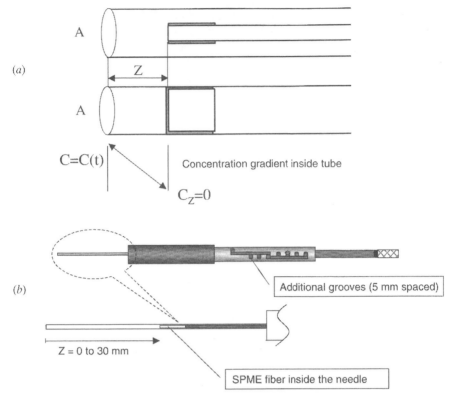

Figure 14.12 Use of SPME for in-needle time-weighted average sampling: (a) schematic; (b) adaptation of commercial SPME manual extraction holder.

where $\Delta C(t)/Z$ is an expression of the concentration gradient established in the needle between the needle opening and the position of the extracting phase; $\Delta C(t) = C(t) - C_Z$, where $C(t)$ is the time-dependent concentration of analyte in the sample in the vicinity of the needle opening and C_Z is the concentration of the analyte in the vicinity of the coating, C_z being close to zero for a high extraction phase/matrix distribution constant; then $\Delta C(t) = C(t)$. The concentration of analyte, C_z, at the coating position in the needle will increase with integration time, but it will remain low compared with the sample concentration, because of the presence of the extraction phase. The amount of analyte accumulated over time can therefore be calculated as

$$n = D_m \frac{A}{Z} \int C(t)\, dt \qquad (14.12)$$

As expected, the amount of analyte extracted is proportional to the integral of sample concentration over time, the diffusion coefficient of analyte in the matrix filling the needle, D_m, and the area of the needle opening, A, and inversely proportional to the distance Z of the coating from the needle opening. It should be emphasized that Equation (14.8) is valid only when the amount of analyte extracted onto the sorbent is a small fraction (below the RSD of the measurement, typically 5%) of the equilibrium amount for the lowest concentration in the sample. To extend integration times, the coating can be placed further into the needle (larger Z), the opening can be reduced by placing an additional orifice over the needle (smaller A), or a higher capacity sorbent can be used. The first two solutions will result in low measurement sensitivity. Increasing the sorbent capacity is a more attractive proposition. It can be achieved either by increasing the volume of the coating or by changing its affinity for the analyte. Because increasing the coating volume would require an increase in the size of the device, the optimum approach to increasing the integration time is to use sorbents characterized by large coating/gas distribution constants. If the matrix filling the needle is something other than the sample matrix, an appropriate diffusion coefficient should be used in Equations (14.11) and (14.12).

In the system described, the length of the diffusion channel can be adjusted to ensure that mass transfer in the narrow channel of the needle controls overall mass transfer to the extraction phase, irrespective of convection conditions.[25] This is a very desirable feature of TWA sampling, because the performance of this device is independent of the flow conditions in the system investigated. This is difficult to ensure for high-surface-area membrane permeation-based TWA devices, for example, passive diffusive badges[26] and semipermeable membrane devices.[27] For analytes characterized by moderate to high distribution constants, mass transport is controlled by diffusive transport in the boundary layer. The performance of these devices therefore depends on the convection conditions in the investigated system.[28]

14.3.4 Derivatization

The main challenge in organic analysis is polar compounds. These are difficult to extract from environmental and biological matrices and difficult to separate on the chromatographic column. Derivatization approaches are frequently used to address this challenge. Figure 14.13 summarizes various derivatization techniques that can be implemented in combination with SPME.[29] Some of the techniques, such as direct derivatization in the sample matrix, are analogous to well-established approaches used in solvent extraction. In the direct technique, the derivatizing agent is first added to the vial containing the sample. The derivatives are then extracted by SPME and introduced into the analytical instrument.

Because of the availability of polar coatings, the extraction efficiency for polar underivatized compounds is frequently sufficient to reach the sensitivity required. Occasionally, however, there are problems associated with the separation of these analytes. Good chromatographic performance and detection can be facilitated by in-coating derivatization following extraction. In addition, selective derivatization to analogues containing high detector response groups will result in enhancement in sensitivity and selectivity of detection. Derivatization in the GC injector is an analogous approach, but it is performed at high injection port temperatures.

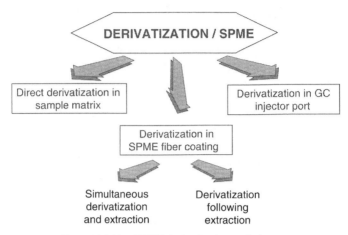

Figure 14.13 SPME derivatization techniques.

The most interesting and potentially very useful technique is simultaneous derivatization and extraction, performed directly in the coating. This approach allows high efficiencies and can be used in remote field applications. The simplest way to execute the process is to dope the fiber with a derivatization reagent and subsequently expose it to the sample. Then the analytes are extracted and simultaneously converted to analogues having high affinity for the coating. This is no longer an equilibrium process as derivatized analytes are collected in the coating as long as extraction continues. This approach, which is used for low-molecular-mass carboxylic acids, results in exhaustive extraction of gaseous samples.[30]

A similar approach is used for the analysis of formaldehyde from gaseous samples.[31] The derivatizing agent, o- (2,3,4,5,6-pentafluorobenzyl) hydroxylamine, is first doped onto the fiber by room temperature headspace extraction from an aqueous solution. Formaldehyde is subsequently extracted from an unknown sample and converted to the oxime derivative in the fiber. In this case the kinetics of the derivatizing reaction are fast, and uptake is controlled by mass transport. The process is similar to the mass transport controlled extraction of volatile organic compounds (VOCs) from air, where uptake rates for short sampling times are proportional to diffusion coefficients. The significant difference between the two approaches is that the formaldehyde analysis approach has a much higher capacity.

This simple but powerful procedure, as described above for this analysis, is limited to low-volatility reagents. The approach can be made more general by chemically attaching the reagent directly to the coating. The chemically bound product can then be released from the coating by, for example, high temperature in the injector or light illumination or change of an applied potential. The feasibility of this approach was recently demonstrated by synthesizing standards bonded to silica gel, which were released during heating. This approach allowed solvent-free calibration of the instrument.[32]

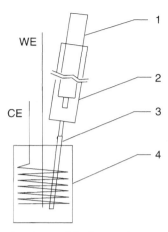

Figure 14.14 SPME electrochemistry minicell: 1, reference electrode; 2, 100-mL plastic syringe; 3, Teflon capillary; 4, 8-mL Teflon vial; 5, SPME fiber working electrode (WE); 6, platinum wire counter electrode (CE).

In addition to using a chemical reagent, electrons can be supplied to produce redox processes in the coating and convert analytes to more favorable derivatives. In this application, the rod as well as the polymeric film must have good electrical conductivity. Figure 14.14 shows the schematic of the three-electrode cell used to deposit mercury species onto a gold-coated metallic fiber. A similar principle has been used to extract amines onto a pencil "lead" electrode.[33] In addition, the SPME approach can be used to study the properties of electrochemical processes occurring on the electrode surface. The use of conductive polymers, such as polypyrole, will introduce additional selectivity of the electrochemical processes associated with coating properties.

14.3.5   Method Development

Developing a new SPME method in most cases requires the following steps:

1. Selection of fiber coating
2. Selection of derivatization reagent, if required
3. Selection of extraction mode
4. Selection of agitation method
5. Selection of separation and/or detection technique
6. Optimization of desorption conditions
7. Optimization of sample volume
8. Determination of extraction time profile in a pure matrix
9. Determination of extraction time
10. Calculation of distribution constant

11. Optimization of extraction conditions (pH, salt, temperature)

12. Determination of linear range for pure matrix at optimum extraction conditions

13. Selection of calibration method

14. Optimization of extraction conditions for heterogeneous samples

15. Verification of equilibration time, sensitivity, and linear dynamic range for complex samples

16. Determination of method precision

17. Determination of method detection limit

18. Validation of method

19. Automation of method

Most SPME methods developed to date are used in combination with GC separation and an appropriate detection method. In most cases, not all the steps have to be performed. Knowledge gained from previous experiments as well as from the literature can often be applied to the problem at hand. Some of the steps might involve additional experimentation, but overall benefits will be better understanding of the extraction process and better performance of the method. It should be emphasized that the optimization process has been evolving since the inception of the technique and may change in the future. Detailed discussion of each step can be found in the literature.[34]

14.4   SOLID-PHASE MICROEXTRACTION DEVICES AND INTERFACES TO ANALYTICAL INSTRUMENTATION

14.4.1   Commercial Devices

14.4.1.1   Fiber Assemblies. The first commercial version of the laboratory SPME device was introduced by Supelco in 1993 (see Fig. 14.3). The device is similar in operation principle to the custom-made device shown in Figures 14.1 and 14.2. An additional improvement is the adjustment for depth of the fiber with respect to the end of the needle, which allows control of the exposure depth in the injector and extraction vessel. The device incorporates such useful features as color marking of the fiber assemblies to distinguish among various coating types. In addition to standard PDMS coatings of various thicknesses and polyacrylate (PA), Supelco developed new mixed phases based on solid/liquid sorption, such as Carbowax–DVB, PDMS–DVB, and Carboxen–PDMS.

14.4.1.2   Autosamplers for GC. Varian has developed an SPME autosampler based on their 8000 GC autosampler system, taking advantage of the fact that the SPME device is analogous to a syringe in its operation and that after desorption the coating is cleaned and ready for reuse.[35] The major challenge is to incorporate agitation and temperature control as well as other enhancements, such as fiber internal cooling or dedicated injectors. One improvement is an SPME system that incorporates an agitation mechanism consisting of a small motor and a cam to vibrate the needle. The fiber in this design works as a stirrer. The vibration causes the vial to shake and the fiber to move with respect to the solution; the result is a substantial decrease of equilibration times compared to a static system. This mode of agitation simplifies fiber handling because it does not require the introduction of foreign objects into the sample prior to extraction.

CTC Analytics (Switzerland) incorporated SPME sampling on their CombiPal autosampler. This is a robotic system with a great deal of flexibility for programming SPME analyses. Samples are loaded onto trays accommodating five vial sizes and are heated and agitated during extraction using a separate sample preparation chamber. To facilitate agitation, the extraction chamber is rotated at a programmable rotation speed during extraction. Additional vials/stations are present to accommodate wash solutions, derivatizing agents, temperature control, derivatization, and fiber conditioning to facilitate operation of SPME at optimal conditions. While the built-in software can be used to perform basic SPME analyses, extra programming flexibility is provided by the Cycle Composer software, available as an accessory. This new instrument with its greatly enhanced flexibility will significantly expand the range of SPME methods amenable to automation. New coatings and devices are expected to follow, as interest in SPME grows along with the unprecedented numbers of new applications appearing in the literature.

14.4.1.3 *Field Portable SPME-FastGC.* SPME coupled to high-speed GC is a good combination to perform rapid and cost-effective investigations in the field, even of complex organic samples. As discussed above, SPME is particularly suited for fast GC, as it is solvent free, and the thin coatings can provide very fast desorption of analytes at high temperatures. Some instrumental modifications were performed recently in order to achieve successful fast separations.[36]

A portable system was optimized for SPME-fast GC field investigations and was commercialized by SRI Instruments (model 8610C, SRI Instruments, Torrance, CA). The instrument was tested in combination with a flame ionization detector (FID), a photoionization detector (PID), and a dry electrolytic conductivity detector (DELCD). A dedicated injector, presented in Figure 14.15, was mounted on the portable system in order to use SPME for high-speed separation. The injector guarantees very fast thermal desorption of the analytes from the SPME fiber.[37]

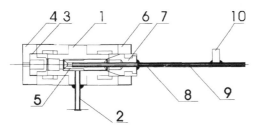

Figure 14.15 Design of dedicated injector: 1, modified Swagelok fitting; 2, stainless steel tubing; 3, molded septum; 4, nut; 5, needle guide; 6, nut; 7, blind ferrule; 8, stainless steel tubing; 9, 0.53-mm-i.d. fused silica capillary; 10, contact.

The injector for high-speed GC should produce as narrow an injection band as possible. Internal volumes of regular injector ports are too large (e.g., split/splitless injector), since they have been designed to accommodate large volumes of gaseous samples or vapors produced by solvent injection. Thermal focusing for separation improvement is not convenient for fast separations, since column temperature programming is impractical for high-speed GC. Hence, an injector port with a small internal volume was required for this application. Also, very fast thermal desorption from the SPME fiber was required to produce a narrow injection band and achieve effective separation. In the dedicated injector for SPME-fast GC, the injector port was maintained cold during needle introduction and was rapidly heated only when the fiber was exposed to the carrier gas stream. The desorption area of the injector was heated by capacitive discharge that allowed heating rates as fast as 4000°C/s, and very narrow injection bands were observed, as required by fast GC.

14.4.1.4 Fiber Conditioners. SRI also has a dedicated fiber conditioner available for its systems, and CTC Analytics has introduced a fiber cleaning station for its SPME autosampler. New SPME fibers require initial conditioning at manufacturer-recommended temperatures ranging from 210 to 320°C for time periods ranging from 0.5 to 4 h. Conditioning is also recommended at the beginning of the workday and between runs when analyte carryover is a possibility. In addition, many SPME-fast GC applications, such as field sampling, require additional fiber cleaning in order to reduce desorption time in the GC injector and eliminate carryover of nontarget analytes.[38]

Typically, fiber conditioning is performed by desorption in a temperature-controlled GC injector for the recommended conditioning time. However, this procedure reduces available GC time by occupying an injector and in many cases prevents analytical work from being performed on the instrument. It may also load a GC column with unwanted products of desorption, which in turn may require additional column conditioning and column "blank" determinations.

To address the aforementioned concerns, a stand-alone SPME fiber conditioner was designed, built, and tested.[39] This new device allows for fast conditioning of SPME fibers using high temperature and gas flow for the desorption and subsequent purging of fiber contaminants. This device, intended for laboratory and field sampling applications, was based on a modified commercial syringe cleaner. The performance of the new fiber conditioner was tested for several types of commercially available SPME fibers and compared with the traditional GC injector fiber conditioning method. The new device performed equal to or better than GC injectors for both new fiber conditioning and the desorption of *n*-alkanes, representing a wide range of boiling points, in addition to being significantly less expensive.

14.4.2 Interfaces to Analytical Instrumentation

Because of its solvent-free nature and small fiber or capillary size, SPME can be interfaced conveniently to analytical instruments of various types. Only extracted analytes are introduced into the instrument, since the extracting phase is nonvolatile.

Thus there is no need for complex injectors designed to deal with large amounts of solvent vapor, and these components can be simplified for use with SPME. Although in most cases the entire complement of analytes is not extracted from the sample, all material that is extracted is transferred to the analytical instrument, resulting in good performance. Also, the solvent-free process results in narrow bands reaching the instrument, giving taller, narrower peaks, and better quantification.

14.4.2.1   SPME–GC Interface.   The analytical instrument used most frequently with SPME has been the gas chromatograph. Standard GC injectors, such as split/ splitless, can be applied to SPME as long as a narrow insert with an inside diameter close to the outside diameter of the needle is used. The narrow inserts are required to increase the linear flow around the fiber, resulting in efficient removal of desorbed analytes. The split should be turned off during SPME injection. Under these conditions, the desorption of analytes from the fiber is very rapid, not only because the coatings are thin but also because the high injector temperatures produce a dramatic decrease in the coating/gas distribution constant and an increase in the diffusion coefficients. The speed of desorption in many cases is limited by the time required to introduce the fiber into the heated zone.

One way to facilitate sharper injection zones and faster separation times is to use rapid-injection autosampling devices. An alternative solution is to use a dedicated injector, which should be cold during needle introduction but which heats up very rapidly after exposure of the fiber to the carrier gas stream. A schematic of such an injector is presented in Figure 14.16. During desorption, the fiber is located inside

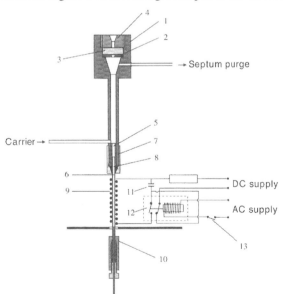

Figure 14.16   Schematic of flash SPME injector: 1, injector body; 2, washer; 3, septum; 4, nut; 5, needle guide; 6, 0.53-mm-i.d. fused silica capillary; 7, nut; 8, ferrule; 9, heater; 10, butt connector; 11, relay; 12, capacitor; 13, switch.

the heated part of the fused silica capillary, its end being close to the bottom of the heated zone. The distance between the fiber and the capillary wall is approximately 0.15 mm. A close match between the inner diameter of the capillary and the outer diameter of the fiber assures effective heat transfer from the heater to the fiber and a high linear flow rate of the carrier gas along the fiber. The injector is rapidly heated via a capacitive discharge. Heating rates of 1000°C/s have been determined experimentally.[36]

The fiber can also be designed to contain the heating element, as shown in Figure 14.17. In this case, no injector is necessary. The modified fiber can be introduced directly into the front of the column, and analytes can be desorbed rapidly by heating with a capacitive discharge current after the fiber has been exposed from the needle.

Flash desorption injectors can be alternatively designed by passing a current directly through the fiber. This is possible if the rod is made of conductive material, as in electrochemical SPME devices. Figure 14.18 illustrates such an interface.[31] when the electrical connection is made at the bottom of the interface, the fiber is rapidly heated by the discharging current. The other option is to use laser energy to desorb analytes from the surface of the fused silica optical fiber, as discussed previously.

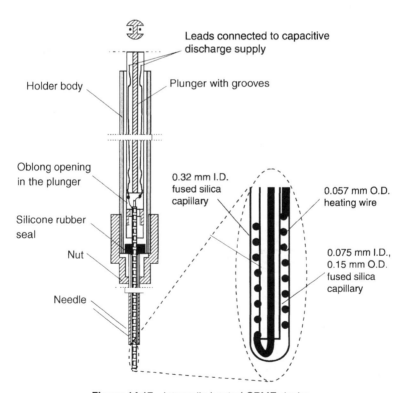

Figure 14.17   Internally heated SPME device.

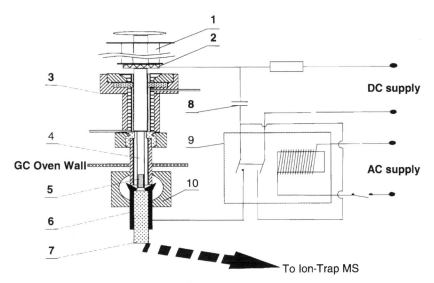

Figure 14.18 Direct capacitive discharge desorption system: 1, SPME syringe; 2, electric connection I; 3, injector body; 4, steel wire; 5, gold coating; 6, electric connection II; 7, transfer line; 8, capacitor; 9, relay; 10, butt connector.

Flash desorption injectors can be applied to directly interface SPME to a range of detection devices such as mass spectrometers and atomic emission devices. The sharp bands obtained during the desorption process result in very sensitive detection. To facilitate proper quantification, the extraction needs to be very clean, which puts an additional demand on the coating selectivity. Some help in proper quantification can be obtained if the apparatus for tandem mass spectrometry (MS–MS) is available.

14.4.2.2 Other Interfaces. Smaller injection volumes for applications of SPME to micro-HPLC (high-performance liquid chromatography) and capillary electrophoresis can be accomplished by modifying microinjector designs. The sliding injector developed for capillary isotachophoresis is one example.[40] The other approach is to design an appropriate sample introduction system based on guides to introduce the fiber with the extracted components directly into the capillary (see Fig. 14.19). Using this method, very efficient separation was achieved.[41]

The SPME method can be directly combined to optical detection based on reflectometric interference spectrometry.[42–44] A light beam passing through an optically transparent fiber coated with transparent sorbing material interacts with absorbed substances through internal reflection. Therefore, if any of the extracted analytes strongly absorb the transmitted light, there is a loss in intensity that can be detected with a simple optical sensor. These devices demonstrate poor sensitivity primarily because it is difficult to find light wavelengths that are specifically adsorbed by the analytes and not by the coating or interferences. In an alternative design, the light can be passed directly through the absorbing polymer that is then

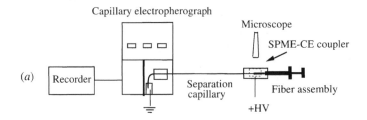

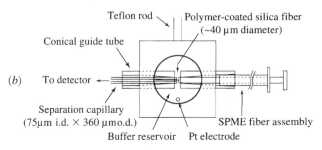

Figure 14.19 Schematic of (*a*) SPME CE system and (*b*) interface.

cooled to facilitate high sensitivity of determination.[45] Fluorescence can be used to detect analytes in the coating. The selectivity of the extraction process and spectroscopy can be combined with selectivity of the electrochemical process, resulting in a spectroelectrochemical sensor.[46]

14.5 APPLICATIONS OF SOLID-PHASE MICROEXTRACTION

This section discusses the more fundamental concepts related to the application of SPME for analysis of real matrices. The applications of the technique are covered in more detail in the recently published book *Applications of Solid Phase Microextraction*.[47] An updated list of SPME publications is posted in a database at the University of Waterloo website (www.spme.uwaterloo.ca). A basic understanding is emerging regarding approaches which can be taken to the basic matrix types: gaseous, liquid, and solid. General approaches to facilitate successful extraction and quantification in these systems are emphasized. The objective of this section is not to demonstrate a comprehensive review but rather to discuss fundamental principles facilitating successful quantification.

14.5.1 Gaseous Matrices

All SPME methods developed to date for gas analysis are directed toward air analysis. However, the approaches described are also suitable for the analysis of other gas mixtures.

Model studies on extraction from air have been performed by two methods: static and (the more realistic) dynamic. In the static method, the target compounds are introduced to a glass bulb equipped with a septum. After the analytes vaporize completely, SPME fiber is introduced to the bulb through a septum. In practical ambient air measurements, the system is not static, since convection is always present. It is more appropriate, therefore, to use dynamic gas chambers for the modeling studies.[48] Equilibration times for extraction of trace contaminants from moving air are short, as predicted from the theory. Increased air flow decreases the equilibration times for less volatile analytes.

The PDMS coating is generally not affected by major air components. A small decrease in the response for organics was observed only for samples of relative humidity close to 100%. The primary experimental parameter which controls the response is temperature. However, the effect of temperature on the distribution constant can be easily predicted, since $\log K_{fg}$ is linearly related to $1/T$ and the heat of vaporization of the pure solute. Figure 14.20 illustrates the above relationship graphically for a range of compounds varying in volatility. The linear relationship is clearly illustrated. In addition, the heat of vaporization and activity coefficient are related to the retention time of a compound on the GC column using the same coating material as the SPME fiber. Thus, the appropriate distribution constants can be determined directly from the retention indices of the target analytes. For PDMS coating, the following relationship between the gas/coating distribution constant and the linear temperature programmed retention index (LTPRI) of a compound has been found[49]:

$$\log_{10}K_{fg} = 0.00415 \times \text{LPTRI} - 0.188 \qquad (14.13)$$

$$\log K = a\left(\frac{1}{T}\right) + b$$

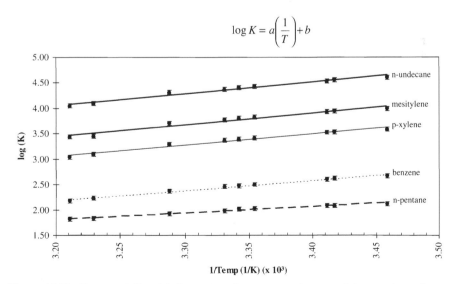

Figure 14.20 Representative plots for pentane, benzene, *p*-xylene, mesitylene, and *n*-undecane demostrating linear relationship between $\log K_{fg}$ and $1/T$.

Thus, the LTPRI system permits interpolation of the K_{fg} values from the plot of log K_{fg} versus retention index. The differences between the values of distribution constants calculated from the retention indices and determined directly were generally found to be within experimental error. This approach can be extended to other coatings as long as columns with appropriate coatings are available. In addition, using this approach, it is possible to determine the distribution constants of unidentified compounds; therefore it is possible to use SPME for the determination of parameters such as total petroleum hydrocarbons (TPHs) in air. A comparison of SPME and PDMS with the standard charcoal tube technique obtained for an air sample drawn near a gas station resulted in 262 and 247 µg/L, respectively.[5] The values show a very good agreement indicating that the approach described above can be used to quantify analytes in gaseous samples without the need for identification. It should be pointed out, however, that this approach could be used only with detectors whose response is independent of the nature of the analyte (e.g., FID for the analysis of hydrocarbons).

The SPME fibers can also be used at high temperatures. Particulate matter can be collected at the same time and characterized using spectroscopic means.[50] Alternatively, the particulate collection device can also be constructed as a packed needle (see Fig. 14.21). The needle can then be introduced to the analytical instrument for desorption and analysis.

The solid-phase microextractor can also be used as an integrated sampling device. As mentioned previously, the simplest way to accomplish this task is to retract the fiber into the needle. An alternative approach is to use a chemical reaction in the coating. This approach is particularly useful for integrated sampling of very low analyte concentrations in air. An important application of SPME coupled to an on-fiber derivatization step was demonstrated for the analysis of aldehydes in air.[51]

Frequently, after the extraction, the fibers with analytes in the coating cannot be analyzed immediately. This is particularly true in field applications, where an appropriate portable instrument might not be available. In such situations, the fibers have to be stored in such a way as to ensure that analyte losses are minimal. This can be simply accomplished by retracting the fiber into the needle and sealing its open end with a piece of a septum. Cooling the needle provides additional protection against losses. For PDMS fibers, even with such precautions, losses of the most volatile analytes might occur when the fiber is stored for more than a few hours.

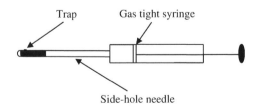

Trap   Gas tight syringe

Side-hole needle

Figure 14.21 Alternative embodiment of SPME: in-needle trap.

When properly protected (retracted fiber, needle sealed with a septum or other sealing approach,[52] low temperature), fibers coated with PDMS/DVB can be safely stored for much longer periods of time.

14.5.2 Liquid Matrices

Most methods for liquid matrices reported in the literature were developed for aqueous samples. The fiber coating/water distribution constant can be calculated from the equation

$$K_{fw} = K_{fg}K_{aw} \qquad (14.14)$$

where K_{fg} can be determined from chromatographic data as discussed above and K_{aw} is the air–water distribution constant for a given analyte which can be found in the tables of Henry's constant values. For example, the equation for a PDMS coating and aqueous matrix can be calculated from the equation

$$\log_{10}K_{fw} = 0.00415 \times \text{LPTRI} - 0.188 + \log_{10}K_{aw} \qquad (14.15)$$

This equation is obtained after substituting the expression for K_{fg} given by Equation (14.14) into Equation (14.13). The K_{fw} values found using this equation agree very well with experimental data considering that typically the errors in Henry's constant determination are larger than 10%. In addition, Henry's constants are similar for closely related compounds, as illustrated in Figure 14.22. As can be

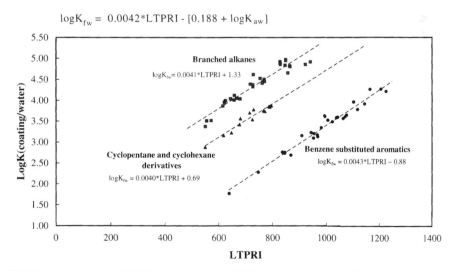

Figure 14.22 Log K_{fw} (PDMS) as function of LTPRI for isoparaffins, substituted benzenes, and cycloalkanes at 25°C.

predicted from Equation (14.15), the slopes of the lines are similar, but their intercepts vary due to differences in the average values of Henry's constant for the different classes of compounds. This relationship enables quantification of unknown analytes in water provided that the class to which a compound belongs is known.

Several reports indicated the existence of a linear relationship between the coating/water distribution constant and the octanol–water distribution constant, K_{ow}.[53] Considering the discussion above, such relationships are expected to exist only for individual classes of compounds, such as aromatic hydrocarbons. Because the activity coefficients of various classes of analytes in octanol are expected to be different than in PDMS or other fiber coatings, the relationship between the two extracting phases will vary with the change of chemical properties of the analytes. It is possible, however, to predict the general trends in K_{fw} within a group of related compounds by using the corresponding values of K_{ow}.

The above discussion pertains to pure water as the sample matrix. The presence of other components in water might affect the distribution constants of the analytes. Also, liquid chromatographic experience gives some clues about the trends in the distribution constant change with modification of the matrix. For example, addition of salt would generally result in an increase in the distribution constant for neutral organics, but the change is expected to be noticeable only if the concentration of the salt exceeds 1%. Also, the presence of water-miscible polar organic solvents may change the properties of the matrix by reducing its polarity. In addition, swelling of the polymer with the solvent might occur for polar coatings, resulting in a change of the coating volume and possibly also of K_{fs}. However, the change is not expected to be substantial when the concentration of the solvent is below 1%.[54] When samples contain higher amounts of salt and/or dissolved organics but are well defined so that pure matrix can be prepared, external calibration using standards in the matrix may still be appropriate. Otherwise, standard addition (preferentially of isotopically labeled analytes) should be used to compensate for variations in the matrix composition.

A major analytical challenge is always associated with the analysis of samples containing solids, such as sludge. Several approaches can be implemented with SPME. Sometimes modification of the extraction conditions, such as temperature, pH, salt, and other additives, facilitates the displacement of analytes into the aqueous phase or the headspace, resulting in distribution constants similar to those for pure water. In certain cases, direct extraction is not possible because of a very dirty matrix or pH conditions which damage the fiber (above pH 12 for the PDMS coating). In such situations, the headspace mode is more suitable for many applications. Even semivolatiles can be analyzed by this method as long as the extraction temperature is sufficiently high and good agitation conditions are provided.

14.5.3 Solid Matrices

Accurate quantification of target analytes in solids represents a very significant challenge to the analytical community. Although SPME cannot be used directly to extract analytes from solids, several approaches can be taken to facilitate simple sample preparation. For volatiles, the typical approach is to perform headspace analysis. To quantitatively release the analytes from the matrix, the temperature needs to be increased. This facilitates extraction and improves the kinetics of the process, but it also decreases the distribution constant. Typically, a maximum is observed for the relationship between the amount of analyte extracted by the fiber and the temperature. Initially, the loss of sensitivity due to the decreased distribution ratio is more than compensated for by the increased concentration of the analyte in the headspace. At even higher temperatures, the loss of sensitivity becomes predominant and the amount of analyte extracted by the fiber decreases. The inherent loss of sensitivity associated with the decrease of the distribution constant at high temperatures can be compensated for by cooling the fiber. However, this approach has not yet been commercialized. Alternatively, a matrix modifier can be added to the solid sample. Water has been found to be very effective in displacing the analytes from the solid surfaces for many types of samples, especially at elevated temperatures.

Another successful approach to SPME analysis of solids involves the use of water or a polar organic solvent, such as methanol, for the extraction of the analytes from the solid matrix prior to SPME analysis. When a polar solvent is used, the extract is spiked into pure water. Low-temperature water extraction followed by SPME was found to be a very useful approach for polar compounds, such as herbicides.[55] Application of methanol with water spiking, on the other hand, has been found to be useful for the analysis of volatile hydrocarbons.[56] An interesting modification to the above procedure involves volatizing the extract followed by fiber extraction from the gaseous phase.[57] Quantification of analytes in the gas phase is easier, as discussed above.

For less volatile analytes, the methanol approach described above can give good results. In addition, high-pressure hot-water extraction is a very suitable solvent-free alternative.[58] Both static and dynamic hot-water extraction have their advantages. The static method is very simple and inexpensive since it does not use high-pressure pumps. The dynamic approach, on the other hand, provides an extract which is much cleaner than the original matrix. However, it might be possible to extract many semivolatile target analytes from very dirty matrices with the high-temperature static system when using the headspace mode of SPME.

LIST OF SYMBOLS

b	Fiber-coating outer radius
C_0	Initial concentration of analyte in sample
D_g	Diffusion coefficient of analyte in gas
D_m	Diffusion coefficient of analytes in sample matrix
DELCD	Dry electrolytic conductivity detector
DVB	Divinylbenzene
FID	Flame ionization detector, commonly used in a gas chromatograph

GC	Gas chromatography
HPLC	High-performance liquid chromatography
i.d.	Inside diameter
K_{aw}	Air/water distribution constant
K_{fg}	Fiber/gas distribution constant
K_{fs}	Fiber/sample matrix distribution constant
K_{fw}	Fiber/water distribution constant
K_{ow}	Octanol/water distribution constant
L	Fiber-coating length
LTPRI	Linear temperature programmed retention index
MS	Mass spectrometry
n	Amount of analyte extracted onto coating
o.d.	Outside diameter
PA	Poly(acrylate)
PAH	Polynuclear aromatic hydrocarbon
PDMS	Poly(dimethylsiloxane)
SPE	Solid-phase extraction
SPME	Solid-phase microextraction
t_e	Equilibration time
TWA	Time-weighted average
V_f	Volume of fiber coating
VOC	Volatile organic compounds
δ	Boundary layer thickness

REFERENCES

1. J. Pawliszyn and S. Liu, *Anal. Chem.* **59**, 1475 (1987).

2. R. G. Belardi and J. Pawliszyn, *Water Pollut. Res. J. Can.* **24**, 179 (1989).

3. C. L. Arthur and J. Pawliszyn, *Anal. Chem.* **62**, 2145 (1990).

4. M. A. Nickerson, Sample Screening and Preparation Within a Collection Vessel. U.S. Patent No. 5,827,944 (1998).

5. E. Baltussen, P. Sandra, F. David, H-G. Janssen, and C. Cramers, *Anal. Chem.* **71**, 5213–5216 (1999).

6. I. Bruheim, X. Liu, and J. Pawliszyn, *Anal. Chem.* **75**, 1002–1010 (2003).

7. P. Popp and M. Moeder, in *ExTech'2001*, Barcelona, Spain, September 2001.

8. L. Müller, Field Analysis by SPME, in J. Pawliszyn (Ed.), *Applications of Solid Phase Microextraction*, RSC Chromatography Monographs, Royal Society of Chemistry, Cambridge, 1999.

9. D. Louch, S. Motlagh, and J. Pawliszyn, *Anal. Chem.* **64**, 1187 (1992).

10. J. Al-Hawari, in *78th Canadian Society for Chemistry Conference and Exhibition*, Memorial University of Newfoundland, St. Johns, NF, 1996.

11. J. Poerschmann, F-D Kopinke, and J. Pawliszyn, *Environ. Sci. Technol.* **31**, 3629 (1997).

12. Z. Mester and J. Pawliszyn, *Rapid Commun. Mass Spectrom.* **13**, 1999 (1999).

13. S. Motlagh and J. Pawliszyn, *Anal. Chim. Acta* **284**, 265 (1993).

14. Z. Zhang, J. Poerschmann, and J. Pawliszyn, *Anal. Commun.* **33**, 219 (1996).

15. Z. Zhang and J. Pawliszyn, *Anal. Chem.* **65**, 1843 (1993).

16. D. Ruthven, *Principles of Absorption and Adsorption Processes*, Wiley, New York, 1984.

17. H. Carslaw and J. Jaeger, *Conduction of Heat in Solid*, Clarendon, Oxford, 1986.

18. J. Koziel, M. Jia, and J. Pawliszyn, *Anal. Chem.* **72**, 5178–5186 (2000).

19. F. Augusto, J. Koziel, and J. Pawliszyn, *Anal. Chem.* **73**, 481–486 (2001).

20. K. Sukola, J. Koziel, F. Augusto, and J. Pawliszyn, Diffusion-Based Calibration for Fast SPME Analysis of VOCs in Aqueous Samples with Carboxen/PDMS Coating, unpublished work (University of Waterloo, Canada, 2002).

21. Y. Chen, J. Koziel, and J. Pawliszyn, *Anal. Chem.* **75**, 6485–6493 (2003).

22. R. Hilpert, *Forsch. Geb. Ingenieurwes.* **4**, 215 (1933).

23. J. G. Knudsen and D. L. Katz, *Fluid Dynamics and Heat Transfer*, Mcgraw-Hill, New York, 1958.

24. M. Chai and J. Pawliszyn, *Environ. Sci. Technol.* **29**, 693–701 (1995).

25. Y. Chen and J. Pawliszyn, *Anal. Chem.* **75**, 2004 (2003).

26. J. Koziel, Sampling and Sample Preparation for Indoor Air Analysis, in J. Pawliszyn (Ed.), *Sampling and Sampling Preparation for Field and Laboratory*, Elsevier, Amsterdam, 2002.

27. J. Petty, C. Orazion, J. Huckins, R. Gale, J. Lebo, K. Echols, and W. Cranor, *J. Chromatogr. A* **879**, 83 (2000).

28. B. Vrana and G. Schuurmann, *Environ. Sci. Technol.* **36**, 290 (2002).

29. L. Pan and J. Pawliszyn, *Anal. Chem.* **69**, 196 (1997).

30. L. Pan and J. Pawliszyn, *Anal. Chem.* **67**, 4396 (1995).

31. P. Martos and J. Pawliszyn, *Anal. Chem.* **70**, 2311 (1998).

32. P. Konieczka, L. Wolska, E. Luboch, J. Namiesnik, A. Przyjazny, and J. Biernat, *J. Chromatogr. A* **742**, 175 (1996).

33. E. D. Conte and D. W. Miller, *J. High Resolut. Chromatogr.* **19**, 294 (1996).

34. J. Pawliszyn, *Solid-Phase Microextraction*, Wiley-VCH, New York, 1997.

35. C. Arthur, L. Killam, K. Buchholz, J. Berg, and J. Pawliszyn, *Anal. Chem.* **64**, 1960 (1992).

36. T. Górecki and J. Pawliszyn, *J. High Resolut. Chromatogr.* **18**, 161 (1995).

37. T. Górecki and J. Pawliszyn, *Field Analyt. Chem. Technol.* **1**(5), 277–284 (1997).

38. J. A. Koziel, M. Jia, A. Khaled, J. Noah, and J. Pawliszyn, *Anal. Chim. Acta* **400**, 153 (1999).

39. J. Koziel, B. Shurmer, and J. Pawliszyn, *J. High Resol. Chromatogr.* **23**, 343 (2000).

40. T. McDonnell and J. Pawliszyn, *Anal. Chem.* **63**, 1884 (1991).

41. C-W. Whang and J. Pawliszyn, *Anal. Commun.* **35**, 353 (1998).

42. B. L. Wittkamp and D. C. Tilotta, *Anal. Chem.* **67**, 600 (1995).

43. G. L. Klunder and R. E. Russo, *Anal. Chem.* **49**, 379 (1995).

44. H. M. Yan, G. Kraus, and G. Gauglitz, *Anal. Chim. Acta* **312**, 1 (1995).

45. J. Pawliszyn, Device and Process for Increasing Analyte Concentration in a Sorbent, U.S. Patent No. 5,496,741.

46. W. R. Heineman, Personal communication, University of Cincinnati, Cincinnati, OH.

47. J. Pawliszyn (Ed.), *Applications of Solid Phase Microextraction*, RSC Chromatography Monographs, Royal Society of Chemistry, Cambridge, 1999.

48. P. Martos and J. Pawliszyn, *Anal. Chem.* **69**, 206 (1997).

49. P. Martos and J. Pawliszyn, *Anal. Chem.* **69**, 402 (1997).

50. M. Odziemkowski, J. Koziel, D. Irish, and J. Pawliszyn, *Anal. Chem.* **73**, 3131(2001).

51. P. Martos and J. Pawliszyn, *Anal. Chem.* **70**, 2311 (1998).

52. L. Muller, T. Gorecki, and J. Pawliszyn, *Fresenius J. Anal. Chem.* **364**, 610 (1999).

53. J. Dean, W. Tomlinson, V. Makovskaya, R. Cumming, M. Hetheridge, and M. Comber, *Anal. Chem.* **68**, 130 (1996).

54. C. Arthur, L. Killam, K. Buchholz, and J. Pawliszyn, *Anal. Chem.* **64**, 1960 (1992).

55. A. Boyd-Boland and J. Pawliszyn, *J. Chromatogr.* **704**, 163 (1995).

56. B. MacGillivray, Analysis of Substituted Benzenes in Environmental Samples by Headspace Solid Phase Microextraction, M.Sc. Thesis, University of Waterloo, Waterloo, 1996.

57. A. Saraullo, Determination of Petroleum Hydrocarbons in the Environment by SPME, M.Sc. Thesis, University of Waterloo, 1996.

58. H. Daimon, Analysis of Solid Samples by Hot Water Extraction-SPME, in J. Pawliszyn (Ed.), *Applications of Solid Phase Microextraction*, RSC Chromatography Monographs, Royal Society of Chemistry, Cambridge, 1999.

15

CONTINUOUS PARTICULATE MONITORING

William J. Averdieck, C. Eng., IMechE
PCME
Cambridgeshire, United Kingdom

Environmental Instrumentation and Analysis Handbook, by Randy D. Down and Jay H. Lehr
ISBN 0-471-46354-X   Copyright © 2005 John Wiley & Sons, Inc.

15.1   OVERVIEW

The subject of continuous particulate emission monitoring to satisfy regulatory requirements is of relatively new interest as a result of recent changes in legislation. Historically regulators were concerned with the visual impact of the discharge from a stack, and therefore emission limits were expressed in terms of color or opacity. However, with the advent of emission limits for a process being defined in terms of mass concentration (expressed in mg/m^3 or grains/dscf), the issue of continuous particulate monitoring has become a new and growing regulatory requirement.

Particulate emission monitoring is a challenging technical field, not only because of the application-specific accuracy and performance of particulate monitors but also due to the harsh environment in which they must continuously operate. By definition, the stack environment includes particulate or dust which will test the robustness of any instrument.

At the cornerstone of all absolute measurements (in mg/m^3 or grains/scf) is isokinetic or gravimetric sampling in which a sample of flue gas is collected and weighed. This testing provides the calibration that gives most in situ continuous emission monitors the ability to monitor particulate in absolute terms. Most instruments can also provide qualitative indication of changes and trends in dust levels should they not be calibrated by isokinetic testing. This provides a solution to a growing requirement of environmental legislation for indicative, parametric, or qualitative monitoring.

15.2   WHY PARTICULATE EMISSIONS ARE CONTINUOUSLY MONITORED

Operators of industrial stacks use continuous particulate monitoring instrumentation for a variety of process and environmental purposes, whether it be to provide better feedback on a process, satisfy environmental legislation, or provide positive proof that stack emissions are under continuous control. Continuous monitoring of particulate is complementary to isokinetic sampling (the method used to periodically assess absolute particulate levels) in that it gives visibility to the dynamics of a process. These dynamic data are crucial in many industrial process or particulate control applications.

Continuous particulate monitoring is sometimes wrongly confused with opacity or color monitoring. Opacity is often a surrogate for particulate monitoring in combustion processes; however, it is really a measurement of the visual impact of an emission. Significantly, environmental emission regulations are changing to specifically limit the particulate concentration in milligrams per cubic meter rather than the color, since in modern industrial processes many emissions, while still being finite, are colorless. Also, the environmental impact of a process is much more related to the quantity of particulate emissions rather than its color. Continuous particle monitoring has therefore become an important regulatory tool for enforcing particulate emission limits.

15.3   TYPES OF CONTINUOUS PARTICULATE MONITORING

Particulate emission monitoring is normally divided into three distinct areas categorized by the quality and type of information provided:

- Concentration measurement (in mg/m^3)
- Qualitative monitoring of the trend in particulate levels
- Gross failure detection

15.3.1   Concentration Measurement

At the top end of continuous particulate emission monitoring is continuous measurement. Such an instrument provides an output calibrated to show particulate levels (in mg/m^3) on a continuous basis. This output is generally recorded for reporting and analysis purposes. Figure 15.1 shows a typical report.

With the exception of a vibrating tapered element, none of the instruments in commercial use actually measures mass concentration; all work by inferring mass concentration from measurement of a different property of the particulate (i.e., attenuation of a light beam by the particles) and are calibrated by comparison to a reference isokinetic sampling method.

The important challenge of particulate monitoring is to measure a property that is only effected by mass and not other particle or process conditions which change in that particular stack. This explains why there are a number of different

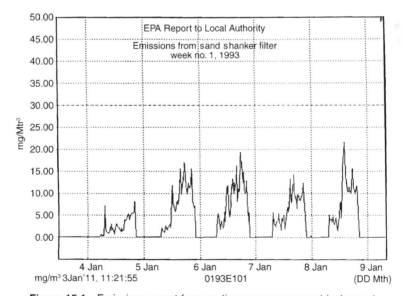

Figure 15.1   Emissions report from continuous measurement instrument.

measurement techniques (each with their own cross-sensitivities) suitable for different types of application.

The performance of concentration monitors is defined by how well the instrument output relates to dust concentration determined by isokinetic sampling over a range of dust concentrations and over time. Of course, this is application specific. Also reflecting that there are systematic errors in isokinetic sampling (typically 10%), the correlation between instrument output and isokinetic result is assessed in terms of a statistical analysis. For example, the International Organization for Standardization ISO-10155 (the international performance standard for particulate monitors) defines a minimum correlation coefficient of 0.95 and limits on other statistical limits such as confidence and tolerance interval. The performance curve as defined by ISO-10155 is shown in Figure 15.2.

Concentration monitors are typically recalibrated at least once a year by reference to an isokinetic test, since this is the only way of checking the calibration

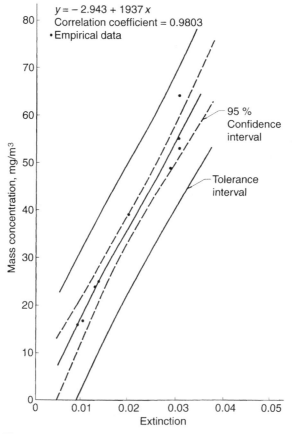

Figure 15.2 Correlation requirement defined in ISO-10155.

(in mg/m^3). More frequent quality control checks are performed to satisfy more advanced regulatory requirements to ensure the instrument is working properly. These are typically implemented as automatic zero or span checks which simulate the parameter being measured (e.g., a reference object which scatters light is inserted in the measurement volume of the light-scattering instrument).

15.3.2 Qualitative Monitoring

In many applications, the absolute level of particulate is not the issue of critical importance and does not justify the cost of isokinetic sampling. Of much more relevance is the trend in particulate levels and monitoring changes in levels over time. This is particularly true for bagfilter applications with high emission limits (50 mg/m^3) where, provided the bagfilter is working properly, emissions will be predictably below the emission limit, in the range 5–10 mg/m^3. The monitoring objective therefore is to provide feedback on the performance of the bagfilter, and a pragmatic, cost-effective approach is qualitative monitoring. Likewise, many process applications, such as measuring particle loss from a drying process, are driven more by a desire to monitor changes rather than to know exact levels.

Qualitative monitoring is the same as concentration measurement with two important exceptions:

- There is no need to calibrate the instrument with isokinetic sampling since the output is in terms of a relative dust output rather than an absolute level. Sometimes approximate calibrations (based on an engineering estimate of the emissions) are applied to qualitative instruments, but this is for approximate referencing rather than to satisfy any desire to make measurements.

- The limits on statistical correlation between dust concentration and instrument output can be relaxed. At one level a 50% accuracy in measurement may be sufficient if an instrument is being used to monitor changes over a 100 : 1 range and there is no critical emission limit. At another, the same performance standards as for concentration measurement may be necessary. Regulators usually define what quality of correlation they accept between instrument response and dust concentration if they are using qualitative monitoring as a basis for regulatory control. Instruments suitable for concentration measurement in an application can always be used to satisfy qualitative measurement requirements. Of course, they require no isokinetic calibration. A response curve from TUV evaluation of a qualitative dust monitor is given in Figure 15.3.

Qualitative instruments provide a relative output (4–20 mA or serial data) of relative dust levels. Units of measurement are usually percentage of full scale or a factor of normal emissions. A typical output from a qualitative instrument is given in Figure 15.4.

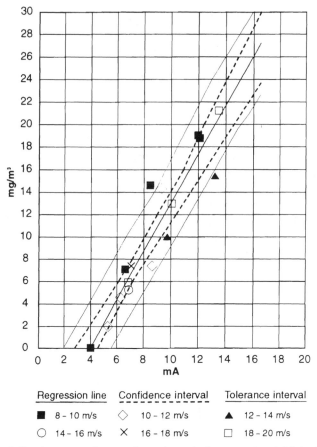

Regression line   Confidence interval   Tolerance interval

■ 8 – 10 m/s   ◇ 10 – 12 m/s   ▲ 12 – 14 m/s

○ 14 – 16 m/s   ✕ 16 – 18 m/s   □ 18 – 20 m/s

Figure 15.3 Calibration curve for an electrodynamic instrument established during TUV field test.

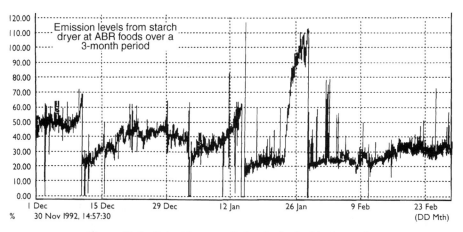

Figure 15.4 Output from qualitative (indicative) instrument.

15.3.3   Gross Failure Detection

Gross failure detection or broken-bag detection is the simplest form of particulate monitoring. An alarm is activated should a significant increase of particulate loading be detected, indicating a failure of the pollution arrestment plant (e.g., filter gross failure or cyclone overflow). Instruments used for filter failure detection do not necessarily need to be accurate or have the sensitivity to measure dust levels in the unfailed conditions; of more importance is a relatively repetitive response to an increased particle level.

Features present in more sophisticated broken-bag detectors includethe following:

- Dual alarm levels allow an early-warning alarm to be supplied in addition to gross failure.
- Alarm delay and averaging differentiate between dust pulses from bag cleaning and proper failure conditions.
- Pulse-tracking capability compare the magnitude of dust pulses resulting from bag cleaning. This permits the development of broken bags to be anticipated and their position in the baghouse to be located. The broken-bag location response from such an instrument is shown in Figure 15.5.

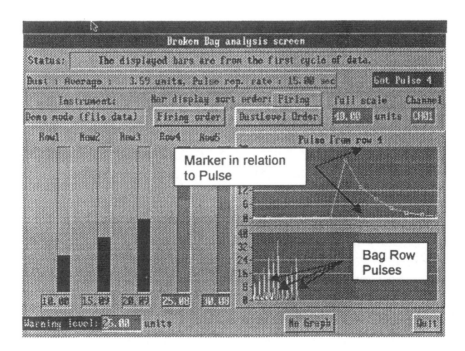

Figure 15.5   Software permits location of failed bagrow to be diagnosed.

15.4   OVERVIEW OF LEGISLATIVE REQUIREMENTS

Environmental regulators worldwide require continuous particulate monitoring for the same two reasons:

- To enforce particulate emission limits in milligrams per cubic meter
- To provide feedback that pollution abatement equipment is working correctly

The overall trend is that continuous particulate monitoring is becoming required in the large and most environmentally sensitive industrial stacks. Other types of particulate monitoring can be required in smaller processes.

Historically many utility and larger stacks were initially regulated in terms of an opacity (color) limit, and therefore opacity instruments are fitted to these stacks. However, as the regulations have moved to specifying limits in milligrams per cubic meter instead of or in addition to color, other types of instruments have often become more suitable for making this measurement.

There is growing demand for continuous particulate monitoring in most parts of the industrial world, reflecting the increased regulatory focus on continuous particulate measurement. The situations in the United Kingdom, Germany, and the United States are worth special mention since each country has adopted an approval scheme for particulate monitors reflecting national regulatory demands.

15.4.1   United Kingdom

Continuous monitoring of particulate is widely implemented in the United Kingdom with divisions between the types of monitoring often being decided on stack air flow. Only in combustion processes do additional Ringlemann limits remain. With the implementation of the Environmental Protection Act in 1990, continuous monitoring of particulate was required in the majority of industrial stacks since it was considered the BATNEEC (best available technique not encurring excessive cost).

The Part B less polluting processes (e.g., Roadstone plant, foundries, animal feed plant, combustion plant < 50 MW) regulated for air emissions by local authorities generally require:

- Continuous measurement in stacks with air flow greater than 300 m^3/min
- Qualitative monitoring and broken-bag detection in stacks with air flow greater than 50 m^3/min
- No monitoring in stacks below 50 m^3/m

The division between the different types of monitoring on larger Part A processes (e.g., chemical plants, steel mills, cement industry, utility boilers) regulated for integrated pollution by the U.K. Environment Agency is subject to individual inspector discretion. Typically continuous measurement is required, but for smaller emission points broken-bag detection is sufficient. Very few industrial stacks have no continuous monitor.

The Environment Agency has historically given little guidance to industrial operators on the quality of instrumentation required to satisfy continuous monitoring requirements. This is likely to change with the introduction of MCERTS, a new approval scheme for continuous emission monitors adopted by the Environment Agency in 1999. This scheme defines standards to which continuous monitors must perform. Instruments obtain a certificate for specific processes and measurement ranges based on a laboratory and 3-month field test overseen by an independent test body (SIRA). ISO-10155 is used as a basis for the test standards against which particulate monitors are tested. Measurement and qualitative instruments are covered by this scheme.

15.4.2 Germany

Regulatory limits are based on particulates rather than opacity, although there are requirements for continuous opacity instruments as a qualitative measurement in certain combustion processes. Continuous measurement is required based on local air pollution issues (i.e., stack is close to residential area) or on stacks when the total mass emissions of particulate are likely to exceed defined limits. These limits depend on the toxicity of the particulate.

Specific regulations impacting the use of continuous particulate monitors are:

Bimsch 13: combustion plant >50 MW
Bimsch 17: incineration plant
Bimsch 27: qualitative monitoring of particulate after filter plant

A type approval scheme exists in Germany. This scheme is widely respected, due to the importance placed on field testing and quality assurance issues (such as instrument checks). Particulate monitors are tested by independent test authorities (e.g., TUV) against standards and for measurement ranges defined by each of the above regulations. Test certificates note any restrictions on the use of an instrument.

15.4.3 United States

Historically there has been little use for particulate emission monitors in the United States for regulatory purposes, since emission limits and monitoring methods have been specified in terms of opacity (color). Interest in particulate monitoring is growing as a result of the following regulatory changes resulting from the Clean Air Act of 1990 amendments:

- The Environmental Protection Agency (EPA) is seriously considering requiring *continuous particulate measurement or PM CEMs* (particulate matter continuous emission monitors) in a number of industrial processes, including incinerator and cement kiln applications. A new standard PS-11 is being developed to define the performance of particulate monitors. This standard uses the same performance approach as ISO-10155 and as such is similar to the U.K. and German type approval scheme. However, a significant difference is that each PM CEM will require validation in the specific stack in which it is being used.

- Some of the new MACT standards for processes with HAPs (hazardous air pollutants) require *broken-bag detectors* to be fitted to baghouses.
- Title V plants (the major metals, chemical, mineral, and combustion processes) are required under the new CAM (compliance assurance monitoring) regulations to develop a method to ensure the continuous compliance of their particulate arrestment plant (e.g., baghouses and electrostatic precipitators). It is likely *qualitative particulate monitoring* and *gross failure detection* will be chosen by many sites as pragmatic solutions to this new requirement.

15.5 TECHNOLOGIES FOR CONTINUOUS PARTICULATE MONITORING

There are at least eight different technologies used in commercially available particulate emission monitors. No one technology has ideal characteristics in all applications, and therefore different technologies are used in different applications depending on exact requirements. The techniques most commonly used for particulate monitoring are as follows:

1. Opacity
2. Dynamic opacity (scintillation)
3. Back/side scatter
4. Forward scatter
5. Oscillating filter (vibrating tapered element)
6. Beta attenuation
7. Triboelectric
8. Electrodynamic

In the main, the adoption of each technology has been driven by its effectiveness and value in a particular application. However, the regulatory approval process has also had an effect on the use of certain technologies. An overview of each technique and its characteristics follows.

15.5.1 Optical Attenuation (Opacity)

15.5.1.1 Principle of Operation. Light is transmitted across the stack and attenuated by particles in the stack. The amount of light received after it has been attenuated by the particles is measured and can be related to the dust concentration. A schematic of an opacity monitor is shown in Figure 15.6.

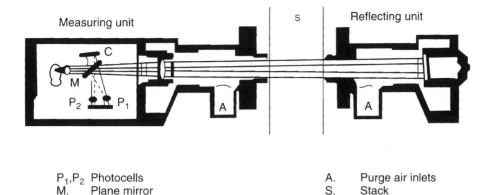

P$_1$,P$_2$ Photocells   A.   Purge air inlets
M.   Plane mirror   S.   Stack
C.   Concave mirror

Figure 15.6 Schematic of opacity instrument.

As light passes through particles, it is scattered and absorbed by the particles. For the dust concentrations found in industrial stacks (0–500 mg/m^3) and small particles (<30 μm) the relationship between transmittance of light T and number of particles is predicted by the Beer–Lambert law, which states

$$\ln\left(\frac{1}{T}\right) = \text{extinction} = EnaL$$

where E = extinction or calibration coefficient for type of particles
         n = number of particles per unit path length
         a = projected surface area of each particle
         L = measurement path length

Simple instruments monitoring pure opacity measure and calculate the transmittance; however, more advanced versions used for particulate monitoring measure and calculate extinction. Provided the physical properties of the particles which affect light absorption and scattering do not change, extinction is proportional to dust concentration, and therefore the output of such an instrument can be correlated to dust concentration and is calibrated by reference to an isokinetic gravimetric sample.

15.5.1.2 *Practical Considerations.* In field instruments the following techniques are often used to improve performance:

1. The light beam is transmitted across the stack and reflected back to a receiver at the same location as the transmitter (double pass). This is done to:

- Increase the path length and hence improve the minimum detection level of the instrument
- Enable a parallel reference light path to be split in the transmitter to compensate for changes in light intensity characteristics of the source or receiver

2. The light beam is modulated or chopped to eliminate stray background and ambient light effects. Techniques for modulation are:

- A rotating slotted disc
- Pulsed light-emitting diode (LED) or laser
- Electronic shutter

3. The source of the light is selected to give sufficient intensity and stability for the application. Filament bulbs have been replaced by lasers and LEDs in more recent instruments.

4. The wavelength of the light is selected to be consistent with the human eye response (the peak response is specified as between 500 and 600 nm by the EPA when used to measure opacity). It is not possible to select a wavelength that does not suffer interference from water vapor.

5. Automatic instrument zero and span checks involve the insertion of a filter of a known optical density into the light path and measuring the change in response of the instrument. This checks the ability of the instrument to measure the intensity of light, but as with other particulate emission monitors, it does not check the calibration in milligrams per cubic meters.

15.5.1.3 Use and Limitations of Measurement.

Opacity monitors are used extensively worldwide to monitor opacity. This is particularly true in the utility and power generation industries in the United States, where regulations have traditionally required an opacity measurement. Their adoption as particulate monitors is less universal due to their inapplicability to the lower levels of particulate now found in industrial processes. Industries where opacity monitors are still well accepted are the power, cement, and steel industries due to their historical experience in satisfying opacity requirements. There are a number of opacity instruments with TUV approvals for particulate measurement. Their limitations are widely accepted as follows:

1. They cannot monitor particulate levels below 25 mg/m^3 per meter path length since at low concentrations the reduction in the light beam caused by the particles is indistinguishable from the zero drift of the source/detector (i.e., variation in the intensity of the receiver with no dust conditions). This fundamental limitation, which makes the instrument unsuitable for many well-abated emission applications, is as a result of the instrument requiring to measure a small reduction in the large baseline signal.

2. The system is sensitive to dust contamination on the lens surfaces since it is not possible to distinguish between the reduction in light caused by dust in the stack and dust on the lenses. In practice, a curtain of air (provided by a 40-cfm blower) is injected into the transmitter and receiver heads to keep the lens surfaces clean.

3. Systems without retroreflectors (i.e., not double pass) are sensitive to misalignment between the transmitter and receiver.

4. The calibration of the instrument changes with changes in the following particle properties:
   - Particle type and refractive index (mainly changes in the amount of light scattering)
   - Particle color (mainly changes in the amount of light absorbed)
   - Particle size and shape (changes in the amount of light scattering)

5. Water vapor absorbs light over the light frequency range used by opacity monitors, and therefore they are not really suitable for drier stacks.

15.5.2 Dynamic Opacity (Scintillation)

15.5.2.1 Principle of Operation. Like opacity monitors, dynamic opacity monitors are based on a light beam passing across the air flow. The essential difference is that they measure not only the beam intensity as such but also the ratio of the temporal variation in intensity to the intensity. This intensity variation derives from the statistical variations in the distribution of particles in the air stream and is depicted in Figure 15.7. The higher the concentration of particles, the greater the

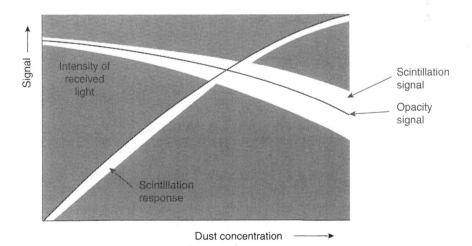

Figure 15.7 Response curve from dynamic opacity (scintillation) instument.

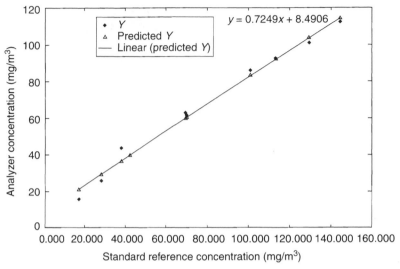

Figure 15.8 A calibration curve for a dynamic opacity instrument.

range of variation. Empirical results as shown in Figure 15.8 confirm a simple linear relationship between scintillation and dust concentration and show that with zero dust there is no scintillation (i.e., the instrument has a true zero, unlike opacity devices). The term *dynamic opacity* or *scintillation* is related to the dust concentration as follows:

$$\text{Dynamic opacity} = \frac{\text{variation in intensity}}{\text{intensity}} = KC$$

where C is the dust concentration and k is an empirical constant for the particle's physical properties.

15.5.2.2 Practical Considerations

1. One of the most important characteristics of the dynamic opacity technique is its tolerance to instrument contamination. A dynamic opacity monitor will continue to function without error even when its lenses are heavily coated with dust. As a result, there is no need to fit large air purge blowers to the system. In high-dust loading applications, the transmitter and receiver are connected to a 1-cfm supply of instrument air to stop the light beam being completely obscured by catastrophic buildup. Provided sufficient light is getting through for the instrument to make a measurement, its response is unaffected by any contamination since the numerator and denominator of the ratio are affected by the same amount. As shown in Figure 15.9, a particular dust concentration produces a variation x in a light intensity I when the lenses are clean. If the light intensity is reduced by a contaminant coating on the

Lens condition	Light intensity	Variation	Dynamic opacity
100% transmission	i	x	x/i
90% transmission	$0.9i$	$0.9x$	$0.9x/0.9i = x/i$
50% transmission	$0.5i$	$0.5x$	$0.5x/0.5i = x/i$

Figure 15.9   Tolerance of dynamic opacity technique to dust contamination.

lenses, the variation in intensity will be affected in the same proportion, giving no net effect.

2. The scintillation instrument is not significantly affected by the absolute alignment between the transmitter and receiver, since, like contamination, this affects the numerator and denominator of the ratio by the same amount. As a result, the adjustable mounting alignment of the instrument can be set by eyesight on first setting up the instrument.

3. The instruments are generally single pass since there is no need to increase the path length due to concerns on detection level. Unlike the opacity technique, the instrument measures no signal when there is no dust, and therefore it is possible to increase the signal-to-noise ratio. In practice, this means the instrument can detect low dust concentrations as low as 2.5 mg/m^3 m (at least 10 times better than an opacity device).

4. Relevant light and electronic automatic zero and span checks can be built into the instrument to check for instrument integrity. As with all types of in situ devices, these do not check for changes in particulate calibration.

5. There are also certain applications in which measurements must be made in terms of opacity and Ringelmann (color) characteristics as well as milligrams per cubic meter. To allow for such applications, dynamic opacity monitors can be switched into the opacity measurement mode.

15.5.2.3   Use and Limitations of Measurement. Dynamic opacity instruments can be a more reliable alternative to opacity instruments where regulations require particulate measurement as opposed to opacity. They are becoming accepted in the utility and cement industries and in recent years have received regulatory approval in the United Kingdom and Germany as particulate monitors. They are often used in combustion applications and other industrial processes with large stack discharges from an electrostatic precipitator or bagfilter. Their limitations are as follows:

1. The calibration of a dynamic opacity instrument will shift if there are changes in parameters affecting the attenuation of light by a particle. These include:

   • Particle size and shape
   • Particle material

In many applications, these parameters normally remain constant. If, however, changes occur for any reason, the instrument must be recalibrated.

2. The calibration is also effected by changes in process conditions affecting the statistical distribution of the particles. These include:
   • Velocity (this effect is really only significant below 5 m/s)
   • Conditions in which short-term process changes cause variations in particle distribution greater than the statistical variations

   In practice, this means that the start-up of certain processes may not be accurately monitored by dynamic opacity.

3. The scintillation response is affected by water vapor, and therefore this is an offset in measurement.
4. An offset in measurement is created by heat haze effects. This can increase the minimum detection level to above 2.5 mg/m^3.

15.5.3   Beta Attenuation

15.5.3.1   Principle of Operation. The moving gas stream is sampled isokinetically, and the sample of particulate is collected on a filter. The filter is advanced periodically (typically every 15 min) into a measurement chamber so that radioactive beta particles can be passed through the sample and the amount of beta particles transmitted through the sample is measured (see Figure 15.10). The amount of transmitted signal is related to the amount of particles by the Beer–Lambert law. The main advantage of this technique is that the absorption of radioactivity is not significantly affected by the type of particle (although particles with different nucleonic density have different responses). It is therefore often used in municipal incinerator applications where the type of particulate changes with the feedstock added to the incinerator.

15.5.3.2   Practical Issues. The major practical issues relate to the difficulties of sampling the particulate:

1. It is expensive and complicated to maintain true isokinetic conditions in the sampling train.
2. High maintenance is required for the sampling train and the mechanical filter advancing system.

15.5.3.3   Use and Limitations of Technology. Beta systems are extensively used in older incinerator plants due to their ability to cope with wet stack conditions and changing particulate type. Regulatory approvals exist for these types of instruments; however, due to their high purchase and running costs, they are rarely used

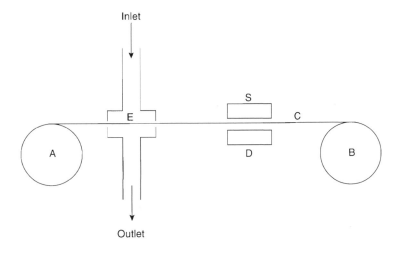

A. Feed spool
B. Takeup spool
C. Filter tape
D. Detector
E. Dust spot
S. Source of beta radiation

Figure 15.10 Schematic of beta attenuation instrument.

in less demanding and more standard applications. In addition to maintenance the principal limitations are:

1. Dust losses in the sample-handling train reduce measurement accuracy and repeatability.
2. Measurement is only at a single point and therefore is not always representative.

15.5.4 Light Scattering

Light-scattering particulate monitors make use of the physical effect that small particles similar in size to the wavelength of light used (i.e., 0.5 µm) scatter or reflect a light beam in all directions. This phenomenon is observed by the human eye when observing sun rays illuminating suspended dust in a darkened room. Light-scattering instruments are not concerned with transmitted light, and the minimum detection limit is significantly better than opacity instruments. The receiver measures the amount of light reflected (scattered) from the direction of incidence by the particles. Importantly, instruments have different characteristics depending upon the specific scatter or reflection angle chosen for measurement. This behaviour is well predicted by Mie and Raleigh scattering theory.

In situ instruments are generally catagorized as either back/side-scatter instruments or forward-scatter instruments according to their scattering angle.

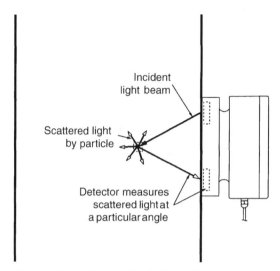

Figure 15.11 Schematic of side scatter instument.

15.5.4.1 *Back- and Side-Scatter Devices.* The transmitter and receiver are mounted on the same side of the stack relatively close to each other. Light is transmitted into the particle stream and the receiver measures the light that is scattered backward or sideways (depending on whether it is a back- or side-scatter instrument). Figure 15.11 shows the layout of a side-scatter instrument. For a given type of particle type, particle size, and particle shape, the intensity of the received (i.e., scattered) light is proportional to dust concentration, and therefore the output can be calibrated in absolute terms by comparison to results from an isokinetic test.

15.5.4.1.1 Practical Issues

1. A light trap is required at the far side of the stack to eliminate light reflected back from the opposite stack wall.
2. Since the transmitter and receiver are mounted on the same side of the stack, alignment problems are avoided.
3. An air purge is required to minimize dust contamination on the lens since otherwise this is a source of error. Side-scatter devices have optical span checks in which the transmitter and receiver lenses are mechanically rotated to face each other so that dust contamination on the lens can be measured and compensated for.
4. Correct location of the instrument is important to ensure representative measurement. For side-scatter instruments the measurement volume is relatively small, and for back-scatter devices the responsiveness to dust is defined by an inverse law depending on the distance of the dust from the transmitter.

15.5.4.1.2 Use and Limitations of Technology. Side-scatter and Back-scatter instruments are used in low-dust-concentration applications such as found for lead smelters and incinerators. In Germany they are widely used due to regulatory approval. Use outside Germany has been less significant due to their higher purchase and installation costs. Their technical limitations are as follows:

1. The calibration is affected by changes in particle size and type of particle. For example, with absorbing particles the response of a back-scatter device is reduced by a factor of 20% from peak response when particle size changes by 0.1 μm. The peak response for nonabsorbing particles is three times greater than for absorbing particles.

2. Back- and side-scatter devices are less sensitive than forward-scatter devices and therefore require more powerful light sources (e.g., lasers) to provide sensitivity of less than 1 mg/m^3.

3. In situ light-scattering instruments cannot differentiate between water aerosols and solid particles.

15.5.4.2 Forward-Scatter Devices. A forward-scatter device measures the light that is scattered at a small angle from the angle of incidence, as shown in Figure 15.12. This angle is typically in the range of 5° to 15°. The response is proportional to dust concentration for a given set of particle properties and can therefore be calibrated in milligrams per cubic meter.

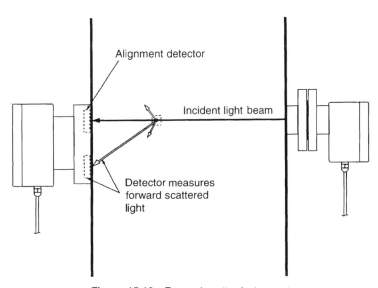

Figure 15.12 Forward scatter instrument.

15.5.4.2.1 Practical Issues

1. Forward-scatter devices have better sensitivity than back-scatter devices so it is possible to use LEDs and diode lasers as sources of light.
2. The receiver can be mounted either on the far side of the stack or in the case of a probe-based instrument at the far end of the measurement cell embedded in the probe.
3. Since measurement is occurring at small angles to the incidence, it is very important to shield the receiver properly from directly transmitted light.
4. Air purges are required for optical surfaces, although compensation for dust accumulation can be made by separately measuring changes in directly transmitted light.
5. Instruments with multiple detectors at different scattering angles could be used to determine particle size. These types of instrument are currently in development.

15.5.4.2.2 Use and Limitations of Technology. Forward-scatter instruments are playing an increasing role in the particulate measurement market. Importantly, their calibration is unaffected by changing particle type or shape. Currently, few instruments have regulatory approval, although this will no doubt change in the future. Limitations of forward-scatter instruments are:

1. Calibration is dependent on particle size, although cross-sensitivities are less than with back-scatter devices.
2. In situ light-scattering instruments cannot differentiate between water aerosols and solid particles.
3. The measurement volume is relatively small, especially in probe-type designs. Appropriate instrument location is therefore important.

There are also extractive light-scattering instruments in which a sample of flue gas is sampled under isokinetic conditions and then passed into an external light-scattering chamber. A forward light-scattering technique is normally used in the chamber, and a heater unit can be fitted before the chamber to evaporate and eliminate water from a wet or humid application. The same problematic issues of sample handling are as relevant with these instruments as for beta attenuation devices. Extractive light-scattering instruments with a heated chamber have German regulatory approval for wet applications.

15.5.5 Vibrating Tapered Element

15.5.5.1 Principle of Operation. A sample of particulate is collected onto a filter mounted on the end of a resonating or oscillating tapered element. The change in frequency of oscillation is measured since this depends on the mass of the

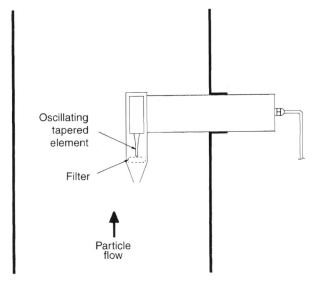

Figure 15.13 Schematic of vibrating tapered element instrument.

combined filter and collected particulate. This technique is currently only semi-continuous since the filter must be periodically replaced (every 1–3 days), but it has the advantage of being mass dependent. A schematic of such a device is shown in Figure 15.13.

15.5.5.2 Practical Considerations

1. The tapered element is generally mounted on an assembly mounted in the stack to minimize particle losses in any extractive line.
2. Air flow into the measurement assembly is controlled to give isokinetic conditions. Therefore, when filters are replaced, they can be used for a traditional isokinetic gravimetric analysis and an analysis of particle content.
3. Interference from stack vibrations is minimized by using a heavy instrument assembly.
4. The measured oscillating frequency is compensated for changes in stack temperature.

15.5.5.3 Use and Limitation of Technology. This technique, while widely accepted in the ambient particulate field, is relatively new to emission measurement. Being an absolute measurement technique with appropriate quality control features, it is beginning to be adopted in incinerator and high-profile emission applications, especially in the United States. Adoption in other applications, except as a periodic measurement, is currently limited due to the novelty of the technique, the

need to regularly replace the filter, and the relatively high purchase costs. Other limitations are:

1. Losses from evaporation of volatiles in the particulate on the filter over time and the effects of moisture on the sample weight
2. Single-point measurement unless manually traversed across the stack
3. Supervision required for systems currently available

15.5.6 Charge Measurement Techniques

Triboelectric and electrodynamic instruments are both charge measurement techniques. They measure a current produced by the interactions of moving particles and a grounded sensor rod installed across the duct of the stack. These techniques are highly sensitive and can be used at concentrations below 0.1 mg/m^3. However, triboelectric and electrodynamic instruments use different processing and analysis of the current, resulting in different characteristics.

15.5.6.1 Triboelectric Instruments

15.5.6.1.1 Principle of Operation. A triboelectric instrument measures the total amount of current produced by particles impacting and passing by the sensor rod. The essential parts of a triboelectric system are shown in Figure 15.14. In many dry baghouse applications the charge transfer per collision is repeatable, and therefore the total current is proportional to the concentration of particulate for a fixed particle size. The system is simple to install and maintain, requiring single-point access on the stack.

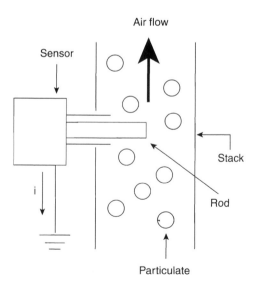

Figure 15.14 Principal components of triboelectric instrument.

15.5.6.1.2   Practical Considerations

1. A rod length of at least half the stack size is used to ensure representative measurement.
2. Amplification of the small triboelectric signal (10 pA) is done in the sensor head to maximize instrument signal-to-noise ratio.
3. The insulator at the base of the sensor rod must be kept clean to avoid false signals from ground loops and stack currents. An air purge can be used, but this is not sufficient in humid applications.
4. Relevant instrument operation checks include zero and span checks, injecting known currents into the sensor rod, and an insulator contamination check.

15.5.6.1.3   Use and Limitations of Technology. Triboelectric instruments are widely used in bagfilter applications in the metals, mineral, and process industries. In the main their use has been confined to process, qualitative, or broken-bag monitoring due to perceived performance limitations and in the United States due to lack of regulatory requirements and approvals. Triboelectric instruments have recently received regulatory approval for measurement in specific applications in Germany. Technical limitations are as follows:

1. The charge transfer resulting from a particle collision is related to the change in its kinetic energy on impact. Calibration is therefore dependent on the particle velocity. Use of triboelectric instruments is generally restricted to constant-velocity applications unless velocity measurement and compensation systems are incorporated.
2. The calibration is affected by contamination and polarization effects on the rod. This reduces the absolute accuracy and reliability and has resulted in their use as qualitative instruments in many applications.
3. The calibration is dependent on particle properties, such as:
   • Particle type
   • Particle size
4. The systems do not work in wet applications due to the problems of measuring small electrical currents in such applications.

15.5.6.2   Electrodynamic Instruments

15.5.6.2.1   Principle of Operation. In such a system the sensing probe is installed across part of the stack and the DC produced by particle collisions is eliminated by AC filtering techniques. The instrument measures the remaining alternating signal produced by charged particles inducing charge flow in the sensor rod as they pass it. Since the signal is not dependent on particle collisions (unlike triboelectric), the related problems of rod contamination and velocity dependence are minimized.

In applications where the particle charge, particle size, and particle distribution remain constant, the resulting AC is proportional to dust concentration. These

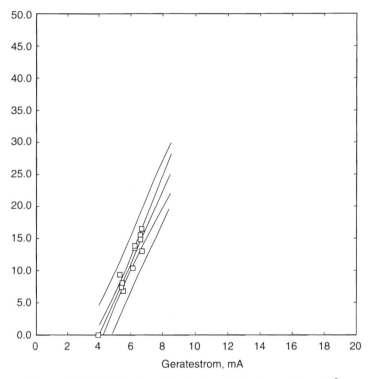

Figure 15.15 Calibration of electrodynamic instrument in mg/m^3.

instruments can be calibrated in milligrams per cubic meter, by comparison to the results of an isokinetic test shown in Figure 15.15.

15.5.6.2.2   Practical Considerations

1. The sensor rod can be completely insulated to extend operation into humid (drier) applications.

2. Analysis of the frequency components of the AC signal is used in certain instruments to improve accuracy.

3. Automatic zero and span checks are performed by injection of known AC into front-end electronics. An insulator contamination check can also be performed automatically.

4. The sensor rod can tolerate contamination without reduction in performance since the measurement signal derives from induction rather than collision.

5. The sensor can be incorporated in an in situ heating chamber assembly to permit measurement in wet stacks. This provides a pragmatic alternative to beta systems.

15.5.6.2.3 Use and Limitations of Technology. Electrodynamic instruments are used to satisfy qualitative and measurement requirements on bagfilters in the metal, mineral, and chemical industries. Their adoption in the United Kingdom is extensive, and their use in Europe, Japan, and Australia is widespread. Regulatory approvals exist for qualitative and measurement instruments in the United Kingdom and Germany. Technical limitations are as follows:

1. The use of electrodynamic technology for particulate measurement requires applications with predictable particle type and precharge, noncondensing conditions and a minimum velocity of 5 m/s. There are only minor effects of changing velocity if the velocity is greater than 8 m/s.
2. The instrument cannot be used for measurement with the presence of water droplets; however, it is often used in noncondensing humid applications (after driers) since it can discriminate between solid particles and water vapor.
3. The technology is only suitable for indicative monitoring in applications in which the precharge on the particle is likely to change. In practice, this covers electrostatic precipitator (EP) and combustion applications where charge on the particle may be changed by EP condition and flame ionization effects, respectively.

15.6 CHECKLIST FOR PARTICULATE MONITORING

It is always worth checking the suitability of a specific instrument and technology for a process due to the application-specific performance of all the measurement techniques. In general, a particulate monitor must satisfy the following requirements:

- It must be suitable for measuring the expected range of particulate (under normal and limit conditions).
- It should be insensitive to changes in non-mass-related properties of the particles, such as particle size, particle material, and particle charge, which are likely to change in the particular application in which it is to be used.
- It should have a defined and repeatable response in relation to dust concentration. This should preferably be linear to facilitate calibration against isokinetic sampling.
- It must make a representative measurement without sampling errors.
- It must be capable of operating reliably for long periods without the need for maintenance.

Any compromise in these characteristics should be made with caution, and costs of purchase and ownership should also be considered to ensure an effective and pragmatic solution. Regulatory approvals are generally designed to assess the above characteristics and provide the best assessment of an instrument's performance in a particular application.

16

GAS SURVEY INSTRUMENTS

Randy D. Down, P.E.
Forensic Analysis & Engineering Corp.
Raleigh, North Carolina

Environmental Instrumentation and Analysis Handbook, by Randy D. Down and Jay H. Lehr
ISBN 0-471-46354-X   Copyright © 2005 John Wiley & Sons, Inc.

16.1 INTRODUCTION

Each year, industrial operations in the United States emit nearly 100 million tons of pollutants into the air. Many of the sources of this air pollution are in facilities such as petroleum refineries and chemical plants, which can potentially have thousands of emission points.

The 1990 amendments to Title 1 of the Clean Air Act require industries and the government to gather data and develop control strategies for compliance with the National Ambient Air Quality Standards. Title III of the act requires programs to assess, reduce, and control hazardous air pollutants. This act, when first enacted, imposed challenging air-monitoring requirements, and traditional technologies were not entirely adequate to meet those requirements. The Clean Air Act amendments renewed interest in the development of monitoring methods with enhanced capabilities. There was a clear need for a small, portable instrument that could characterize stack emissions and fugitive gas leaks.

As regulatory requirements regarding fugitive emissions became more stringent, newer, more accurate and reliable hand-held gas analyzers were developed to detect and measure them. This chapter addresses their purpose, their functionality, and applications utilizing these instruments.

Sometimes referred to as gas survey instruments, or "sniffers," these hand-held analyzers can aid a technician in determining the type of vapor that is leaking and its airborne concentration. It may also aid in tracing the leak to its source (such as a leaking fitting or failed seal). This type of survey is necessary in order to comply with the U.S. Environmental Protection Agency (EPA) Title V and applicable state requirements requiring facilities to monitor and quantify fugitive air emissions. In the case of toxic gas leaks, proper use of these instruments can result in early detection and potentially avoid serious risks to human health and the environment.

16.2 INSTRUMENT SELECTION

When selecting the proper hand-held or portable analytical instrument, you should consider the following:

- To what type of environment will the device be exposed?
- Is it needed just to determine the presence of a gas or also to quantify it?
- How much accuracy is required to meet your objective?
- Does it need to be capable of detecting a broad range of gases or a specific gas?
- What levels of accuracy and resolution are required?
- What are the requirements to maintain proper calibration?
- What is the hazard classification of the area in which the analyzer will be used?

- Has calibration of the device been verified (using a calibration gas)?
- Is good technical support readily available should you encounter problems with the instrument?

The type of analyzer needed will depend largely upon the type of application and the type of gas or vapor to be monitored. A gas survey instrument is used to do the following:

(a) *Locate Source of Leak.* This can often be accomplished by means of a fairly unsophisticated hand held-gas analyzer that is capable of detecting chemical vapors in a small, concentrated area, where quantification (providing an actual measurement of the airborne concentration) is not required.

(b) *Measure Level of Human Exposure to Leak (Airborne Concentrations).* Usually used within the breathing zone, this instrument provides a measurement displayed as a percentage (by volume) of a measured quantity of air. If only the incident of exposure, and not a determination of the level of exposure, is needed, then a device as unsophisticated as a clip-on card that reacts chemically in the presence of sensed gas or vapor is needed to alert the wearer of exposure.

(c) *Chemical Analysis of Leaking Vapor or Gas.* Necessary to determine the type of gas or vapor that is leaking in order to protect people, the environment, and equipment.

(d) *Data Collection.* Recorded measurements of the fugitive emissions must be maintained and submitted to the state government in order to comply with federal regulations under Title V. It is also valuable in quantifying the losses that occur in a process due to leakage.

The type of sensor technology to be used is dependent upon the type of gases to be detected and measured. The two primary types of sensor technologies used in gas survey instruments are electrochemical and infrared (IR). Both are accurate and reliable methods of gas measurement.

16.2.1 Electrochemical Sensors

Electrochemical sensor technology offers highly specific gas sensing in the low-ppm (parts per million) range for many gases. Functionally, the electrochemical sensor comes into direct contact with the tested gas or vapor. The sensor is constructed of a material that changes its electrical properties due to a chemical reaction that occurs at the sensor when in the presence of a specific chemical vapor. The change in electrical properties is amplified and displayed as an alarm (indicating that the specific gas or gases being tested for are present), in ppm, or as a percentage of the lower explosive limit (LEL) of the measured gas. Over a period of time, the electrochemical sensor will be depleted (due to the chemical reaction with sensed gases) and will need to be replaced. Refer to the manufacturer's recommendations regarding the frequency of replacement.

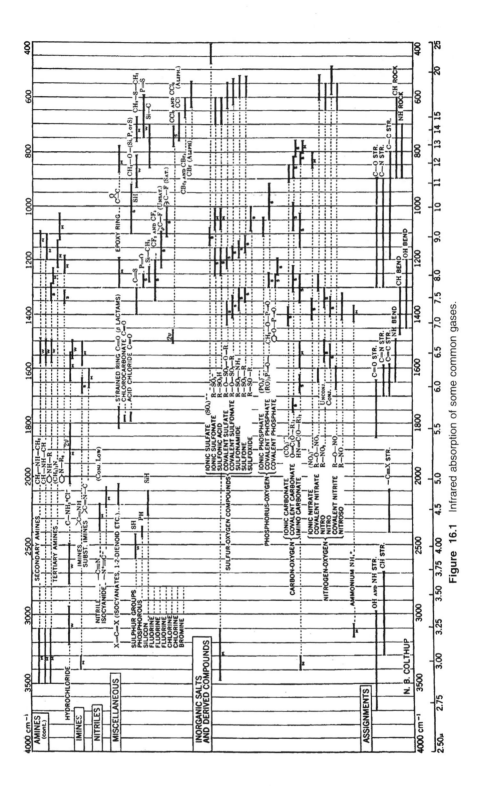

Figure 16.1 Infrared absorption of some common gases.

N. B. COLTHUP

16.2.2 Infrared Sensors

Infrared gas sensors work on the principle that most gases have unique IR signatures within the 2–14-μm wavelength region. An example of the IR signature of some common gases is shown in Figure 16.1.

Optical gas sensors measure the transmission of light at a different wavelength for each gas. The particular wavelength identifies the gas, and the amount of light absorbed by the gas determines the gas concentration.

Because each gas has a unique IR absorption line, IR gas sensors can provide conclusive identification and measurement of the target gas with little interference from other gases. Infrared gas sensors are highly accurate, responsive, and reliable. In IR analyzers the sensor does not touch the gas—only IR light comes in contact with the gas. Therefore it is not poisoned by contact with the environment.

16.2.3 Ultrasonic Detectors

Ultrasonic analyzers find potential leaks and sources of energy waste in containers that are too large to make pressurizing practical (such as large vessels or heat exchangers). Analysis is performed by placing a transmitter inside the container that generates ultrasonic sound waves. The sound waves escape through leaks in much the same way that a gas or fluid would. The sound waves are then detected by the hand-held analyzer probe, which is located outside of the test chamber.

16.3 TYPICALLY MEASURED GASES

A broad range of gases can be detected and measured using hand-held portable instruments. These instruments are selected based on the type of gas or multiple gases that you need to detect the presence of or quantify.

Types of leaking gaseous vapors that are commonly detected and/or measured using hand-held measurement instruments include:

- Refrigerants (chlorofluorocarbons, or CFCs)
- Carbon monoxide (CO)
- Combustible gases
    Natural gas
    Propane
    Methane
    Butane
    Benzene
    Isopentane
    Acetylene
    Hydrogen
    Toluene
    Paint thinners
    Naphtha

A list of less common gases and vapors that are also detectable using commercially available hand-held gas survey instruments is given in Table 16.1.

Applications of hand-held gas survey instruments include:

Power Plants. The use of ammonia is increasing at power plants for both NO_X reduction and turbine inlet cooling. Gas survey instruments using electrochemical sensors can monitor ammonia at low-ppm levels.

16.1 Gases/Vapors Less Commonly Detected Using Hand-Held Gas Survey Instruments

Acetaldehyde	Dinitrobenzene	Ketone(hexone)
Acetone	Dinitroluene	Methyl butyl ketone
Acetonitrile	Dipropylene glycol methyl	Methyl mercaptan
Alcohol	Ether	Methyl chloroform
Allyl Alcohol	Disobutyl ketone	Methyl alcohol
Ammonia	Epichlorhydrin	Methyl chloride
Benzoyl chloride	Ether	Methyl-*n*-amyl ketone
Benzoyl peroxide	Ethoxyethanol	Methyl ethyl ketone
Butanone (MEK)	Ethyl butyl ketone	Methylal
Butoxyethanol	Ethyl Ether	Methylamine
Butyl alcohol	Ethyl bromide	Methylcyclohexanol
Butyl acetate	Ethyl formate	Methylcyclohexane
c-Allylglycidylether	Ethyl benzene	Methylene chloride
c-Chloroform	Ethyl alcohol	Napthalene
c-Dichloroethyl ether	Ethyl chloride	Nitrobenzene
Camphor	Ethylamine	Nitrochlorobenzene
Carbon tetrachloride	Ethylene dichloride	Nitroethane
Chloro-1-nitropropane	Ethylene oxide	Nitroglycerin
Chloroacetaldehyde	Ethylenediamine	Nitromethane
Chlorobenzene	Formaledehyde	Nitrotoluene
Chloropicrin	Furfuryl alcohol	Pentane
Chloroprene	Gasoline	Pentanone
cl-1-Dichloro-1-nitroethane	Glycol monoethyl ether	Perchloroethylene
Cumene	Heptane	Petroleum distillates
Cyclohexane	Hexachloroethane	Phenylether
Cyclohexanol	Hexane	Propargyl alcohol
Cyclohexanone	Hexanone	Propylene oxide
Cyclopentadiene	Hydrogen sulfide	Propyne
DDT	Hydrogen chloride	Sulfur dioxide
Diacetone alcohol	Hydrogen bromide	Tetrachloronaphthalene
Diasomethane	Isoamyl alcohol	Tetranitromethane
Diborane	Isobutyl alcohol	1,1,1-Trichloroethane
1,1-Dichloroethane	Isopropyl alcohol	1,1,2-Trichloroethane
1,2-Dichloroethane	Isopropyl glycidel ether	Trichloroethylene
Diethylamine	Ketone	Trichloronaphthalene
Diethylamino ethanol	L.P. gas	Trichloropropane
Dimethylamine	Lacquer thinners	Trinitrotoluene
Dimethylaniline	Methyl acetylene	Turpentine
Dimethylformamide	Methyl cellosolve	Xylene
Dimethylhydrazine	Methyl isobutyl	

Chemical Plants. Electrochemical sensor technology offers "gas-specific" monitoring for a long list of gases in a choice of ranges, including ammonia, bromine, carbon monoxide, chlorine, chlorine dioxide, hydrogen, hydrogen chloride, hydrogen fluoride, hydrogen sulfide, nitric oxide, nitrogen dioxide, oxygen, ozone, sulfur dioxide, and others.

Confined-Space Entry. Oxygen, combustibles, refrigerants, carbon monoxide, and other gases can be monitored to determine if a space is safe to enter.

16.4 LEAK TESTING

Leak testing is a nondestructive test procedure that is concerned with the leakage (escape) of liquids, vacuum, or gases from (or into) sealed components or systems. Like other forms of nondestructive testing, leak testing can have a great impact on safety as well as an environmental impact. Reliable leak testing in a manufacturing environment can also save cost by detecting system problems before they result in damaged equipment or process downtime.

The three most common reasons for performing a leak test are:

Material Loss. With the high cost of energy and strong foreign competition, material loss is increasingly important. By leak testing, energy is saved not only directly, through the conservation of fuels such as gasoline and natural gas, but also indirectly, through the saving of expensive chemicals and even compressed air.

Contamination. With stricter Occupational Safety and Health Administration (OSHA) exposure limits and stricter environmental regulations, the need for leak testing and leak detection has rapidly grown. Leakage of dangerous gases or liquids can create a serious personnel hazard as well as contribute to environmental pollution.

Reliability. Component reliability has long been a major reason for leakage testing. Leak tests operate directly to assure serviceability of critical parts.

Leaks can result from poor seals and connections as well as from poor workmanship such as inadequate welds. Technically, a leak consists of a hole in an enclosure that is capable of passing a fluid from the higher pressure side to the lower pressure side.

Leaks do not always operate in the same manner and may tend to increase over time. Some leaks are open only intermittently due to temporary clogging by liquids. High-temperature or high-vacuum leaks can "disappear" when operating conditions are removed.

Simulating actual operating conditions, when possible, is the best approach whenever testing for critical leaks.

Although leakage typically occurs as a result of a pressure differential across the hole, capillary effects can also occur. When a fluid flows through a small leak, its rate of flow depends upon the geometry of the leak, the nature of the leaking fluid, the imposed pressure differential, and temperature conditions.

The flow characteristics of a leak are often referred to as the "conductance" of the leak. Because the hole cannot always be seen with the naked eye or be readily measured, the quantity used to define the leak is the conductance, or leakage rate, of a given fluid through the leak under given conditions.

The two most commonly used units of leakage rates are standard cubic centimeters per second (std. cm^3/s) for pressure leaks and torr-liters per second for vacuum leaks. Adoption of the International System of Units (SI units) has aided in standardizing the units of measure.

16.5 GAS FLOW IN LEAKS

Gas flow is an integral part of leak testing. A working knowledge of gas flows is important for comparing the flows of liquids to air as well as for determining the best conditions to perform a leak test. For laminar flow leaks, an increase in the pressure difference across the leak causes the flow rate through the leak to increase. Therefore, the easiest method of increasing detection sensitivity is to supply increased pressure. The leakage rate is inversely proportional to the test gas viscosity. Therefore, the less viscous the tracer gas, the more sensitive the test. The leaking tracer gas is then detected and measured by the gas survey instrument.

16.5.1 Method Selection

There are a great number of leak-testing methods. Each method has its own advantages and disadvantages as well as its own optimum sensitivity range. However, not all methods are useful for every application. By applying a number of selection criteria, the choice can often be narrowed to two or three methods with the final choice being determined by special circumstances or cost effectiveness.

Obviously, the first question to ask is, "What is the purpose of the test?" Is it to locate every leak of a certain size or is it to measure the total amount of leakage from the test object without regard to leak number or location? After the purpose of the test has been defined, one must determine whether the test object (such as a pipe fitting, valve, vessel, tank, or duct) is under pressure or under a vacuum. Another important criterion is the test object's size. When the size of the test object is quite large, electronic methods for locating leaks in pressurized units become increasingly impractical due to a stratifying of the tracer gas and the slowness with which the detector probe must be moved during the test procedure.

16.6 CAUTIONARY MEASURES

Caution must be used when selecting any hand-held electronic device to be used within an area having a hazardous area classification (such as Class I, Division I or Class I, Division II). Areas or rooms that are deemed Class I, Division I are areas that, under normal conditions, can have a flammable or explosive air–fuel mixture present that will ignite if a source of ignition is present. An example of a Class I, Division I area would be a room with open tanks that contain a flammable solvent. An area or room with a classification of Class I, Division II does not have an ignitable air–fuel mixture present unless an abnormal condition exists, such as a leak. An example of a Class I, Division II area would be a room with chemical reactors and piping that are under a negative pressure and contain a flammable material. Here, there is little risk of a fire or explosion unless a vessel or pipe fitting leaks.

Hazardous areas will typically have stationary gas analyzers in strategic locations of the area that produce an audible and remote alarm if a combustible gas or vapor is detected. The location of the fixed sensors is dependent upon the nature of the gas to be detected. If it is a relatively heavy gas, like chlorine, the detectors will be positioned near the floor level, where the gas is most likely to collect. If the gas is light, such as hydrogen, the detectors will be mounted in a high location where the gas is most likely to collect.

Much like a cellular phone or a calculator, hand-held analytical instruments can be a potential source of ignition if explosive, airborne concentrations of a gas or dust are present and the device is not intrinsically safe. A small spark from a battery-powered device and even a static discharge are sufficient to ignite a serious fire or catastrophic explosion if a flammable or explosive mixture of air and accelerant is present. The device must be rated by its manufacturer for use in that type of environment to avoid the possibility of acting as a source of ignition for a flammable or explosive gas or dust mixture.

Third-party-approved devices must be approved for the specific hazardous area (class, group, and division) where they are used. Acceptable approval generally means listing or approval by a national recognized testing laboratory such as Underwriters Laboratories (UL), Factory Mutual (FM), ETL Testing Laboratories (ETL), or Canadian Standards Association (CSA). All approved equipment must be marked with its approval ratings and in some cases its temperature rating. When it is stated in the National Electrical Code (NEC) that equipment is required to be approved for a particular class, then the equipment must be approved for both Division I and Division II. Do not assume the safety classification of an area without confirming it with the plant safety official or other knowledgeable person at the site.

16.7 PERSONAL EXPOSURE MONITORING

Personal exposure monitors are generally simple gas detectors with audible alarms or a chemically reactive card that changes color in the presence of a particular type of gas or vapor. These devices serve as a precautionary measure for work in the presence of potentially hazardous substances. Their advantages over more sophisticated analyzers include:

- Simplicity—They can be used by anyone (no hygienists or technicians are required).
- Light weight—A reusable badge may weigh as little as $\frac{1}{4}$ ounce.
- Badges and monitors are immediately ready for use. No activation or calibration is needed.

16.8 GAS SURVEY INSTRUMENT CLASSIFICATIONS

Portable gas survey instruments generally fall under one of three categories of gas-specific analyzers:

- "Sniffers"—to pinpoint the locations of leaks.
- Ambient analyzers—to measure ambient concentrations of a gas.
- Personal exposure monitors—to quantify the level of exposure to a specific gas.

Hand-held analyzers range in price from $200 to $2000 depending upon the range of gases they will analyze and their accuracy.

16.9 SNIFFERS

Sniffers typically have a flexible sampling tube or "wand" (rigid tube) or a very small chemical sensor located at the end of an attached rod (Fig. 16.2). The wand or tube serves as a probe that allows the sensor to be positioned very close to the test object and detect very small quantities (milliliters) of vapor at a pinpoint location (at the end of the tube). The gas or vapor at the tip of the wand is then electronically analyzed and scaled through an attached hand-held component.

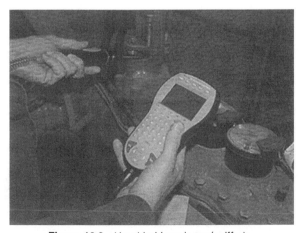

Figure 16.2 Hand-held analyzer (sniffer).

Electrochemical sensor technology offers highly specific gas sensing in the low-ppm range for many gases. Functionally, the electrochemical sensor comes in direct contact with the tested gas or vapor. The sensor is constructed of a material that changes its electrical properties due to a chemical reaction that occurs at the sensor when in the presence of a specific chemical vapor. The change in electrical properties is amplified and displayed as an alarm (indicating that the specific gas or gases being tested for are present) in ppm or as a percentage of the LEL of the measured gas.

An example of an alarm-type device is the TIF 8800 combustible gas detection instrument, which is typically used by home inspectors and HVAC (heating, ventilation, and air conditioning) contractors to detect the presence of hydrocarbons, alcohols, ethers, ketones, and other chemicals and compounds. The device has a 15-in.-long probe attached to a hand-held analyzer that clicks much like a Geiger counter. The clicking frequency increases as the probe gets closer to the source of the leak (a higher gas concentration). This device weighs approximately 16 ounces.

16.10   AMBIENT ANALYZERS

Ambient analyzers are used to determine the general concentration of a gas within a defined area, such as a room. An example of an ambient air analyzer is the Bacharach Snifit CO (carbon monoxide) analyzer, which can accurately measure CO concentration levels in room air to a resolution of 1 ppm over a range of 0–1999 ppm (Fig. 16.3). It weighs approximately 8 ounces, fits in a shirt pocket, and has a back-lit liquid-crystal display (LCD). The unit is powered by a standard 9-V battery. Other similar instruments are on the market and vary in size, cost, and accuracy.

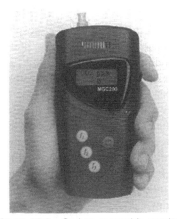

Figure 16.3   Carbon monoxide monitor.

16.11   GAS SURVEY INSTRUMENT CALIBRATION

Portable gas survey analyzers can pose a potential risk to human health and safety if the device is improperly calibrated and inaccurate readings occur. They should be periodically calibrated per the manufacturer's recommendations.

Typically, testing of a gas survey instrument is accomplished by zeroing the measurement and then testing it using a calibration gas (a gas of known concentration in clean air). The measurement reading is compared to the known gas concentration. If the readings differ, then the analyzer calibration is adjusted until the reading agrees with the test gas concentration, and the unit is rezeroed and then retested with the calibration gas. Distributors who sell the gas survey instruments can often provide the calibration gas cylinders needed to calibrate the unit.

A standard calibration kit will typically contain two 17-L disposable cylinders of calibration gas, a calibration adapter, a flowmeter, a valve assembly, and a length of tubing in a carrying case. After the initial purchase of a kit, replacement gas cylinders are ordered on an "as-needed" basis. The calibration gas standards are traceable to the National Institute of Standards and Technology (NIST).

Gas survey instruments should be calibrated to NIST-traceable standards if they are to be used for critical measurement applications. Critical applications would include such things as measuring exposure levels, determining the percentage of LEL, or recording measurements for regulatory compliance.

To help meet the measurement and standards needs of U.S. industry and the nation, NIST provides calibrations, standard reference materials, standard reference data, test methods, proficiency evaluation materials, measurement quality assurance programs, and laboratory accreditation services that assist a customer in establishing traceability of results of measurements or values of standards.

16.12   FUGITIVE EMISSIONS MONITORING

Fugitive emissions are those emissions to the atmosphere resulting from leaking pipe fittings and equipment such as valves, flanges, pump seals, connections, compressor seals, open lines, and pressure-relief valves. In general, these emissions are not visually observable but can be measured in relatively low-ppm concentrations at each source using the proper instrument. Although the emission of a single source can be relatively small, the cumulative effect of a large number of these leaking sources can result in significant emission. The acknowledgment of losses of raw materials, the potential danger of explosions, and the environmental impact has created awareness that companies and organizations need to improve their monitoring programs. In plants that lack thorough preventative maintenance programs, fugitive emissions are determined to be the most significant cause of atmospheric pollution resulting from plant activities.

16.13 GLOSSARY

ACGIH	American Conference of Governmental Industrial Hygienists.
Adsorption	The condensation of gas or vapor onto the surface of a solid.
Aerosol	A suspension in air (or gas) of minute particles of a liquid or a solid.
Ambient air	Air to which the sensing element is normally exposed.
Analyzer	An instrument that can determine qualitatively and quantitatively the components in a mixture.
ANSI	American National Standards Institute.
Area Monitor	Gas-monitoring sensors that are installed throughout an area requiring monitoring.
ASNT	American Society for Non-Destructive Testing.
ASTM	American Society for Testing of Materials.
Background	The environment against which an indication must be evaluated.
Blocking	Certain conditions can cause a sensor not to function. When this happens, normal gas sensing is blocked until the conditions are removed. The most common "block" is due to a lack of oxygen.
Calibration	The procedure used to adjust the instrument for proper response.
Calibration Gas	A gas of known concentration(s) used to set the instrument span or alarm level(s) to a corresponding value.
Capillary Action	The tendency of certain liquids to travel, climb, or draw into a tight crack (e.g., interface areas) due to such properties as surface tension, wetting, cohesion, adhesion, and viscosity.
Colorimetric Detectors	Rapid and inexpensive leak detectors that react chemically with minute leaks, causing a visible color change in the developer.
Combustion	The rapid oxidation of a material evolving heat and generally light.
Consumables	Those materials or components that are depleted or require periodic replacement through normal use of the instrument.
Diffusion	A process by which the atmosphere being monitored is transported to the gas-sensing element by natural random molecular movement.

Explosion

An uncontrolled chemical reaction that generates a large amount of heat and gas within a very short period of time.

Fixed Installation

A gas monitor that is permanently installed, such as in the control panel of a control room. Occasionally gas monitors are mounted in vehicles, such as fire trucks or tankers. These are also generally referred to as fixed-installation monitors.

Flammable (Explosive) Limits

Gases or vapors that form flammable mixtures with air or oxygen have a minimum concentration of vapor in air or oxygen below which the propagation of a flame does not occur on contact with the source of ignition. There is also a maximum mixture of vapor (or gas) and air, above which the propagation of flame does not occur. These boundary line mixtures of vapor (or gas) with air, which if ignited will propagate flame, are known as the lower and upper flammable limits (LFL and UFL) or the lower and upper explosive limits (LEL and UEL) and are usually expressed in terms of percentage by volume of gas or vapor in air. In internal combustion terms, a mixture below the lower flammable limit is too "lean" to burn or explode and a mixture above the upper flammable limit too "rich" to burn or explode.

Flash Point

The minimum temperature at which a liquid emits sufficient vapor to reach 100% LEL (sufficient vapor to form an ignitable mixture with the air near the surface of the liquid).

Gas

A phase of matter that expands indefinitely to fill a containment vessel. Gases are characterized by their low density.

Gas Detection

The process of detecting a gas.

Gas Detection Instrument

An assembly of electrical, mechanical, and chemical components comprising a system that senses and responds to the presence of gas in air mixtures.

Indication

A test response that requires interpretation and evaluation.

Interferent

Any gas other than the target gas that will cause a gas-detecting sensor to give a signal.

Lag Time

The time in a test between input and observable meter response.

Leak

Technically, a hole in an enclosure capable of passing a fluid from the higher pressure side to the lower pressure side.

Leakage Rate	Quantity or measure of leakage per unit time (leak rate).
Leak Detection	The process of detecting a leak.
Liquid	A phase of matter that is free to conform to the internal shape of a vessel but having a fixed volume and a greater density than a gas.
Lower Explosive or Lower Flammable Limit (LFL)	The smallest amount of the gas that will support a self-propagating flame when mixed with air (or oxygen) and ignited. In gas detection systems, the amount of gas present is specified in terms of % LEL: 0% LEL being a combustible gas-free atmosphere and 100% LEL being an atmosphere in which the gas is at its lower flammable limit. The relationship between % LEL and percent by volume differs from gas to gas.
NIOSH	National Institute for Occupational Safety and Health.
Nondestructive Testing	The examination of an object or material with technology that does not affect the object's future usefulness.
Nonincentive	Devices that can spark under normal operating conditions but cannot release enough energy to cause ignition.
OSHA	Occupational Safety and Health Administration, a government agency.
Peak	Maximum one-time exposure (usually 10 min) per OSHA standards.
Permeation Leak	A leak through a barrier that has no hole or discrete passage.
Permissible Exposure Limit (PEL)	The cumulative average concentration over an 8-hour day, 40-hour week to which a worker can be safely exposed per OSHA standards.
Poisons	Gas-detecting sensors can be quickly destroyed (poisoned) by certain materials that chemically attack the sensor.
Portable	A self-contained, battery-operated or transportable gas monitor worn or carried by the person using it (i.e., a gas detector that can be carried).
ppm	Parts per million (1% exposure = 10,000 ppm).
Quantitative Leak Measurement	The overall leakage measurement for a complete component or system but with no location.
Range	The series of outputs corresponding to values of concentrations of the gas of interest over which accuracy is ensured by calibration.

Sensitivity of Leak Test	The smallest leakage rate that the gas survey instrument is capable of detecting under a specified set of conditions (e.g., pressure, temperature).
Sensor	A gas-detecting sensor converts the presence of a gas or vapor into an electrically measurable signal.
Soak Time	The time it takes for a tracer gas to fully penetrate the boundary walls.
Stationary	A gas detection instrument intended for permanent installation in a fixed location.
Stoichiometric	The exact percentage of two or more substances that will react completely with each other, leaving no unreacted residue.
Test Gas	A known amount of the gas to be detected diluted with a known amount of a clean, inert gas.
Toxic Gas or Vapor	Any substance that causes illness or death when inhaled or absorbed by the body in relatively small quantities.
Tracer Gas	A gas that, passing through a leak, can be detected by a specific leak detector and thus disclose the presence of the leak.
Ultrasonic Leak Detector	A detector that translates inaudible ultrasonic frequencies into a variety of recognizable sounds and meter readings.
Vapor	The gaseous state of a substance below its boiling point.
Vapor Density	A relationship between the molecular weight of a gas and the molecular weight of air.
Zero Gas	A clean air source that ensures a small release of gas is not near the sensor while zeroing the sensor signal and during calibration.

17

ION CHROMATOGRAPHY FOR THE ANALYSIS OF INORGANIC ANIONS IN WATER

Peter E. Jackson, Ph.D.

Dionex Corporation
Sunnyvale, California

Environmental Instrumentation and Analysis Handbook, by Randy D. Down and Jay H. Lehr
ISBN 0-471-46354-X   Copyright © 2005 John Wiley & Sons, Inc.

17.1 INTRODUCTION

17.1.1 Background

The technique of ion chromatography (IC) was originally developed by Hamish Small and co-workers at Dow Chemical in the mid 1970s. They employed a low-capacity, ion exchange stationary phase for the separation of analyte ions in conjunction with a second ion exchange column and conductivity detector that allowed for continuous monitoring of the eluted ions.[1] The second column was called a "stripper" column (later termed a *suppressor*) and served to reduce the background conductance of the mobile phase (or eluent) and enhance the detectability of the eluted ions. The term *ion chromatography* was subsequently introduced when this technology was licensed to the Dionex Corporation for commercial development.[2]

The introduction of IC stimulated renewed interest in the determination of ionic solutes by liquid chromatography and prompted much investigation into alternative separation and detection approaches for the analysis of inorganic compounds. In 1979, Fritz and co-workers showed that a suppressor was not essential to achieving sensitive conductivity detection provided that appropriate low-capacity stationary phases and low-conductance eluents were used.[3,4] Since that time, a considerable variety of separation and detection approaches have been employed for the determination of ionic species by IC.

17.1.2 Definition and Scope

This diversity has led to the point where IC is now perhaps best defined by the range of solutes to which it is applied rather than by any specific combination of separation and detection modes.[5] Hence, IC can be considered to encompass liquid chromatographic techniques that can be used for the determination of inorganic anions; inorganic cations, including alkali metal, alkaline earth, transition metal, and rare earth ions; low-molecular-weight (water-soluble) carboxylic acids plus organic phosphonic and sulfonic acids, including detergents; carbohydrates; low-molecular-weight organic bases; and ionic metal complexes.[5–8]

The growth of IC was very rapid because it provided a reliable and accurate method for the simultaneous determination of many common inorganic ions. In the early stages of its development, IC was viewed as a tool for the determination of simple inorganic species in environmental samples. Most early applications of IC were for the analysis of inorganic anions and cations in samples such as air filter extracts, soil extracts, and drinking water and natural water samples.[9] As the range of solutes that could be determined by IC continued to expand, so did the application areas in which the technique was applied. In addition to environmental applications, IC is now routinely used for the analysis of ionic compounds in diverse areas, such as chemical and petrochemical industries, semiconductor and high-purity water applications, food and beverage applications, clinical and pharmaceutical industries, and mining and metallurgical applications.

Ion chromatography can now be considered a well-established, mature technique for the analysis of ionic species.[10] Many organizations, such as the American Society for Testing and Materials (ASTM), Association of Analytical Communities (AOAC), and U.S. Environmental Protection Agency (EPA), have standard or regulatory methods of analysis based upon IC. Despite the diverse range of solutes and sample types currently analyzed by IC, environmental analysis continues to be the largest application area of IC in terms of new instrument sales and the total number of samples analyzed. In terms of the solutes analyzed in environmental applications of IC, inorganic anions are by far the most important. The primary reason for this is the lack of alternative methods for anion analysis, which is not the case for cations, where many other instrumental methods are available.

17.1.3 Basic Principles

Ion chromatography is a liquid chromatographic technique applied specifically to the determination of ionic solutes. The instrumentation used for IC is similar to that employed for conventional high-performance liquid chromatography (HPLC), although the wetted surfaces of the chromatographic system are typically made of an inert polymer, such as poly(tetrafluoroethylene) (PTFE) or (more commonly) poly(ether ether ketone) (PEEK), rather than stainless steel. This is due to the fact that the corrosive eluents and regenerant solutions used in IC, such as hydrochloric or sulfuric acids, can contribute to corrosion of stainless steel components, which ultimately leads to problems due to contamination from metal ions.[8] The components of a typical IC system are shown schematically in Figure 17.1.

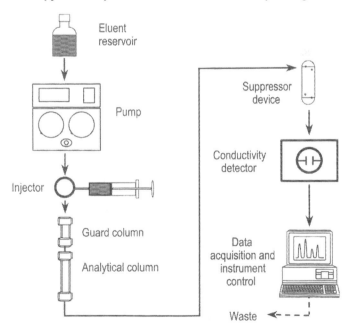

Figure 17.1 Basic components of typical IC system.

An inert pump, which can be either single or dual piston, delivers the mobile phase (or eluent) throught the chromatographic system. As is the case with HPLC, the use of dual piston pumps is preferred as they provide the most pulse-free operation, which typically translates into lower baseline noise.[11] The injector can be either a manually operated loop injector based upon a six-port switching valve or a more complex, automated injector that allows unattended operation. Injection volumes used in IC are typically on the order of 10–200 μL.

Ion exchange is the primary separation mode used in IC, as opposed to traditional HPLC, where reversed-phase chromatography is the dominant separation mechanism. The first commercially available stationary phases used in IC were agglomerated ion exchange resins.[5–7] These agglomerated, or pellicular, materials consist of a monolayer of charged latex particles that are electrostatically attached to a surface-functionalized, internal core particle. This inner particle is totally covered by the fully functionalized latex; consequently the properties of the outer latex determine the ion exchange selectivity of the composite material. These pellicular ion exchange resins have higher efficiencies than conventional microporous ion exchangers as a result of faster kinetics and greater permeability of the pellicular layer.[12]

In addition to ion exchange, other approaches used for the separation of inorganic ions include ion interaction, ion exclusion, and chelation chromatography, while reversed-phase separations can be used for the analysis of neutral metal complexes. Agglomerated ion exchange resins are still the most commonly used stationary phases, although both surface-functionalized and grafted ion exchange resins are also routinely used for IC separations.

Ion chromatography also differs from HPLC in that conductivity is the primary method of detection, as opposed to ultraviolet/visible (UV/VIS) in conventional HPLC. Conductivity is a bulk property detector and provides universal (nonselective) response for charged, ionic compounds.[11] Conductivity detection can be operated in the direct (or nonsuppressed) mode or with the use of an ion exchange–based device, termed a suppressor, which is inserted between the ion exchange column and the conductivity detector. The suppressor is a post–column reaction device unique to IC that greatly improves the signal-to-noise ratio for conductivity detection by (a) reducing the background conductance of the eluent and (b) enhancing the detectability of the eluted ions.

The suppression process is illustrated schematically for anion analysis using the membrane-based anion self-regenerating suppressor (ASRS) shown in Figure 17.2. Hydronium ions, which are generated at the anode by electrolysis of water, are exchanged for sodium ions in the eluent chamber prior to the measurement of conductance. This exchange, which occurs via two cation exchange membranes, results in conversion of the eluent anion (hydroxide) to its much less conductive weak-acid form (water in this case). In addition, the conductance of the analyte ion pair is enhanced as a result of replacing the less conductive co-cation (sodium) with the significantly more conductive hydronium ion. The reverse situation can be described for cation analysis, where the suppressor is an anion exchange, membrane-based device.

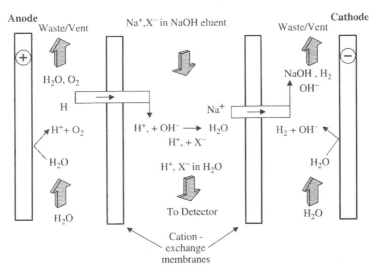

Figure 17.2   Ion movement within a membrane-based ASRS.

Special requirements exist for both the anionic and cationic eluent components to be used for the suppressed ion chromatographic determination of anions. The eluent cation must be able to displace hydronium ions from the suppressor, the eluent anion must be readily protonated to form a weak acid, and the eluent anion must have a sufficient ion exchange affinity to elute the solute anions within a reasonable time.[5] Solutions of sodium carbonate–sodium bicarbonate, sodium borate, or sodium hydroxide are most commonly used as eluents for anion analysis, while sulfuric acid or methanesulfonic acid are most commonly used for cation analysis. The suppressor device itself has undergone considerable improvement since the introduction of IC. Suppressors have evolved from packed-bed columns that required periodic offline regeneration to low-dead-volume, membrane-based devices that use the electrolysis of water (or even recycled eluent) to continuously provide regenerant ions for the suppression neutralization reaction.[13]

Both UV/VIS and RI detection, when operated in the indirect mode, can be employed as alternatives to conductivity for universal detection, although the sensitivity obtained using these methods is typically inferior to suppressed conductivity detection.[5] Direct detection methods, such as UV/VIS or amperometry, have proven to be highly sensitive for certain UV-absorbing or electroactive species, while post–column derivatization followed by UV/VIS absorption or fluorescence is an important detection approach for transition metals, lanthanides, and actinides. Additionally, the use of more advanced detection techniques for IC, such as mass spectrometry (MS) and inductively coupled plasma (ICP) MS, continue to be explored.

Despite the wide variety of separation and detection modes available, the combination of ion exchange separation with suppressed conductivity detection, as originally described by Small and co-workers, remains the most commonly used approach in IC. This is particularly true for the analysis of inorganic anions in environmental water samples. Anion exchange chromatography provides the most appropriate separation selectivity for inorganic anions, while suppressed conductivity detection permits improved detection sensitivity compared to nonsuppressed detection for these solutes.[7]

17.2 ALTERNATIVE ANALYTICAL TECHNIQES

Ion chromatography is well established as a regulatory method for the analysis of inorganic anions in environmental samples. Acceptance of IC for the analysis of anionic solutes was very rapid, primarily due to the lack of alternative methods that could determine multiple anions in a single analysis. A variety of methods have been employed for the analysis of inorganic anions, including traditional spectroscopic techniques, such colorimetry; wet chemical methods, such as gravimetric analysis, turbidimetry, and titrimetry; and electrochemical techniques, such the use of ion-selective electrodes (ISEs) and amperometric titrations.[14] Many of these methods suffer from interferences and limited sensitivity, can be labor intensive, and are often difficult to automate. The use of flow injection analysis (FIA) enables the automation of certain colorimetric and ISE methods of analysis, although still only for one analyte at a time. Multiple analytes can be determined by adding additional "channels" to an FIA system; however this adds complexity and cost to the instrument.

During the early development of IC, many comparisons between wet chemistry methods and IC were performed in order to validate the latter technique. For instance, in a 1984 comparison, IC was found to be equivalent to conventional wet chemistry methods for the determination of common anions, such as chloride, nitrate, and sulfate, in drinking water samples.[15] Table 17.1 details the approved conventional methods used (at the time) for the analysis of the inorganic anions commonly found in drinking water. Considering that these six individual test procedures could be replaced by one 30-min chromatographic separation, it is

Table 17.1 Conventional Methods for Analysis of Inorganic Anions in Drinking Water

Analyte	Conventional Analytical Method	Method Number
Fluoride	Ion-selective electrode	U.S. EPA 340.2
Chloride	Potentiometric titration	APHA 407C
Nitrite	Automated (FIA) sulfanamide[a]	U.S. EPA 354
Nitrate	Automated (FIA) cadmium reduction[a]	U.S. EPA 365.1
Phosphate	Automated (FIA) ascorbic acid[a]	U.S. EPA 353.2
Sulfate	Turbidimetric	U.S. EPA 375.1

Source: Ref. 15.

[a] FIA methods used spectrophotometry for quantification after appropriate color formation.

not surprising that IC quickly became accepted by regulatory bodies worldwide for the analysis of anions in drinking water.

However, the situation regarding the analysis of cations is quite different than the case for anions. Many rapid and sensitive spectroscopic methods, such as atomic absorption spectroscopy (AAS), ICP atomic emission spectroscopy (AES), ICP MS, and electrochemical methods, such as polarography and anodic stripping voltammetry, are available for cation analysis. Many of these are multielement techniques and hence duplicate one of the major attractions of chromatographic methods. Regulatory methods for cation (metal) analysis in environmental samples tend to be based primarily upon AAS and ICP instrumentation,[14] although IC does offer an advantage over spectroscopic techniques in the area of metal speciation.[5]

17.3 SAMPLE HANDLING AND PREPARATION

The primary concerns when collecting environmental samples for analysis using any measurement technique are that the sample collected is representative of the total sample matrix and no contamination occurs during the sampling process.[5] Also, appropriate storage and preservation of the sample is required in order that the final sample analysis is representative of the analyte concentrations present when the sample was originally taken from the field.

17.3.1 Sample Storage and Preservation

Water samples collected for analysis by IC ideally should be collected in plastic containers, such as polystyrene or polypropylene bottles, as glass bottles can contribute ionic contamination when performing trace analysis.[16] The bottles should be thoroughly rinsed with reagent-grade water before use. Sample preservation requirements and holding times for anions typically determined by IC are listed in Table 17.2.

17.3.2 Sample Dissolution

Most water samples collected for IC analysis require little or no sample pretreatment. Drinking water samples, for instance, typically require no pretreatment other than filtration through a 0.45-μm filter to remove particulates. Higher ionic strength water samples (e.g., wastewater) often only require dilution (and filtration) to bring the analytes of interest into the working range of the method. In fact, this so-called dilute-and-shoot approach to sample preparation is one of the advantages of IC when it comes to the practical application of this technique.[20] However, solid samples, such as soils, sludge, plant matter, and geologic materials, are not directly amenable to IC analysis and require additional sample pretreatment. Strategies employed for the dissolution of such samples include extraction (most commonly into aqueous media), acid digestion, alkali fusion, or sample combustion.[5]

Table 17.2 Sample Preservation and Holding Times for Anions Commonly Determined Using IC

Analyte	Preservation	Holding Time (days)
Acetate	Cool to 4°C	2
Bromate[a]	Add 50 mg/L EDA[b]	28
Bromide	None required	28
Chlorate[a]	Add 50 mg/L EDA	28
Chloride	None required	28
Chlorite[a]	Add 50 mg/L EDA, cool to 4°C	14
Chromate	Adjust pH to 9–9.5 with eluent[c]	1
Cyanide	Adjust pH to >12 with NaOH, cool to 4°C	14
Fluoride	None required	28
Formate	Cool to 4°C	2
Nitrate[d]	Cool to 4°C	2
Nitrite[d]	Cool to 4°C	2
o-Phosphate	Cool to 4°C	2
Sulfate	None required	28

Sources: Refs. 14, 17–19.

[a] Samples collected for oxyhalide analysis should be immediately sparged with an inert gas (e.g., nitrogen, argon, or helium) for 5 min to remove active gases such as chlorine dioxide or ozone. Samples for chlorite should be stored in amber containers.

[b] EDA = ethylenediamine.

[c] Eluent = 250 mM ammonium sulfate/100 mM ammonium hydroxide.

[d] Holding times can be increased by adjusting to pH 12 with sodium hydroxide.

17.3.3 Sample Cleanup

Once in solution, it is typically necessary to perform some degree of sample pretreatment or cleanup prior to injection into the ion chromatograph. This pretreatment may be as simple as filtration or a complicated, time-consuming matrix elimination step. The typical intent of sample cleanup is to achieve one or more of the following goals: (1) removal of particulates that could cause blockages or damage to the instrument, (2) reduction of the overall sample loading on the column, (3) concentration or dilution of the target analytes, and (4) removal of matrix interferences.[5]

17.3.3.1 Filtration. As is the case with all liquid chromatographic methods, samples analyzed by IC should be free of particulates to avoid blockages or damage to connecting tubing, column end frits, and other hardware components. Samples are typically filtered through a 0.45-μm (or less) membrane-based filter. Disposable "syringe" filters are commercially available and their use greatly simplifies sample filtration. Also, certain types of autoinjectors will automatically filter the sample before injection into the IC instrument. However, the possibility of sample contamination from these devices can be a concern, particularly when performing trace analysis. Rinsing the filters with 20 mL of deionized water prior to filtration of the sample has been shown to remove most inorganic contaminants.[21]

17.3.3.2 Matrix Elimination. Complex aqueous samples, such as wastewaters and solid leachates, often require further chemical modification (cleanup) of the sample in order to eliminate matrix interferences. Solid-phase extraction (SPE) cartridges represent the most convenient means of removing interferences prior to ion chromatographic analysis.[22] These commercially available, disposable cartridges enable rapid sample pretreatment and require only small volumes of sample. Solid-phase extraction cartridges are available with many different chromatographic packing materials, including silica, alumina, C_{18}, anion exchange resins (OH^- form), cation exchange resins (H^+, Ag^+, Ba^+ forms), neutral polymer, amino, and activated carbon.[5] These cartridges can be employed with IC analyses in a number of different modes of operation, as discussed below.

(**a**) Hydrophobic SPE cartridges (e.g., C_{18} and neutral polymers) can be used to remove neutral organic compounds while allowing inorganic ions to pass through unretained. Hydrophobic organic compounds do not typically interfere during an IC separation; however they can be strongly retained on the stationary-phase material, which can lead to decreased column lifetimes. This approach is typically required when using IC for the analysis of anions in wastewaters and soil leachates, which contain high levels of organic material (e.g., humic acids).[18,23]

(**b**) Cation and anion exchange SPE cartridges in the H^+ and OH^- forms, respectively, can be used to adjust sample pH and reduce total ionic strength without adding a potentially interfering co-anion (for acids) or co-cation (for bases) to the sample.[22]

(**c**) Cation exchange SPE cartridges in the H^+ form can be used to remove carbonate and cationic species, such as Fe(III) and aluminum, which may precipitate under alkaline eluent conditions.[18] Cation exchange cartridges in the Ag^+ and Ba^+ forms can be used to selectively remove halides and sulfate from samples by precipitating insoluble silver halides or barium sulfate, respectively.[22] The Ag^+ form cartridges are widely used to selectively remove chloride from environmental waters to allow trace analysis of anions, such as bromate in ozonated waters, which would otherwise be masked by the excess chloride.[24]

Recently, SPE disks have become available as an alternative to the cartridge configuration. These disks are available with many of the same packings as the cartridges, although their geometry allows the use of higher sample loading flow rates.[25] The use of dialysis across membranes offers another means of reducing sample interferences, and this approach has been used to reduce sample acidity,[26] basicity,[27] and chloride in brine samples prior to IC analysis.[28]

17.3.3.3 Online Matrix Elimination. Many sample preparation techniques can also be performed "on line" using instrumental methods. This approach offers the benefits of greater precision, and the process can often be automated, although the instrumentation is usually more complex. The most common application of online matrix elimination in IC is sample preconcentration. This approach is most frequently used for the determination of trace ions in high-purity waters, e.g., steam and boiler feed water for power station generators.[29] However, sample preconcentration can be sucessfully applied to high-ionic-strength samples, particularly if the solutes of interest are strongly retained on the concentrator column. The strongly retained solutes can be trapped on the concentrator column while weakly retained solutes and water pass through to waste. This approach, which effectively combines sample preconcentration with matrix elimination, has been used for the determination of anions in natural waters and low-ppb levels of anionic metal cyanide complexes in gold tailing solutions.[30,31]

Automated matrix elimination can also be performed using "heart-cut" techniques. This involves loading the sample onto one analytical column, then carefully switching only the fraction of eluent containing the solute(s) of interest toward a second separator column. This approach, which typically performs best when the sample matrix has a consistent composition, has been used for the determination of trace anions in samples containing high levels of chloride.[20]

17.4 REGULATORY METHODS OF ANALYSIS

Ion chromatography has been approved by many standard or regulatory organizations in numerous countries for the analysis of both anions and cations in environmental samples. While a complete listing of approved IC methods worldwide is beyond the scope of this chapter, a list of the most important regulatory IC methods used for the analysis of anions in environmental waters in the Unites States is given in Table 17.3.[32]

Table 17.3 Regulatory IC Methods approved in U.S. for the Analysis of Anions in Environmental Waters and Waste

Method	Analytes	Matrices[a]
EPA 300.0 (A)	F, Cl, NO_2, Br, NO_3, PO_4, SO_4	rw, dw, sw, ww, gw, se
EPA 300.0 (B)	BrO_3, ClO_3, ClO_2	Raw water, dw
EPA 300.1 (A)	F, Cl, NO_2, Br, NO_3, PO_4, SO_4	rw, dw, sw, gw
EPA 300.1 (B)	BrO_3, Br, ClO_3, ClO_2	rw, dw, sw, gw
EPA SW-846 9056	F, Cl, Br, NO_3, PO_4, SO_4	Combustion extracts, all waters
ASTM D 4327-97	F, Cl, NO_2, Br, NO_3, PO_4, SO_4	dw, ww
Standard methods 4110	Cl, NO_2, Br, NO_3, PO_4, SO_4	rw, dw, ww
EPA 300.6	Cl, NO_3, PO_4, SO_4	Wet deposition, rain, snow, dew, sleet, hail
ASTM D 5085-90	Cl, NO_3, SO_4	Wet deposition, rain, snow, sleet, hail
EPA B-1011	NO_2, NO_3	rw, dw
EPA SW-846 9058	ClO_4	rw, dw, gw
ASTM D 2036-97	CN	dw, sw, ww
EPA 218.6	Hexavalent chromium (CrO_4^{2-})	dw, gw, ww
EPA SW-846 7199	Hexavalent chromium (CrO_4^{2-})	dw, gw, ww
ASTM D 5257-93	Hexavalent chromium (CrO_4^{2-})	dw, sw, ww

[a] Matrices: rw = reagent water; dw = drinking water; sw = surface water; ww = wastewater (mixed domestic and industrial); gw = groundwater; se = solid extracts; mw = marine water; ew = esturine water.

Many regulatory agencies promulgate what are essentially similar methods, as Table 17.3 illustrates. For instance, ASTM D 4327-97 uses the same methodology as EPA Method 300.0(A); however, each agency has a unique method format and writing style. Also, differences exist between the methods in the area of quality control (QC); for example, QC is mandated in almost all EPA methods while it is optional in most ASTM methods. Different regulatory agencies exist even within the EPA. Hence Method 300.0 is applicable to the analysis of inorganic anions in drinking water and wastewater under direction of the Office of Ground Water and Drinking Water, while Method 9056 is applicable to the analysis of inorganic anions in all water types and combustion bomb extracts under direction of the Office of Solid Waste and Emergency Response.

While Table 17.3 details only methods approved in the United States, many industrial countries around the world have similar health and environmental standards. Consequently, a considerable number of regulatory IC methods exist worldwide. For instance, German National Standards DIN 38 405 (D 20) and DIN ISO 10 304-1 are similar to EPA Method 300.0, while DIN 38 405 (D 22) is applicable to the determination of chromate, iodide, sulfite, thiocyanate, and thiosulfate in water matrices. French method AFNOR T90-042 is again similar to EPA method 300.0, as is the Italian method UNICHIM 926 (1991). Several IC methods are approved in Japan for the analysis of anions, including matrices such as industrial waters (K0101), industrial wastewater (K0102), and mine water and wastewater (M0202).[33] Japan also has a method detailing general rules for IC analysis (K0127), while Standards Australia has a recommended practice for chemical analysis by IC (AS 3741-1990).[34]

17.5 WATER AND WASTEWATER ANALYSIS

17.5.1 Drinking Water and Wastewater

Water quality in the United States is legislated through the Safe Drinking Water Act (SDWA) and the Clean Water Act (CWA). The SWDA ensures the integrity and saftey of U.S. drinking water, while the goal of the CWA is to reduce the discharge of pollutants into U.S. waters. Most regulatory methods of analysis that use IC are validated for both drinking water and wastewater; hence these matrices will be considered together.

17.5.1.1 Common Inorganic Anions. The U.S. National Primary Drinking Water Standards specify a maximum contaminant level (MCL) for a number of common inorganic anions, including fluoride, nitrite, and nitrate. The MCLs are specified to minimize potential health effects arising from the ingestion of these anions in drinking water. For instance, high levels of fluoride cause skeletal and dental fluorosis, while nitrite and nitrate can cause methemoglobulinemia, which can be fatal to infants.[23] Consequently, the analysis of these anions in drinking waters is mandated, as are the analytical methods that can be used for their quantification. Other common anions, such as chloride and sulfate, are considered secondary contaminants. The Secondary Drinking Water Standards are guidelines regarding taste, odor, color, and certain aesthetic effects that are not federally enforced. However, they are recommended to all states as reasonable goals, and many of states adopt their own enforceable regulations governing these contaminants.[35]

Ion chromatography has been approved for compliance monitoring of these common inorganic anions in drinking water in the United States since the mid-1980s, as described in EPA method 300.0. This same method received interim approval for the analysis of inorganic anions in wastewater under the National Pollution Discharge Elimination System (NPDES) in 1992. Method 300.0 specifies the use of a Dionex AS4A anion exchange column with an eluent of 1.7 mM sodium bicarbonate–1.8 mM sodium carbonate for the separation of common anions. An optional column may be substituted provided comparable resolution of peaks is obtained and the QC requirements of the method can be met.[36] Conductivity detection is used for quantification after suppression of the eluent conductance with an anion micromembrane suppressor (AMMS) or similar device.

Figure 17.3 shows a chromatogram of a standard containing low-ppm levels of common anions obtained using the conditions described in EPA Method 300.0. All the anions are well resolved within a total run time of less than 8 min. The application range and method detection limits (MDLs) that can be achieved for these anions using method 300.0 are shown in Table 17.4. Similar methods, such as ASTM D 4327-97 and Standard Methods 4110, provide comparable performance. Advances in column and suppressor technology continue to improve the methodology for determination of these common anions. The IonPac AS14 column provides

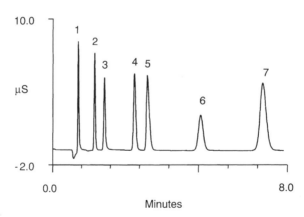

Figure 17.3 Separation of low-ppm anion standard using EPA method 300.0. Conditions: column, Dionex IonPac AS4A-SC; eluent, 1.8 mM sodium carbonate–1.7 mM sodium bicarbonate; flow rate, 2.0 mL/min; detection, suppressed conductivity with an ASRS operated at 50 mA in recycle mode; injection volume, 25 μL; solutes, 1—fluoride (2 mg/L), 2—chloride (3 mg/L), 3—nitrite (5 mg/L), 4—bromide (10 mg/L), 5—nitrate (10 mg/L), 6—phosphate (15 mg/L), 7—sulfate (15 mg/L).

Table 17.4   U.S. EPA Method 300.0 Application Ranges and Detection Limits

Analyte	Application Range (mg/L)	MDL (mg/L)
Fluoride	0.26–8.49	0.01
Chloride	0.78–26.0	0.02
Nitrite-N	0.36–12.0	0.004
Bromide	0.63–21.0	0.01
Nitrate-N	0.42–14.0	0.002
Orthophosphate-P	0.69–23.1	0.003
Sulfate	2.85–95.0	0.02

Source: Ref. 36.

complete resolution of fluoride and acetate and also improved resolution of fluoride ·from the void peak compared to the AS4A column. Figure 17.4 shows a chromatogram of a typical drinking water sample obtained using an AS14 column with a 3.5 mM bicarbonate–1.0 mM carbonate eluent and suppressed conductivity detection.

EPA Method 300.0 is also validated for wastewater analysis, although these samples often require sample pretreatment before injection into the IC. Dilution into the application range followed by filtration is frequently required, while pretreatment with SPE cartridges to remove hydrophobic organic material is recommended to prolong column lifetimes. Figure 17.5 show a chromatogram of a typical domestic wastewater sample obtained using an AS4A column with a carbonate–bicarbonate eluent and suppressed conductivity detection. The performance of environmental

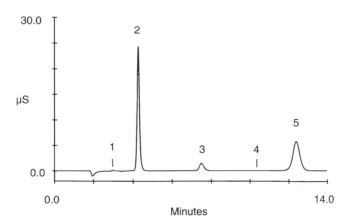

Figure 17.4   Determination of inorganic anions in drinking water. Conditions: as for Figure 17.3, except: column, Dionex IonPac AS14; eluent, 3.5 mM sodium carbonate–1.0 mM sodium bicarbonate; injection volume, 50 µL; solutes, 1—fluoride (0.03 mg/L), 2—chloride (10.1 mg/L), 3—nitrate (3.7 mg/L), 4—phosphate (0.04 mg/L), 5—sulfate (12.2 mg/L).

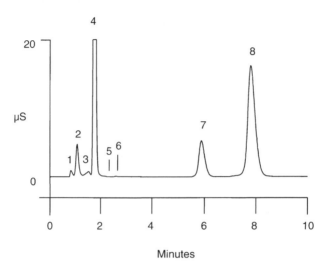

Figure 17.5 Determination of anions in domestic wastewater. Conditions: as for Figure 17.3, except: injection volume, 50 μL; sample preparation, SPE cleanup with a Waters C_{18} Sep-Pak; solutes, 1—injection peak, 2—acetate (4.2 mg/L), 3—bicarbonate (not quantitated), 4—chloride (49 mg/L), 5—bromide (0.04 mg/L), 6—nitrate (0.03 mg/L), 7—phosphate (28 mg/L), 8—sulfate (35 mg/L).

methods, such as EPA Method 300.0, is typically validated through single- and multioperator precision and bias studies on spiked samples. Table 17.5 shows single-operator precision and bias data obtained using Method 300.0 for common anions spiked into reagent water, drinking water, and mixed domestic and industrial wastewaters. Acceptable precision and bias data were obtained for the determination of common anions in all three matrix types when using IC.

17.5.1.2 *Disinfection By-Product Anions.* U.S. EPA Method 300.0 was revised in 1993 to include determination of the disinfection by-product (DBP) anions—bromate, chlorite, and chlorate. Bromate is a DBP produced from the ozonation of source water that contains naturally occurring bromide, while chlorite and chlorate are produced by using chlorine dioxide as a disinfectant. These DBP anions pose significant health risks, even at low-μg/L levels. Bromate has been judged as a potential carcinogen, and the EPA has estimated a potential cancer risk equivalent to 1 in 10^4 for a lifetime exposure to drinking water containing bromate at 5 μg/L.[37]

The occurrence of bromate and other DBPs in U.S. drinking water was recently studied by the EPA through the comprehensive collection of data mandated by the Information Collection Rule (ICR). The EPA proposed a MCL of 10 μg/L bromate in finished drinking water, which was promulgated in stage I of the D/DBP rule.[38] EPA Method 300.0(B) specifies the use of a Dionex AS9-SC anion exchange column with a 1.7 mM sodium bicarbonate–1.8 mM sodium carbonate eluent and

Table 17.5 U.S. EPA Method 300.0 Single-Operator Precision and Bias

Analyte	Matrices[a]	Added Concentration (mg/L)	Mean Recovery[b] (%)	Standard Deviation (mg/L)
Fluoride	RW	2.0	91	0.05
	DW	1.0	92	0.06
	WW	1.0	87	0.07
Chloride	RW	20.0	96	0.35
	DW	20.0	108	1.19
	WW	20.0	101	5.2
Nitrite-N	RW	10.0	91	0.14
	DW	10.0	121	0.25
	WW	5.0	91	0.50
Bromide	RW	5.0	99	0.08
	DW	5.0	105	0.10
	WW	5.0	105	0.34
Nitrate-N	RW	10.0	103	0.21
	DW	10.0	104	0.27
	WW	10.0	101	0.82
Orthophosphate-P	RW	10.0	99	0.17
	DW	10.0	99	0.26
	WW	10.0	106	0.85
Sulfate	RW	20.0	99	0.40
	DW	50.0	105	3.35
	WW	40.0	102	6.4

Source: Ref. 36.

[a]Matrices: RW = reagent water; DW = drinking water; WW = wastewater (mixed domestic and industrial).

[b]Average of seven replicates.

suppressed conductivity detection for the analysis of bromate, chlorite, and chlorate. However, Method 300.0(B) could not meet the quantitation requirements specified by the ICR, and method modifications, including the use of a weaker borate eluent to improve bromate and chloride resolution and the use of sample pretreatment to minimize chloride interference, are required.[39]

Prior to the publication of stage I of the D/DBP rule, the EPA developed Method 300.1 for the determination of inorganic and DBP anions. Method 300.1 specifies a Dionex AS9-HC anion exchange column with a 9.0 mM sodium carbonate eluent and suppressed conductivity detection. The AS9-HC column differs from the column specified in Method 300.0(B) in that it has higher capacity and improved separation of the key oxyhalide anions from potential interferences. Method 300.1(A) is applicable to common inorganic anions in drinking water, while 300.1(B) is applicable to the determination of DBP anions and bromide in drinking water. Methods 300.1(A) and (B) use the same column but different injection

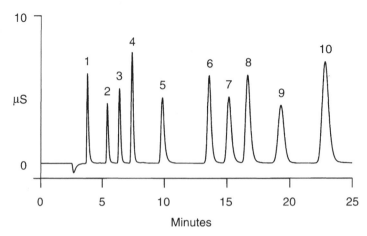

Figure 17.6 *Separation of oxyhalides plus common inorganic anions using EPA Method 300.1. Conditions: column, Dionex IonPac AS9-HC (4 mm i.d.); eluent, 9.0 mM sodium carbonate; flow rate, 1.0 mL/min; detection, suppressed conductivity with an ASRS operated at 100 mA in external water mode; injection volume, 25 µL; solutes, 1—fluoride (3 mg/L), 2—chlorite (10 mg/L), 3—bromate (20 mg/L), 4—chloride (6 mg/L), 5—nitrite (15 mg/L), 6—bromide (25 mg/L), 7—chlorate (25 mg/L), 8—nitrate (25 mg/L), 9—phosphate (40 mg/L), 10—sulfate (30 mg/L).*

volumes to achieve different MDLs. Method 300.1(A) requires 10 µL while 300.1(B) requires 50 µL when using a 2-mm-i.d. column or 50- and 200-µL injections, respectively, with a 4-mm-i.d. column.[17] The relatively large injection volume is necessary for 300.1(B) in order to achieve the required detection limits when analyzing for DBP anions.

Figure 17.6 shows the separation of chlorite, bromate, and chlorate in addition to the common inorganic anions using an AS9-HC column. The MDLs for Method 300.1(B) obtained with a 50-µL injection and 2-mm-i.d. column are 0.89, 1.44, 1.44, and 1.31 µg/L for chlorite, bromate, bromide, and chlorate, respectively.[17] Figure 17.7 shows the application of Method 300.1(B) to the determination of DBP anions in drinking water from Sunnyvale, California. The water in this municipality is disinfected using hypochlorite; hence chlorate appears in the drinking water matrix, shown in Figure 17.7a. The same drinking water spiked with 10 µg/L each of chlorite, bromate, bromide, and chlorate is shown in Figure 17.7b, indicating that the DBP anions are clearly resolved from the common inorganic anions (and bromide) present in drinking water. Despite using a 200-µL injection, no column overloading occurs and bromate can be determined at 12 µg/L in the presence of a 1000-fold excess of chloride. In fact, Method 300.1(B) allows the determination of bromate at the 5-µg/L level in the presence of as much as 200 µg/mL of chloride, a 40,000-fold concentration difference. Table 17.6 shows single-operator precision and bias data obtained using Method 300.1 for the DBP anions spiked into reagent water, drinking water, and high-ionic-strength drinking water.

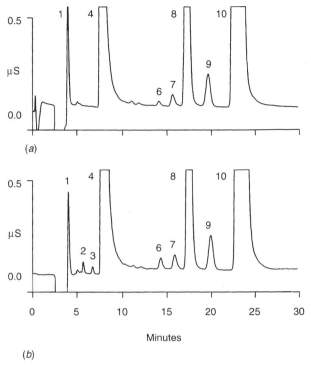

Figure 17.7 Determination of oxyhalides and common anions in Sunnyvale drinking water. Conditions: as for Figure 17.6, except: sample, (a) drinking water and (b) drinking water spiked with 0.01 mg/L of oxyhalide anions and bromide; solutes, (a) 1—fluoride (0.05 mg/L), 4—chloride (19 mg/L), 6—bromide (0.004 mg/L), 7—chlorate (0.03 mg/L), 8—nitrate (1.7 mg/L), 9—phosphate (0.25 mg/L), 10—sulfate (30 mg/L) and (b) 1—fluoride (0.05 mg/L), 2—chlorite (0.008 g/L), 3—bromate (0.012 mg/L), 4—chloride (19 mg/L), 6—bromide (0.013 mg/L), 7—chlorate (0.041 mg/L), 8—nitrate (1.7 mg/L), 9—phosphate (0.25 mg/L), 10—sulfate (30 mg/L).

17.5.1.3 *Hexavalent Chromium.*

U.S. EPA Methods 300.0 and 300.1 and equivalent methods in the United States and other countries represent the most important and widely used applications of IC in environmental analysis. However, a number of other regulatory methods based on IC are applicable to drinking water and wastewater analysis. Inorganic chromium is a primary drinking water contaminant with an MCL of 0.1 mg/L. Hexavalent chromium is the most toxic form of the metal, in addition to being a suspected carcinogen. Hexavalent chromium exists in solution primarily as the chromate anion, which can be separated using a high-capacity Dionex AS7 anion exchange column, as specified in EPA Method 218.6.[19] In this case, detection is achieved using a UV/VIS detector after postcolumn reaction with diphenylcarbohydrazide, as this color-forming reagent provides a more sensitive and selective means for determining chromate than suppressed conductivity.

U.S. EPA Method 218.6 and ASTM D 5257-93 are validated for the determination of hexavalent chromium in drinking water, groundwaters, and industrial

Table 17.6 U.S. EPA Method 300.1 Single-Operator Precision and Bias

Analyte	Matrices[a]	Added Concentration (μg/L)	Mean Recovery[b] (%)	Standard Deviation (μg/L)
Chlorite	RW	100	96.2	0.95
	O3W	100	84.4	0.46
	HIW	100	102	2.19
Bromate	RW	5.00	101	0.45
	O3W	5.00	80.9	0.61
	HIW	5.00	97.5	0.95
Bromide	RW	20.0	104	0.80
	O3W	20.0	—[c]	3.67
	HIW	20.0	92.5	0.79
Chlorate	RW	100	98.3	0.80
	O3W	100	100	1.20
	HIW	100	86.1	1.47

Source: Ref. 17.

[a] Matrices: RW = reagent water; O3W = ozonated drinking water; HIW = high-ionic-strength simulated drinking water.

[b] Average of nine replicates.

[c] Not calculated as added amount was less than the unfortified amount.

wastewaters. An MDL of 0.3 μg/L for Cr(V) in drinking water and wastewater can be achieved using a 250-μL injection. Average recoveries on the order of 98–105% were obtained for 100-μg/L Cr(VI) solutions spiked into reagent, drinking, ground, primary sewage waste, and electroplating wastewaters.[19] Figure 17.8 shows a chromatogram of a spiked wastewater sample obtained using the conditions described in method 218.6. No interfering peaks are observed when using this very specific detection approach for Cr(VI) analysis.

17.5.1.4 Cyanide.
The highly toxic cyanide anion is a primary drinking water contaminant that has an MCL of 0.2 mg/L. Sources of cyanide contamination in drinking water include effluents from electroplating, steel, plastics, and mining industries in addition to certain fertilizers.[35] Cyanide is classified according to its "availability" in the presence of complexing metals. *Total cyanide* refers to CN that can be released as hydrocyanic acid (HCN) from both the aqueous and particulate portions of a sample under total reflux distillation conditions. *Free cyanide* refers to CN that can be released as HCN from the aqueous portion of a sample by direct cyanide determination without reflux distillation. *Cyanide amenable to chlorination* refers to CN determined after chlorinating a portion of sample, while *weak-acid dissociable cyanide* refers to CN determined after distillation with a weak acid.[40]

In practice, the majority of CN determinations, particularly in wastewater samples, involve measurement of total CN, which is determined after reflux distillation of an alkaline sample in the presence of sulfuric acid and a magnesium chloride

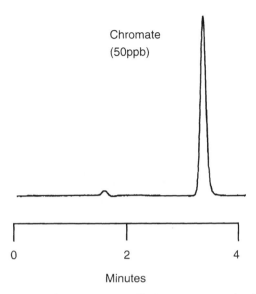

Figure 17.8 Determination of chromate in spiked wastewater using EPA Method 218.6. Conditions: column, IonPac Dionex AS7; eluent, 250 mM ammonium sulfate–100 mM ammonium hydroxide; flow rate, 1.5 mL/min; detection, UV/VIS at 530 nm after postcolumn reaction with 2 mM diphenylcarbazide–10% methanol–1.0 N sulfuric acid delivered at 0.5 mL/min; injection volume, 100 μL; sample, filtered wastewater spiked with 50 μg/L chromate.

catalyst.[40] The released HCN is absorbed into a sodium hydroxide scrubber solution, and the cyanide in this solution can be measured colorimetrically by IC, FIA, titration, or ion-selective electrodes. An IC separation is recommended with electrochemical detection when sulfur, thiocyanate, or other sulfur containing compounds are present in the sample, as H_2S codistills with HCN and can interfere with the FIA determination when using electrochemical detection.[40]

ASTM D 2036-97 is validated for the determination of total cyanide in drinking water, groundwater, surface water, and both domestic and industrial wastes using IC, in addition to other analytical measurement techniques. Cyanide is separated on a Dionex AS7 anion exchange column using an eluent of 100 mM sodium hydroxide–500 mM sodium acetate–0.5% (v/v) ethylenediamine. The CN is then detected using an amperometric detector with a silver working electrode operated at −0.05 V.[24] This very sensitive detection approach provides an MDL of 2 μg/L when using 1.0 mL injection volume and can tolerate sulfur concentrations up to 100 times the cyanide level without degradation of method performance. Figure 17.9 shows a chromatogram of cyanide and sulfide in a spiked wastewater sample obtained using an AS7 column and amperometric detection. The applicable range of this method is from 10 μg/L to 10 mg/L cyanide, and mean recoveries on the order of 85–98% were obtained for samples spiked with CN over the range of 40–1000 μg/L.

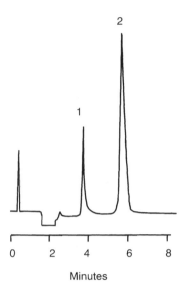

Figure 17.9 Determination of sulfide and cyanide in spiked wastewater using ASTM Method D2036-97. Conditions: column, Dionex IonPac AS7; eluent, 500 mM sodium acetate–100 mM sodium hydroxide; flow rate, 1.0 mL/min; detection, amperometry using a silver working electrode operated at −0.05 V versus Ag/AgCl reference; injection volume, 100 μL; sample, filtered spiked wastewater; solutes, 1—sulfide (25 μg/L), 2—cyanide (100 μg/L).

17.5.1.5 Perchlorate. Ammonium perchlorate is a key ingredient in solid rocket propellant that has recently been found in groundwaters in regions of the United States where aerospace material, munitions, and fireworks were developed, tested, or manufactured. Perchlorate has been found in groundwater and surface water in California, Nevada, Utah, and West Virgina.[41] Perchlorate poses a human health risk, and preliminary data from the EPA report that exposure to less than 4–18 μg/L provides adequate health protection.[42] Perchlorate is listed on the EPA Contaminant Candidate List as a research priority, although it is not currently regulated under the federal SDWA.[37] Perchlorate contamination of public drinking water wells is becoming a serious problem in some western states, and the California Department of Health Services (CDHS) has adopted an action level for perchlorate in drinking water of 18 μg/L. To date, perchlorate has been detected in over 100 public drinking water wells in California, with more than 20 wells being closed due to contamination.[42]

The CDHS developed an IC method based on the use of an hydrophilic IonPac AS5 column, large loop injection, and suppressed conductivity detection to quantify perchlorate at low-μg/L levels.[43] However, the use of an IonPac AS11 or AS16 column with an eluent of 65–100 mM hydroxide, 1000-μL injection, and suppressed conductivity detection provides an MDL for perchlorate of <0.3 μg/L without the need for an organic modifier in the mobile phase.[41] Figure 17.10 shows a chromatogram of perchlorate standard at 20 μg/L, while Figure 17.11 shows a drinking

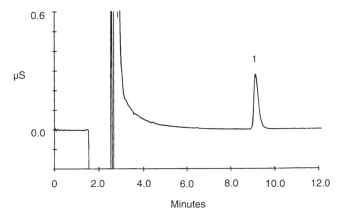

Figure 17.10 Chromatogram of 20 μg/L perchlorate standard. Conditions: column, Dionex IonPac AS11; eluent, 100 mM sodium hydroxide; flow rate, 1.0 mL/min; detection, suppressed conductivity with an ASRS-Ultra operated at 300 mA in external water mode; injection volume, 1000 μL; solutes, 1—perchlorate.

water sample spiked with 6.0 μg/L perchlorate. The applicable range of this method is from 2.0 to 100 μg/L perchlorate. The method is free of interferences from common anions, and quantitative recoveries were obtained for low-μg/L levels of perchlorate in spiked drinking and groundwater samples.[41]

17.5.2 Natural Waters

In addition to being approved for a number of drinking water and wastewater analyses, IC is also widely used for analysis of natural water samples. Many of

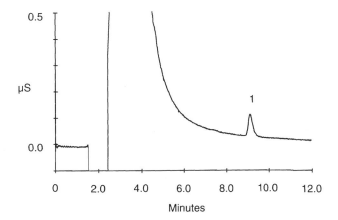

Figure 17.11 Analysis of perchlorate in spiked drinking water. Conditions: as for Figure 17.10, except: sample, filtered Sunnyvale, CA, tapwater spiked with 6 μg/L perchlorate.

the regulatory methods described in Section 17.5.1 are also applicable to natural waters, such as groundwater and surface water. Natural waters encompass a wide variety of sample matrices, including rain and acid rain; groundwaters; surface waters (such as river, stream, lake, and pond waters); soil pore waters; run-off waters; snow, hail, and sleet; ice and ice cores; well and bore waters; and so on. There are few regulations governing the analysis of such samples; hence a great diversity of IC methods are applied to a much wider range of analytes than in the highly regulated area of drinking water and wastewater analysis. The application of IC to the analysis of natural waters and other environmental samples was comprehensively reviewed in 1990.[5,44]

The analysis of rain water and acid rain is one of the more important applications of IC. The determination of ionic components in rain waters by IC is frequently used to estimate the effects of acidification on the natural and urban environments caused by acid rain.[45] The major ionic components of acid rain consist of the hydronium ion, sodium, ammonium, potassium, calcium, magnesium, chloride, nitrate, and sulfate. U.S. EPA Method 300.6, which is similar to EPA Method 300.0, although now somewhat dated, was validated for the determination of inorganic anions in acid rain and rain water.[46]

Many researchers have worked to develop methods that allow the simultaneous determination of both anions and cations in rain waters and atmospheric aerosols. Tanaka and co-workers developed a simple, elegant approach that enables the simultaneous determination of anions and cations in acid rain based upon a simultaneous ion exclusion–cation exchange separation and nonsuppressed conductimetric detection.[45] Figure 17.12 shows an example of a chromatogram of anions and cations in rain water obtained using this approach.

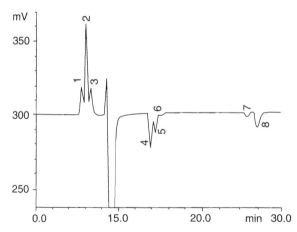

Figure 17.12 Simultaneous determination of anions and cations in acid rain. Conditions: column, Tosoh TSKgel OA-PAK; eluent, 5 mM tartaric acid–7.5% methanol; flow rate, 1.2 mL/min; detection, nonsuppressed conductivity; injection volume, 100 µL; sample, acid rain from Nagoya, Japan; solutes, 1—sulfate, 2—chloride, 3—nitrate, 4—sodium, 5—ammonium, 6—potassium (10 mg/L), 7—magnesium, 8—calcium. Reproduced with permission from Ref. 45.

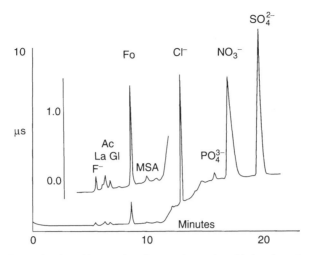

Figure 17.13 Determination of inorganic anions and organic acids in rain water using a borate gradient. Conditions: column, Dionex IonPac AS11; eluent, DI water/sodium tetraborate gradient; flow rate, 1.0 mL/min; injection volume, 25 μL; detection, suppressed conductivity using an AMMS. Reproduced with permission from Ref. 47.

Gradient elution is typically required in order to quantify all of the minor acid components of rain water samples, as organic acid anions, such as formate, acetate, and methanesulfonate, are often present at low levels in rain water samples.[47] Figure 17.13 shows the separation of inorganic anions and organic acids in rain water obtained using a borate gradient with a Dionex AS11 column and suppressed conductivity detection. Gradient separations can be used to quantify a wide range of anionic solutes in complex water samples, such as hazardous waste leachates. In addition to rain water analysis, IC has also been used to analyze terrestrial waters and ice cores from pristine environments, such as Antartica, in order to establish "baseline" levels of pollutants.[48,49]

The analysis of groundwater and surface water is another common application of IC. The determination of inorganic solutes in waters from rivers, streams, lakes, and ponds is similar in complexity to the analysis of typical wastewater samples. Filtration followed by pretreatment with SPE cartridges to remove hydrophobic organic material is recommended when analyzing most groundwater and surface water. Figure 17.14 shows the analysis of inorganic anions in lake water from Salt Lake in Utah using an AS4A column and suppressed conductivity detection.

17.5.3 Brines

The analysis of brines by IC is complicated by the high ionic strength and excess sodium chloride in the sample. Nevertheless, IC is frequently used for the analysis of inorganic solutes in natural water brines, which include seawater, subsurface brines, geothermal brines, and very high salinity groundwaters. The analysis of major components (e.g., chloride and sulfate) in brines is relatively straightforward,

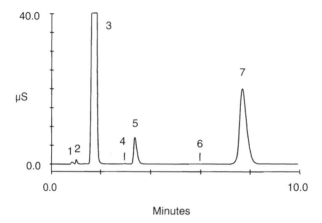

Figure 17.14 Determination of anions in lake water. Conditions: as for Figure 17.3, except: sample, water from Salt Lake, Utah; solutes, 1—injection peak, 2—fluoride (0.28 mg/L), 3—chloride (142 mg/L), 4—bromide (0.18 mg/L), 5—nitrate (11.2 mg/L), 6—phosphate (0.28 mg/L), 7—sulfate (44 mg/L).

only requiring a dilution before direct injection into the IC.[50] However, the analysis of minor components in brines typically requires careful selection of analytical conditions, and a wide variety of approaches have been used for the analysis of these minor components. The use of IC for the analysis of anions in high-salt-concentration environmental waters has been reviewed by Singh and co-workers.[51] In addition to the use of conventional suppressed IC, other approaches include using sodium chloride as the eluent; sample pretreatment with Ag form membranes and SPE cartridges; the use of heart-cut matrix elimination or preconcentration combined with matrix elimination; and the use of more selective detection methods, such as amperometry or post–column derivatization.[22,25,51–56]

The use of sodium chloride as an eluent combined with low-wavelength UV detection allows the determination of UV-absorbing anions, such as nitrate, iodide, and molybdate, in samples containing up to 20,000 mg/L of chloride without loss of chromatographic performance.[52] A similar approach with an eluent containing chloride and an ion pairing reagent has been used for the analysis of thiosulfate and polythionates in natural saline waters.[54] Sample pretreatment with Ag form SPE cartridges or membranes to reduce chloride is a commonly used approach to allow determination of minor anionic components in brines. Chloride precipitates as AgCl, although other halides are also removed to a significant extent using this approach and poor recoveries can be obtained for nitrite.[22,25] Selective detection can be applied to the determination of specific ions; for instance, iodide has been determined in brine using pulsed amperometry at a silver working electrode after separation on a Dionex AS11 column with an eluent of 50 mM nitric acid. Iodide could be quantified at 16 μg/L in 30% NaCl after a 10-fold dilution and calibration using standard addition.[55]

Automated matrix elimination techniques have also been used to determine anions and metals in seawater samples. Nitrite, bromide, nitrate, and sulfate have been determined in brine using a heart-cut and recycling system. A unresolved "cut" of the sample containing chloride and nitrite is trapped in a sample loop and reinjected onto an anion exchange column using four- and six-port switching valves. This method permits a detection limit of 0.5 mg/L for nitrite in spiked seawater samples when using IC with suppressed conductivity detection.[56]

17.6 QUALITY CONTROL FOR ION CHROMATOGRAPHIC ANALYSIS

Quality control is an essential part of environmental water analysis when it comes to generating reliable results using IC or any analytical method. A great deal of literature is available in the area of quality management, and the specifics of quality assurance in environmental analysis have been described in considerable detail elsewhere (e.g., Ref. 57). In addition, U.S. EPA methods contain detailed (and mandatory) QC sections specific to the method of analysis. For instance, EPA methods 300.0 and 300.1 provide detailed instructions on QC procedures to be implemented

Table 17.7 Typical QC Section of IC Method Used for Environmental Analysis

Quality Control Procedure	Intent of Procedure
Initial demonstration of performance (IDP)	Analysis of seven replicates of IDP solution to demonstrate laboratory (or operator) proficiency using the test method
Initial calibration verification using calibration verification standard (CVS)	Run CVS to check calibration standards and acceptable instrument performance.
Run one CVS with each sample batch. A batch is typically defined as somewhere from 10 up to a maximum of 20 samples	Ongoing verification of previously established calibration curves; analyte concentrations to fall within acceptable limits (typically $\pm 15\%$ known value)
Run one reagent blank with each sample batch.	Ongoing check for contamination introduced by laboratory or method
Run one quality control sample (QCS) with each sample batch.	The analyte recoveries of the QCS should fall within control limits of $x \pm 3S$; where x is the mean recovery and S is the standard deviation.
Run one matrix spike (MS) with each sample batch.	Ongoing test of method recovery
Run one matrix duplicate (MD) with each sample batch.	Ongoing test of method precision
Additional quality control (QC)	Any laboratory may perform additional QC as desired or appropriate to its internal quality program

Source: Ref. 57.

when analyzing inorganic anions and disinfection byproduct anions in environmental samples using IC.[36,17] An example of the requirements of a typical QC section contained in IC intended for environmental analysis is detailed in Table 17.7.[58]

ACKNOWLEDGMENTS

All chromatograms are courtesy of the Dionex Corporation unless otherwise stated. The author thanks the Dionex Corporation for permission to use the chromatograms presented in this chapter.

REFERENCES

1. H. Small, T. S. Stevens, and W. C. Bauman, Novel Ion Exchange Chromatographic Method Using Conductometric Detection, *Anal. Chem.* **47**, 1801–1809 (1975).

2. H. Small and B. Bowman, Ion Chromatography: A Historical Perspective, *Am. Lab.* **30**, 56C–62C (1998).

3. D. T. Gjerde, J. S. Fritz, and G. Schmuckler, Anion Chromatography with Low-Conductivity Eluents, *J. Chromatogr.* **186**, 509–519 (1979).

4. D. T. Gjerde, G. Schmuckler, and J. S. Fritz, Anion Chromatography with Low-Conductivity Eluents, *J. Chromatogr.* **187**, 35–45 (1980).

5. P. R. Haddad and P. E. Jackson, *Ion Chromatography: Principles and Applications*, *J. Chromatogr. Library*, Vol. 46, Elsevier, Amsterdam, 1990.

6. C. A. Pohl, J. R. Stillian, and P. E. Jackson, Factors Controlling Ion-Exchange Selectivity in Suppressed Ion Chromatography, *J. Chromatogr.* **789**, 29–41 (1997).

7. H. Small, *Ion Chromatography*, Plenum, New York, 1989.

8. J. Weiss, *Ion Chromatography*, 2nd ed., VCH, Weinheim, 1995.

9. E. Sawicki, J. D. Mulik, and E. Wittgenstein (Eds.), *Ion Chromatographic Analysis of Environmental Pollutants*, Vol. 1, Ann Arbor Science, Ann Arbor, MI, 1978.

10. C. A. Lucy, Recent Advances in Ion Chromatography: A Perspective, *J. Chromatogr.* **739**, 3–13 (1996).

11. K. Robards, P. R. Haddad, and P. E. Jackson, *Principles and Practice of Modern Chromatographic Methods*, Academic, London, 1994.

12. R. W. Slingsby and C. A. Pohl, Anion-Exchange Selectivity in Latex-Based Columns for Ion Chromatography, *J. Chromatogr.* **458**, 241–253 (1988).

13. S. Rabin, J. Stillian, V. Barreto, K. Friedman, and M. Toofan, New Membrane-Based Electrolytic Suppressor Device for Suppressed Conductivity Detection in Ion Chromatography, *J. Chromatogr.* **640**, 97–109 (1993).

14. A. E. Greenberg, L. S. Clesceri, and A. D. Eaton (Eds.), *Standard Methods for the Examination of Water and Wastewater*, 18th ed., American Public Health Association, Washington, DC, 1992.

15. A. Eaton, M. Carter, A. Fitchett, J. Oppenheimer, M. Bollinger, and J. Sepikas, Comparability of Ion Chromatography and Conventional Methods for Drinking Water Analysis, *Proc. AWWA Water Qual. Tech. Conf.*, 175–188 (1984).

16. ASTM D 5542-94, *Standard Test Methods for Trace Anions in High Purity Water by Ion Chromatography*, Vol. 11.01, American Society for Testing and Materials, Philadelphia, PA, 1995, pp. 696–703.

17. U.S. EPA Method 300.1, *The Determination of Inorganic Anions in Water by Ion Chromatography*, U.S. Environmental Protection Agency, Cincinnati, OH, 1997.

18. C. A. Lucy, Practical Aspects of Ion Chromatographic Determinations, *LC.GC* **14**, 406–415 (1996).

19. U.S. EPA Method 218.6, *Determination of Dissolved Hexavalent Chromium in Drinking Water, Groundwater and Industrial Wastewater Effluents by Ion Chromatography*, U.S. Environmental Protection Agency, Cincinnati, OH, 1991.

20. S. R. Villasenor, Matrix Elimination in Ion Chromatography by Heart-Cut Column Switching Techniques, *J. Chromatogr.* **602**, 155–161 (1992).

21. R. Bagchi and P. R. Haddad, Contamination Sources in the Clean-Up of Samples for Inorganic Ion Analysis, *J. Chromatogr.* **351**, 541–547 (1986).

22. I. K. Henderson, R. Saari-Nordhaus, and J. M. Anderson, Sample Preparation for Ion Chromatography by Solid Phase Extraction, *J. Chromatogr.* **546**, 61–71 (1991).

23. P. E. Jackson, P. R. Haddad, and S. Dilli, The Determination of Nitrate and Nitrite in Cured Meats using High Performance Liquid Chromatography, *J. Chromatogr.* **295**, 471–478 (1984).

24. R. J. Joyce and H. S. Dhillon, Trace Level Determination of Bromate in Ozonated Drinking Water Using Ion Chromatography, *J. Chromatogr.* **671**, 165–171 (1994).

25. R. Saari-Nordhaus, L. M. Nair, and J. M. Anderson, Elimination of Matrix Interferences in Ion Chromatography by the Use of Solid Phase Extraction Disks, *J. Chromatogr.* **671**, 159–163 (1994).

26. P. E. Jackson and W. R. Jones, Membrane-Based Sample Preparation Device for the Pretreatment of Acidic Samples Prior to Cation Analysis by Ion Chromatography, *J. Chromatogr.* **586**, 283–289 (1991).

27. P. R. Haddad and S. Laksana, On-Line Analysis of Alkaline Samples with a Flow-Through Electrodialysis Device Coupled to an Ion Chromatograph, *J. Chromatogr.* **671**, 131–139 (1994).

28. P. E. Jackson and W. R. Jones, Hollow-Fiber Membrane-Based Sample Preparation Device for the Clean-up of Brine Samples Prior to Ion Chromatographic Analysis, *J. Chromatogr.* **538**, 497–503 (1991).

29. D. F. Pensenstadler and M. A. Fulmer, Pure Steam Ahead, *Anal. Chem.* **53**, 859A–868A (1981).

30. P. E. Jackson and P. R. Haddad, Studies on Sample Preconcentration in Ion Chromatography VII. Review of Methodology and Applications of Anion Preconcentration, *J. Chromatogr.* **439**, 37–48 (1988).

31. P. R. Haddad and N. E. Rochester, Ion-Interaction Reversed-Phase Chromatographic Method for the Determination of Gold(I) Cyanide in Mine Process Liquors Using Automated Sample Preconcentration, *J. Chromatogr.* **439**, 23–36 (1988).

32. Dionex, *Official U. S. Analytical Methods Specifying Ion Chromatography*, Dionex, Sunnyvale, CA, 1996.

33. Dionex, *Environmental Analysis Seminar*, Dionex, Sunnyvale, CA, 1995.

34. Australian Standard AS 3741–1990, *Recommended Practice for Chemical Analysis by Ion Chromatography*, Standards Australia, North Sydney, NSW, 1990.

35. U.S. EPA, *Fed. Reg.*, 62 FR 52193 (1997).

36. U.S. EPA Method 300.0, *The Determination of Inorganic Anions in Water by Ion Chromatography*, U.S. Environmental Protection Agency, Cincinnati, OH, 1993.

37. U.S. EPA, *Fed. Reg.*, 59 (145) FR 38709 (1994).

38. R. Roehl, R. Slingsby, N. Avdalovic, and P. E. Jackson, Applications of Ion Chromatography with Electrospray Mass Spectrometric Detection for the Determination of Environmental Contaminants in Water, *J. Chromatogr. A* **956**, 245–254 (2002).

39. U.S. EPA, *DBP/ICR Analytical Methods Manual*, U.S. Environmental Protection Agency, Cincinnati, OH, 1996.

40. ASTM D 2036-97, *Standard Test Methods for Cyanide in Water*, American Society for Testing and Materials, Philadelphia, PA, 1997.

41. P. E. Jackson, S. Gokhale, T. Streib, J. S. Rohrer, and C. A. Pohl, An Improved Method for the Determination of Trace Perchlorate in Ground and Drinking Waters by Ion Chromatography, *J. Chromatogr. A* **888**, 151–158 (2000).

42. California Department of Health Services, California's Experience with Perchlorate in Drinking Water, update September 5, 2001, http://www.dhs.cahwnet.gov/ps/ddwem/chemicals/perchl.htm.

43. California Department of Health Services (CDHS), *Determination of Perchlorate by Ion Chromatography*, CDHS, Berkeley, CA, 1997.

44. W. T. Frankenberger, H. C. Mehra, and D. T. Gjerde, Environmental Applications of Ion Chromatography, *J. Chromatogr.* **504**, 211–245 (1990).

45. K. Tanaka, K. Ohta, J. S. Fritz, S. Matsushita, and A. Miyanaga, Simultaneous Ion-Exclusion Chromatography Cation Exchange Chromatography with Conductometric Detection of Anions and Cations in Acid Rain Waters, *J. Chromatogr.* **671**, 239–248 (1994).

46. U. S. EPA Method 300.6, *Chloride, Orthophosphate, Nitrate, and Sulfate in Wet Deposition by Chemically Suppressed Ion Chromatography*, U.S. Environmental Protection Agency, Cincinnati, OH, 1986.

47. A. A. Ammann and T. B. Ruttimann, Simultaneous Determination of Small Organic and Inorganic Anions in Environmental Water Samples by Ion-Exchange Chromatography, *J. Chromatogr.* **706**, 259–269 (1995).

48. K. A. Welch, W. B. Lyons, E. Graham, K. Neumann, J. M. Thomas, and D. Mikesell, Determination of Major Elements in Terrestrial Waters from Antartica by Ion Chromatography, *J. Chromatogr.* **739**, 257–263 (1996).

49. J. Ivask and J. Pentchuk, Analysis of Ions in Polar Ice Core Samples by Use of Large Injection Volumes in Ion Chromatography, *J. Chromatogr.* **770**, 125–127 (1997).

50. R. Kadnar and J. Rieder, Determination of Anions in Oilfield Waters by Ion Chromatography, *J. Chromatogr.* **706**, 301–305 (1996).

51. R. P. Singh, N. M. Abbas, and S. A. Smesko, Suppressed Ion Chromatographic Analysis of Anions in Environmental Waters Containing High Salt Concentrations, *J. Chromatogr.* **733**, 73–91 (1996).

52. Marheni, P. R. Haddad, and A. R. McTaggart, On-Column Matrix Elimination of High Levels of Chloride and Sulfate in Non-Suppressed Ion Chromatography, *J. Chromatogr.* **546**, 221–228 (1991).

53. A. C. M. Brandao, W. W. Buchburger, E. C. V. Butler, P. A. Fagan, and P. R. Haddad, Matrix-Elimination Ion Chromatography with Post-Column Reaction Detection for the Determination of Iodide in Saline Waters, *J. Chromatogr.* **706**, 271–275 (1996).

54. S. I. Weir, E. C. V. Butler, and P. R. Haddad, Ion Chromatography with UV Detection for the Determination of Thiosulfate and Polythionates in Saline Waters, *J. Chromatogr.* **671**, 197–203 (1994).

55. Dionex Application Update 122, *Determination of Iodide in Brine*, Dionex, Sunnyvale, CA, 1998.

56. P. F. Subosa, K. Kihara, S. Rokushika, H. Hatano, T. Murayama, T. Kuboto, and Y. Hanoaka, Ion Chromatography of Inorganic Anions in Brine Samples, *J. Chrom. Sci.* **27**, 680–684 (1989).

57. W. P. Cofino, Quality Assurance in Environmental Analysis, in D. Barcelo (Ed)., *Environmental Analysis—Techniques, Applications and Quality Assurance* Elsevier, Amsterdam, 1993.

58. ASTM 6581-00, *Standard Test Method for Bromate, Bromide, Chlorate, and Chlorite in Drinking Water by Chemically Suppressed Ion Chromatography*, American Society for Testing and Materials, Philadelphia, PA, 2000.

18

ULTRAVIOLET–VISIBLE ANALYSIS OF WATER AND WASTEWATER

Bernard J. Beemster

Applied Spectrometry Associates, Inc.
Waukasha, Wisconsin

Over the past few years, photometric and spectrometric analyzers have been introduced for numerous online analysis applications. Although commonplace in the laboratory, spectrometry was not considered to be a reliable technology for field

Environmental Instrumentation and Analysis Handbook, by Randy D. Down and Jay H. Lehr
ISBN 0-471-46354-X   Copyright © 2005 John Wiley & Sons, Inc.

use prior to the mid-1980s due to the need for a controlled environment, expert operators, and frequent maintenance for complex, mechanically adjusted optical systems. The availability of low-cost standardized optical components such as light sources, solid-state wavelength separation filters and gratings, improved photomultiplier and diode array detectors, fiber optics, and embedded computation systems made it possible to construct rugged, reliable systems using light absorbance measurements for analysis. These systems may use single or multiple (sometimes numerous) wavelengths in the ultraviolet (UV), visible, or near-infrared wavelength range, with or without assistance from reagents, for environmental analysis of water and wastewater.

18.1   INTRODUCTION: SPECTROMETRIC METHODS OF ANALYSIS

Although a wide variety of models are available to describe interactions between matter and electromagnetic radiation, this section addresses a subset of optical analysis where the energy level of the matter to be analyzed is altered through the application of radiation in the form of light and where changes in the energy level of the matter is subsequently measured as a function of wavelength. The radiation source may be monochromatic, such as a lamp followed by a filter or a laser, which will result in the application of radiation at a single wavelength. Or the radiation source may be polychromatic and result in either the simultaneous application of all radiated wavelengths or the sequential application of a range of wavelengths. Detection may also be at a single wavelength or over a range of wavelengths using a photoelectric detector.

For purposes of this chapter, we will refer to any method that measures a change in energy anywhere in the optical range of the electromagnetic spectrum (from the far infrared through the vacuum UV) as a *photometric* method. We will reserve the term *spectrophotometric* for those methods that measure changes in energy over a *range* of wavelengths in the optical range (a spectrum).

In practical terms, photometric methods can be classified by the number of wavelengths used to measure energy changes (single or multiple), the source of the radiation (arc, spark, flame, laser, lamp, etc.), and the range of the radiation source (continuous or discrete) and the orientation of the detector in relation to the source and the sample matrix (transmissive or reflective.) Methods frequently classified as photometric measure other effects of optical radiation, such as the angle of refraction or amount of defraction from the sample or the time interval between the application of radiation and the measurement of a change in energy.

18.2   BASIC PRINCIPLES OF UV–VISIBLE ABSORPTION SPECTROMETRY

Radiation of light in the UV and visible wavelength range will result in a change in energy by certain molecules resulting from the absorption of energy. When a molecule absorbs radiation, its energy increases. In molecules responsive to UV and visible light, the increase in energy is attributable to changes in electronic, vibrational, and rotational energy in outer shell electrons. The electrons most likely to gain energy in molecules are loosely bound valence electrons, electrons participating in double or triple bonds, and the unpaired electrons in certain radicals or ions. Under normal conditions a molecule occupies the lowest allowable energy level (called the ground state), but after energy is gained from an external source, the energy level will jump from the ground state to a higher energy level (called the excited state).

All molecules subject to absorption will return to the ground state following removal of the radiation. Most molecules will return to the ground state without emitting a measurable product, but in certain molecules the return from the excited state to the ground state is accompanied by the emission of light from the molecule. Although the change in energy due to absorption is the object of analysis in absorption spectrometry, it should be noted that measurement of the light emission resulting from return to the ground state is the basis for fluorescence spectrometry.

18.2.1 Beer's Law

Beer's law describes the relationship between the amount of light absorbance occurring across a given path length of a sample matrix and the concentration of matter in the matrix. This law applies to a simple matrix consisting of one absorbing component, measured using the intensity change of a single wavelength of light transmitted through the sample matrix. If the absorbing component in the matrix conforms to Beer's law, the concentration of the absorbing component will be proportionate to the change in absorbance. This relationship is illustrated in Figure 18.1.

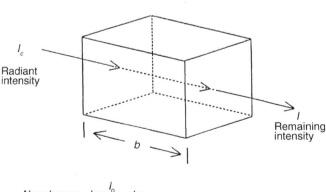

$$\text{Absorbance} = \log \frac{I_0}{I} = \varepsilon b c$$

where
  ε = molar absorbtivity constant
  b = path length through sample
  c = concenentration of absorbing substance

Figure 18.1 Beer's law.

18.2.2 Detection of Absorbance Spectra

Although numerous chemicals will absorb light in the UV or visible wavelength range (including many nutrients, heavy metals, halogens and organics), many other chemicals of interest for environmental analysis do not have the bond structures or free electrons that allow absorption of UV or visible light. When light absorbance measurements can be used for direct detection of a substance using natural absorbance spectra, this is called *primary* absorbance analysis. When chemical reagents must be used to induce a detectable absorbance spectra, this is called *secondary* absorbance analysis.

18.2.2.1 *Primary Absorbance.* In a simple matrix, where only one absorbing component is present, any wavelength conforming to Beer's law can be used for analysis. In the absence of issues arising from other components in the matrix, the wavelength providing the greatest dynamic range will be used for analysis. This is typically the peak wavelength of the absorbance spectra. In a more complex matrix, a number of issues need to be addressed before primary analysis can be considered. If the absorbance of the target substance is at unique wavelengths not affected by absorbances from other components, primary absorbance can be considered for analysis. This implies that the analyst has access to information concerning the identity and absorbance characteristics of the other components in the matrix. In the absence of this information, the sample may need to be screened for the presence of substances that have absorbance spectra known to interfere (overlap) with the wavelength intended for analysis of the target compound. These interferences either must be removed prior to primary analysis or must be separately measured so that primary absorbance measurements can be adjusted to compensate for their presence.

18.2.2.2 *Secondary Absorbance.* The detection of UV absorbance or color in a sample following the addition of a reagent is fundamental to the field of chemical analysis. Many of the standard analytical methods available for use in environmental laboratories are applications of secondary absorbance analysis at UV or visible wavelengths. Such tests typically rely on reactions between a reagent and a target substance to form a new compound that will possess light absorbance characteristics that conform to Beer's law at a specific wavelength and can therefore be used to calculate the concentration of the target substance. Secondary absorbance analysis is often used as a strategy to circumvent issues resulting from a mixed matrix, which contains several components. The absorbance created after addition of the reagent is intended to be at wavelengths that are unaffected by the presence of the other components in the matrix.

18.2.3 Interpretation of UV-Visible Absorbance Spectra

Detection and measurement of absorbance spectra from a sample can provide sufficient information for analysis only if the measurement represents the light absorbance attributable to matrix components. When light is transmitted through a sample matrix, Beer's law calculations are based on the change in intensity at a given wavelength. Thus, a starting intensity of the light must be known as well as the resulting intensity following transmittance of the light through the sample matrix. The observed change in intensity may not be entirely attributable to absorbance within the matrix, especially if the matrix is contained within a cell or cuvette that may offer reflective or transmissive light intensity losses independent from the absorbance losses within the matrix. These losses must be eliminated from any subsequent calculations using a technique called *zeroing*. This technique makes a measurement in an identical cell or cuvette that contains a nonabsorbing matrix (a blank). This allows a measurement of all losses not attributable to the matrix components to be performed. The blank measurement can then be subtracted from any subsequent sample measurements, leaving only the absorbance attributable to matrix components.

A calibration is essentially an equation in the form $Y = a + bX$, where a is the Y intercept and b is the slope of a line that defines the relationship between absorbance measured in a matrix and concentration of the absorbing component in the matrix. (Calibration curves can also be in the form $\log Y = a + bX$.) This relationship can be derived experimentally by preparing a minimum of two standards with known concentrations of the component to be measured and obtaining the absorbance of these samples across a given path length. A larger number of samples is always preferable, however, in order to verify that the relationship between absorbance and concentration is truly linear (i.e., that Beer's law is obeyed) across a given range of concentrations. The additional standards may be used either as test samples, where the concentration value predicted by the calibration equation is compared to the actual concentration of the standard, or to derive the equation using a statistical regression.

18.2.3.1 Single-Wavelength Primary Analysis.

In a simple matrix, such as one containing a single absorbing component in transparent media, the number of wavelengths used for analysis is not significant provided that the selected wavelength is within a range that obeys Beer's law. In the absence of interference from a second component, one wavelength will yield a satisfactory result. An example is the analysis of nitrate in deionized water, where any wavelength in the range of 220–250 nm will provide excellent correlation for nitrate concentrations between 0.1 and 20.0 ppm.

If a matrix contains more than one absorbing component, the absorbance spectra detected will be a combined function of all absorbing components. In liquid media, the overlap of spectra for the individual components will result in a smooth absorbance signature, as illustrated in Figure 18.2. If the individual component spectra partially overlap, it may be possible to identify some portion of the matrix signature where wavelengths are well correlated with the concentration of one individual component, but in the absence of automated tools that can search for such correlations, special knowledge of the components and their absorbance spectra are required. If two or more individual component spectra fully overlap, no one

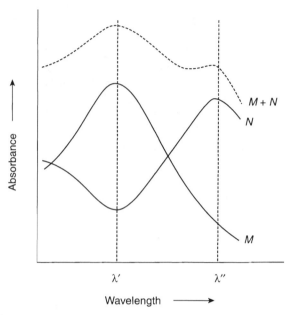

Figure 18.2 Absorption spectrum of a two-component mixture.

wavelength will yield an acceptable correlation with any one absorbing component because the absorbance at any one wavelength will actually be the sum of the individual absorbances from multiple substances.

18.2.3.2 Multiple-Wavelength Primary Analysis.

When a matrix contains overlapped spectra from multiple absorbing components, the number of wavelengths used for primary analysis becomes significant. As the matrix becomes more complex, a larger number of properly selected wavelengths will improve the correlation between the calculated and actual concentrations for a target component when analyzed using techniques such as multiple linear regression of absorbance values.

One form of this problem is a mixture in which all of the absorbing components are known. A series of simultaneous equations (known as a K-matrix) can be constructed in the form

$$A = KC$$

where A = absorbance of light, $= \log I_0/I = \varepsilon bC$

       $K = \varepsilon b$

       ε = absorption coefficient

       b = path length

       C = concentration

Because there are several absorbing components, total absorbance at any wavelength will be a function of the sum of the individual component absorbances based on their concentrations in the solution. Equations can be defined for each of m wavelengths as

$$A_1 = K_{11}C_1 + K_{12}C_2 + \cdots + K_{1n}C_n$$
$$A_2 = K_{21}C_1 + K_{22}C_2 + \cdots + K_{2n}C_n$$
$$\vdots$$
$$A_m = K_{m1}C_1 + K_{m2}C_2 + \cdots + K_{mn}C_n$$

This analysis requires knowledge of all of the absorbing components in the solution in order to provide information for correlation. This is an unrealistic demand for analysis of most samples. Therefore, an inverse technique that defines concentration as a function of absorbance is a more practical approach. This P-matrix approach does not require a knowledge of all absorbing components. The approach uses a series of simultaneous equations in the form

$$C = PA \qquad \text{where} \qquad P = K^{-1}$$

Equations can be defined for each of n absorbing components as follows:

$$C_1 = P_{11}A_1 + P_{12}A_2 + \cdots$$
$$C_2 = P_{21}A_1 + P_{22}A_2 + \cdots$$
$$\vdots$$
$$C_n = P_{n1}A_1 + P_{n2}A_2 + \cdots$$

The P-matrix approach has the further advantage of allowing the use of various pre-processing and transformation techniques, such as derivatives, *Fourier* transforms, principal-components analysis, and other so-called *chemometric* techniques.

The potential for primary analysis in a complex matrix using multivariate analysis techniques can be seen in Table 18.1, where nitrate was the target component in a simple matrix consisting of nitrate and deionized water as a solvent and in a complex matrix consisting of wastewater containing numerous components. The nitrate was analyzed using multiple linear regression of absorbance values from 1 to 50 wavelengths for each matrix. In the simple matrix no difference in correlation was obtained as the number of wavelengths included in the analysis was increased. In the complex matrix, however, there was improved correlation as the number of wavelengths was increased.

18.2.3.3 Single-Wavelength Secondary Analysis.

Single-wavelength methods are among the most widely used laboratory methods. Many of these

Table 18.1 Nitrate Correlation Coefficient: Calculated Versus Actual Concentration

Number of Wavelengths (Excluding Reference)	Simple Matrix (Deionized Water)	Complex Matrix (Wastewater)
1	0.99	—
2	0.99	0.82
3	0.99	0.88
5	0.99	0.92
10	0.99	0.96
20	0.99	0.98
30	0.99	0.99
40	0.99	0.99
50	0.99	0.99

methods have been adapted for online application. As previously noted, these methods typically rely on reagent additions to form compounds with target components that will produce spectra that obey Beer's law and can be measured at wavelengths that are not influenced by other components in the matrix.

An example of a typical laboratory method using single-wavelength light absorbance measurements for chemical analysis is the cadmium reduction method for nitrate analysis. This method uses a column of cadmium granules to reduce nitrate to nitrite in the sample. The sample must first be prepared by extracting any oil or grease present, removing any chlorine, color, or turbidity in the sample, and adjusting the pH if necessary. Ethylenediaminetetraacetic acid (EDTA) may also need to be added to maintain reduction column efficiency if metals are present in the sample. The cadmium-reduced sample is then treated with a reagent to form an azo dye that can be measured for intensity at 543 nm. The intensity results from the sample are compared to a blank in order to eliminate losses not attributable to the azo complex. Nitrate standards of known concentration are similarly reduced in the cadmium column, treated with reagent and measured at 543 nm in order to develop a calibration equation. The equation is used to calculate the unknown nitrate concentration in the sample.

Thus, with this standard method, a fairly demanding sample conditioning and processing procedure is required in order to measure concentration using a comparatively simple single-wavelength measurement. Following suppression of interferences, the absorbance spectra for the target component is shifted and amplified through the creation of a new compound, as illustrated in Figure 18.3. This procedure can be automated for online application, but only with great difficulty and after excluding some of the sample preprocessing steps.

18.2.3.4 Multiple-Wavelength Secondary Analysis.
Multiple-wavelength methods can be used in conjunction with reagent additions in order to avoid

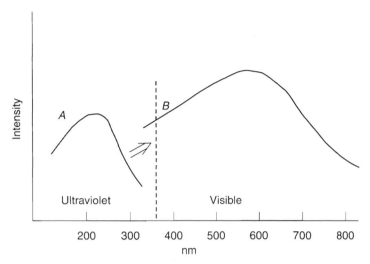

Figure 18.3   *A*—Natural absorption spectra. *B*—Shift in wavelength and intensity after analyte is complexed with an auxochromatic reagent. Reagents are often used in peak-wavelength analysis to shift absorption spectra of analyte from one waveband to another (often from ultraviolet to visible) or to increase intensity of spectra by changing position or shape of signature.

some of the sample conditioning procedures that would otherwise be required for analysis. Following addition of a reagent, spectra from the new compound may be overlapped with spectra from other components, from excess quantities of the reagent, from the solvent, of from intercomponent reactions. In these instances multiple-wavelength analysis techniques offer the same benefits that are available for primary analysis, and these techniques are similarly applied.

Multiple-wavelength analysis is an especially useful method for monitoring the change in absorbance for one or more substances involved in a titration procedure provided the substance obeys Beer's law.

Spectrophotometric analysis can also be used for compound identification, which requires a single-component matrix, a full absorbance spectrum for examination, and knowledge of the relationship between absorbance structure and compound composition.

18.3   APPARATUS FOR ONLINE UV–VISIBLE ANALYSIS

In a laboratory environment, the analyst must prepare a sample and perform each of the individual steps of an established analytical procedure. If the procedure happens to use photometric analysis, a laboratory photometer or spectrophotometer may be employed as one of the several tools that will be used by the analyst during the course of the procedure. Other tools may be required to filter or remove contaminants, dissolve or remove suspended matter, measure appropriate amounts of sample and reagent, mix the sample and the reagent, time the reaction, apply heat or agitation at different stages of the procedure, prepare blanks and standards, record data from the analysis step, derive calibration equations, and calculate a concentration for the target compound based on the data from the analysis and the calibration equation. Even if certain steps in the procedure benefit from automated apparatus, it is the responsibility of the analyst to supervise the process, initiate each individual step, and assure that the complete analytical procedure is performed correctly.

An *automated* analyzer is designed not only to perform the final analytical step but also to perform all required sample and reagent preparation, conditioning, and handling necessary for analysis of the sample and for computation of concentration in the sample or in reference solutions or blanks. Although the analyzer automates the performance of a specific sequential procedure, the analyzer may need to be prepared by an analyst to perform the specific procedure, stocked with a supply of the necessary chemicals and commodities used in the analysis, and provided with the sample or samples to be analyzed. The analyst must initiate the start of the automatic analysis sequence after ensuring that the appropriate settings and materials have been provided.

An *online* analyzer is more than a device that will perform an automated sequence for a specific analytical procedure. An online analyzer is a device that will *automatically perform an analysis procedure communicate the analysis result to an observer or to an external device at established intervals on samples that can be automatically delivered to the analyzer from one or more fixed sample points.* An important additional requirement is the expectation that an online analyzer may be required to perform these functions in a location outside the laboratory where the apparatus may be subject to a variable operating environment. Thus, all online analyzers are automated analyzers, but not all automated analyzers are online analyzers.

18.3.1 Online Filter Photometers

An online filter photometer is an online analyzer designed to perform single-wavelength analysis procedures. Many of these procedures are automated versions of established colorimetric analysis methods originally developed for use in a laboratory. The original laboratory procedures typically require manipulation of the sample to remove interferences and develop a colored compound for analysis.

The photometer is a simple device consisting of a light source and power supply, bandpass- or wavelength-selective filter, fixed-path-length cell for the sample, entrance slit, and photodetector. Complications in the design of filter photometers arise from:

1. The need to compensate for losses attributable to absorbance, scattering, and reflection other than from the sample components. The major choices involve single- and double-beam designs and the use of blanks and/or standards to obtain comparative measurements of absorbance or intensity.
2. The need to automate a method for determining and applying the calibration equation, which may involve the use of reference standards and/or standard additions.
3. The need to automate all required sample conditioning and preparation steps.
4. The need to avoid contamination from prior samples and analysis steps.

The inherent simplicity of single-wavelength photometric measurement for online analysis may be more than offset by complexities arising from the automation of the sample handling and conditioning steps.

18.3.2 Online Spectrometers

If multiple-wavelength spectrometers are to be used for online analysis, they must offer some improvement over automated versions of one-wavelength photometry. Since spectrophotometric detection and analysis are inherently more complex due to the additional wavelengths involved, this complexity can only be justified if precision and accuracy are substantially improved or there is a corresponding reduction in sample handling and conditioning at no significant loss in precision and accuracy compared to photometric or other alternative methods.

There are substantial differences in design among the products offered by various manufacturers of online spectrometers, but one of the most fundamental differences is in the apparatus used for detection at individual wavelengths. A *scanning*-type system contains a monochromator containing a wavelength-selective grating mounted to a cam or stepper motor that allows a sequential series of wavelengths to be transmitted through a sample and onto a detector. The grating must physically move through a series of positions in order to detect light intensity at a range of wavelengths. An alternative design uses a curved grating to separate the information at discrete wavelength intervals and simultaneously project the information onto a detector array. This *array*-type system allows light intensity at a range of wavelengths to be detected without moving parts in the detection system. Information from any, some, or all of the automatically detected wavelengths can be used for analysis, as dictated by the specific analytical method.

18.4 APPLICATIONS FOR WATER AND WASTEWATER ANALYSIS

The use of an online analyzer to detect any substance in water or wastewater implies more than simple curiosity concerning the composition of the sample. If curiosity about the composition of a sample is the only issue, a grab sample analyzed in a laboratory would be sufficient. Online analysis is performed because there is a sustained need for fresh information concerning the presence and concentration of a specific component in a sample that is obtained from a specific sample point. The use of online analysis implies that the location, timing, and frequency of analysis are important considerations that justify the use of special apparatus that is dedicated to the task of automatically detecting and measuring a specific target component at regular intervals.

A major reason for online analysis in water or wastewater is for *monitoring*, where the objective is to ensure that some minimum concentration of a desirable component is present or some maximum concentration for an undesirable component is not exceeded. These monitoring tasks often require the analyzer to initiate an alarm signal upon detection of some minimum or maximum concentration for a target component and may also require that information be communicated to an external device so that a record of the concentration value at each measurement interval can be maintained.

The other major reason for online analysis in water or wastewater is for *control*, where the objective is to provide information concerning the concentration of a variable component in order to automatically provide the control algorithm with reliable data that can be used to control the process. (This does not imply that the analyzer actually performs the control tasks, although the analyzer may be part of an integrated control system.)

Whether the application is for monitoring or control, the use of online analysis allows a sample to be immediately analyzed at or near the point of extraction so that the results from analysis can be immediately available. The term "immediately" is a relative term. Some photometric analyzers can be operated continuously, but in practical terms, measurement of absorbance and calculation of component concentration is a dissect event that requires some finite interval of time to accomplish. Thus, even the most rapid primary analysis method, performed using light absorbance measurements made in situ, will be discontinuous events. If the analysis is performed on a sample that is delivered to the analyzer through a sample line, time will be required to flush the line to ensure the freshest possible sample is available for analysis. If the sample matrix is complex, some type of sample processing may be required prior to analysis. This work will impose a time penalty on the analysis cycle that will limit the time period between analysis intervals. If secondary analysis methods must be used, additional time may be required for sample conditioning, reagent addition, and chemical reactions prior to spectra being available for detection.

18.4.1 Primary Analysis Parameters

Most successful primary analysis methods will, of necessity, be multiple-wavelength procedures. Although single-wavelength methods are possible, they require a target component that has substantial natural absorbance in a matrix that offers little or no interference from other absorbing components. This is rarely seen in nature, although is possible. Groundwater, for example, may contain a component with strong natural absorbance such as iron or nitrate and also contain other components such as calcium and magnesium that contribute background chemistry to the sample matrix but do not contribute background absorbance in the UV or visible wavelength range. Although single-wavelength methods could successfully analyze nitrate or iron in this water, if both nitrate and iron were present, a single-wavelength method might not be successful for analysis of either nitrate or iron.

Single-wavelength methods are most often used to detect nonspecific parameters where the absorbance, transmittance, or light intensity is a meaningful value. An example is turbidity, where the amount of light transmitted through the sample (measured using a Jackson turbiditiy unit scale) is a measure of transmissive light losses attributable to the presence of dissolved and particulate matter, without reference to the specific identity of the matter. Other examples are the measurement of absorbance at 254 nm as a nonspecific measure of organic matter and the measurement of light absorbance at a selected wavelength for measurement of color.

In addition, there is some antedotal evidence that, under certain conditions, absorbance or transmittance measurements in the UV wavelength range (principally at 254 nm) can be used to measure total organic carbon (TOC) or even carbonaceous oxygen demand (COD). This is simply not true. TOC or COD measurements are the result of a very specific procedure that is not duplicated by direct UV absorbance measurements. It is true that some (but not all) organic substances will absorb light in the UV range. Among the organic substances that absorb in the UV range, identical concentrations of different organic substances will have different absorbance intensities at 254 nm or any other individual wavelength used for analysis. Thus, unless the application is such that the ratio between UV-absorbing organics and the total organics at the sample point is stable, UV absorbance will not provide a good correlation with TOC or COD. If nonabsorbing organics such as alcohol, methanol, ethanol, glycol, or other similar substances are present, the use of UV absorbance as a surrogate for TOC or COD analysis may result in substantial inaccuracy.

Table 18.2 Primary Analysis Parameters in Water and Wastewater

Absorbance, specific wavelength	Ascorbic acid
Chlorine, free	Chromium, hexavalent
Color	Copper II
Fluoride	Fulvic acid
Humic acid	Iron, ferrous
Molybdate	Monochloramine
Nickel	Natural organic matter (NOM)
Nitrate	Nitrite
Ozone	Permanganate
Polyacrylate	Sulfite
Tannic acid	Transmittance, specific wavelength
Triazole	Turbidity

Table 18.2 provides a list of parameters that are being analyzed using primary analysis methods based on a review of the literature published by leading manufacturers of online photometers and spectrometers. All chemical substances listed are the dissolved fractions as found in the sample presented for analysis. (If a digestion, acidification, pH adjustment, or filtration is required, the parameter would be listed as a secondary analysis parameter.)

Table 18.3 Online Secondary Analysis Parameters and Methods for Water and Wastewater

Parameter	Sample Preparation	Reagents	Detection
Aluminum	—	Pyrocathechol violet	Color 570
	—	Eriochrome cyanine-R	Color 535
Ammonia	—	Chlorine, salicylate	Color 660
Ammonia, free	—	Chlorine, hydroxide	UV 220–450
Calcium	Dialysis	8-Hydroxyquinoline, cresolphthalein	Color 580
Carbonate	—	Acid, pH indicator	Color 550
Chloride	—	Thiocyanate, ferric iron	Color 490
Chlorine, total	—	N,N-diethyl-p-phenylenediamine	Color 530
Chromium VI	—	Acid, diphynylcarbazide	Color 550
Cobalt	—	Acid, nitroso-R	Color 500
Copper	—	Cuproindisulfonic acid, ascorbic acid	Color 480
Cyanide (total)	Digest	Chloramine-T, iso-nicotinic acid	Color 600
Cyanide (free)	Distill	Chloramine-T	Color 600
Ethanol	—	NAD, alcohol dehydrogenase	UV 340
Formaldehyde	—	Ammonium acetate, acetyl acetone	UV 410
Hardness (total) 520,669	—	EDTA, calmagite	Color
Hydrazine	—	Acid, p-dimethyl-aminobenzaldehyde	Color 460
Hydrogen sulfide	—	Carboxymethylcellulose, zinc chloride	Color 660
Iron II	—	Hydroxylammonium chloride	Color 590
Magnesium	Dialysis	Color reagent	Color 505
Manganese	—	Sodium periodate, leucomalachite	Color 620
Methanol	Dialysis	Acid, permanganate, acetylacetone	UV 410
Nitrate	Dialysis	Hydrazinium sulfate, sulfanilamide, azo	Color 540
Nitrogen	Digest	Hydrazinium sulfate, sulfanilamide, azo	Color 540
Nitrite	—	Hydrazinium sulfate, azo	Color 540
Phenol, free	—	Ferricyanide, 4-aminoantipyrine	Color 505
Phenol, ortho	Distill	Ferricyanide, 4-aminoantipyrine	Color 505
Phosphate, total	—	Molybdate, antimony, ascorbic acid	NIR 880
Phosphate, ortho	—	Molybdophosphoric acid, vanadate	Color 420
Silica	—	Silicamolbdate, heteropoly blue	Color 452
Silicate	—	Ammonium molybdate, ascorbic acid	NIR 810
Urea	—	Acid, diacetylmonoxim	Color 520
Zinc	—	Buffer, zircon, cyanide, chloralhydrate	Color 600

18.4.2   Secondary Analysis Parameters

The majority of the photometric and spectrophotometric methods in use for online chemical analysis in water and wastewater are classified as secondary analysis methods. Table 18.3 shows a list of parameters available for analysis by leading manufacturers of online photometers and spectrometers. The table identifies the major reagents and the detection range or specific wavelength used for analysis. One procedure not listed but gaining in popularity uses online photometry as a part of an automated titration procedure for alkalinity analysis. The key aspects of the titration procedure is the addition of indicator solutions followed by a titrant, which results in a color change in the sample that can identify the end points for hydroxide, carbonate, and bicarbonate alkalinity

18.5   MULTIPLE-PARAMETER SYSTEMS

Automatic photometer systems that detect information at one specific wavelength are typically used to analyze one specific parameter. Scanning or array-based spectrometer systems, however, can be used to detect multiple parameters, sometimes using a combination of primary and secondary analysis methods. The actual methods are not substantially different whether applied individually or in sequence with other methods. When used for analysis of multiple parameters, the online analyzer must be designed to avoid contamination of samples and apparatus used for each successive analysis.

II

WATER QUALITY
PARAMETERS

19

THERMAL CONDUCTIVITY DETECTORS

John M. Hiller[*] and Nancy M. Baldwin

Lockheed Martin Energy Systems, Inc.
Oak Ridge, Tennessee

[*]Corresponding author. Current affiliation is OI Analytical, CMS Field Products, Pelham, Alabama.

Environmental Instrumentation and Analysis Handbook, by Randy D. Down and Jay H. Lehr
ISBN 0-471-46354-X   Copyright © 2005 John Wiley & Sons, Inc.

A number of analytical instruments perform measurement functions by sensing the loss of heat from a transducer and relating that heat loss, through an intermediate measurable parameter such as current, voltage, or resistance, to the property that the instrument is designed to report. These instruments include vacuum gauges, flow sensors, leak detectors, and the thermal conductivity detectors (TCDs) found on many gas chromatographs.

Heat can be lost (actually transferred to a different object) by radiative, conductive, and convective processes. The radiative process is simply energy transferred without physical contact with any medium; the standard example is light or heat transferred through empty space. The conductive and convective processes require physical contact with the medium through which heat is transferred. In the case of conductive transfer, heat generally travels along a solid. In convective transfer, heat is generally transferred from a solid to a liquid or gas. The sensors described in this chapter principally rely on the convective (or forced convective) mechanisms to sense gases.

19.1 THERMAL CONDUCTIVITY

Thermal energy, or heat, moves from a region of higher temperature to regions of lower temperature. This observation was formalized by Fourier over a hundred years ago and is incorporated in the equation that now bears his name[1]:

$$\mathbf{q}' = -\lambda \, \nabla T$$

where $\mathbf{q}'$ is the heat flux vector, λ is the thermal conductivity of the material, and ∇T is a representation of gradients of temperature over space. Here, heat loss is attributed to the random motion of molecules.

If we impose a moving fluid (such as a gas) over a solid object, we allow heat to be transferred in bulk as well as by the random process described above. This brings us into the realm governing convection: Newton's law of cooling. Here,

$$\mathbf{q}' = k(T_{\text{solid}} - T_{\text{gas}})$$

where k is the convection coefficient, which is a composite "constant" representing a number of factors that affect heat transfer.

19.1.1 As a General Property of Materials

Thermal conductivity is a property of any real material. In the case of solids, it depends only slightly on temperature and is generally treated as a constant over ranges encountered on thermal conductivity–based analytical instruments. In the case of gas sensors, most instruments are designed such that heat loss by conduction is minimized and kept as constant as possible.

19.1.2 As a Specific Property of Gases

The thermal conductivity of gases is a complex function of the gas itself, other gases in the gas stream, temperature, and pressure. For pure gases, tables of thermal conductivities are generally available[2] (Ref. 3, pp. 6–251, Thermal Conductivity of Gases, and 8–65, Properties of Carrier Gases for Gas Chromatography). In general, thermal conductivity is largely a function of molecular velocity, which is inversely related to the square root of molecular mass.

A theoretical derivation of thermal conductivity (based on the kinetic theory of gases) appears in most physical chemistry[4] and chemical engineering textbooks. The derivation also appears in engineering handbooks.[5] However, most authors are quick to indicate that kinetic theory is best described as an ideal approximation that might not be correct for any particular gas. The thermal conductivity of a mixture of two gases has likewise been estimated from kinetic theory.[2]

A number of units for thermal conductivity are used in the literature; the most common unit is W/K · m (watts per kelvin meter). The *CRC Handbook of Chemistry and Physics*[3] (pp. 1–38, Conversion Factors for Thermal Conductivity Units) provides an entry point for the engineering units commonly employed in the United States.

An aeronautics view of gases is available[5] that should be consulted if one wants to design or use any of the instruments described later under extreme (altitude, pressure, or gas speed) conditions.

19.2 SENSOR PRINCIPLES OF OPERATION

Thermal conductivity is a property that cannot be sensed or used directly in determining a physical or chemical property of a gas. Instead, temperature is sensed. The principal transducers for temperature are based on the voltage generated by a thermocouple and the change in resistance generated by a thermistor or hot wire. In each case, the measured response is small; hence, the transducer's signal is directed into a Wheatstone bridge (or equivalent). From the bridge, one typically extracts voltage, which is ultimately related to the bulk property for which the instrument was designed.

The raw sensors (thermocouple, thermistor, and hot-wire detectors) and the Wheatstone bridge are considered a mature technology.

19.2.1 Thermocouple

The thermocouple is a voltaic device that is based on creating a voltage at the junction of two dissimilar metals (the thermoelectric effect). The signal generated is

$$v = s \, \Delta T$$

where v is the measured voltage across the device, s is the sensitivity factor for the particular metal couple used in the construction of the thermocouple (nominally about $3\,\mu V/K$ for many metal couples), and ΔT is the temperature difference in kelvins. The thermocouple is used in some forms of vacuum gages that are described later.

19.2.2 Thermistor

The thermistor is a solid-state device that functions as a thermal transistor or thermally sensitive resistor. These devices usually have two leads and a bead that contains a transition metal oxide. Like the conventional semiconductors, the thermistor changes resistance with temperature in a predictable fashion. The thermistor is commercially available in various configurations: The vendor of the instrument containing the thermistor (a leak detector or TCD) will usually select a thermistor with a thin coating that will confer chemical resistance while still providing a fast analytical response.

The thermistor is usually operated at near room temperature. The principal reason for ambient temperature operation is that the sensitivity (change in resistance per unit temperature change) is quite large; hence, the "ΔT" generated between the "reference" and "specimen" need not be very large. Depending on the composition of the thermistor, the sensitivity can vary from -4 to -7% per degree Celsius. The thermistor has a short working (or linear) range because the response is quite nonlinear. It is common practice to extend the working range by conditioning the raw signal with the support electronics.

As indicated earlier, the thermistor resistance decreases with increased temperature. The negative response function introduces a potential failure mechanism called thermal runaway. In this failure, increased temperature leads to decreased resistance, decreased resistance leads to more current, more current leads to more heat, and more heat leads to increased temperature, which brings us back to the forward-feeding loop, ultimately resulting in the destruction of the thermistor and associated circuits (destruction can be caused by the current and/or the temperature). The standard engineering practice to avoid the thermal runaway situation is to place a current-limiting resistor in series with the thermistor.

The principal support circuit for using a thermistor is usually based on a Wheatstone bridge. The use of a thermistor in a Wheatstone bridge largely parallels the discussion of using a hot wire-detector with the principal (and critical) difference being the negative response to increasing temperature. A good overview of thermistor technology is available elsewhere.[6]

19.2.3 Hot Wire

The hot-wire detector is based on the property that a resistor increases resistance with increasing temperature. These detectors are usually operated at temperatures substantially above ambient temperature because the sensitivity (change in resistance with temperature) is small (on the order of $3\ m\Omega/K$). The hot-wire resistors are generally made from tungsten/rhenium, nickel, platinum, or gold-coated tungsten.[7]

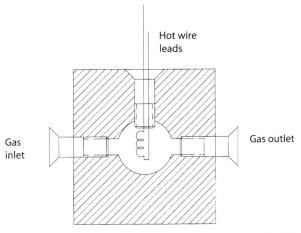

Hot wire
leads

Gas
inlet

Gas outlet

Figure 19.1 Thermal conductivity cell. Basic elements of thermal conductivity cell as used in leak detector or TCD used in gas chromatography. The individual detector units are usually used in pairs or quads. The block is usually kept at a constant temperature. The filament leads are thermally and electrically insulated from the block.

Because the hot-wire detector is used at elevated temperature, some thought must be given to chemical reactivity with the leak gas. Vendor data provide some guidance.[8] In general, the nickel wire has poor chemical resistance to aqua regia (HCl/HNO$_3$) and damp amine/ammonia vapors; the tungsten/rhenium filaments have poor chemical resistance to fluorine, hydrogen fluoride, and nitric acid vapors. Also note that the working range (linearity of response) of a hot wire is also a function of its elemental composition.

In most instruments and other engineered systems, the hot-wire detector elements are designed for ease of replacement since they are considered a consumable component. A drawing of a "typical" hot-wire detector element is presented as Figure 19.1. The elements are often used in pairs or quads for insertion into a Wheatstone bridge. In a two-hot-wire system, one detector senses the sample gas; the second senses a reference gas. In a four-hot-wire system, two sensors are plumbed together (usually in parallel) to the sample side; the other two sensors are similarly plumbed to the reference side. To improve the analytical signal, the hot wire is usually presented as a coil (the electrical resistance is proportional to the length of the wire; also, the amount of heat conducted away from the device is a function of surface area).

In an engineered system, one can usually improve the sensitivity by increasing the temperature of the filament or the amplification of the signal. In general, sensitivity will increase four-eightfold for a doubling of filament current.[9] Given a choice, it is generally better to operate a hot-wire detector with higher electronic amplification than with higher filament current since the latter tends to decrease filament life, especially with corrosive gases. One should read the vendor literature carefully to determine if the instrument "sensitivity" settings are adjustments on the filament current or amplifier gain.

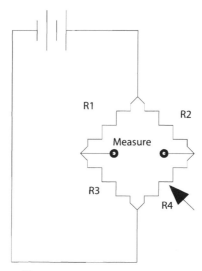

Figure 19.2   Wheatstone bridge.

19.2.4   Wheatstone Bridge

A "classical" Wheatstone bridge is illustrated in Figure 19.2. This device is used primarily to measure small, differential resistances. In the classical (freshman physics) form, R_1 and R_2 are fixed and identical resistors. Resistor R_3 is the resistive element under test, and R_4 is a variable resistor adjusted such that the voltage measured across the bridge (at the points identified as "measure") drops to zero.

The classical Wheatstone bridge has been modified to operate in several ways. The earliest form is based on nulling the measured voltage across the bridge. This could be done by hand (e.g., the freshman physics experiment) or by a servomechanism. Many earlier strip chart recorders operated with a nulling servomechanism.

Later applications have used the bridge in the unbalanced mode (without fulfilling the null condition). In this mode, instead of balancing resistance, one relies on the measured voltage (or current) across the bridge for an analytical result. For sensitivity as well as signal conditioning, an operational amplifier is usually connected to the measure points. A more complete discussion of variations on the Wheatstone bridge can be found in the literature.[10] The continued utility of the Wheatstone bridge in extracting low-level signals is evident from the continual traffic in U.S. patents.

19.3   INSTRUMENT EMBODIMENTS

The property of thermal conductivity is ultimately used in instruments to measure residual gas pressure, flow, and leak rates. It is also used to determine the concentration of any compound or element that is amenable to gas chromatography

(GC) and has a thermal conductivity different that that of the carrier gas. These devices are best described as a mature technology that the vendor should have optimized for the intended application.

This section describes how thermal conductivity is sensed in the instruments and includes some applications in environmental sensing, calibration, maintenance, and other analytical considerations.

19.3.1 Vacuum Gauges

The vacuum gauge is a device that measures the pressure of residual gases in a container. For a vacuum gauge based on thermal conductivity, one places a (usually heated) sensing element in contact with the "vacuum" and monitors the temperature change as heat is lost from the sensor to the ambient atmosphere (vacuum). Alternatively, the feedback system is designed to keep the sensor at constant temperature and monitor the current necessary to do so. In designing a vacuum gauge based on thermal conductivity, one creates a design that will maximize heat loss due to convection to the surrounding atmosphere (vacuum) while minimizing other heat losses. Vacuum gauges of this class are generally based on thermistors and thermocouples. These devices sense pressure in the rough vacuum range (1 mtorr–760 torr; 760 torr is 1 atm).

The amount of heat lost is related to the concentration of the ambient gases in an extremely nonlinear fashion. At high (near-atmospheric) pressures, the net change in heat transfer with change in pressure is small. Thus, the instruments are generally imprecise at the upper end of the working range. This is easily seen by looking at the face of an analog thermal conductivity vacuum gauge—the scale is usually logarithmic.

At the lower end of the working range, the number of gas molecules available to transfer heat is small, resulting in very few collisions with the sensor. This results in essentially no temperature change in the sensor and hence no change in output.

For measurement purposes, a pair of thermocouples (or thermisors) is often placed in a bridge circuit (such as the Wheatstone bridge described earlier) to minimize errors caused by changes in ambient temperature; one "arm" of the circuit is kept at a fixed pressure or referenced to ambient atmosphere.

In general, these types of vacuum gauges are not calibrated themselves, but the entire apparatus (including the support electronics and readout device) is calibrated as a whole at two points. The high-pressure calibration point is usually ambient atmospheric pressure. For the low calibration point, the instrument is set at a pressure below the working range of the vacuum gauge.

Vacuum gauges that operate on the basis of thermal conductivity are commonly called (1) *thermocouple gauges*, which naturally use thermocouples, and (2) *Pirani gauges*, which usually are based on the use of thermistors. One often finds a Pirani gauge combined with a much more sensitive ionization vacuum gauge to increase the ranges of pressure sensed. In the combination units, the Pirani gauge also functions to protect the ionization gauge (at high pressures, the ionization gauge could burn out from arcing and excess current flow).

As indicated earlier, the thermal conductivity of a gas changes with temperature, pressure, and the atomic composition of the gas. Hence, the heat transfer coefficient is not constant. Because of these constraints, one must specify the temperature, pressure, and gases to be measured when purchasing a vacuum gauge. Lacking a specification, these vacuum gauges are often available off-the-shelf and are designed to be calibrated at two points, with dry air or dry nitrogen.

An analytical accuracy of 2–3% is possible with these vacuum gauges if one works in the middle of the working range. Often that level of accuracy is not achieved (or desired). Frequent operator errors include using *standard atmospheric pressure* of 760 torr rather than ambient pressure when setting the high-pressure point. Also, such gages are incorrectly calibrated by using ambient pressure (obtained from the weather services or weather-monitoring barometer) without correcting for altitude.

For better accuracy at higher pressures (>1 torr) one generally uses a sensor based on mechanical distortion of a flexible membrane (capacitance manometer). To measure pressures below 10^{-4} torr, one usually uses a sensor based on ionization of the residual gas and measurement of the resultant current (such as with an ionization, Penning, or cold-cathode gauge).

19.3.2 Flow Sensors and Controllers

Flow sensors based on thermal conductivity are designed to maximize heat loss through forced convection. These are among the oldest instruments based on thermal conductivity, dating back to the 1880s, when they were used for process control in petrochemical refineries.

In operation, the gas passes over a sensor, losing heat. The temperature drop, if large, is sensed directly and output to a suitable recorder. If the temperature drop is small, the output of the sensor feeds back to a current-supplying circuit that restores the sensor to its initial temperature; the current necessary to restore the temperature to its initial value is related to flow.

In a flow controller, an output from the sensor (directly or indirectly) controls a valve that admits fluid over the sensor. This feedback mechanism keeps the sensor at constant temperature by adjusting a valve that regulates the flow rate of the fluid.

In purchasing a flow sensor or controller, expect to specify the gas(es), temperature range, pressure ranges (both upstream and downstream from the sensor), and flow rate range. Each of these parameters affects the heat transfer "constant." Also, one must specify the output of the sensor, plumbing connections, and type of hazard of the atmosphere (flammable gases, dust, water, etc.) in which the sensor/controller will reside.

A good troubleshooting guide for flow sensors (as well as TCDs used in GC) is available in the literature.[9]

19.3.3 Leak Detectors

Leak detectors based on thermal conductivity are based on measuring the heat loss from a heated wire or thermistor. When a gas passes the sensor, the heat loss changes (relative to a reference gas), the temperature changes, resistance changes, and current flow into the Wheatstone bridge changes. The output of the bridge feeds back to the circuit modifying the applied current, stopping when the original temperature has been reestablished. In the end, current change is related to the flow of the gas corresponding to the leak. For a leak detector based on the hot wire, current is increased to restore heat. For a leak detector based on thermistors, current is reduced. These opposite responses to heat loss are required because resistors and thermistors have opposite temperature–resistance characteristics.

Leak detectors based on thermal conductivity occupy an intermediate point in leak detection technology. Larger leaks are detected by soaping the region of the leak and watching for any soap bubbles. The very smallest leaks are detected by helium leak detectors, instruments based on a mass spectrometer optimized for the 4-dalton mass of helium.[*]

One vendor's specification for a portable thermal conductivity–based leak detector gave a 10% full-scale response for helium of 1×10^{-5} cm^3/s (about 20 in.^{3}/ year). This device uses air as the reference gas (for balancing the bridge circuit). Sensitivity is about 10-fold higher (less favorable) for gases with thermal conductivities near that of air (e.g., Ar, CO_2, and common fluorinated refrigerants). The device uses thermistors because of their sensitivity and operation at ambient temperature. It is important to note that many leak detectors cannot be used with flammable gases.

Leak detectors can be calibrated; however, they are often used as "indication-only" troubleshooting tools.

19.3.4 Gas Chromatography Detectors

The thermal conductivity detector is among the most common detectors used in GC. It is often described as a universal detector since it responds to most materials of potential interest. The most common application is for the analysis of permanent gases such as nitrogen, oxygen, methane, argon, and carbon dioxide and water. The TCD is also used in series (upstream) with a destructive detector, such as the flame ionization detector. In general, an analysis based on GC with a TCD is good for many compounds at the 10–100-ppm level.[11] The working range for a GC TCD is about 10^5.

19.3.4.1 Hardware. The TCD built for GC usually consists of a heated block with hot-wire sensor elements. The block is heated (and precisely controlled) above ambient temperature to ensure maintaining a constant temperature (to avoid instrumental drift). The temperature is set slightly higher than the highest temperature reached by the analytical column in the course of an analysis to avoid condensation of less volatile analytes. The filaments in the detector block are usually 200 K hotter

---

[*]In operating a helium leak detector, the operator connects the instrument to the vessel under investigation and "sprays" helium on the outside; any helium that leaks into the vessel will ultimately be detected by the mass spectrometer.

Table 19.1 Connections of GC Column(s) to TCD

Number of Columns	Number of Sensors	Routing of GC Gases
2	2	Effluent from analytical column over R_3, effluent from reference column over R_4
2	4	Effluent from analytical column split; half over R_3, half over R_1 Effluent from reference column split; half over R_4, half over R_2
1	2	Carrier gas split ahead of injection system. Some gas directed to the analytical column, the effluent of which is directed over R_3. The balance of the carrier gas is directed over R_4; gas flow is reduced by a small orifice, capillary, or needle valve.
1	4	Carrier gas split ahead of injection system. Some gas directed to analytical column, the effluent of which is split and directed over R_3 and R_1. The balance of the carrier gas is split and directed over R_4 and R_2; gas flow is reduced by one or more small orifices, capillaries, or needle valves

Note: References to resistors (R) refer to Wheatstone bridge (Fig. 19.2).

than the block itself. The hot wires are typically part of a Wheatstone bride. Helium and hydrogen are the preferred carrier gases because of their higher thermal conductivity relative to other gases.

A number of methods of "plumbing" the gas chromatograph to the TCD have been used. These are summarized in Table 19.1. The principal design consideration is to balance the Wheatstone bridge and reduce instrumental drift due to variations in the environment, carrier gas pressure, and carrier gas flow rates. A good collection of photographs and wiring/plumbing diagrams of a number of TCDs is available elsewhere.[7]

A number of reviews on the design of TCDs for gas chromatography have been written.[12–15]

One can expect hydrogen to become the predominant carrier gas in portable instruments and capillary systems. Hydrogen offers lower cost (especially in locations where helium is not readily available), can be generated electrochemically with a simple apparatus, is available at higher purity, and allows faster and/or superior chromatography. Using an electrochemical hydrogen generator obviates the need for "dealing with" gas cylinders (weight, demurrage, regulator and building ventilation, and gas inventory hazard). There is an extensive safety database for using hydrogen in chromatography, largely generated in areas where helium is generally not available (e.g., Europe).

For best accuracy, it is necessary to precisely control the volumetric flow of carrier gas. The use of electronic flow controllers is now becoming widespread.

A thorough review of gas chromatography thermal conductivity detectors is available elsewhere.[16]

19.3.4.2   Applications. The TCD, being a universal detector, is often used for screening to prevent fouling of a more sensitive detector (such as a flame ionization detector, electron capture detector, or mass spectrometer) and possibly the analytical column.

It is routinely used in process control analyzers, particularly in the petroleum-refining industry. The detector can also be used to monitor petroleum refinery emissions (e.g., sulfur, hydrocarbons). A number of other industries use a GC TCD for process control and emissions monitoring.

The TCD has been miniaturized[17] using micromachining for use with portable and space-borne instruments. This miniature unit was used to analyze products from an experiment package designed to study combustion gases in zero gravity. The MTI GC was constructed of four modules, with each module containing an injector, a column, and a TCD. One of the modules used argon as the carrier gas to determine CH_4, CO, O_2, H_2, and N_2. A second module used helium as the carrier gas to determine CH_4, CO_2, and SF_6. The modules operated simultaneously, producing an analysis time of 160 s. A variation of this device is available commercially. A review article on micromachining (including the specific technology used in the space-borne and commercial GC TCDs) has been written.[18]

Other uses for GC TCDs have included the analysis of medical oxygen and breathing air for oxygen content; the analysis of combustion products for use in dedicated analyzers for carbon, hydrogen, nitrogen, and sulfur; the analysis of off-gases from transformers (used to identify the need for maintenance before failure of the transformer dielectric fluid); the analysis for hydrogen isotopes;[19] and the analysis of evolved gases (H_2, CO, CO_2, CH_4, N_2O) from high-level nuclear waste.[20]

19.3.4.3   Other Information. The American Society for Testing and Materials (ASTM)[21] has a specification for testing the performance of TCDs used as GC detectors. The standard provides guidance for determining sensitivity, minimum detectable quantity (or concentration), linear range (nominally ±5% deviation from a straight line), dynamic range, noise and drift, and response time (time to reach 63% of the eventual steady-state signal).

ACKNOWLEDGMENT

This research was sponsored by the U.S. Department of Energy and performed at the Oak Ridge Y-12 Plant, managed by Lockheed Martin Energy Systems, Inc., for the U.S. Department of Energy under Contract DC-AC05-84OR21400.

REFERENCES

1. F. P. Incropera, Heat Transfer, in B. D. Tapley (Ed.), *Eshback's Handbook of Engineering Fundamentals*, 4th ed., Wiley, New York, 1990, Chapter 10.2.

2. I. M. Kolthoff, E. B. Sandell, E. J. Meehan, and S. Bruckenstein, *Quantitative Chemical Analysis*, 4th ed., Macmillan, New York, 1969, pp. 1016–1017.

3. D. R. Lide, *CRC Handbook of Chemistry and Physics*, CRC Press, Boca Raton, FL, 1995.

4. W. J. Moore, Kinetic Theory, in *Physical Chemistry*, 4th ed., Prentice-Hall, Englewood Cliffs, NJ, 1972.

5. E. J. Jumper, C. A. Forbrich, Jr., and L. M. Nicolai, Atomic and Molecular Theory of Gases, in B. D. Tapley (Ed.), *Eshback's Handbook of Engineering Fundamentals*, 4th ed., Wiley, New York, 1990, Chapter 6.1.

6. A. J. Diefenderfer, *Principles of Electronic Instrumentation*, W. B. Saunders, Philadelphia, PA, 1975.

7. Gow-Mac Instrument Co., *Gow-Mac Detectors for Gas Analysis and Gas Chromatography, SB-10*, Gow-Mac Instrument Co., Bethlehem, PA, 1992.

8. Gow-Mac Instrument Co., *Gow-Mac Thermal Conductivity Detector Elements for Gas Analysis, SB-13*, Gow-Mac Instrument Co., Bethlehem, PA, 1993.

9. Gow-Mac Instrument Co., *General Service Bulletin for Repair of Thermal Conductivity Cells*, Gow-Mac Instrument Co., Bethlehem, PA, 1996.

10. G. W. Ewing, Sir Charles' Bridge. *J. Chem. Ed.* **52**(4), A239–A242 (1975).

11. J. D. Waters, Gas Chromatography, in H. A. Strobel and W. R. Heineman (Eds.), *Chemical Instrumentation: A Systematic Approach*, 3rd ed., Wiley, New York, 1989, Chapter 25.

12. C. H. Lochüller, B. M. Gordon, P. M. Gross, A. E. Lawson, and R. J. Mathieu, A Systematic Approach to Thermal Conductivity Detector Design I. Use of Computer-Aided Data Acquisition for the Evaluation of Cell Contributions to Peak Shape, *J. Chromat. Sci.* **15**(8), 285–289 (1977).

13. C. H. Lochüller, B. M. Gordon, P. M. Gross, A. E. Lawson, and R. J. Mathieu, A Systematic Approach to Thermal Conductivity Detector Design II. Flow Rate Dependence of Non-column Contributions to Band Volume, *J. Chromat. Sci.* **16**(4), 141–144 (1978).

14. C. H. Lochüller, B. M. Gordon, P. M. Gross, A. E. Lawson, and R. J. Mathieu, A Systematic Approach to Thermal Conductivity Detector Design III. Applications and Analysis Using Small-Diameter, Open-Tubular Columns, *J. Chromat. Sci.* **16**(11), 523–527 (1978).

15. A. B. Richmond, Thermal Conductivity Detectors for Gas Chromatography: A Literature Survey, *J. Chromat. Sci.* **9**(2), 92–98 (1971).

16. R. L. Grob, *Modern Practice of Gas Chromatography*, Wiley, New York, 1995.

17. MTI Analytical Instruments, Chemical Diagnostics in Space, *MTI's QuickPeak*, February 1998, p. 2.

18. J. B. Angell, S. C. Terry, and P. W. Barth, Silicon Micromechanical Devices, *Sci. Am.* **248**(4), 44–55 (1983).

19. R. H. Kolloff, Hydrogen, Deuterium Thermal Conductivity Detector, U.S. Patent No. 4,464,925, 1984.

20. K. Silvers, *Tank Vapor Analysis: A Wide Range of Methods*, U.S. Department of Energy Report PNNL-SA-28107.

21. American Society for Testing and Materials (ASTM), *Standard Practice for Testing Thermal Conductivity Detectors Used in Gas Chromatography*, Designation E516-95a, ASTM, Philadelphia, PA, 1995.

DISCLAIMER

20

OPACITY MONITORS

Julian Saltz

Datatest Industries, Div. of Redkoh Industries
Levittown, Pennsylvania

Environmental Instrumentation and Analysis Handbook, by Randy D. Down and Jay H. Lehr
ISBN 0-471-46354-X   Copyright © 2005 John Wiley & Sons, Inc.

20.1 INTRODUCTION

The term *opacity* is typically used for measuring particulate matter (smoke particles) in a smokestack.

By definition, *opacity* is the obscuration of a light beam by the particles in the beam. It is a percentage measurement. For example, *0% opacity* means that there are no particulates and *100% opacity* means that the light beam is completely blocked by the particulates. It is a linear relationship.

20.2 BASIC MEASUREMENTS

Other terms that are sometimes used in opacity monitoring, such as transmittance, optical density, and Ringleman number (see the basic specifications for a USEPA compliance opacity meter in Table 20.1).

20.2.1 Transmittance and Opacity

The relationship between transmittance and opacity is

$$\text{Opacity} = (100 - \text{transmittance})$$

where *transmittance* is the amount of light that passes through the particulates. For example, in Figure 20.1, I_0 is the light transmitted from a light source and I is the light received.

Table 20.1 Performance Specification

Parameter	Specifications
Calibration error[a]	$\leq 3\%$ opacity
Response time	≤ 10 s
Conditioning period[b]	≤ 168 h
Operational test period[b]	≤ 168 h
Zero drift (24-h)[a]	$\leq 2\%$ opacity
Calibration drift (24-h)[a]	$\leq 2\%$ opacity
Data recorder resolution	$\leq 0.5\%$ opacity

[a] Expressed as the sum of the absolute value of the mean and the absolute value of the confidence coefficient.

[b] During the conditioning and operational test periods, the CEMS must not require any corrective maintenance, repair, replacement, or adjustment other than that clearly specified as routine and required in the operation and maintenance manuals.

Figure 20.1   Transmittance of a light beam.

The transmittance (T) is

$$T = \frac{I}{I_0} \times 100\% \tag{20.1}$$

and the opacity (Op) is

$$Op = \left[1 - \frac{I}{I_0}\right] \times 100\% \tag{20.2}$$

Opacity is the complement of transmittance.

The term *transmittance* is rarely used in the United States but is occasionally used in other countries.

20.2.2 Optical Density

The relationship between optical density (OD) and opacity is

$$OD = -\log T = -\log\left[1 - \frac{I}{I_0}\right] \tag{20.3}$$

Optical density is sometime used for determining mass flow such as grams per cubic meter.

20.2.3 Ringleman Number

The term *Ringleman number* has been used since the beginning of the twentieth century by observers to measure the density of smoke from a smokestack. It is calibrated by using a gray scale next to the observer for comparison between the smokestack emission and the scale. The calibration is in steps of 20% opacity; thus Ringleman 1 is 20%, Ringleman 2 is 40%, and so on up to Ringleman 5, which is 100% (Table 20.2).

Table 20.2   Relationship Between Transmittance, Opacity, and Optical Density (OD)

Transmittance (%)	Opacity (%)	OD
100	0	0.0
90	10	0.046
80	20	0.097
70	30	0.16
60	40	0.22
50	50	0.3
0	100	Infinity

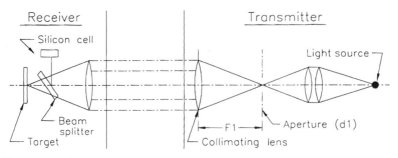

Figure 20.2 Single-pass opacity meter—light passes through the smokestack once.

20.3 DESIGN CONSIDERATIONS FOR OPACITY MONITORS

20.3.1 Single-Pass Opacity Monitor

As shown in Figure 20.1, a simple opacity meter consists of a well-regulated light source and receiver. The particles in the smokestack block the light. The resultant light is measured at the receiver. In Figure 20.2 the optical system is shown in more detail.

In Figure 20.2 the light source could be any of the following:

- Tungsten lamp
- Light-emitting diode (LED)
- Laser

If a laser were used, the lens system in the transmitter would be omitted since the light beam would be well collimated.

20.3.2 Projection and Viewing Angles

20.3.2.1 Angle of Projection. It is important to have a well-collimated beam. The United States Environmental Protection Agency (USEPA) specification for collimation, known as the *angle of projection*, is that it be less than 4°.

The angle of projection Figure 20.3 may be calculated from the dimensions shown in Figure 20.2:

$$\text{Angle of projection (Op)} = \frac{d_1}{f_1} \times \frac{360°}{2\pi} \tag{20.4}$$

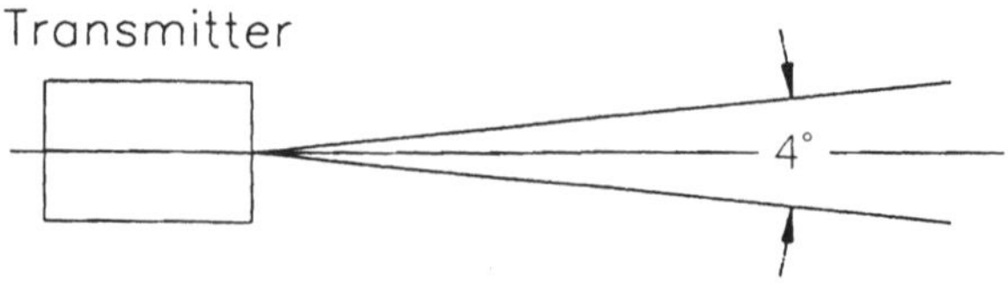

Transmitter

4°

Figure 20.3 Angle of projection.

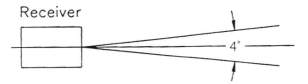

Figure 20.4 Angle of view.

where d_1 is the diameter of the aperture and f_1 is the focal length of the collimating lens.

20.3.2.2   Angle of View. The angle of view is the angle seen by the receiver; any light outside this angle would not be detected (see Fig. 20.4).

20.3.3   Response Time

The response time is the time it takes the opacity signal to reach 95% of its final value. The USEPA specification is 10 s, maximum.

20.3.4   Double-Pass Opacity Meter

The double-pass opacity meter is similar to the single-pass model except the light beam makes two passes through the smokestack.

20.3.5   Photopic Light

The light beam through the smokestack should be in the visible spectrum, specifically between 500 and 600 nm. The overall system response should be within 400–700 nm. The effect of wavelength outside this spectrum is shown in Figures 20.5 and 20.6.

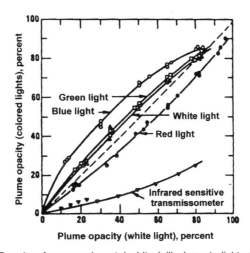

Figure 20.5 Opacity of an experimental white (oil) plume to light of various colors.

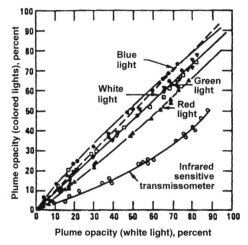

Figure 20.6 Opacity of an experimental black (carbon) plume to light of various colors.

For example, in Figure 20.5 an opacity of 20% would be 15% if the wavelength of the beam were 700 nm (in the near-infrared region).

If a red laser were used as the light source and the wavelength were 670 nm, the curve in Figure 20.5 could be compensated for this change in opacity. Most opacity meters use a microprocessor, so the modification to the curve of Figure 20.5 could be stored in an erasable programmable read-only memory (EPROM).

The light beam is reflected back to the receiver after it passes through the stack. In the double-pass system the opacity is a nonlinear function:

$$\text{Opacity} = \left[1 - 10 \exp \frac{1}{2}\log\left(\frac{I}{I_0}\right)\right] \times 100\% \qquad (20.5)$$

where I_0 is transmitted light (see Fig. 20.7) and I is reflected light. The $\frac{1}{2}$ in Equation (20.5) relates to the light beam passing through the particles twice.

In Figure 20.7 the light from the light source is focused and the focal point is at the aperture. The beam then passes through a 50/50 beamsplitter and then to the collimating lens. The collimating lens will now form almost parallel beams of light. The USEPA standard is a maximum divergence of 4°.

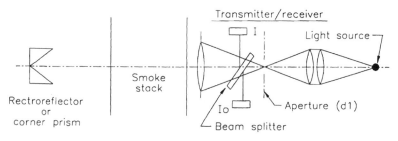

Figure 20.7 Double-pass opacity system.

The retroflector can be either a glass corner prism or a plastic reflector. Angular rotation of the retroflector will not affect the direction of the reflected beam. If a plain mirror were used, the angle of incidence would equal the reflected angle. This certainly is not desirable.

Various detectors may be used, such as silicon, gallium arsenide, vacuum diodes, and photomultiplier tubes (PMTs). In most cases, silicon diodes are used since they have a low-temperature coefficient and are relatively inexpensive.

20.3.6 Single- Versus Two-Cell Detector

In Figure 20.7 the measurement of I and I_0 is made using two cells. The problem with this type of system is unequal temperature drift of the two cells, thereby causing an opacity error. The temperature swing from night to day or summer to winter can be large; 100°F is not uncommon, thereby causing a differential error between the two cells. The typical temperature coefficient for a silicon cell is 0.3%/°C.

An alternative method is to use a single-cell detector and timeshare it for I and I_0. This can be done with a rotating chopper disk as shown in Figure 20.8.

In Figure 20.8 the light beam is divided by beamsplitter BS_1 into two beams. Beam A goes through the open hole of the rotating disk and the smokestack, and is then reflected back by the retroflector to BS_2 and then to the silicon cell. Beam B goes through the rotating disk when the disk has rotated 180°, reflected by mirror 2 through BS_2 and onto the Si cell.

In this manner, the silicon cell now shares beams A and B. Since both beams impinge on a single silicon detector cell, ambient-temperature changes do not cause any change in the opacity output. In actual practice, the light may be directed using glass fiberoptics rather than beamsplitters and mirrors.

In Figure 20.9 the Datatest Model 1000MPD uses a single beamsplitter and glass fiberoptics to achieve the same purpose.

Other signals are also measured on the same silicon cell. An optical filter is placed on the rotating disk for the span measurement and the dust accumulation is sampled at the cover glass that is exposed to the flue gas. Ambient light in the smokestack is measured when the rotating disk blocks both beams A and B as shown in Figure 20.8.

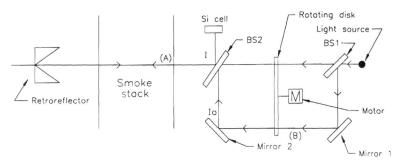

Figure 20.8 Optical system using a single-cell detector.

Figure 20.9   Model 1000MPD transmitter/receiver.

20.3.7   Zero and Span Measurements

USEPA requires a simulated zero and span measurement at least once every 24 h to ensure that the opacity meter is in calibration. It is simulated since the smokestack has flue gas and the only way to get an absolutely true reading is to operate the zero and span in a clear stack. This is seldom not possible (clear stack) in most operating plants.

Simulated measurements are made in a variety of ways, as follows:

1. In a double-pass system a reflector is interposed at the transmitter/receiver so that the light is reflected right back as shown in Figure 20.10. The span is measured by putting some dots over another mirrored surface. The disk can either rotate continuously or can be put in position on command. Another method is to use a glass fiberoptic cable to measure the light from the source after it has gone through the optical system.

2. In a single-pass system the light beam is directed through a glass fiberoptic cable that goes around the smokestack (on the outside). The light through this cable has the same intensity as the same light beam through a clear stack. Again it is a simulated measurement.

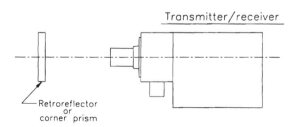

Transmitter/receiver

Retroreflector
or
corner prism

Figure 20.10   Measurement of the simulated zero.

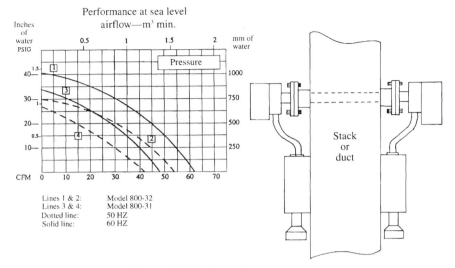

Figure 20.11 Purge air pressure curve.

In both the single- double-pass systems, if the light beam shifts as a result of structural changes in the stack or duct, the instrument will go out of calibration. The opacity meter must have an optical alignment device so that the shift can be corrected. The 24-hr zero–span calibration will not pick up this shift.

Air purge is used on the optical surfaces that are exposed to the flue gas. In almost all cases air purge is an absolute necessity for keeping the lenses clean. The backpressure of the smokestack should be overcome by the purge so that the flue gases do not deposit a film on the optical surfaces of the opacity meter. A typical purge air pressure curve is shown in Figure 20.11.

In most cases 15–25 acfm (absolute cubic feet per minute) is sufficient to keep the optical surfaces clean.

20.3.8 Slotted Pipes

The slotted pipe (see Fig. 20.12) is used in a smokestack or duct to maintain optical alignment between the light transmitter on one side and the receiver or retroreflector on the other side. It is usually less than 12 ft long for mechanical stability.

The pipe is typically made of steel, but it can be made of other materials such as stainless steel, poly(vinyl chloride) (PVC), or fiberglass. In accordance with USEPA specifications (40 CFR 60, PSI, paragraph 5.1.8; see Ref. 1), the clear area where the flue gases pass should be 90% minimum. The diameter should be large enough so that the light beam does not hit the sides of the pipe.

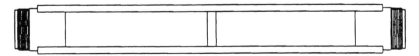

Figure 20.12 Typical slotted pipe.

20.4   LIGHT SOURCES

Essentially three different types of light sources are used in opacity meters.

20.4.1   Incandescent Lamps

One light source is a tungsten filament filled with a halogen gas, typically 10 W or greater. The lamp envelope may be glass or quartz. A small, concentrated filament is best since it makes focusing easier. The lamp voltage should be regulated, and the electronics of the system should be designed so that the system is insensitive to lamp intensity changes. Most lamps should be operated below their rated voltage to prolong life. As the lamp ages, there is a tungsten loss and the envelope blackens. Lamp life under continuous operation should be at least 3 years.

20.4.2   Light-Emitting Diodes (LEDs)

These may be used as the light source of the opacity meter. The emitted light should have wavelengths between 500 and 600 nm. Acceptable LEDs should be green or amber (yellow). The green LED is typically 530–560 nm, where as the amber (yellow) is about 590 nm.

20.4.3   Laser Light Source

The gas or solid-state laser may be used as a light source for the opacity meter, but the wavelength must be between 500 and 600 nm. Typically, a green laser is used with a wavelength of about 550 nm.

The problem with the laser is that the beam diameter is only several millimeters; consequently, the receiver or retroreflector must be perfectly aligned at all times.

It is also quite possible to use a red laser with a wavelength of about 670 nm if the overall response is adjusted to the response curve shown in Figures 20.5 and 20.6. This can be easily accomplished using a computer program that would store the corrected response curve.

20.5   USEPA OPACITY SPECIFICATIONS

The USEPA specifications for opacity are given in the *Federal Register* (40 CFR 60 AP B, PSI). The most important parts are abstracted from the USEPA document. They will be listed by paragraph number as they appear in the *Federal Register*.

20.5.1   Peak and Mean Spectral Responses

The peak and mean spectral responses[2] must occur between 500 and 600 nm. The response at any wavelength below 400 nm or above 700 nm shall be less than 10% of the peak spectral response.

20.5.2 Viewing and Projection Angles

20.5.2.1 Angle of View. The total angle of view[2] shall be no greater than $4°$.

20.5.2.2 Angle of Projection. The total angle of projection[2] shall be no greater than $4°$.

20.5.3 Optical Alignment Sight

The optical alignment sight[2] is an essential part of an opacity meter since the metal of a duct or smokestack will be distorted because of temperature change and wind conditions.

Each analyzer must provide some method for visually determining that the instrument is optically aligned. The method provided must be capable of indicating that the unit is misaligned when an error of $+2\%$ opacity occurs as a result of misalignment at a monitor pathlength of 8 m. Instruments that are capable of providing an absolute zero check while in operation on a stack or duct with effluent present, and while maintaining the same optical alignment during measurement and calibration, need not meet this requirement (e.g., some "zero pipe" units).

20.5.4 Simulated Zero and Upscale Calibration System

Since it is not possible to obtain an absolute measurement of zero and span, this is the next best compromise.

Each analyzer must include a calibration system for simulating a zero (or no greater than 10%) opacity and an upscale opacity value for the purpose of performing periodic checks of the transmissometer calibration while on an operating stack or duct. This calibration system will provide, as a minimum, a system check of the analyzer internal optics and all electronic circuitry, including the lamp and photo-detector assembly.

Table 20.3 lists the neutral density attenuator values for calibrating the opacity instrument. A revision of these values has been proposed. For example, for an

Table 20.3   Required Calibration Attenuator Values (Nominal)

Span Value (% Opacity)	Calibrated Attenuator Optical Density (Equivalent Opacity in Parentheses) O_2		
	Low-Range	Midrange	High-Range
40	0.05 (11)	0.1 (20)	0.2 (37)
50	0.1 (20)	0.2 (37)	0.3 (50)
60	0.1 (20)	0.2 (37)	0.3 (50)
70	0.1 (20)	0.3 (50)	0.4 (60)
80	0.1 (20)	0.3 (50)	0.6 (75)
90	0.1 (20)	0.4 (60)	0.7 (80)
100	0.1 (20)	0.4 (60)	0.9 (87.5)

Figure 20.13 The transmitter/receiver module on a smokestack.

instrument span of 0–100% opacity, the calibration filters are 20%, 60%, and 87.5%. The USEPA maximum allowable emission is 20%. This is not practical. The new proposed specification is 16%, 24%, and 40%.

20.6 EXAMPLE OF AN INSTALLATION IN A CHEMICAL PLANT

Opacity meters are rigidly mounted on smokestacks or on the ducts leading to the stack.

Figure 20.13 a transmitter/receiver module is firmly attached to the smokestack using companion 4-in ANSI flanges.

The optical alignment is accomplished using three external alignment screws.

REFERENCES

1. 40 CFR 60, App. B, Spec. 1, USEPA.
2. *Measurement of the Opacity and Mass Concentration of Particulate Emissions by Transmissometry*, EPA 650/2-74-128, USEPA
3. Datatest opacity meters.

21

TEMPERATURE MEASUREMENT

Randy D. Down, P.E.

Forensic Analysis & Engineering Corp.
Raleigh, North Carolina

Environmental Instrumentation and Analysis Handbook, by Randy D. Down and Jay H. Lehr
ISBN 0-471-46354-X   Copyright © 2005 John Wiley & Sons, Inc.

21.1 INTRODUCTION

Although used extensively in many other applications, there are many environmental applications that require the use of instruments capable of accurate and reliable temperature measurement.

Many techniques have evolved over the years for measuring temperature. In 1714, Gabriel D. Fahrenheit invented a liquid-filled thermometer, in which liquid, such as mercury or alcohol, expands with temperature and moves up a glass tube. His measurement system had a fixed volume, with the vapor pressure above the liquid reduced to near zero, allowing the liquid freedom to expand. Graduation marks along the glass tube gave an indication of the temperature.

There are numerous temperature measurement instruments on the market today produced by a variety of manufacturers. The instruments we will focus on in this chapter are those commonly used in environmental measurement and control applications. We will discuss how they are selected and applied to various applications and what potential problems or disadvantages can arise from misapplying or improperly installing these devices.

21.2 INSTRUMENT SELECTION

Selection of the appropriate temperature measurement device requires answering the following questions:

- How precise a measurement is required?
- How quickly must it respond to track the temperature accurately?
- How repeatable is its measurement?
- What range of measurement is needed?
- How much resolution is required?
- What conditions will the instrument be exposed to, such as dust, moisture, corrosive substances, heat, and vibration?
- Can the sensor come in contact with the measured media?

By answering these questions, many pitfalls associated with temperature measurement can be avoided.

21.3 AVOIDING ERRORS

The most common mistakes made in selecting a temperature sensor are as follows:

- The sensor range does not match the range of the display device or transmitter.
- The sensor gets exposed to temperatures that exceed its range or can damage it.
- Poor electrical connections exist at termination points (terminal blocks or splices).
- The sensor is poorly installed (sensor location does not pick up actual temperature to be measured), typically due to probe not being extended far enough into the measured fluid stream.

21.4 MECHANICAL AND ELECTROMECHANICAL TEMPERATURE SENSING

Mechanical temperature measurement is accomplished by measuring a change in the position of a component of the instrument that changes its shape or moves in response to a change in sensed temperature. Examples of these components are bimetallic elements and gas- or liquid-filled capillaries and bellows.

21.4.1 Bimetallic Elements

Bimetallic elements are typically comprised of two dissimilar metal strips that are bonded together. Since different metals have a different coefficient of expansion, they each grow in length (expand) at a different rate for a given change in temperature. This thermal expansion forces the bonded metals to bend (deform) to one side (toward the metal strip with the lower thermal coefficient). This characteristic makes them a highly reliable method of thermal overload protection of equipment such as thermal oxidizers and other equipment that needs to be protected from damage from exposure to excessively high temperature.

Electric room thermostats, such as the common Honeywell round-wall thermostat that used to be found in many homes and apartments, used a helical-shaped bimetallic strip attached to a Microswitch. As a change in the bimetallic element's shape occurred, in response to a change in measured room temperature, the movement actuated the Microswitch, which was wired to the controls of a furnace or air conditioning unit to turn it on and off. A second bimetallic coil mounted inside the thermostat cover was attached to a temperature indicator on the face of the thermostat.

21.4.2 Remote Bulb Thermometers and Thermostats

Remote bulb thermometers and thermostats use a gas- or liquid-filled bulb connected by a thin capillary tube to a remote bellows. The bellows expands or contracts as the temperature at the bulb increases or decreases, expanding the gas in the bulb according to the ideal gas law. This is due to the expansion of the fluid (gas or liquid) within the bulb and capillary. The bellows is attached to a dial or switch to provide temperature indication or control.

The simplicity of these devices makes them highly reliable and relatively inexpensive. They are also somewhat fragile and have a limited temperature range. The capillary conveying the fluid to the remote indicator or sensor must be protected so that it is not kinked, pinched, or subjected to abrasion (from vibration) that could eventually wear a hole through the capillary wall.

21.5 ELECTRONIC TEMPERATURE SENSORS

An electronic temperature sensor, or "sensing element," consists of a simple device that converts a change in temperature into some type of physical change—typically a change in position (such as a bimetallic strip or diaphragm), a change in electrical current, or a change in electrical resistance. With the declining cost of solid-state electronics, the accuracy, repeatability, durability, and compact size of electronic sensors have made them the preferred type to use on most applications. The lower current-handling requirements of solid-state devices such as integrated circuits used for signal amplification have reduced concerns of packaging these components to keep the physical size of a field transmitter compact and still not overheat the internal components.

Three basic types of noncontact electronic temperature sensors are used in environmental measurement and control applications: thermocouples, thermistors, and RTDs (resistance–temperature detectors).

Resistance–temperature detectors, thermocouples, and thermistors are typically encapsulated in an epoxy enclosure (to protect them) or within some type of metal sheath to protect the delicate sensor. The sheath material is dependent upon the environment to which it will be exposed.

21.6 THERMOCOUPLES

First developed in 1821 by T. J. Seebeck, a thermocouple is a temperature sensor that consists of a junction between two dissimilar metals, such as copper and nickel. A typical schematic of a thermocouple junction and electrical terminations is shown in Figure 21.1.

When heat energy is introduced at a thermocouple junction, a difference in temperature (due to the difference in thermal conductivity of the differing metals)

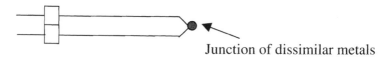

Junction of dissimilar metals

Figure 21.1 Typical thermocouple.

Table 21.1   Temperature in Thermocouples

Thermocouple Type	Thermocouple Wire Materials	Temperature Range (°F)
T	Copper and constantan	−328–662
J	Iron and constantan	32–1382
E	Chromel and constantan	−328–1652
K	Chromel and alumel	328–2282
R	Platinum and platinum with 13% rhodium	32–2642
S	Platinum and platinum with 10% rhodium	32–2642
B	Platinum and platinum with 30% rhodium	32–3092

occurs. This induces a small electrical current flow (measured in millivolts), the magnitude of which changes as the temperature at the junction changes. As long as a temperature difference exists at the junction, current will continue to flow through the wire.

Junctions formed from different pairs of dissimilar metals allow sensors with different temperature ranges to be created. Typical sensors include type J, K, and L thermocouples. A comparison of temperature ranges is provided in Table 21.1.

Thermocouples require the use of thermocouple wire (made of the same materials—two dissimilar metals—as the sensor). All termination points between the sensor and display device, transducer, or transmitter (that converts the signal to a 4–20-mA DC signal) must also be made of the same metals. They are given the same letter designation as the sensor. For example, to work properly, a type J thermocouple would require type J thermocouple wire and type J terminal blocks. Selecting the wire size used in a thermocouple sensor depends upon the application requirements. Generally, when longer life is required and you are dealing with higher temperatures, a larger wire size should be chosen. When sensitivity is the primary concern, a smaller wire should be used.

There are several basic thermocouple configurations—*exposed-junction, grounded-junction*, and *ungrounded-junction* thermocouples:

Exposed-Junction Thermocouple. An exposed-junction thermocouple has the quickest response time of the three basic types because the junction is in direct contact with the media being sensed. This also exposes the unprotected junction to potential damage.

Ground-Junction Thermocouple. A ground-junction thermocouple has a slower response time than that of an exposed-junction thermocouple and is electrically grounded to its protective metal sheath. The sheath is typically provided in the form of a narrow metallic tube enclosed at the junction end to protect the sensor from physical damage. Carbon steel, 316 or 304 stainless steel, and cast-iron tubes are used for temperatures below 1000°F. For higher temperature applications, ceramic tubes and metals such as Inconel are used.

Ungrounded-Junction Thermocouples. Ungrounded-junction thermocouples have the slowest response time but are the most commonly used thermocouples.

Figure 21.2 Typical thermistor.

They are not electrically insulated from their protective metal sheath (to prevent potential ground loops and pickup of electrostatic noise). This makes them the most reliable of the three types in terms of life expectancy and performance.

21.7 THERMISTORS

Thermistors are small, two-wire semiconductor devices with a junction between conductive materials that changes electrical resistance as the temperature changes (Fig. 21.2). The semiconductor junction is typically encapsulated in an epoxy bead. A small change in temperature results in a measurable change in resistance. Because of their simplicity, thermistors are relatively inexpensive and reliable. They can be purchased in a wide range of sizes and shapes to fit numerous applications. For harsh environments they are installed in a protective metal sheath in the same fashion as a thermocouple or RTD.

21.8 RESISTANCE–TEMPERATURE DETECTORS

In the 1860s, William Siemens, based on earlier research conducted by Sir Humphrey Davy, discovered that a relationship existed between the temperature of a metal and its electrical resistance. Further analysis by Siemens established platinum as the preferred element for the resistance thermometer. As the temperature of a platinum element increases, its resistance also increases. Modern RTDs are made using several different metals, including platinum, copper, and nickel. For example, a 1000-Ω nickel RTD will typically exhibit a change in its electrical resistance of approximately 3 Ω per degree Fahrenheit change in sensed temperature. Each material has an acceptable measurement range.

- Copper RTDs have a useful temperature range of from -150 to $500°F$.

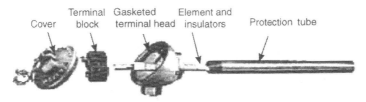

Figure 21.3 Exploded view of sensor assembly.

- Nickel RTDs have a useful temperature range of from −150 to 400°F.
- Platinum RTDs have a useful temperature range of −325 to 1200°F.

Resistance–temperature detectors consist of a hair-thin strand of nickel or platinum wire wound around an electrical insulator form. The resistance wire element is then encapsulated within a protective layer of epoxy. Depending on the type of application, the epoxy-coated element may then be enclosed within a protective metal jacket. The wire resistance changes almost proportionally to a change in resistance.

A RTD is a passive device. It is typically connected as one leg of a bridge circuit, requiring a small amount of current to produce a measurable resistance change that is proportional to the change in temperature. The power supply used to generate this small current typically resides in the measuring or control device, such as a temperature controller or data logger.

Care must be taken to ensure that the wires that connect the RTD to the measuring device do not affect the reading. These extension wires can add resistance, which will change the measured value. For this reason, RTDs often are provided in a three-wire version. The third (extra) lead is used to offset any potential for error by balancing the measurement bridge circuit.

21.8.1   RTDs Versus Thermocouples

Resistance–temperature detectors are generally more expensive than thermocouples but will tend to be more stable and more repeatable.

In most cases, RTDs are slower to respond to temperature changes than thermocouples, and they require an external power source and a third (and sometimes a fourth) conductor to offset additional error.

21.9   THERMOWELL ASSEMBLIES

If it is undesirable to have the sensor probe (even when enclosed in a protective sheath) directly in contact with a measured media, then a thermowell assembly is installed (Fig. 21.4). This well assembly is designed so that the thermocouple can be inserted into it in such a fashion that the junction has good physical contact with

Figure 21.4 Standard thermowell.

the bottom of the well. This ensures good thermal conductivity between the sensor, the well, and the measured media. Typically, wells are used in pipelines so that they can be removed and inspected (or replaced) without having to drain the pipeline or allowing the liquid or gas to leak from the resulting pipe opening.

The drawback of any thermowell is that it further slows the response time of the sensor because thermodynamically the temperature change must travel through the wall of the well and the sensor sheath before detection can occur at the element or junction. This inherently dampens the sensor's reaction time to a change in media temperature.

21.10 INFRARED DETECTION

A fourth type of electronic sensor senses infrared radiation emitted from the heat source. This "noncontact" method of temperature measurement is highly accurate but very expensive compared to other "contact"-type measurement methods. It is typically used where accessibility for servicing or sensor replacement is a problem or the surface temperature of the measured media is very high and will damage sensors in direct contact with the media.

21.11 COMPONENTS OF A TEMPERATURE MEASUREMENT LOOP

Temperature measurement systems typically consist of a sensor, a signal conditioner or transmitter (to convert the sensor signal to an industry standard signal, such as 4–20 mA DC), and a device to record or display the measured value (Fig. 21.5). The sensor (or sensing element) detects a change in the temperature of the measured media and converts it to a change in electrical value (millivoltage, current, or resistance), which in turn is converted to an industry-standard 4–20-mA DC signal and scaled over the accurate range of the sensor and then displayed

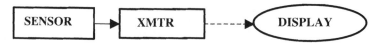

Figure 21.5 Temperature loop block diagram.

at the display device, recorder, or controller in standard engineering units (degrees Fahrenheit or degrees Celsius).

21.12 CALIBRATION

As mentioned in other chapters regarding instrumentation, traceability to NIST (National Institute of Standards and Technology) standards is highly important when dealing with critical applications that require precise measurement.

According to NIST, traceability to NIST temperature standards is addressed in the Measurement Assurance Program (MAP), as implemented using standard platinum resistance thermometers (SPRTs) in the range of 190–660°C.

In a MAP, NIST sends to a customer (who utilizes the ITS-90 or an approximation to the ITS-90) a set of three SPRTs that have previously been calibrated by NIST on the ITS-90. The customer calibrates the set of SPRTs and then returns the SPRTs to NIST for a final calibration. NIST analyzes the raw data, uncertainty statements, methods of data analysis, methods of realization of the ITS-90, and calibration report results and format. A full descriptive report of this analysis is sent to the customer with suggestions for improvement and a quantitative statement of the present level of agreement of the customer calibrations with the NIST calibrations.

A MAP is a quantitative test of the customer's ability to calibrate SPRTs and the validity of the claimed uncertainties of the calibrations. When conducted at regular intervals, a MAP supports the customer's claim of continuous traceability to NIST at a specified uncertainty.

21.13 TEMPERATURE TRANSMITTERS

Temperature transmitters serve several purposes. They amplify and filter the sensor signal and electronically isolate it from other electronic devices. They also establish a scalable signal range over which the transmitter output signal will operate (e.g., from 4 mA DC at 50°F to 20 mA DC at 300°F).

If the environment in which they will be installed is classified as a Class I hazardous environment, then the transmitters must be housed in explosion-proof enclosures (Fig. 21.6). These typically consist of an air-tight, cast-aluminum or stainless

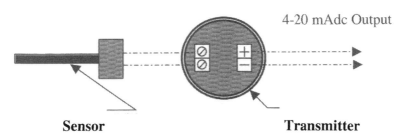

Sensor   **Transmitter**

Figure 21.6 Temperature transmitter wiring.

steel enclosure to prevent explosive gases or dust from entering. They are designed to be strong enough to contain a blast should the gas or dust ignite within.

21.14 TYPICAL APPLICATIONS

A few select examples of environmental applications that the author has encountered when required to apply a temperature measurement include the following:

- Exhaust air or gas temperatures needed for temperature correction of flow measurements—to determine the true volumetric flow rate of plant exhaust air emissions for permitting purposes.
- Water discharge temperature needed to satisfy wastewater permit applications—in some cases there are maximum temperature limits at which wastewater can be discharged to a sanitary drain.
- Exhaust air temperature entering a carbon adsorption system—to determine if precooling the exhaust air is needed to maintain proper removal efficiency. Activated charcoal used in the adsorption process is most effective within certain temperature and humidity conditions.
- Exhaust air temperature entering a regenerative thermal oxidizer (RTO). If the entering air temperature is too high (above the maximum design condition), the oxidizer will start to overheat and shut down on a high-limit thermostat.
- Sensing motor bearing temperatures on critical exhaust fans. As a bearing begins to fail, the friction factor increases, generating heat that can potentially damage the motor shaft and result in expensive repairs or motor replacement and possible production or occupancy downtime.
- Sensing water temperature in a sampling system to maintain stasis during storage of the biological grab samples. Undesirable bacterial growth can occur above stasis (40°Fahrenheit).

21.15 TEMPERATURE MEASUREMENT TO ESTABLISH FLOW RATES

Another function of temperature measurement is to derive air flow velocities using a hot-wire anemometer (Fig. 21.7). Hot-wire anemometers are typically used to measure velocities in flows where a Pitot tube is inadequate (turbulent or rapidly varying flows). There are two types of hot-wire anemometers: constant current and constant temperature. Constant-temperature anemometers (CTAs) are more common.

A CTA consists of a fine wire (or coated cylinder in the case of a hot film) which is exposed to the fluid flow (exhaust air stream). The wire is heated electrically to some constant temperature much higher than the fluid temperature. As the fluid

Figure 21.7 Hot-wire anemometer.

passes over the wire, the wire is cooled by forced convection. Electrical current flow to the wire is adjusted to keep the wire temperature constant. The current required to keep the wire at a constant temperature then becomes a measure of velocity. Convective cooling is balanced by electrical energy input.

Convective heat transfer away from a body in a moving air stream is a primarily a function of the fluid's Reynolds number (Re). Overall heat transfer increases as Re increases.

For measurement purposes, the hot wire is connected to a Wheatstone bridge circuit, with one section of the bridge being current flow through the hot film. Since the hot film is a metal, its resistance increases as the temperature increases and vice versa. As the measured fluid temperature changes, the bridge becomes unbalanced. The unbalance is sensed by a DC amplifier which then produces a change in voltage (and thus current) through the hot-wire sensor to keep the bridge balanced. Resistance of the controlling resistor determines the operating temperature of the hot film.

21.16 ADVANTAGES AND DISADVANTAGES OF SENSOR TYPES

Each type of detection device has distinct advantages and disadvantages over the other types. They are summarized below.

Thermocouples   *Advantages*: very wide sensing range, low cost, high durability, self-powered
Disadvantages: poor resolution, drift, thermocouple wire and thermocouple terminals required for field wiring

Thermistors	*Advantages*: high resolution (large change in resistance per degree change in temperature), low cost
	Disadvantages: lower reliability and susceptibility to ambient conditions
RTDs	*Advantages*: high resolution (change in resistance per degree change in temperature), high repeatability
	Disadvantages: high cost, susceptibility to ambient conditions
Infrared	*Advantages*: noncontact sensing (which allows it to be used for very high temperature conditions), high resolution, high repeatability
	Disadvantages: very high cost compared to other types of temperature sensors

21.17 CONCLUSION

Selection of the proper temperature sensor, compatible transmitter, and indication device is necessary to produce an accurate temperature measurement. Work closely with the equipment manufacturer or manufacturer's representative to ensure that the proper devices are purchased and installed. Field verify the accuracy of the measurement after initial installation as part of a startup and commissioning procedure.

21.18 GLOSSARY

Accuracy The closeness of an indication or reading of a measurement device to the actual value of the quantity being measured.

Ambient Temperature The average or mean temperature of the surrounding air which comes in contact with the equipment and instruments under test.

Cold-Junction Compensation Measures the ambient temperature at the connection of the thermocouple wire to the measuring device. This allows for accurate computation of the temperature at the hot junction by the measuring device.

Cold Junction (Reference Junction) The junction generally at the measuring device that is held at a relatively constant temperature.

Exposed Junction The thermocouple junction or measuring point is exposed (no outer well assembly or tube). Such thermocouples have the fastest response time.

Extension Wire Wires which connect the thermocouple to a reference junction (controller, receiver, recorder). Extension wire must be of the same type (material) as the thermocouple. Special plugs and jacks made of the same alloys as the thermocouple should be used if a quick connect is needed.

Grounded Junction The internal conductor of a thermocouple is welded directly to the surrounding sheath material, forming a completely sealed integral junction.

Head   An enclosure attached to the end of a thermocouple within which the electrical terminations are made.

Hot Junction   The measuring junction. Also referred to as the *measurement junction*.

Immersion Length   The portion of the thermocouple which is subject to the temperature being measured.

Protection Tube   A tubelike assembly in which the thermocouple, RTD, or thermistor is installed in order to protect the element from harsh environments.

RTD   Abbreviation for resistance–temperature detector. A sensor that operates on the principle that electrical resistance increases with an increase in temperature at a specific rate. Commonly manufactured using a platinum or nickel resistance element. More accurate and more linear than most thermocouples. Generally much more costly and slower responding.

Thermal Coefficient of Resistance   The change in resistance of a semiconductor per unit change in temperature over a specific range of temperature.

Thermal Conductivity   The ability of a material to conduct heat in the form of thermal energy.

Thermocouple   A junction of two dissimilar metals that induces (through thermal conductance) a voltage output proportional to the difference in temperature between the hot junction and the lead wire (cold junction).

Transmitter (Two-Wire)   A device used to transmit temperature data from either a thermocouple or RTD via a two-wire current loop.

Ungrounded Junction   Although the internal thermocouple conductors are welded together, they are electrically insulated from the external sheath material and are not connected to the sheath in any way. Ungrounded-junction thermocouples are ideal for use in conductive solutions or wherever circuit isolation is required. Ungrounded junctions are required where the measuring instrumentation does not provide channel-to-channel isolation.

Thermowell   A threaded or flanged closed-end tube which is mounted directly to the process or vessel; designed to protect the thermocouple from the process surroundings.

Transducer   A device (or medium) that converts energy from one form to another, such as a physical media (pressure, temperature, humidity, flow), and converts it to an electrical signal.

REFERENCES

1. *Annual Book of ASTM Standards, Vol. 14.03: Temperature Measurement*, American Society for Testing and Materials, W. Conshohocken, PA, 2003.
2. R. D. Down, *Environmental Control Systems*, ISA Publications, 1992.

3. T. W. Kirlin, *Practical Thermocouple Thermometry*, ISA Publications, 1999.

4. L. Michalski, K. Eckersdorf, and J. McGhee, *Temperature Measurement*, Wiley, New York, 1991.

5. J. V. Nicholas and D. R. White, *Traceable Temperatures*, 2nd ed., Wiley, New York, 2001.

6. G. F. Strouse, *NIST Measurement Assurance Program for The ITS-90*, Test and Calibration Symposium, Arlington, VA, 1994.

22

pH ANALYZERS AND THEIR APPLICATION

James R. Gray

Rosemount Analytical Inc.
Irvine, California

Environmental Instrumentation and Analysis Handbook, by Randy D. Down and Jay H. Lehr
ISBN 0-471-46354-X   Copyright © 2005 John Wiley & Sons, Inc.

22.1 THEORY OF pH MEASUREMENT

22.1.1 Definition of pH

pH is a measure of the acidity or alkalinity of a water solution. Acidity or alkalinity is determined by the relative number a hydrogen ions (H^+) or hydroxyl (OH^-) present. Acidic solutions have a higher relative number of hydrogen ions, while alkaline (also called basic) solutions have a higher relative number of hydroxyl ions. *Acids* are substances that either dissociate (split apart) in water to release hydrogen ions, or react with water to form hydrogen ions. *Bases* are substances that dissociate to release hydroxyl ions, or react with water to form hydroxyl ions.

In the strict definition, pH is defined as the negative logarithm of the hydrogen ion activity, $pH = -\log a_H$. Activity is the molar concentration of hydrogen ion, multiplied by an activity coefficient, γ, which takes into account the effects of the ion–ion interactions in solution: $a_H = \gamma[H^+]$.

In general usage, the term *p*H is, for the most part, shorthand, which allows the concentrations of hydrogen ions over a tremendously wide range to be conveniently expressed. The activity coefficient is assumed to be constant and is eliminated by calibration, or is simply ignored. Under these assumptions, pH is taken to be equal to the negative logarithm of hydrogen ion concentration: $pH = -\log [H^+]$ (Fig. 22.1). This simplified expression of pH is sufficient for most applications, and only runs into trouble when pH is being precisely related to a hydrogen ion concentration.

22.1.2 Acid–Base Chemistry

The following discussion addresses the key points of acid–base chemistry. The chemical equations corresponding to topics that follow are summarized in Figure 22.2. In this discussion, water is assumed to be the solvent with, at most, a trace of nonaqueous solvents, such as alcohols. A later section will address the influence of nonaqueous solvents on the measurement of pH.

22.1.2.1 *Hydrogen Ion and Hydroxyl Ion Equilibrium.* In water solutions, the product of the molar concentrations of hydrogen and hydroxyl ions is equal to a

ACIDIC

NEUTRAL

BASIC

pH	Hydrogen ion concentration	Hydroxyl ion concentration
25 C	moles / liter	moles / liter
0	1.0	0.00000000000001
1	0.1	0.0000000000001
2	0.01	0.000000000001
3	0.001	0.00000000001
4	0.0001	0.0000000001
5	0.00001	0.000000001
6	0.000001	0.00000001
7	0.0000001	0.0000001
8	0.00000001	0.000001
9	0.000000001	0.00001
10	0.0000000001	0.0001
11	0.00000000001	0.001
12	0.000000000001	0.01
13	0.0000000000001	0.1
14	0.00000000000001	1.0

Figure 22.1 Hydrogen and hydroxyl ion concentrations.

dissociation constant (K_w): $[H^+]$ $[OH^-] = K_w$. This equilibrium can also be expressed using the "p" notation, where p is an operator meaning the negative logarithm: $pH + pOH = pK_w$.

The dissociation constant of water is temperature-dependent and can be used to calculate the hydroxyl ion concentration from the hydrogen ion concentration, and vice versa, at a given temperature.

In acidic solutions, the hydrogen ion concentration is greater than the hydroxyl concentration ($[H^+] > [OH^-]$). In alkaline solutions, the reverse is true ($[H^+] < [OH^-]$). In neutral solutions, the concentrations are equal ($[H^+] = [OH^-]$).

Acid and Base Chemistry

ACIDS:

General:	MA	$\longrightarrow$	M^+	+	A^-	
Strong Acid:	HCl	$\longrightarrow$	H^+	+	Cl^-	(Complete Dissociation)
Weak Acid:	HAc	$\longrightarrow$	H^+	+	Ac^-	(Partial Dissociation)

BASES:

General:	MOH	$\longrightarrow$	M^+	+	OH^-	
Strong Base:	NaOH	$\longrightarrow$	Na^+	+	OH^-	(Complete Dissociation)
Weak Base:	NH_4OH	$\longrightarrow$	NH_4^+	+	OH^-	(Partial Dissociation)

ACIDS/BASES:

General:	ACID	+	BASE	$\longrightarrow$	SALT	+	WATER
Strong Acid/Strong Base:	HCl	+	NaOH	$\longrightarrow$	NaCl	+	H_2O
Strong Acid/Weak Base:	HCl	+	NH_4OH	$\longrightarrow$	NH_4Cl	+	H_2O
Weak Acid/Strong Base:	HAc	+	NaOH	$\longrightarrow$	NaAc	+	H_2O

HYDROLYSIS:

Weak Acid/Strong Base Salt	Ac^-	+	H_2O	$\rightleftharpoons$	HAc	+ OH^-	(Increase pH)
Strong Acid/Weak Base Salt	NH_4^-	+	H_2O	$\rightleftharpoons$	NH_4OH	+ H^-	(Decrease pH)

Figure 22.2 Acid–base chemistry formulas.

At 25°C, the value of K_w is 10^{-14} ($pK_w = 14.0$). Acidic solutions will have a pH less than 7.0 pH, alkaline solutions will have a pH greater than 7.0 pH, and neutral solution will have a pH of 7.0 pH. It is useful to note that these relations are valid only at 25°C, and although they are also commonly thought to hold true at all temperatures.

22.1.2.2 Acids and Bases.

Acids and bases can be classified as either strong or weak, depending on the degree to which they dissociate. Strong acids, such as hydrochloric acid or nitric acid, are completely dissociated in water solutions, while weak acids, such as acetic acid or citric acid, are only partially dissociated. Similarly, strong bases such as sodium or potassium hydroxides completely dissociate in water solutions, while weak bases such as ammonia only partially dissociate.

The degree of dissociation for weak acids is given by a dissociation constant K_A for weak acids and K_B for weak bases. These constants can be used to calculate the concentration of hydrogen ion or hydroxyl ion for a given concentration of weak acid or base. Some weak acids have more than one acidic hydrogen. Their dissociation proceeds in steps, and the equilibrium for each step can be described by its own dissociation constant.

22.1.2.3 Neutralization and Salts.

Acids and bases neutralize one another. They react to form water and a salt, which is ionic compound, consisting of the negative ion (anion) from the acid and the positive ion (cation) from the base.

22.1.2.4 Hydrolysis.

When a salt from the reaction of a strong acid and strong base is dissolved in water and dissociates the anions and cations have no tendency to associate with hydrogen or hydroxyl ions in solution. A case in point is sodium chloride, which is the salt from the reaction of hydrochloric acid and sodium hydroxide, a strong acid and a strong base. When dissolved in water, they form sodium and chloride ions, neither of which reacts with the hydrogen or hydroxyl ions present in the solution. They have no effect on the solution pH.

However, when the salt of a strong base and a weak acid such as sodium acetate is dissolved in water, the acetate ions will partially combine with hydrogen ions in solution, in accordance with the dissociation constant of acetic acid, to form acetic acid. This reaction reduces the hydrogen ion concentration, and raises the pH. This effect is known as *hydrolysis*.

In the same way, the salt of a weak base and a strong acid, such as ammonium chloride, will lower the pH, by hydrolysis, through the association of ammonium ion with hydroxyl ion, which lowers the hydroxyl ion concentration.

Adding the salts of weak acids and bases to a solution will influence its pH.

22.1.2.5 Buffers and Buffering.

Buffers are solutions designed to maintain a constant pH in spite of the addition of small amounts of acids and bases. They consist of a mixture of a weak acid and its salt, which will have a pH given by the following equation: $pH = pK_A + \log([A^-]/[HA])$, where $[A^-]$ and $[HA]$ are the concentrations of the salt and acid, respectively.

When buffers are formulated with equal concentrations of salt and acid, the pH of the buffer is equal to pK_A. When a base is added to the buffer, the acid component of a buffer neutralizes it, while hydrogen ion from the addition of an acid is consumed by the salt component through hydrolysis. As long as the ratio of salt to acid remains fairly constant, the pH of the buffer will not change.

Buffering capacity is a measure of the how much acid or base a buffer can tolerate without significantly changing pH. Higher concentrations of acid and salt used to prepare the buffer increase its capacity.

Buffering is not only observed in prepared solutions, but also in process samples because of their composition. Simply knowing the pH of a process sample says nothing about the buffering capacity of the sample. A small amount of acid or base reagent will product a significant change in a sample with no buffering capacity, while a sample with a high buffering capacity will show little or no change with the same amount of reagent. The buffering capacity of a sample can be known only from knowledge of its composition or its titration curve.

22.1.2.6 Titration Curves, Acidity, and Alkalinity.

If the pH of a sample solution is plotted against the amount of titrant added, that is, an acid or base, a characteristic titration curve for the solution is obtained. The neutralization point for the sample is defined as the amount of acid or base that has to be added to reach 7.0 pH. The equivalence point is the point where an equal (equivalent) concentration of acid or base has been added to the sample.

If a strong acid and a strong base are used in a titration, the equivalence point and the neutralization point will be the same: 7.0 pH (Fig. 22.3). If either the acid or base is weak, the equivalence point will be greater or less than 7.0 pH because of hydrolysis (Fig. 22.4). In general, the titration curve shows how much titrant is required to adjust the sample to any given pH.

Acidity and alkalinity are used to describe the buffering capacity of a water sample or solution. The acidity or alkalinity values are based on the amount of a reagent

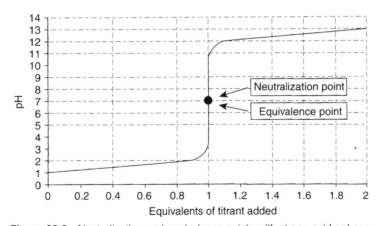

Figure 22.3 Neutralization and equivalence points with strong acid or base.

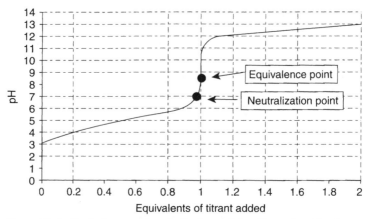

Figure 22.4   Equivalence and neutralization points with weak acid or base.

required to adjust the pH of the sample to a particular pH. The amount is then reported in terms of a standard acid or base. For example, in the case of *m* alkalinity, the sample is titrated with an acid to 4.0 pH. The alkalinity is reported in terms of parts per million as calcium carbonate, based on the equivalents of acid required.

22.1.3   Theory of pH and Reference Electrodes

pH measurement is based on the use of a pH-sensitive electrode (usually glass), a reference electrode, and a temperature element to provide a temperature signal to the pH analyzer.

   The pH electrode uses a specially formulated, pH sensitive glass in contact with the solution, which develops a potential (voltage) proportional to the pH of the solution. The reference electrode is designed to maintain a constant potential at any given temperature, and serves to complete the pH measuring circuit within the solution, and provide a known reference potential for the pH electrode. The difference in the potentials of the pH and reference electrodes provides a millivolt signal proportional to pH, to the pH analyzer, which relates the millivolt signal to pH.

22.1.3.1   The Glass pH Electrode.   Glass pH electrodes use a specially formulated pH glass surface welded to the end of inert glass tube. The silicon–oxygen matrix of pH glass is slowly corroded by water to form a thin pH sensitive region, called the *gel layer*. This region develops a millivolt potential, proportional to the pH of the solution, as the result of electrochemical equilibrium between it and the hydrogen ions in the solution. The pH-sensitive glass is typically formed into a hemispherical or cylindrical bulb, which contains a specially formulated fill solution of known pH. The overall millivolt potential of the pH glass is the difference in the response of the gel layer on the exterior of the electrode in the sample solution, and the response of the interior gel layer to the fill solution.

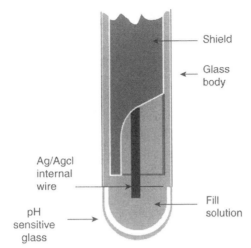

Figure 22.5 Theoretical response of a pH electrode.

A silver wire with a silver chloride coating is immersed in the fill solution of the pH electrode and makes contact with the lead to the electrode. When silver chloride is in contact with chloride ion–containing solution, silver chloride dissolves into the solution in the form of a number of silver chloride complexes, as well as silver ion. This results in a potential on the silver chloride proportional to the concentration of dissolved silver in the fill solution. The overall potential of the glass electrode is the sum of the potential on the silver chloride and the potential difference of the pH glass. The theoretical response of a pH electrode is −59.16 mV/pH, although in practice, this response is somewhat less than the theoretical value. (Fig. 22.5)

22.1.3.2 *The Reference Electrode.* The purpose of the reference electrode is to maintain a reproducible potential at any given temperature and complete the pH measuring circuit through electrolytic conduction with the sample solution. The potential of the typical reference electrode is due to a silver–silver chloride wire immersed in a chloride-containing fill solution, almost always potassium chloride, as was the case with the interior of the pH electrode. Since the potential on the silver–silver chloride wire is due to the concentration of silver ion in the fill solution, the chemical environment within the reference electrode must be maintained so that the concentration of silver ion is reproducible at any given temperature. Moreover, because the potential of the internal silver–silver chloride wire is temperature-dependent, changes in the chemical makeup of the fill solution will affect the temperature behavior of the reference electrode.

For a reference electrode to complete the measuring circuit with the pH electrode through the process solution, its fill solution must be brought into contact with the process solution. This can be accomplished by an actual flow of the reference fill solution, or by diffusion of ions from the fill solution into the process.

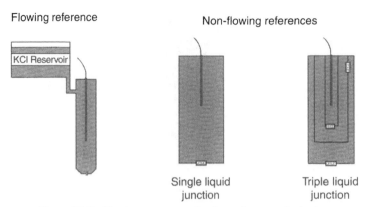

Flowing reference

Non-flowing references

KCl Reservoir

Single liquid junction

Triple liquid junction

Figure 22.6 Flowing versus nonflowing reference electrodes.

These two approaches define the two fundamental types of reference electrodes: flowing and nonflowing reference electrodes (Fig. 22.6).

A flowing reference electrode makes contact with the sample solution by flowing the fill solution into the sample through a narrow orifice or ceramic frit. For ambient-pressure samples, a reservoir of fill solution is elevated with the hydrostatic pressure providing the motive force. When the sample is at elevated pressures, the fill solution reservoir must be pressurized to ensure flow against sample pressure. Although the flow rate of the fill solution is not large, the reservoir must be periodically refilled as a routine maintenance task. Flowing references have largely been replaced by nonflowing references because of their lower initial cost and reduced maintenance costs, by eliminating the need for periodic refilling.

Nonflowing references rely on the diffusion of fill solution ions to maintain electrolytic contact with the fill solution. Diffusion occurs through a liquid junction, which can be a porous ceramic frit, wood, or polymeric material. Potassium chloride is used as a reference fill solution, not only because it provide chloride ion to maintain reproducible reference potential but also because it is equitransferent.

Equitransference is a property of ionic solutions in which the positive ions (cations) and negative ions (anions) diffuse at the same rate. In the case of a nonflowing reference, this means that the potassium ions and chloride ions diffuse through the liquid junction at same rate, so that there is no charge imbalance from the interior to the exterior of the liquid junction. Were this not the case, a faster diffusion rate of either ion would lead to a charge imbalance across the liquid junction, which would result in an opposing potential, called the *liquid junction potential.*

By relying on diffusion to maintain electrolytic contact, the nonflowing reference is subject to the effects of the diffusion of ions from the sample solution, which can lead to an upset of the chemical environment within the reference electrode and problems within the liquid junction. This is the greatest drawback to choosing a nonflowing over a flowing reference. These effects and the means for overcoming them will be examined later.

22.1.3.3 *The Need for Temperature Measurement.* A glass pH electrode and a reference electrode, in combination with a pH analyzer, provide a complete pH measuring circuit. In combination, they provide a millivolt signal that is proportional to pH. However, the millivolt potentials produced by the pH and reference electrodes are temperature-dependent. The Nernst equation describes the behavior of a pH and reference combination:

$$E = E^0 + \frac{dE^0}{dT}(T - 298.15\,\text{K}) - \frac{RT}{F}(2.303)\log\text{pH}$$

where E = the potential of the pH and reference electrode pair
E^0 = standard potential of the pH and reference electrode pair
dE^0/dT = change in the standard potential with temperature
T = absolute temperature
R = universal gas constant
F = Faraday's constant

This equation can be manipulated by taking the first derivative of the potential, setting it to zero, and solving for a pH and potential at which the potential is constant with temperature. This pH and potential is called the *isopotential point* (Fig. 22.7).

By choice of the pH electrode fill solution, pH sensors are almost always designed to have an isopotential pH of 7.0 pH and an isopotential voltage of 0 mV. Using the isopotential point with a theoretical knowledge of electrode behavior makes it possible to compensate (correct) the pH measurement at any

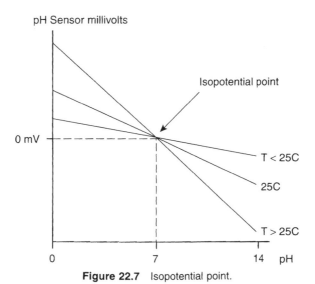

Figure 22.7 Isopotential point.

temperature to a reference temperature (usually 25°C), using a temperature signal from the temperature element. This makes the pH measurement independent of changes in the electrodes' output with temperature. It is important to note that this temperature compensation corrects only for electrode potential changes, and not for actual solution pH changes with temperature, which can occur.

The linear equations used to calculate pH involve different manipulations of the Nernst equation, and can result in different terms for the expressing the zero and slope of the calibration line relating millivolt potential to pH.[1] The millivolt response of the pH electrode can be expressed as the Nernstian slope at 25°C, with units of mV/pH, or it can be expressed as a percentage of the theoretical Nernstian slope of −59.16 mV/pH (PTS). In either case, the slope is calculated by the analyzer from a two-point calibration, and provides a measure of the pH electrode response to pH, which is an important diagnostic tool.

The zero of the line can be expressed as a millivolt offset to the theoretical isopotential voltage (usually 0 mV), or it can be in terms of pH as the pH at 0 mV and 25°C. Both expressions provide an indication of changes in the pH electrode or reference electrode, and are calculated by the pH analyzer from single- and two-point calibrations.

22.1.3.4   pH Sensor Design

22.1.3.4.1   Preamplifiers. pH sensors consist of a pH electrode, a reference electrode, and a temperature element. They may also contain a preamplifier, which lowers the impedance of the pH measuring circuit. The impedance of the circuit formed by the pH electrode and the reference electrode in solution is in the tens to hundreds of megohms, which can make the signal from the pH sensor prone to interference. Preamplification lowers the impedance of the signal, allowing the sensor to be located hundreds of meters from the analyzer.

The preamplifier is, in a sense, the first stage of the input circuit to the analyzer, and so has to be matched with the analyzer in terms of the manufacturer and model; a sensor containing a preamplifier from one manufacturer cannot, in general, be used with another manufacture's analyzer. Since it is an electronic device, a sensor-mounted preamplifier should not be exposed to process temperatures in excess of 85°C. In these cases the preamplifier can be mounted in a junction box along the cable run from the sensor to the analyzer.

22.1.3.4.2   Rebuildable and Combination pH Sensors. There are two major types of pH sensors: rebuildable sensors and combination sensors. The rebuildable sensor is constructed with the pH electrode, reference electrode, and temperature element as discrete components, which allows each to be replaced on failure, as well as other components such as O-rings and seals. In contrast, the combination sensor provides all three in a unitized construction. Failure of any one component necessitates replacement of the entire sensor. Combination sensors have replaced rebuildable sensors in most applications where the labor costs associated with maintenance

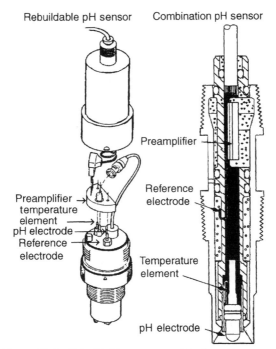

Rebuildable pH sensor   Combination pH sensor

Preamplifier

Preamplifier
temperature
element
pH electrode
Reference
electrode

Reference
electrode

Temperature
element

pH electrode

Figure 22.8   Rebuildable versus combination pH sensors.

are a major issue, although rebuildable sensors may be favored in certain applications, where process conditions drastically shorten the life of a particular component (Fig. 22.8).

22.2   PROCESS EFFECTS ON pH MEASUREMENTS

22.2.1   Effect of Temperature on Solution pH

The pH of a solution can change with temperature, due to the effect of temperature on the dissociation of weak acids and bases, and the dissociation of water itself. In fact, any solution with a pH of 7 or above will have some degree of temperature dependence, due to the temperature dependence of the dissociation constant of water (K_w) (Fig. 22.9). The pH of process solutions with sparingly soluble components can also be temperature-dependent because of higher solubility at elevated temperature. The magnitude of solution pH change with temperature will depend on the composition of the solution and the process temperature. Solution pH temperature dependence is frequently the explanation for discrepancies between laboratory and online measurements.

The temperature dependence of solution pH can lead to errors in reagent addition. For example, a sample consisting of a strong base, with a pH of 8 or above,

Temperature C	Dissociation constant of water (pK_w)	Neutral water	Water with strong base pH = 9.00 pH
10	14.54	7.27 pH	9.54 pH
20	14.17	7.08 pH	9.17 pH
25	14.00	7.00 pH	9.00 pH
30	13.83	6.92 pH	8.88 pH
40	13.54	6.77 pH	8.54 pH
50	13.26	6.63 pH	8.26 pH
60	13.02	6.51 pH	8.02 pH

Figure 22.9 Effect of temperature on solution pH.

will decrease by 1 pH from 25 to 60°C, so the pH will appear to be 1 pH lower than it actually is at room temperature. If the pH were adjusted at 60°C to raise the pH by 1 pH, to overcome what appears to be a discrepancy, the pH on cooling to room temperature would be 1 pH high. This would result in a waste of reagent and a possibly off-specification sample.

The most straightforward approach to determining the temperature dependence of a sample is to measure its pH at various temperatures in the laboratory. Because of the logarithmic nature of pH and the error associated with pH measurement, a straight-line approximation is usually sufficient. The slope of this line is termed the *solution temperature coefficient*, with units of pH per degree Celsius. Since temperature dependence is a function of composition, samples used for the determination, from processes undergoing a chemical reaction should represent the final product composition. Likewise, the temperature dependence of a process with widely varying composition or pH, such as a wastestream, can exhibit a variety of temperature-dependent behaviors.

The standard temperature compensation used in pH analyzers only compensates for temperature changes in the millivolt output of the pH sensor and does nothing to correct for solution pH changes. However, most microprocessor-based pH analyzers allow their temperature compensation routines to be modified to compensate for both the electrodes and the solution. Simply entering the temperature coefficient of the solution into the analyzer is typically all that is required. With this temperature compensation in place, changes in solution pH with temperature do not affect the final reading.

22.2.2 Effects of Mixed Solvents on pH Measurement

The conventional pH analyzer and sensor are designed to measure the pH of water solutions. When a nonwater solvent is present in appreciable quantities, the pH reading will be shifted from the expected value by effects of the nonwater solvent on the pH and reference electrodes, and by effects on the activity of hydrogen ion itself. Since pH sensor components are designed for use in water, a nonwater solvent may attack seals and O-rings, resulting leakage of process into the sensor and short-circuiting it.

A few mixed-solvent cases can be considered:

1. A solvent that is miscible with water may only cause a shift in the pH reading. If the solvent's concentration is above 15% by weight, it can cause a loss of water from the gel layer of the pH electrode, and, after a period of time the electrode will cease functioning. In these applications a periodic rewetting of the sensor by a spray of water is necessary.

2. A solvent that is nonmiscible with water will likely not have an appreciable concentration of hydrogen ions; most will be in the water phase. It is better to make the measurement in a place in the process where the water and solvent phases separate and keep the pH sensor in the water phase.

3. Completely nonaqueous processes may be possible only with periodic rewetting of the pH sensor with water and may also require a specially designed reference electrode to eliminate solvent effects on the liquid junction.

In all of these cases, pH measurement of a mixed-solvent solution should be studied in the controlled conditions of the laboratory, before going on line. This can help avoid unpleasant surprises on line. The study should include prolonged exposure of the pH sensor to the process sample, as some effects take time to appear.

22.2.3   Process Effects on the Glass pH Electrode

22.2.3.1   Temperature Effects. Elevated temperature accelerates the aging of the electrode, as evidenced by a decrease in the electrode's slope, while low temperature can lead to sluggish response. Extremely high or low temperatures can alternatively boil or freeze the fill solution, causing the electrode tip to break or crack. Elevated temperatures can also affect the interior and exterior of the pH electrode differently, giving rise to asymmetry potential. Asymmetry potential is caused by an imbalance between the inner and outer gel layers of the pH electrode. This creates a shift in the zero point of the pH measurement and changes the temperature behavior of the pH electrode, which leads to temperature compensation errors.

The temperature specifications for the pH sensor should be examined to ensure that the sensor can accommodate the elevated (or low) temperature being considered. But even a pH sensor within its specifications will have a shorter life at elevated temperature. In processes with extremely high temperatures, a sample line with cooling must be used.

22.2.3.2   Sodium Error. More correctly called *alkali ion error*, sodium ion error occurs at high pH, where hydrogen ion concentration is lower than to sodium ion by orders of magnitude. The sodium ion concentration can be so high relative to hydrogen ion that the pH electrode response to sodium ions becomes significant. As a result, the pH reading is lower than the actual pH. Depending on the pH glass formulation, this can occur at pH as low as 10. When an accurate high pH reading is required, pH electrode specifications should be checked and, if necessary, a specially formulated, "high pH" glass electrode should be used. With regard to the other alkali ions, lithium ions will produce a larger error than will sodium, while the effect of potassium ions is negligible.

22.2.3.3 Solution Components Attacking pH Glass

22.2.3.3.1 Concentrated Caustic Solutions. High concentrations of hydroxyl ions shorten the life of pH electrodes. In general, the more alkaline the solution, the shorter the glass electrode life will be. Solutions that can reach a pH in excess of 14 (equivalent to 4% caustic soda) can destroy a pH electrode in a matter of hours. Nothing can be done to prevent this, short of simply avoiding pH measurements in these solutions, and using another technique, such as conductivity.

22.2.3.3.2 Hydrofluoric Acid. Hydrofluoric acid (HF) quickly dissolves most sensitive pH glasses, but there are pH glass formulations designed to resist destruction by HF, which, when used within their limits, can give satisfactory electrode life.

It is important to note that, while only hydrofluoric acid (HF), and not fluoride ion (F^-), attacks glass, hydrofluoric acid is a weak acid. Therefore a solution can contain a relatively high concentration of fluoride ion at a high pH and do no damage to the electrode. But if the pH of the solution decreases, the fluoride ion will combine with hydrogen ion to form HF, which will damage the electrode. So it is important to look at the extremes of the pH measurement (even upset conditions) and the total concentration of fluoride in the solution to determine what the maximum concentration of HF will be.

22.2.3.3.3 Abrasion. Abrasives in the process solution can wear away the glass electrode, and larger particles can crack or break the electrode. While glass electrode designs with thicker glass have been tried to minimize the effects of abrasion, the best course of action is to protect the electrode from direct impact of an abrasive solution, with a baffle or by judiciously locating the sensor away from the abrasive flow.

22.2.4 Process Effects on Nonflowing Reference Electrodes

To function properly, a reference electrode must maintain a reproducible concentration of silver ions in the fill solution, which in turn results in a reproducible potential (voltage) on the silver–silver chloride wire. For the reference electrode to maintain a reproducible potential, the fill solution must remain relatively uncontaminated by certain components in the process solution. At the same time, the reference electrode must be in electrical contact with the pH electrode through the process solution. For a nonflowing reference, diffusion of ions between the reference electrode and the process is necessary to maintain electrical contact, but also creates the potential for contamination of the reference fill solution by components in the process solution.

22.2.4.1 *Reference Poisoning.*

The mechanism of reference poisoning is a conversion of the reference from a silver–silver chloride–based electrode to an electrode based on a different silver compound. The anions that typically cause this form less soluble salts with silver than does chloride. They include bromide, iodide, and sulfide ions. When these ions enter the fill solution, they form insoluble precipitates with the silver ions in the fill solution. But there is no initial effect on the potential of the reference, because the silver ions lost to precipitation are replenished by silver ions dissolving off the silver chloride coating of the silver wire. It is not until the silver chloride coating is completely lost that a large change in the potential and temperature behavior of the reference occurs. This is evidenced by a large shift in the zero point of the measurement. At this point, the reference electrode must be replaced.

The reference can also be poisoned by reducing agents, such as bisulfite or complexing agents (ammonia), which reduce the concentration of silver ion in the fill solution by reducing it to silver metal or complexing it.

To counteract this effect, multiple junction reference electrodes are used, which consist of two or more liquid junctions and fill solutions to slow the progress of the poisoning ions. Gelling of the reference fill solution is also used to prevent the transport of poisoning ions by convection. A new approach has been to use a reference with a long tortuous path to the silver–silver chloride wire, along with gelling of the fill solution.

22.2.4.2 *Plugging of the Liquid Junction.*

To maintain electrical contact between the reference electrode and the pH electrode, there must be a relatively free diffusion of ions between the reference fill solution and the process solution. In some cases, large concentrations of an ion that forms an insoluble precipitate with silver ion (most notably sulfide ion) will precipitate within the liquid junction and plug it. Metal ions that form insoluble salts with chloride ion (typically the heavy metals silver, lead, and mercury) will also precipitate in the liquid junction.

When the liquid junction is closed, the pH reading will drift aimlessly. To avoid these effects, multiple junction reference electrodes have been used with the outermost fill solution containing potassium nitrate, rather than potassium chloride. This lowers the concentration of chloride available for precipitation by heavy metals, as well as lowering the concentration of free silver ion, which can precipitate with sulfide.

22.2.4.3 *Liquid Junction Potential.*

As was discussed, potassium chloride is chosen for the reference fill solution not only for its ability to solubilize silver ion but also because it is *equitransferent*, which means that potassium and chloride ions diffuse at the same rate. However, not all solutions are equitransferent, and when the process solution is not, a liquid junction potential can result. When a positive ion diffuses through the liquid junction faster than does a negative ion, or vice versa, a charge imbalance will result. This gives rise to an opposing potential, which is a liquid junction potential. The magnitude of this potential depends on the composition and concentrations in the process solution, and the design of the liquid junction.

This potential is added to the potentials of the pH and reference electrodes and causes an offset to the pH measurement of typically a few tenths of a pH. Buffer solutions used to calibrate pH measurements are by and large equitransferent, and so this phenomenon is not observed until 15–20 min after the sensor has been put on line. By the same token, the liquid junction will take some time to dissipate, after the pH sensor is first put into a buffer for calibration. The remedy for liquid junction potential is to standardize the pH measurement after sensor has been put on line and has stabilized.

22.2.4.4 *Low-Conductivity Water Samples.* Samples with a conductivity of 50 μS/cm (microsiemens per centimeter) or less can exhibit large, unstable liquid junction potentials, leading to a drift, sample flow dependence, and a generally unsatisfactory pH measurement. The lower the sample conductivity, the worse the effects are. In very low-conductivity samples, there can be static buildup on nonmetal sensor components, which adds to the errors. The best approach to low-conductivity samples is to use a pH sensor specially designed for this application.

22.2.5 pH Sensor Coating and Cleaning

Undissolved solids and liquids in a process can coat a pH sensor, drastically increasing its response time, and plug the liquid junction of the reference electrode. Depending on rate of fouling, the sensor can be cleaned either manually or on line. A sample flow rate of 5 ft/s or greater can reduce the accumulation of fouling materials, through the turbulence created.

Manually cleaning the sensor should use the mildest but most effective cleaning solution available. Alkaline deposits can be removed with weak-acid solutions (5% HCl or vinegar), while acidic deposits may be removed with mild caustic (1% caustic soda). Organic deposits (oil and grease) can often be removed with a detergent solution, but more tenacious coating may require the use of a solvent. The solvent used should be carefully chosen to avoid attack on the sensor's O-rings and seals.

In all cases, the exposure to the cleaning solution should be minimized, to limit the amount of cleaning solution entering the liquid junction. Components from the cleaning solution that enter the liquid junction can create a liquid junction potential, which will persist until the cleaning solution components diffuse out of the liquid junction.

Online cleaning reduces maintenance in processes that quickly foul the sensor. Ultrasonic cleaning was an early approach to online cleaning, but was ineffective on soft or gelatinous coating, and only marginally effective with hard crystalline coatings, which it was designed to address. It has, for the most part, been abandoned as a cleaning method. One of the most fruitful methods for online cleaning has been the jet spray cleaner. This method directs a spray of water or cleaning solution at the face of the pH sensor, at timed intervals, or when sensor diagnostics are available, based on a high-reference-impedance alarm.

In many cases, liquid junction designs that feature a large surface area and very small pore size or a nonwettable surface have proved effective in avoiding the effects of severe coating. When a coating problem is anticipated in a new application, the best approach is to first use one of these junctions without online cleaning, and then add online cleaning, if the frequency of manual cleaning requires it. Use of these junction designs can drastically reduce the required frequency of manual cleaning, and in many cases, eliminates the need for online cleaning.

22.3 pH SENSOR MOUNTING

22.3.1 Sensor and Analyzer Location

While a pH sensor with a preamplifier can be located hundreds of meters from the analyzer, it must be remembered that maintenance tasks, especially buffer calibration, require the user to access both the sensor and the analyzer at the same time. Long distances between the sensor and the analyzer may make maintenance inconvenient for a single operator, or require two operators, and some type of remote communication. The analyzer itself should be mounted in an easily accessible location with the sensor nearby. pH installations that are difficult to access tend not to be properly maintained.

22.3.1.1 General Sensor Mounting Considerations. pH sensors may be mounted in a number of ways, as will be seen, but in all cases the following points should be considered:

1. The sensor should be mounted so as to receive a continuous flow of a sample representative of the process solution.

2. The sensor (and analyzer) should be mounted in a location that allows convenient access for maintenance and calibration.

3. If there is reagent being added to the process, the sample should be well mixed before reaching the sensor. If the sensor is mounted in a vessel, good agitation is necessary to avoid large lag times in the measurement.

4. The pressure limitations of the pH sensor should not be exceeded. Locate the sensor at a point of lower pressure, or if this is not possible, a sidestream should be taken off the process with pressure reduction. This is also a safety issue.

5. Likewise, temperature limitations should not be exceeded. It should be remembered that lower temperatures promote longer glass electrode life. Processes with very high temperatures or temperatures in excess of the limits of the sensor should be provided with a sample sidestream and cooling. This is also a safety issue.

6. Sensors with preamplifiers should not be exposed to temperatures above 85°C. If preamplification is necessary, the preamplifier should be located nearby in a junction box along the sensor to analyzer cable run.

7. For installations in hazardous areas, it is best to use an analyzer (usually a two-wire transmitter) designed for use in the hazardous area, rather than a long cable run from the sensor to an analyzer in a nonhazardous area. If more information than simply a pH measurement is required from the analyzer, use a smart pH analyzer.

8. In processes where there is a potential for coating, the sensor should be exposed to a high flow rate of sample. In pipes, closer to the center is better. In vessels, the sensor should be located to avoid regions where there is settling of precipitates or debris.

9. If the sensor will be exposed to the process only intermittently, it should be mounted in a low point in the piping to keep it wet. After prolonged drying, a pH electrode may take some time to rehydrate. Reference fill solution can crystallize on the face of the liquid junction and may be slow to dissolve. Dissolved or suspended solids, which dry on the pH sensor, can form a tenacious coating that is difficult to remove.

10. Many pH sensors must be mounted at an angle of at least 15° above horizontal to keep an air bubble in the glass electrode at the top of the electrode.

22.3.2 Mounting Configurations

The mounting configuration of a pH sensor is determined by the requirements of the process piping and vessels, and other considerations such as the need for sample cooling or pressure reduction. Some mounting configurations are more convenient to the operator than others, and the choice of a configuration should always default to the most convenient given the requirements of the process (Fig. 22.10).

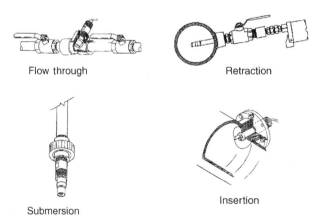

Flow through   Retraction

Submersion

Insertion

Figure 22.10 Mounting configurations for a pH sensor.

22.3.2.1   *Submersion Mounting.* Submersion mounting is immersing the sensor in a process vessel or pond on a length of conduit. It is typically used when the only access to the process is from the top. It is the most inconvenient mounting configuration for the operator, because both the sensor and its conduit must be removed for maintenance and calibration. Care has to be taken to ensure that a watertight seal is made between the sensor and the conduit, and backfilling the conduit is advisable to avoid leakage of the process into the cable end of the sensor, which causes short circuits.

22.3.2.2   *Insertion Mounting.* This configuration generally involves mounting the pH sensor into a tee or in through the side of a vessel. If a threaded sensor is screwed into the process, installation and removal for maintenance can require the sensor to be disconnected from the analyzer to avoid twisting the cable. For this reason it is often advantageous to use a flange or some manner of insertion fitting to allow the sensor to withdrawn without unscrewing it. The sensor should be mounted in a pipe run, where it will always remain fully immersed and air entrapment can be avoided.

22.3.2.3   *Retraction Mounting.* One shortcoming of insertion mounting is the need to shut down the process or drain a vessel in order to remove the sensor for maintenance. This shortcoming is overcome using retraction mounting, which is an insertion type of mounting using a ball valve or some other isolating valve that allows the sensor to be removed while the process is running. The sensor insertion and retraction can be done manually or by pneumatic action. With this configuration, the sensor can be quickly removed for cleaning, or replacement in the event of a catastrophic failure, minimizing the downtime of the pH measurement during the process.

The process pressure is an important safety consideration in retraction installations. With ball-valve-type insertion devices using manual insertion and retraction, the maximum rating for the device should be strictly observed, to avoid the sensor from being ejected from the device and the operator being sprayed with the process. Inserting the sensor into the process against the force of a pressurized sample can be difficult. In cases where excessive pressure is possible but access to the sensor during running is necessary, a pneumatically operated ball valve, a retractable device other than a ball valve, or a sample sidestream with pressure reduction should be chosen over a manually operated ball valve.

22.3.2.4   *Flowthrough Mounting.* A flowthrough configuration can be a sensor specially designed for flowthrough mounting with flanges or female threads to mate with process piping, or it can simply be a insertion or retractable sensor mounted in a tee. The term *flowthrough* is often taken to mean a pH sensor mounted in a sample sidestream. When provided with isolation valves, this mounting configuration is the most convenient because it allows the sample flow to be shut off and the sample line to be drained at will for maintenance and calibration. When sample conditioning is required, such as cooling, pressure reduction, or filtration and separation, this is the only configuration possible.

The major drawbacks to using a flowthrough sensor in a sample sidestream are the losses of process sample, unless sample return is provided, plugging of the sample line by solids or precipitates, and the possible lag time in getting the sample from the process to the sensor.

In general, choice of sensor mounting will be determined largely by the requirements of the process vessels and piping, as in the case of a pond or a sump, where the only mounting option is submersion. In other cases, there can be greater flexibility. Insertion mounting is most often the option of choice, due to its simplicity and low cost, but if sample cooling or pressure reduction is required, flowthrough mounting is the only option. Retraction or flowthrough mounting should be used in cases where it is necessary to access the pH sensor, while the process is running for cleaning or maintenance, especially when a reliable pH measurement is critical to the process. The added hardware and piping necessary for retraction and flowthrough mounting makes the up-front cost greater than with insertion or submersion.

22.4   CALIBRATION

22.4.1   Buffer Calibration

Buffer calibrations use two buffer solutions, usually at least 3 pH units apart, which allows the pH analyzer to calculate a new slope and zero value. The slope and zero derived from a buffer calibration provide an indication of the condition of the glass electrode from the magnitude of its slope, while the zero value gives an indication of reference poisoning or asymmetry potential. Overall, the buffer calibration can demonstrate how well the pH sensor responds to pH.

Buffer calibrations serve to recalculate the slope used by the analyzer to calculate pH. A buffer calibration is required only when the slope of the pH electrode has changed, which depends on the process temperature and the pH range, among other things. Buffer calibrations are more costly than a simple standardization in terms of the maintenance time required. The pH measurement is also lost during a buffer calibration since the sensor has to be removed from the process.

22.4.1.1   *Buffer Calibration Frequency.* The proper frequency for a buffer calibration in a new installation can be determined by initially buffer calibrating weekly, to determine the how fast the slope changes and then decreasing the frequency accordingly. With the advent of glass electrode impedance diagnostic, to be discussed later, a broken glass electrode can be detected, and so the usual rate of slope decrease can be relied on to determine the buffer calibration frequency.

In cases where a daily buffer calibration appears necessary because of an apparent rapid rate of slope decrease, the application should be reviewed for components that attack the glass electrode. Highly concentrated ionic solutions can create large liquid junction potentials, and can also affect the gel layer of the glass electrode, resulting in a reading that can take over an hour to stabilize in buffer solution and the process. This effect can lead to difficulties doing a buffer calibration, and can give the false impression that the electrode slope is changing rapidly.

22.4.1.2 Buffer Calibration Errors. Buffer solutions have a stated pH value at 25°C, but the actual pH of the buffer will change with temperature, especially when the stated value of the buffer is 7 pH or above. The values of commercially available buffer solutions at temperature other than 25°C are usually listed on the bottle. The pH value at the calibration temperature should be used, or errors in the slope and zero values, calculated by the calibration, will result. An alternative is to use the "buffer temperature compensation" feature on microprocessor-based pH analyzers, which automatically corrects the buffer value used by the analyzer for the temperature.

Buffer calibrations can be done in haste, which may not allow the pH sensor to fully respond to the buffer solution. A common error is failure to allow a warm pH sensor enough time to cool down to the temperature of the buffer solution. The temperature compensation of a pH analyzer is based on the assumption that the sample and all the components of the pH sensor are at the same temperature. If the sensor is warmer than the buffer solution, the millivolt and temperature signals read by the analyzer will be in error, as will be the buffer temperature correction if used. Current pH analyzers have a "buffer stabilization" feature, which prevents the analyzer from accepting a buffer pH reading that has not reached a prescribed level of stabilization based on a minimum rate of change of the pH reading with time.

22.4.2 Single-Point Standardization

Standardization is a simple zero adjustment of a pH analyzer to match the reading of a sample of the process solution made, using a laboratory or portable pH analyzer. It is most useful for zeroing out liquid junction potential, but some caution should be used when using the zero adjustment. A simple standardization does not demonstrate that the pH sensor is responding to pH, as does a buffer calibration. In some cases, a broken pH electrode can result in a believable pH reading, which can then be standardized to a grab sample value. The following guidelines should be observed for grab sample standardization:

1. The grab sample should be taken near the online pH sensor, so the sample represents what the pH sensor is actually measuring.

2. The grab sample should be taken when the process pH and temperature are stable. Otherwise discrepancies can result from the grab sample not matching the process sample at the sensor. The online reading may not have fully responded to a change in pH, or the pH sensor may not have reached the process temperature.

3. Solution pH temperature dependence may show up as a discrepancy between the online and laboratory measurements, if solution temperature dependence is not compensated for in the pH analyzer.

4. If a chemical reaction is taking place in the process, the reaction may not have gone to completion at the online sensor, but will have by the time a grab sample reaches the laboratory, resulting in discrepancies.

5. The stability of the process sample is an important consideration. Some samples can react with air, or precipitate on cooling and change pH.

Finally, it must be remembered that the laboratory or portable analyzer used to adjust the online measurement is not a primary pH standard, as is a buffer solution. Errors in the laboratory or portable analyzer will result in an incorrect standardization of the online unit.

22.5 DIAGNOSTICS AND TROUBLESHOOTING

22.5.1 Traditional Error Detection and Troubleshooting

There are two categories of pH diagnostics, online and offline diagnostics. Online diagnostics alert the user to a problem in real time, during operation of the analyzer. Offline diagnostics are indications of a problem during calibration or maintenance of the sensor and analyzer.

The traditional pH analyzer was a basic pH analyzer, which is an analog device that simply transmits and displays pH. The only online indication of a problem is a pH reading that is obviously incorrect, based on what is known about the process. In fact, a broken glass electrode can go undetected until routine maintenance is performed, if the normal process pH is near 6. The only offline indication of a sensor problem is the inability to buffer calibrate the sensor, which indicates that the glass electrode slope is too low, or that the zero had drifted beyond the analyzer's ability to correct for it. The actual slope and zero can, in some cases, be calculated from a laborious measurement of the sensor millivolt output and the buffer pH values. Troubleshooting consists of following a table of tests based on a description of the behavior of the pH signal online and during offline tests.

22.5.2 Analyzer Self-Diagnostics

Along with microprocessor-based pH analyzers came the ability to do more online diagnostic measurements and the ability show them on digital displays. These included self-checks of the microprocessor itself for malfunctions in the CPU (central processing unit) and memory, but more importantly, included some checking of the input signals from the pH sensor. Input signal checking was able to detect open- or short-circuited temperature elements, and open or short circuits in the sensor cable.

Offline diagnostics received a boost from the ability of the analyzer to calculate and display the pH sensor's slope and zero value from buffer calibrations, which had become an automated routine within the analyzer. It was now much easier to keep records of these key indicators of pH sensor health and plan maintenance accordingly. The sensor's millivolt output could also be easily displayed, which aided in troubleshooting.

22.5.3   Glass pH Electrode Impedance Diagnostics

The pH sensor is the most maintenance-intensive part of the pH measurement, and with the widespread use of combination pH sensors, sensor replacement is necessary at intervals of 6 months to 1 year, or more frequently in severe applications. Failures can occurs slowly, such as the gradual loss of pH electrode slope, or poisoning of the reference electrode, both of which can usually be tracked with offline diagnostics. But failures can also occur suddenly or catastrophically, as in the case of pH electrode breakage or coating and plugging of a reference electrode. In either case, failure might go undetected until the next routine buffer calibration. The purpose of impedance diagnostics is to detect these problems in real time.

22.5.3.1   Glass pH Electrode Impedance.   Glass pH electrode impedance is in the range of tens to hundreds of megohms, and is strongly temperature-dependent. Studies of glass electrode impedance have shown these electrodes to be composed of a set of resistances in series, some of which correspond to the gel layer of the glass electrode, while others have their origins in the bulk pH glass.[2] Moreover, each component of the overall resistance behaves like an *RC* (resistance × capacitance) circuit, by virtue the fact that each has its own temperature-dependent time constant. So, measuring the full electrode impedance by applying a voltage or using a parallel resistance as in the IEC method,[1] requires sufficient time for all of the resistance components to fully respond. In cold samples this can take up to an hour.

Online impedance measurements, which must be much shorter in duration, cannot measure the full impedance of the electrode. As a result, the impedance values measured on line are lower than those measured in the laboratory. But even this more limited measurement allows a broken or cracked glass electrode to be detected, which is its greatest utility.

Cracks and breakage can create a short circuit across the glass electrode, resulting in a large decrease in impedance. An impedance decrease can also occur as a result of a temperature increase; glass electrode impedance decreases by approximately one-half for every 8°C increase in temperature. Detecting broken or cracked pH electrodes requires the impedance measurement to include temperature compensation circuitry so as to distinguish between temperature effects and cracking. It is important to note that high temperature can cause the impedance of the glass electrode to drop below impedance measuring the circuit's ability to make a reliable measurement. In these cases, the analyzer generally turns its impedance alarm off.

To implement glass impedance diagnostics, the user simply sets a low alarm point for the glass impedance. Alarms based on high-glass-impedance measurements are not generally as useful, as the low alarm, but can in some instances detect a pH sensor that is no longer wetted by the process.

22.5.3.2 *Reference Electrode Impedance.* The overall impedance of a reference electrode is a sum of the resistances of its components; the largest is the liquid junction. This is due to the limited volume of current carrying electrolyte within the liquid junction. Reference electrode impedance is not very temperature-dependent, as was the case with glass electrode impedance, and so no temperature compensation is used.

The usefulness of reference electrode impedance is in the detection of coating or blockage of the liquid junction, which increases the impedance of the liquid junction. A reference electrode not wetted by the process solution, or that has run out of reference fill solution, can also exhibit high impedance. Detection of coating or blockage involves setting a high-reference-impedance alarm. Low-reference-impedance alarms are, in reality, glass electrode impedance alarms and are used only with special pH sensors using a second glass electrode as a reference.

22.6 ANALYZER–USER INTERFACE AND OUTPUT SIGNAL

The user interface and output signals of a pH analyzer enable the analyzer to communicate with the user or a supervisory device. In describing each type of analyzer, the feature set of the analyzer determines what information is available, and the display and output signal determines where and how that information is made available. Although the platforms described below roughly correspond to the historical development of pH analyzers, they are nevertheless all still currently commercially available (Table 22.1).

22.6.1 The Basic Analog pH Analyzer

The basic pH analyzer is an analog device with zero and span adjustments (typically potentiometers) for doing buffer calibration and standardization. They have a very limited feature set, typically providing only a 4–20-mA output proportional to a pH range and a display (sometime also optional in two-wire transmitters) either digital or analog, for displaying pH only. Line-powered units can include high- and low-pH alarm contacts. These units typically lack the ability to do solution pH temperature compensation. As discussed earlier, there is a minimum of diagnostic information available, which takes considerable effort on the part of the user to extract. The major advantage of basic analog pH analyzers is their lower cost.

22.6.2 The Microprocessor-Based pH Analyzer

Microprocessor-based pH analyzers offer a marked improvement over the basic analog devices. In terms of features, they typically include the ability to do solution pH temperature compensation, automatic buffer calibration features (buffer stabilization and buffer pH temperature correction), and a full complement of diagnostic features, including analyzer self-checking and pH and reference electrode diagnostics. A major improvement is also the ease of use. The analyzer can be queried for diagnostic and error information, configured and calibrated using software routines.

Table 22.1   pH Analyzers

	Feature Set	User Interface	Remote Communication
Basic analog analyzer	Zero, span, and range adjustment pH sensor temperature compensation only	Analog or digital display of pH Potentiometers / switches for zero, span, and range adjustment	4–20-mA signal proportional to pH pH alarm contacts
Microprocessor-based analyzer	Solution pH temperature compensation Buffer calibration with buffer temperature compensation and stabilization Self-diagnostics pH and reference electrode impedance diagnostics	Digital display of pH, calibration constants, configuration parameters, and error messages Software routines for calibration and configuration Keypad for entering numerical values and navigating among software routines	4–20-mA signal proportional to pH range pH alarm contacts Fault alarm via alarm contact or overange 4–20-mA signal
Smart pH analyzer	Solution pH temperature compensation Buffer calibration with buffer temperature compensation and stabilization Self-diagnostics pH and reference electrode impedance diagnostics	Digital display of pH, calibration constants, configuration parameters, and error messages Software routines for calibration and configuration Keypad for entering numerical values and navigating among software routines	Digital pH signal (also 4–20-mA signal in some cases) Digital pH alarms (also pH alarm contacts in some cases) Digital fault messages (also fault alarm contact in some cases) Full access to the instrument for calibration and configuration via a handheld remote terminal, PC, or host control system

In all of these analyzers, while the typical user interface consists of a keypad and display, the display types can vary from small LED or LCDs (liquid crystal displays), which display nonnumerical information in mnemonics, to larger displays capable of spelling out complete messages. To address the world market, the larger displays have become multilingual, although there is also an emerging trend toward symbolic, language-independent displays.

The output signal of these analyzers remains the analog 4–20-mA signal proportional to pH, with alarm contacts in the line-powered devices. Line-powered devices may also include a second 4–20-mA output for a second pH range or a temperature. Line-powered units can use a dedicated alarm contact for remote indication of a diagnostic alarm, while two-wire transmitters might use an overange 4–20-mA signal (21 or 22 mA) as an indication of a problem. Since all the diagnostic information resides in the analyzer, these remote signals are "come read" alerts to the user, who has to then go and interrogate the analyzer.

22.6.3 Smart pH Analyzers

Smart pH analyzers combine the features of the microprocessor-based analyzer with "smart," digital communications. This allows not only diagnostic information to be transmitted to a remote location along with the pH measurement but also makes it possible to interrogate, configure, and even calibrate the analyzer remotely. There are a number of protocols available for smart communications, many of which are manufacturer-specific. No protocol has, as yet, become a standard in the same way that the 4–20-mA analog signal has. The most widely used protocols for analyzers are HART, Modbus, FOUNDATION Fieldbus, and Profibus. Smart analyzers are more costly than the general microprocessor-based analyzers.

To access the digital information provided by the analyzer, the user needs a communication device, which can take the form of a handheld keypad display for configuration and calibration in the field, a PC, or a host control system. The use of a PC for communication with smart pH analyzers has led to the development maintenance software packages that provide a software environment for configuration, troubleshooting, and calibration. These software packages can include a database for recording timestamped error messages and calibration data, thus automating the task of keeping maintenance records. The information provided by the smart pH analyzers, in combination with maintenance software packages, offers the potential of reducing the instances of unnecessary routine maintenance.

22.7 THE APPLICATION OF pH

As with all liquid analytical measurements, the application of pH measurements can roughly be divided into three tasks: leak detection, concentration measurement, and reaction monitoring/endpoint detection. But it is first necessary to examine the range of acid and base concentrations, which can be measured using pH.

22.7.1 Applicable pH Measurement Range

Since pH measures the acidity or alkalinity of a solution, it is often assumed that pH is applicable to any concentration of an acid or base. This is not the case, and although the pH range of an analyzer is often stated as 0–14 pH, solutions with acid and base concentrations near the extremes of this range are often better measured using another analytical technique.

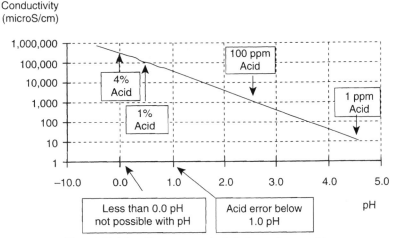

Figure 22.11 pH versus conductivity for a strong acid.

At pH 1.0 and below, there can be a large liquid junction potential due to the high concentration of hydrogen ion, and the pH electrode can be subject to acid errors, which may make the reading appear higher than it actually is. Sensor materials can also be subject to chemical attack. pH measurements normally in the range of 1.0 pH and below should be avoided. A solution with a pH of 1.0 pH contains a strong acid approaching the percentage range in concentration, which becomes even stronger as the pH decreases. Strong-acid solutions with this strength can often be measured very successfully with a conductivity determination (Fig. 22.11).

At the high end of the pH scale, the argument for avoiding pH measurements is even more compelling, since highly alkaline solutions drastically shorten the life of a pH electrode. In general, the higher the alkalinity, the shorter the life of the glass electrode will be. The caustic soda (NaOH) concentrations at pH 13 and 14 are 0.4% and 4.0% by weight, respectively, both of which can often be successfully measured with conductivity. In applications where pH simply has to be measured above pH 13, the user has to anticipate a short pH electrode life. Applications at or above pH 14 can destroy the glass electrode in less than 1 day and must be avoided at all costs (Fig. 22.12).

In general, pH should be applied to solutions with a pH in the 1–13 range.

22.7.2 Leak Detection with pH

Leak detection with a pH measurement involves detecting the leakage of an acid or base contaminant into a sample by the pH response of the sample. How well pH can detect a leak depends on the magnitude of the pH response for a given volume of contaminant. The pH response of the sample to a contaminant can be determined by looking at a titration curve of the sample, where the titrant is the acid or base contaminant. Considering a general titration curve, a few conclusions can be reached regarding the sample pH and the pH of the contaminant:

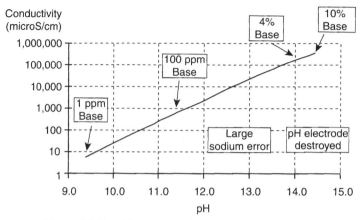

Figure 22.12   pH versus conductivity for a strong base.

1. The best sensitivity for leak detection with pH is for a sample, at its neutralization point, with little or no buffering capacity, and a contaminant, which is a strong acid or base. In this case, the sensitivity of leak detection decreases as the buffering capacity of the sample increases, and as the concentration of the acid or base contaminant decreases.

2. Sensitivity decreases for samples, the further their pH is from the neutralization point.

3. Small volumes of acids leaking into strongly alkaline solutions produce a very small pH response, as do strong bases leaking into strongly acidic solutions.

When evaluating a sample for leak detection by pH, the sensitivity of the pH response at the extremes of the normal pH range of the sample must be evaluated. Good sensitivity at one extreme of the pH range might completely disappear at the other extreme. In some cases, pH measurements before and after the source of the leak can help cancel out the effects of normal changes in the sample pH. The transit time of the sample through the source of the leak should be shorter than the frequency of normal pH variations to avoid false leak alarms. Leak detection applications with a single pH measurement use a high/low-pH alarm to signal the detection of a leak. When two pH measurements are used, the difference between the two measurements is used for alarming.

22.7.3   Concentration Measurement with pH

Although pH is the negative logarithm of the hydrogen ion concentration, or more correctly, the hydrogen ion activity, in most applications, pH is considered to be a thing unto itself, and no correlation to concentration is drawn from the pH measurement. In some cases, however, the purpose of the pH measurement is to derive the concentration of an acid or base from the pH reading. To gauge how well this can be done, it is necessary to examine the mathematics behind deriving concentration from pH and the general errors associated with a pH measurement.

A simple case can be examined with the assumptions that pH is the negative logarithm of hydrogen ion concentration, and that the acid or base under consideration is a strong acid or strong base at 25°C. The molar concentrations of the acid [HA] and the base [MOH] can be calculated from the pH as follows:

$$[HA] = 10^{-pH}$$
$$[MOH] = 10^{pH-14}$$

The important thing to note in the calculation of concentration from pH is that the pH appears as an exponent of 10. As a result, small changes in pH result in large changes in the calculated concentrations. It also means that any errors in the pH measurement have a significant effect on the derived concentration. For example, an error of ± 0.01 pH, which is the typical readability of a pH analyzer, results in a relative error in concentration of $\pm 2\%$. The errors in a pH measurement can come from a variety of sources. These errors include not only the error in the analyzer's millivolt measurement, which is typically better than ± 1 mV, but also errors in the temperature measurement, calibration errors, small liquid junction potentials, and the errors resulting from the pH sensor not having reached temperature equilibrium with the process.

It is useful to look at pH errors and derived concentration errors in terms of equivalent millivolt measurement errors. If there is a deviation of only ± 3.0 mV, due to a combination of these errors, a pH error of ± 0.05 pH will result, which yields a relative error in concentration of $\pm 11\%$. A millivolt deviation of ± 6 mV, which is not at all unreasonable in a process application, is equivalent to a pH error of ± 0.1 pH and a relative error in concentration of $\pm 22\%$. It has been claimed that in general, pH measurements are accurate only to ± 0.3 pH. If this were the case, the derived concentration could be one-half or double the actual concentration.

As can be seen, deriving acid and base concentrations from pH measurements is subject to large errors due to the logarithmic nature of pH. There can also be additional errors in derived concentration resulting from factors affecting the activity of the hydrogen ion.

If an online analysis technique is required, with an accuracy comparable to a laboratory titration, conductivity can often be applied to samples with a simple component matrix, while samples with a more complex component matrix will require an online titrator. pH measurements are useful because of their ability to cover a wide dynamic range of concentrations. When the target pH in an application involves a range of only pH 2, it must be remembered that two orders of magnitude of hydrogen ion concentration are involved.

22.7.4 Reaction Monitoring and Endpoint Detection

Many chemical and biological reactions are accompanied by a change in pH of the reaction solution. An example of this is the production of a lactobacilli culture used in the cheesemaking process. As the lactobacilli culture grows, lactose is converted to lactic acid, which lowers the pH. During production anhydrous ammonia is injected to bring the pH back up to a value optimal for culture production. When the culture is ready for further use to inoculate a cheese batch, the lactose has been consumed and the pH of the batch ceases to decrease. The role of pH in this application is primarily monitoring a reaction.

Another example is the precipitation of heavy metals through the addition of caustic soda (NaOH) to form insoluble metal hydroxides. If the pH is maintained above a certain minimum value, the concentration of heavy metals remaining in the solution is known to be below the desired concentration limit for further processing of the solution. The pH measured in the solution can even be used to control the addition of caustic soda, which in this case is relatively easy to do.

In many cases, the pH of a process needs to be controlled to promote the desired chemical reaction, protect process equipment from corrosion, and, finally, to condition wastestreams for discharge into the environment. While pH control itself is the subject of entire texts, the general issues involved can be brought to light from the following considerations.

The success of a pH control application depends on a number factors, and the starting point for evaluating a pH control application begins with the titration curve of the process sample where the reagent is used to make the adjustment. The first thing the titration curve shows is that the amount of reagent required to adjust the sample by 1 pH decreases roughly tenfold as the neutralization point of the titration curve is approached. This is due to the logarithmic nature of pH.

The shape of the titration curve is also important. The titration curve of a strong acid and strong base will have a sharp change in pH near the neutralization point, while if either the acid or base is weak, the change will be more gradual near the neutralization point. This means that when enough reagent has been added to be in the vicinity of the neutralization point, the remainder of the reagent to be added is a relatively small volume, which has to be measured very precisely to avoid overshooting the target pH. In the case of a strong-acid/strong-base neutralization, the last portion of reagent has to be added much more accurately than when weak acids and bases are involved.

How accurately does reagent need to be added to adjust the pH to between 6.0 and 8.0 pH in the strong-acid/strong-base case? This can be calculated from the titration curve, and the accuracy of addition depends on how far the starting pH is from the neutralization point. If the starting point is three pH units, above or below the neutralization point, reagent has to be added with a volumetric accuracy of $\pm 1.0\%$. Even higher accuracy is required if the starting point is further away from the neutralization point (e.g., 0.1%, 0.01%, and 0.001% for four, five, and six units, respectively, above or below the neutralization point).

Two rules of thumb for pH control are based on the preceding analysis. The first is that it is best not to attempt pH adjustments of more than three pH units. In cases where adjustments have to be made over a wider range, the adjustment should be made in smaller steps, usually in separate vessels. The second is to use weak acids to adjust strong bases, and vice versa. The change in pH near the neutralization point will have a milder slope. The analysis also highlights the need for accurate reagent addition, and the final control elements, that is, the values or proportioning pumps, must be evaluated as to whether they can deliver reagent with required accuracy.

A further consideration is sample and reagent mixing. While acid and base reactions are among the fastest reactions known, the reagents must be brought into contact for the reaction to occur. The sample and reagent must be well mixed in the vessel or pipe, or else the response of the sample to a reagent additions will be too slow, or may appear faster than it really is, due to reagent that short-circuits the mixing process.

When evaluating a pH control application, the initial pH or initial pH range of the sample must be considered, in conjunction with the target pH range, and the titration curve of the sample using the reagent. How accurately must reagent be delivered to hit the target pH range from the range of initial pH range? In some cases, the target pH range is beyond the neutralization point, like the application for the precipitation of heavy metals, and a lot of variation in reagent addition can be tolerated, which is the case anytime a pH adjustment is made away from the neutralization point. In cases where there may be a need to alternately neutralize acidic or basic components, such as in a wastestream, a more detailed analysis of the application is needed, and the reader is referred to texts dealing with pH control in far greater detail.

22.8 SUMMARY

The first thing to consider when applying a pH measurement is the purpose of the measurement. Is the goal of applying pH leak detection, concentration measurement, and simple monitoring or pH control? Evaluating pH in terms of the goal of the measurement will give a good indication of the probable success of the application. To evaluate leak detection and control applications, it is necessary to have a titration curve of the sample using the reagent or contaminant. It must be kept in mind that pH may not be the best technique for achieving the goal of the application.

Even in cases of simple pH monitoring, the sample pressure, temperature, and composition must be considered, including the sample conditions under upset conditions. The acid or base concentrations should be within the range addressed by pH, or another technique should be chosen. Are components present in the sample, which attack the pH sensor or adversely affect the reference electrode? Is the sample conductivity low? Are nonaqueous solvents, abrasives, or undissolved materials present that can coat the sensor? Does the sample pH change with temperature? If consideration of any of these factors is neglected and their potentials for problems not addressed, the application may have only limited success and may become costly in terms of the maintenance required.

Analyzer location and sensor mounting can facilitate maintenance and calibration. The maintenance of analyzers and sensors, which are difficult to access, is often neglected, and in extreme cases, abandoned. The increased accessibility of retraction and flowthrough mounting can easily justify the higher cost in a critical pH application.

Unless up-front cost is of paramount importance, a microprocessor-based pH analyzer should be chosen over a basic analog analyzer. The feature set and greater ease of use of the microprocessor-based analyzer give it a lower overall cost of ownership in terms of reduced maintenance costs. The choice of smart pH analyzers is usually determined by the presence of other smart instruments on site, and the handheld configurators, control system, or PCs that communicate with them. In multiple, smart instrument installations, maintenance software can help reduce emergency and unneeded routine instrument maintenance.

REFERENCES

1. IEC Publication 746-2, 1st ed. 1982, *Expression of Performance of Electrochemical Analyzers*, Part 2: pH value.
2. A. Wikby and G. Johansson, *J. Electroanal. Chem.* **23**: 23–24 (1969).

FURTHER READING

1. G. K. McMillan, *pH Control*, Instrument Society of America, Research Triangle Park, NC, 1984
2. H. Galster, *pH Measurement*, VCH Publishers, New York, 1991

23

CONDUCTIVITY ANALYZERS AND THEIR APPLICATION

James R. Gray
Rosemount Analytical Inc.
Irvine, California

Environmental Instrumentation and Analysis Handbook, by Randy D. Down and Jay H. Lehr
ISBN 0-471-46354-X   Copyright © 2005 John Wiley & Sons, Inc.

23.1 THEORY OF ELECTROLYTIC CONDUCTIVITY

23.1.1 Background

Conductivity measurement has widespread use in industrial applications, which involve the detection of ionic contaminants in water and concentration measurements. Conductivity measures how well a solution conducts electricity.

The units of conductivity are siemens per centimeter (S/cm), which is identical to the older unit of mho/cm. Conductivity measurements cover a wide range of solution conductivity from pure water at less than 1×10^{-7} S/cm to values in excess of 1 S/cm for concentrated solutions (Fig. 23.1). For convenience, conductivity is usually expressed in the units of microSiemens/cm (μS/cm, one-millionth of a S/cm) or mS/cm (mS/cm, one-thousandth of a S/cm).

In certain industries, the conductivity of water approaching purity is measured in terms of its resistivity. The units used are megohms-centimeters ($M\Omega \cdot cm$), where 18.3 $M\Omega \cdot cm$ is the resistivity of pure water at 25°C. In many of these applications, a minimum resistivity value is used as a criterion for acceptability of the water sample.

23.2 CONDUCTIVE SOLUTIONS

Conductivity is measured primarily in aqueous solutions of electrolytes. Electrolytes are solutions containing ions, which are charged particles formed by

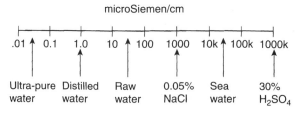

Figure 23.1 Conductivity of common samples.

the dissociation of acids, bases, salts, and certain gases such as carbon dioxide, hydrogen chloride, and ammonia. Water is the solvent in virtually all electrolyte solutions, because water is one of the rare solvents that has the capability of stabilizing the ions formed, by a process called *solvation*. Pure water itself can conduct electricity, because it dissociates to form hydrogen ion (H^+) and hydroxyl ion (OH^-). In the range of common industrial conductivity measurements, solutions made with organic solvents and water solutions made with nonelectrolytes (like alcohol) are only sparingly conductive and are, for the most part, considered nonconductive.

Conductivity is nonspecific. Electric current can be carried in a solution by any ions present. As a result, conductivity measurements respond to any and all ions present in a solution. All that can be said about a conductive solution is that it contains ions. An electrolyte in a solution cannot be identified, or its concentration known from conductivity alone. To determine concentration with a conductivity measurement, the user must have additional information about the solution.

23.3 CONCENTRATION DEPENDENCE OF CONDUCTIVITY

Since ions are the charge carriers in a conductive solution, it would be expected that the higher the concentration of ions in a solution, the more conductive it would be. While this is true for dilute solutions and also for certain electrolytes up to their saturation point, ion–ion interactions in the solution can cause the conductivity response to deviate considerably from what would be expected to be a linear response.

In very dilute electrolyte solutions, with conductivity below 10 μS/cm, ions behave independently, so the conductivity response with increasing concentration is essentially linear. In this low range, the conductivity of a mixture of electrolytes is pretty much the additive sum of their individual conductivities, barring any reaction between them.

Above 10 μS/cm, ion–ion interactions become significant greater as the concentration increases. The conductivity response with increasing conductivity is progressively nonlinear. Some electrolyte solutions can even reach a maximum conductivity in the percent by weight concentration range, after which there is a decrease in conductivity with increasing concentration. This makes the task of deriving concentration from conductivity measurements highly dependent on the conductivity behavior of the electrolyte of interest.

Mixtures of electrolytes, in concentrated solutions, will exhibit conductivity less than the sum of the conductivities of their components in pure solution. How much the conductivity deviates from a simple sum depends on the electrolytes in the mixture, so no simple, general model can be developed for mixtures of electrolytes. As a result, meaningful concentration information for electrolyte mixtures cannot be obtained from conductivity measurements.

23.4 CONDUCTIVITY MEASUREMENT TECHNIQUES

23.4.1 Electrode Conductivity

Electrode conductivity uses a sensor with two metal or graphite electrodes in contact with the electrolyte solution. An alternating-current (AC) voltage is applied to the electrodes by the conductivity analyzer, and the resulting AC current flowing between the electrodes through a volume of solution is used to determine the conductance of the sensor (Fig. 23.2)

23.4.1.1 Probe Constant. The amount of current that flows depends not only on the solution conductivity but also on the length, surface area, and geometry of the sensor electrodes. The probe constant (also called "sensor constant" or "cell constant") is a measure of the current response of a sensor to a conductive solution, due to its dimensions and geometry. Its units are cm^{-1} (reciprocal centimeters; length divided by area). The probe constant scales the response to a given conductivity range to the input circuit requirements of a particular conductivity analyzer. Probe constants can vary from 0.01 to 50 cm^{-1} and, in general, the higher the conductivity, the greater the probe constant necessary.

23.4.1.2 Characteristics of Electrode Conductivity. By choosing the appropriate cell constant, electrode conductivity can measure down to pure water conductivity, and even below in some special applications with organic solvents. By choosing a large cell constant, contacting conductivity can be used to measure highly conductive solutions. But there are drawbacks to using contacting conductivity in highly conductive solutions.

The main drawbacks are that the electrode sensor's susceptibility to coating and corrosion and the fact that many strong electrolyte solutions are corrosive and tend to contain undissolved material that can coat the sensor. Coating or corrosion on the

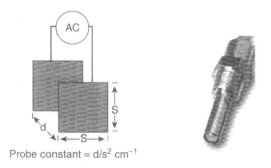

Probe constant = d/s^2 cm^{-1}

Figure 23.2 Apparatus for measuring electrode conductivity.

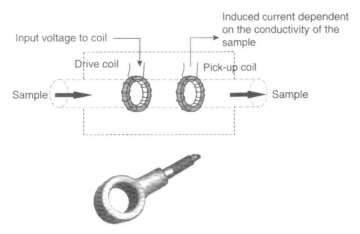

Figure 23.3 Apparatus for measuring toroidal (inductive) conductivity.

surface of an electrode lowers its effective surface area, decreasing its probe constant, and lowering the reading. The conductivity reading will decrease in proportion to the fraction of the surface area of the sensor that is coated or corroded. In strongly conductive solutions, there can also be polarization effects, which result in nonlinearity in the measurement.

23.4.2 Toroidal (Inductive) Conductivity

Toroidal conductivity measurements are made by passing an AC voltage through a toroidal coil, which induces a current in the electrolyte solution. This induced solution current, in turn, induces a current in a second toroidal coil in the sensor. The amount of coupling between the two coils is proportional to the solution conductivity (Fig. 23.3).

23.4.2.1 Characteristics of Toroidal Conductivity. The major advantage of toroidal conductivity is that the toroidal coils are not in contact with the solution. They are either encased in a polymeric material or external to the solution, as is the case with a flowthrough sensor. The toroidal sensor can be completely coated by a solid or oily contaminant in the process, with essentially no decrease in the reading until the coating builds up to a thickness of 1 cm. The polymeric material housing the toroids, or the pipe material in a flowthrough configuration, can be chosen to be resistant to corrosive solutions, which would quickly corrode contacting sensors with metal electrodes.

The major drawback to the toroidal conductivity measurement has been its lower sensitivity compared with contacting conductivity, and, although some recently (as of 2003) developed toroidal conductivity analyzers can measure low conductivity, their application has been restricted to those rare applications where the process has low conductivity and is fouling.

Toroidal sensors are also typically larger than contacting sensors and the solution current induced by the toroid occupies a volume around the sensor. So toroidal sensors need to be mounted in a larger pipe than a contacting sensor. If the toroidal sensor is mounted less than a sensor diameter from a conducting pipe, there can be an upscale reading from conduction through the pipe. Near a nonconductive (plastic) pipe, the readings can be low because of a reduction in the conductive solution volume near the sensor. Both of these effects can usually be corrected for, by zeroing the measurement loop with the sensor mounted in the dry process pipe.

23.4.3 Criteria for Choosing the Conductivity Measurement Technique

General rules for choosing the conductivity measurement technique can be outlined as follows:

- If the process is dirty or corrosive, toroidal conductivity is the only choice. An electrode conductivity sensor will require frequent maintenance, and may require frequent replacement if chemically attacked (Fig. 23.4).
- If a conductivity measurement below 10 µS/cm is required, electrode conductivity is the best choice, not only for its sensitivity but also because electrode conductivity analyzers have the features that support conductivity measurements in this range. Toroidal conductivity should only be considered for use in this range, if the potential for fouling is a major concern. The toroidal analyzer may not have the features to provide a highly accurate reading in very low-conductivity samples.
- Toroidal conductivity is generally favored for applications above 20,000 µS/cm, and it is favored, by some for applications above 5000 µS/cm, because of its lower maintenance requirements.
- Below 5000 µS/cm either toroidal or electrode conductivity can be used. Toroidal sensors are favored because they rarely require cleaning in a reasonably clean application. Electrode sensors are typically smaller and easier to install than toroidal sensors.

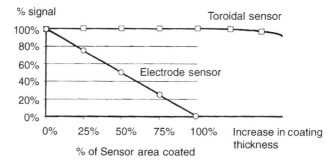

Figure 23.4 Toroidal versus electrode sensors in conductivity measurement.

23.5 EFFECTS OF TEMPERATURE ON SOLUTION CONDUCTIVITY

The conductivity of a solution always increases with temperature, due to the higher mobility of the ions, which carry the electric current. However, a solution's change in conductivity with temperature depends on the conductivity range and the identity and concentration of the electrolyte.

Online measurement of conductivity seeks to determine the concentration of a particular electrolyte, or measure the bulk concentration of electrolytes in the solution. The conductivity, reported by a conductivity analyzer, must be independent of temperature, or else a temperature increase could be mistaken for a concentration increase. As a result, all conductivity analyzers include a temperature compensation routine, which corrects raw conductivity for temperature, using a temperature element in the conductivity sensor. In most cases, the conductivity is corrected to the value it would have at 25°C, but in certain applications a higher reference temperature is used. For accuracy, the temperature compensation routine must match the temperature behavior of the solution. The three common types of temperature compensation will be examined.

23.5.1 Temperature Compensation of Moderately Conductive Solutions

In moderately ($>10\,\mu$S/cm) and highly conductive solutions with a constant electrolyte concentration, the increase in conductivity with temperature is assumed to be linear, and can be compensated for using a linear equation. This temperature compensation is termed *straight-line* temperature compensation and uses a temperature coefficient (z), which is the percent increase in conductivity per degree centigrade: $C(25°C) = C(T)/[1 - z(T-25)]$, where $C(25°C)$ is the conductivity at 25°C, $C(T)$ is the conductivity at some temperature T, and z is expressed as a fraction.

The temperature coefficients of the following electrolytes generally fall in the ranges shown below:

Acids	1.0–1.6%/°C
Bases	1.8–2.2%/°C
Salts	2.2–3.0%/°C
Fresh water	2.0 %/°C

In some conductivity analyzers, the temperature coefficient is assumed to be 2.0%, and only that value is provided.

Finding the temperature coefficient for a particular electrolyte in the published literature can be difficult, simply because there is not a great deal of published data on temperature coefficients. The easiest alternative is to measure the change in conductivity from 25°C to the process temperature in the laboratory, and backcalculate the temperature coefficient from the equation in the preceding paragraph.

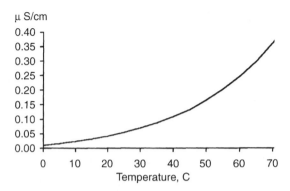

Figure 23.5 Exponential increase in conductivity with increase in temperature.

Alternately, the temperature coefficient in the analyzer can be adjusted until the conductivity reading of the warm solution is the same as it was at 25°C. Some microprocessor-based analyzers have a built-in software routine for doing this determination.

23.5.2 Temperature Compensation in High-Purity Water

In solutions with a conductivity of $1 \mu S/cm$ or less, the conductivity increase with temperature is highly nonlinear. This occurs because the conductivity of water itself is a large fraction of the overall conductivity. The ions carrying charge in water are hydrogen (H^+) and hydroxyl (OH^-) ions from the dissociation of water. With an increase in temperature, not only are the hydrogen and hydroxyl ions more mobile, but water dissociates more, so there are more ions to carry the charge. This double effect causes the conductivity to increase exponentially (Fig. 23.5). Large errors can result from using linear temperature compensation in these applications (Fig. 23.6).

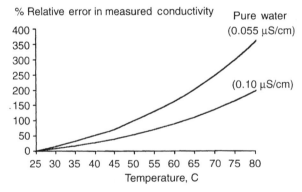

Figure 23.6 Error resulting from linear temperature compensation.

Temperature compensation for these solutions must take into account not only the increase in the conductivity of water but also the increase in conductivity of the solute (dissolved electrolyte). From pure water at 25°C (0.055 µS/cm) to 1.0 µS/cm, the fraction of the total conductivity due to the solute increases, and so the temperature compensation must also take into account the accompanying change in temperature behavior. The increase in the conductivity due to the solute will also depend on what type of electrolyte is present (i.e., acid, base, or salt).

Two major types of temperature compensation address the bulk of high-purity water applications. The first, salt temperature compensation, assumes that the small concentration of solute is salt (NaCl) and is used for most high-purity water applications. The second, more specialized compensation, termed *cation* temperature compensation, addresses applications where the solute is acidic. It gets its name from the fact that the effluent from a cation demineralizer bed, which replaces cations with hydrogen ions, is acid. This type of temperature compensation assumes that hydrochloric acid (HCl) is the solute. Other temperature compensation routines have been developed for more specialized applications, such as pure water containing a sizeable fraction of isopropyl alcohol for use in the electronics industry.

Both temperature compensations involve highly nonlinear mathematical routines, and thus are available only in microprocessor-based analyzers. Before the advent of microprocessors, the temperature of the sample was conditioned to 25°C or a portion of the compensation curve was approximated using resistor networks.

23.5.3 Temperature Compensation in Strongly Conductive Solutions

The temperature coefficient not only varies among electrolytes but can also vary over a wide concentration range of a single electrolyte (Fig. 23.7). This behavior is especially apparent when a concentrated solution of a single electrolyte is being measured. Linear temperature compensation cannot accurately address the

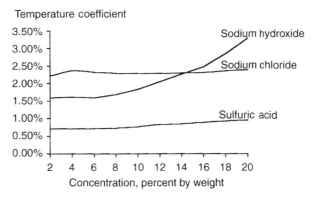

Figure 23.7 Variation of temperature with electrolyte concentration.

temperature behavior over the entire concentration range, and so a compromise is the only solution. In most cases, this means using the temperature coefficient of the concentration at the control point or alarm point, so as to maximize the accuracy at and near this point.

Changes in the temperature coefficient with concentration pose a special problem for conductivity analyzers designed to read out the concentration of a particular electrolyte. For an accurate reading over the full concentration range, temperature compensation at each concentration has to be addressed. This is usually accomplished by building the temperature compensation into the mathematical routine used to derive concentration from conductivity.

23.6 CONDUCTIVITY SENSOR MOUNTING

23.6.1 Sensor and Analyzer Locations

The conductivity analyzer should be mounted in an easily accessible location with the sensor nearby. Conductivity sensors can generally be can be located 100 m or more from the analyzer, but long distances between the sensor and the analyzer make maintenance and calibration inconvenient for a single operator, or require two operators, and some type of remote communication.

23.6.1.1 General Sensor Mounting Considerations. Conductivity sensors may be mounted in a number of ways, as will be seen, but in all cases the following points should be considered:

1. The sensor should be mounted so as to receive a continuous flow of a sample representative of the process solution. Air entrapment in the piping in the vicinity of the sensor must be avoided, as it can cause low or noisy readings.
2. The sensor (and analyzer) should be mounted in a location that allows convenient access for maintenance and calibration.
3. If there is reagent being added to the process, the sample should be well mixed before reaching the sensor. If the sensor is mounted in a vessel, good agitation is necessary to avoid large lag times in the measurement.
4. The pressure limitations of the conductivity sensor should not be exceeded. Locate the sensor at a point of lower pressure, or if this is not possible, a sidestream should be taken off the process with pressure reduction. This is also a safety issue.
5. Likewise, temperature limitations should not be exceeded. Processes with very high temperatures or temperatures in excess of the limits of the sensor should be provided with a sample sidestream and cooling. This is also a safety issue.
6. For installations in hazardous areas, it is best to use an analyzer (usually a two-wire transmitter) designed for use in the hazardous area, rather than a

long cable run from the sensor to an analyzer in a nonhazardous area. If more information than simply a conductivity measurement is required from the analyzer, use a smart conductivity analyzer.

7. In processes where there is a potential for coating, a toroidal sensor should be used. In vessels, the sensor should be located to avoid regions where there is settling of precipitates or debris.

23.6.2 Mounting Configurations

The mounting configuration of a conductivity sensor is determined by the requirements of the process piping and vessels, and other considerations such as the need for sample cooling or pressure reduction. Regardless of the mounting configuration used, air entrapment in the process near the sensor is a major source of error and has to be avoided at all costs (Fig. 23.8).

23.6.2.1 Submersion Mounting. Submersion mounting involves immerson of the sensor in a process vessel or pond on a length of conduit. It is typically used only when only access to the process is from the top. It is the most inconvenient mounting configuration for the operator, because both the sensor and its conduit must be removed for maintenance and calibration. Care has to be taken to ensure that a watertight seal is made between the sensor and the conduit, and backfilling the conduit is advisable to avoid leakage of the process into the cable end of the sensor, which causes short circuits. Because of the difficulties accessing the sensor, submersion applications are best served by toroidal conductivity, due to their lower maintenance requirements.

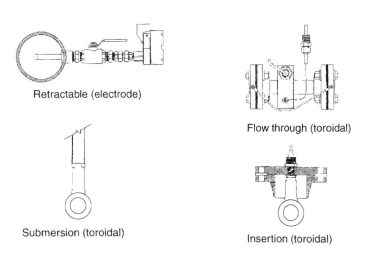

Retractable (electrode)

Flow through (toroidal)

Submersion (toroidal)

Insertion (toroidal)

Figure 23.8 Mounting configurations for a conductivity sensor.

23.6.2.2 *Insertion Mounting.* This configuration generally involves mounting the conductivity sensor in a tee or in through the side of a vessel. A threaded electrode sensor can simply be screwed into a pipe tee, without the need for additional hardware. Toroidal sensors are most often inserted using a flange or some manner of insertion fitting because of their larger size.

23.6.2.3 *Retraction Mounting.* One shortcoming of insertion mounting is the need to shut down the process or drain a vessel in order to remove the sensor for maintenance. This shortcoming is overcome using retraction mounting, which is an insertion type of mounting using a ball valve or some other isolating valve that allows the sensor to be removed while the process is running. With this configuration, the sensor can be quickly removed for cleaning, or replacement in the event of a catastrophic failure, minimizing the downtime of the conductivity measurement during the process. But it should be noted that if the conductivity range permits, using a toroidal sensor with insertion mounting is often preferable to using a contacting sensor with retraction mounting, if the purpose of the retraction mounting is simply to facilitate cleaning.

The process pressure is an important safety consideration in retraction installations. With ball-valve-type insertion devices using manual insertion and retraction, the maximum rating for the device should be strictly observed, to avoid the sensor from being ejected from the device and the operator being sprayed with the process. Inserting the sensor into the process against the force of a pressurized sample can be difficult. In cases where excessive pressure is possible but access to the sensor during running is necessary, a retractable device other than a ball valve, or a sample sidestream with pressure reduction should be chosen over a manually operated ball valve.

23.6.2.4 *Flowthrough Mounting.* A flowthrough configuration can be a sensor specially designed for flowthrough mounting with flanges or female threads to mate with process piping, or it can simply be a insertion or retractable sensor mounted in a tee. The term flowthrough is often taken to mean a conductivity sensor mounted in a sample sidestream. When provided with isolation valves, this mounting configuration is the most convenient because it allows the sample flow to be shut off and the sample line to be drained at will for maintenance and calibration. When sample conditioning is required, such as cooling, pressure reduction, or filtration and separation, this is the only configuration possible.

The major drawbacks to using a flowthrough sensor in a sample sidestream are the losses of process sample, unless sample return is provided, plugging of the sample line by solids or precipitates, and the possible lag time in getting the sample from the process to the sensor.

In general, choice of sensor mounting will be determined largely by the requirements of the process vessels and piping, as in the case of a pond or a sump, where the only mounting option is submersion. In other cases, there can be greater flexibility. Insertion mounting is most often the option of choice, because of its simplicity and low cost, but if sample cooling or pressure reduction is required, flowthrough mounting is the only option. Retraction or flowthrough mounting should be used in cases where it is necessary to access the conductivity sensor, while the process is running for cleaning or maintenance, especially when a reliable conductivity measurement is critical to the process. The added hardware and piping necessary for retraction and flowthrough mounting make the up-front cost greater than with insertion or submersion.

23.7 CONDUCTIVITY CALIBRATION

23.7.1 Calibration in Moderate to Highly Conductive Solutions

For conductivity measurements in excess of 50 μS/cm, a conductivity standard may be used to perform a single-point calibration on a conductivity loop, which is usually preceded by zeroing the analyzer with the sensor in air. The conductivity measurement may also be calibrated using a grab sample of the process. The following guidelines should be observed for grab sample standardization:

1. The grab sample should be taken near the online conductivity sensor, so the sample represents what the sensor is actually measuring.
2. The grab sample should be taken when the process conductivity and temperature are stable. Otherwise discrepancies can result from the grab sample not matching the process sample at the sensor. The online reading may not have fully responded to a change in conductivity, or the conductivity sensor may not have reached the process temperature.
3. Incorrect temperature compensation in the analyzer may show up as a discrepancy between the online and laboratory measurement.
4. If a chemical reaction is taking place in the process, the reaction may not have gone to completion at the online sensor, but will have by the time a grab sample reaches the laboratory.
5. The stability of the process sample is an important consideration. Some samples can react with air, or precipitate on cooling and change conductivity.
6. Grab sample standardization should never be used for conductivity measurements under 50 μS/cm. They are easily contaminated, especially in the case of high-purity water, which is almost immediately contaminated by carbon dioxide in ambient air.

Finally, it must be remembered that the laboratory or portable analyzer used to adjust the online measurement is not a primary conductivity standard. Errors in the laboratory or portable analyzer will result in an incorrect standardization of the online unit. Both the laboratory and online measurements can be checked with a primary conductivity standard, however.

23.7.2 High-Purity Water Measurements

Low-conductivity samples are highly susceptible to contamination by trace contaminants in containers and by CO_2 in air. As a result, calibration with a conventional standard is not possible. Any sample or grab sample taken would also be subject to the same contamination, and would be rendered unusable within seconds.

The alternative used by conductivity analyzers designed for high-purity water measurements is to calibrate the sensor input to the analyzer, using precision resistors. A conductivity sensor is used, which has an accurately known probe constant determined in a primary conductivity standard. Calibration of the analyzer and sensor combination is a simple matter of entering the probe constant into the analyzer. This feature is available only with microprocessor-based analyzers.

23.8 DIAGNOSTICS AND TROUBLESHOOTING

23.8.1 Traditional Error Detection and Troubleshooting

There are two categories of diagnostics: online and offline diagnostics. Online diagnostics alert the user to a problem in real time, during operation of the analyzer. Offline diagnostics are indications of a problem during calibration or maintenance of the sensor and analyzer.

A basic conductivity analyzer, is an analog device that simply transmits and displays conductivity. The only online indication of a problem is a conductivity reading that is obviously incorrect, based on what is known about the process. Troubleshooting is based on following a table of tests based on a description of the behavior of the conductivity signal online and during offline tests.

23.8.2 Analyzer Self-Diagnostics

Along with microprocessor-based conductivity analyzers came the ability to do more online diagnostic measurements and to show them on digital displays. These included self-checks of the microprocessor itself for malfunctions in the CPU and memory, but more importantly, included some checking of the input signals from the conductivity sensor. Input signal checking was able to detect open- or short-circuited temperature elements, and open circuits or short circuits in the sensor cable.

Offline diagnostics received a boost from the ability of the analyzer to calculate and display the probe constant on the basis of a calibration that had become an automated routine within the analyzer. It was now possible to note decreases in the probe constant, which indicate a coating problem. In fact, some analyzers calculate the probe constant from the first calibration, and calculate a correction factor, which gives an indication of coating or corrosion.

The most common catastrophic failure of a contacting or toroidal conductivity sensor is an open- or short-circuited temperature element, and, as noted, the analyzer checks this fault. Other failures that can occur are short circuits due to leakage of process into the sensor, and a sudden, abnormal amount of coating.

23.9   ANALYZER–USER INTERFACE AND OUTPUT SIGNAL

The user interface and output signals of a conductivity analyzer are the means by which the analyzer communicates with the user or a supervisory device. In describing each type of analyzer, the feature set of the analyzer determines what information is available, and the display and output signal determines where and how the information is made available. While the platforms described below roughly correspond to the historical development of conductivity analyzers, they are nevertheless all still currently commercially available (Table 23.1).

23.9.1   The Basic Analog Conductivity Analyzer

The basic conductivity analyzer is an analog device with zero and span adjustments (typically potentiometers) for zeroing and calibration. They have a very limited feature set, typically providing only a 4–20-mA output proportional to a conductivity range and a display (sometimes also optional in two-wire transmitters), either digital or analog, for displaying conductivity only. Line-powered units can include high/low-conductivity alarm contacts. These units, in most cases, lack the ability to accurately temperature compensate high-purity water. The major advantage of basic analog conductivity analyzers is their lower cost.

23.9.2   The Microprocessor-Based Conductivity Analyzer

Microprocessor-based conductivity analyzers offer a marked improvement over the basic analog devices. In terms of features, they typically include the ability to do high-purity water temperature compensation, and be calibrated for high-purity water, and may also include the ability to read out and be calibrated in concentration units. Software routines are included for configuration and calibration, which make the analyzer easier to use than the basic analyzer.

While the typical user interface consists of a keypad and display in all of these analyzers, the display types can vary from small LED or LCDs, which display non-numerical information in three-character mnemonics, to larger displays capable of spelling out complete messages. To address the world market, the larger displays have become multilingual, although there is also an emerging trend toward symbolic, language-independent displays.

The output signal of these analyzers remains the analog 4–20-mA signal proportional to conductivity, with alarm contacts in the line powered devices. Line-powered devices may also include a second 4–20-mA output for a second conductivity range or a temperature. Line-powered units can use a dedicated alarm contact for remote indication of a diagnostic alarm, while two wire transmitters might use an overange 4–20-mA signal (21 or 22 mA) as an indication of a problem. Since all the diagnostic information resides in the analyzer, these remote signals are "come read" alerts to the user, who has to then go and interrogate the analyzer.

Table 23.1 Conductivity Analyzers

	Feature Set	User Interface	Remote Communication
Basic analog analyzer	Zero, span, and range adjustment Straight-line temperature compensation only	Analog or digital display of conductivity Potentiometers/ switches for zero, span, and range adjustment	4–20-mA signal proportional to conductivity Conductivity alarm contacts
Microprocessor-based analyzer	Temperature compensation for high-purity water and concentration measurements Special calibration routines for high-purity water and concentration measurements Calculation of concentration from conductivity Self-diagnostics	Digital display of conductivity, calibration constants, configuration parameters, and error messages Software routines for calibration and configuration Keypad for entering numerical values and navigating among software routines	4–20-A signal proportional to conductivity Conductivity alarm contacts Fault alarm via alarm contact or overange 4–20-mA signal
Smart conductivity analyzer	Temperature compensation for high-purity water and concentration measurements Special calibration routines for high-purity water and concentration measurements Calculation of concentration from conductivity Self-diagnostics	Digital display of conductivity, calibration constants, configuration parameters, and error messages Software routines for calibration and configuration Keypad for entering numerical values and navigating among software routines	Digital conductivity signal (also 4–20-mA signal in some cases) Digital conductivity alarms (also pH alarm contacts in some cases) Digital fault messages (also fault alarm contact in some cases) Full access to the instrument for calibration and configuration via a handheld remote terminal, PC, or host control system

23.9.3 Smart Conductivity Analyzers

Smart conductivity analyzers combine the features of the microprocessor-based analyzer with "smart," digital communications. This not only allows diagnostic information to be transmitted to a remote location along with the conductivity measurement but also makes it possible to interrogate, configure, and even calibrate the analyzer remotely. There are a number of protocols available for smart communications, a number of which are manufacturer specific. No protocol has, as yet, become a standard in the same way that the 4–20-mA analog signal has. The most widely used protocols for analyzer are HART, Modbus, FOUNDATION Fieldbus, and Profibus. Smart analyzers are more costly than the general microprocessor-driven analyzers.

To access the digital information provided by the analyzer, the user needs a communication device, which can take the form of a handheld keypad display for configuration and calibration in the field, a PC, or a host control system. The use of a PC for communication with smart conductivity analyzers has led to the development of maintenance software packages that provide a software environment for configuration, troubleshooting, and calibration. These software packages can include a database for recording timestamped error messages and calibration data, thus automating the task of keeping maintenance records.

23.10 APPLICATION OF CONDUCTIVITY

23.10.1 Nonspecific Applications

Nonspecific applications involve simply measuring conductivity to detect the presence of a certain bulk concentration of electrolytes, that is, there is no attempt to derive the concentration of a particular electrolyte from the measurement. The majority of conductivity applications fall within this category. They include monitoring and control of demineralization, leak detection, and monitoring to a prescribed conductivity specification, as in the case of power generation equipment. In most instances, there is a maximum acceptable concentration of electrolyte, which is related to a conductivity value, and that conductivity value is used as an alarm point.

23.10.1.1 Interface Detection. An often over looked application of conductivity is in detecting the interface between a conductive solution and a nonconductive solution. The nonconductive solution can be a solvent or hydrocarbon, and the conductive solution can be virtually any water sample. The principle is simple: There is some upscale conductivity reading in the water solution and no conductivity reading in the nonconductive solution. The conductivity output is used to provide an on/off signal to indicate which phase the sensor is in. This application almost always employs a toroidal conductivity sensor, because solutions of solvents and hydrocarbons can foul an electrode sensor.

23.10.2 Leak Detection with Conductivity

Conductivity can be used to detect the leak of more conductive contaminant into a less conductive sample. The greater the difference in conductivity is between the contaminant and the sample, the greater the sensitivity of leak detection will be.

In general, unless the contaminant is at least 100 times more conductive than the sample, the sensitivity of leak detection will be poor.

An approximation of the fractional change in conductivity accompanying a leak can be calculated as follows: $F = (V_C / V_S) (C_C / C_S)$, where F is the fractional change in conductivity; V_C and V_S are the volumes (or flow rates) of the contaminant and the sample, respectively; and C_C and C_S are the conductivities of the contaminant and the sample, respectively. The minimum leak detectable will be the volume or flow rate of contaminant, which gives a conductivity change over the normal sample conductivity that can be reliably measured by the analyzer. The minimum leak detectable will depend on the accuracy of the conductivity measurement and the extent to which the conductivity of the sample normally varies.

Where the sample conductivity varies over time, the maximum conductivity of the sample should be used for estimating the sensitivity of leak detection. In some cases, two conductivity measurements can be used before and after the source of a potential leak and the difference between the two reading used to detect a leak. The accuracy of the conductivity measurements and the normal changes in sample conductivity over time will determine the minimum conductivity difference, which can be used for alarming.

23.10.3   Concentration Measurements

Conductivity is nonspecific, but it can sometimes be applied to concentration measurements if the composition of the solution and its conductivity behavior are known. The first step is to know the conductivity of the solution as a function of the concentration of the electrolyte of interest. These data can come from published conductivity–concentration curves for electrolytes, or from laboratory measurements. The conductivity of mixtures usually requires a laboratory measurement due to the scarcity of published conductivity data on mixed electrolytes.

Over large concentration ranges, conductivity will increase with concentration, but may then reach a maximum and then decrease with increasing concentration. It is important to use conductivity data over the temperature range of the process, because the shape of the conductivity–concentration curve will change with temperature, and a concentration measurement may be possible at one temperature but not another.

23.10.3.1 Necessary Conditions. With the conductivity versus concentration data in hand, it can be reviewed for the two necessary conditions for a concentration measurement:

1. There must be a measurable change in conductivity over the desired concentration range.
2. Conductivity must either strictly increase or decrease with increasing concentration. If a maximum or a minimum is present in the of conductivity–concentration curve, in the desired concentration range, concentration cannot be measured over concentration ranges containing these extreme points (Fig. 23.9).

Measurable concentration ranges

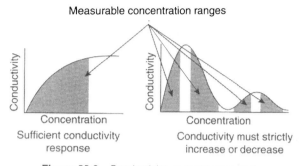

Figure 23.9   Conductivity–concentration plot.

23.10.3.2 Cases Where Concentration is Measurable. If the two necessary conditions are met, concentration can be measured in each the following cases:

1. The solution is a single electrolyte in solution, or the conductivity of other solution components is very low.
2. The conductivity due to other electrolytes is large relative to the electrolyte of interest, but their concentration remains constant.
3. The concentration of a mixture of electrolytes and any one or all of its components can be measured, if their ratio remains constant (e.g., a detergent solution).

23.10.3.3 Concentration Analyzers Based on Conductivity. Many conductivity analyzers include an output option for displaying and transmitting the concentration of a particular electrolyte over a specified concentration range. This is great convenience to users over having to develop their own concentration applications. As these concentration measurements are conductivity-based, the principles outlined above apply to them as well. The concentration ranges chosen are those where the conductivity strictly increases or decreases with increasing concentration, within specified temperature limits. Samples with concentrations outside the specified concentration or temperature range will give erroneous readings, which can appear believable. For this reason, it is important to apply these analyzers only within the specified ranges.

The algorithms used in concentration analyzers are based on the conductivity curve of the pure electrolyte in solution. Therefore they are applicable only to samples consisting of the pure electrolyte, or samples with negligible background conductivity. A common error is to apply these analyzers to samples that have a significant concentration of other electrolytes. The concentration reading will be erroneous, but can appear to be the true concentration. The sample composition must be known and evaluated before these analyzers are applied.

23.10.4   Reaction Monitoring and Endpoint Detection

Chemical reactions, typically batch reactions, in electrolyte solutions can be accompanied by conductivity change sufficient to allow their progress to be followed by conductivity. Potential applications can be evaluated by measuring the conductivity at various stages in the reaction and relating it to a laboratory analysis of the reaction mixture at each stage. Conductivity can be related to the percent completion of the reaction or to the concentration of any reactant or product.

23.11   SUMMARY

When properly applied, online conductivity is probably the most straightforward and lowest maintenance of all online analytical measurements. While conductivity can be applied to the full range of water solutions, from high-purity water to the most conductive solutions known, the key to a successful application is knowing the sample's conductivity characteristics and applying the correct technology. These can be outlined as follows:

- Use toroidal conductivity for dirty, corrosive, or high-conductivity applications.
- For accuracy in applications approaching 1 μS/cm, use electrode conductivity and an analyzer with high-purity water temperature compensation.
- Use contacting conductivity for low-conductivity applications in clean process streams.

In many cases, conductivity is used as a nonspecific measurement, with a high alarm point based on a prespecified conductivity limit. Conductivity may also be applied with a specific analysis goal in mind. If leak detection is the goal of the application, the conductivity range of the sample and contaminant must be known, as well as an estimate of the overall measurement accuracy. Applying conductivity to concentration measurements requires the conductivity of the sample over the concentration and temperature range to be known. Following a chemical reaction with conductivity must be based on a laboratory analysis of the sample composition and its conductivity at various stages in the reaction.

24

TURBIDITY MONITORING

John Downing, Ph.D.
D & A Instrument Company
Port Townsend, Washington

Environmental Instrumentation and Analysis Handbook, by Randy D. Down and Jay H. Lehr
ISBN 0-471-46354-X   Copyright © 2005 John Wiley & Sons, Inc.

24.1   INTRODUCTION AND SCOPE

Suspended particulate matter (SPM) and dissolved light-absorbing material affect light and near infrared (NIR) transfer through surface waters (lakes, rivers, and the oceans). SPM does so mainly by scattering, causing radiation to follow a zigzag rather than a straight path while absorption attenuates light intensity without changing its direction. In the environment, SPM is primarily suspended clay, silt, and sand (sediment), organic particles, plankton, and other microscopic organisms. These materials give water a hazy or cloudy appearance called turbidity. Optical instruments that measure turbidity are called turbidimeters, and although particle detection is their main purpose, dissolved radiation-absorbing materials affect turbidity by controlling the availability of radiation scattered to detectors. The presence of dissolved radiation-absorbing materials in water is commonly indicated by its color.

Turbidity has practical utility in sanitation, industrial, and environmental sectors because it indicates the presence of SPM in water, and in many cases, its relative concentration volume. It is convenient to measure with simple instrumentation that demands only modest training. Turbidity is used as a surrogate for suspended sediment by engineers and scientists to evaluate the effects of logging, agriculture, and mining on sediment loads in surface water. It indicates water suitability for biological productivity, fisheries, drinking, as well as industrial and recreational uses. Groundwater turbidity is a measure of progress in well development.

The U.S. Clean Water Act of 1972 (CWA) and amendments distinguish between monitoring low-level turbidity in drinking water and the monitoring of wastewater, groundwater, and surface water (lakes, rivers, and the ocean) for environmental purposes. The scope of this chapter is confined to the latter three monitoring tasks and focuses on submersible and battery-powered, field-portable instruments that work at remote sites. Some important applications within this scope include:

- Compliance with permits, water quality guidelines, and regulations
- Determination of transport and fate of waterborne particles and contaminants

- Conservation, protection, and restoration of surface waters
- Measure performance of water- and land-use management
- Monitor waterside construction, mining, and dredging operations
- Characterization of wastewater and energy production effluents
- Tracking water-well completion

Although standard methods for turbidity measurements have existed for more than 30 years, they permit wide latitude in optical components and their arrangement in turbidimeters. As a consequence, the turbidity of a water sample depends on how and with what apparatus it is measured. Different turbidity values can therefore be reported for an individual sample by multiple turbidimeters meeting the same standard method, challenging the integrity of a monitoring program. It is a goal of this chapter to help analysts and managers overcome this challenge and conduct effective monitoring programs in the environment. To meet this goal, the following objectives will be achieved: (1) facilitate better understanding and refinement of technical requirements; (2) provide guidance for selecting turbidimeters and standards; (3) introduce field sampling, measurement, and quality assurance procedures; (4) develop relationships between turbidity and suspended matter; and (5) review U.S. regulations that deal with turbidity.

24.2 ORIGINS OF TURBIDIMETERS AND TURBIDITY STANDARDS

More than a century ago, Whipple and Jackson[1] described an apparatus for measuring the extinction of light from a candle flame by scattering and absorption in a water sample in Jackson turbidity units (JTU), where 1 ppm of Jackson's silica standard equaled 1 JTU. Although it was a practical analytical tool, the range of the candle turbidimeter was limited to 25–1000 JTU and measurements depended on visual observations. Moreover, the JTU refers to a ppm of Jackson's silica suspension, which was not traceable to a physical standard and is no longer available.[2] In 1926, Kingsbury et al.[3] described a cloudy, polymer suspension in water that mimicked the light extinction properties of silica in a candle turbidimeter with ±1% precision. They made their suspension from analytical-grade reagents and called it formazin. It provided a synthetic, light-scattering proxy for diatomaceous earth that could be used in any turbidimeter. When combined with developments in optoelectronics, this was a leap forward in turbidimeter technology. An example of this was the turbidimeter developed in the early 1930's by Exton,[4] which consisted of an incandescent lamp, optics, a water sample tube, and photovoltaic detectors. An early backscatter turbidimeter was invented by Hach in 1959.[5]

Forty-five years after Exton's patent, the 13th edition of *Standard Methods for the Examination of Water and Wastewater*[6] recognized formazin as the primary turbidity standard. The U.S. Environmental Protection Agency (EPA) specified Method 180.1, EPA 180.1, for turbidity measurements in Methods for Chemical

Analysis of Water and Wastes[6a] and described a turbidimeter that complies with the method. EPA 180.1 is known as the nephelometric method (*nephelos* is the Greek word for *cloud*). It measures light scattered at about 90° by a sample and assigns a formazin-equivalent turbidity value in nephelometric turbidity units (NTU). Submersible sensors from Partech and optical backscatter (OBS) probes appeared in the mid-1980s. Continuous improvements in photometry, solid-state light sources, and microcontrolled electronics have led to the long list of instruments currently available.

24.3 LIGHT-SCATTERING THEORY FOR TURBIDITY

The word *light* is used for both visible light [wavelengths (λ) from 390–760 nm] and near-infrared radiation (NIR), 760–3000 nm, in the remainder of this chapter; however, the wavelength dependence of optical phenomena must not be overlooked. When wavelength dependence is important in a particular context, it is signified by the λ symbol.

Light is an ensemble of photons that are absorbed and scattered as they travel through a medium like water. The absorption coefficient, $a(\lambda)$, is a measure of the conversion of radiant energy to heat, and the scattering coefficient, $b(\lambda)$, is a measure of radiant energy dispersion without changes in wavelength.[7] The units for these coefficients are reciprocal length (m^{-1}) where larger values indicate stronger effects. For example, water with $b(\lambda) = 10\,\text{m}^{-1}$ will scatter 1/e of the energy out of a light beam over a distance of 10 cm, whereas another sample with $b(\lambda) = 1\,\text{m}^{-1}$ will scatter the same proportion of energy in 100 cm. The attenuation coefficient, $c(\lambda)$, is a measure of the loss of light from a beam resulting from the combined effects of scattering and absorption. These processes are illustrated by the overhead projection of two petri dishes, one containing inky water and the other containing skim milk. The images will be gray because both water samples attenuate light; however, the milky sample does so mainly by scattering, whereas the inky one mainly absorbs light.

At 600 nm, particle-free water absorbs about 250 times more light than it scatters,[8,9] and it has a turbidity of only 0.01 NTU, which is 1/30th the allowable limit for finished drinking water.[10] In some spring-fed water bodies, mountain streams, and clear lakes, light absorption is the dominant optical effect, and turbidity is of no environmental consequence. Drinking water measurement systems are appropriate for monitoring such waters. Sediment particles are ubiquitous in most surface water, however, and even minute amounts will cause scattering to be the major cause of turbidity and light attenuation.

The angle between incident and scattered light rays is called the scattering angle, α. Forward-scattered radiation occupies the hemisphere surrounding a light beam, and, oriented away from the source, backscattered radiation fills the opposite hemisphere. The angular distribution of scattered light intensity around a water parcel is called the volume scattering function (VSF). In general, VSFs depend on particle and fluid refractive indices and the ratio of particle diameter to wavelength, D/λ. SPM scatters light in proportion to λ^{-1} to λ^{-2}. Small particles, $D \sim \lambda/10$, scatter about the same amount of light in the forward and backward directions, whereas intermediate-size particles ($D \sim \frac{1}{4}\lambda$) scatter most of it in the forward direction. Particles with $D > \lambda$ scatter nearly all light in a small, forward cone. The VSFs and their integral effects indicated by $b(\lambda)$ have been measured in coastal ocean waters by Petzold.[11] His measurements showed that about half of the scattered light is contained in a $10°$ cone oriented in the forward direction, and less than 2.5% of the light is backscattered. Together, the aforementioned effects result in wide variability in the angular distribution and intensity of light scattering from a water sample, which means that different combinations of particle size, operating wavelength, color, shape, and surface quality can produce similar turbidity readings. They also mean that control of light wavelength and consistent light transmission in turbidimeters is essential for obtaining reproducible measurements of surface water samples; see Section 24.4.

24.4   TURBIDIMETER DESIGNS

Turbidimeters respond mainly to light scattering and to a lesser degree to absorption by a water sample, dissolved material, and SPM; however, only research instruments can measure the inherent optical properties, VSF, $a(\lambda)$ and $c(\lambda)$, of a sample as defined in radiation transfer theory. Governmental agencies and standards organizations have defined methods for turbidity measurements and described how an apparatus must be configured to measure turbidity in compliance with standard methods. Two classes of turbidimeters are discussed in this chapter. They include portable, battery-powered meters for measuring grab samples as well as submersible meters that can measure turbidity in situ. Before reviewing the basic turbidimeter designs, it must be stressed that these standards do not approve or disapprove of a particular make and model of turbidimeter. They only provide guidance on how a compliant meter should be constructed. In other words, a manufacturer's certification that his meter complies with a particular method does not mean that the originator of that method has approved his meter. As will soon become clear, there is a lot of latitude in how turbidimeters can be designed while still meeting the specifications of an approved method. For this reason, it is a rare coincidence when two meters made by different manufacturers report exactly the same turbidity value for a sample of surface water.

A good example of the variation in turbidity one can expect from a turbid water sample measured by different turbidimeters comes from Lander's[12] blind comparison. The turbidities of three samples of suspended sediment containing different proportions of fine sediment (< 63 μm diameter), or mud, and sand were determined with turbidimeters made by several different manufacturers. The reported turbidities ranged from 11 to 53 NTU for a sample with a sediment concentration (SSC) of 150 mg/L and composed of 100% mud. The other ranges were 112–268

NTU for a sample with SSC = 600 mg/L (7% sand plus 93% mud) and 85–221 NTU for a sample with SSC = 600 mg/L (20% sand plus 80% mud). In addition to design, inconsistent meter turbidimeter operation also contributed to imprecision in reported values.

Two messages are illustrated by these examples. First, the physical characteristics of the suspended sediment that determine its light-scattering and light-absorbing properties vary widely in the environment. Second, variance in the design of turbidimeters that comply with the same standard or method cause them to read different turbidity values on the same sample. Therefore, it is very important to use the same model of turbidimeter and method for a complete study to ensure that the data are internally consistent. There is a current strategy, exemplified by the ongoing effort by the ASTM, to categorize turbidimeter designs and units of measure rather than to set "standard" designs. The goals are to: 1) reduce the range of values measured in natural water samples by a turbidimeter category, 2) improve means for tracing variability to turbidimeter designs, and 3) heighten awareness of the vagaries and shortfalls inherent in turbidity measurements. Finally, consistent light transmission and a stable, operating wavelength are critical. General factors are (1) a low blank value (turbidity reading on turbidity-free water in a clean scratch-free, properly oiled sample vial, and (2) service and maintenance (sample compartment of portable is dust free and submersible optics are clean and scratch free).

24.5 EPA 180.1

This method requires a turbidimeter with a tungsten-filament lamp (TFL), operated at a color temperature between 2200 and 3000 K and a detector plus bandpass filter if used with peak spectral response between 400 and 600 nm (Fig. 24.1). The acceptance cone of the detector must be no wider than 60° and oriented 90° to the incident light beam. TFLs have peak emittances in the NIR band near 850 nm. They are rugged and inexpensive; however, tungsten condensation on the bulb causes them to get dimmer with age, necessitating recalibration or optical feedback for stable

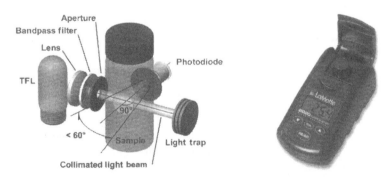

Figure 24.1   EPA Method 180.1 (*left*) and portable EPA 180.1 turbidimeter (*right*).

operation. TFLs are impractical for submersible turbidity probes because they consume a large amount of power, over 95% of which expended making heat, and they cannot be switched electronically to reject ambient light. Silicon photodiodes are used almost exclusively to detect light in turbidimeters because they are: (1) stable and respond linearly over a 10^6 range of light levels, (2) well matched spectrally to infrared emitting diodes (IRED) and filtered TFLs, and (3) compact and inexpensive.

The EPA 180.1 method is for low-level measurements (< 40 NTU). Samples with turbidities greater than 40 NTU must be diluted down to the range 30–40 NTU with turbidity-free water[13] before measurement by the EPA 180 method. Because sample dilutions made in the field can result in loss of material due to settling, incomplete mixing, and measurement errors, the method is recommended only for grab samples with turbidity less than 40 NTU. Colored water and particles can produce different turbidity readings from EPA 180.1 turbidimeters because of the loose tolerances for the TFL operating wavelengths, bandpass filter, and detector spectral response.[10] Variance in wavelength and particle diameter also affects the ratio, D/λ, and will result in additional meter-to-meter variations.[14]

Standard Method (SM) 2130 B is nearly identical to EPA 180.1. SM-2130 B meters measure in NTU and have the same limitations for high-level measurements and unattended operations. The photoelectric nephelometer in American Society for Testing and Materials (ASTM) Standard D 1889-00[15] is also similar to EPA 180.1 except the detector and filter spectral responses are unspecified, making the design specifications looser than the prior two methods. The D 1889-00 meter measures in NTU and has the same limitations for field use as EPA 180.1. Some turbidimeters meeting these standards are listed in Table 24.1. ASTMD 1889-00 will be superceded by a new standard in 2005.

24.6 GLI-2 METHOD

The GLI-2 method (Great Lakes Instruments, Inc. 1992. Turbidity, GLI method 2 Milwaukee, Wisconsin) incorporates two source-detector pairs arranged orthogonally,

Table 24.1 Turbidimeters That Comply with EPA 180.1 and GLI-2 Methods

Manufacturer	Model	Type	Range (NTU)	Method	Logs Data
Hach Company	2100P	P	2000	180.1/R	○
HANNA Instruments	C102	P	50	180.1	✓
HydroLab (Hach) Inc.	DataSonde4	S	1000	GLI-2	✓
HF Scientific Inc.	DRT-15CE	P	1000	180.1/R	○
Industrial Chemical Measurements	11520	P	200	180.1	○
Lamotte	2020	P	1100	180.1	○

Note: R = ratiometric; P = portable; S = submersible. Measurements above 40 NTU do not meet EPA reporting or regulatory requirements.

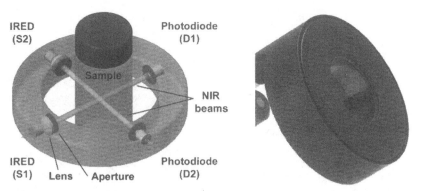

Figure 24.2 GLI-2 design standard (*left*) and submersible GLI-2 probe (*right*). (Photo courtesy of HACH/HydroLab.)

as shown in Figure 24.2, and is approved for drinking water measurements (0–40 NTU). The first infrared (IR) emitting diode, S1, is turned on and detector D1 measures the intensity of the beam transmitted through the sample while detector D2 measures scattered-light intensity at 90°, giving two numbers. A second measurement is then made with S2 producing two more numbers. A microcontroller performs a calculation with the two resulting ratios of transmitted to scattered intensities to determine turbidity. The advantage of this approach is that errors resulting from fluctuating IRED power, photodiode sensitivity, and uniform window fouling are nearly eliminated. When fouling is not uniform, however, the effects will not cancel in a ratio and measurement errors will occur. The probability of uneven fouling in natural waters is significant, so caution is advised when analyzing turbidity data from environments where this can occur. GLI-2 meters measure in NTU and the EPA 180.1 limitations for field use apply. Only the HydroLab four-beam turbidimeter meets GLI-2 (Table 24.1) and is suited for field measurements in locations where fouling is not a big concern. In biologically active water, regular service will be necessary to clean the sensor.

24.7 ISO 7027

The International Standards Organization (ISO) 7027[15a] apparatus (also known as EN 27027 and DIN 38404) has an 860-nm IRED or IR laser diode (LD) light source (Fig. 24.3). Like EPA 180.1, the axes of the emitter and detector acceptance cones must form an angle of $90 \pm 2.5°$. The emitter beam must be collimated to better than $\pm 1.5°$. The acceptance angle of the detector must be 20–30°. IREDs and LDs efficiently convert electrical current to light energy, making them ideal for battery-powered turbidimeters; they can be switched rapidly to reject ambient light; they have narrow emission spectra and age slowly. When used for drinking water testing, sample dilution is required when turbidity exceeds 40 NTU. A TFL with

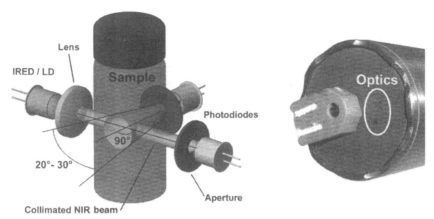

Figure 24.3 ISO 7027 standard design (*left*) and fiber-optic version of the standard (*right*). (Courtesy of McVan Instruments Pty)

bandpass filter is permitted; however, ISO 7027 is not approved for EPA regulatory or reporting purposes. Colored dissolved matter in high concentrations can negatively bias turbidity readings when the dissolved matter absorbs light in the source or detector spectral bands.[16] The turbidity of runoff from acid mine tailings, for example, can be only 80% of what it would be were the same particles suspended in clean water. This is a rare condition, however, since environmental concentrations of organic and inorganic radiation-absorbing matter are normally too low to measurably affect turbidity.[16] Particle color will also change ISO 7027 turbidity by as much as a factor of 10; see backscatter designs for a discussion of color effects. ISO 7027 meters measure in FNU, formazin nephelometric units, to distinguish the standard from EPA 180.1. An FNU nominally equals an NTU; however, a water sample that produces a 100-NTU reading from an EPA 180.1 or GLI-2 turbidimeter may not read 100 FNU on an ISO 7027 meter.

Fiber optics (FO) are used to transmit light to and from a sample in some ISO 7027 meters, making them well suited for in situ measurements (grab samples are not needed) and for wiper mechanisms (Figs. 24.3 and 24.4). In particle

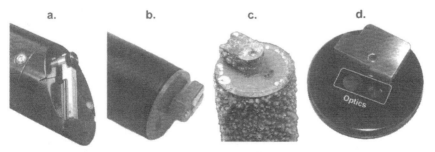

Figure 24.4 (*a*) Neoprene wiper. (Photo courtesy of FTS Inc.) (*b*) Pad wiper. (Photo courtesy of YSI, Inc.) (*c*) ISO 7027 probe with wiper after 38 days in coastal waters. (Photo courtesy of YSI, Inc.) and (*d*) Copper shutter. (Photo courtesy of WEB Labs, Inc.)

measurements, uncertainty is proportional to $1/\sqrt{n}$, where n is the number of particles sampled. The number of particles measured by a turbidimeter is proportional to the sample volume. FO-based meters illuminate a small sample volume compared to other types and therefore larger measurement uncertainties, or noise, can be expected. Longer measurement times will be needed to average out this noise. Another drawback of FOs is their susceptibility to dirt and scratches on FO tips, which can affect light transmission and cause apparent shifts in turbidity. Corrective actions include cleaning the fiber tips and recalibration.

In summary, there are several advantages to the ISO 7027 standard. These include: (1) NIR versions are less prone to negative bias resulting from water color than EPA 180.1; (2) compact, low-power, submersible probes meet the standard; (3) ambient-light rejection is possible; and (4) fiber-optic versions can be fitted with wiper mechanisms. Table 24.2 lists turbidimeters meeting the ISO 7027 standard.

Table 24.2   Turbidimeters Meeting ISO 7027 Standard

Manufacturer	Model	Type	Range	Cleaning	Logs Data
Eureka Environmental Eng.	Manta	SD	1000 NTU	Wiper	✓
HANNA Instruments	HI93703	P	1000 NTU	○	✓
HF Scientific Inc.	Micro TPI	PR	1100 NTU	○	○
HydroLab (Hach) Inc.	DataSonde4	SD	3000 NTU	Wiper	✓
Innovative Sensors Inc.	41	S	9999 FTU	Wiper	○
McVan Instruments Pty Ltd.	Analite 160-3	SD	3000 NTU	○	✓
McVan Instruments Pty Ltd.	Analite 390	SD	1000 NTU	○	○
McVan Instruments Pty Ltd.	Analite 391	SD	1000 NTU	○	○
McVan Instruments Pty Ltd.	Analite 395	SD	1000 NTU	Wiper	○
McVan Instruments Pty Ltd.	Analite 396	SD	1000 NTU	Wiper	○
McVan Instruments Pty Ltd.	Analite 495	SD	1000 NTU	Wiper	✓
McVan Instruments Pty Ltd.	Analite 9000	S	1000 NTU	○	○
McVan Instruments Pty Ltd.	Analite 9500	S	1000 NTU	Wiper	○
Tyco Greenspan	TS100	S	1000 NTU	○	○
Tyco Greenspan	TS300	SD	1000 NTU	○	✓
WTW Measurement Sys. Inc.	Turb 350 IR	P	1100 NTU	○	○
YSI Inc.	YSI 6136	SD	1000 NTU	Wiper	✓

Note: S = submersible; SD = submersible with RS-232, RS-485, or SDI-12 output; P = portable, PR = portable/ratiometric; Measurements above 40 NTU will not meet EPA requirements.

24.8 RATIOMETRIC DESIGNS

Turbidimeters that incorporate two or more detectors and process their signals in weighted ratios have a ratiometric feature (indicated by R in Tables 24.1 and 24.3). The basic arrangement has a detector at $\alpha = 90°$ to measure scattered radiation and a second one at $\alpha = 0°$ to detect transmitted radiation (Fig. 24.3). The prime advantage of ratiometric processing is that it partially compensates for light absorption by water and particle color and multiple scattering effects, thereby extending the linear range and alleviating the need for sample dilution. Some portable turbidimeters have this feature, but users should confirm that they comply with applicable regulations before using them. Most ratiometric turbidimeters also have detectors that satisfy either EPA 180.1 or ISO 7027, so they can measure both low- and high-level turbidities.

24.9 ATTENUATION METERS AND SPECTROPHOTOMETERS

These meters have collimated TFL, light-emitting diodes (LED), or IRED sources that illuminate a sample and detect the light transmitted through it at $\alpha = 0°$. The measured quantity is either transmittance T, expressed as the ratio of received power, to incident power, or absorbance A, which equals $1 - T$ and are exponential functions of turbidity. When particles are present in the water sample, the measurement includes the combined effects of light attenuation caused by absorption and scattering and small-angle, forward scattering. To avoid the positive error resulting from including forward-scattered light in its measurements, an attenuation meter must have a narrow (<1 mm), well-collimated beam. Only research-grade instruments, called transmissometers, meet this criterion. However, some portable spectrophotometers have software allowing them to be calibrated with formazin and used for turbidity measurements. The unit of measure of such spectrophotometers is the formazin absorption unit (FAU), named to distinguish it from EPA 180.1, ISO 7027, and GLI-2. One FAU nominally equals one NTU; however, a water sample that produces a 100-NTU reading from an EPA 180.1 or GLI-2 turbidimeter may not read 100 FAU on a spectrophotomer. Such instruments are suited for measuring grab samples with turbidities > 5 FAU.

24.10 CLEANING SYSTEMS: WIPERS AND SHUTTERS

Several turbidimeters have mechanisms for clearing sediment, bubbles, and chemical and biological foulants off optical surfaces. The effectiveness of such mechanisms is illustrated in Figure 24.4, which shows a sensor after being deployed for 38 days in biologically active coastal waters. During this time, the unwiped areas were heavily fouled with barnacles while the optics under the wiper continued to function properly. Wiper cleaning parts consist of brushes, sponges, and plastic/neoprene wiper blades. Shutters made of biocides like copper keep biological

Table 24.3 Turbidimeters That Do Not Conform to a Published Standards

Manufacturer	Model	Method	Type	Range	Cleaning	Logs Data
Aanderaa	3200	NIR BS	S	500 NTU	◯	◯
Alec Electronics Inc.	ATU5-8M	NIR BS	SD	2000 JIS	Wiper	✓
Alec Electronics Inc.	ACLW-CMP	NIR BS	SD	2000 JIS	Wiper	✓
Alec Electronics Inc.	ATU40-D	NIR BS	SD	2000 JIS	◯	✓
AMEL Instrument s.r.l.	345	NIR Scattering	SD	200 NTU	◯	◯
Analyticon Instruments Corp	WQC-24	NIR Scattering	SD	800 NTU	◯	✓
Chelsea Instruments Ltd.	Aquatracka III	540-nm Scattering	SD	200 FTU	◯	✓
D & A Instrument Company	OBS-3	NIR BS	S	4,000 NTU	◯	◯
D & A Instrument Company	OBS-3A	NIR BS	SD	4,000 NTU	◯	✓
D & A Instrument Company	OBS-5	NIR BS	SD	2000 NTU	◯	✓
ES & S Pty. Inc.	2600	NIR BS	S	2000 NTU	S Shutter	◯
FTS Inc.	DTS-12	NIR Scattering	SD	1500 NTU	Wiper	✓
Global Water Inc.	WQ710	NIR Scattering	S	1000 NTU	◯	✓
Global Water Inc.	WQ770	NIR Scattering	SD	1000 NTU	◯	✓
HANNA Instruments	C114	527-nm Scattering/R	P	50 NTU	◯	◯
Horiba Instruments	U-10	NIR Scattering/R	SD	800 NTU	◯	◯
Horiba Instruments	U-20XD	NIR Scattering/R	SD	800 NTU	◯	◯
In-Situ, Inc.	TROLL 9000	NIR Scattering	SD	2000 NTU	◯	✓
McVan Instruments Pty Ltd.	Analite 160-1	NIR BS	SD	30,000 NTU	◯	✓
McVan Instruments Pty Ltd.	Analite 180	NIR BS	S	30,000 NTU	◯	✓
McVan Instruments Pty Ltd.	Analite 185	NIR BS	S	30,000 NTU	◯	✓
Partech Instruments Ltd.	Turbi-Tech 2000LS	NIR Scattering	SD	1000 NTU	Wiper	◯
Seapoint Inc.	Seapoint	NIR Scattering	S	2500 NTU	◯	◯
WET Labs Inc.	FLNTU	NIR BS	SD	1000 NTU	Shutter	✓

Note: NIR = near IR; BS = backscatter; P = portable; S = submersible (SD with RS-232, RS-485, or SDI-12 output); R = ratiometric.

growth off the optics between measurement times. These instruments are the best choice when: (1) EPA 180.1 or GLI-2 methods are not required; (2) long deployments must be made in biologically active water; and (3) frequent service is not practical. Before investing in a turbidimeter with moving parts, investigate how reliable the mechanism is and how it will work in an environment similar to the one you will be monitoring [e.g., freshwater/saltwater, sediment/forest debris, or biological activity (barnacle/algae infestation)]. Some turbidimeter probes having wipers or shutters are indicated in Tables 24.2 and 24.3.

24.11   BACKSCATTER SENSORS

Backscatter sensors measure radiation scattered by more than 90°. Their response to turbidity is nearly linear up to 4000 NTU. Figure 24.5 shows an exploded view of an OBS sensor that measures scattered light in the range from $\alpha = 140-165°$. It consists of an IRED surrounded by four photodiodes, a light baffle, and a temperature sensor for temperature compensation enclosed in a NIR-absorbing cavity, the "Black Box".[17] Currently, no design standard exists for backscatter sensors, although ASTM is in the process of developing one.

When particle colors change during monitoring, turbidity measurements with backscatter sensors will also change independently of other factors such as SPM size, particle shape, and concentration. This occurs because color can indicate darkness level, which makes particles more or less reflective in the NIR band and more or less efficient at scattering NIR. Sutherland et al.[18] determined that the sensitivity of a backscatter sensor can change by a factor of 10 from color

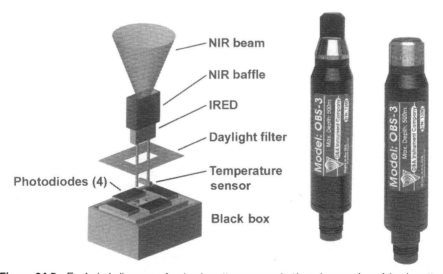

Figure 24.5 Exploded diagram of a backscatter sensor (*left*) and examples of backscatter turbidity meters (*right*).

effects alone. For example, a suspension of white particles will produce a turbidity reading about 10 times higher than suspension of black particles with the same size distribution and concentration. In rare situations, such as in runoff from acid mine tailings, dissolved, inorganic, NIR-absorbing matter can reduced turbidity by as much as 20%. Nearly everywhere else in the environment, however, such effects are immeasurable.[16]

Some of the advantages of backscatter sensors include: (1) wide linear range for turbidity and SPM, (2) low power, (3) compact, and (4) well suited for unattended, subsurface, continuous turbidity monitoring. Disadvantages include: (1) susceptible to effects of particle darkness value (indicated by color); (2) not the best choice for low-level measurements (< 5 NTU); (3) susceptible to NIR absorption by dissolved, colored matter when present in high concentrations; and (4) can "see" long distances in clear water so sensors are susceptible to debris and require special mounting structures.

24.12 FORMAZIN—THE PRIMARY STANDARD

Standard Methods for the Examination of Water and Wastewater[6] defines a primary standard for turbidity to be one prepared by the user from traceable raw materials, using precise methodologies, under controlled environmental conditions. Formazin prepared by the user from scratch ("scratch formazin") is the only material that meets this definition. Turbidimeter range and accuracy specifications, calibration, and alternative standards must be traceable to scratch formazin. To make formazin: (1) dissolve 10.0 g hydrazine sulfate in 1.0 L of water and 100 g of hexamethylenetetramine in 1.0 L of water; (2) combine equal volumes of these solutions; and (3) let combined solution stand at 25°C for 24 h. The resulting milky liquid has a turbidity of 4000 NTU by definition. Obtaining 2% precision requires conformity with the specifications of the Committee on Analytical Reagents of the American Chemical Society, scrupulously clean glassware, and strict adherence to laboratory procedures. For this reason, most analysts purchase premixed formazin which is

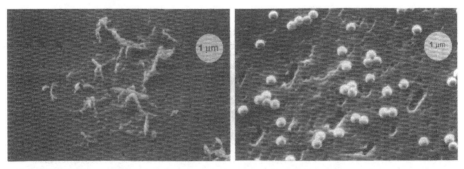

Figure 24.6 SEM images of formazin particles (*left*) and SDVB spheres (*right*). (Courtesy of GFS Chemicals, Inc.)

Table 24.4 Recommended Formazin Storage Times and Tolerances for Reporting Turbidity

Turbidity (NTU)	Maximum Storage Time	Tolerance (NTU)
1–10	1 hour	0.1
2–20	1 day	—
10–40	1 day	1
20–400	1 month	—
40–100	1 month	2
100–400	1 month	5
400–000	1 year	20
> 1000	1 year	50

approved for calibration but is not a primary standard. Strict adherence to the recipe does not guarantee optical consistency because the VSF of formazin will vary from batch to batch causing different readings in the same turbidimeters. Working standards are prepared by volumetric dilution of 4000-NTU stock formazin.[19] The procedures for diluting formazin are given in SM 2130 and ASTM D 1889.

Figure 24.6 shows a scanning electron microscope (SEM) image of formazin particles and illustrates how substantially they can vary in size and shape. The mean particle size is 2.48 μm, with a standard deviation of 0.57 μm.[20] The refractive index cannot be measured because the bulk material would be chemically and physically altered by the measurement of its index. Zaneveld et al.[21] determined $c(650 \text{ nm})$ to be $0.47(\text{NTU}) \text{ m}^{-1}$ and the VSF (45°) at 650 nm to be $0.045(\text{NTU}) \text{ m}^{-1} \text{ Sr}^{-1}$ for the range 0–40 NTU. The preparation, storage, and handling of formazin will affect its accuracy and stability. Recommended formazin storage times and turbidity reporting tolerances[50] are listed in Table 24.4.

Besides being the only primary standard, formazin has two other advantages. It can be made from scratch (hence the coined name, "scratch" formazin), and it is also the least-expensive, premixed, commercially available turbidity standard. There are several disadvantages that include: (1) formazin is toxic (see MSDS); (2) turbidity varies 2% from batch-to-batch; (3) its physical properties, size, shape, and aggregation, change with temperature, time, and dilution; (4) it settles in storage and must be mixed immediately prior to use; (5) dilute formazin standards have a storage life as short as one hour (see Table 24.4); and (6) field disposal of formazin may be prohibited by local water quality regulations.

24.13 ALTERNATIVE STANDARDS

An alternative standard is approved for turbidimeter calibration, but it must be traceable to formazin. AMCO Clear (formerly AEPA-1), supplied by GFS Chemicals, is a suspension of styrene divinylbenzene (SDVB) microspheres in ultrapure water containing a trace of the biocide sodium azide. It is one of two EPA-approved alternative standards. SDVB spheres have a mean size and standard deviation of 0.12 μm (1/20 the size of an average formazin particle) and 0.10 μm, respectively, and a refractive index of 1.56. As shown on an SEM image (Fig. 24.6), they are dimensionally uniform. The size distribution of SDVB spheres is adjusted to produce a formazin-equivalent response from a particular turbidimeter. SDVB standards formulated for one make and model of turbidimeter, therefore, cannot be used with one from a different manufacturer or model even though it conforms to the same standard method. It is therefore critical for users to purchase SVDB standards from a vendor who guarantees that they have carried out the formazin traceability of the standard using the same turbididimeter as the user will be calibrating with the SDVB standard.

The consistency of new SDVB standard batches is verified by three independent methods. First, absorbance is measured with a spectrophometer calibrated with National Institute of Standards and Technology (NIST) standard reference materials SRM 2034 and 930e.[22] Absorbance of a scattering media provides an estimate of the combined effects of $c(\lambda)$ and forward scattering. Second, the size distribution and shape are measured by SEM image analysis. Third, new material is compared to retained batches to verify time stability of size and absorbance. These quality control measures result in a physically well-characterized light-scattering material. Better precision of both replicate measurements and measurements made with similar turbidimeters can be obtained with SDVB than with formazin.[20] Their tests showed lower standard errors in replicates, less than 0.1% for SDVB compared to 2.1% for formazin, and interinstrument variability of 1.6% for SDVB compared to 1.9% for formazin. The linearity of SDVB standards determined from deviations of measurements from a best-fit, straight line is also better than formazin, 0.15 NTU compared to 0.32 NTU (lower deviations indicate smaller, random, measurement errors and a more linear standard). These results reflect the superior overall physical consistency of SDVB.

The key benefits of SDVB standards are: (1) lot-to-lot variation in turbidity < 1%; (2) optical properties are constant from 10 to 30°C; (3) one-year stability is guaranteed; (4) mixing and dilution are not required; (5) they are not toxic; and (6) an EPA 180.1 SDVB standard can be used to calibrate any ISO 7027 or EPA 180.1 below 40 NTU. Two drawbacks are that SDVB standards can only be used with the turbidimeter model for which they are made to measure turbidities greater than 40 NTU and they are expensive. For example, 1 L of 4000 NTU standard costs about two times more than an equivalent amount of 4000 NTU formazin. Submersible systems can require large volumes of standard, special calibration containers, and the standard may have to be discarded after each calibration.

StablCal, a stable formazin standard from Hach Company, is another EPA-approved, alternative standard. StablCal has the same physical properties as formazin, but Hach claims it maintains its original NTU value within 5% for more than 2 years. Like SDVB, StablCal cannot be made by the user and must be purchased premixed. With the exception of stability it has all the disadvantages of formazin.

24.14   RELATIONSHIPS BETWEEN TURBIDITY AND SEDIMENT

There is a lot of emphasis in this chapter on SPM and sediment because they are the primary cause for surface water turbidity, and excessive sediment is the leading cause of water quality violations in the United States.[23] Moreover, most water-borne metal contaminants, cause number 6, and insecticides, cause number 7, are sediment bound.[24–26] The use of turbidimeters to monitor sediment and sediment-bound contaminants is therefore a potentially powerful tool for environmental applications despite the mismatched physical properties of turbidity standards and sediment. It is a useful tool because turbidimeters respond in proportion to the intensity of scattered light. More important for environmental monitoring is that continuous SSC records can be calculated from locally specific SSC/turbidity relationships and records with submersible turbidimeters.[27] The key advantages of using turbidimeters to monitor suspended sediments are: (1) continuous records can be obtained with submersible meters, (2) detailed, high-frequency measurements of events (storms, dredging, construction, etc.) are feasible, and (3) lower labor costs for sampling and laboratory analyses.

It is often assumed that turbidity standards and SPM scatter and absorb light the same way in turbidimeters. They can only do this when the light-scattering characteristics of the sediment, resulting from size, color, refractive index, and shape, are similar to the standard. This is infrequently the case. Of these characteristics, particle size is the most important factor because it varies over about three orders of magnitude in the environment. Sediment color can also be important because it

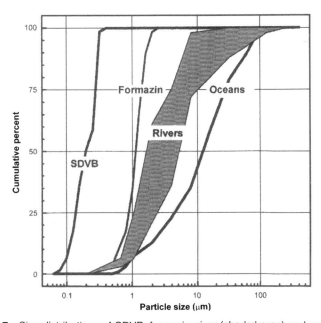

Figure 24.7   Size distributions of SDVB, formazin, river (shaded area) and oceanic SPM.

can indicate reflectivity for the operating wavelength of a turbidimeter. Therefore formazin and SDVB, being translucent materials, can only mimic light-colored sediments. Mineral refractive indices, on the other hand, fall in a narrow range from 1.5 to 1.9, and they have a smaller influence on light scattering than size and color. The refractive index of SDVB, however, falls within the range.

Figure 24.7 shows size distributions for river[28] and oceanic[29] SPM and SDVB and formazin turbidity standards.[20] The median sizes, D_{50}, for the materials are 0.2 μm (SDVB), 1.5 μm (formazin), 1.6–5 μm (river SPM) and 12 μm (oceanic SPM). The formazin size distribution overlaps river SPM in the clay/colloidal size range (< 2 μm), which comprises 40% of the sediment by mass, on average. The size distributions of formazin and oceanic SPM overlap only over the lowest 14% of the SPM size distribution curve. SDVB spheres are finer than about 95% of river and oceanic SPM and are not a good size proxy for SPM in surface water. Of these two turbidity standards, formazin better represents SPM on the basis of size; however, it is not a good size proxy for SPM larger than 2 μm and such material is prevalent in surface waters.

The effects of SPM shape on turbidimeter sensitivity have not been quantified; however, shape and particle flocculation also influence how SPM scatters light. Matches among physical properties of SPM and turbidity standards in the environment are rare. Figure 24.8 shows SEM images of SPM from the Colorado and Columbia Rivers.[30] Material from the Columbia River is a mix of plankton shells and clay-mineral flocs in roughly equal proportions, and the Colorado River material is nearly all clay flocs. The SEM images shown in Figures 24.6 and 24.8 illustrate the diversity of shapes and degrees of flocculation that can exist in turbidity standards, and SPM. Their effects on light scattering and absorption in a water sample can degrade the accuracy of a turbidimeter.

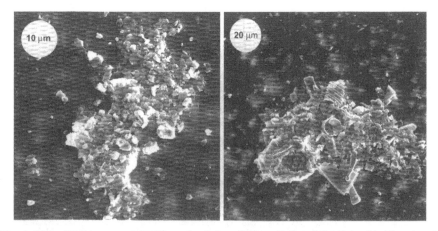

Figure 24.8   SEM images of SPM from the Colorado River (*left*) and the Columbia River (*right*).

24.15   SECONDARY (CHECK) STANDARDS

Secondary, "check," turbidity standards are used to check instrument calibration; they cannot be used for calibration. The NTU-equivalent value of a secondary standard is defined by measuring it with a turbidimeter that has been calibrated with formazin, StablCal, or SVDB standards. They are handy for checking turbidimeter calibrations and very convenient for field use where handling primary standards is impractical. Secondary standards, "check standards," include devices made from mirrors, glass rods, frosted cubes, and plastic cups that produce known NTU values from turbidimeters for which they were made. They also include standards such as GELEX (Hach Company), which consist of metal oxide particles immobilized in a clear silicone material and Quick Cubes for use with the Hydrolab/Hach GLI-2 meter. Secondary standards are instrument specific, so a standard made for one model cannot be used in a different model. Moreover, every time a turbidimeter is recalibrated with a primary standard, secondary standards must be remeasured and assigned new NTU values.

24.16   GENERAL GUIDANCE

Turbidity monitoring is conducted to support a description of water quality, to measure changes in water quality, and to understand how a water systems functions. Whichever core reasons apply in a monitoring program, its objectives should be specific, measurable, and focused on water quality problems and the processes that affect water clarity. It is important to select a turbidimeter that fulfills the applicable standard method while meeting these objectives. Cost is always a consideration in the planning phase, but compromises in performance to cut costs will frequently obligate a program to added maintenance, calibration, and service costs later in the program. Also, lost data production resulting from equipment malfunctions can cost more than equipment, and suppliers are rarely liable for the cost of repeating measurements. Finally, ask associates experienced with the chosen method about their preferences for turbidimeter models and recommendations for ancillary equipment.

General questions to answer in the planning phase of a turbidity-monitoring program include:

1. Will the selected instrument and method satisfy the applicable regulations? Instrumentation must be suited for the intended operation, submersible (i.e., ISO 7027, GLI-2, etc.), regulatory compliance (EPA 180.1), discrete samples, etc.
2. Does the instrument have all essential capabilities, especially the range and accuracy needed for the task? Measurement range must include turbidity levels excepted under all conditions.
3. Are operational capabilities of the chosen equipment consistent with the planned use, project budget, and personnel?

4. Can the data be readily converted to information upon which management can take action?

5. Will the QA standards for duplicate samples, replicate analyses, and calibration be met?

Reporting excess data can result in nonessential questions from regulatory officials. Report only required data, retaining additional data for internal purposes, QA, and QC.

24.17 PORTABLE TURBIDIMETERS

Portable meters require grab samples to be measured in the field. This means that an operator must be on-site at regular intervals for sampling and continuously for event-based sampling. Some portable turbidimeters can record measurements for later retrieval with a personal computer while the data from most must be logged in a field notebook or on a printed form. Some portable meters have plumbing for continuously streaming samples through the meter to make groundwater measurements. Ancillary data and information will aid the exploration of the causes for turbidity fluctuations. These frequently include: (1) water temperature, level, and flow or discharge; (2) local weather conditions; (3) land-use activities such as construction, logging, and dredging; and (4) natural processes such as land sliding, erosion, and deposition.

Activities to complete before going to the field are:

1. Prepare standard operating procedures (SOPs) for field activities and test them under controlled conditions.

2. Check the field equipment inventory and replenish consumables and spares as necessary.

3. Verify that calibrations are current and check meters with secondary standards before and after field trips.

4. Service equipment and test turbidimeter batteries.

5. Clean sample vials with warm tap water and laboratory glassware detergent. Triple rinse with filtered, deionized water.[13] Package them for travel in clean aluminum foil or polyfilm.

6. Use clean, scratch-free vials. Discard damaged ones.

7. Turbidity readings depend on sample vial orientation in the meter. Mark the vial position that produces the minimum reading for clean water and use the same orientation for sample measurements.

8. Determine stray-light correction by manufacturer's recommendations or measure a sample of turbidity-free water in a clean, new sample vial (item 5). This blank turbidity should be less than the desired precision level.

9. Document calibration, maintenance, checks with secondary standards, and other critical preparations.

In the field analyze samples as soon after they are collected as possible. Turbidity begins to change immediately after a sample is removed from the source water because of: (1) particle settling, (2) dissolution and precipitation, (3) flocculation, and (4) biological production and consumption of particles. When samples cannot be analyzed immediately, store them in clean, capped, amber bottles, out of direct light; refrigerate them to 1–4°C; and analyze them within 24 h.

The general field guidelines for portable meters include:

1. Avoid water transfers by using sample vials to get grab samples.
2. Cap sample vials to prevent spillage and evaporation when immediate measurements are not possible.
3. Let aerated samples stand for 10 min and tap them on a hard surface to knock bubbles off vial walls.
4. Do not shake sample vials. Aeration raises turbidity. Gently tumble sample vial end-over-end to keep sediment uniformly suspended.
5. Coat sample vials with silicone oil in accordance with manufacturer's instructions.
6. Measure samples < 5 s after tumbling.
7. Insert vials in the same orientation every time and close the sample lid to keep light out of the measurement compartment.
8. Note evidence of evaporation (increases turbidity), material settling and dissolution (reduces turbidity), and biological activity (can increase or reduce turbidity) for samples that are not analyzed immediately after collection.
9. Document or photograph unusual conditions that influence sample integrity, measurement quality, and the physical conditions affecting turbidity.

Any portable meter could have one or more of the following problems: (1) erratic readings due to settling or bubbles in a sample; (2) dirty, scratched, or improperly oriented vials; (3) condensation; (4) dead batteries or lamp failure; (5) low readings due to particle or dissolved color; and (6) calibration drift. Follow the manufacturers troubleshooting procedures for the identification of faults and remedies specific to an individual meter and check the calibration (Section 24.15) every time you use it.

24.18 SUBMERSIBLE TURBIDIMETERS

Submersible probes are used for spot measurements at accessible locations and to make continuous turbidity records at underwater locations for extended time periods. In many situations, field operators will have to modify commercial equipment and adopt field procedures to meet unique measurement challenges. Selection of the best turbidimeter for a project will require the evaluation of many factors, including: (1) the objectives, for example, compliance with water-quality regulations, establishment of ambient turbidity levels, monitoring before and after projects, confirmation of compliance, or evaluation of the effectiveness of management practices; (2) range and accuracy requirements, for example, compliance and background measurements might require high accuracy while a wide range would facilitate monitoring trends. (3) space–time sampling requirements; (4) equipment and manpower costs; (5) safety, training, supervision; and (6) data reporting. Similarly, before-after, control-impact (BACI) studies will usually require higher accuracy than event-based sampling. Automatic water sampling equipment is frequently used in conjunction with submersible meters to develop relationships between sediment concentration and turbidity, and in streams, turbidity and sediment load.

24.18.1   Before Going to the Field

Unique field problems require practice runs in the field. Basic preparations before heading to the field include:

1. Prepare field SOPs and test them in the laboratory to ensure that they work under controlled conditions.
2. Maintain a calibration log recording the time, adjustments, and relevant notes about turbidimeter performance, drift, noise, and the like.
3. Post lists of field items such as turbidimeter, secondary standards, sample vials, spares, and other accessories and consumables in the field cases.
4. Recalibrate the turbidimeter if the calibration has lapsed and record the action in a calibration log. Quarterly calibrations are recommended.
5. Restock depleted equipment and supplies (list from item 4) to ensure that you have everything for the operation.
6. Check cleanliness and working condition of all equipment and perform necessary maintenance.

24.18.2   Sampling and Data Considerations

The turbidity changes continuously in natural systems. To get an idea about how much it can change at a location, measure 5–10 grab samples taken over a period of 15 min and compute the standard error for the values (see Section 24.22). This is a measure of the natural variability in the waters you are monitoring at the time of the observation. For stream monitoring, it is useful to measure turbidity at several locations and depths to assess its variability across a channel. A sampling plan should include enough replicate measurements and time averaging to produce a statistically representative turbidity measurement for the sampling time and location. Take replicates under a variety of conditions, for example, low and high stream flows, upstream and downstream from ground disturbances, before and after storms to get an idea how many replicates and what measurement frequency may be required to get representative samples over the study region and program duration.

One serious consequence of not doing this includes the possibility of making a type I error, which is to falsely detect an environmental impact or disturbance (a false alarm), causing an operation to be shut down. Another possible consequence is a type II error, which is not detecting an impact or disturbance that has occurred, resulting in a false sense of security. This type of error might allow an activity to continue that should be stopped.

Maintenance of submersible turbidimeters is essential for their successful field use. Abrupt changes or gradual drift in readings that appear unrelated to environmental conditions usually indicate that service is required. Sensor optics get dirty, scratched, and sometimes buried in sediment. Their position can shift due to debris and ice accumulation, causing them to "see" a nearby object or structure and report erroneously high or low turbidity. Batteries expire, data memory can fill up, and automated cleaning mechanisms can get stuck producing data loss and gaps in a record. All of these conditions warrant attention in planning maintenance and service activities.

Besides spot measurements with submersible turbidimeters, other data handling techniques may be appropriate. An internally recording meter is useful for applications when there is no host system to log data or when telemetry is impractical because the study area is not served by satellites, a cell net, or line-of-sight radio frequency (RF) links. Turbidimeters having this feature are listed in Tables 24.2 and 24.3. Digital communication of turbidity data over a wire is useful when a host logger services other equipment and logs data in a central storage medium. RS-232, SDI-12 (serial data interface, 1200 baud), and RS-485 are the most common serial communication protocols. The RS-232 protocol is for short-haul transmissions (< 15 m), and the latter two share the attractive feature of sensor addressing, meaning that a host controller can get turbidity and ancillary data from one sensor at a time on a programmed schedule.

Wireless transmission of turbidity data (telemetry) is by far the most versatile and useful means to get them where they are needed for resource management and operational requirements. Telemetry is also the most expensive means for handling data. Wireless telemetry enables: (1) timely intervention in response to system changes, (2) feedback on instrument performance and maintenance requirements, and (3) better information from BACI studies. The main types of telemetry include: (1) ultrahigh frequency (UHF) and very high frequency (VHF), spread-spectrum, radio-frequency links; (2) cell and landline telephony; and (3) GOES, ARGOS, and Inmarsat satellite communication. Selection of a particular telemetry system will depend on site terrain, vegetation (forested versus barren ground), and distance between the field site and data host. Whenever possible, get recommendations from someone experienced with telemetry applications at a site like yours.

24.19   ESTIMATING SUSPENDED SEDIMENT FROM TURBIDITY

Suspended sediment concentration is determined by weighing the dry material separated from a water sample with a 1.5-μm filter and calculating suspended-sediment concentration (SSC) or total suspended solids (TSS). The preferred unit of measure is mg/L.[31,32] Assuming water and sediment densities are 1.00 and 2.65 g/mL, ppm can be calculated from mg/L by the multiplicative factor $[1.0-(mg/L)(622 \times 10^{-9})]^{-1}$. The SSC and TSS methods should give the same result for a water sample, but they usually do not.[33] The reason is the SSC method uses the entire sample, whereas the TSS analysis uses only part of it and material can be lost to settling during processing. SSC values can exceed TSS values by 30%,[34] so select the SSC method whenever possible.

The following section gives some examples of predictive relationships between turbidity and SSC and potential difficulties in data interpretation. The first example is turbidity and coincidentally measured sediment concentration in runoff from a Massachusetts freeway (Fig. 24.9).[35] These data were acquired with a variety of turbidimeters, and the size of the sediment in the runoff varied from place to place. The scatter plot shows that although turbidity is not an accurate SSC predictor, it is useful for investigating the relative magnitude and timing of sediment production by runoff. The quality of the relationship between SPM and turbidity is reflected in the coefficient, r^2, which is 0.7 (see Section 24.22). This coefficient indicates that about 70% of the change in SSC can be attributed to changing turbidity. An r^2 value equal to 1 would mean that turbidity is a perfect predictor of SSC.

Other sites have tighter correlations between SSC and turbidity, as is shown by the data from a site in San Francisco Bay (Fig. 24.9). Turbidity in this example was measured with an OBS sensor, and water samples were drawn from time to time at a nearby location and analyzed for SSC.[36] The r^2 for these data is 0.97, indicating that turbidity is an accurate predictor of SSC at the site.

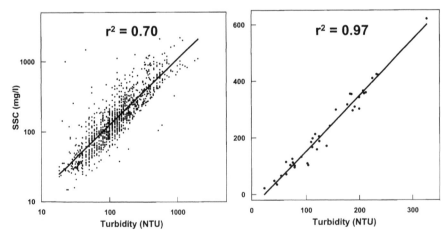

Figure 24.9 Correlation between turbidity and SSC in highway runoff (*left*) and in San Francisco Bay (*right*).

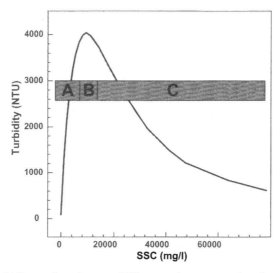

Figure 24.10 Turbidity readings from an OBS sensor for a range of sediment concentrations.

A final example illustrates that caution must be exercised when using turbidity as an SSC surrogate. Figure 24.10 shows the response of an OBS sensor to a range of SSC values. In region A, the relationship between NTU and mg/L is nearly linear because the area of the light-scattering particles increases in direct proportion to SSC. Here, the OBS gives excellent estimates of SSC. At about 6000 mg/L, multiple scattering and light absorption cause the indicated turbidity to peak sharply and then to decline with increasing SSC, indicated by region B where the dominant effect changes from scattering to attenuation. In region C, the OBS sensor behaves like an attenuation meter and received light diminishes exponentially with increasing SSC. In the example, an OBS reading of 2000 NTU could either mean SSC is equal to 2000 or 32,000 mg/L. Similar effects can influence ISO 7027 and GLI-2 meters leading to errors in data interpretation when high SSC occurs and an analyst is unaware of the response characteristics of his turbidimeter. So be sure to characterize the SPM-NTU relationship over the full range of expected SSC values.

It is essential that any program intending to use turbidity to estimate SSC rigorously establish a locally specific working relationship between turbidity and SSC by direct sampling, as shown in Figure 24.9. This process may require several dozen to a few hundred SSC samples. Correlations will be valid for specific sites and turbidimeters, and they may depend on conditions such as water flow, depth, time of year, and land-use activities as well. It is essential that the same instrument be used at a site for the entire monitoring period. Switching turbidimeters, even the same model, could require a new sediment calibration.

Coordinate the location and frequency of surface water sampling and related measurements with the water level, flow, or discharge (the hydrograph). Sampling frequency should generally be higher by a factor of 2–5 during the growth and decline of events than between events when conditions are changing slowly. The primary reason for this is that most of the transport of sediments and sediment-bound contaminants occurs during events when water flow and SPM concentrations are high. While point-for-point correlations are important in lakes, reservoirs, and the ocean, the goal in stream monitoring is usually to develop a relationship between turbidity and the sediment load.

24.20 WATER SAMPLING

Three types of samplers are used to collect water for SSC and turbidity measurements.[37] They include: (1) grab samplers, containers for collecting instantaneous water samples; (2) area- and time-integrating samplers, consisting of a container, tube, and intake structure for withdrawing water over a time interval at a fixed location or over a range of locations in a cross section; and (3) systems with a pump above water to draw water into a sample container. Manual grab samples are impractical in water deeper than about one arm length. Integrating samplers can withdraw water from a range of depths or a fixed depth to characterize suspended material and turbidity in time or in space.[38] They should be isokinetic so as to withdraw water through an orifice at the same velocity as the water surrounding the sampler. Otherwise, inertial effects will cause more or less SPM to be collected than is in the source water, causing erroneous turbidity and SSC values. Pump samplers operate automatically on a programmed schedule keyed to stage, time, or turbidity.[39] They are very useful for establishing a relationship between turbidity and SSC. A number of handbooks and standard methods provide detailed descriptions of how to select and use water samplers; see: Keith,[40] ASTM D 4411-1998,[41] USDA,[42] Water Survey of Canada,[43] USEPA,[44] ISO 4365,[45] and ANZECC.[45] The recommended one is the National Field Manual for the collection of Water-Quality Data.[37]

Some of the recommendations for handling turbidity standards also apply to suspended sediment samples. General recommendations for sampling in conjunction with turbidity monitors are:

1. Maintain a field log of times, dates, locations, water stage, and other environmental conditions relevant to samples.

2. Take sediment samples close to turbidity-monitoring instruments because turbidity and SSC can vary widely between closely spaced stations and over short periods. In streams, it will be necessary to sample at several locations and depths to average out cross-channel variations in turbidity and SSC.

3. Take paired sediment and turbidity water samples using identical methods, samplers, containers, and other materials.

4. Avoid sample transfers because sediment can settle and be lost. When transfers are necessary, mix samples immediately before they are poured from one container to another.

5. Select the standard or methods prescribed by local regulations.

Table 24.5 QA/QC Activities

Quality Control	Quality Assurance
1. Duplicate samples and replicate analysis	1. Sample tracking and chain of custody
2. Stray-light and turbidity-free water analyses	2. Sample size and frequency consistent with desired precision
3. Turbidimeter calibration and maintenance	3. Clear personnel assignments and training
4. Statistical measures of precision	4. Field-operations and data audits

24.21 QUALITY CONTROL AND ASSURANCE

Good quality control (QC) is essential in turbidity monitoring programs to maximize the integrity and value of data for its intended purpose. Quality assurance (QA) includes the steps taken to confirm that QC is successful.[46] Effective QA incorporates performance measures into day-to-day procedures to improve data accuracy, precision, and compatibility with external data. In this way, potential errors can be anticipated and avoided. The design of a specific QC/QA plan must satisfy applicable regulations and standards. Some key elements of QC/QA for turbidity monitoring programs are given in Table 24.5.

To the greatest extent possible, acquire and record turbidity in the field so that sample storage and transportation are not required. Usually a water sample taken and measured in the field can be returned to the water from which it was taken, whether contaminated or not. Once samples arrive at a laboratory, they will have to be disposed of in an approved manner. Sometimes the chain of custody (COC), establishing the movement of water samples from their acquisition until they are analyzed and disposed of must be documented. The COC document is an essential element of QA and the eventual use of data to meet the objectives of the monitoring program. Additional costs must be budgeted to create COC documentation and for QA testing by an outside laboratory.

24.22 STATISTICS

Statistics are an integral part of turbidity analyses and reporting.[47] At a minimum, statistics should: (1) describe turbidity trends over time and across a study area, (2) measure variation attributable to measurement errors and natural turbidity fluctuations, and (3) provide confidence limits for internal and external data comparisons. The mean and median provide estimates of the expected, or most probable value from a large number of measurements made under similar conditions. The mean turbidity, $\overline{NTU}$, equals the sum of measured values divided by the number of measurements. The median turbidity value, NTU_{50}, exceeds 50% of the values in a sample. It is a more reliable measure of the expected value for a sample of measurements than the mean because it is less influenced by outliers, which are prevalent in records from submersible turbidimeters. Measures of variability or spread indicate how much values can be expected to change over time, from one location to another, and because of measurement imprecision. They provide insights about processes, data integrity, and quality. Measures of spread include the sample standard deviation σ_{NTU} and quartiles such as NTU_{25} and NTU_{75}. Quartile measures are similar to the median. They are the values that exceed 25 and 75% of values in the sample, respectively. A third useful measure of spread is the standard error, which is the sample standard deviation divided by the sample mean expressed in percent, $100[\sigma_{NTU} / \overline{NTU}]$. It indicates how large the spread is relative to the sample mean. A similar measure for data containing outliers is the value, $100 (NTU_{75} - NTU_{25}) / NTU_{50}$. When a measure of spread changes suddenly or for no apparent physical reason, it can indicate an equipment malfunction, a service requirement, or a change in data quality that should be promptly investigated.

More complex statistics may be required to meet program objectives, including, for example, regression analyses to show relationships between turbidity and suspended sediment or analyses of variance (ANOVA) to test "cause-and-effect models." In the former method, the mathematical relationship between two measured values, say NTU and SSC (called a regression of SSC on NTU), is computed. The coefficient of variation, r^2, for the regression indicates the proportion of the variance of a dependent variable (SSC) that can be attributed to change in the independent variable, NTU. A perfect correlation between SSC and NTU would result in $r^2 = 1$ and turbidity would be a perfect predictor of suspended sediment concentration. An r^2 value less than about 0.6 usually indicates that the independent variable is a poor predictor of the dependent one.

Consult a statistician during program development to ensure that sampling, and data processing and analysis activities, will meet program objectives. The statistical analysis and graphing tools in spreadsheet programs such as EXCEL and LOTUS will meet all routine needs for handling turbidity data. More advanced analyses may require a statistical program such as SAS, MINITAB, STATISTCAL, SYSTAT, and S-PLUS.

24.23 U.S. REGULATIONS

Internationally, water quality regulations and requirements for monitoring turbidity vary greatly from one country to another. They will depend on what stage of economic development a country is in, its water uses, and proximity to other nations. It is not possible to cover these regulations comprehensively, but many countries have modeled their water quality regulations after those developed in the United States. Accordingly, the U.S. regulations are described in some detail to indicate what elements of a turbidity-monitoring program may need to comply with environment and water quality laws.

The cornerstone of surface water protection in the United States is the Clean Water Act. Section 402 of the CWA established the National Pollution Discharge

Elimination System (NPDES) to govern most environmental applications and was placed under the control of the EPA. NPDES initially focused on wastewater treatment facilities, industrial sources, and mining and petroleum-drilling operations, but since the late 1980s, NPDES has been expanded to include municipal stormwater runoff, feedlots and aquaculture facilities, agricultural lands, and construction operations larger than one acre. In general, if an activity produces turbid water that will end up in a river, lake, or the ocean, it will likely fall under the NPDES and require some sort of state or federal permit. In water bodies impaired by excessive nonpoint loads of suspended sediment, Section 303(d) of the CWA requires the development of total maximum daily loads (TMDL),[48] for sediment. Sediment TMDLs are being determined with the aid of submersible turbidimeters to improve sediment control and the efficiency of watershed management. CWA Sections 319, 401, and 404, respectively, relate to non–point sources, water quality in wetlands and tide lands, and licensing of facilities such as dams. Any of these CWA sections could affect your project. Groundwater is regulated by state authorities.

In addition to expanding the scope of NPDES, the EPA delegates increasing authority for permitting, compliance monitoring, and enforcement to individual states. As of 2003, Oklahoma, South Dakota, Texas, Utah, and Wisconsin were fully certified by the EPA for such activities. Idaho, Massachusetts, New Mexico, and New Hampshire, on the other hand, have no NPDES authority. The remaining 41 states fall somewhere in between. Because water quality regulations continue to evolve and state authorities are changing, it is wise to confirm with state and federal officials what regulations apply to your monitoring program. Delegation of authority has led to regulations tailored to the environmental needs, water uses, and surface geology of individual states. Pruitt[49] surveyed the usage of turbidity in water quality monitoring and observed turbidity ranges in the United States. Observed turbidities were found to vary widely from mountain streams with NTU values from 1 to 50 NTU, to moderately high levels (several hundred NTU) in muddy rivers such as the Mississippi, to extreme values exceeding several thousand NTU for barren watersheds subject to flash flooding. The quartiles (25, 50, and 75%) of the maximum observed turbidities are: 115, 550, and 957 NTU, respectively. Because acceptable turbidimeters vary from state to state, interstate comparisons of turbidity data can be inconclusive.

Diverse surface water conditions, land-use practices, and geology across the United States have resulted in wide range of state turbidity limits. Most of them can be placed in one of the following five categories:

1. One turbidity limit above background for all water types
2. Individual limits for specific water types or uses, and geographical regions
3. Discharge-specific limits, for example, runoff from highways, feedlots, mining operations
4. Percent increase above background specified by region, water type, or use
5. Thirty-day-average turbidity limits with instantaneous caps

The turbidity levels within these categories will vary from state to state. Consult the state agency with jurisdiction over your project to determine what the limits are and to ensure that a monitoring plan will document compliance with them.

24.24 SUMMARY

Turbidity is a measure of water cloudiness that is most frequently caused in surface waters by suspended sediment. Excess sediment is the leading cause of water quality violations in the United States, and sediment-bound heavy metals and insecticides are among the top 10 causes of violations.[23] Although potential linkages between turbidity, physicochemical, and environmental factors are alluring, turbidity measurements do not always provide meaningful indicators of water quality. This arises from factors under the analyst's control such as water-sampling and storage methods, turbidimeter choices, and calibration procedures as well as from others that are outside his control. Optical effects of suspended matter size, shape, and color and how a particular turbidimeter was assembled are in the latter group.

As a consequence of these factors, one can get similar turbidity readings on water samples even when physical characteristics of the suspended material are different. Moreover, turbidimeters made by different manufacturers and complying with a standard method will nearly always indicate different turbidity values on the same water sample. For these reasons it very important to use the same model of turbidimeter, measurement approach, and calibration standards for a complete study to ensure that turbidity data are internally consistent. Choosing an appropriate meter, standards, and ancillary equipment will enable meaningful turbidity monitoring. These choices are driven by the study goals and accuracy requirements. In the United States and several other countries that have modeled their water quality regulations after U.S. ones, allowable turbidity levels in waterways can be simply a percentage increase over a background level.[23] An appropriate turbidimeter must therefore have a range that includes the background turbidity as well as the expected excursion above it. In other environments turbidity can depend in a more complicated way on geographic location, water use, or time of year. A wider measurement range as well as several meters and continuous monitoring may be required. Selecting a turbidimeter is usually the most important first step as it will drive other important choices.

24.24.1 Choosing a Turbidimeter

A selection matrix summarizing applications and general recommendations is presented in Table 24.6 to provide a starting point for the selection process. The primary selection criteria in descending order of importance are: 1) satisfaction of regulations, 2) measure grab or in situ samples, 3) turbidity range, 4) accuracy, and 5) automatic cleaning. Once the type of instrument has been selected, refer to Tables 24.1, 24.2, or 24.3 to identify suppliers of the equipment you need.

Table 24.6 Turbidimeter Selection Guide

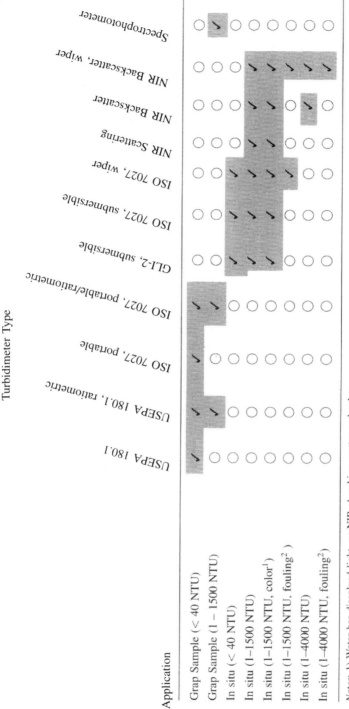

Application	USEPA 180.1	USEPA 180.1, ratiometric	ISO 7027, portable	ISO 7027, portable/rationmetric	GLI-2, submersible	ISO 7027, submersible	ISO 7027, wiper	NIR Scattering	NIR Backscatter	NIR Backscatter, wiper	Spectrophotometer
Grap Sample (< 40 NTU)	✓	✓	✓	✓	○	○	○	○	○	○	○
Grap Sample (1 – 1500 NTU)	○	✓	○	✓	○	○	○	○	○	○	✓
In situ (< 40 NTU)	○	○	○	○	○	○	○	○	○	○	○
In situ (1–1500 NTU)	○	○	○	○	✓	✓	✓	✓	✓	✓	○
In situ (1–1500 NTU, color[1])	○	○	○	○	✓	✓	✓	✓	✓	✓	○
In situ (1–1500 NTU, fouling[2])	○	○	○	○	○	○	✓	○	○	✓	○
In situ (1–4000 NTU)	○	○	○	○	○	○	○	○	✓	✓	○
In situ (1–4000 NTU, fouling[2])	○	○	○	○	○	○	○	○	○	✓	○

Notes: 1) Water has dissolved light- or NIR-absorbing matter and color.
2) Severe biological growth or chemical fouling.
3) Several submersible turbidity sensors perform well below 40-NTU; review manufacturers specifications.

Turbidimeters conforming to EPA 180.1 and similar methods (SM 2130 B and ASTM D 1889-00) are recommended only for grab samples with turbidity less than 40 NTU and for locations with water having no visible color. Portable ISO 7027 turbidimeters can also be used for grab samples and for spot checks of submersible instruments, but the data may not satisfy EPA requirements. Because they operate in the NIR band, ISO 7027 meters are less affected by dissolved colored substances than meters with TFL sources. Meters with ratiometric optics are less prone to color effects and should be used when permitted. GLI-2 is suitable for field measurements where turbidity is less than 1000 NTU and where fouling rates are low and regular cleaning is feasible. Submersible sensors of the ISO 7027 design are recommended for environments with low to moderate turbidities (1 to ~1500 NTU) and for installations with difficult access. Use probes with wipers and shutters when long deployments must be made in biologically active water where frequent service is not practical. Several nonstandard turbidimeters perform as well as or better than ISO 7027 designs; see Table 24.3. Some or them have cleaning mechanisms making them ideal for field use, and they are frequently permitted by water quality regulations so long as they are calibrated with approved turbidity standards. Finally, backscatter designs have the widest linear range for turbidity and SPM and are a good choice for unattended, subsurface, continuous turbidity monitoring where turbidities as high as 4000 NTU are expected. Care must be taken, however, when mounting backscatter sensors because they can "see" long distances and debris accumulation around sensors can produce erroneous high readings.

Once turbidimeter candidates have been chosen and before developing a final monitoring plan, consult state and/or federal agencies with jurisdiction over your project to determine if the chosen equipment, standards, and calibration protocols will comply with applicable regulations.

24.24.2 Selecting Standards

Four criteria drive the selection of calibration standards. In descending order of importance they are: (1) regulatory requirements, (2) accuracy and precision, (3) operational needs (e.g., convenience), and (4) cost. Use formazin when regulations, cost, or the need for large quantities are prerequisites. SDVB standards are good choices when high initial accuracy and precision, ease of use (no mixing required), and long storage times are prevailing factors, however, they are about four times more expensive than formazin. Other selection factors being equal, always select turbidimeters that come with secondary, "check", standards. These are handy for checking turbidimeter calibrations in the field and when primary standards are impractical.

24.24.3 Sediment–Turbidity Relationships

Turbidity is a useful indicator of suspended sediment in most waterways and can be used to quantify sediment concentrations continuously over time. The establishment of relationships between turbidity and SSC by direct sampling at individual sites is vital for success in sediment gauging applications. Sediment–turbidity relationships will be valid for specific sites, turbidimeters, and may depend on water flow, depth, time of year, and land-use activities. It is therefore important to use the same model turbidimeter and sediment-sampling protocol consistently to avoid sampling and calibration errors. Grab samplers obtain useful spot checks at monitoring locations; however, integrating samplers provide superior time and space averages of suspended sediment at fixed locations or over a range of locations. Pump samplers are best for making co-located time-integrating simultaneous SSC and NTU measurements. The advantages of simultaneous turbidity and sediment monitoring include: (1) continuous SSC and sediment transport records are possible, (2) labor costs will be lower than when just water sampling a laboratory analysis are undertaken, and (3) instruments can gather data when manual sampling is impossible.

ACKNOWLEDGMENTS

I thank Kemon Papacosta, Doug Glysson and Chauncey Anderson (USGS), and John McDonald (YSI, Inc.) for their insight and technical reviews.

REFERENCES

1. G. C. Whipple and D. D. Jackson, A Comparative Study of the Methods Used for the Measurement of Turbidity in Water, *M.I.T. Quarterly* **13**, 274 (1900).

2. R. W. Austin, Problems in Measuring Turbidity as a Water Quality Parameter, U.S. Environmental Protection Agency Seminar on Methodology for Monitoring the Marine Environment, Seattle, Washington, 1973, pp. 40–51.

3. Kingsbury, Clark, Williams, and Post, *J. Lab. Clin. Med.* **11**, 981 (1926).

4. W. G. Exton, Electrooptical Method and Means for Measuring Concentration, Colors, Dispersions, Etc. of Fluids and Similar Substances, U.S. Patent Number 1,971,443, 1934.

5. C. C. Hach, Turbidimeter for Sensing the Turbidity of a Continuously Flowing Sample, U.S. Patent Number 3,306,157, 1967.

6. American Public Health Association (APHA) (AWWA, WPCF), *Standard Methods for the Examination of Water and Wastewater*, 13th edition, 1971.

6. (a) U.S. Environmental Protection Agency (EPA), *Method for Determining the Turbidity of a Water Sample*, EPA 79–97, Washington D.C. Office of Water, EPA, Washington, DC, 1979.

7. C. F. Bohren and D. R. Huffman, *Absorption and Scattering of Light by Small Particles*, Wiley, New York, 1983.

8. A. Morel, Optical Properties of Pure Water and Pure Sea Water, in N. Jerlov and E. Steemann Nielsen (Eds.), *Optical Aspects of Oceanography*, Academic, New York, 1974, pp. 1–24.

9. R. C. Smith and K. S. Baker, Optical Properties of the Clearest Natural Waters (200–800 nm), *Appl. Opt.* **20**(177) (1981).

10. E. Huber, Light Scattering by Small Particles, *AQUA* **47**(2), 87–94 (1998).

11. T. J. Petzold, *Volume Scattering Functions for Selected Ocean Waters*, SIO 72-28, Scripps Institution of Oceanography, La Jolla, CA, 1972.

12. M. A. Landers, Summary of Blind Sediment Reference Sample Measurement Session, in J. R. Gray, and G. D. Glysson, (Eds.), *Proceedings of the Federal Interagency Workshop on Turbidity and Other Sediment Surrogates*, April 30–May 2, 2002, Reno, Nevada, Circular 1250, U.S. Geological Survey, Washington, DC, 2003, pp. 29–30

13. National Committee for Clinical Laboratory Standards (NCCLS), *Preparation and Testing of Water in the Clinical Laboratory*, 2nd Ed., Vol. 11(13).

14. V. Sethi, P. Patanaik, P. Biswas, R. M. Clark, and E. W. Rice, Evaluation of Optical Detection Methods for Waterborne Suspensions, *J. AWWA* **89**(2), 98–112 (1997).

15. American Society for Testing and Materials (ASTM), *Standard Test Method for Turbidity of Water*, ASTM Designation D 1889-00, ASTM, West Conshohocken, PA, 2000.

15. (a) International Organization for Standardization (ISO), *Water Quality—Determination of Turbidity*, ISO 7027, ISO, Geneva, 1993. 10 pages.

16. J. Downing and W. E. Asher, The Effects of Colored, Dissolved Matter on OBS Measurements, paper presented at the Fall Meeting of the American Geophysical Society, San Francisco, CA, 1997.

17. J. Downing, Optical Backscatter Turbidimeter Sensor, U.S. Patent Number 4,841,157, 1989.

18. T. F. Sutherland, P. M. Lane, C. L. Amos, and J. Downing, The Calibration of Optical Backscatter Sensors for Suspended Sediment of Varying Darkness Level, *Marine Geol.*, **162**, 587–597 (2000).

19. M. Sadar, *Turbidity Standards*, Technical Information Series, Booklet No. 12, Hach Company, 1998.

20. K. Papacosta and M. Katz, The Rationale for the Establishment of a Certified Reference Standard for Nephelometric Instruments, Technical Conference Proceedings, paper presented at the American Waterworks Association Water Quality Technical Conference, Paper Number ST6-4, 1990, pp. 1299–1333.

21. J. R. V. Zaneveld, R. W. Spinrad, and R. Bartz, *Optical Properties of Turbidity Standards*, SPIE Volume 208 Ocean Optics VI, Bellingham, WA, 1979, pp. 159–158.

22. K. Papacosta, NIST-Traceable, Ready-to-Use Spectrophotomer Standards, *Am. Lab. News*, October 1996, pp. 39–51.

23. http:// www.epa.gov/watertrain/cwa/cwa27.htm.

24. C. A. Ruhl and D. H. Schoellhamer, *Time Series of SSC, Salinity, Temperature, and Total Mercury Concentration in San Francisco Bay during Water Year 1998*, Regional Monitoring Program for Trace Substances Contribution Number 44, San Francisco Estuary Institute, Richmond, CA, 2001.

25. D. C. Whyte and J. W. Kirchner, Assessing Water Quality Impacts and Cleanup Effectiveness in Streams Dominated by Episodic Mercury Discharges, *Science of the Total Environment*, **260**, 1–9 (2000).

26. J. L. Domagalski and K. M. Kuivila, Distribution of Pesticides and Organic Contaminants between Water and Suspended Sediment, San Francisco Bay, California, *Estuaries* **16**(3A), 416–426 (1993).

27. C. J. Gippel, Potential of Turbidity Monitoring for Measuring the Transport of Suspended Solids in Streams, *Hydrol. Process.* **9**, 83–97 (1995).

28. U.S. Geological Survey (USGS), Summary of the U.S. Geological Survey On-line Instantaneous Fluvial Sediment and Ancillary Data, in *Proceedings of the 7th Federal Interagency Sedimentation Conference*, Reno, NV, USGS, Washington, DC, 2001.

29. R. W. Sheldon, A. Prakask, and W. H. Sutcliffe, Jr., The Size Distribution of Particles in the Ocean, *Limnol.* Oceanogr. **17**(3), 327–340 (1972).

30. T. Chisholm, *SEM Images of River SPM*, Oregon Graduate Institute, 2003.

31. American Public Health Association (APHA) (AWWA, WEF), *Standard Methods for the Examination of Water and Wastewater*, 19th ed., 1998.

32. American Society for Testing and Materials (ASTM), *Standard Test Method for Determining Sediment Concentration in Water Samples*, ASTM Designation D 3977-97, ASTM, West Conshohocken, PA, 2000 pp. 395–400.

33. G. C. Brent, J. R. Gray, K. P. Smith, and G. D. Glysson, *A Synopsis of Technical Issues for Monitoring Sediment in Highway and Urban Runoff*, Open-File Report 00-497, U.S. Geological Survey, Washington, DC, 2001.

34. J. R. Gray, G. D. Glysson, L. M. Turcios, and G. E. Schwarz, *Comparability of Suspended-Sediment Concentration and Total Suspended Solids Data*, Water-Resources Investigations Report 00-4191, U.S. Geological Survey, Washington, DC, 2000.

35. G. C. Brent, J. R. Gray, K. P. Smith, and G. D. Glysson, *A Synopsis of Technical Issues for Monitoring Sediment in Highway and Urban Runoff*, Open-File Report 00-497, U.S. Geological Survey, Washington, DC, 2001.

36. P. A. Buchanan and C. A. Ruhl, *Summary of Suspended-Solids Concentration Data, San Francisco Bay, California, Water Year 1998*, Open-File Report 00-88, U.S. Geological Survey, Washington, DC, 2000.

37. U.S. Geological Survey (USGS) *National Field Manual of the Collection of Water-Quality Data*, Book 9, Handbooks for Water-Resources Investigations, USGS, Washington, DC, 2003.

38. T. K. Edwards and G. D. Glysson, *Field Methods for Measurement of Fluvial Sediment*, Techniques of Water-Resources Investigations of the U.S. Geological Survey Book 3, Chapter C2, 1999.

39. J. Lewis, Turbidity-Controlled Suspended Sediment Sampling for Runoff-Event Load Estimation, *Water Resour. Res.* **32**(7), 2299–2310 (1996).

40. L. H. Keith, *Principles of Environmental Sampling*, 2nd ed., ACS Professional Reference Book, American Chemical Society, Washington, DC, 1996.

41. American Society for Testing and Materials (ASTM), *Standard Guide for Sampling Fluvial Sediment in Motion*, ASTM Designation D 4411-98, ASTM, West Conshohocken, PA, 1998.

42. U.S. Department of Agriculture, *National Handbook of Water Quality Monitoring*, Part 600, USDA SCS, Washington, DC, 1994.

43. Water Survey of Canada, *Field Procedures for Sediment Data Collection*, Vol. 1: *Suspended Sediment*, National Weather Services Directorate, Ottawa, Canada, 1993.

44. U.S. Environmental Protection Agency (EPA), *Handbook for Sampling and Sample Preservation of Water and Wastewater*, EPA-600/4-82-029, Environmental Monitoring and Support Laboratory, Cincinnati, OH, 1982.

45. Australian and New Zealand Environment and Conservation Council (ANZECC), National Water Quality Management Strategy Paper No. 7, ANZECC, Canberra, 2000.

46. T. A. Ratliff, *The Laboratory Quality Assurance System—A manual of Qualtiy Procedures and Forms*, 3rd ed., Wiley-Interscience, New York, 2003.

47. D. R. Helsel and R. M. Hirsch, *Statistical Methods in Water Resources*, Studies in Environmental Science, Vol. 49, Elsevier, New York, 1992.

48. U.S. Environmental Protection Agency (EPA), *Protocol for Determining Sediment TMDLs*, EPA 841-B-99-004, Washington D.C. Office of Water (4503F), EPA, Washington, DC, 1999.

49. B. A. Pruitt, Uses of Turbidity by States and Tribes, in J. R. Gray, and G. D. Glysson (Eds.), *Proceedings of the Federal Interagency Workshop on Turbidity and Other Sediment Surrogates*, April 30–May 2, 2002, Reno, Nevada, Circular 1250, US Geological Survey, Washington, DC, 2003, pp. 31–46

50. International Organization for Standardization (ISO), *Liquid Flow in Open Channels—Sediment in Streams and Canals—Determination of Concentration, Particle Size Distribution and Relative Density*, 1st ed., ISO 4365, ISO, Geneva, 1997.

25

WATERSHED SCALE, WATER QUALITY MONITORING–WATER SAMPLE COLLECTION

Randy A. Dahlgren, Kenneth W. Tate, and Dylan S. Ahearn

University of California
Davis, California

Environmental Instrumentation and Analysis Handbook, by Randy D. Down and Jay H. Lehr
ISBN 0-471-46354-X   Copyright © 2005 John Wiley & Sons, Inc.

25.1 INTRODUCTION

Water quality monitoring at the watershed scale is an important component of water resources management, conservation, and protection. The federal Clean Water Act establishes water quality monitoring as an important tool for the identification and restoration of water bodies that are impaired by pollutants.[1] Watershed-scale strategies known as total maximum daily loads (TMDLs) must be developed and implemented to restore every impaired water body by controlling both point and nonpoint sources of the pollutant(s) causing the impairment. Comprehensive monitoring is required to determine the location and magnitude of point and nonpoint pollutant sources. While point sources are relatively easy to quantify, addressing nonpoint source water pollution in large watersheds is extremely challenging due to the diversity of the physical environment (e.g., climate, geology, soils, topography, vegetation, hydrology) and land use (e.g., urban, mining, forestry, agriculture) occurring within a given watershed. Water quality data are often required for development, calibration, and validation of mechanistic models used to evaluate alternative watershed management scenarios proposed for water quality improvement during TMDL development. Once management plans and practices to improve water quality have been implemented, continued monitoring is necessary to document the effectiveness of these activities. With more than 21,000 impaired water bodies identified in the United States[1] and an average cost estimated at more than $1 million per TMDL, there is a critical need to control monitoring costs while generating credible, defensible, and quality data that are useful for addressing water resource objectives.

A well-designed monitoring program can help reduce the cost of data collection and optimize information return on monitoring investment. The most important step in formulating an efficient and informative water quality monitoring program is the initial specification of objectives. For example, a monitoring project that is attempting to detect a relatively small change with a high degree of certainty will be more rigorous and costly than a monitoring program with a lower standard for identifying a statistically significant change. MacDonald et al.[2] define seven types of water quality monitoring activities that are useful for formulating monitoring objectives: baseline, trend, implementation, effectiveness, project, validation, and compliance. A typical first step in watershed monitoring is characterization of ambient water quality conditions across the entire watershed. *Baseline monitoring* is used to characterize existing water quality conditions and to establish a reference set of data for planning or future comparisons. The intent of baseline monitoring is to capture a representation of the temporal variability of a particular water quality parameter for a given site. In turn, *trend monitoring* implies that measurement will be made at regular, well-spaced time intervals in order to determine the long-term trend in a particular water quality parameter. There is no explicit end point at which continued baseline monitoring becomes trend monitoring. Given the focus on utilizing TMDLs to address nonpoint source water impairments, it is critical to establish baseline monitoring programs to serve as the basis for evaluating the watershed-wide effectiveness of best management practices (BMPs) applied to address nonpoint source pollution. This latter type of monitoring is termed *effectiveness monitoring*. Monitoring efforts designed to determine whether specific water quality criteria are being met is termed *compliance monitoring*. Because the objectives of a water quality monitoring program are constantly evolving, it is prudent to consider potential future objectives that might be asked of a water quality monitoring program.

The purpose of this chapter is to provide a general overview of the complexities involved in designing a water quality monitoring program at the large-watershed scale and to provide a detailed discussion of how automatic pump samplers can be deployed to enhance the value of data collected as part of a monitoring program. Technological advances in automatic pump samplers allow for cost-effective collection of water quality data in a manner that can fully characterize temporal water quality dynamics.

25.2 THEORETICAL CONSIDERATIONS

25.2.1 Temporal Scales

An effective watershed monitoring program must account for the inherent temporal characteristics of various water quality parameters to accurately characterize their fluxes. Water quality parameters may show strong temporal variability on the interannual, seasonal, storm event, and diel time steps (Fig. 25.1).[3] For example, mineral nitrogen ($NH_4 + NO_3$) fluxes in California oak woodland watersheds display more than an order-of-magnitude variation on the interannual time step as a result of differences in total precipitation, precipitation intensity, and seasonal distribution of precipitation events (Fig. 25.1a). In many cases, regulatory and management decisions must be made without a long-term data set to provide perspective. Figure 25.1a demonstrates the importance of investing in the development of long-term water quality records for future assessments of water resources impairment.

Large differences in stream water nitrate concentrations are prevalent on a seasonal basis in the Mediterranean climate of California where there is a temporal asynchrony between hydrologic fluxes (wet winters–dry summers) and plant nitrogen demand (primarily spring) (Fig. 25.1b).[4] Therefore, more intensive sampling frequency is required during the earlier winter to adequately characterize the peak nitrate concentrations and allow for accurate flux calculations.

For several constituents, the majority of the annual constituent flux (or load) is transported during a few large storm events each year. Storm and snowmelt events almost always result in large changes in most water quality parameters (Fig. 25.1c). Changing hydrologic flow paths occur during storm events, resulting in a transition from a dominance of groundwater stream flow generation sources during base-flow conditions to interflow through the soil zone and surface contributions from variable source areas during peak flows.[4] Within storm event hysteresis effects, due to pollutant supply limitations and hydrologic flushing, typically result in higher

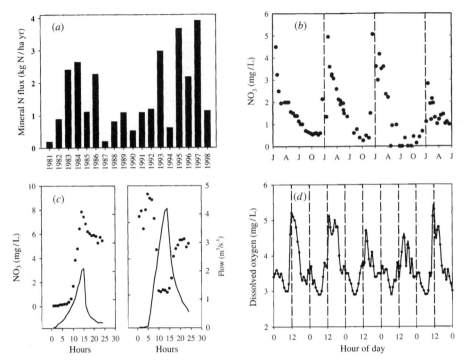

Figure 25.1 Examples of temporal changes in water quality parameters occurring at (a) interannual, (b) seasonal, (c) storm event, and (d) diel time scale.

concentrations at a given discharge on the rising limb of the storm hydrograph as compared to the same discharge on the falling limb. Given the large fluctuations in concentrations during a storm event, it is necessary to collect samples intensively throughout a storm event to accurately characterize constituent fluxes. In California oak woodlands, nitrate concentrations increase with increasing flow during early rainfall season storms due to flushing of nitrates from the soil zone. However, concentrations decrease during storms later in the water year due to dilution and reduced nitrate concentrations in the soil zone.

For water quality parameters regulated strongly by biological processes (e.g., dissolved oxygen, pH, and chlorophyll a), diel changes may be relatively large (Fig. 25.1d). Dissolved oxygen levels reach a maximum concentration during the early afternoon as photosynthesis by algae and aquatic plants produces oxygen that results in supersaturated dissolved oxygen levels. In contrast, minimum dissolved oxygen concentrations occur just before sunrise due to the lack of photosynthesis during the nighttime hours and continued stream respiration throughout the dark period.

To accurately determine water quality fluxes and account for inherent fluctuations in water quality parameters, it is critical to capture the temporal variability for the parameters of interest before devising a monitoring protocol. At the large-watershed scale, temporal dynamics may vary spatially due to differences in water quality drivers, such as climate, atmospheric deposition, geology/soils, vegetation, land use, and population density. As a result, the temporal sampling design may require independent optimization for different sites within a given watershed. In many cases water quality parameters display similar temporal dynamics at the seasonal and storm event scales, which allows for a similar sampling protocol to be utilized for multiple parameters.

Given the short timeline for TMDL development (8–13-year deadline) and lack of sufficient resources (people and money) for comprehensive water quality monitoring, several states have adopted a rotating-basin approach. As part of a rotating-basin approach, individual water bodies are assessed at differing levels of intensity each year. For example, intensive baseline monitoring may occur once every 5 years while a less intensive monitoring frequency is employed in the interim years.

25.2.2 Spatial Scales

An important goal of water quality monitoring at the large-watershed scale is to adequately characterize source area contributions with the fewest number of sample locations, therefore minimizing monitoring cost. A source-search protocol is an efficient initial step to help identify spatial patterns in water quality parameters. The source-search technique involves periodic sampling of representative sub-watersheds to determine concentrations/fluxes from the range of physical environments (e.g., climate, vegetation, geology/soils, land use) contained within the large watershed. Samples from all the subwatersheds should be collected over a short period of time, typically within a day or two. Repeated samples must be collected to address the temporal considerations discussed above (e.g., seasonal, storm event, diel). While samples collected during base flow (representing groundwater contributions) may have similar water quality values throughout the entire watershed, samples collected during storm events may differ appreciably among subwatersheds. Therefore, it is necessary to stratify sampling to characterize base-flow, rising-limb, peak-flow, and falling-limb water quality characteristics. Source-search protocols are generally effective for isolating the subwatersheds that contribute disproportionately to water quality impairment. Once water quality impaired "hot spots" are identified, further upstream sampling may be able to isolate the specific source of the impairment. Continued refinement of sampling sites using the source-search methodolgy can be utilized to optimize the final selection of water quality monitoring sites for a more rigorous whole-watershed investigation. This strategy is being employed throughout the United States to identify impaired water bodies for TMDL development.

The source-search technique was employed in the 1700-km^2 Mokelumne River watershed in California to elucidate the source of nitrogen contributing to periodic eutrophication of downstream reservoirs.[5] The initial hypothesis was that forest harvest practices in the high-elevation coniferous forests were responsible for elevated nutrient concentrations in stream water. From an initial survey of 37 sites, water samples were collected biweekly and during selected storm events for 4 years at

25 sites throughout the watershed. Selection of these sites effectively divided the upper Mokelumne River watershed into a series of subwatersheds for the purpose of identifying the contribution of each subwatershed (i.e., cumulative effects) to the overall water quality entering the downstream reservoirs. At high elevation sites, sampling was limited during the winter to periodic helicopter access due to the deep snowpack. As the watershed is oriented in an east–west direction, we established several north–south transects across the Mokelumne mainstem and its major tributaries to follow the evolution of water quality as it flows from high to low elevation. This approach allowed us to examine the major climate, land use, and geologic/soil associations occurring throughout the watershed. A factor that typically drives site selection is the availability of flow data from an established and regularly maintained guaging station. Another major consideration for selection of sampling sites is long-term accessibility to the site. Thus, all samples were collected from bridges (on right-of-ways) of public roads. For consistency, grab samples were collected from the center of the stream at the 0.6 depth.[6]

Maximum nitrate concentrations over the 4-year period showed a general increase with decreasing elevation and very high concentrations in subwatersheds immediately above the reservoirs (Fig. 25.2). Combining water fluxes with nitrate concentrations indicated that about 90% of the nitrate–nitrogen flux originated from about 10% of the watershed area. This analysis allowed us to reject the hypothesis that forest harvest practices were the primary source of stream water nitrates in the Mokelumne River watershed. We therefore focused our efforts on determining the source of the high nitrate concentrations in the lower elevation sites. Process level

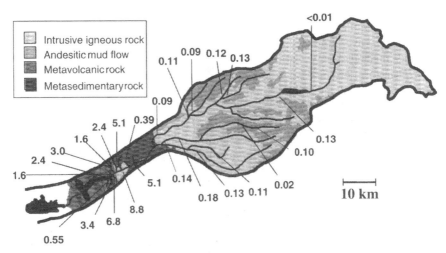

Figure 25.2 Example of using source-search methodology to isolate dominant sources of pollutants in a large watershed. The 1700-km² Mokelumne River watershed is located in the western Sierra Nevada of central California. Values represent the maximum nitrate–nitrogen concentrations (mg N/L) observed over a 4-year monitoring period. High concentrations of nitrate in the lower watershed were associated with geologic nitrogen (nitrogen contained in bedrock).

investigations demonstrated that bedrock nitrogen (i.e., geologic nitrogen) contributed to nitrogen saturation of these ecosystems and the elevated nitrate concentrations. The source-search technique was then expanded to a rigorous temporal sampling regime in these watersheds to determine accurate fluxes for these sub-watersheds.

25.2.3 Concentration Versus Load

While most water quality standards are based on concentrations (in milligrams per liter) determination of loads (or fluxes, in kilograms) is required to assess and compare source contributions and load allocations. Therefore, it is necessary to simultaneously determine river discharge (in cubic meters per second) as well as constituent concentrations in order to calculate loads (flux = concentration × discharge). Because of variable flows in many rivers, it is necessary to provide a continuous record of river discharge using a calibrated stage–discharge relationship to calculate accurate loads. While gauging stations that provide flow information may be available along major rivers, their density on tributary streams is often inadequate to allow accurate load calculations for these streams. In lieu of continuous discharge measurements, it may be possible to determine instantaneous discharge estimates using the area–velocity method based upon stream cross-sectional area (in square meters) and stream velocity (meters per second) at the time of sample collection. Hydrologic models may also be employed to estimate discharge in the absence of measured data. When utilizing discharge values from hydrologic models, it is critical to provide measures of uncertainty so that discharge data are interpreted within their limitations. The lack of reliable river discharge measurements is a major uncertainty in the calculation of constituent loads.

25.3 WATER SAMPLE COLLECTION METHODOLOGIES

25.3.1 Grab Samples

An effective stream water sampling strategy requires a basic understanding of stream characteristics. Flowing streams are dynamic in nature and discharge volumes and water quality are seldom constant in time. Low-frequency data collection (weekly, monthly, seasonal) is typically accomplished by collecting grab (discrete) samples taken by a field crew. A single grab sample represents a snapshot in time and can easily misrepresent pollutant loadings unless sufficient samples are taken to characterize all aspects of a pollutant's temporal dynamics. It is virtually impossible to capture the large fluctuations that occur in pollutant concentrations during storm events using a grab-sampling protocol. Depending on when a grab sample is collected during a storm event, it may under- or overestimate the average condition by more than an order of magnitude (Fig. 25.1c).

There are several logistical aspects to consider in collecting a grab sample, the most important being collecting a single sample that is representative of the entire volume of water within the river channel. In well-mixed rivers the dissolved (<0.45 µm) constituents (e.g., nitrate, dissolved organic carbon, electrical conductivity) are generally uniform in concentration throughout the river cross section. Thus, collecting a sample from the middepth and middle cross section of the channel using a weighted bottle might well provide a representative sample. In contrast, water quality constituents associated with particulate (>0.45 µm) matter (e.g., suspended sediment, particulate nitrogen and phosphorus, pesticides, mercury) are highly variable depending on the water velocity and turbulence, which varies vertically and horizontally within the water column. In smaller streams with numerous pools, riffles, and run sections, the run portion of the stream provides the most representative characterization of overall stream water quality. Care must be taken not to stir up bottom sediments when taking samples from shallow streams.

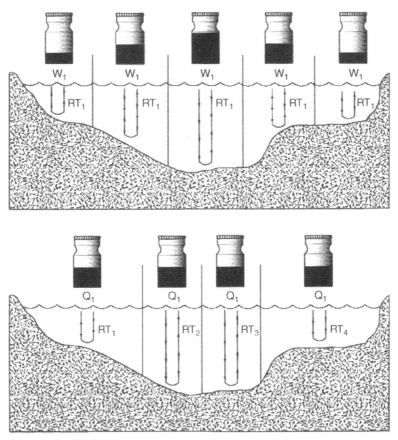

Figure 25.3 Example of equal-width-increment method for collection of depth-width integrated water quality samples (upper panel). Cross section is divided into equal-width intervals and a vertical sample is collected using a constant transit rate (RT) for each interval. The equal-discharge-increment method divides the stream cross section into intervals with equal discharge (lower panel). Transit rate (RT) is adjusted to collect a full sample bottle from each-sectional interval.

Collecting a representative sample for any constituent in larger river systems generally requires compositing several depth-integrated samples collected along the cross section of the channel. The *equal-width-interval method* divides the cross section of the stream into a number of intervals having equal widths (Fig. 25.3). Generally, a minimum of 10 measurement points across a large stream or river is sufficient to establish the horizontal cross-sectional variability of water quality constituents.[7] Samples are collected by lowering and raising a weighted sampler through the water column at the center of each cross-sectional increment. The combination of the same constant transit rate used to sample each increment and the isokinetic property of the sampler results in a discharge-weighted sample that is proportional to total stream flow. All cross-sectional samples are composited and a representative sample is obtained using an appropriate sample-splitter.[8]

The *equal-discharge-increment method* delineates a number of cross-sectional increments having equal discharge volumes as calculated from the stream cross-sectional area and velocity (Fig. 25.3). Generally, a minimum of five increments across a large stream or river is sufficient to provide a discharge-weighted mean and establish the horizontal cross-sectional variability of the measured water quality constituents.[7] An equal volume of sample is collected over the depth interval for each increment and composited to obtain a representative sample. While application of either method improves the sample's representation of the entire stream cross section at a point in time, application of these methods greatly increases monitoring costs.

25.3.2 Automated Pump Samplers

25.3.2.1 General Characteristics. Automated pump samplers have revolutionized our ability to effectively characterize temporal dynamics in water quality constituents. There are more than 150 different types of automatic sample collection devices commercially available. The samplers differ in several ways, such as pump type, intake design, transport velocity, microprocessor capabilities, refrigeration options, and sample volumes. The majority of the samplers consist of a peristaltic pump that draws water through a flexible sample tube from the stream and into a collection bottle (Fig. 25.4). The electric motor is typically powered by batteries, which can be attached to a solar panel for remote operation. Most samplers have microprocessors that can be programmed for a variety of sample collection routines. The microprocessors can also serve as a datalogger for storing data from external sensors. External sensors can be integrated into the system to automatically turn on the sampler during a rainfall event (e.g., liquid level detector or automatic rain gauge) or to drive the sample collection regime (e.g., flow or stage proportional sampling schemes, turbidity threshold sampling). Programs can be paused and resumed for intermittent discharge flow monitoring. An optional telephone modem allows programming changes and data collection to be performed remotely though a cellular telephone connection.

Depending on the water quality constituent of interest, samplers can be fitted with glass, polyethylene, or Teflon bottles with capacities typically ranging from

Figure 25.4   (*a*) Automatic pump sampler deployed in a locked, steel box chained to a tree to prevent theft and loss during high-flow events. A flowmeter was utilized to activate pump the sampling sequence during storm events. (*b*) Sample tube inlet attached to a floating boom anchored to the stream bottom. Tube inlet was located at a position along the boom to maintain inlet at 0.6 depth. Sampling tube was enclosed in a rigid PVC tube anchored to bottom to prevent debris from collecting and disturbing sampling tube.

0.5 to 30 L. Most samplers have a full-bottle shutoff valve that turns off the pump when the bottle is full to prevent cross-contamination between sample bottles. Most samplers can hold a total of 24 bottles having volumes of 0.5–1 L. If larger volumes are required, the number of bottles is proportionally reduced. The system can be programmed to perform multiple rinses between samples and for an air purge before and after each sample so that the collection tube remains empty between samples. These rinse-and-purging steps greatly reduce the potential for cross-contamination between successive samples. To help preserve sample integrity, some samplers have a refrigeration unit that can be employed when AC power is available. Alternatively, most samplers have the capacity to be packed with ice in an insulated base to preserve samples for extended periods. The automated sampler is enclosed in an environmentally sealed housing that protects the samples from outside sources of contamination.

The suction produced by peristaltic pumps can strip volatile organic compounds (VOCs) from the water sample, and storage in open collection bottles can further result in loss of VOCs. As a result, collecting water samples for VOC analysis has traditionally been a manual process. New designs employ a bladder pump to gently push the liquid to a sealed sample bottle.

A critical factor in using automatic pump samplers for obtaining a representative sample is to maintain a sufficient transport (intake) velocity so that larger particles do not preferentially settle out during collection. The U.S. Environmental Protection Agency (EPA)[9] recommends that sample transport velocities should exceed 2 ft/s (0.61 m/s). Transport velocity is not a factor for dissolved constituents but is critical for water quality constituents associated with particulate matter. Based on a minimum transport velocity of 2 ft/s, the maximum head that most automatic pump samplers can draw against is about 20–26 ft (6.1–7.9 m). Horizontal pumping distances are generally not limiting as the pump is not pulling against the increasing weight of the water in the vertical direction.

25.3.2.2 Site Selection. Selection of a water quality monitoring site is determined by the data quality objectives. The best location for a site is often one that is optimal for measuring surface water discharge. Ideally, the site should be located in a straight portion of the stream, flow should be confined to a single channel, the streambed should not be subject to scour or fill, the site should not be affected by the confluence with another stream or tidal effects, nor should it be susceptible to nearby point source discharge, and the site should be readily accessible and safe for instrument access and maintenance.

Sampling at midstream and middepth in the stream channel is standard practice because water quality at this point is commonly more representative of the whole stream. It is a prudent step to determine the cross-sectional variability in water quality constituents before a monitoring site is installed to properly locate the most representative measurement point in the stream cross section and to determine if a cross-sectional correction is necessary. In most cases, placement of the inlet tube at the horizontal and vertical center of the stream is best. To obtain a representative water sample in smaller streams, the autosampler is best placed in a "run" section rather than "pool" or "riffle" sections. In larger streams, one must ensure that the autosampler is placed in a well-mixed segment of the stream. Lateral mixing of tributary streams into large rivers is not complete for several miles downstream of their confluence. As a result, sampling near the streambank may be more representative of local runoff or tributary streams, whereas a location near the center of the channel may be more representative of areas farther upstream. While it is best to place an autosampler in a straight-stream reach, if it is necessary to place it in a bend, the cut bank side is preferred rather than the inside edge where deposition may bury the sample collection tube.

25.3.2.3 Intake Placement. The measurement point in the vertical dimension also needs to be appropriate for the primary purpose of the monitoring installation. The vertical measurement point may be different for low-, medium-, or high-flow conditions. In shallow streams that can be accessed by wading, an anchoring post can be driven into the streambed. Ideally, the top of the post will be below the water surface to minimize collection and disturbance by debris. The entire length of tubing between the inlet and the autosampler should be staked to the river bottom to prevent the tube from getting snagged by debris. It is recommended that the collection tube inlet be placed at about the 0.6 depth of the stream.[6] Anchoring an inlet tube at a fixed depth results in collection of waters from a fixed depth during storm events in which river stage may increase and decrease appreciably in a short time period. It is critical that the inlet be high enough above the streambed to prevent collection of bottom sediments and bed load.

In some instances it is possible to collect from variable depths during storm events by anchoring the tube inlet to a floating boom anchored either in the stream bottom or from an overhanging structure (Fig. 25.4*b*). By anchoring the floating boom to the river bottom, it is possible to place the inlet at a position along the boom to maintain the sampling depth at a fixed relative depth in the stream (e.g., 0.6 depth). A limitation to this method is that debris can dislodge the boom from its anchor during high-flow events. For deeper water streams, a floating boom can be attached to a structure, such as a bridge pylon, which will allow sampling of water at a fixed depth from the surface. By placing the boom immediately downstream of a bridge pylon, it is possible to avoid, to some extent, disturbance by floating debris.

25.3.2.4 Sample Collection Frequency. Automated samplers can be programmed to initiate a sampling schedule based on input from external sensors, such as input from an electronic rain gauge or a liquid level detector that detects an increase in stream stage. It is often critical to catch the earlier portion of a storm hydrograph as it may contain extremely high concentrations of constituents due to flushing of entrainable materials from the system. By relying on manual initiation of the sampler, the operator must be prepared to visit the site at anytime and predictions of precipitation may not materialize, resulting in a wasted trip to the site to turn the sampler on and then again to turn the sampler off and replace contaminated sample bottles.

Automated samplers can be programmed to collect samples at regular intervals that are determined either by time (time weighted) or by flow volumes (flow weighted) determined using an external sensor. Samplers also have a random-interval sample collection program that can provide random samples for statistical purposes. Flow-proportional sampling requires an external sensor that measures stream discharge. The user defines a predetermined discharge volume (e.g., one sample every 2000 m^3) and a sample is collected each time this volume is reached. While base-flow conditions typically display only small temporal differences in water quality constituents, storm events can cause order-of-magnitude differences in constituent concentrations. Because of the high water volumes during storm events, these events typically transport the largest constituent fluxes (or loads).

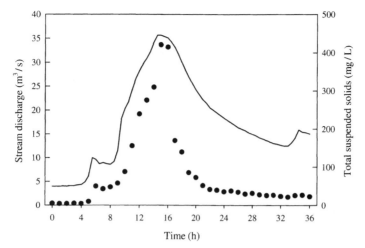

Figure 25.5 Example of 36-h storm hydrograph with TSS samples analyzed at a 1-h time step. This example was used to calculate TSS loads for the storm event assuming various sample collection strategies.

With an idealized goal of accurately characterizing water quality concentrations and constituent fluxes, a number of sample interval strategies may be employed.

To illustrate differences in calculated loads from various sampling routines, we selected a 36-hr storm event in which stream discharge was measured at a 30-min interval and total suspended solids (TSS) were measured at a 1-hr interval using an automatic pump sampler (Fig. 25.5). For fixed-interval sampling, a period-weighting method was utilized and consisted of multiplying the water volume (in cubic

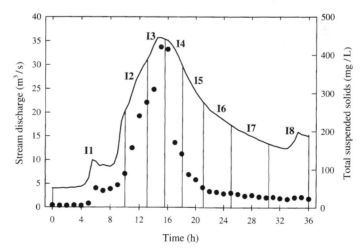

Figure 25.6 Example of dividing the storm hydrograph into eight increments having identical total discharge volumes (286,411 m³). These intervals were used to calculate flow-proportional TSS loads for the storm hydrograph.

meters) over the sample interval (e.g., 1 or 4 h) times the mean TSS value (mean value from the beginning and end of the sample interval). The sample technique was used to calculate loads for composite samples in which a single bottle was filled over a 4- or 12-h period with equal fractions collected each hour. A flow-proportional calculation was performed by dividing the total flow into eight increments of 286,411 m³ of water along the hydrograph (Fig. 25.6). The period-weighting method was used to determine the average TSS concentration based on the initial and final TSS values associated with the flow increment. Finally, a simple discharge–log TSS rating curve was established using the 1- and 4-h data to construct simple regression equations predicting log TSS from the discharge (Fig. 25.7). The simple linear regression equation was then applied for each 30-min discharge value and the 30-min loads summed for the entire storm event.

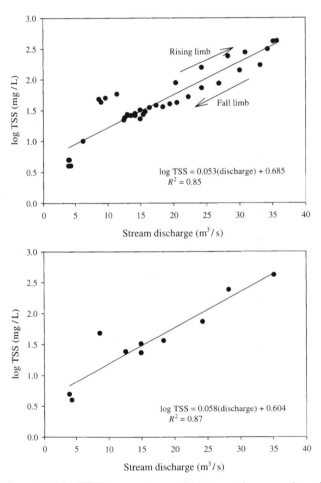

Figure 25.7 Example of log(TSS) versus stream discharge rating curve for calculating TSS loads. Upper curve is constructed using all 36 pairs of data collected on a 1-h time step (from data in Figure 5), while the bottom curve is constructed using 9 pairs representing data collected on a 4-h step (from data in Fig. 5).

Table 25.1   Comparison of TSS Fluxes Based on Various Sample Collection Routines

Flux Calculation Method	Calculated Flux (Mg)	Deviation from 1-h Calculation (%)
1-h sampling interval	308.0	—
2-h sampling interval	314.9	2.2
4-h sampling interval	340.6	10.6
8-h sampling interval	363.0	17.9
12-h sampling interval	241.5	−21.6
4-h composite, 1-h interval	296.0	−3.9
8-h composite, 1-h interval	268.5	−12.8
Flow proportional, 8 segments	287.4	−6.7
Rating curve, 1-h sampling interval	302.2	−1.9
Rating curve, 4-h sampling interval	349.5	13.5

Note: The percent deviation from the most rigorously defined flux value (1-h TSS sampling interval) is reported as a comparative measure.

The results of these various load calculations are reported in Table 25.1. Because the most rigorous determination of TSS load should result from the 1-h time step collection of TSS values, we use a TSS load of 308.0 Mg calculated from the 1-h time step sample collection as the reference for comparison to the alternative sampling routines. As the time step for sampling increases from 1 to 12 h, the deviation from the 1-h sampling frequency increases from 2 to 22%. Compositing samples over a 4- and 8-h period with hourly samples resulted in a substantial improvement compared to the fixed-interval independent sampling for the same time periods. Flow-proportional sampling resulted in a load similar to that obtained by 1-h fixed-interval sampling. To program a flow-proportional sample collection program, you must anticipate the magnitude of the storm event and how large the storm hydrograph will be. In many cases, predicting the size of the storm hydrograph will be very difficult, resulting in too few samples collected for smaller than anticipated storms and possible too many samples collected for larger than anticipated storms and thus filling all sample bottles prior to the end of the storm.

The rating curve method constructed from the 36 hourly pairs of TSS–discharge data provides a nearly identical load calculation to the 1-h fixed-interval sampling. The simple regression equation obtained from using only 9 pairs of data (4-h sampling interval) was similar in form and fit to the simple regression equation developed using 36 pairs of data (1-h sampling interval); however, due to the log transformation, slight differences in model coefficients generated a final TSS load calculation that was nearly 14% higher than the 1-h fixed-sampling routine. The hysteresis in TSS concentrations is clearly observed in the data with higher TSS levels at a given discharge on the rising limb of the hydrograph compared to similar discharge values on the falling limb (Fig. 25.7). Potential complications of using the simple regression based rating curve approach include (1) the need to collect sufficient data to develop a significant TSS–discharge relationship, (2) hysteresis; (3) changing storm event TSS–discharge relationships on a seasonal basis, and

(4) variability in TSS–discharge relationships among subwatersheds. These complications can all be overcome with adequate sample collection from a representative suite of subwatersheds and application of more complex multiple-regression techniques. However, the cost savings and accuracy of such an approach should be evaluated on a case-by-case basis.

Assuming a tolerance of 10% for calculated loads, the 4-h composite sample consisting of hourly subsamples provides optimal balance between accuracy (3.9%) and minimum number of samples (9 samples for the 36-h storm event). The flow-proportional sampling with 8 samples resulted in 6.7% deviation, while the 2-h fixed-interval sampling with 18 samples provided a 2.2% deviation. Therefore, from an accuracy-and-cost basis, a composite sampling strategy provides the best sampling routine for this particular example. For constituents that display a strong rating curve relationship over the course of the monitoring period, it may be possible to make good estimates of flux from stream discharge data and minimal sample collection for validation.

Water sampling design can also be driven by results from external sensors, such as an in situ turbidity probe. Turbidity threshold sampling collects physical samples using an automatic pump sampler that are distributed over a range of rising and falling turbidities, independent of stream discharge.[10,11] An algorithm is used to distribute samples over the entire turbidity range during each storm event. A few (less than 10) data pairs (TSS–turbidity) spanning the range of concentrations are generally sufficient to reliably establish the relationship between TSS and turbidity during each storm event. Then TSS loads can be calculated from the continuous measurements of turbidity acquired from the in situ probe.

25.4 OVERVIEW OF COMPLEMENTARY FIELD INSTRUMENTATION

Technological advancements in the field of in situ water quality monitoring may one day replace the need for automated pump samplers and many costly laboratory analyses.[12,13] While present-day in situ probes can only measure a select few water quality constituents, the data resolution and storage capabilities of these instruments make them invaluable monitoring tools. Common in situ probes employed for water quality analyses include pH, electrical conductivity, temperature, chlorophyll, dissolved oxygen, turbidity, oxidation–reduction potential, and rhodamine for dye tracer studies. Major limitations of using in situ probes are the need to verify calibration and fouling of the probes by contaminants in stream water. Some sensors have self-cleaning devices that greatly reduce the need for maintenance. In some cases, such as ammonium, the probe sensitivity may be too poor for use in many natural waters; however, these probes may work well for enriched systems like wastewater or industrial discharge.

The combination of automated pump samplers and continuous monitoring equipment can provide a powerful approach to understanding changes in water quality over short time steps. Samples collected by the automated pump sampler can also be used to calibrate and/or validate the results obtained from the in situ probes. Similarly, results from the in situ probes can document how representative discrete samples are of water quality along the storm hydrograph. In situ probes can be used to trigger automated pump samplers at any preprogrammed threshold level, but the real value of these probes is in the data they collect between sampling events. Time steps on in situ probes can be set as low as 1 s, so any fluctuation in stream chemistry which may have been missed by an automated sampler is recorded by the data logger associated with the probe. Although in situ probes are limited to measuring only a few water quality constituents, regression analysis with other constituents may yield accurate estimates of values which otherwise would have to be obtained through analysis in a laboratory. The most common use of this technique is with in situ turbidity probes. Samples are collected through a full range of flow and suspended sediment conditions. The suspended sediment calculations determined in the laboratory are regressed against the in situ turbidity probe reading at the time of collection. Once the regression is established, the probe alone (with occasional recalibration and verification) can calculate suspended sediment concentrations at a high resolution resulting in very accurate and cost-effective estimates of suspended sediment loading. Laboratory analysis of water samples is expensive and time consuming; the many applications of in situ probes revolve around these factors. Whether data are needed quickly during a field experiment or cheaply during periods that require intensive sampling, in situ probes are a viable and often necessary tool.

REFERENCES

1. National Research Council, *Assessing the TMDL Approach to Water Quality Management*, National Academy Press, Washington, DC, 2001.

2. L. H. MacDonald, A. W. Smart and R. C. Wissmar, *Monitoring Guidelines to Evaluate Effects of Forestry Activities on Streams in the Pacific Northwest and Alaska*, EPA Publication 910/9-91-001, U.S. Environmental Protection Agency, Washington, DC, 1991.

3. K. W. Tate, R. A. Dahlgren, M. J. Singer, B. Allen-Diaz, and E. R. Atwill, Timing, Frequency of Sampling Affect Accuracy of Water-Quality Monitoring, *Calif. Agric.* **53**, 44–48 (1999).

4. J. M. Holloway and R. A. Dahlgren, Seasonal and Event-Scale Variations in Solute Chemistry for Four Sierra Nevada Catchments, *J. Hydrol.* **250**, 106–121 (2001).

5. J. M. Holloway, R. A. Dahlgren, B. Hansen, and W. H. Casey, Contribution of Bedrock Nitrogen to High Nitrate Concentrations in Stream Water, *Nature*, **395**, 785–788 (1998).

6. S. L. Ponce, *Water Quality Monitoring Programs*. USDA Forest Service Technical Paper WSDG-TP-00002, USDA Forest Service, Fort Collins, CO, 1980.

7. U.S. Geological Survey, *National Field Manual for the Collection of Water-Quality Data*, U.S. Geological Survey Techniques of Water-Resources Investigations, Book 9, 1998. Available: http://pubs.water.usgs.gov/twri9A.

8. A. J. Horowitz, J. J. Smith, and K. A. Elrick, *Selected Laboratory Evaluations of the Whole-Water Sample-Splitting Capabilities of a Prototype Fourteen-Liter Teflon* ^{R} *Churn Splitter*, Open-File Report 01-386, U.S. Geological Survey, Atlanta, GA, 2001.

9. U.S. EPA, *National Pollutant Discharge Elimination System (NPDES) Compliance Sampling Inspection Manual*, U.S. Environmental Protection Agency, Section 5, Washington, DC, 1994.

10. J. Lewis, Turbidity-Controlled Suspended Sediment Sampling for Runoff-Event Load Estimation, *Water Resources Res.* **32**, 2299–2310 (1996).

11. J. Lewis and R. Eads, Turbidity Threshold Sampling for Suspended Sediment Load Estimation, in *Proceedings of the Seventh Federal Interagency Sedimentation Conference*, March 25–29, 2001, Reno, NV, 2000.

12. R. J. Wagner, H. C. Mattraw, G. F. Ritz, and B. A. Smith, *Guidelines and Standard Procedures for Continuous Water-Quality Monitors: Site Selection, Field Operation, Calibration, Record Computation, and Reporting*, Water-Resources Investigations Report 09-4252, U.S. Geological Survey, Reston, VA, 2000.

13. W. M. Jarrell, *Water Monitoring for Watershed Planning and TMDLs*, YSI Incorporated, 2003; available: (www.ysi.com/watershed-tmdl).

GROUND WATER
MONITORING

26

LEVEL MEASUREMENTS IN GROUNDWATER MONITORING WELLS

Willis Weight

Montana Tech. of the University of Montana
Butte, Montana

Environmental Instrumentation and Analysis Handbook, by Randy D. Down and Jay H. Lehr
ISBN 0-471-46354-X   Copyright © 2005 John Wiley & Sons, Inc.

26.1   LEVEL MEASUREMENTS

Determining the direction of groundwater flow is based upon static water-level data. Level measurements in groundwater monitoring wells are one of the most fundamental tasks performed at any site. It is one of the first things done when arriving to a site and performed on a routine basis. Waterlevels can be collected manually with a portable field device or rigged to collect continuous data with datalogger sentinel. It is possible to telemetrically send these data back to the office, powered by a solar panel (Fig. 26.1). Water-level data are of little value without knowing the well-completion information and what the relative elevations

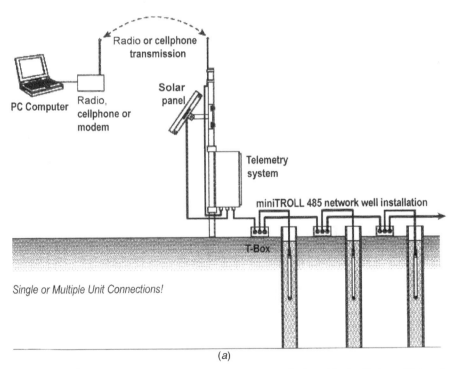

(a)

Figure 26.1   Schematic of an enhanced telemetry system designed by in-situ Inc. of Laramie, Wyoming.

(*b*)

Figure 26.1   (*Continued*)

are. Grave errors can result if one assumes that all water-level data for a given area are collected from the same aquifer unit, ignores recharge and discharge areas, and fails to survey the measuring points!

The purpose of this chapter to present the practical aspects of obtaining level measurement data and to describe the most commonly used devises to obtain them. It is instructive to present the most common sources of error from field mistakes and from idiosyncracies of the equipment to give the reader a better understanding of the pitfalls one can fall into. It is beyond the scope of this chapter to discuss the interpretation of level measurements in wells.

26.1.1   Defining Level Measurements

For a given project area or data set, it is imperative that there be some consistency among those who collect the data. One of the first considerations is where the common datum is. Many use mean sea level (MSL) as a datum. This would mean that all data would be reduced to elevations above MSL. For some projects, an arbitrary relative datum may become the base datum. For example, you are out in the field and there is no convenient way to tie into a bench mark of known elevation. You still need to evaluate the relative elevations or may have security reasons for keeping your database as arbitrary elevations. Keeping exact elevations from being known by others may be important if a public presentation is being conducted in a politically sensitive area. Whatever ends up being used as the base station needs to be a relatively permanent feature that is not likely to be disturbed. It would be disastrous to select a large rock or drive a stake that is later excavated and removed during construction or by vandalism.

Surveying the well locations and measuring points is critical to proper interpretation of the data. Surveying can be done by estimating on a topographic map, by tape and measure, by simple level surveys, total station systems that give northing, easting, and elevation, or by the Global Positioning System (GPS). For example, level surveys provide a relative vertical positioning but are not capable of providing northing and easting positions.

The GPS is very useful for widely spaced monitoring locations, distributed farther than is practical with traditional surveying equipment. The handheld GPS units are widely used by drillers and geologists with accuracies within 10 ft (3 m) if the receivers have wide Area Augmentation System (WAAS) capabilities. The WAAS provides greater accuracy and precision by incorporating ground stations in the United States to correct for ionospheric disturbances timing of the z4 Department of Defense satellites and satellite orbit errors (Garmin, 2003). Without the WAAS, GPS units are typically accurate to within 50 ft (15 m). If a more expensive GPS system is employed, one with a base station and a rover unit on a tripod, accuracies within a fraction of an inch or centimeter can be achieved.

Figure 26.2 Pumping well and monitoring well with flush mount security plate.

26.1.2   Access to Wells

Taking level measurements in wells appears to be straightforward, provided the access to wells is planned for in advance. At a site where water levels are being measured from wells constructed by a variety of contractors, there may be an assortment of security devices and locks. Each may require its own key or require some special way to remove the cap. For example, monitoring wells that are constructed on playgrounds or in paved roadways may have a flush mounting security plate (Fig. 26.2). In this case, a socket or wrench is required to remove the surface metal plate before being able to reach a subsequent lock wisk brooms and screwdrivers are also helpful. This flat completion method allows vehicles to run over the well locations or individuals to run around them without tripping.

26.1.3   Removing a Well Cap

To save on costs for more regional studies, monitoring networks can be expanded by including domestic, stock-water, or other existing wells. Nothing is more frustrating in field work than not to have access to a well. This may be something as simple as forgetting the keys to the well caps or the gate to get in. Many stock or domestic wells typically have a well cap secured by three bolts (Fig. 26.3). These are usually more than finger tight to prohibit animals from knocking off the well

Figure 26.3   Domestic well cap secured with three bolts.

caps and discourage the casual passer by from intruding. An indispensable tool is the handy multitool device purchased at most sporting goods or department stores. These should be equipped with pliers that can be used to loosen the bolts and free the cap. This is a good excuse to tell your spouse that you need one, or it makes a great safety award.

Well caps that become wedged over a long season from grit, changes in air temperature, or moisture can usually be loosened by taking a flat stick or your multitool and tapping upward on the sides of the well cap until it loosens up. This process will eventually allow access to the well. Tight-fitting polyvinyl chloride (PVC) well caps can be modified with slits on the sides of the well cap to allow more flexibility. Another important point to remember is to place the well cap back on and resecure any locking device the way you found it. Well owners can get really angry about such neglect and may not allow a second opportunity for access.

In areas where gas migration takes place (coal mines or landfill areas) gas pressures may buildup under well caps if they are not perforated with a breathe hole. One field hydrogeologist was measuring water levels in a coal mining region and had an idea to kick the well cap. When the cap blew 150 ft into the air, he contemplated what would have happened to his head. This resulted in a company policy of drilling breathe holes in well caps or in the casing to bleed off methane gas pressures. Pressure changes in any monitoring well from changing weather conditions may affect the static water level.

26.1.4 Negotiating Fences

Another obstacle in gaining access to wells in the western United States is from passing through barbed-wire fences. After one obtains permission to obtain level measurements in rural areas, it is important to leave all fences and gates the way they were found. Some gates are tough to open and are tougher to close. To close a tight gate, insert the gate post into the bottom loop of the adjoining fence post and pull the upper part of the gate toward the top loop (sort of like a head lock in the sport of wrestling), with the fence post braced against your shoulder. This allows more leverage.

Some fences may have an electrically hot top wire to keep livestock in. In this case if you touch the wire, you will get shocked. The wire looks distinctively different from the other wires and is located near the top of the fence. Test the wire by gently tossing something metal at the suspected wire. Having rubber boots on can be helpful. Hot wires can be especially problematic for fences keeping buffalo in. Network mogul Ted Turner has a fence in Montana that puts out 5000 V (although the amperage is low) to the unwary animal. It is shocking enough to keep a buffalo away but also strong enough to knock a person down and stop his or her heart from beating for a couple of minutes.

26.1.5 Measuring Points

Once a datum has been established, data collected in a field book is brought back to the office. All persons collecting data should clearly indicate whether data were

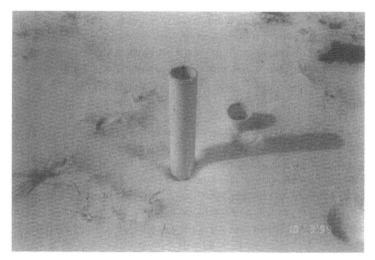

Figure 26.4 Two-inch PVC well with measuring point marked inside the casing.

collected from the top of casing (TOC) or from ground surface (GS), and so forth. Are the data recorded as depth from TOC or depth from GS or from some other point of reference? For example, a well casing cut off below the floor of a shed may be recorded from the shed floor surface. The distance that the casing extends above the ground surface is referred to as the "stick up." All these become important later on as a level survey is conducted and relative elevations need to be established.

It is suggested that markings be clearly made on the casing indicating where to take a measurement. This is especially important for PVC casing cut off at an angle or steel casing cut off with a torch. On steel casing, it is helpful to take a hacksaw or file and make three nick marks close together. A dark permanent maker can also be used to make an upward arrow pointing to the three marks. Marking MP for measuring point is also helpful. A similar arrangement can be used on PVC casing; however, the dark permanent marker works well enough that hacksaw or file markings are usually not necessary (Fig. 26.4). This works best if all markings are made inside the casing. Markings on the outside, even with bright spray paint, tend to fade quickly in the weather. If a name plate is not present, it is also helpful to indicate well name and completion information, such as total depth drilled (TDD) and screen interval on the inside of the well cap.

26.2   HOW TO TAKE A LEVEL MEASUREMENT

The basic procedure of taking a level measurement involves gaining access to the well, removing the well cap, and lowering a device down the well to obtain a reading. Normally, this procedure only takes a matter of a few minutes depending on the well depth, device used, and whether the well is a dedicated monitoring well or whether it is a domestic well being used as a monitoring well. As simple as this procedure sounds, if one is not aware of the weaknesses and idiosyncrasies of the equipment, serious errors in readings can result that affect the interpretation of the data. Each piece of equipment has its own shortcomings that may affect the quality of data. These are discussed in this section and in Section 26.2.6.

Most level measurements are under static conditions. Dynamic conditions occur during a pumping test. If the intention is to take a static water level measurement and the well is in use or was recently used, it is important to allow recovery to equilibrium conditions before taking the measurement. For example, a stock well may be on all the time or the pump activated by a timer. The pump can be shut off for awhile (until there are no changes between readings). Then take a reading and turn the pump back on or reset the timer. Once again, if timers are not reset or pumps turned back on, then well owners may not allow a second opportunity.

26.2.1   Water Level Devices

There are a host of water-level devices used to collect level measurement data. Do not be fooled into thinking one is best, for each of the different devices has its own unique applications. Manual devices such as steel tapes and electric tapes or E-tapes are discussed first followed by those with continuous recording capabilities, such as chart recorders and transducer data-logger combinations.

26.2.2   Steel Tapes

Most data collected before the late 1970s were likely collected with a steel tape. Steel tapes represent the "tried-and-true" method that still has many important applications today. Water levels from the 1940s upward were all measured this way until electric tapes (E-tapes) became the norm. The proper way to use a steel tape is to apply a carpenter's chalk to approximately 5–10 ft (1.5–3 m) of tape before lowering it into the well. The tape is lowered to some exact number next to the measuring point, for example, 50 ft (15 m). The tape is retrieved to the surface, where the "wet" mark on the tape, accentuated by the carpenter's chalk is recorded (Fig. 26.5). A reading such as 50 − 4.32 ft would yield a reading of 45.68 ft depth to water below MP. There is usually historical knowledge of the approximate depth to water to guide this process. Otherwise, it is done by trial and error. Usually, a good reading can be made even on a hot day, with the wet chalk mark, otherwise the tape dries quickly. Getting a clean marking may require more than one trip with the tape, hence its loss in popularity.

Sometimes the only way to get a water level *is* with a steel tape. In many pumping wells, with riser pipe and electrical lines, or wells with only a small access port, the only way to get a water level is with a thin steel tape. Steel tapes are rigid and tend to not get hung up on equipment down the hole. A 300-ft (100-m) steel tape is an important part of any basic field equipment. It can also be used to measure distances between wells or other useful tasks.

Figure 26.5 Level measurement taken with a steel tape. The "wet" mark is accentuated by a carpenter's chalk.

26.2.3 Electrical Tapes (E-tapes)

By the early 1980s, E-tapes began to dominate the market and are currently considered as basic equipment. The first devices were usually marked off in some color-coded fashion every 5 ft (1.5 m), with 10-ft markings and 50- or 100-ft markings in some different color. E-tapes now are usually marked every 1/100th of a foot (3 mm) (Fig. 26.6). All E-tapes function with the same basic principal. A probe is lowered into the water that completes an electrical circuit, indicated by a buzzer sound or an activated light or both. The signal from the water level is transmitted up the electronic cable of the E-tape to the reel where the signal occurs. Each has its own appropriate function. During a pumping test with a generator running, it may be hard to hear a buzzer sound, or you may have a hearing impairment from hanging around drill rigs too long without wearing ear protection. In this case a light is helpful. In very bright sunlight, your light indicator may be hard to detect and the buzzer is more helpful. It should be noted that if oils or other floating organics are present in the well, they will not be detected and may provide a false water level reading.

Most E-tapes have a sensitivity knob from the conductivity sensor in the probe. For example, the sensitivity knob may be a turn dial that ranges from 1 to 10. Probe

Figure 26.6 E-tapes are marked every 1/100th of a foot.

sensitivity is needed for a variety of water qualities. Low sensitivity (1–3) will give a clear signal in high total dissolved solids (TDS) waters; such as those found in Cretaceous marine shales. The higher the TDS in the water, the lower the sensitivity needed to detect a water level. Relatively pristine, or low-TDS waters require that the turn dial be adjusted to a higher sensitivity setting (8–10) to get a clear signal. A good general rule is to put the dial in midrange (4–6) and lower the probe until a buzz/light signal is indicated. If little attention is payed to the sensitivity setting, the hydrogeologist may obtain false readings. This may be one of the first questions asked to a field technician when trying to explain why a water-level reading does not make sense.

To take a water-level reading using an E-tape, turn the device on by moving the sensitivity knob dial to midrange. Lower the probe until the light goes on or the buzzer goes off. This gets you close to where the reading will be. At this point, lift the E-tape line above the depth that activates the buzzer and gradually lower the probe once again until the buzz sound is repeated. Hold this spot on the E-tape line with your thumb nail or with a pointer, like a pencil (Fig. 26.7). The E-tape line should be held away from the measuring point toward the center of the well and shaken lightly to remove the excess water and the process repeated. If the light signal is clear and the buzz is crisp, the reading is probably accurate.

False readings are fairly common if one does not pay attention to the sensitivity knob setting. In some wells, condensation inside the casing can trigger a false reading. The sound of the buzz may not be clear or the light may flicker instead of providing a clear, bright reading. In this case, the sensitivity should be turned down to the lowest setting that yields a good signal. This will be variable, depending on the water quality involved. Cascading water from perforated sections above the actual water level can also be problematic. This can occur during a pumping test, where one is trying to get manual readings in the pumping well that has been perforated at

(a)

(b)

Figure 26.7 Marking the location of the water level on an E-tape.

various depths. In pristine waters, it may be difficult to get a reading at all. In this case, the highest sensitivity knob settings may not be discerning enough to detect when the water table has been reached. Detection may or may not occur until one has submerged the probe well under the water surface. In this case, turn the dial to the high sensitivity range (8–10), and the buzzer or light should give a clearer signal. It is always a good idea to lightly shake the line and repeat the process until a clear reading is obtained.

E-tapes come in a variety of designs. Some have round or flat marked electronic cable lines, and probes come in a variety of sizes. Some are easier to use, but most used by hydrogeologists and field technicians today tend to be marked every 1/100th of a foot (3 mm). Many drillers still use devices marked every 1 or 5 ft. Common spool lengths range from 100 (30.5 m) to 700 ft (213.2 m), although lengths of 2500 ft (762 m) are routinely used in mining applications. It is usually helpful to have a couple of different depth capabilities. There are a number of multiparameter probes that will measure other water quality parameters in addition to static water levels.

Water levels taken in monitoring wells tend to be uninhibited by riser pipe and electrical lines. These pose the fewest concerns as to probe size. Pumping wells, however, may present some challenges. Here, probes may wind around or get hung up in wiring or spacers placed down the well to hold the riser pipe in the central part of the well. Getting hung up can be a real problem, not to mention costly. Here is where steel tapes or probes with a very narrow size can be helpful. In some cases a service call to a well driller may be cheaper than paying for a new E-tape.

26.2.4 Chart Recorders

Many times it is advisable to perform a continuous recording of level measurements. One of the traditional devices that has been used for many years is the chart recorder, such as the Steven's recorder. A device like this uses a drum system onto which a chart is placed. The position of the water level in the well is tracked by having a weighted float connected to a beaded cable that passes over the drum and is connected to a counter weight (Fig. 26.8). As the drum turns forward or rolls backward with the movement of the water level, a stationary ink pen marks the chart. The pen is only allowed to move horizontally to correspond with time set by a timer. Timers can be set for 1 month or up to 3 months. Chart paper is gridded where each column line usually represents 8 hours. The row lines mark the vertical water level changes.

Since the turning of a drum can be significant, control of movement is through setting the ratio of float movement to drum movement of the chart paper. This is

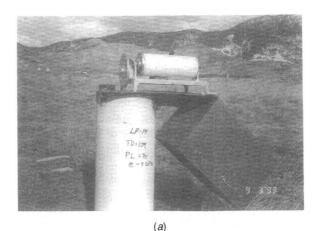

(a)

Figure 26.8 A chart recorder mounted on top of a monitoring well.

(b)

Figure 26.8 (Continued)

known as setting the gear ratio. For example, a 4:1 gear ratio means that the float will move four times the distance the drum would turn forward or backward. This helps keep the pen on the chart paper.

Once the chart is replaced with new paper, the information from the chart has to be reduced or converted into numbers. A technician often logs numbers from a chart into a file for data analysis or hydrograph plotting. Corrections for drift are made by noting the time and date at the end of the chart and then applying a linear correction from the beginning time. For example, if the final marking on the chart is 12 h short (because the timer setting is off) of the supposed time, a correction factor is applied to "stretch" the data to match up with real time. Reduced data can be quickly adjusted on a spreadsheet.

Frequently, wells located near surface water streams are fitted with a chart recorder to compare the groundwater and surface water elevations or in an area where water level fluctuations are significant. Other reasons may be for placement during winter months or remote areas when access is difficult.

The chart recorder is positioned directly above the well on some kind of constructed platform. The platform sits above the well casing, with the float attached to a beaded cable lowered down to the water surface. A hole cut in the platform allows the beaded cable to pass up over the drum and then through another hole cut in the platform for the counter weight. The top of the well casing and the recorder can be encased in a 55-gal drum that can be locked.

Errors and malfunctions can occur in a variety of ways. For example, the batteries can go dead on the timer or the pen can become clogged. The beaded cable occasionally comes off of the drum from lack of maintenance or from rapid water level changes. Replacement floats can be jury-rigged with a plastic [1- or 2-L] soda pop bottle weighted by sand, if one cannot find a replacement float in a timely way. The chart recorder has been phased out and replaced by enhanced telemetry systems or well sentinels.

26.2.5 Enhanced Telemetry Systems

When monitoring conditions change rapidly, wells are located in remote sites, or require real-time data, enhanced telemetry systems (ETS) are becoming a cost-effective way to go. These systems provide wireless signal technology, which provides a greater control over the data, and has the benefit of reducing the number of site visits. The basic layout was shown in Figure 26.1.

Telemetry options include very high and ultrahigh frequency (VHF/UHF) radio, cell-phone modem, license free spectrum radio, or telephone line modem, while power options include either 100–240 VAC line power or solar-charged battery power (In-Situ Inc., 2003). Wireless technology is currently (2003) limited to a distance of tens of miles; therefore, a phone jack is preferred. The technology is changing rapidly, so performing a cost analysis on purchasing such a system is worth looking into.

26.2.6 Transducers and Data Loggers

Transducers are pressure-sensitive devices that "sense" the amount of water above them. The "sensing" is performed by a strain gauge located near the end of the transducer that measures the water pressure. The water pressure can be converted into feet of water above the strain gauge by the following simple relationship:

$$1 \text{ psi} = 2.31 \text{ ft} \tag{26.1}$$

The relationship in Equation (26.1) comes from the pressure head component of the Bernoulli equation:

$$h_p = \frac{P}{\rho^* g} \tag{26.2}$$

where h_p = pressure head (L, ft)
P = pressure (M/LT^2)
ρ = fluid density (M/L^3)
g = gravity (L/T^2)

Equation (26.1) is valid if the following values are put into Equation (26.2): fluid density = 62.4 lb/ feet^3, gravity is 32.2 ft/s^2, and pressure is measured in pounds/ inch^2 and converted into pounds per feet^2.

The above relationship is important in knowing the maximum pressure or how deep the transducer can be placed below the static water surface. Typical ranges of pressure transducers are 5, 10, 15, 20, 30, 50, 100, 200, 300, and 400 psi. For example, a 10-psi transducer should not be placed more than 23 ft below the static water surface. If it is, the transducer may become damaged. Another problem is that the data logger will not record any meaningful data. Personal experience, indicates that the data logger will give level readings of zero. Some 400-psi transducers have been used in deep mine shafts to gauge recovering water levels.

During a pumping test, one should estimate the range of water-level changes that will take place during a test. The pumping well is usually equipped with the highest psi range followed by lower psi ranges for observation wells located at various distances away. If the transducer is to be placed in an observation well, it is a good idea to lower the transducer to an appropriate depth, and let it hang and straighten out before defining the initial level measurement. A common error is to forget to establish the base level in each well, as a last setting in the data logger before

Figure 26.9 Placement of electrician's tape around riser hose and electrical lines.

the beginning of the test. If this is done immediately as the transducer is placed into the well, especially with the placement of the pump, the water levels may not be equilibrated. It is also a good idea to select a base level other than zero. Fluctuations of the water level may occur above the initial base reading, resulting in negative numbers for drawdown. This can be problematic when trying to plot the data in a log cycle. Placement of the transducer in the pumping well should be just above the top of the pump. The transducer should be secured with electricians tape, with additional tape being placed approximately every 5 ft (1.5 m) as the pump and riser pipe are being lowered into the well (Fig. 26.9).

Data loggers record level measurement readings from pressure transducers and store these values into memory. The stored data can be retrieved for analysis on a PC. Transducers and data loggers are indispensable equipment for taking rapid succession readings from multiple wells during a pumping test (Fig. 26.10) or a slug test. Loggers can record data in log-cycle mode, which allows hundreds of readings within the first few seconds of turning the pump on or during the initial recovery phase. Data loggers can also record data on a linear timescale determined by the user. In this way, a data logger is similar to a chart recorder; however, the information stored is already in digital format and does not need to be corrected or reduced.

Care must be taken in the field to make sure pressure transducers are connected to the data loggers. Sometimes even though a proper connection is made, the cable line from the transducer to the data logger may be kinked or damaged, thus prohibiting accurate data collection. Most cables have a hollow tube that allows the strain gauge to sense changes in air pressure. The typical field procedure is to make a loop greater than 1 in. (2.5 cm) with duct tape and attach this to the well casing also with duct tape. The duct-tape loop may slip in hot weather, so a dowel or similar object should be placed. The cable can be protected in heavy-traffic areas by placing

(a)

Figure 26.10 Pumping tests conducted in (a) west-central Montana and (b) near mine-impacted soils.

(b)

Figure 26.10   (*Continued*)

boards on the ground, with the cable located between the boards. Most field technicians are encouraged to obtain manual backup readings to the data-logger readings using E-tapes in case the automatic system fails.

Well sentinels are more commonly being placed into wells to record data over a long time period similar. For example, In Situ Inc. has one called the Troll. Trolls are stainless steel probe-like devices with their own built-in pressure transducers and data loggers. Solinst. Inc. has one called the Levellogger. These can be programmed directly via communications software loaded on a personal computer or a handheld computer (Fig. 26.11). Well sentinels can also be equipped with a variety of water quality probes in addition to a pressure transducer. Measurements of specific conductance, pH, dissolved oxygen, and temperature are also popular options. These can be very helpful in monitoring changes in water quality over time. For example, during a pumping test, changes in water quality may indicate information that is helpful in interpretation. In a case example during a pumping test in mining-impacted waters near Butte, Montana, a steady-state head condition was being approached during the test. It was clear that a recharge source was contributing cleaner water to the pumping well, indicated by a higher pH and lower specific conductance, possibly from a nearby stream or from a gravel channel connected to a nearby stream. The water-quality data nicely augmented the interpretation of the pumping test.

26.3   PRACTICAL DESIGN OF LEVEL MEASUREMENT DEVICES

Based upon the above discussion there are some practical considerations one should be aware of when purchasing level measurement devices. This section will go into

(a)

(b)

Figure 26.11 (a) Programming a mini-troll with a handheld pocket computer. (b) Levellogger well sentinel.

more detail on the designs of probes, cables, and battery placement. Most of these have implications for the popular manual E-tapes, although others will be addressed.

26.3.1 Probe Design

The purpose of the probe is to detect the static water level. Some probes have their detection sensor near the middle, while others are located near the bottom of the probe (Fig. 26.12). With an E-tape, when the static-water surface is encountered,

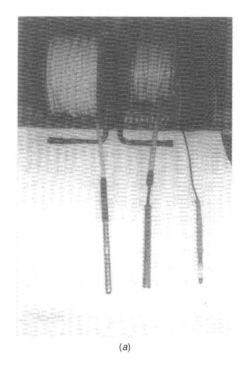

(a)

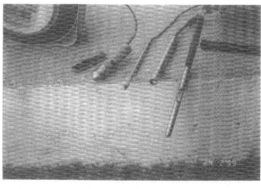

(b)

Figure 26.12 Probe design showing various locations of the sensor.

an electrical circuit is completed that activates a light or audible buzzer. Both are desirable. The sensor may be made of copper or some other metal. Probes are mostly made of stainless steel. The sensor may be accessible on the outside of the probe or through a hole in the probe with the sensor in the middle or the bottom inside of the probe. In the latter case, the bottom part is somewhat hollow like a bell housing with a wire sensor that extends across the hollow section. In the case of a transducer, the size needed depends on the application. Fairly standard are the $\frac{5}{8}$-in. (1.6-cm) models, although $\frac{3}{8}$-in. (0.95 cm) or $\frac{5}{16}$-in. (0.79 cm) are made by most manufacturers. These are sufficient to be used in 1-in. piezometers, but if mini-peizometers of smaller diameters are being measured, significant displacement can occur. This is also one reason for having the sensor at the bottom of the probe.

It was mentioned that sometimes false readings are made when the E-tape probe scrapes off moisture down the side of the casing. Condensation is especially problematic in hot weather or very cold weather. The hole in the middle and bell-like probe housing with the sensor near the bottom is designed to avoid this problem. Whether this is a problem or not depends on the application. For monitoring wells, probe diameters are not a problem down to a well diameter of about 1 in. (2.5 cm), however for minipiezometers or well plates with small access ports probe diameters may be too large. The shape of older probes also tended to be more cylinder and "blocky-like" with abrupt edges, which provided places for them to become hung up on protrusions in the well, like centralizers, or wiring in pumping wells. Newer designs have probes that are more sleek, tapered, and have greater depth capabilities.

Smaller well probes, usually equipped with an outside sensor, tend to be tapered and smaller diameter. The minimum size is down to $\frac{1}{4}$ of an inch (6.4 mm), which easily allows measurement in a $\frac{1}{2}$-in. (1.3-cm) minipiezometer, or access through a port in the top of a well plate is designed by Solinst Inc. These can be very effective for shallow wells with depth capacities usually less than 100 ft (30.4 m). Another example of this kind of design is by Slope Indicator, which has a $\frac{5}{16}$-in. (8-mm) probe with depth capacity up to 700 ft (213 m).

Sometimes you may wish to get a water level in a production well that does not have convenient access. In this instance, instead of having a rigid probe, a string of brass or stainless steel beads covers the water level sensor (this design is carried by drillers more than anyone else). These devices will snake through a difficult access port better than anything else. They are marked every foot (30.5 cm) or every 5 ft (1.5 m) with a brass plate located on the cable for each marking. Once again, these may give false readings from scraping along the casing.

26.3.2 Electronic Cable Design

Another variation in level measuring devices is in the cable and reel design. Most E-tapes have a crank handle that allows the cable to spool onto a reel. Some are free standing while others require the person to handle the whole thing to maneuver it. Some manufacturers provide both capabilities, with the free-standing design requiring an extra charge. Having some kind of free-standing capability is desirable for taking repeated measurements and wanting to set the device down in between readings. This helps keep the reel, probe, and cable cleaner. Another useful design is a simple braking stop that keeps the reel from turning when it is helpful to "hold" a position. This is usually a plastic or rubber stop that can be twisted in or out or a lever arm that is moved half of a rotation or so to keep the reel from turning.

Some reels and spooling designs have too small a housing for the cable. When the probe is fully reeled in, some of the wraps may slip off creating a tangled mess. Whenever too much cable is placed on too small a reel, there will be problems. Another aspect of level measuring devices is the cable design itself. Electronic cables come in a flat, tape-like design, round, or variations in between. Most of the tape designs have the electronics fully encapsulated in an unbreakable ultraviolet radiation-resistant plastic, with two reinforcing wires or thin metal cables that pass through the outer edges for added strength and resistence to cable stretch. This is a good idea. Markings occur on both sides in hundredths of a foot and fractions of a meter. A drawback to this cable design is in the spooling on the reel. The tape-like design allows easier slippage on the reel. Newer models allow for replacement lines in the field.

Round electronic cable designs are also marked to 1/100th of a foot (3 mm) but have no capability for having metric markings without another separate E-tape. The choice is either in feet or meters, not both. The round electronic cable has a reinforced mesh design around the transmitting signal wire for strength. Round cables tend to not slip like the flat tape-like cables do, but one worries more about cable stretch with round-cable measurements, especially in measurements that are deep.

26.3.3 Other Errors in Level Measurements

Another aspect of any measurement is whether the kinks and bends in the electronic cable lead to inaccurate readings. This can also be argued with steel tapes too. Generally, as one uses the same device for each measurement, the results tend to be consistent. If minor changes occur in the data that are significant to a project, one should investigate whether there was a change in field technicians or equipment or both.

Other discrepancies could be the result of improper reading of the measuring device. There are less errors associated with taking readings from E-tapes marked every 1/100th of a foot (3 mm) than with those marked every foot (30.5 cm) or every 5 ft (1.5 m). Typically errors amount to reading a 6 or a 9 upside down. If one is always consistent at reading from the marking below or above, it is easier to detect errors later on. If one arbitrarily changes how to follow protocol in the field, serious interpretation problems may result. For example, in a high transmissivity aquifer 1/10th (0.03 m) of a foot can make a big difference in interpreting the direction or rate of groundwater flow.

26.3.4 Battery Location

As obvious as it sounds, the location of where batteries are housed can be a big factor on level measurement device maintenance. Some newer models and many of the earlier models required that all of the cable be unreeled before having access

Figure 26.13 Batteries located on the reel housing provide good design.

to the battery compartment. Some allow the reel housing to become separated to have access to the batteries. A good design is where the batteries are quickly accessible through a plate cover on the reel housing (Fig. 26.13). This a simple remove-the-plate and change-the-batteries scenario. When purchasing a level measuring device, accessibility and easy maintenance can be a deciding factor if all other features are the same. Most level measuring devices are powered with a 9-V battery.

Level measuring devices also have a battery or light tester button. One merely presses a test button and the light or buzzer goes off or both. Some allow the user to choose whether they want to have the light active or the buzzer active and can test each separately. If the buzzer signal or light is weak, it is a good idea to replace the batteries. Having extra batteries on hand is always a good idea. The multitool is helpful for removing a few screws, should changes of batteries be required.

26.4 OTHER PRACTICAL APPLICATIONS

Some of the problems associated with level measurements come from well completion rather than device design. For example, pumping wells arranged with spacers and centralizers are more likely to encounter a hangup of the equipment than a monitoring well dedicated only to water level measurements. Once a level measurement device becomes stuck, how can it be retrieved from the well? Should E-tapes be used to sound the depth of wells? Additionally, what should be done about decontamination of equipment before lowering into other wells?

26.4.1 Retrieving Lost Equipment

When pumping wells are being used to obtain level measurement readings, there is always a risk of getting the equipment stuck. This may be during a pumping test, where an operator is trying to follow the pumping level in a well, with an E-tape or simply trying to obtain a level measurement from a domestic well. Extending upward from the top of the pump is the riser pipe and electrical lines to power

Figure 26.14 Riser pipe and electrical lines extending upward from pump.

the motor (Fig. 26.14). If the pump is fairly deep, plastic centralizers are placed to keep the riser pipe and wiring near the middle of the casing.

In cold climates, pitless adapters are used to make a connection from a water line pipe placed in a trench from the house to the well. An access port is welded to the casing, approximately 6–8 ft below the ground surface to prevent freezing of water lines. An elbow connector from the riser pipe fits into the pitless adapter. This way the riser pipe elbow can be removed from the pitless adapter and the pump can be pulled from the well for servicing.

It is amazing how a level measuring probe can become stuck down a well. When probes become lodged, it is usually because the probe has wound around the riser pipe and electrical lines. Sometimes the probe will slip through the electrical lines and become wedged as it is being retrieved or hung up on a pitless adapter. Some well owners will have a rope or cable attached to the pump that extends up to the surface for pump retrieval. If an E-tape probe threads through a gap between the cable and another obstruction, it will become stuck. There are other cases where the probe actually forms a slip knot during the process of obtaining a level measurement. The electronic cable line loops outward as an obstruction is encountered, and the probe ends up falling back through the loop.

When an E-tape becomes stuck, the first task is to try and free the probe. Depending on how far down the probe is, one can try different methods. Sometimes success occurs from loosening the electronic cable and striving to jiggle the probe free. If this does not work and the probe is within 20 ft (6.1 m), a "fishing" device can be made from one or more sections of 1-in. PVC electrical conduit pipe, typically 10 ft (3 m) long. In the bottom piece, a slit or long notch about 6 in. (15.2 cm) long is cut. The level measuring cable is hooked into the notch. By sliding the conduit pipe toward the probe, there is more leverage available to move the probe around. In the event the well probe really becomes stuck, it may be necessary to pull the pump and riser pipe to retrieve the E-tape. This can be a real pain, particularly in a production well that has a lot of heavy piping. If you want to save the E-tape, it is cost effective to call in a service truck with a cable-and-hoist capability to pull up the piping and pump. The cost of service is a minor percentage of the cost of a new E-tape.

Sometimes if someone pulls really hard on an electronic cable, the line breaks and the cable and probe fall down the well. In this case, after one pulls out the pump and riser pipe, it is possible to "fish out" the offending object with a treble hook attached to 30- to 50-lb test monofilament fishing line. Treble hooks may be helpful in hooking onto things that unwittingly fall down a well.

26.4.2   Using E-tapes to Sound Wells

A common stress placed on E-tape probes is to use them for sounding well depths. One turns off the electronics and lowers the probe until the bottom is "felt". The problem with this is that many probes are not designed to endure the stress of pressures placed on a sensor from sounding tens of feet below the surface. Someone may wonder why his or her probe does not work very well. A question asked by the manufacturer or vender is if you used this probe to sound well depths? If yes, then the vender may balk at providing replacement equipment.

26.4.3   Detangling Equipment

When an E-tape cable line slips off of a reel, a tangled mess can occur. Rather than trying to fight the reel, it is worthwhile to unspool the reel by having someone walk the probe and cable out until the problem area is resolved. Sometimes untangling becomes akin to untangling kite strings or fishing lines. Time and patience can be well worth the effort.

In some cases the electronic cable line may become damaged. It is possible to reattach the probe to a convenient marking on the cable (such as at 100 ft). In this case the repaired device can be used as a backup E-tape. Most manufacturers also have replacement cables. One must compare the economics of purchasing a new probe over the time and effort to repair or replace a new electronic cable line.

26.4.4   Decontamination of Equipment

As an operator performs a level measurement survey of a series of monitoring wells, it is important to decontaminate, or DECON, the equipment. Merely going from one well to the next with the same device may cause cross-contamination. DECON practices will vary from company to company. Something as simple as rinsing with an Alconox–water solution to steam cleaning or a submergence bath may be required. A small amount of Wooltite into a bucket of water makes a great DECON solution. It depends on the nature of the site. If the monitoring is for trends in water levels, a simpler DECON step would be taken, compared to monitoring wells completed in a plume of toxic organics.

Newer level measuring devices are designed to have the electronic reel built as a module that can be removed for DECON purposes or the electronics are encased in an epoxy coating for easier decontaminating. It is significant to realize that certain chemicals may adsorb onto probes. Most probes are constructed of stainless steel because this tends to be less reactive than other materials, except in mining-impacted waters where pHs are low and dissolved metals are high. It is also helpful to have clean looking equipment when you go knock on a landowner's door.

26.4.5 Summary

Level measurement devices are necessary to obtain static water level data. Each has its own application and design. Generally, it may be helpful to have a variety of designs and depth capabilities.

26.5 SUMMARY OF MANUAL METHODS

1. Establish a common datum. It may be worth defining a particular elevation above MSL so that maps of different scales can be constructed later on.
2. Make sure that the base station is a relatively permanent feature, unlikely to be moved or removed later on.
3. Do not forget your keys, wrenches, or multitool to remove the well cap.
4. Establish a measuring point (MP) physically on the well casing or the location from which all water level measurements will take place.
5. To prohibit affects from weathering, make all markings on the inside of the well or under the well cap.
6. Use a carpenter's chalk with a steel tape for a clearer reading.
7. E-tapes should have sensitivity dials adjusted appropriate to the water quality to get a crisp, clear buzz or light indication.
8. Always move the E-tape cable line to the middle of the well, shake lightly, and repeat to make sure the reading is the same.
9. Record the depth to water from TOC or GS. The "stick up" should also be measured and recorded. Remember to put the well cap back on after the reading and secure the well.
10. Do not use your E-tape to sound well depths.
11. Make sure your field notes are clear because getting back to some wells again may be difficult.

If wells are locked or inside locked gates, one must remember the appropriate keys or tools necessary to gain access. Multitools are handy to remove tightened bolts and loosen wedged well caps. Steel tapes are helpful to gain access to small openings in well caps and can be used to measure field distances. E-tapes are the most common level measuring device and come in a range of designs. Probe size is important for small piezometers and pumping wells, and proper care must be taken to ensure the readings are accurate. Flat electronic cable designs are easy to read, with English and metric markings and resist stretching, but may slip on the reel. Round cable designs do not slip but have markings on one side and may stretch on deeper readings. Other design features, such as the location of the battery housing, may be important for choosing between devises and maintenance capabilities.

Chart recorders have been helpful in monitoring continuous hydrograph data in the past. Chart recorder systems are being replaced by enhanced telemetry systems or well sentinels. Phone jacks provide the greatest distances for sending informations although "wireless" technologies are now available for monitoring sites within tens of miles of the office. These are the way to go for obtaining "real-time" data, while well sentinels can be up-loaded at the end of a season.

Pressure transducers and data loggers are indispensable for rapid succession readings needed during pumping and slug tests. These are more expensive but are easier to manipulate during the data reduction and analysis stage.

26.6 SUMMARY OF AUTOMATED METHODS

1. Lower each transducer down the well early on to allow the cable to straighten and then spool each transducer cable to the data logger. Allow a 1-in. loop in the cable (to sense air pressure changes), secured with duct tape and then taped to the well so that it does not move. Place a dowel in the loop.

2. Check the psi range of the transducer and make sure the transducer is not lowered into a water depth that exceeds its capacity.

3. Attach the transducer in the pumping well above the pump with electricians tape and secure the cable to the riser pipe every 5 ft (1.5 m) to the surface.

4. Make sure each transducer is connected to the data logger. (This can be checked from the data logger.)

5. Establish the base level as a last item before starting the pumping test and use a value other than zero.

6. Back up the automated system with manual E-tape level measurements.

Getting level measurement devices stuck down a well is a real possibility in pumping wells. Fishing techniques may be used as a first step, but it may be necessary to call a service rig to free an E-tape probe. This would only be a percentage of the cost of a new probe. Decontamination of level measuring devises is essential to prevent cross-contamination. Simple rinsing with an Alconox–water solution or steam cleaning can do a good job. Having clean equipment is important when showing up at the door of a landowner.

27

LABORATORY ANALYSIS OF WASTEWATER AND GROUNDWATER SAMPLES

Lawrence R. Keefe

Terracon
Colorado Springs, Colorado

Environmental Instrumentation and Analysis Handbook, by Randy D. Down and Jay H. Lehr
ISBN 0-471-46354-X   Copyright © 2005 John Wiley & Sons, Inc.

27.1   INTRODUCTION

The intent of this chapter is provide a overview of U.S. Environmental Protection Agency (USEPA)–approved laboratory methods for the analysis of wastewater, groundwater, soil, sediment, and sludge. The focus of this chapter is on the most common types of laboratory instrumentation and identification of the various analytical methods used to analyze environmental samples. Because of length limitations, the theory behind the chemistry and methods is not discussed in detail.

The information presented herein is based primarily on USEPA guidance documents, including SW-846;[1] EPA 200, 500, and 600 Series, Standard Methods, and the Contract Laboratory Program (CLP). Cross-reference tables identifying available methods for common compounds and analytes are provided at the end of this chapter.

27.1.1   Overview of Development of Analytical Methods

The Comprehensive Environmental Response, Compensation, and Liability Act of 1980 (CERCLA) and the Superfund Amendments and Reauthorization Act of 1986 (SARA) granted legislative authority granted to USEPA to develop standardized analytical methods for the measurement of various pollutants in environmental samples from known or suspected hazardous-waste sites.

Under this authority, USEPA developed standardized analytical methods for the measurement of various pollutants in environmental samples.[2,3] These include volatile, semivolatile, pesticide, and Aroclor [polychlorinated biphenyl (PCB)] compounds that are analyzed using gas chromatography (GC) and gas chromatography coupled with mass spectrometry (GC/MS). Inorganic analytes, including various heavy metals, are analyzed using atomic absorption (AA) and inductively coupled plasma (ICP) techniques.

Through use of these standardized procedures, the chemical analyses can produce data of known and documented quality.

27.1.2   Selection of Analytical Methods

The selection of the appropriate analytical method is based on several factors including the compounds (either singularly or by class) or analytes of interest; intended use of the data, the required detection limits or practical quantification limits, the media being analyzed (i.e., drinking water, wastewater, sludge, or soil), and cost. The following sections provide a brief description of most commonly used instruments used in laboratory for analysis of environmental samples. Following these descriptions, Tables 27.1 and 27.2 (in Section 27.3) provide a cross-references for common test methods and analytes.

27.1.3   Available Resources

A partial listing of available resources is provided in the references and further reading lists. In addition, users should consult with the analytical laboratory to best determine the appropriate methods based on their data objectives and requirements.

27.2 OVERVIEW OF TYPICAL INSTRUMENTATION

Provided below is a brief overview of typical instrumentation that are used in the laboratory for analysis of environmental samples. Because of length limitations, theories and principles of operation of these instruments are not described. If interested, the reader should refer to the references and further additional reading lists for this information.

27.2.1 Gas Chromatography

Chromatography is a process used to separate and/or identify similar compounds by allowing a solution of the compounds to migrate through or along a substance that selectively adsorbs the compounds in such a way that materials are separated into zones. An example of this process is the separation of colors in a drop of ink placed on a piece of paper and then dampening the paper. As the water moves up through the paper, it separates the different dye colors in the ink.[4] A mixture of compounds can be separated using the differences in physical or chemical properties of the individual components. Some useful chemical properties by which compounds can be separated are solubility, boiling point, and vapor pressure.

Gas chromatography (GC) is a chromatographic technique that can be used to separate organic compounds to provide both qualitative and quantitative information about the compounds. The GC is often used for the analysis of samples containing volatile and semivolatile organic compounds, chlorinated herbicides and pesticides, and PCBs.

A GC consists of a flowing mobile phase, an injection port, a separation column, and a detector. The compounds are separated as a result of differences in their partitioning behavior between the mobile gas phase and the stationary phase in the column. A variety of detectors can be used depending on the compounds of interest. These include flame ionization detectors (FIDs), photoionization detectors (PIDs), electron capture, nitrogen–phosphorus, and hall electrolytic conductivity detectors (HECDs).

The GC is considered a selective instrument in that its detector is selected for the compounds of interest. For this reason, the GC can be used to detect known target compounds at lower practical quantification limits (PQLs) and lower cost than the other methods. However, this can lead to false-positive detection, a disadvantage of the GC method.

27.2.2 Gas Chromatography/Mass Spectroscopy

Gas chromatography/mass spectroscopy (GC/MS) is a GC with a mass spectrometer (MS) detector. Similar to GC, GC/MS is often used to analyze samples containing unknown volatile and semivolatile organic compounds, chlorinated herbicides and pesticides, and PCBs. It is often the workhorse of organic analysis.[5]

The sample is injected into the GC/MS where the organic compounds are fragmented by a stream of electrons producing positive ions. The MS detector relies on the difference in mass : charge ratio of ionized molecules to separate them from each other. The MS detector responds to each ion and records a corresponding peak. The height of each peak is proportional to concentration of each element present in the compound. The pattern of the series of peaks, called the *mass spectrum*, is compared to the mass spectrum of known compounds to identify the unknown compounds present in the sample.

Because of its MS detector, the GC/MS provides dual confirmation of the presence and concentrations of target compounds with less frequency of false positives as compared to the GC. However, because the MS is nonselective, the PQLs can be higher than the GC.

27.2.3   Flame and Graphite Furnace Atomic Absorption Spectroscopy

Atomic absorption (AA) spectroscopy is an analytical technique commonly used to identify and quantify the concentration of metals in an environmental sample. AA spectroscopy is based on measurement of the absorbance of the wavelength of light emitted from metals from within the sample. AA requires vaporization of the sample using a high-temperature source such as a flame or graphite furnace. In AA, the sample is subjected to temperatures high enough to cause dissociation into atoms and ionization of the sample atoms to take place. Once the atoms or ions are in their excited states, they decay to lower states, resulting in the emission of light wavelengths specific to the elements present in the sample. In AA, these wavelengths are measured and used to determine the concentrations of the elements of interest.

Flame AA uses a flame burner to vaporize the sample. A sample solution is usually drawn into a mixing chamber using a gas flow to form small droplets before entering the flame. The use of flame AA is generally limited to solutions.

A graphite furnace is an AA spectrophometer that uses an electric current to vaporize the sample. The graphite furnace is more efficient than flame AA since it can directly accept very small quantities of solutions, slurries, or solid samples. Environmental samples are placed directly in the graphite furnace, and the furnace is electrically heated in several steps to dry the sample, ash organic matter, and vaporize the atoms.

27.2.4   Inductively Coupled Plasma

An inductively coupled plasma (ICP) is a very high-temperature (7000–8000-K) excitation source that efficiently vaporizes, excites, and ionizes atoms. Molecular interferences are greatly reduced with this excitation source but are not eliminated completely. ICP sources are used to excite atoms for atomic emission spectroscopy and to ionize atoms for mass spectrometry.

Samples in solution are pumped into a nebulizer that produces a fine mist, entrained in argon. The argon gas is inductively heated to about 10,000°C, forming a plasma (a type of electrical discharge) containing electrons and positively charged atoms. A portion of the plasma is introduced into a mass spectrometer, where positive ions in the plasma are separated. Because each element produces a unique mass spectrum, the identification and concentration of each element can be measured from the intensity and uniqueness of the signal.

ICP is a versatile technique, due not only to the large number of elements that can be simultaneously and quickly determined at trace levels but also to the wide variety of sample types that can be analyzed.

27.2.5 Cold-Vapor Atomic Absorption

The presence an accuracy of the mercury in environmental samples is one of the most difficult metals to determine, primarily because of its volatility under ambient temperature and conditions. Cold-vapor atomic absorption (CVAA) utilizes the volatility of mercury at room temperature. Ironically, this volatility is what makes mercury difficult to determine by other absorption and emission techniques such as atomic absorption spectroscopy (AAS) or ICP.

27.3 SUMMARY OF ANALYTICAL METHODS

A brief summary of laboratory methods for analysis of environmental samples is provided below. Although numerous analytical methods are available, those discussed below are based on common USEPA-approved methodologies. As mentioned above, the laboratory should be consulted to determine the most appropriate method(s) to meet the data and project objectives.

27.3.1 Analytical Methods for Organic Constituents

One of the most common groups of organic compounds of interest in the environment is the Target Compound List (TCL). These compounds were originally derived from the USEPA Priority Pollutant List, which consisted of 129 compounds. Since development of the priority pollutant list in the early 1980s, compounds have been added to and deleted to the TCL, based on advances in analytical methods, evaluation of method performance data, and the needs of programs regulated by USEPA. Currently, the TCL includes 125 organic compounds consisting of 33 volatile, 64 semivolatile, and 28 pesticide/ PCB target compounds in water and soil/sediment environmental samples.

A partial listing of available analytical methods for analysis of organic compounds in environmental samples are summarized in Table 27.1, which presents the analytical techniques and analysis methods based on current EPA-approved test procedures.[1]

Table 27.1 Partial Listing of Analytical Methods for Analysis of Organic Compounds in Environmental Samples

Instrumentation Method / EPA Method	GC 8010	GC 8015	GC 8020	GC 8021	GC/MS 8040	GC/MS 8060	GC 8080	GC 8100	GC 8120	GC 8150A	GC/MS 8240	GC/MS 8260	GC/MS 8270	HPLC 8310
Volatile Organic Compounds														
1,1,1-Trichloroethane	•		•	•							•	•		
1,1,2,2-Tetrachloroethane	•		•	•							•	•		
1,1,2-Trichloroethane	•		•	•							•	•		
1,1-Dichloroethane	•		•	•							•	•		
1,1-Dichloroethene	•		•	•							•	•		
1,2-Dichloroethane	•		•	•							•	•		
1,2-Dichloroethene (total)											•			
1,2-Dichloropropane	•		•	•							•	•		
2-Butanone		•												
2-Hexanone											•			
4-Methyl-2-pentanone		•									•			
Acetone											•			
Benzene			•	•							•	•		
Bromodichloromethane	•		•	•							•	•		
Bromoform	•		•	•							•	•		
Bromomethane	•		•	•							•	•		
Carbon disulfide											•			
Carbon tetrachloride	•		•	•							•	•		
Chlorobenzene	•		•	•							•	•		
Chloroethane	•		•	•							•	•		
Chloroform	•		•	•							•	•		
Chloromethane	•		•	•							•	•		
cis-1,3-Dichloropropene											•	•		
Dibromochloromethane	•		•	•							•	•		
Ethylbenzene			•	•							•	•		
Methylene chloride	•		•	•							•	•		

Table 27.1 (*Continued*)

Instrumentation Method	GC	GC	GC	GC	GC/MS	GC/MS	GC	GC	GC	GC	GC/MS	GC/MS	GC/MS	HPLC
EPA Method	8010	8015	8020	8021	8040	8060	8080	8100	8120	8150A	8240	8260	8270	8310
Styrene	•													
Tetrachloroethene			•	•							•	•		
Toluene			•	•							•	•		
m-Xylene				•							•	•		
o-Xylene			•	•							•	•		
p-Xylene			•	•							•	•		
Total xylene														
trans-1,3-Dichloropropene	•										•	•		
Trichloroethene	•		•	•							•	•		
Vinyl chloride	•		•	•							•	•		

Semivolatile Organic Compounds

	GC	GC	GC	GC	GC/MS	GC/MS	GC	GC	GC	GC	GC/MS	GC/MS	GC/MS	HPLC
	8010	8015	8020	8021	8040	8060	8080	8100	8120	8150A	8240	8260	8270	8310
1,2,4-Trichlorobenzene													•	
1,2-Dichlorobenzene									•				•	
1,3-Dichlorobenzene									•				•	
1,4-Dichlorobenzene									•				•	
2,4,5-Trichlorophenol					•								•	
2,4,6-Trichlorophenol					•								•	
2,4-Dichlorophenol					•								•	
2,4-Dimethylphenol					•								•	
2,4-Dinitrophenol					•								•	
2,4-Dinitrotoluene													•	
2,6-Dinitrotoluene													•	
2-Chloronaphthalene									•				•	
2-Chlorophenol					•								•	
2-Methylnaphthalene													•	
2-Methylphenol													•	
2-Nitroaniline													•	

599

Table 27.1 (*Continued*)

Instrumentation Method EPA Method	GC 8010	GC 8015	GC 8020	GC 8021	GC/MS 8040	GC/MS 8060	GC 8080	GC 8100	GC 8120	GC 8150A	GC/MS 8240	GC/MS 8260	GC/MS 8270	HPLC 8310
2-Nitrophenol					•								•	
3,3'-Dichlorobenzidine													•	
3-Nitroaniline													•	
4,6-Dinitro-2-methylphenol					•								•	
4-Bromophenyl phenyl ether													•	
4-Chloro-3-methylphenol					•								•	
4-Chloroaniline													•	
4-Chlorodiphenylether													•	
4-Methylphenol													•	
4-Nitroaniline													•	
4-Nitrophenol					•								•	
Acenaphthene									•				•	•
Acenaphthylene									•				•	•
Anthracene									•				•	•
Benzo(*a*)anthracene									•				•	•
Benzo(*a*)pyrene									•				•	•
Benzo(*b*)fluoranthene									•				•	•
Benzo(*ghi*)perylene									•				•	•
Benzo(*k*)fluoranthene									•				•	•
Bis(2-chloroethoxy) methane													•	
Bis(2-chloroethyl) ether													•	
Bis(2-chloroisopropyl) ether													•	
Bis(2-ethylhexyl) phthalate						•							•	
Butyl benzyl phthalate						•							•	
Carbazole													•	
Chrysene								•	•				•	•
Di-*n*-butyl phthalate						•							•	•

Table 27.1 (*Continued*)

Instrumentation Method	GC	GC	GC	GC	GC/MS	GC/MS	GC	GC	GC	GC	GC/MS	GC/MS	GC/MS	HPLC
EPA Method	8010	8015	8020	8021	8040	8060	8080	8100	8120	8150A	8240	8260	8270	8310
Di-*n*-octyl phthalate						•							•	
Dibenzo(*a,h*)anthracene								•					•	•
Dibenzofuran													•	
Diethyl phthalate						•							•	
Dimethyl phthalate						•							•	
Fluoranthene								•					•	•
Fluorene								•					•	•
Hexachlorobenzene									•				•	
Hexachlorobutadiene									•				•	
Hexachlorocyclopentadiene									•				•	
Hexachloroethane									•				•	
Indeno(1,2,3-*cd*)pyrene								•					•	•
Isophorone													•	
N-Nitroso-di-*n*-propylamine													•	
N-nitrosodiphenylamine													•	
Naphthalene				•				•					•	•
Nitrobenzene													•	
Pentachlorophenol					•								•	
Phenanthrene								•					•	•
Phenol					•								•	
Pyrene								•					•	

Pesticides/Herbicides/PCBs

	GC	GC	GC	GC	GC/MS	GC/MS	GC	GC	GC	GC	GC/MS	GC/MS	GC/MS	HPLC
	8010	8015	8020	8021	8040	8060	8080	8100	8120	8150A	8240	8260	8270	8310
4,4'-DDD							•							
4,4'-DDE							•							
4,4'-DDT							•							
Aldrin							•						•	
α-BHC							•						•	
α-Chlordane							•						•	

Table 27.1 (Continued)

Instrumentation Method EPA Method	GC 8010	GC 8015	GC 8020	GC 8021	GC/MS 8040	GC/MS 8060	GC 8080	GC 8100	GC 8120	GC 8150A	GC/MS 8240	GC/MS 8260	GC/MS 8270	HPLC 8310
Aroclor 1016							●						●	
Aroclor 1221							●						●	
Aroclor 1232							●						●	
Aroclor 1242							●						●	
Aroclor 1248							●						●	
Aroclor 1254							●						●	
Aroclor 1260							●						●	
β-BHC							●						●	
δ-BHC							●						●	
Dieldrin							●						●	
Endosulfan I							●						●	
Endosulfan II							●						●	
Endosulfan sulfate							●						●	
Endrin							●						●	
Endrin aldehyde							●						●	
Endrin ketone							●						●	
γ-BHC (Lindane)							●						●	
γ-Chlordane							●						●	
Heptachlor							●						●	
Heptachlor epoxide							●						●	
Methoxychlor							●						●	
Toxaphene							●						●	
2,4-D										●				
2,4-DB										●				
Dalapon										●				
Dicamba										●				
Dinoseb										●				
2,4,5-T										●				
2,4-TP (Silvex)										●				

Table 27.2 Summary of Analytical Techniques for Inorganic Analytes

	AAS Methods		ICP Methods	
Element	SW-846 700 Series	EPA 200 Series	SW-846 6010A	EPA 200 Series
Aluminum	●	●	●	●
Antimony	●	●	●	●
Arsenic	●	●	●	●
Barium	●	●	●	●
Beryllium	●	●	●	●
Cadmium	●	●	●	●
Calcium	●	●	●	●
Chromium	●	●	●	●
Cobalt	●	●	●	●
Copper	●	●	●	●
Iron	●	●	●	●
Lead	●	●	●	●
Magnesium	●	●	●	●
Manganese	●	●	●	●
Mercury	●	●		
Nickel	●	●	●	●
Potassium	●	●	●	●
Selenium	●	●	●	●
Silver	●	●	●	●
Sodium	●	●	●	●
Thallium	●	●	●	●
Vanadium	●	●	●	●
Zinc	●	●	●	●

Source: 1993 EPA, *methods for determination of inorganic substances in environmental samples*, EPA-600R-93-100, Aug. 1993.

27.3.2 Analytical Methods for Inorganic Constituents

Similar to the TCL, one of the most common list of inorganics of interest in the environment is Target Analyte List (TAL). As with the TCL, these analytes were also originally derived from the USEPA Priority Pollutant List and currently includes 23 analytes. Available analytical methods for analysis of inorganic compounds in environmental samples are summarized in Table 27.2, which presents the analytical techniques and analysis methods.

REFERENCES

1. *Test Methods for Evaluating Solid Waste*, USEPA, SW-846, 1994.
2. *Multi-Media, Multi-Concentration, Inorganic Analytical Service for Superfund*, USEPA, Publication 9240.0-09FSA, Nov. 1996.

3. *Multi-Media, Multi-Concentration, Organic Analytical Service for Superfund*, USEPA, Publication 9240.0-09FSA, Nov. 1996.

4. L. H. Stevenson and B. C. Wyman, *The Facts on File Dictionary of Environmental Science*, Facts on File, Inc., New York, 1990.

5. RECRA Environmental, Inc., Quality Assurance Management Program, RECRA Environmental, Buffalo, New York. 1997.

FURTHER READING

1. APHA, *Standard Methods for the Examination of Water and Wastewater, 18th ed.*, American Public Health Association, American Water Works Association, Water Pollution Control Federation, Washington, DC, 1992.

2. USEPA, *Methods for Chemical Analysis of Water and Wastes, 3rd ed.*, USEPA, 600/4-79-020, 1983.

3. USEPA, W. Mueller, and D. Smith, *Compilation of EPA's Sampling and Analytical Methods*, Lewis Publishers, Chelsea, MI, 1992.

28

TECHNIQUES FOR GROUNDWATER SAMPLING

Robert M. Powell

Powell & Associates Science Services
Las Vegas, Nevada

Environmental Instrumentation and Analysis Handbook, by Randy D. Down and Jay H. Lehr
ISBN 0-471-46354-X   Copyright © 2005 John Wiley & Sons, Inc.

605

28.1   INTRODUCTION

Groundwater sampling techniques have evolved from the determination of water yield and general groundwater quality to techniques that must delineate extremely low concentrations of contaminant(s) in three dimensions and the changes in these concentrations over time. This evolution has resulted in the development of ground water contaminant monitoring methods referred to as *low-flow purging and sampling* or *passive sampling* techniques. These techniques are distinct from the procedures that have traditionally been used to sample water wells, namely high–flow rate purging and sampling or well bailing, which incorporate the removal of several borehole or casing volumes of water before collecting the water samples. Low-flow techniques seek to minimize disturbance to the well water and to reduce the volume of water that must be removed from the well before collecting the sample. The objectives of withdrawing water from monitoring wells (to sample contaminants and delineate their flow paths with high sensitivity) and water supply wells (to provide high water yield while minimizing contaminant concentrations and particulates in the supply via filter packs, dilution, etc.) are quite different. Fortunately, this difference is being recognized by most practitioners and state and federal regulators. Low-flow and passive sampling techniques are being widely adopted as the groundwater monitoring methodologies of choice.

The techniques of low-flow and passive purging and sampling of groundwater have not arisen suddenly but have developed gradually during the last two decades of environmental investigation. The development of these concepts required numerous observations by investigators and analysts over the course of countless sampling events, combined with the ongoing accumulation of knowledge about subsurface processes and the factors that influence these processes. The development of databases of groundwater analytical information and the ease with which data sets could be compared also contributed to our awareness by making data inconsistencies obvious. It gradually became apparent that something was wrong with the way groundwater was being sampled.

Sample reproducibility using high–flow rate pumping and bailing was an obvious problem, especially when two or more sampling devices were compared or when the same samples were handled differently in the field. Samples were often laden with sediment, although the initial water removed from the screened interval might be clear. Filtration visually clarified the samples, but secondary effects were often manifested by this procedure. When sample results came back from the laboratory, one was often confronted with a choice of which well data to believe, the acid digestion of the sediment-bearing sample or the field-filtered acidified solution. In fact, which of a pair of duplicate samples could be trusted was often in question since they often carried differing sediment loads and yielded different analytical results.

Fortunately, the solution to most of these problems was very simple and, in hindsight, fairly obvious: Slow the pumps. Do not "plunger" the well with a bailer or with a pump. Take care not to disrupt the water, the sand pack, or the outlying aquifer materials when collecting the sample. The reward for this approach is better data quality in terms of both accuracy and reproducibility. A bonus is generally the production of less purge water for expensive transport and disposal.

28.2 GROUNDWATER SAMPLING OBJECTIVES

Many problems associated with groundwater sampling can be eliminated simply by establishing the goals and objectives of the monitoring program and its constraints before implementation. Sampling of groundwater for contaminants is typically done to achieve one of the following goals, several of which are interrelated:

- To investigate the presence or absence of contaminants
- To delineate a plume
- To determine the concentrations of contaminants at specific points in a plume at a given time
- To understand the transport and fate of contaminants in the system
- To carry out regulatory compliance monitoring
- To evaluate a treatment system through remediation performance monitoring

The common factor in achieving these objectives is that analytical data resulting from groundwater samples must accurately represent the contaminant concentrations and geochemistry of the subsurface at the points in space and time where the samples were acquired. The usefulness of such accurate groundwater data is illustrated in Figure 28.1 for monitoring wells M_1 and M_2 at times t_1 and t_2 and pump locations M_{1xyz} and M_{2xyz}. Proper sampling at these x, y, and z spatial coordinates

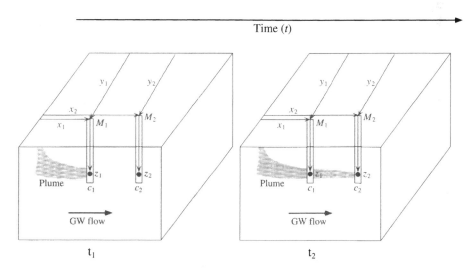

Figure 28.1 Usefulness of accurate groundwater data.

provides information about the presence or absence of the contaminant at those points, its concentrations (c_1 and c_2) at those coordinates, whether these wells are in regulatory compliance, and how these concentrations are changing over time (e.g., $t_1 c_2$ vs. $t_2 c_2$). This information increases understanding and allows calculation of the transport and fate of the contaminant in the subsurface (t_1 concentrations vs. t_2 concentrations at M_{1xyz} and M_{2xyz}). Unfortunately, obtaining accurate unbiased data to achieve sampling objectives is not always readily accomplished and is highly dependent upon the sampling methodologies used in the wells.

28.3 GROUNDWATER UNCERTAINTY PRINCIPLE

There is a principle in physics that is applicable to groundwater sampling, with some rethinking and rewording. The uncertainty principle was discovered by Werner Heisenberg in 1927. It is "a quantum-mechanical law which states that, if one measures the position of an object or the strength of a field with high precision, one's measurement must necessarily perturb the object's velocity or the field's rate of change by an unpredictable amount" (Ref. 1, p.).

Reworded for groundwater it could be stated as follows:

If one samples groundwater to measure the location of a contaminant or the relationship of the contaminant to the overall geochemical milieu of the subsurface, one's measurement must necessarily perturb the contaminant's location, concentration, and rate of change in its relationship with the other components of the subsurface environment by an unpredictable amount.

This means that:

- All impacts on the sample before, during, and after collection degrade the sample and diminish our ability to understand the subsurface.
- "Perfect" samples can never be obtained but must be approximated to achieve our sampling objectives.
- We must eliminate avoidable subsurface perturbations and sample alterations when collecting samples.

28.4 SCHOOLS OF THOUGHT ON GROUNDWATER SAMPLING

During the period that groundwater has been sampled for contaminants, several schools of thought have developed. It was realized that changes in sampling protocols were needed to improve sample quality. In a sense, these schools of thought on sampling procedures have developed in a manner that could be perceived as the evolution of a technique. In much the same manner as evolution might proceed, we now have several "species" of groundwater sampling techniques that coexist.

That is, at this point in time, none of the techniques has driven any of the others to extinction, although it is clearly time for some of them to go. The following list simplistically delineates these schools of thought and the general order in which they came into being, although there is broad overlap in both technique and time frame:

1. Traditional sampling: Purge a specified number of casing volumes (three to five) with high–flow rate pumping or bailing and collect the samples.
2. Determine formation hydrogeology and calculate a purge rate and volume.
3. Purge until temperature, pH, and conductivity are stable.
4. Purge with low flow; monitor dissolved oxygen (DO) redox, turbidity, and the contaminant if possible.
5. Do not purge the well, but purge the sampling device, and sample using extremely low flow, that is, passive, sampling.

Although there are numerous permutations and combinations of these techniques, this division breaks them down into categories that can be easily discussed. The primary focus of this chapter will be to compare and contrast traditional sampling with low-flow and passive purging and sampling techniques.

28.5   TRADITIONAL GROUNDWATER SAMPLING

Having derived from water supply investigations and, frankly, dropping the bucket down the well to get drinking water, early methods of groundwater sampling for contaminants (referred to here as "traditional") did not consider the significance of many of the potential sampling impacts on the quality of the water sample. The same techniques were generally used for all sampling objectives and depended upon the available equipment, familiarity of the site personnel with the sampling technique and equipment, flow rate at which the wells could be sampled (faster = cheaper = better), and any access or infrastructure limitations of the site (e.g., availability of electricity). These techniques generally included rapid purging of multiple borehole volumes of water carried out using either bailers (high-technology buckets) or high-flow pumps. Speaking of these techniques in the past tense is somewhat misleading, however, since these traditional approaches are still being used by many site investigators. A typical traditional sampling approach can include the following steps:

1. Remove the well cap and measure the water level.
2. Drop in the pump, tubing, or bailer.
3. Lower the device the chosen depth for purging.
4. Pump or bail at the highest rate possible.
5. Remove three to five casing volumes of purge water.

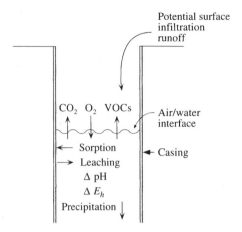

Figure 28.2 Standing well water.

6. Adjust the depth (perhaps) and collect the sample or change the sampling device and collect the sample.
7. Decontaminate the devices and move to the next well.

Purging before sample collection is done because the standing well water is considered stagnant due to a variety of interactions with the well materials and the physical environment of the well (Fig. 28.2), including:

- The air interface at the top of the water column (gas exchange, e.g., O_2 intrusion, CO_2 loss)
- Loss of volatiles up the column (loss of volatile contaminants to the atmosphere)
- Leaching from or sorption to the casing material (of elements and compounds)
- Sorption to the sand pack (of elements and compounds)
- Chemical changes due to clay/cement seals or backfills (e.g., alterations, leaching of metals)
- Surface infiltration (runoff entering the well and introducing artifactual elements, compounds, and particulates)

Although most investigators would agree that the standing water column in the casing of a well is stagnant and should not be sampled, several studies indicate that the water in the screened interval of many wells (Fig. 28.3) is representative of the formation water.[2,3] This water can be sampled without excessive purging, provided care is taken not to mix the overlying casing water with the screened interval water during the sampling procedure.

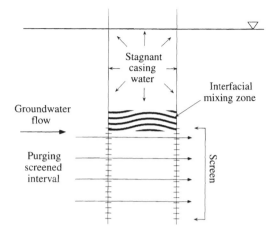

Figure 28.3 Water in the screened interval of a well.

Unfortunately, traditional high–flow rate/bailer purging and sampling can have several negative effects,[4] including:

- Device insertion that mixes the stagnant casing water into the screened interval
- The generation of large volumes of purged waters for disposal
- Potential loss of volatile contaminants and dissolved gases from the samples
- Shifts in chemical equilibria that can impact the analytes
- Dewatering of the well and aeration of both samples and the sediments around the screened interval
- Confounding hydrogeologic effects
- Particle entrainment due to turbulence and induced stress, that is, artificial turbidity

Due to these effects, it can be said, and is supported by data, that traditional sampling techniques influence the groundwater sample. As was stated by the groundwater uncertainty principle, any change in the sample from its condition in the aquifer degrades the quality of the sample and our ability to understand the subsurface. The effects of traditional sample collection on sample quality are worth consideration. It is important to note that these effects are interdependent; that is, one can result in the other. It is also important to note that disruption of the well and its waters by even the best sampling approaches, improperly applied, can result in the effects discussed below.

28.5.1 Turbulence Effects on Volatiles and Dissolved Gases

Turbulence of the water in the well screen and casing is caused both by high–flow rate pumping and by bailing. The *American Heritage Dictionary* (1993) describes a turbulent fluid as:

1. Violently agitated or disturbed; tumultuous
2. Having a chaotic or restless character or tendency

Turbulence can begin with the insertion of the sampling device and persist throughout the purging and sampling. Repeated bailer insertion and removal result in a "plunger effect" in the well bore. Turbulence can cause mixing of atmospheric gases into the well, outgassing of the dissolved gases, shifts in chemical equilibria, and generation of artificial turbidity due to disruption of the sand pack and aquifer materials. Sample turbulence and aeration can also occur when bailers are emptied into sample containers by pouring.

Traditional high-rate pumping and bailing methods offer no real solution for turbulence effects on samples. Turbulent effects can be avoided by the careful application of low-flow and passive sampling techniques.

28.5.2 Shifts in Chemical Equilibria of Sample

Shifts in the chemical equilibria of the groundwater sample can result from all types of disturbances to the sample or to the well. Changes can occur in Eh, pH, dissolved gas concentrations, solute speciation, mineral dissolution, mineral precipitation, complexation, adsorption, desorption, and so on. These effects are not discrete but are interrelated and interactive. When such changes occur in the fundamental chemistry of the sample, it is difficult or impossible to accurately quantify the constituents of interest.

Traditional purging and sampling techniques have largely ignored problems with shifts in chemical equilibria of the sample during the collection process, although many techniques have been investigated and developed for preserving the samples for analysis after collection. Again, the goal must be to minimize disturbances to the well and water to be sampled.

28.5.3 Dewatering and Aeration of the Well

Dewatering and aeration occur most commonly in low-yield wells evacuated by pumping during purging and allowed to recover for sampling but may also occur in higher yielding wells with long screens when the pumping rate is excessive (Fig. 28.4). In the latter situation, the drawdown might dewater some upper portion of the screen, if not the entire screened interval. Dewatering can cause significant effects on aquifer minerals, including the intrusion of air, some drying, oxidation of mineral surfaces, and so on. These changes in the aquifer solids can influence the chemistry of the aquifer water as it flows across the mineral surfaces in the affected zone through processes such as oxidation/reduction, adsorption, and dissolution. In a low-yielding well, water can slowly seep into the screen, resulting in significant atmospheric exposure. In the higher yielding scenario, aquifer waters can cascade through the screened interval and down the well bore, creating turbulence (Fig. 28.4). Effects on water chemistry can result from air contact with the water

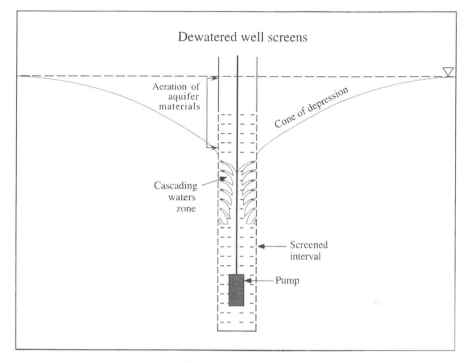

Figure 28.4 Dewatered well screens.

entering the screen either by seepage or turbulence, including the oxidation of analytes and dissolved minerals, the uptake of atmospheric gases, loss of dissolved gases and volatile contaminants, and other potential impacts, such as precipitate formation.

For extremely low-yield wells, it may be difficult or impossible to obtain samples that are truly representative of the water in the formation. The traditional approach for sampling such wells is to pump them to dryness, allow them to recover, and then sample the water. The problem with aeration of the water as it seeps back into the well during this approach should be obvious. A potentially better solution is to use passive sampling techniques to collect the water that is present in the well without pumping it to dryness.

28.5.4 Confounding Hydrogeologic Effects

Improper groundwater sampling can cause hydrogeologic effects around and within the well that can be incorporated into the collected sample. These hydrogeologic disturbances can result in a mixing of contaminated and uncontaminated waters when both purging and sampling. This mixing confounds a true understanding of contaminant distributions and concentrations at the point of the well in space and time. These effects are exacerbated by:

- Long well screens interacting with
- Variable stratigraphy and flow across the screened interval ("short-circuiting")
- Excessively high pump rates or bailing

Long well screens often intersect multiple sedimentary or hydrogeologic units that have varying characteristics and properties. These differences can include factors that affect the flow of groundwater, such as permeability and porosity. This can result in preferential flow of both water and contaminants in some units relative to the others. When water is pumped from the well, it will typically flow into the well in greater volume from these units (Fig. 28.5a). Therefore, if an investigator is expecting to collect a sample that is volume integrated to some radius equidistant from all points on the well screen (Fig. 28.5b), the pump, or the sandpack, the sample will not accurately represent this expected volume. The actual integrated volume sampled will extend further into the more permeable zones, such as sands, and less into the tighter formations, such as silts and clays. This scenario can be further complicated by multiple high-flow zones, only one or a few of which actually contain and are transporting the contaminants (Fig. 28.5c). In this scenario, the contaminant from the high-concentration zone is not accurately represented since it is being diluted by water from the other permeable zones. The vertical location of the plume is also not accurately known. In addition, high pumping rates, drawing water significantly faster than the natural flow velocity through the stratigraphic units, can pull water from a much greater distance than anticipated by the sampling design and objectives. This can cause significant errors in estimating the true locations of contaminants and their concentrations. In fact, high pumping rates can actually pull contaminants into zones that were previously uncontaminated, thus spreading the contamination throughout a greater volume of the aquifer.

Confounding hydrogeologic effects can be reduced by using discrete screen lengths and locating the screens in relation to the sampling zone of interest based on stratigraphy and zones of contamination (when known). Nested or clustered wells should be used for sampling multiple depths at the same surface location whenever possible. Pump rates in the screened zone of interest should be set to attain water only from that zone and minimize induced effects.

28.5.5 Artificially Entrained Turbidity

Turbidity in well water or a water sample results from solid particulate matter that is present in the water. These particulates either can represent a colloidal phase, that is, remain suspended in the water due to their small particle size, or they may be larger particles that will eventually settle to the bottom. *Webster's New World Dictionary* defines *turbidity* as:

1. Muddy or cloudy from having the sediment stirred up
2. Thick, dense, or dark, as clouds
3. Confused

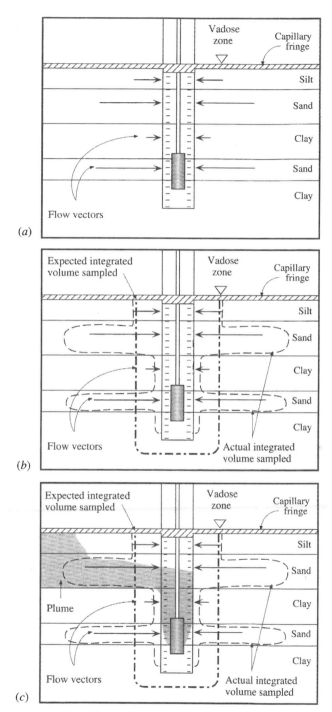

Figure 28.5 Long well screens.

Turbidity in a water sample is very adequately described by all three of these definitions, the first two due to sample appearance, the third because data resulting from the analysis of turbid samples can be difficult or impossible to understand. There are several possible causes of turbidity in a water sample. Among these are:

- Poor well construction or improper well development
- Surging with bailers or excessive rates of pumping or purging relative to local hydrogeologic conditions, resulting in
- Grain flow in the sand pack surrounding the well screen and casing
- Disruption of aquifer mineral cementation, mobilizing particulates
- Significant transport of mobile colloidal particles naturally present in the aquifer

Entrained artificial turbidity alters the physicochemical system of the water and can result in several effects due to the advent of freshly exposed reactive mineral surfaces in the solution, including:

- Surface/mineral matrix dissolution, with metals and major ion release to the solution that will now be measured in the analyses
- Desorption of previously immobilized metals or other contaminants (formerly of no concern to the environment) into the solution that will now be measured as though they were mobile contaminants
- Surface precipitation (with metals and major ion removal), lowering the apparent concentrations of contaminants in the solution
- Sorption/adsorption of dissolved contaminants of interest, removing them from solution
- Adsorption of colloidal contaminants of interest, removing them from suspension

If any of these effects, or some combination thereof, impacts the water sample being collected, it is unlikely that the laboratory analyses will provide information on the sample that represents the true condition of the aquifer water quality and/or contaminant concentrations.

The presence of solids in water samples has long been recognized as an analytical problem, primarily due to acidification of the sample (to preserve metals for analysis) releasing major ions such as Si, Al, Ca, and Fe into the solution, artificially increasing their concentrations. The traditional answer to this problem has been to collect two samples and handle them differently; for example:

- Field filter a raw sample through a 0.45-μm-pore-diameter filter for "dissolved" metals; then acidity to $pH < 2$ (to maintain the dissolved metals in solution until analysis).
- Field acidify and/or laboratory digest a raw sample for "total" metals. Numerous approaches to the acidification/digestion procedure have been applied, often with differing results.

Filtration of water samples intended for metals analysis through 0.45-μm filters has become a standard practice. Unfortunately, the filter pore diameter of 0.45 μm has no physical relevance for discriminating dissolved versus particulate materials, and the process of filtration itself can have large impacts on the water sample and the subsequent analytical data. Some concerns created by filtration are:

- Colloidal particles from 0.003 to 10 μm in diameter can be mobile in certain environments; therefore filtration using 0.45-μm filters does not truly separate particulate from dissolved material.
- Particles <0.45 μm may or may not be naturally mobile and may or may not contain the contaminant of interest but will usually pass the filter.
- Particles >0.45 μm may be mobile, carry adsorbed contaminant, and be excluded from analysis by filtering, thereby making the mobile contaminant load appear to be less than it actually is.
- Most filters have been shown to selectively remove dissolved metals from solution and some have been known to leach metals to the solution.
- As filtration proceeds, the mean pore diameter of the filter changes due to blockage by particle accumulation, causing the removal of particles smaller than 0.45 μm.

Due to the irrelevant filter pore diameter and the altered physicochemical system due to the initial presence of solids, contaminant values may be either too high or too low versus the actual dissolved and mobile colloidal fractions present in the aquifer. The problem is, once artificial sample turbidity has occurred, there is no analytical method or sample treatment that can restore confidence in the data quality. Indeed, if filtration is routinely done on water samples that are properly collected (i.e., no artificially induced particulate turbidity), it is possible for naturally mobile colloidal particles bearing contaminants to be removed and not measured.

To avoid the entrainment of artificial turbidity, methods should be used that do not disrupt the well during sampling. This can usually be accomplished by using low-flow purging and sampling. When low-flow sampling is used, there is generally no filtration of the samples for metals analysis. This provides a better estimate of the total mobile contaminant concentration (total mobile = dissolved + mobile colloidal) at that point in the aquifer. There are special cases when filtration might be used after low-flow and passive sampling—for example, when data are needed for transport modeling of dissolved versus mobile particulate. Recommended filter choices for this scenario have pore diameters of 0.1 and 10 μm, intended to represent:

- Generally dissolved materials if <0.1 μm
- Possibly mobile colloidal if 0.1–10 μm
- Generally not mobile if entrained if >10 μm

Although these are also somewhat arbitrary choices, they are better approximations than the traditional 0.45-μm-filter choice.

28.6 CONCEPTS AND TECHNIQUES OF LOW-FLOW AND PASSIVE PURGING AND SAMPLING

Although no current sampling methodology or equipment is perfect, many of the problems encountered during traditional high-rate pumping or bailing can be minimized or eliminated by using low-flow sampling techniques. The goal of low-flow sampling techniques is to reduce disruption to the well, the surrounding aquifer, and the samples during the collection process. The developmental research and use of low-flow techniques have been well documented during the past decade (see References). These techniques are even more effective when coupled with clear sampling objectives and proper monitoring well design and installation.

28.6.1 Monitoring Wells

Monitoring well installations always disrupt the formation to be sampled, especially during the drilling process. These disruptions can affect the hydrodynamic properties of the well and influence the aquifer geochemistry in the well vicinity. Biasing effects of drilling can include vertical mixing of contaminated and uncontaminated waters in the borehole, smearing of clay layers across porous zones of higher hydraulic conductivity, introduction of drilling mud or fluids into the formation, cross-contamination between boreholes, and aeration and/or degassing of zones in the formation that are not at atmospheric equilibrium.[5]

Many studies have also shown effects on samples due to construction materials, such as the casing and screen, sand pack, and grouting or sealing materials.[5-11] To avoid long-term impacts on sample collection and quality, drilling methods and construction materials should be carefully considered in the context of the groundwater sampling objectives.

Following construction, the well should be immediately developed. This helps to remove the particulates and residues from the well construction activities. It also unblocks the more permeable stratigraphic flow zones adjacent to the well screen that may have been smeared by clay layers during drilling and had their flow restricted. Pumping should be used for well development, but the wells should not be pumped to dryness. Surging the well with plungers or swabs should also be avoided, since this forces the borehole water, its particulates, and drilling contaminants into the aquifer during the downward stroke. This intrusion could result in potential long-term chemical effects on the aquifer materials and groundwater in this zone. Before sampling, the well should be allowed to stabilize and approach equilibrium with respect to both groundwater flow and geochemistry. The time that this requires will vary from site to site and well to well and will depend largely on the system hydrodynamics and chemical characteristics. Stabilization will probably not occur during the first few weeks following the well installation. Data from samples collected during this period should be used with caution.

Wells should be designed, located, and installed with the objectives of the site sampling and the sampling methods and equipment to be used in mind. Screen lengths should be chosen to encompass only the stratigraphic unit of interest at a given depth. Ideally, the screened interval would acquire a sample from only one stratigraphic unit; that is, stratigraphy with multiple flow zones would not intersect a single screen as it does in Figure 28.5. In addition, it is usually preferable to limit the diameter of a well to the minimum that will accommodate the sampling device to be used. This decreases the distance from any point within the well to the formation water, minimizes the cross-sectional area of the air–water interface in the casing, and reduces the total volume of water in the well bore that can be considered stagnant.

To obtain groundwater data from multiple depths at a single surface location, nested or clustered wells, each of which is screened and/or completed only to a single depth, are very effective and recommended. For such wells at depths within suction lift range (≈ 25 ft or less), tubing bundles with short screens (e.g., 1-ft screens) have been successfully used following installation in a single completed borehole with no sandpack (i.e., the formation was allowed to collapse upon the bundle). These installations are typically sampled by low-flow or passive methods using peristaltic pumps. Sampling of depths below suction lift range requires the use of in situ pumps, such as pneumatic bladder pumps or low-flow centrifugal submersible pumps. These in situ pumps are typically useful to depths of about 1000 ft for the bladder pump and about 300 ft for the low-flow centrifugal pump. Sampling at greater depths with such pumps will usually require specialized equipment due to the height of the water column being pumped to the surface and the pressure it exerts (e.g., potential for tubing leaks at fittings).

Unfortunately, meeting the objectives of a groundwater monitoring program is often constrained by physical limitations. These physical limitations can include both technological and fiscal constraints. The cost of the drilling and sampling program is always a consideration and has become even more important during the past few years due to increasing concerns about cost versus benefit and the development of risk assessment technologies to evaluate this trade-off. For example, in a water-supplying formation deep below or within fractured rock—for example, an aquifer 1000 ft below ground surface (bgs)—drilling and emplacing a single monitoring well can cost as much as $750,000. This is clearly a significant constraint on the number of monitoring wells that can be drilled. This cost must, of course, be weighed versus the potential risk of providing contaminated water to the users of the water supply. In addition to the cost, monitoring wells at such great depth push the limits of our current monitoring well technology for both the drilling and the sampling of such wells. Due to the need to sample throughout the thickness of such a deep aquifer, while only being able to drill one or a few such monitoring wells, these installations often have extremely long screens. These long screens are, of course, subject to the previously discussed problems when sampling and can act as a conduit from one stratigraphic flow zone to another, potentially spreading contamination. Accurately sampling such deep, long-screened wells presents a technological challenge. Devices such as multilevel port sampling systems are capable of collecting discrete-interval samples from a single deep borehole when installed as the well completion technology.

28.6.2   Low-Flow Purging and Sampling

Low-flow purging and sampling consist of a variety of concepts and processes designed to minimize disruption to the well, outlying aquifer, and collected samples. These techniques are also generally designed to provide confirmation that the water being collected is representative of the formation water in the aquifer through the observation of sensitive indicator parameters. These processes, concepts, and techniques include:

- Low pump rates, usually 0.1–0.5 L/min, with no bailers are allowed.
- Purging and sampling are always performed in the screened interval when standard monitoring wells are used.
- Samples collect in the formation immediately adjacent to the well and pump rather than in outlying waters.
- Sampling follows stabilization of the most sensitive purging indicator parameters.
- Dedicated pumps or tubes are desirable but not required.
- Short screened intervals are preferred but longer screens can be sampled.

The low pumping rates and elimination of the use of bailers minimize artificial turbidity, aeration, hydrogeologic mixing, volatile organic compound (VOC) loss, and outgassing and reduce equilibrium shift in the water being collected. Since waters are collected from the aquifer in the immediate vicinity of the well, better concentration data at that point are obtained. This localization of water withdrawal is extremely useful for monitoring the new passive remediation techniques, such as permeable reactive barriers (PRBs). For example, monitoring might be needed immediately down gradient of a PRB, but forcibly drawing contaminated up-gradient water through the PRB and into the well must be avoided.

Low-flow purging and sampling expand the list of parameters that are routinely monitored during purging—that is, temperature, pH, and conductivity—to include DO, Eh, turbidity, and the contaminant of concern if possible. Field research has shown that temperature, pH, and conductivity are relatively insensitive parameters for indicating continuity with formation water compared with those added to the list.[5] Flow-through cells are required for the DO and Eh measurements because these parameters immediately change upon exposure to the atmosphere. Figure 28.6 shows the stabilization of indicator parameters in a monitoring well near Elizabeth City, North Carolina, and illustrates the potential differences in parameter sensitivity.

Although not required for low-flow techniques, dedicated pumps or tubes reduce purge time relative to portable pumps. This is because inserting the pump into the well and lowering it to the zone to be sampled mixes stagnant casing water with screen water. When dedicated pumps are used, this disruption does not occur except when the pump is initially inserted. Dedicated pumps also have the considerable advantage of eliminating cross-contamination between wells and of not requiring the extensive decontamination procedures that are necessary when a portable pump is moved from well to well. This reduces sampling time and eliminates the

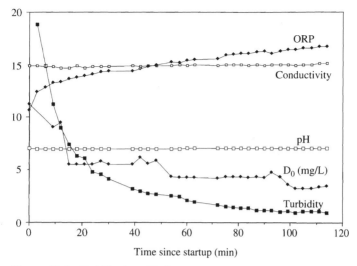

Figure 28.6 Stabilization of indicator parameters in a monitoring well.

need for disposal of the wash solutions. If a portable pump is used, it should be lowered into the well gently to minimize disruption to the extent possible. Portable pumps should also be used to sample the wells in a sequence from the least to the most contaminated, if this sequence is known, to reduce the possibility of contaminant carryover.

Short screened intervals are preferred for low-flow sampling. Shorter screens reduce "short-circuiting" between high-flow and low-flow stratigraphy. Low-flow techniques have been used to successfully sample long screened intervals, however. Irrespective of the screen length, low-flow techniques are preferable to high-rate pumping or bailing for all the reasons detailed in the earlier sections of this chapter.

28.6.3 Passive Sampling

Passive sampling depends solely on the natural flow of groundwater to purge the well, or hydrogeologic zone, where the sampler is located. That is, the well or zone is not purged, although the sampling device itself is evacuated to remove possible stagnant water from within the device and tubing before sample collection. However, passive sampling should not be confused with the recent proposals for "no-purge sampling," in which petroleum hydrocarbons or other contaminants are sampled in wells using bailers, rapid pumping, and so on, with no purging. Such techniques are seriously flawed and should not be used.

Passive sampling techniques should have the following characteristics:

1. Sampling devices are dedicated, whether they are simply down-hole tubing (within suction lift limits), down-hole pumps, or retrievable devices such as membrane cell samplers. Bailers are not allowed.
2. When the devices are emplaced in standard monitoring well installations, they must be located within the screened interval.

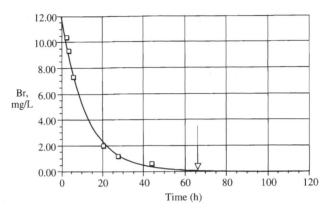

Figure 28.7 Results of a borehole flushing test in a monitoring well using bromide as a tracer.

3. Following emplacement, time must be allowed for the screened interval of the well to recover from the mixing with the casing waters. This time will vary with the well and formation hydraulics but should be fairly rapid (days to weeks) for transmissive aquifers and can be estimated or determined. Figure 28.7 shows the results of a borehole flushing test in a monitoring well using bromide as a tracer. This test shows that about 60–65 h was needed for the tracer to be completely removed from the screened interval[3]; that is, the tracer-bearing water was completely replaced by formation water in 60–65 h.

4. Purging of the standing water column, or to attain continuity with the formation water, is not done before sample collection. For this technique, continuity of the screened interval water with the formation water is assumed or has been tested (preferable).

5. Samples are collected at low flow rates, preferably 100 mL/min or less.

Passive sampling benefits both monitoring process logistics and the quality of the samples in many of the same ways as low-flow sampling but should provide even less disruption to the water being sampled. It almost completely eliminates disposal of contaminated purge waters as hazardous waste, reduces the time spent at each monitoring well, eliminates or greatly reduces decontamination requirements, results in no disruption of the sand pack and aquifer around the screen, eliminates turbulence and particle loading effects on sample chemistry, and avoids artificially induced flow and zone mixing. Passive sampling is probably best in short-screen, small-diameter ($\leq$2-in.), collapsed formation wells with no sandpack. This reduces the distance of the device from the formation water and eliminates one material (the sandpack) that could influence the results. In longer screen or sand-packed wells, passive data should be confirmed initially with low–flow rate purging and sampling, when possible, to determine that the results are comparable. Passive sampling techniques are potentially the best currently available methods for low-yielding, long–recovery period wells that could be pumped to dryness by traditional sampling or even by low-flow methods.

REFERENCES

1. K. S. Thorne, *Black Holes and Time Warps, Einstein's Outrageous Legacy*, W. W. Norton & Company, New York, 1994.
2. M. J. L. Robin and R. W. Gillham, Field Evaluation of Well Purging Procedures, *Ground Water Monitoring Rev.*, Fall 1987, pp. 85–93.
3. R. M. Powell and R. W. Puls, Passive Sampling of Groundwater Monitoring Wells without Purging. Multilevel Well Chemistry and Tracer Disappearance, *J. Contam. Hydrol.* **12**, 51–77 (1993).
4. R. M. Powell and R. W. Puls, Hitting the Bull's-Eye in Groundwater Sampling, *Pollution Eng.* June 1997, pp. 50–54.
5. R. W. Puls and R. M. Powell, Acquisition of Representative Ground Water Quality Samples for Metals, *Ground Water Monitoring Rev.*, Summer 1992, pp. 167–176.
6. J. M. Marsh and J. W. Lloyd, Details of Hydrochemical Variations in Flowing Wells, *Ground Water* **18**(4), 366–373 (1980).
7. G. D. Miller, Uptake and Release of Lead, Chromium and Trade Level Volatile Organics Exposed to Synthetic Well Casings, paper presented at the Second National Symposium on Aquifer Restoration and Ground Water Monitoring, May 1982, Columbus, OH, 1982.
8. M. J. Barcelona, J. P. Gibb, and R. A. Miller, *A Guide to the Selection of Materials for Monitoring Well Construction and Ground-Water Sampling*, Illinois State Water Survey Contract Report No. 327, USEPA-RSKERL, EPA-600/52-84-024, U.S. Environmental Protection Agency, Washington, DC, 1983.
9. R. W. Gillham, M. J. L. Robin, J. F. Barker, and J. A. Cherry, *Ground-Water Monitoring and Sample Bias*, Department of Earth Sciences, University of Waterloo, Ontario, API Publication 4367, American Petroleum Institute, 1983.
10. M. J. Barcelona, J. P. Gibb, J. A. Helfrich, and E. E. Garske, *Practical Guide for Ground-Water Sampling*, Publication No. EPA/600/2-85/104, U.S. EPA Office of Research and Development, R. S. Kerr Environmental Research Laboratory, Ada, OK, 1985.
11. M. J. Barcelona, and J. A. Helfrich, Well Construction and Purging Effects on Groundwater Samples, *Environ. Sci. Technol.* **20**(11), 1179–1184 (1986).

BIBLIOGRAPHY

M. J. Barcelona, H. A. Wehrmann, and M. D. Varljen, Reproducible Well-Purging Procedures and VOC Stabilization Criteria for Ground-Water Sampling, *Ground Water* **32**(1), 12–22 (1994).

T. Giddings, Bore-Volume Purging to Improve Monitoring Well Performance: An Often-Mandated Myth paper presented at NWWA 3rd Nat. Symp. & Aquifer Rest. & Ground Water Monitoring, Columbus, OH, May 1983, pp. 253–256.

D. T. Heidlauf and T. R. Bartlett, Effects of Monitoring Well Purge and Sample Techniques on the Concentration of Metal Analytes in Unfiltered Groundwater Samples, paper presented at NGWA Outdoor Action Conf., Las Vegas, May 1993, p. 437.

D. B. Kaminski, MicroPurge: Cost Savings and Method Control for Ground-Water Sampling Programs, paper presented at SWANA Third Annual Landfill Symposium, Palm Beach Gardens, FL, June 1998, pp. 65–72.

D. B. Kaminski, Alternative to High-Volume Well Purging Reduces Costs, *Pollution Prevention*, February 1993, pp. 53–57.

P. M. Kearl, N. E. Korte, M. Stites, and J. Baker, Field Comparison of Micropurging vs. Traditional Ground Water Sampling, *Ground Water Monitoring and Remediation*, Fall 1994, pp. 183–190.

P. M. Kearl, N. E. Korte, and T. A. Cronk, Suggested Modifications to Ground Water Sampling Procedures Based on Observations from the Colloidal Borescope, *Ground Water Monitoring Rev.*, Spring 1992, pp. 155–161.

K. F. Pohlmann, G. A. Icopini, and R. D. McArthur, *Evaluation of Sampling and Field-Filtration Methods for the Analysis of Trace Metals in Ground water*, Publication No. EPA/600/R/94/119, U.S. Environmental Protection Agency, Washington, DC, October 1994.

R. W. Puls, A New Approach to Purging Monitoring Wells, *Ground Water Age*, January 1994, pp. 18–19.

R. W. Puls and M. J. Barcelona, *Ground Water Sampling for Metals Analysis*, Publication No. EPA/540/4-89/001, U.S. Environmental Protection Agency, Washington, DC, March 1989.

R. W. Puls and M. J. Barcelona, *Low-Flow (Minimal Drawdown) Ground-water Sampling Procedures*, Ground Water Issue, Publication No. EPA/540/S-95/504, U.S. Environmental Protection Agency, Washington, DC, April 1996.

R. W. Puls, D. A. Clark, B. Bledsoe, R. M. Powell, and C. J. Paul, Metals in Ground Water: Sampling Artifacts and Reproducibility, *Haz. Waste Haz. Mater.* **9**(2), 149–162 (1992).

R. W. Puls, J. H. Eychaner, and R. M. Powell, *Colloidal-Facilitated Transport of Inorganic Contaminants in Groundwater: Part 1. Sampling Considerations*, EPA/600/M-90/023, U.S. Environmental Protection Agency, Washington, DC, 1990.

R. W. Puls and C. J. Paul, Low-Flow Purging and Sampling of Ground Water Monitoring Wells with Dedicated Systems, *Ground Water Monitoring and Remediation*, Winter 1995, pp. 116–123.

K. Schilling, Low-Flow Purging Reduces Management of Contaminated Groundwater, *Environ. Protection* **6**(12), 24–26 (1995).

C. L. Serlin and L. M. Kaplan, Field Comparison of MicroPurge and Traditional Ground Water Sampling for Volatile Organic Compounds, in *Proceedings of the NGWA Petroleum Hydrocarbons and Organic Chemicals in Ground Water Conference*, 1996, pp. 177–190.

D. E. Shanklin, W. C. Sidle, and M. E. Ferguson, Micro-Purge Low-Flow Sampling of Uranium-Contaminated Ground Water at the Fernald Environmental Management Project, *GWMR*, Summer 1995, pp. 168–176.

U.S. EPA, *Use of Low-Flow Methods for Ground Water Purging and Sampling: An Overview*, Region 9, Quick Reference Advisory, U.S. Environmental Protection Agency, Washington, DC, December 1995.

U.S. EPA, *Ground Water Sampling—A Workshop Summary*, Office of Research and Development, EPA Publication No. EPA/600/R-94/205, U.S. Environmental Protection Agency, Washington, DC, 1995.

29

SOIL PERMEABILITY AND DISPERSION ANALYSIS

Aziz Amoozegar, Ph.D.

North Carolina State University
Raleigh, North Carolina

Environmental Instrumentation and Analysis Handbook, by Randy D. Down and Jay H. Lehr
ISBN 0-471-46354-X Copyright © 2005 John Wiley & Sons, Inc.

29.1 INTRODUCTION

A major portion of the earth is covered with unconsolidated porous materials commonly referred to as soil. The term *soil* has different meaning to different people. To an agronomist, soil is the medium in which plants can grow and provide food and fiber. To a civil engineer, on the other hand, soil is the material that can be moved around to support buildings and other man-made structures. In general, soils are formed as a result of physical and chemical weathering of rocks, and their properties are impacted by the soil-forming factors.[1]

29.1.1 Soil Texture and Structure

Soil is considered to be a three-phase system composed of solids, liquids, and gases.[2,3] The solid portion of the soil is composed of mineral and organic particles that can be less than 1 μm. In general, the mineral particles that are ≤ 2 mm in diameter are grouped into three major categories of sand, silt, and clay.[4] Figure 29.1 presents the categorization of these particles based on their size (i.e., diameter) by various organizations. The proportion of sand-, silt-, and clay-sized particles, referred to as soil texture, is used to classify soils into textural classes. One such classification system is the U.S. Department of Agriculture (USDA) classification in which soils are grouped into 12 different soil textural classes (Fig. 29.2).

The arrangement and organization of soil particles in the soil into larger units called soil aggregates is referred to as soil structure. Unlike soil texture, which cannot be changed easily, soil structure is inherently unstable and can be impacted by factors such as climate, soil management (e.g., deep plowing), chemicals, and biological activities. Soil structure plays an important role in retention and movement of water, solute transport, aeration, soil strength, compactability, temperature, and heat flux, among other factors. This is because any alteration in soil structure will have a direct impact on the geometry of the pores (size, shape, and turtousity) as well as on the total amount of pore spaces in the soil. The volume of pore spaces per unit volume of bulk soil is called porosity, and the ratio of pore spaces and solids in the soil is called soil void ratio. The distribution of various size pores in a given volume of soil is referred to as pore size distribution.

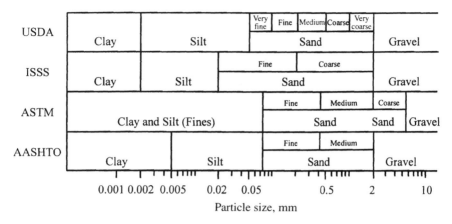

USDA — U.S. Department of Agriculture
ISSS — International Soil Science Society
ASTM (Unified) — American Society for Testing and Materials
AASHTO — American Association of State Highway and Transportation Officials

Figure 29.1 Classification of soil mineral particles by various organizations.

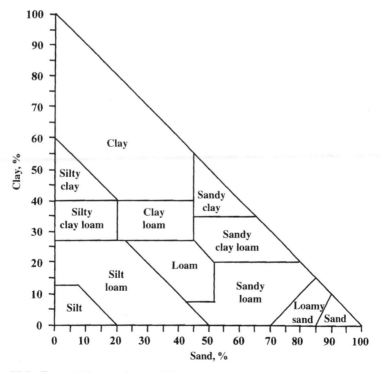

Figure 29.2 Textural triangle showing 12 textural classes based on sand and clay content.

The liquid phase of soil is generally water that contains various amounts of different chemicals. Because of the presence of these chemicals, the water in the soil is referred to as soil water or soil solution. The gas phase of the soil, referred to as soil air, generally has a composition similar to the atmosphere with a lower level of oxygen and a higher level of carbon dioxide.[5] In addition to gases commonly found in the atmosphere, soil air may also contain various levels of other gases, such as volatile organic compounds.

29.1.2 Soil Water and Soil Water Flow

Soil water and soil air occupy the pore spaces in the soil. For nonswelling soils (i.e., soils that do not expand and shrink with the addition or removal of water, respectively), the soil porosity remains constant, and the water-filled porosity is inversely related to the air-filled porosity with small pores being occupied by water and large pores being occupied by air. The amount of water in a given volume of soil is called soil water content and can be expressed on the basis of the dry mass of the soil, the wet mass of the soil, or the bulk volume of the soil. The amount of water expressed on a mass basis is often called gravimetric soil water content, and the water content on a volume basis is referred to as volumetric soil water content.

Water in the soil is held by soil particles and in pores of various shapes and dimensions. Similar to other objects in nature, soil water contains energy (for a more complete description of soil water energy see Refs. 3, 6, and 7). Because of the tortuous path and narrowness of soil pores, the velocity of water moving through soil is relatively low. Therefore, the kinetic energy of soil water is considered to be negligible, and only its potential energy, which is related to the relative position and internal condition of soil water, is of primary importance. The soil water potential expressed on a volume basis has dimensions of pressure and can be expressed as dynes per square centimeter (or other units of pressure such as pounds per square inch). For assessing soil water movement, however, the most convenient way of expressing total soil water potential (and its components) is on a weight basis. Soil water potential expressed on a weight basis, referred to as potential head or head, is similar to expressing it in terms of the height of a vertical column of water corresponding to the given pressure. For most practical applications related to soil water movement, the main components of total (potential) head H (dimensional units of length L), referred to as hydraulic head, are gravitational (potential) head h_g and pressure (potential) head h_p. In general, the pressure head in the saturated zone (related to the submergence potential) is positive ($h_p > 0$) and in the unsaturated zone (related to matric potential) is negative ($h_p < 0$). At the boundary between the saturated and unsaturated zones (referred to as the water table when considering groundwater), the pressure head is similar to the atmospheric pressure and is taken to be zero ($h_p = 0$). The matric potential of the unsaturated zone may also be expressed as tension or suction to avoid the use of negative pressure. Soil water tension or suction is numerically the same as the absolute value of the pressure head in the unsaturated zone.

As the water content decreases in an unsaturated soil, soil water tension increases (i.e., soil water pressure head decreases and becomes more negative). The relationship between the soil water content and pressure head is referred to as soil water or soil moisture characteristics. This relationship is hysteretic and varies with wetting and drying of the soil.

29.1.2.1 *Darcy's Law.*

Soil water flow is multidimensional, but the direction of water flow is always from a point at higher total soil water potential (i.e., higher hydraulic head) to a point at lower soil water potential (i.e., lower hydraulic head). The rate of change of the hydraulic head along the flow path, $\Delta H/L$, called hydraulic gradient (a vector, dimensional units $L \cdot L^{-1}$), is the driving force behind water movement. Henri Darcy, a French engineer, discovered over a century ago that the quantity of water $Q(L^3)$ flowing through water-saturated sand filters of length L and cross-sectional area $A(L^2)$ during a time period t (dimensional units of time T) was proportionally related to the hydraulic gradient $\Delta H/L$ by

$$\frac{Q}{At} = v = K_{sat} \times \frac{\Delta H}{L} \qquad (29.1)$$

where $K_{sat}(L \cdot T^{-1})$ is the saturated hydraulic conductivity of the medium and v is called the flux density, flux, or Darcian velocity. The above equation, known as Darcy's law, was later modified and presented in a differential form to describe water flow through both saturated and unsaturated soils. For one-dimensional flow, Eq. (29.2) can be written as

$$v = -K\frac{dH}{dx} \qquad (29.2)$$

where K is the soil hydraulic conductivity and dH/dx is the gradient. Darcy's law is mathematically similar to other linear transport equations describing electrical current (Ohm's law), heat transport (Fourier's law), and gaseous movement by diffusion (Fick's law). (For a more comprehensive discussion on Darcy's law see Ref. 3 or 7.)

29.1.3 Hydraulic Conductivity and Permeability

Hydraulic conductivity is defined as a measure of the ability of soil to transmit water.[8] Under saturated flow conditions (e.g., groundwater flow into a well), the hydraulic conductivity is referred to as saturated hydraulic conductivity and is generally denoted as K_{sat} or K_s. Under unsaturated conditions, the soil hydraulic conductivity (referred to as K_{unsat}) is a function of soil water content $\theta(L^3 \cdot L^{-3})$ or pressure head h_p and can be presented as $K(\theta)$ or $K(h_p)$, respectively. In groundwater flow analysis, the product of the saturated hydraulic conductivity and thickness of the aquifer is called transmisivity $(L^2 \cdot T^{-1})$ and is generally denoted as T.[9]

In a rigid porous medium such as sandstone and in soils with stable structure (i.e., a soil in which the geometry of pores and void ratio remain constant with time), the saturated hydraulic conductivity at a given location can be taken as a constant value at any given time. Since the flux across a cross-sectional area of a soil depends on the rate of water flow through individual voids conducting water and the soil porosity and pore geometry depend on soil texture and structure, soil hydraulic conductivity is undoubtedly affected by variations in soil texture and structure. In general, clayey soils have a larger total porosity than sandy soils, but sandy soils have a significantly higher saturated hydraulic conductivity than clayey soils. Also, macropores (such as cracks, worm holes, and decaying root channels) impact water flow. According to Vepraskas et al.[10], only a small fraction of flow occurs through interparticle (i.e., matrix) pores while tubular and planar voids (mostly macropores) conduct the bulk of soil water.

As the water content decreases in a soil, the large pores drain, and only small pores (which hold water more tightly than the larger pores) conduct water. As a result, as the water content decreases (as in drainage), the relatively high hydraulic conductivity of a saturated sandy soil initially decreases at a faster rate than the hydraulic conductivity of a clayey soil (Fig. 29.3). At higher tension, the hydraulic

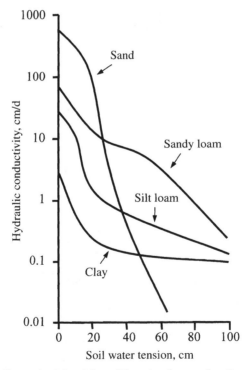

Figure 29.3 Hydraulic conductivity of four different soils as a function of soil water tension. (Adapted from Ref. 11.)

conductivity of the sandy soil may be much smaller than the K_{unsat} of the clayey soil. Similar to the soil water characteristics, the relationship between unsaturated hydraulic conductivity and soil water pressure head or water content is not unique and changes with wetting and drying (draining) of the soil (i.e., K_{unsat} is subject to hysteresis).

Hydraulic conductivity of a soil not only is impacted by viscosity and density of water (which are dependent on temperature) but also is influenced by the chemical properties of the water (or other fluids) going through it. This is because any change in the chemical composition or concentration of solutes in soil water will impact the balance of exchangeable ions on soil particles, which in turn may impact the soil structure. Biological activities can also influence soil hydraulic conductivity. For example, application of sewage effluent to a soil may result in the formation of biological byproducts, which can physically block soil pores.[12] Another factor that may result in significant reduction in the soil hydraulic conductivity is the translocation of colloidal particles through soil.[13] Translocated colloidal particles can physically block soil pores and reduce K_{sat}. The impact of the narrowing of soil pores on K_{unsat}, however, is not clear. As pores narrow, they retain water at higher tension and may provide continuity among small pores to conduct water.

As stated above, soil hydraulic conductivity not only depends on the soil characteristics but is also impacted by the properties of the liquid going through it. Therefore, we may express soil hydraulic conductivity in terms of the soil intrinsic permeability $k(L^2)$, which is an exclusive property of the solid portion of the soil and the fluidity of the liquid going through it. The fluid properties that affect hydraulic conductivity are fluid density ρ_f (dimensional units of mass over volume $M \cdot L^{-3}$) and viscosity $\eta(M \cdot L^{-1} T^{-1})$. The relationship between intrinsic permeability and soil saturated hydraulic conductivity is

$$k = \frac{K_{sat}\eta}{g\,\rho_f} \tag{29.3}$$

where $g(L \cdot T^{-2})$ is the acceleration due to gravity.

Intrinsic permeability depends mainly on the soil porosity and pore geometry. As long as the liquid that moves through the soil does not interact with the soil solid phase to impact soil porosity and pore geometry, the intrinsic permeability of the soil should remain the same. That is, for a stable soil, the intrinsic permeability should remain constant for different liquids (e.g., water, gasoline) or gases (e.g., air, methane) going through it.

The term *permeability* is used to express intrinsic permeability or specific permeability (see Refs. 3 and 9) and is defined by the Soil Science Society of America (SSSA)[14] as "the property of a porous medium itself that expresses the ease with which gases, liquids, or other substances can flow through it." In some texts, however, permeability is at least partly associated with the rate of water flow through porous media.[15–18] For example, the Natural Resources and Conservation Service (NRCS, formerly the Soil Conservation Service, or SCS) of the USDA classifies soils into different permeability classes to express qualitatively the ease with which water can move through soils (see Ref. 4).

29.1.4 Soil Dispersion

Individual soil particles (sand, silt, and clay) join together with other materials (e.g., organic matter) to form soil aggregates called peds. Soil structure is basically described based on the size, shape, and strength of these aggregates.[4,19] The pore spaces in the soil can be classified based on the size and shape of pores.[20,21] The pores can also be classified in general terms into interparticle or matrix pores, tubular (channel) pores, and planar voids.[22,23] Interparticle pores refer to spaces between individual soil particles, and they are the same as the matrix pores.[19] Tubular pores, on the other hand, are voids that are usually created by animals boring through the soil (e.g., worm holes) and decaying plant roots. These pores are generally round and continuously follow a smooth path for distances ranging from a few millimeters to a few meters. The third type of pores are referred to as planar voids due to rather large width and length as opposed to thickness. These are pore spaces that are associated with cracks as well as spaces between stable soil aggregates or peds. Any change in the arrangement of soil particles that results in the alteration of soil structure will have an impact on the soil porosity and pore geometry. As a result, water flow and specifically soil hydraulic conductivity and permeability can be severely impacted by changes in soil structure. The soil particles within soil aggregates can become loose or dispersed, resulting in the destruction of soil structure. As a result, dispersion of soil particles will negatively impact the soil's ability to transmit water; that is, soil dispersion can significantly reduce soil hydraulic conductivity. Near the soil surface, the soil structure can be physically altered by our activities, such as plowing the soil or compacting it, but at deeper depths soil aggregates can be altered by chemical reactions between the soil particles and the fluid passing through it. For example, sodium applied to a soil through irrigation, land application of waste, or other means replaces other cations on the exchange sites of clay particles. If the amount of Na on the exchange sites exceeds a certain value for a given soil, the soil particles repel each other, causing dispersion when the soil becomes wet. As a result of dispersion, the pore geometry changes and the hydraulic conductivity decreases.[24,25] Calcium, on the other hand, is known to cause soil particles to flocculate (i.e., join together) to form stable aggregates.[26] Stabilization of soil aggregates generally increases the water-conducting pores, which in turn increases soil hydraulic conductivity.

29.1.5 Chemical Diffusion/Dispersion

Liquid chemicals, including pollutants (e.g., gasoline), can move through the soil directly as a fluid or with water as dissolved or suspended particles (emulsified). Chemicals in solid form, on the other hand, can only move as dissolved or suspended particles in the liquid phase of a soil. When a liquid, solution, or suspension moves through the soil, it does not displace the soil water already present in the soil

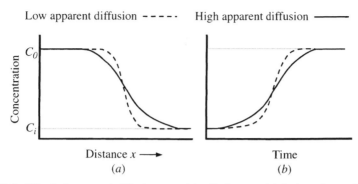

Figure 29.4 Effect of apparent diffusion on solute displacement (a) along flow path and (b) with time.

uniformly. As a result, the soil water concentration of the chemical under consideration at any given time along the flow path (Fig. 29.4a) or at a location with time (Fig. 29.4b) changes gradually from the concentration in the original soil solution, C_i, to the concentration of the incoming liquid, C_0. This gradual change in solute concentration (Fig. 29.4) can be attributed to hydrodynamic dispersion, which is related to the mechanical mixing of soil solution as well as to the molecular diffusion of the solute under consideration in soil water. The hydrodynamic or mechanical dispersion of soil solution $D_{lh}(L^2 \cdot T^{-1})$ can be expressed in terms of the average soil water convective velocity $v(L \cdot T^{-1})$ by

$$D_{lh} = \alpha_l v \tag{29.4}$$

where α_l, the dynamic dispersivity or dispersivity (L), is a property of the soil. The collective effects of the hydrodynamic (or mechanical) dispersion D_{lh} and molecular diffusion D_l^s can be expressed as "apparent diffusion" by

$$D_a = D_{lh} + D_l^s = \alpha_l v + D_l^s \tag{29.5}$$

For small soil water velocities, the apparent diffusion is dominated by molecular diffusion, whereas for large velocities hydrodynamic dispersion becomes important. For a more complete discussion of diffusion and dispersivity see Refs. 6, 7, and 27.

29.1.6 Theoretical Considerations and Practical Applications

Measurement of dynamic soil properties (e.g., hydraulic conductivity) as well as assessment of the movement of water and transport of pollutants through soils may require theoretical modeling. Although these models may be based on physical and chemical principles, in most cases, they are also based on assumptions that may not hold true under most field conditions. For example, many mathematical models assume the soil to be homogeneous and isotropic, while in general soils are heterogeneous and anisotropic by nature. Therefore, in consideration of models for assessing movement of water or transport of pollutants, particularly as related to measurement of soil properties, such as K_{sat}, one must consider practical applications of the concept. In some cases, simplifying assumptions resulting in a less elaborate and less costly procedure may be advantageous over a more complicated and laborious procedure that is based on a sophisticated model with unrealistic assumptions.

As indicated above, the term *permeability* is sometimes used synonymously as *saturated hydraulic conductivity* to express the ability of a soil to transmit water. In other cases, the term implies intrinsic permeability, which is a property of the soil alone (i.e., independent of the property of the liquid under consideration). Some may numerically equate air permeability of a dry soil to its intrinsic permeability.

There is no direct method to measure soil intrinsic permeability. However, there are field and laboratory procedures for measuring soil saturated hydraulic conductivity, unsaturated hydraulic conductivity, and air permeability. Soil permeability can then be obtained from measured saturated hydraulic conductivity.

29.2 MEASUREMENT OF SATURATED HYDRAULIC CONDUCTIVITY IN VADOSE ZONE

Many applications require knowledge of the saturated hydraulic conductivity of the vadose zone. Examples include designing drainage systems for water table management, designing infiltration galleries for bioremediation of contaminated soils and groundwater, and estimating K_{sat} for modeling soil water flow in the unsaturated zone. Here, two field procedures for measuring K_{sat} below the ground surface and determining infiltration rate and K_{sat} of the surface layers of the soil and one procedure for determining unsaturated hydraulic conductivity of the soil surface under low tension will be presented.

Because soil hydraulic conductivity is impacted by the quality of the water moving through it, the chemical composition of the water used for measuring conductivity in the vadose zone should be close to the quality of the soil water in the area. In most cases, however, it is impractical to determine the chemical composition of the soil water and prepare an adequate volume of water for hydraulic conductivity measurements. As an alternative, water without undesirable characteristics (e.g., high solute content, turbid water) from shallow wells or streams in the area can be obtained for measurements. In some areas, the quality of tap water may be comparable to the quality of soil water (or groundwater). Distilled or deionized water is not recommended for measuring hydraulic conductivity, but chemicals could be added to these waters to simulate the soil water or to minimize the adverse impact of the water quality on measurement. Klute and Dirkesen[28] suggested using a 0.005 M $CaSO_4$ solution with thymol (or another biological activity inhibitor) for laboratory measurement of K_{sat}. Since biological activities are part of the natural processes in the soil, the use of biological inhibitors may not be advisable for field measurements.

Dissolved gases in water may also be released, causing reduction in the rate of water movement through soil during K_{sat} measurement. The use of deaerated water has been recommended for laboratory measurement of K_{sat}. However, the use of deaerated water may not be practical for in situ measurements. Instead, care should be taken to avoid the use of freshly drawn tap water (particularly when air is introduced to the water at the faucet) and to minimize aeration of water or solution (e.g., avoiding shaking of the container) that will be used for conductivity measurement. Water that will be used can be stored at room temperature in a large container with a spigot at the bottom. If air bubbles appear on the side of the container, vacuum can be applied to the container to remove the excess air from the container.

Temperature of water infiltrating the soil may also impact the measured hydraulic conductivity. According to Youngs,[29] the measured K_{sat} increases nearly 3% per degree Celsius as the viscosity of water decreases with increasing temperature. Since it is not practical to conduct field measurements only during times when ambient temperature is the same as soil water temperature at the desired depth, a measured K_{sat} value can be adjusted to a desired temperature by multiplying it by the ratio of the viscosity of water at the measured temperature to that of the desired temperature. For example, the measured K_{sat} at 35°C (water viscosity 0.7194 mPa·s)[30] can be corrected to 25°C (water viscosity 0.8904 mPa·s) by multiplying it by 0.7194/0.8904. Although the measured K_{sat} values can be adjusted for temperature, it is best to avoid field measurements of this property during extreme temperatures. Also, we should keep in mind that if the temperature of the water infiltrating the soil decreases, entrapped gases in the soil may dissolve in water, resulting in an increase in the flow rate of water, which increases the measured K_{sat} values.[31,32] Therefore, temperature correction should be used cautiously.

29.2.1 Constant-Head Well Permeameter Technique

The constant-head well permeamater method, also known as shallow-well pump-in technique and borehole permeameter technique, is perhaps the most versatile and practical procedure for in situ measurement of the K_{sat} of the vadose zone from the soil surface to depths exceeding a few meters. In this procedure, the steady-state rate of water flow under a constant depth of water at the bottom of a cylindrical hole, dug to the desired depth, is measured for calculating K_{sat}.

The constant-head well permeameter method was developed more than 50 years ago.[33,34] However, it was not until the early 1980s that this procedure became more common. Originally, it was recommended that a float be used for maintaining a constant depth of water at the bottom of a relatively large diameter hole (on the order of 10 cm or more).[32,35] In addition, it was believed that a steady-state would be achieved after a relatively long time (over 24 h) and 2–6 days would be required to complete a test.[36,37] As a result, a large quantity of water was required to fill the hole and to measure the steady-state rate of water flow for measuring K_{sat}. Theoretical assessment of three-dimensional water flow and field measurement of K_{sat} by this procedure have shown that steady-state flow rate from a small-diameter (e.g., 6-cm) hole may be achieved in a relatively short time (e.g., a few hours).[38–41]

Because the volume of water required to fill a small-diameter hole to the desired depth is relatively low, measurement of K_{sat} for most practical applications can be accomplished with a few liters of water in 2–3 h.

Talsma and Hallam[41] conducted a series of field measurement of K_{sat} of the soils within four different forest catchments in Australia by this procedure. They developed a simple Marriott bottle device for maintaining a constant head of water at the bottom of 6.4-cm-diameter auger holes and completed each measurement using a few liters of water. Later, Reynolds et al.[39] presented a constant-head device, called Guelph Permeameter, for maintaining a constant depth of water and measuring the rate of water flow at the bottom of a relatively small hole (≥ 4 cm in diameter). The Guelph permeameter (hereafter referred to as GP) is basically a long Mariotte bottle (Fig. 29.5) capable of maintaining a constant depth of water at the bottom of a hole and allowing measurement of the rate of water flow into the hole. Currently, the GP is commercially available from Soilmoisture (Santa Barbara, CA).

Amoozegar[38,43] presented yet another permeameter, called Compact Constant-Head Permeameter (also known as Amoozemeter), for measuring K_{sat} by this procedure (Fig. 29.6). The compact constant-head permeameter (hereafter referred

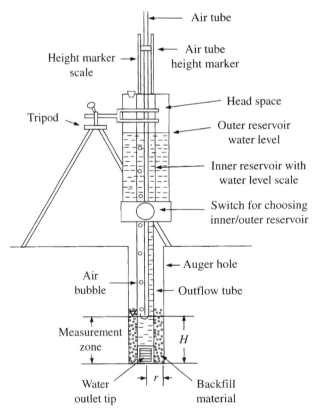

Figure 29.5 Schematic diagram of Guelph permeameter. (Adapted from Ref. 42.)

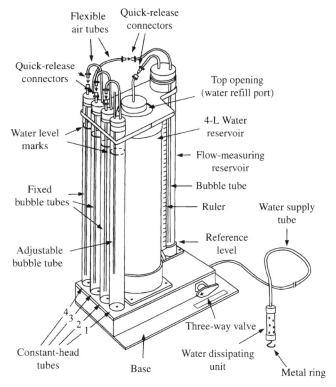

Figure 29.6 Schematic diagram of compact constant-head permeameter. (Courtesy of K_{sat}, Inc.)

to as CCHP) is portable, versatile, and easy to operate. It uses four Mariotte bottles (called constant-head tubes) for maintaining a constant depth of water at the bottom of a small-diameter hole ($\geq$4 cm) from the soil surface to 2 m depth. Using four additional constant-head tubes, the depth of measurement can be extended to 4 m. In addition, Amoozegar[43] introduced an accessory equipped with a pressure transducer for allowing measurements below 4 m depth. The CCHP is commercially available from Ksat (Raleigh, NC).

29.2.1.1 Materials and Equipment.
Materials and equipment needed for measuring K_{sat} include a set of augers for excavating a cylindrical hole, a constant-head device for maintaining a constant head and measuring the rate of water flow from the bottom of the hole into the soil, a measuring tape or long ruler for determining the depth of the hole and the depth of water in the hole, a timer or stopwatch, a data sheet for recording field measured information (Fig. 29.7), a marking pen and pencil, more than 5 L of water, and a small shovel or other devices for cleaning a small area where K_{sat} measurements will be conducted. To excavate the hole, a set of augers, including a hand auger, a planer auger (also called hole cleaner), a nylon brush, and an adequate length of auger extension with a handle, is

Measurement no. _____   Conducted by _____   Page ___ of ___

Location _____   Date _____

Weather condition _____   Temperature _____

Soil Horizon _____   Source of water _____

Hole Depth	_____ cm	Measured (actual) water level in hole
Distance between reference level		Initial _____ cm
and soil surface	_____ cm	Final _____ cm
Distance from the hole bottom to		
the reference level (D)	_____ cm	Clock time
Desired water depth in hole (H)	_____ cm	Start saturation _____
Constant-head tube setting (d)	_____ cm	Steady-state reading _____

Reservoirs used for measurement of the steady-state flow rate
  Flow-measuring reservoir only   _____   Conversion factor (C.F.) = 20 cm²
  Both flow-measuring and main reservoir _____   Conversion factor (C.F.) = 105 cm²

(To obtain flow volume multiply change in water level by the appropriate C.F. from above.)

Clock time h:min	Reservoir reading cm	Δt min	Change in water level cm	Flow volume cm³	Q cm³/min	Q cm³/h	K_{sat} cm/h
___	___	___	___	___	___	___	___
___	___	___	___	___	___	___	___
___	___	___	___	___	___	___	___
___	___	___	___	___	___	___	___
___	___	___	___	___	___	___	___
___	___	___	___	___	___	___	___
___	___	___	___	___	___	___	___

Average of last three measurements: K_{sat} = _____ cm/h _____   (other units)

Comments: _____

_____

_____

_____

Figure 29.7 Sample data sheet for constant-head well permeameter method using a compact constant-head permeameter. (Courtesy of K_{sat}, Inc.)

needed. Various sizes and types of augers are commercially available from AMS (American Falls, ID), Terra Systems (Washington, NC), and other vendors and supply companies (e.g., Forestry Supplies). A permeameter (e.g., CCHP, GP), a float,[36,37] or other devices can be used to maintain a constant head of water at the bottom of the hole and to measure the volume of water entering the hole. In this publication the CCHP will be introduced for collecting field data. The readers interested in the GP or other equipment are referred to the corresponding articles cited above.

The CCHP is composed of a set of four constant-head tubes, a 4-L reservoir, a 1-L flow-measuring reservoir, a water-dissipating unit, and a base with a three-way valve that connects the reservoirs to the water-dissipating unit (Fig. 29.6). The four constant-head tubes, filled with water, can be connected in series and provide from 0 to −50 cm of water pressure (vacuum) with the first constant-head tube and up to −150 cm of additional water pressure with the other three tubes for maintaining a constant level of water in a hole from the soil surface to approximately 200 cm below the device. The 4-L reservoir is located in the middle part of the unit and maintains the center of the gravity above and in the middle portion of the base. The 1-L flow-measuring reservoir is constructed of 2-in. clear (transparent) polycarbonate tube, which allows measurement of the water level in the device. A bubble tube that will be connected to one of the constant-head tubes is set at a fixed level that is used as the reference level for the permeameter. The water-dissipating unit, connected to a 2.5-m-long flexible water supply tube for carrying water to the bottom of the hole, allows uniform water flow from the permeameter into the hole to prevent scouring the hole wall. (*Note:* When additional constant-head tubes are used to increase the depth of measurement to 4 m, a water-dissipating unit with a longer flexible tubing must be used.) The other end of the flexible tube is attached to the outlet of a three-way valve that is connected to the 4- and 1-L reservoirs. By turning the valve, either the 1-L reservoir alone or both reservoirs can be used to supply water to the hole.

29.2.1.2 *Field Data Collection.*

Clean a small area in the field of plant residues and debris. Bore an auger hole of radius *r* to the desired depth using a hand auger. For most practical applications, a 6-cm diameter hole is recommended, but holes in the range of 4–12 cm in diameter may also be used depending on the application. For measurements at deep depths, a larger auger or a mechanical drill can be used for digging the hole to within 50 cm of the depth where conductivity measurement will be conducted. The bottom 50 cm of the hole, however, must be dug with a hand auger of known diameter. Using a planer auger of the same size as the auger, clean the bottom of the hole (shave it) to form a flat-base cylindrical hole. (*Note:* The diameter of the bottom part of the hole must be known. It should be mentioned that the diameter of an auger hole dug with a hand auger is generally larger than the nominal diameter of the auger. The diameter of the hole generally corresponds to the distance between the cutting edges of the auger blades.) To minimize the effect of smearing of the auger hole side wall, use a round nylon brush or other devices (see Ref. 44) to scrape off the hole side wall. However, it should be noted that if smearing is the result of soil being too wet, brushing the auger hole side wall may not reduce the smearing on the wall of the auger hole.

Place the CCHP next to the hole and record the depth of the hole and the distance between the bottom of the hole and the reference level on the CCHP on the data sheet. Fill the flexible tubing attached to the water-dissipating unit with water from both reservoirs and serially connect an adequate number of constant-head tubes. Adjust the bubble tube in the first constant-head tube to maintain the desired depth of water in the hole (referred to as head of water, H). Lower the water-dissipating unit into the hole and turn the three-way valve to allow water to enter the hole from both reservoirs. In the system depicted in Fig. 29.8a for measuring K_{sat} in the upper 50 cm of the soil, a constant depth of water equal to H is maintained at the bottom of a hole by adjusting the location of the tip of the air tube

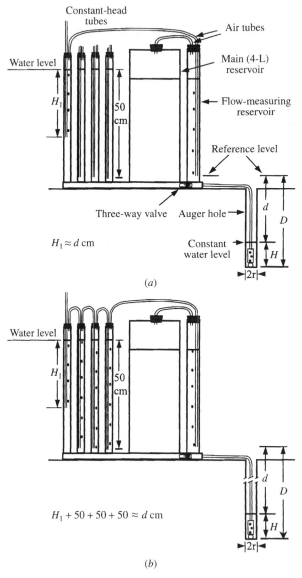

Figure 29.8 Schematic diagram of compact constant-head permeameter showing proper connection of constant-head tubes for (a) measurements in upper 50 cm and (b) measurements between 150- and 200-cm depths

inside the first constant-head tube such that the distance d from the top of the desired depth of water in the hole to the reference location on the permeameter corresponds to the distance H_1 from the tip of the air tube to the water level inside the constant-head tube. For measurements at deeper depths the distance from the reference location on the device to the water level in the hole can be increased by serially connecting a series of constant-head tubes as shown in Fig. 29.8b for measurements between 150 and 200 cm depths. It should be noted that the distance d may not be exactly the same as distance H_1 plus 50 × number of fixed constant-head tubes (as depicted in Fig. 29.8b). The depth H can be adjusted by moving the bubble tube in the first constant-head tube up or down as needed.

After establishing a constant head of water at the bottom of the hole, allow water to move from the permeameter into the soil through the hole until a steady-state condition is achieved (i.e., the rate of water flow into the soil Q under a constant head of water becomes constant). For most practical applications, we can assume a steady-state rate is achieved when three consecutively measured flow rates are equal. To ensure that a constant head of water is maintained, measure the depth of the water in the hole a few times during flow rate measurements. The depth of the water in the hole can be determined by accurately measuring the distance between the water level in the hole and the reference location (distance d shown in Figs. 29.8a and b) and subtracting it from the distance between the bottom of the hole and the reference location on the permeameter (distance D in Figs. 29.8a and b).

29.2.1.3 *Calculation.*

After determining the steady-state flow rate, calculate the saturated hydraulic conductivity by the Glover model,

$$K_{sat} = \frac{CQ}{2\pi H^2} \tag{29.6}$$

where

$$C = \sinh^{-1}\left(\frac{H}{r}\right) - \left[\left(\frac{r}{H}\right)^2 + 1\right]^{1/2} + \frac{r}{H} \tag{29.7}$$

The above model is valid when the distance between the bottom of the hole and an impermeable layer below the hole, s, is greater than or equal to $2H (s \geq 2H)$. When using the Glover model, Amoozegar[45] has suggested that H be at least five times greater than the radius of the hole (i.e., $H \geq 5r$). For cases where $s < 2H$, K_{sat} can be calculated using the model

$$K_{sat} = \frac{3Q \ln(H/r)}{\pi H(3H + 2s)} \tag{29.8}$$

The Glover model presented above is one of the original models for calculating K_{sat}.[33,34] This model is suggested by the author because it is simple and reliable.

The Glover model was developed by only considering the saturated flow of water around the auger hole. Several other models and approaches presented since the early 1980s consider both saturated and unsaturated water flow but use the same field measured data for Q, r, and H.[32,39,46–49] In these models, the saturated flow around the auger hole is parameterized by K_{sat} and the unsaturated flow component is represented by an empirical parameter that must be determined independently, estimated based on selected soil properties, or calculated simultaneously. To develop their analytical models, Reynolds and colleagues[39,46] and Philip[47] used the matric flux potential[50]

$$\phi = \int_{h_i}^{h} K(h)\, dh \qquad (29.9)$$

and the hydraulic conductivity function

$$K(h) = K_{sat} \exp(\alpha h) \qquad (29.10)$$

where $h(L)$ is the soil water pressure head, h_i is the initial soil water pressure head, and $\alpha(L^{-1})$ is an empirical constant. The models by Stephens and Neuman[49] and Stephens et al.[32] are regression based. They used the unsaturated hydraulic conductivity and soil water pressure head relationship (i.e., K versus h) to represent the unknown parameter related to the unsaturated flow in their regression models.

Reynolds and Elrick[51] presented the simultaneous equations approach for obtaining K_{sat} and the matric flux potential ϕ simultaneously. The simultaneous equations approach, however, may result in a negative value for K_{sat} or ϕ (see Ref. 45). To alleviate the problem of negative values for K_{sat} or ϕ, Elrick et al.[46] offered fixed values of α for four different types of soils. The fixed-α approach of Elrick et al.[46] does not result in a negative value for K_{sat}; however, this parameter must be estimated based on soil texture and structure.

There is no agreement among investigators on the use of a specific model for calculating K_{sat}. Amoozegar[38,45,52] compared the Glover model with the other models that consider both saturated and unsaturated flow. The author's conclusion has been that for all practical applications the K_{sat} values calculated for various flow rates from identical cylindrical holes by the Glover and other models are not significantly different.

29.2.1.4 Comments. Smearing of the auger hole side wall during its construction may reduce the rate of infiltration of water from the hole into the soil. If the soil under investigation is not excessively wet, the smearing may be removed by inserting a nylon brush into the hole and moving it gently up and down with a slight twist a few times. Wilson et al.[53] have discussed the possibility of allowing the bottom section of the hole to dry out for removing smearing before applying water to the hole. In a research study, Campbell and Fritton[54] used an ice-pick to remove a thin layer of smeared soil at the bottom of the hole. Their technique, however, may not be practical for removing smearing from deep holes and also may enlarge the hole and increase the diameter of the bottom section of the hole significantly. Another factor that must be considered is air entrapment within the wetted zone around the auger hole.[31,32] Variations in the temperature of water at the bottom of the hole may also impact the measured flow rate and the calculated K_{sat} value. The calculated K_{sat} value can be adjusted to any desired temperature, as discussed earlier. Finally, to prevent collapse of the auger hole side wall in sandy or otherwise unstable soils, a section of perforated pipe with an outside diameter equal to that of the hole (e.g., a section of 2-in. well casing for 6-cm-diameter holes) can be inserted inside the hole immediately after the hole construction.

29.2.2 Double-Cylinder Infiltrometer Method

Knowledge of the infiltration rate may be required for designing irrigation systems (including effluent disposal), surface drainage systems, or storm water management systems. Infiltration rate (often denoted as i) is defined as the volume of water entering a unit area of the soil surface per unit time. The surface under consideration here could be the land surface or any infiltrative surface such as the bottom of a trench or the dispersal area for a subsurface infiltration gallery. The initial rate at which water enters the soil depends on antecedent soil water content of the surface layer under consideration (Fig. 29.9). In the absence of any impermeable or slowly permeable layer that can impede vertical flow, the final infiltration rate for any given event generally is a constant that is assumed to be equal to the saturated hydraulic conductivity of the surface layer where measurements are conducted.

Single- and double-cylinder infiltrometers[55] are used for measuring infiltration rate. A single-cylinder infiltrometer is basically a thin-wall cylinder constructed from a rigid material such as aluminum or steel that can be vertically inserted into the soil. A double-cylinder infiltrometer is composed of a relatively small cylinder (e.g., 30 cm diameter) that is installed vertically in the middle of a larger cylinder (e.g., 60 cm diameter) of the same length (Fig. 29.10). While water is

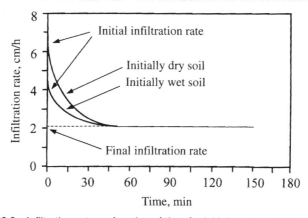

Figure 29.9 Infiltration rate as function of time for initially wet and initially dry soil.

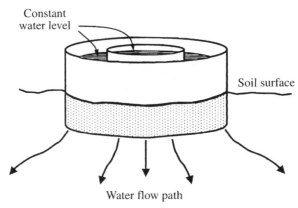

Constant
water level

Soil surface

Water flow path

Figure 29.10 Schematic diagram of double-cylinder infiltrometer showing potential flow path below infiltrometer.

applied to both inner and outer cylinders, only infiltration rate from the inner cylinder is measured. The purpose for this is to minimize the effect of the lateral divergence of water flow under the cylinders (see Fig. 29.10). Bouwer[55] and Reynolds et al.[56], however, stated that a double-cylinder infiltrometer is not effective in reducing the lateral flow divergence. Bouwer[55] suggested using a single-cylinder infiltrometer with a diameter larger than 100 cm, and Reynolds et al.[56] suggested the cylinder diameter be as large as possible while the depth of water ponded in the cylinder be as small as possible. For some applications, such as measuring hydraulic conductivity of landfill liners, a relatively large single- or double-cylindrical or rectangular infiltrometer is used to measure the very low infiltration rate over a long period of time (e.g., >24 h). For description of open and sealed single- and double-cylinder (ring) infiltrometers see Daniels.[57]

29.2.2.1 Materials and Equipment.
Materials and equipment needed for measuring infiltration rate and estimating K_{sat} include a double-cylinder infiltrometer with 25–30 and 50–60 cm diameter, two Mariotte reservoirs or other devices for maintaining a constant depth of water in the cylinders, measuring tape or ruler, shovel or other tools for clearing the soil surface, a few sections of 70-cm-long 2×4 or 4×4 lumber, a sledge hammer, a stopwatch, a marking pen, a data sheet, and water. A double-cylinder infiltromerter with 30.5 and 61 cm diameter for the inner and outer cylinders, respectively, is commercially available from ELE International (Loveland, CO). For some applications, two CCHPs, described in the previous section, can be used for maintaining a constant depth of water in the inner and outer cylinders. As an alternative to the CCHP or other Mariotte-type reservoirs, two small floats can be used for maintaining a constant depth of water in each of the inner and outer cylinders.

29.2.2.2 Field Data Collection.
Clean a 1-m² flat area of debris and loose plant materials. If a crust is present at the surface, you need to remove it prior to the installation of the infiltrometer unless the crust itself must be evaluated. Also,

for measurements below the soil surface or in sloping areas, you must remove some soil to create a level area for this measurement. After cleaning and leveling the soil surface, push the outer cylinder vertically into the soil at least 5–10 cm. For this purpose, place a 2×4 or 4×4 piece of lumber over the top of the cylinder and, using the sledge hammer, pound (or hydraulically push) the board on the two sides of the cylinder alternately. To prevent formation of any gap between the soil and the cylinder wall, rotate the board frequently to force the cylinder vertically into the soil. After installing the outer cylinder, tamp the soil around the outside wall (and if necessary inside wall) of the cylinder to avoid leakage. Then, place the inner cylinder at the center of the outer cylinder and push it at least 5 cm into the soil using the sledge hammer and a piece of board. To prevent leakage between the two cylinders, lightly tamp a small strip around the inner and outer walls of the inner cylinder. To prevent scouring of the soil during water application, place a piece of cheese-cloth, some straw, or another material with high conductivity on the soil surface inside each of the cylinders.

Apply water to both the inner and outer cylinders under a constant head (i.e., depth of water in the cylinders) by two constant-head devices (e.g., floats). To optimize one-dimensional (vertical) flow from the inner cylinder, the water levels inside the inner and outer cylinders should be the same. Also, to minimize divergence of flow, the depth of water in the cylinders should be kept at a minimum (say 1–5 cm). Care should be taken to keep all areas of the soil surface in the cylinders under water at all times during measurements. Immediately after application of water to the cylinders, measure the amount of water entering the soil within the inner cylinder with time to determine the initial rate of infiltration. Continue measurement until a steady-state condition is achieved (i.e., the volume of water entering the soil in fixed time intervals becomes constant). Matula and Dirksen[58] introduced an automatic system for controlling the depth of water and measuring the amount of water infiltrating the soil. Based on their claim, the volume of water entering the inner cylinder can be measured within ± 1 mL. As an alternative to using constant-head devices, small quantities of water can be frequently added to both the inner and outer cylinders to maintain approximately the same level of water in both cylinders. The infiltration rate is then measured by determining the rate of fall of water level in the inner cylinder using small time intervals.

29.2.2.3 *Calculation.*

Infiltration rate is calculated by dividing the volume of water entering the inner cylinder per unit time by the surface area of the soil inside the inner cylinder (i.e., cross-sectional area of the inner cylinder). Plot the volume of water entering the soil per unit surface area against time to determine the initial and final infiltration rates. Plot the cumulative volume of water entering a unit surface area of the soil against time to obtain the cumulative infiltration (often denoted as I). The rate of infiltration into the soil immediately after water application is relatively high and depends on the antecedent soil water content (see Fig. 29.9). With time, the rate of infiltration decreases and reaches a quasi-steady-state value. This quasi-steady-state infiltration rate has been equated to the K_{sat} of the surface layer where infiltration takes place.[59] However, Bouwer[55] has stated that, based on limited observations, the final infiltration rate is approximately 50% of the saturated hydraulic conductivity of the infiltrative layer. Due to incomplete saturation and air entrapment, the final infiltration rate may be lower than K_{sat} of the infiltrative layer. Reynolds et al.[56] described a model to calculate K_{sat} using the infiltration rate. This model, however, contains empirical parameters (including the α parameter discussed in the previous section) that must be independently measured or estimated. The readers interested in the model are referred to the above reference as well as to Reynolds and Elrick.[60]

29.2.2.4 Comments. Measurement of infiltration rate by the cylinder infiltrometer technique is rather simple and can be accomplished without requiring expensive equipment. The procedure, however, requires a flat area and is not suitable for measuring infiltration on sloping grounds under natural conditions. Also, this procedure can only be used to measure infiltration rate and perhaps vertical K_{sat} at the soil surface or at the bottom of a pit. As a result, it is not a useful method for measuring K_{sat} at deep depths.

29.3 MEASUREMENT OF UNSATURATED HYDRAULIC CONDUCTIVITY IN VADOSE ZONE: TENSION INFILTROMETER METHOD

The tension infiltrometer (also known as disk infiltrometer), as we know it today, was initially designed and used by Australian and other scientists until Perroux and White[61] described it in detail. This tension infiltrometer has been modified by others for measuring soil hydraulic properties.[62,63] Figure 29.11 presents a schematic diagram of a tension infiltrometer.

The tension infiltrometer is used to measure K_{sat} under a small positive pressure and K_{unsat} between zero and up to 30 cm of tension (equivalent to -30 cm of soil water pressure head). Measurements can be made by forcing a one-dimensional flow within a cylinder inserted into the soil (referred to as confined technique)[64,65] or by establishing a three-dimensional flow (referred to as unconfined technique).[66,67] Since measurements are made under small tensions, the tension infiltrometer method is suitable for assessing water flow by excluding macropore flow during infiltration.

Different procedures have been suggested for determining K_{unsat} by the unconfined technique. In one method, the infiltration rate into soil is measured at one location under different tensions,[68] and in another infiltration rate is measured under one tension at a given location.[69] In the later, soil water content under the infiltrative surface must be determined before and immediately after measurement,[70] and a new location must be used for each tension. Soil water content can be determined outside the measuring area before measurements and under the infiltrometer immediately after measurements are completed. Both gravimetric and in situ techniques have been suggested for this purpose.[70,71] For procedures to measure soil water content, see Ferré and Topp[72] and Topp and Ferré.[73]

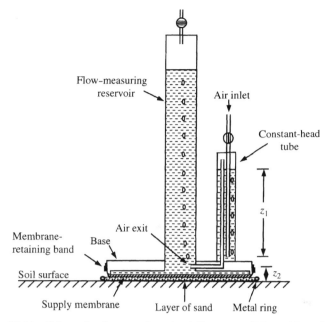

Figure 29.11 Schematic diagram of tension infiltrometer. (Adapted from Ref. 69.)

29.3.1 Materials and Equipment

A tension infiltrometer, a shovel or spade to clean and level a small area, fine sand, a 3–4 mm-tall ring or a double-cylinder infiltrometer (and equipment for their installation as described previously), a data sheet, a stop watch, and water are needed for collecting field data. In addition, equipment for measuring soil water content in situ or for collecting and determining soil water content by the gravimetric method (as will be discussed later) may be needed. The sand is used for making a solid contact between the soil and the infiltrometer. The air entry value for the contact materials must be greater than the highest tension considered for measuring K_{unsat} (e.g., 30 cm). Therefore, when compacted, the contact materials should have adequate fine pores that remain saturated under the highest tension used. Perroux and White[61] have listed the criteria for the contact materials and suggested a fine sand eolian deposit for the purpose. Fine-grain silica sand with a hydraulic conductivity greater than the conductivity of the soil under consideration is also suitable (see Ref. 64). The inside diameter of the inner cylinder of the infiltrometer or the ring must be slightly larger than the diameter of the tension infiltrometer. Different sizes and types of tension infiltrometers are available commercially from Soil Measurement Systems (Tucson, AZ) and Soilmoisture (Santa Barbara, CA).

29.3.2 Field Data Collection

For the confined technique (one-dimensional flow), clean and level a small area at the soil surface (or at the desired depth) and install the double-cylinder infiltrometer as described earlier. For measurements at the soil surface, remove the surface crust unless the surface crust itself must be analyzed. After measuring the final infiltration rate under a small positive head (set equal to K_{sat} as described by Watson and Luxmoore[64]), cease water application to the cylinders and remove the excess water from the cylinders (or alternatively allow all the water to infiltrate the soil). Immediately after removing the water from both cylinders, apply and level a thin layer (2–3 mm) of wet sand inside the inner cylinder and place the tension infiltrometer on top of the sand inside the inner cylinder. Following the manufacturer's instructions, apply water to the inner cylinder under the smallest tension desired (e.g., $z_1 - z_2$ in Fig. 29.11 equal to 3 cm corresponding to -3 of soil water pressure head) and measure the volume of water moving from the infiltrometer into the soil in regular time intervals until a steady-state rate is achieved. Then, increase the tension to the next level (e.g., 6 cm) and measure the rate of water flow into the soil again. After reaching the steady-state rate of infiltration into the soil, repeat the steps by successively increasing the tension on the infiltrometer (by increasing the distance z_1 shown in Fig. 29.11). Make sure that a solid contact is maintained between the soil and the bottom surface of the tension infiltrometer to prevent air entering the system, as this will break the tension.

For the unconfined technique (three-dimensional flow), clean and level a small area where measurements are needed as described above. Place the ring on top of the soil and push it slightly into the soil. Then, apply a thin layer of sand, and using the edge of the ring, level the sand inside the ring for full contact with the base of the infiltrometer. Prepare the tension infiltrometer based on the manufacturer's instructions and place it on top of the sand in the ring. Apply water to the soil inside the ring under a desired tension and measure the rate of water flow into the soil until a steady-state rate is achieved. Then, change the tension on the infiltrometer to the next desired level for measuring infiltration rate. Since measurements start under natural field conditions, it is best to start with the highest tension (e.g., 20 cm) and then successively decrease tension for subsequent measurements. Different tension values for measuring infiltration rates have been suggested by various investigators. Ankney[66] suggested measurements under tensions of 3, 6, and 15 cm. If soil water content measurements are needed, collect at least two soil samples from the upper part of the soil on opposite sides outside of the ring before installing the infiltrometer. At termination of the steady-state flow rate measurement, remove the infiltrometer and the sand contact layer and collect two samples from the upper 2–5 mm depth of the soil. Determine the soil water content of these samples by the gravimetric method.[73] Alternatively, you can determine soil water content indirectly by time domain reflectometry (TDR)[72,74] and/or tensiometry[75] using the soil water characteristic curve for the soil under consideration.[70]

29.3.3 Calculation

For the confined method, assume a unit hydraulic gradient within the wetted zone under the infiltrometer and set the unsaturated hydraulic conductivity equal to the steady-state (final) infiltration rate for each tension by

$$K(h) = \frac{q(h)}{\pi r^2} \tag{29.11}$$

where $q(h)$ is the total volume of water entering the soil per unit time under soil water pressure head h (equal to the negative of the tension set on the infiltrometer, $z_2 - z_1$) and r is the radius of the disk (base) of the infiltrometer.[64]

For the unconfined method, plot the cumulative infiltration I obtained under the desired tension versus time t and $t^{1/2}$. Determine the sorptivity S, defined as the slope of I versus $t^{1/2}$ during the early stages of infiltration (i.e., $dI/dt^{1/2}$ as $t \to 0$; see Ref. 3), and calculate K_{unsat} at each tension by[69]

$$K = \frac{dI_\infty}{dt} - \frac{2.2 S^2}{\Delta\theta\,\pi r} \tag{29.12}$$

where dI_∞/dt is the steady-state (final) infiltration rate at tension under consideration, $\Delta\theta$ is the difference in soil water content under the infiltrometer before and after measurements, and r is the radius of the infiltrating surface area (i.e., radius of the disk infiltrometer). Although this procedure is straightforward, it can be used to measure K_{unsat} under only one tension at a location if the gravimetric method is used for water content determination. As a result, due to spatial variability of soil water content and infiltration capacity, erratic K_{unsat} values may be obtained when measurements are conducted at various tensions. As mentioned earlier, measurements at various tensions can be conducted at one location by in situ determination of soil water content with an indirect (nondestructive) method, such as the TDR technique or tensiometry. The procedure, however, suffers from the lack of agreement between the values obtained by the indirect techniques and the actual soil water content of the soil.

Ankney et al.[68] proposed a procedure for determining K_{unsat} of different tensions at one location. They used the Wooding[76] model for water flow from a circular source and developed a system of three equations for calculating K_{unsat} corresponding to a pair of soil water pressure heads h_1 and h_2. The three equations are

$$q(h_1) = K(h_1)\left(\pi r^2 + \frac{4r}{A}\right) \tag{29.13a}$$

$$q(h_2) = K(h_2)\left(\pi r^2 + \frac{4r}{A}\right) \tag{29.13b}$$

$$\frac{K(h_1) - K(h_2)}{A} = \frac{\Delta h[K(h_1) + K(h_2)]}{2} \tag{29.13c}$$

where $q(h_1)$ and $q(h_2)$ are the steady-state volumetric flux under two successive soil water pressure heads (e.g., -3 and -6 cm, -6 and -15 cm), r is the radius of the infiltrometer base, and A (L^{-1}) is an empirical constant set equal to $K(h)/\phi$

[obtained from Equations (29.9) and (29.10)]. Under these equations, each pair of steady-state flow rates results in one constant A and two hydraulic conductivity values at h_1 and h_2. For example, for the pairs $q(-3)$ and $q(-6)$ the corresponding K_{unsat} values are $K(-3)_{3,6}$ and $K(-6)_{3,6}$, and for $q(-6)$ and $K(-15)$ the corresponding K_{unsat} values are $K(-6)_{6,15}$ and $K(-15)_{6,15}$, respectively. It should be noted that the two estimated K_{unsat} values for a given tension [$K(-6)_{3,6}$ and $K(-6)_{6,15}$ in the above example] are not equal. To determine a single K_{unsat} value for each tension, Ankeney et al.[68] suggested to take the average of the two resulting $K(h)$ values obtained by this procedure. For the above example, the final values will be

$$K(-3) = K(-3)_{3,6} \qquad (29.14a)$$

$$K(-6) = \frac{1}{2}[K(-6)_{3,6} + K(-6)_{6,15}] \qquad (29.14b)$$

$$K(-15) = K(-15)_{6,15} \qquad (29.14c)$$

To use the procedure offered by Ankney et al.[68], measure the flow rate (i.e., volume of water per unit time) from the infiltrometer under different tensions (e.g., 0, 3, 6, and 15 cm corresponding to 0, -3, -6, and -15 cm soil water pressure heads, respectively) without moving the infiltrometer. Then use Equations (29.13a)–(29.13c) to obtain the corresponding $K(h)$ values for each pair of neighboring soil water pressure head values. Determine the final $K(h)$ values by averaging the corresponding values as described above.

29.3.4 Comments

The tension infiltrometer technique allows measurement of K_{unsat} only near soil saturation at the soil surface or at the bottom of a pit. Therefore, the K_{unsat} values obtained by this technique may have limited application for assessing water flow under unsaturated conditions. However, the technique is useful for assessing the contribution of macropores to water flow. Also, the procedure can be used to obtain data for determining some of the parameters needed for modeling soil water flow.

29.4 MEASUREMENT OF SATURATED HYDRAULIC CONDUCTIVITY IN SATURATED ZONE

The available field methodologies for measuring saturated hydraulic conductivity K_{sat} and transmissivity T of aquifers can only be used in areas where the saturated zone extends horizontally in all directions. To the knowledge of the author, no field technique is available for measuring K_{sat} of small pockets of saturation that may occur under various field conditions (e.g., perched water table above a small lens of clay layer). Some applications in which knowledge of K_{sat} or T is required are designing drainage systems for lowering a permanent or seasonally high water table, analysis of water flow from recharge wells to discharge wells in groundwater bioremediation analysis, and mounding analysis of groundwater for designing large septic systems or other waste disposal systems.

The procedures for measuring K_{sat} below a water table have the advantage of using the actual groundwater for measurement. In general, there is no need for temperature correction, and the impact of the quality of groundwater on measured K_{sat} value for the location under consideration is minimal. Here, we will present two procedures for measuring K_{sat} of the upper part of an unconfined aquifer. These procedures can be used at various locations within a field to assess the variability of soil hydraulic conductivity.

29.4.1 Auger Hole Method

The auger hole method is perhaps the oldest procedure for in situ measurement of the K_{sat} of the soil below a shallow water table. Basically, the auger hole method involves boring a cylindrical auger hole of known diameter $(2r)$ to the desired depth (D) at least 30 cm below the water table (Fig. 29.12). After preparing the hole for measurements and allowing the water level in the hole to reach equilibrium with the water table, water is removed quickly from the hole and the rate of rise of water level in the hole is measured with time $(\Delta y/\Delta t)$ for calculating a K_{sat} value using one of the available models. The procedural techniques and the models for calculating K_{sat} have been improved by a number of investigators to make the procedure simpler or to extend its use.[77–83] The readers interested in the history of this procedure are referred to Bouwer and Jackson.[84]

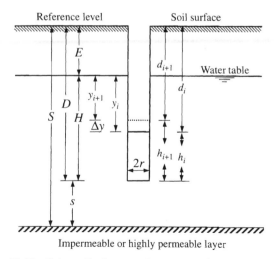

Figure 29.12 Schematic diagram of geometry of auger hole method.

29.4.1.1   Materials and Equipment. Compared to other procedures, the auger hole method may be performed with commonly available materials and equipment. The materials and equipment include an auger set, a device to lower the water level in the hole, a device to measure the changes in the water level in the hole, a timer, a measuring tape or ruler, a data sheet (Fig. 29.13), a marking pen, and a shovel or other devices for cleaning the ground area.

An auger capable of boring a 6- to 20-cm-diameter hole (preferably 8–10 cm diameter), a matching planer auger, and an adequate number of auger extensions

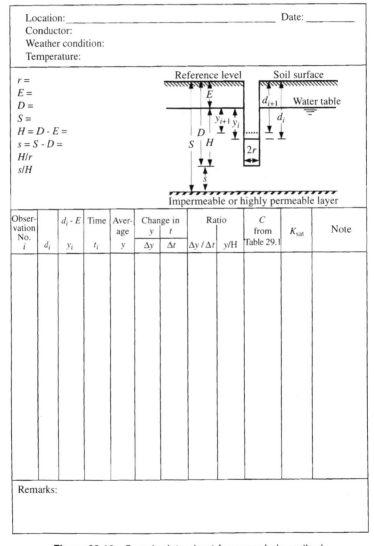

Figure 29.13 Sample data sheet for auger hole method.

are required to bore a cylindrical hole to the desired depth below the water table. Bucket-type augers are preferred because they are more suitable for describing the soil profile than Dutch-type augers. In general, smaller diameter holes are preferred for low-conductivity soils and larger diameter holes are better for high-conductivity soils. As mentioned earlier, various sizes and types of augers are commercially available from different vendors.

A bailer, slug, or a high-capacity pump (e.g., 38 L/min or more) is needed for lowering the water level in the hole quickly. For small-diameter holes, a bailer with a check valve at the bottom end can be constructed from a section of thin-wall aluminum or polyvinyl chloride (PVC) pipe with an outside diameter 2–3 mm smaller than the diameter of the hole under consideration. The bailer should be long enough to reach the bottom of the hole. As an alternative to a bailer or a pump, a slug constructed of a solid section of metal or plastic rod (or a section of a pipe with sealed ends) with a diameter slightly less than the diameter of the hole can be used to raise or lower the water level in the hole quickly.

For low-conductivity soils, a popper,[85] or a battery-operated probe attached to a flexible measuring tape or to a set of long and narrow rulers, can be used to monitor changes in the water level in the hole. For high-conductivity soils, where water level rises rapidly inside the hole, a reel-type recorder equipped with a float[82,83] or a fast-acting pressure transducer (equipped with relatively thin wires) placed at the bottom of the hole can be used for continuously measuring the water level in the hole. Downey et al.[86] presented a computerized probe for rapid recording of the water level in an auger hole or well.

29.4.1.2 *Field Data Collection.*

29.4.1.2 Field Data Collection. Remove loose plant materials and debris to clean a small area where conductivity measurements will be conducted. Using the auger, bore a hole to the desired depth below the water table. Check the soil texture during the boring of the auger hole to determine the degree of layering and the presence of any hydraulically limiting layer below the water table. If needed, describe the soil profile during the process. Because the auger hole cannot be used if the groundwater is under artesian conditions (i.e., confined aquifer), note any impermeable or slowly permeable layer within the depth of the hole. A rapid rise in the water level in the hole during construction may indicate a confined aquifer below an aquitard (i.e., an impermeable or slowly permeable layer). After boring the hole with the auger, use a planer auger to clean the bottom of the hole and make it cylindrical without deepening the hole. For coarse-textured soils a section of thin-wall perforated pipe (e.g., a perforated well casing) with an outside diameter equal to the diameter of the hole can be inserted into the hole immediately after its construction. The perforation of the pipe must extend to above the water table, and its wall must be thin to minimize errors in determining K_{sat}. [*Note:* For some soils (e.g., sandy soils) it may be impractical to extend the hole below the water table because of caving of soil materials into the hole. If caving during the hole construction is a problem, other procedures, such as jetting with water or driving a highly perforated pipe (e.g., a section of well casing) with a cone attached to the bottom end must be used to construct the hole. The thickness of the casing and blocking of the bottom floor of the hole are not considered in the theoretical models for calculating K_{sat}. As a result, the measured K_{sat} values may contain additional errors and should be viewed accordingly.]

Allow adequate time for the water level in the hole to come to equilibrium with the water table, a condition known as *static water level*. After achieving equilibrium, set a reference level at the soil surface next to the hole and measure and record the depth of the hole (D), the depth to the water table (E), and the radius of the bottom section of the hole (r) under the water table (see Fig. 29.12). Estimate the depth to any impermeable or highly permeable layer that may be present at the site (distance S in Fig. 29.12) from existing information or by boring holes at locations a few meters away from the measurement area. As an alternative, the depth to an impermeable or highly permeable layer can be determined by extending the auger hole after completing the conductivity measurements. Using D, E, and S, determine the initial depth of the water in the hole ($H = D - E$) and the distance between the bottom of the hole and the impermeable or highly permeable layer ($s = S - D$).

To minimize the impact of smearing the side wall of the hole during its construction, remove water from the hole and allow the hole to refill with water. Repeat this process several times until the water removed from the hole no longer contains appreciable amounts of fine materials. This is similar to developing a water supply well. Dispose of the water removed from the hole in this process (and during later measurements) away from the measurement area. After developing the well, allow the water level in the hole to reach equilibrium with the water table (i.e., to reach its static level).

Using a pump, slug, or bailer, quickly lower the water level in the hole to a distance y_0 below the static water level (see Fig. 29.12). Remove the intake pipe or the hose from the hole if a pump is used. Immediately after lowering the water level in the hole, at time t_0, measure the distance from the reference level to the water level (d_0), or measure the initial depth of water in the hole (h_0). Then, measure the depth to the water level (d_i) or the depth of water in the hole (h_i) a few times (at $t_i, i = 1, 2, 3, \ldots$). To use the model developed by Boast and Kirkham,[77] more than half the water in the hole is removed and measurements must stop when the depth of water in the hole (h in Fig. 29.12) reaches about one-half of H (i.e., when $y/h < 0.5$). To use the graphs or models presented by van Beers,[83] measurements must be stopped when the rise of the level of water in the hole from the initial measurement ($\sum \Delta y$) reaches 0.25 y_0 (i.e., $\sum \Delta y \leq 0.25\, y_0$). van Beers[83] suggested to measure the water level in the hole five times before the change in the water level reaches 0.25 y_0. To measure the rate of rise of water level in the hole, either measure the water level in the hole at fixed time intervals (e.g., measure the depth to the water level in the hole every 10 s) or record the time for a fixed amount of rise in the water level (e.g., measure the time for every 2-cm rise in the water level in the hole).

29.4.1.3 Calculation. The equation for calculating K_{sat} using field-measured data can be written in the form

$$K_{sat} = \left(\frac{C}{864}\right)\frac{\Delta y}{\Delta t} \tag{29.15}$$

where $C/864$ is a dimensionless shape factor related to r, y, H, and s and $\Delta y/\Delta t$ is the rate of rise of water level in the hole (a positive value with dimensions of $L \cdot T^{-1}$). In the above equation, r, y, H, and s have the same units (e.g., centimeters) and the calculated K_{sat} has the same units as $\Delta y/\Delta t$ (e.g., centimeter per hour).

Based on the original work of E. F. Ernst (published in 1950 in Dutch as cited by Bouwer and Jackson[84]), van Beers[83] and Bouwer and Jackson[84] presented four nomographs for obtaining the shape factor C in Equation (29.15) for r values of 4 and 5 cm, $s = 0$ and $s \geq 0.5\,H$, and various values of H and y (the average value for two consecutive measurements y_i and y_{i+1}). In addition to the nomographs, van Beers[83] and Bouwer and Jackson[84] offered two equations for approximating the shape factor C using r, y, H, and s. These equations are

$$C = \frac{3600r}{y(10 + H/r)(2 - y/H)} \qquad \text{when } s = 0 \tag{29.16}$$

$$\frac{4000r}{y(20 + H/r)(2 - y/H)} \qquad \text{when } s \geq 0.5\,H \tag{29.17}$$

van Beers[83] has suggested that these equations do not give the exact relationship between various parameters and K_{sat}. Therefore, they should only be used when no nomographs can be used for obtaining C (i.e., when r is not equal to 4 or 5 cm and $0 < s < 0.5\,H$). According to van Beers,[83] using Equation (29.17) rather than the graphs to obtain the C factor for calculating K_{sat} has sufficient accuracy (maximum error of 20%) for cases where $20 < H < 200$ cm, $3 < r < 7$, $s > H$, $y > 0.2\,H$, and $\sum \Delta y = 0.25\,y_0$.

Boast and Kirkham[77] developed a model for water entering a partially empty cylindrical hole dug to below a water table, and presented the K_{sat} in the form of Equation (29.15) with the shape factor C related to r, y, H, and s. The C factors developed by Boast and Kirkham[77] for various hole geometries and distances to an impermeable or highly permeable layer below the hole are presented in Table 29.1. The ratio of C values calculated using Equation (29.16) and the corresponding values obtained from Table 29.1 for $s = 0$ are between 0.73 and 0.85 for $H/r \leq 2$ and between 0.87 and 1.04 for $H/r \geq 5$. The ratios of the corresponding C values obtained from Equation (29.17) and Table 29.1 for $s/H \geq 0.5$ are <0.79 for $H/r \leq 2$ and between 0.79 and 1.1 for $H/r \geq 5$. Because the C values presented by Boast and Kirkham[77] are based on a more accurate model of water flow into the auger hole than either of Equations (29.16) and (29.17), it is better to obtain the C values from Table 29.1 to calculate K_{sat} for all cases where $y/H \geq 0.5$.

Table 29.1 Values of Coefficient C in Equation (29.15) for Auger Hole Underlain by Impermeable or Infinitely Permeable Layer

		s/H for Impermeable Layer								s/H	s/H for Infinitely Permeable Layer			
H/r	y/H	0	0.05	0.1	0.2	0.5	1	2	5	∞	5	2	1	0.5
1	1	447	423	404	375	323	286	264	255	254	252	241	213	166
	0.75	469	450	434	408	360	324	303	292	291	289	278	248	198
	0.5	555	537	522	497	449	411	386	380	379	377	359	324	264
2	1	186	176	167	154	134	123	118	116	115	115	113	106	91
	0.75	196	187	180	168	149	138	133	131	131	130	128	121	106
	0.5	234	225	218	207	188	175	169	167	167	166	164	156	139
5	1	51.9	48.6	46.2	42.8	38.7	36.9	36.1		35.8		35.5	34.6	32.4
	0.75	54.8	52.0	49.9	46.8	42.8	41.0	40.2		40.0		39.6	38.6	36.3
	0.5	66.1	63.4	61.3	58.1	53.9	51.9	51.0		50.7		50.3	49.2	46.6
10	1	18.1	16.9	16.1	15.1	14.1	13.6	13.4		13.4		13.3	13.1	12.6
	0.75	19.1	18.1	17.4	16.5	15.5	15.0	14.8		14.8		14.7	14.5	14.0
	0.5	23.3	22.3	21.5	20.6	19.5	19.0	18.8		18.7		18.6	18.4	17.8
20	1	5.91	5.53	5.30	5.06	4.81	4.70	4.66		4.64		4.62	4.58	4.46
	0.75	6.27	5.94	5.73	5.50	5.25	5.15	5.10		5.08		5.07	5.02	4.89
	0.5	7.67	7.34	7.12	6.88	6.60	6.48	6.43		6.41		6.39	6.34	6.19
50	1	1.25	1.18	1.14	1.11	1.07	1.05			1.04			1.03	1.02
	0.75	1.33	1.27	1.23	1.20	1.16	1.14			1.13			1.12	1.11
	0.5	1.64	1.57	1.54	1.50	1.46	1.44			1.43			1.42	1.39
100	1	0.37	0.35	0.34	0.34	0.33	0.32			0.32			0.32	0.31
	0.75	0.40	0.38	0.37	0.36	0.35	0.35			0.35			0.34	0.34
	0.5	0.49	0.47	0.46	0.45	0.44	0.44			0.44			0.43	0.43

Note: r = radius of auger hole; H = depth of water in hole at equilibrium; y = average distance from water level in hole to water table at equilibrium for two consecutive measurements; and s = distance between bottom of hole and impermeable or infinitely permeable layer below hole.

Source: Ref. 77.

To determine K_{sat}, calculate and record on the data sheet the distance between the water level in the hole and the static water level for each step (i.e., $y_i = d_i - E$ for $i = 1, 2, \ldots$). Then, calculate the average $y = (y_i + y_{i+1})/2$, $\Delta y = (y_i - y_{i+1})$, and $\Delta t = (t_{i+1} - t_i)$ for each pair of consecutive measurements. (Note that Δy is made to be positive by reversing the order of its measurements.) For $y/H \geq 0.5$, obtain a C value from Table 29.1 and calculate a K_{sat} value for each of the y/H values and the corresponding H/r and s/H values by Equation (29.15). For the C values missing in columns for $s/H = 2$ and $s/H = 5$, use the appropriate value from column $s/H = \infty$. Also, because the C factor does not change linearly with any of the parameters H/r, y/H, and s/H, it is best to interpolate graphically between the neighboring values for the C factors not presented in Table 29.1. For other cases or to obtain an approximate value for K_{sat}, use Equation (29.16) or (29.17), whichever applies, to obtain the shape factor for each pair of consecutive measurements.

As the water level rises in the hole, the rate of change of the water level ($\Delta y/\Delta t$) decreases and the C factor increases, resulting in a set of consistent K_{sat} values. If the calculated K_{sat} values for a set of measurements vary erratically (e.g., change by more than $\pm 20\%$), all measurements for that set should be repeated until a consistent set of K_{sat} values is obtained. To ensure that the resulting K_{sat} represents the actual K_{sat} of the site under consideration, it is recommended that measurements be repeated until the K_{sat} results for two consecutive sets of measurements are consistent. Care should be taken to allow the water level in the hole to reach the static water level before emptying the hole for measurements.

29.4.1.4 Comments. Boring a hole with an auger in some soils may result in the smearing of the hole side wall. Removing water from the hole and allowing the hole to refill several times may open the pores on the hole wall by forcing particles responsible for smearing into the hole. As indicated earlier, this step should be repeated several times until the water entering the hole becomes clear. If smearing cannot be removed, other field and laboratory methods should be considered for measuring K_{sat}.

The measured K_{sat} by the auger hole method represents the average horizontal saturated hydraulic conductivity of a relatively large volume of the soil, particularly for large H/r values.[83,87] Using a theoretical evaluation, van Bavel and Kirkham[82] estimated the volume of the soil contributing to water flow into the hole to be a pear-shaped volume, with an average radius of five times the radius of the hole. van Beers[83], on the other hand, stated that the volume of the soil influencing the measured value is within a radius of 30–60 cm around the hole between the water table and 10–15 cm below the hole bottom.

According to Massland and Haskew,[88] the K_{sat} values determined by the auger hole method are reproducible to within 10%. For best results, they recommended that measurements be made until the amount of water returned into the hole exceeds 20% of the volume of the water removed from the hole (i.e., $\Sigma \Delta y \leq 0.20 y_0$). As indicated earlier, van Beers[83] suggested that measurements be ceased when $\Sigma \Delta y > 0.25 y_0$. He suggested the hole be extended to 60–70 cm below the water table (30–50 cm for homogenous, high-conductivity soils) and that the water level be lowered more than 40 cm in the hole. Boast and Kirkham's[77] approach for determining C values can be used for shallow holes ($H/r = 1$), and measurements must stop when the hole becomes half full (i.e., when $y < 0.5 H$).

The auger hole method can be used in gravelly soils as long as a cylindrical hole can be constructed below the water table. For high-conductivity soils the rate of water entry into the hole may be too rapid for manual measurement of the rise of water level in the hole. A fast-acting pressure transducer attached to a data logger or another automatic water level recording device may be required to monitor the water level in the hole. Also, care should be taken to use the appropriate model or nomograph for calculating K_{sat} properly.

29.4.2 Slug Test

The slug test is very similar to the piezometer method originally proposed by Kirkham[89] and later modified by Frevert and Kirkham[90] and Luthin and Kirkham.[91] In the piezometer method, a cavity of known length (say h_c) and diameter $2r$ is created at the bottom of a snugly fitted piezometer below the water table. Water is then removed from the piezometer and the rate of rise of water level in the piezometer is measured for calculating K_{sat} by an appropriate procedure.[92,93] In the slug test presented by Bouwer and Rice,[94] a hole is dug below the water table and a section of casing with perforation at the bottom is inserted into the hole. The space between the solid portion of the casing and soil is sealed to prevent water flow along the casing. The slug test can be performed by quickly removing a volume of water from the hole and measuring the rate of rise of water level in the casing. Alternatively, a volume of water can be quickly added for suddenly raising the water level in the casing above the static water level and measuring the rate of fall of water level within the casing. The rate of rise or fall of water level in the casing is then used to calculate a K_{sat} value for the volume of soil surrounding the perforation section of the casing. Because of its simplicity and relative ease of measurements, a slug test, rather than a pumping test, is commonly used for measuring conductivity and transmissivity of aquifers.[16,18]

29.4.2.1 *Materials and Equipment.* Because the procedure for the slug test is similar to that of the auger hole method, the same equipment and materials listed for the auger hole method are needed for this procedure. In addition, a well casing with perforation at the bottom section and packing and sealing materials are needed to construct the hole with a geometry similar to the one shown in Fig. 29.14.

According to Bower,[95] although any diameter hole can be used for measuring K_{sat} by this technique, due to spatial variability of soils and inaccuracies associated with estimating the thickness of the gravel pack (see Fig. 29.14), small-diameter holes (e.g., 5 cm) may not yield representative conductivity values for the zone under consideration. Therefore, select the largest auger set that allows you to conveniently bore a hole to the desired depth below the water table. For information on

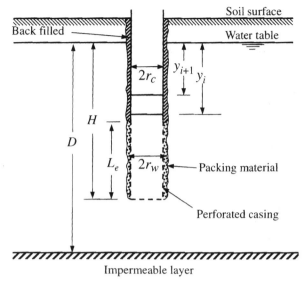

Figure 29.14 Schematic diagram of slug test for partially perforated well in unconfined aquifer. (Adapted from Ref. 94.)

the auger set, bailer or slug, water level measuring device, and other materials, see Section 29.4.1.

A section of perforated well casing (or another type of perforated pipe) with inside radius r_c and length L_e (to be at least four times the radius of the hole, r_w, as shown in Fig. 29.14) attached to a solid section of a pipe of the same diameter is needed to line the hole and isolate a section of the saturated zone. To prevent caving of the hole, gravel or very coarse sand is needed to pack around the perforated section of the casing. To prevent water flowing along the solid section of the casing into the hole, bentonite or another suitable material is needed to seal around the solid section of the casing below the water table.

29.4.2.2 *Field Data Collection.*

After cleaning a small area, use the auger to bore a hole (diameter $2r_w$) to the desired depth H below the water table (at least four times the radius of the hole). After cleaning the bottom section of the hole with a hole cleaner auger, insert the casing (inside diameter $2r_c$) with the perforation (length L_e) into the hole and pack gravel or coarse sand (with a conductivity much greater than the conductivity of the soil) around the perforated section of the casing (see Fig. 29.14). Then, seal around the solid section of the casing with bentonite or other low-permeability materials (e.g., packed clay). Allow the water level in the hole to reach the static water level. Measure the depth of the hole below the water table, H, and measure or estimate the thickness of the aquifer, D, as shown in Figure 29.14. Remove the water from the hole and allow the hole to refill several times until the water removed from the hole does not contain an appreciable amount of suspended particles (developing the well). Dispose of water removed from the hole away from the measurement area. Allow the water level in the hole to reach the static water level at the end of this process.

Use the bailer or high-capacity pump and quickly remove the water from the hole to lower the water level in the hole to a distance y below the water table (see Fig. 29.14). Immediately start measuring the water level in the hole with time. Repeat the measurements a few times until the rate of rise of the water level becomes small. As an alternative to a bailer or pump, a solid section of a rod with a diameter slightly less than the inside diameter of the casing (referred to as a slug) can be used for lowering the water level in the hole quickly. For this purpose, remove the water from the hole and insert the solid rod to the bottom of the hole. Allow the water level in the hole to reach its static level. Then, remove the rod to lower the water level in the hole quickly. If a pressure transducer or a system such as the one presented by Downey et al.[86] is used for measuring the depth of water in the hole, care should be taken not to disturb the transducer during the insertion and removal of the bailer or slug. Alternatively, the pressure transducer can be quickly inserted into the bottom of the hole immediately after lowering the water level in the hole.

29.4.2.3 *Calculation.* Saturated hydraulic conductivity is calculated by

$$K_{\text{sat}} = \frac{r_c^2 \ln(R_e/r_w)}{2L_e} \frac{\ln(y_0/y_t)}{t} \tag{29.18}$$

where y_0 is y at time zero, y_t is the y at time t, R_e is a shape factor that must be determined based on the geometry of the hole, and the other parameters are as defined above. Using an electrical resistance analog, Bouwer and Rice[94] estimated the R_e values and expressed the dimensionless parameter $\ln(R_e/r_w)$ by

$$\ln\left(\frac{R_e}{r_w}\right) = \begin{cases} \left(\dfrac{1.1}{\ln(H/r_w)} + \dfrac{A + B\ln(D-H)/r_w}{L_e/r_w}\right)^{-1} & \text{when} \quad D \gg H \qquad (29.19) \\[4mm] \left(\dfrac{1.1}{\ln(H/r_w)} + \dfrac{C}{L_e/r_w}\right)^{-1} & \text{when} \quad D = H \qquad (29.20) \end{cases}$$

where A, B, and C are obtained for various L_e/r_w values from the graphs presented in Figure 29.15.

Where the perforated section extends above the water table or when measurements of the water level are made while the water level rises within the perforated section with gravel (or coarse sand) packed around it, an equivalent r_c value that is based on the thickness and porosity of the gravel pack must be used in place of r_c in Equation (29.18). Bouwer and Rice[94] (also see Ref. 95) suggested to set the equivalent r_c value equal to $[(1-n)r_c^2 + nr_w^2]^{1/2}$, where n is the porosity of the gravel (or sand) pack around the perforation.

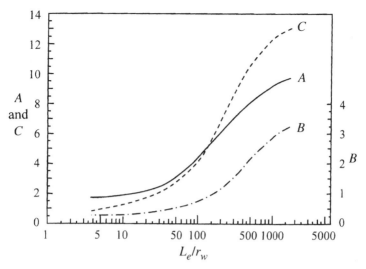

Figure 29.15 Dimensionless parameters A, B, and C of Equations 29.19 and 29.20 as function of L_e/r_w. Parameters A and C are read from left axis and B is read from right axis. (Adapted from Ref. 94.)

The rate of rise of water level in the hole decreases with time. Plot $\ln(y)$ versus t and draw a best-fit line through the points during the early measurement periods. Take the slope of this line as $\ln(y_0/y_t)/t$, and calculate a K_{sat} value from Equation (29.18). In some cases, two straight lines may be fitted through the points for early measurements. Bouwer[95] described this as a "double-straight-line effect" and suggested using the portion of the second line to determine the slope of the line $\ln(y_0/y_t)$.

29.4.2.4 *Comments.* In many respects, the slug test is similar to both the auger hole (as described above) and the piezometer methods (described in Refs. 8, 92, and 93). Saturated hydraulic conductivity of the upper portion of the saturated zone (i.e., aquifer) can be measured by extending the perforation of the well casing above the water table (similar to the auger hole method). On the other hand, by keeping the length of the perforated section as small as possible, K_{sat} of individual layers in a layered soil can be determined (similar to the piezometer method). The procedure described above was for measuring the rise of water level in the hole after lowering it quickly. The slug test can also be used by measuring the fall of water level in the well after raising it quickly provided that the static water level is above the screened portion of the well.[95] According to Bouwer and Rice,[94] the volume of the soil contributing to flow in the slug test is approximately a cylinder of radius equal to R_e and height somewhat larger than the height of the perforated section below the water table.

29.5 LABORATORY MEASUREMENT OF SATURATED HYDRAULIC CONDUCTIVITY

It may be desirable or necessary to measure K_{sat} of a soil by a laboratory procedure. For this purpose an intact or repacked core sample of the soil is saturated with water and the rate of water flow through the core is measured under a constant head, a falling head, or an oscillating head. The saturated hydraulic conductivity of the soil core is then calculated by directly using Darcy's equation [Equation (29.1)]. To minimize the impact of the chemical and physical properties of the water on measured hydraulic conductivity values, care should be taken to prepare an adequate volume of water prior to the initiation of measurements. For a discussion on the quality of water for hydraulic conductivity measurements, see the discussion in Section 29.2.

29.5.1 Sample Collection and Preparation

Because of soil structure and other natural features (e.g., root channels), it is best to use intact core samples for measuring K_{sat} in the laboratory. However, it should be recognized that collection of intact cores from some soils (e.g., sandy textured soils) may be impractical. For these and other cases where repacked samples can be used (e.g., measuring conductivity of field compacted soils such as liners), soil materials must be packed uniformly in appropriately sized columns to a bulk density similar to that found in the field. For this purpose, the use of cylinders with an inside diameter and height of 7 cm or larger is highly recommended. Also, care should be taken to avoid layering of the soil or smearing of the ends of the core during packing. Using a cylinder made from aluminum or other rigid materials, crushed air-dried soil can be packed in layers by tamping the materials to the desired (or practical) bulk density. Klute[96] suggested packing uniformly moist soil materials in a cylinder using a piston. For simulating landfill liners and other field-packed soils, the amount of force for packing may be specified by regulations or other procedures.

An intact core is needed to preserve the structural pores (e.g., planar voids between soil peds) as well as other natural pores (e.g., root channels) in the field. Since both structural and natural pores are important in conducting water, it is necessary not only to preserve these pores but also to minimize induced cracks during the sample collection and preparation. A thin-wall cylinder made from a rigid material (e.g., aluminum, brass, steel) with a sharp edge beveled outward at one end or a double-cylinder sampler such as an Uhland sampler[97] can be driven into the soil by hammering or by pressing it into the soil hydraulically.[98] After evaluating the hydraulic conductivity and bulk density of intact samples obtained by hammering and hydraulically driving Uhland samplers into the soil, Rogers and Carter[99] concluded that hammering the sampler into the soil results in a greater distortion of the soil structure than hydraulically driving the sampler. To minimize induced cracks, Amoozegar[100] offered a procedure for collecting intact cores using a hydraulically driven sampling tube and preparing 8–10 cm long cores for hydraulic conductivity measurements. A 3-in. sampling tube with a 6.5-cm-diameter cutting head is suitable for this technique. Amoozegar's[100] procedure also does not require excavation of a large pit for collection of intact core samples at deep depths.

Samples collected by the above techniques are usually 5–10 cm in diameter and less than 10 cm long. To collect larger samples, a large cylindrical or cubical column can be carved in place, covered with concrete or gypsum, encased in a wooden frame, and then transported to the laboratory for analysis. Regardless of the size of the intact core, some macropores, such as root channels, worm holes, or natural seams and cracks, may extend from the top to the bottom of the core, resulting in abnormally high saturated hydraulic conductivity.[101] Under field conditions, however, these pores are limited in length, and their contribution to high flow rate is somewhat limited. Therefore, sampling locations should be selected carefully to avoid an excessive number of macropores.

Before applying water to the core for conductivity measurement, the sample must also be saturated. A soil is completely saturated when all the pores are filled with water and the soil water pressure head is ≥ 0. Under field conditions, entrapped air often prevents complete saturation of all the pores even when the soil water pressure head is ≥ 0. This degree of saturation is referred to as natural saturation, field saturation, or satiated water content.[96] For laboratory measurements, one should select an appropriate procedure to achieve the desired degree of saturation.

To determine saturated hydraulic conductivity of a given site, collect intact cores from the desired depths at a few locations within the field. To collect the samples, use a double-cylinder sampler (similar to the Uhland sampler) driven by a hammer or hydraulics or use the procedure described by Amoozegar[100] for collecting and preparing intact samples. Cover each sample in a plastic bag with appropriate labeling for identifying the sample and determining its upper end and transport it to the laboratory in a cooler. Avoid prolonged exposure to heat or excessive cold as temperature extremes may result in cracking or drying of the soil. In the laboratory, trim the two ends of the core and remove any smearing by picking or plucking the exposed surface with the tip of a sharp knife. Cover the bottom end of the core with two layers of cheesecloth (or another highly permeable material) to prevent the soil from falling out. To hold the cheesecloth on the core, place the bottom of the core on a square or round piece of cheesecloth, wrap the cheesecloth around the cylinder, and hold it in place with one or two pieces of rubber band. If the sample is encased in a cylinder, securely place another cylinder of the same size on top and seal the gap between the two cylinders (e.g., with a wide rubber band). If the intact sample is encased in paraffin (or other materials) as described by Amoozegar,[100] secure and seal a section of PVC or rigid wall pipe to the top of the core.

Place each core on a piece of plastic or metal mesh (that fits inside a flat bottom funnel) in a tub. To saturate the soil and allow the air in the pores to escape, apply water to the tub so the sample can take up water from its bottom end. To prevent air entrapment, slowly raise the level of water in the tub to slightly above the top level of the soil core. For soils with high to moderate hydraulic conductivity a 24-h period may be adequate. For some soils (e.g., clayey soils), it may be necessary to extend the saturation period to a few days. To enhance the degree of saturation, saturate the core under vacuum (in a vacuum chamber) or flush the soil with CO_2 followed by saturating it with deaerated water (see Ref. 96). Do not apply vacuum to the top of the core alone as this may cause water to move quickly through large pores leaving air entrapped in smaller pores. After saturating the core (as indicated by a thin film of water above the soil inside the cylinder), raise the level of water in the tub to allow water to accumulate above the soil surface in the cylinder. If necessary, add a small amount of water to the top of the core and carefully move the core and place it on a highly permeable base, such as a wire or plastic mesh, inside a flat bottom funnel on the rack where K_{sat} measurements will be conducted.

Initially, apply water to the top of the core (or maintain a constant head above the core) and measure the outflow volume with time. If the flow rate is less than 5 mL/h (under the highest possible gradient), the falling-head procedure should be used. Otherwise, maintain a constant depth of water above the soil core and continue to measure K_{sat} by the constant-head procedure. The oscillating-head technique developed by Child and Poulovassilis[102] can be used if there is a possibility for changes in soil structure as a result of leaching that accompanies water flow through the core. For more information on this procedure, interested readers are referred to Ref. 102.

29.5.2 Constant-Head Method

To measure K_{sat} by this procedure, maintain a constant head of water on top of the core (Fig. 29.16) and measure the volume of outflow with time using a graduated cylinder. Alternatively, weigh the outflow and determine its temperature to find the density of water for converting the weight of the outflow to its volume.

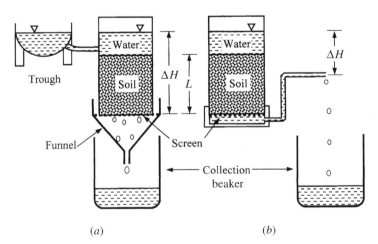

(a) (b)

Figure 29.16 Methods to provide a constant head of water for (a) gradients >1 and (b) gradients ≤ 1 for high-conductivity soils. By adjusting the arm in the setup shown in (b), the gradient can be increased to >1.

Then, directly calculate the saturated hydraulic conductivity of the core by Darcy's equation,

$$K_{sat} = \frac{QL}{At\,\Delta H}$$ (29.21)

where Q is the volume of the outflow collected from the core of length L and cross-sectional area A during the time period t under a hydraulic head difference of ΔH between the top and bottom of the core. Cover the outflow collection to minimize evaporation and loss of outflow volume.

Klute and Dirksen[28] described a number of procedures for maintaining a constant head of water over the cores. A simple Marriott bottle attached to a trough (Fig. 29.17) can be used to maintain a constant depth of water over a number of cores simultaneously. In this setup, water inside the trough is at the same level as the tip of the air tube inside the Marriott bottle, which is at atmospheric pressure. Since the top of the cylinder in each core is directly connected to the bottom of the trough, a constant level of water will be maintained over all the cores connected to the trough. Water can be applied to the top of the cores through an adapter on the side of the upper cylinder (Fig. 29.16a).

If the core sample is prepared based on the above instructions, the hydraulic gradient for the sample will be between 1 and 2. To increase the gradient above 2,

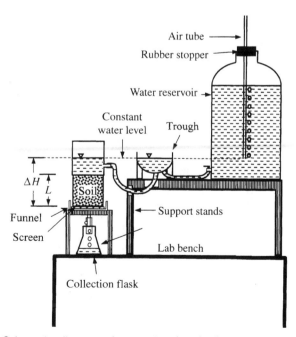

Figure 29.17 Schematic diagram of apparatus for simultaneously measuring saturated hydraulic conductivity of a number of core samples by constant-head method (adapted from Ref. 92).

additional cylinders can be attached to the top cylinder. However, extremely large gradients (i.e., $\Delta H \gg L$) should be avoided because Darcy's law may become invalid if the flow through the core becomes turbulent. For high-conductivity soils, it may become necessary to maintain a gradient below 1. For this purpose, tightly attach a cap equipped with an adapter to the bottom section of the core and raise its outflow tube to the desired elevation so $\Delta H < L$ (Fig. 29.16b). In this setup, water can also be applied through the bottom and collected from the top.

For samples collected in cylinders, the flow between the soil core and the cylinder wall, commonly referred to as wall flow, may be significant. If the soil sample is covered with paraffin or other materials, as described by Amoozegar,[100] no wall flow occurs. If wall flow is suspected, the flow from the center of the core can be collected for conductivity measurement.[103] Collecting the flow from the center portion of the bottom surface, however, will decrease the cross-sectional area of the flow.

29.5.3 Falling-Head Method

Saturated hydraulic conductivity cannot be measured accurately if the outflow is small (e.g., <5 ml/h). To measure K_{sat} under a falling head of water, tightly connect a standpipe constructed of clear glass or plastic with an inside cross-sectional area a to the top of the saturated core of length L and cross-sectional area A (Fig. 29.18). Apply water to the standpipe, and after allowing some water to go

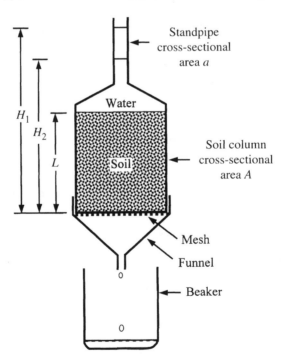

Figure 29.18 Schematic diagram of apparatus for measuring saturated hydraulic conductivity by falling-head method.

through the core, measure the water level in the standpipe at different times. Calculate the saturated hydraulic conductivity by

$$K_{sat} = \left(\frac{aL}{A(t_2 - t_1)}\right) \ln\left(\frac{H_1}{H_2}\right) \tag{29.22}$$

where H_1 and H_2 are the hydraulic heads at times t_1 and t_2, respectively, and A and L are the cross-sectional area and length of the core, respectively. The inside diameter of the standpipe should be large enough to avoid capillary rise but small enough to allow accurate measurement of the drop in hydraulic head during a reasonable time. According to Amoozegar and Wilson,[92] an inside diameter of 2 mm–2 cm and measurement intervals of a few minutes to 1 h are convenient for measuring K_{sat} by this procedure.

29.5.3.1 Comments. In general, saturated hydraulic conductivity is highly variable across a landscape.[104] Therefore, to characterize this parameter with confidence, a relatively large number of samples may be needed. Increasing the sample size reduces the variability among samples (see Ref. 105), but collecting large samples may not be practical for measuring K_{sat}. To obtain an accurate measure of the saturated hydraulic conductivity of the soil by the laboratory techniques, the sample should contain a representative amount of peds (soil structural units) and macropores found in the field under consideration. The optimum sample size for measuring a soil property is referred to as representative elementary volume (REV) (see Ref. 106). Bouma[107] estimated the REV to range between 100 cm^3 for a sandy soil with no peds to 10^5 cm^3 for a clayey soil with large peds and continuous macropores.

Unfortunately, the REV cannot be easily determined for a given location (and soil horizon), and collecting and analyzing large samples for more accurately estimating K_{sat} may not be practical. As an alternative, an adequate number of intact samples with relatively small dimensions (e.g., 7.5 cm diameter, 7.5 cm long) may be sufficient for determining K_{sat} of a soil. The number of samples required to achieve a certain degree of confidence can be estimated from the mean and standard deviation for a limited number of measurements. Also, it has been demonstrated that K_{sat} values for a given field are log-normally distributed.[104] Instead of averaging the measured K_{sat} values of a number of soil cores, it may be more appropriate to determine the geometric mean of the measured values to represent the saturated hydraulic conductivity of a group of samples.

29.6 FIELD MEASUREMENT OF AIR PERMEABILITY

Air permeability k_a can be defined as a measure of the ability of soil to conduct air by convective flow, which is the movement of soil air in response to a gradient of total gas pressure. Since the pores in the soil are generally occupied by air and water, air permeability is inversely related to soil water content and can be expressed as a function of air-filled porosity (or volumetric air content). Also, similar to water conductivity (which depends on water-filled porosity, i.e., soil water content), air permeability is related to the total air-filled porosity as well as to the pore size distribution and the connectivity of the open pores in the soil (which is related to the type of soil). For a completely dry soil, air permeability is considered to be equal to the intrinsic permeability of the soil.

Analogous to Darcy's law for the movement of water, convective flow of air through soil in the direction of the x axis can be expressed by

$$q_v = -\left[\frac{k(\varepsilon)}{\eta}\right]\frac{dP}{dx} \tag{29.23}$$

where q_v is the volumetric air flux density $(L \cdot T^{-1})$, $k(\varepsilon)$ is the air permeability of the soil (L^2) as a function of air-filled porosity $\varepsilon(L^3 \cdot L^{-3})$, η is the air viscosity $(M \cdot L^{-1} T^{-1})$, P is the total air pressure $(M \cdot L^{-1} T^{-2})$, and x is the distance (L) in direction of air flow. The flux density q_v is equal to $V_a/A_s t$, where V_a is the volume of air going through a soil column of cross-sectional area A_s during a time period t. For a soil column of finite length L and cross-sectional area A_s, Equation (29.23) can be rearranged to calculate air permeability by

$$k_a = \frac{V_a\eta L}{\Delta P A_s} \tag{29.24}$$

A number of laboratory and field techniques are available for determining air permeability. In both laboratory and field methods, a bulk volume of soil is confined within a cylinder and the rate of air flow through the soil is measured under a steady-state (constant-pressure) or non-steady-state (falling-pressure) condition. Air permeability can also be estimated based on the total soil porosity and air-filled porosity. Ball and Schjønning[108] present a fairly complete review of field and laboratory techniques as well as theoretical methods for determining air permeability. Here, only a brief description of the steady-state field procedure originally developed by Grover[109] will be presented. The readers interested in other methods are referred to Ball and Schjønning.[108]

29.6.1 Steady-State Field Method

In the steady-state method proposed by Grover[109] a volume of soil is isolated within a cylinder (referred to as inlet tube), as depicted in Figure 29.19, and the volumetric flux of air passing through the soil under the constant pressure produced by the weight of the float is measured. Because the air pressure at the lower end of the inlet tube is not atmospheric and the boundary conditions are not well defined (see Fig. 29.19), Equation (29.24) cannot be used directly to calculate k_a. To overcome this problem, Grover[109] used an empirical shape factor A (as an approximation of the A_s/L quotient) and introduced

$$k_a = \frac{q_v\eta}{\Delta P A} \tag{29.25}$$

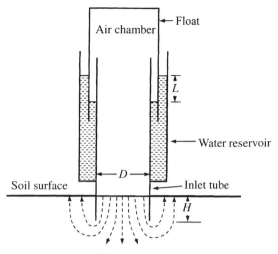

Figure 29.19 Schematic diagram of air permeameter. (Adapted from Ref. 109.)

where ΔP is the air pressure inside the air chamber (shown as L in Fig. 29.19). He also produced a graph for estimating the shape factor A based on the diameter of the inlet tube, D, and the depth of insertion of the inlet tube into the soil, L (see Fig. 29.19). To estimate A Liang et al.[110] developed the equation

$$A = D\left[0.4862\left(\frac{D}{H}\right) - 0.0287\left(\frac{D}{H}\right)^2 + 0.1106\right] \qquad (29.26)$$

by regression analysis.

29.6.1.1 *Materials and Equipment.* An air permeameter similar to the one shown in Figure 29.19 can be constructed from thin-wall steel. The air pressure inside the chamber can be measured by installing a gauge on top of the float or by measuring the distance between the water inside and outside the float in the water reservoir using two small manometers. Other equipment and materials needed for measuring k_a include tools for cleaning and leveling a small area, tools to drive the inlet tube into the soil (see Section 29.2.2), a meter stick, a data sheet, a stop watch, water, and additional weights if higher pressures are required.

29.6.1.2 *Field Data Collection.* Insert the inlet tube (e.g., 20 cm diameter, 20 cm long with beveled edge at the bottom) of the apparatus shown in Figure 29.19 into the soil vertically in a flat area. Care should be taken not to create gaps between the inlet tube wall and the soil (for installation information see Section 29.2.2). Tamp the soil in a small ring around the inside and outside walls of the inlet tube to minimize wall flow. Then, assemble the upper part of the apparatus to the inlet tube as shown in Figure 29.19. The float (inverted bucket) placed in the water around the inside cylinder provides a constant pressure that drives the air

from under the air chamber through the soil within the inlet tube. Additional weight can be added to the float to increase the pressure gradient. Measure the rate of drop of the float in the water reservoir to calculate the volume of air flowing through a unit cross-sectional area of the soil in the inlet tube in unit time (e.g., 1 min, 1 h).

29.6.1.3   Calculation. Determine the volume of air going through soil by multiplying the length of the drop in the level of the float and the cross-sectional area of the inlet tube. Determine the shape factor from Equation (29.26) and calculate the air permeability (or conductivity) by Equation (29.25)

29.6.1.4   Comments. Air permeability is a function of air-filled porosity; hence it depends on soil water content. It was indicated that the value of air permeability for dry soil is equated with the intrinsic permeability. However, even for nonswelling soils, drying a soil may result in shrinkage and formation of cracks, that will impact the soil air permeability. Reeve[111] used the ratio of air permeability to that of the saturated hydraulic conductivity to assess soil stability. For stable soils, both air permeability of dry soil and saturated hydraulic conductivity remain relatively high. On the other hand, for unstable soils, where soil particles disperse upon wetting saturated hydraulic conductivity is significantly lower while the air permeability of the dry soil remains relatively high. The air permeability can therefore be used as a tool to assess soil stability.

29.7   SOIL DISPERSION

Dispersion of soil particles results in the destruction of soil aggregates, which in turn negatively impacts the soil saturated hydraulic conductivity. The stability of soil aggregates depends on many factors, including the soil water content, organic matter content, and ratio of sodium to other ions on the exchange sites of clay particles. As indicated above, Reeve[111] used the intrinsic permeability and saturated hydraulic conductivity ratio as an index of the stability of soil structure. Another measure of expressing the degree of soil dispersion is by measuring aggregate stability.

Soil aggregate is defined by the Soil Science Society of America[14] as "a group of primary soil particles that cohere to each other more strongly than to other surrounding particles." One way to measure aggregate stability is to apply certain disruptive procedures to a volume or mass of soil and determine the relative amounts of various size aggregates that remain intact. Another way of expressing aggregate stability is to determine the amount of force necessary to break an aggregate apart.[112] Aggregate stability measurements are generally performed to assess soil stability in tillage practices in agriculture. The interested readers are referred to Nimmo and Perkins[112] for specific procedures for wet and dry aggregate stability analysis.

ACKNOWLEDGMENT

The use of trade names in this publication does not imply endorsement by the North Carolina Agriculture Research Service of the products named or criticism of similar ones not mentioned.

REFERENCES

1. W. W. Buol, R. J. Southard, R. C. Graham, and P. A. McDaniel, *Soil Genesis and Classification*, 5th ed., Iowa State Press, Ames, IA, 2003.

2. W. H. Fuller and A. W. Warrick, *Soils in Waste Treatment and Utilization*, Vol. I, CRC Press, Boca Raton, FL, 1985.

3. D. Hillel, *Environmental Soil Physics*, Academic, New York, 1998.

4. P. J. Schoeneberger, D. A. Wysocki, E. C. Benham, and W. D. Broderson, *Field Book for Describing and Sampling Soils*, Nat. Resour. Conser. Ser., USDA, National Soil Survey Center, Lincoln, NE, 1998.

5. I. L. Pepper, C. P. Gerba, and M. L. Brusseau, *Pollution Science*, Academic, San Diego, CA, (1996).

6. W. A. Jury, W. R. Gardner, and W. H. Gardner, *Soil Physics*, 5th ed., Wiley, New York, 1991.

7. A. W. Warrick, *Soil Water Dynamics*, Oxford University Press, New York, 2003.

8. A. Amoozegar and A. W. Warrick, Hydraulic Conductivity of Saturated Soils: Field Methods, in A. Klute (Ed.), *Methods of Soil Analysis, Part 1. Physical and Mineralogical Methods*, 2nd ed., Agron. Monogr. 9, American Society of Agronomy and Soil Science Society of America, Madison, WI, 1986, pp. 735–770.

9. R. A. Freeze and J. A. Cherry, *Groundwater*, Prentice-Hall, Englewood Cliffs, NJ, 1979.

10. M. J. Vepraskas, A. G. Jongmans, M. T. Hoover, and J. Bouma, Hydraulic Conductivity of Saprolite as Determined by Channels and Porous Groundmass, *Soil Sci. Soc. Am. J.* **55**(4), 932–938 (1991).

11. J. Bouma, Unsaturated Flow during Soil Treatment of Septic Tank Effluent, *J. Environ. Eng. ASCE.*, **101**, 967–983 (1975).

12. R. Siegrist, Soil Clogging during Subsurface Wastewater Infiltration as Affected by Effluent Composition and Loading Rate, *J. Environ. Qual.* **16**, 181–187 (1987).

13. R. Kretzschmar, W. P. Robarge, and A. Amoozegar, Filter Efficiency of Three Saprolite for Natural Clay and Iron Oxide Colloids, *Environ. Sci. Tech.* **28**, 1907–1915 (1994).

14. Soil Science Society of America (SSSA), *Glossary of Soil Science Terms*, SSSA, Madison, WI, 2001.

15. J. E. Bowles, *Physical and Geotechnical Properties of Soils*, 2nd ed., McGraw-Hill, New York, 1984.

16. P. A. Domenico and F. W. Schwartz, *Physical and Chemical Hydrology*, 2nd ed., Wiley, New York, 1998.

17. R. W. Miller and D. T. Gardiner, *Soils in our Environment.* 8th ed., Prentice-Hall, Upper Saddle River, NJ, 1998.

18. Z. Sen, *Applied Hydrogeology for Scientists and Engineers*, CRC Press, Lewis Publisher, Boca Raton, FL, 1995.

19. Soil Survey Division Staff, *Soil Survey Manual*, Agricultural Handbook No. 18, USDA, U.S. Government Printing Office, Washington, DC, 1993.

20. J. Bouma, A. Jongerius, O. Boersma, A. Jager, and D. Schoonderbeek, The Function of Different Types of Macropores during Saturated Flow through Four Swelling Soil Horizons, *Soil Sci.Soc. Am. J.* **41**, 945–950 (1977).

21. R. J. Luxmoore, Micro-, Meso, and Macroporosity in Soil, *Soil Sci. Soc. Am. J.* **45**, 671–672 (1981).

22. S. W. Buol, A. Amoozegar, and M. J. Vepraskas, *Southeast. Geol.* **39**, 151–160 (2000).

23. M. J. Vepraskas, W. R. Guertal, H. J. Kleiss, and A. Amoozegar, Porosity Factor That Control the Hydraulic Conductivity of Soil-Saprolite Transitional Zones, *Soil Sci. Soc. Am. J.* **60**, 192–196 (1996).

24. J. E. Mace and C. Amrhein, Leaching and Reclamation of a Soil Irrigated with Moderate SAR Waters, *Soil Sci. Soc. Am. J.* **65**, 199–204 (2001).

25. B. L. McNeal and N. T. Coleman, Effect of Solution Composition on Soil Hydraulic Conductivity, *Soil Sci. Soc. Am. Proc.* **30**, 308–312 (1966).

26. I. Lebron, D. L. Suarez, and T. Yoshida, Gypsum Effect on the Aggregate Size and Geometry of Three Sodic Soils under Reclamation, *Soil Sci. Soc. Am. J.* **66**, 92–98 (2002).

27. C. W. Fetter, *Contaminant Hydrology*, 2nd ed., Prentice-Hall, Upper Middle River, NJ, 1999.

28. A. Klute and C. Dirksen, Hydraulic Conductivity and Diffusivity: Laboratory Methods, in A. Klute (Ed.), *Methods of Soil Analysis, Part 1. Physical and Mineralogical Methods*, 2nd ed., Agron. Monogr. 9, American Society of Agronomy, CSSA, and Soil Science Society of America, Madison, WI, 1986, pp. 687–734.

29. E. G. Youngs, Hydraulic Conductivity of Saturated Soils, in K. A. Smith and C. E. Mullins (Eds.), *Soil Analysis—Physical Methods*, Marcel Dekker, New York, 1991, pp. 161–207.

30. D. R. Lide (Ed.), *Handbook of Chemistry and Physics*, 84th ed., Chemical Rubber Company Press, Cleveland, OH, 2003.

31. D. B. Stephens, Application of the Borehole Permeameter, in G. C. Topp et al. (Eds.), *Advances in Measurement of Soil Physical Properties: Bringing Theory into Practice*, Soil Sci. Soc. Am. Spec. Publication No. 30., Soil Science Society of America, Madison, WI, 1992, pp. 43–68.

32. D. B. Stephens, K. Lambert, and D. Watson, Regression Models for Hydraulic Conductivity and Field Test of the Borehole Permeameter, *Water Resour. Res.* **23**, 2207–2214 (1987).

33. C. N. Zangar, *Theory and Problems of Water Percolation*, Engineering Monogr. No. 8, Bureau of Reclamation, U.S. Department of the Interior, Denver, CO, 1953.

34. D. B. Stephens and S. P. Neuman, Vadose Zone Permeability Tests: Summary. *J. Hydraul. Div. ASCE* **108**, 623–639 (1982).

35. L. Boersma, Field Measurement of Hydraulic Conductivity above a Water Table, in C. A. Black et al. (Eds.), *Methods of Soil Analysis, Part 1. Physical and Mineralogical Properties, Including Statistics of Measurement and Sampling*, Agron. Monogr. 9, American Society of Agronomy Madison, WI, 1965, pp. 234–252.

36. J. N. Luthin, *Drainage Engineering*, Robert E. Krieger, Huntington, NY, 1978.

37. R. J. Winger, In-Place Permeability Tests and Their Use in Subsurface Drainage, in *Trans. Int. Congr. Comm. Irrig. Drain., 4th*, Madrid, Spain, 1960, pp. 11.417–11.469.

38. A. Amoozegar, Compact Constant Head Permeameter: A Convenient Device for Measuring Hydraulic Conductivity, in G.C. Topp et al. (Eds.), *Advances in Measurement of Soil Physical Properties: Bringing Theory into Practice*, Soil Sci. Soc. Am. Spec. Publication No. 30., Soil Science Society of America, Madison, WI, 1992, pp. 31–42.

39. W. D. Reynolds, D. E. Elrick, and G. C. Topp, A Reexamination of the Constant Head Well Permeameter Method for Measuring Saturated Hydraulic Conductivity above the Water Table, *Soil Sci.* **136**, 250–268 (1983).

40. T. Talsma, Some Aspects of Three-Dimensional Infiltration, *Aust. J. Soil Res.* **8**, 179–184 (1970).

41. T. Talsma and P. M. Hallam, Hydraulic Conductivity Measurement of Forest Catchments, *Aust. J. Soil Res.* **30**, 139–148 (1980).

42. W. D. Reynolds, Saturated Hydraulic Conductivity: Field Measurement, in M. R. Carter (Ed.), *Soil Sampling and Methods of Soil Analysis*, Lewis, Boca Raton, FL, 1993, pp. 599–613.

43. A. Amoozegar, A Compact Constant-Head Permeameter for Measuring Saturated Hydraulic Conductivity of the Vadose Zone, *Soil Sci. Soc. Am. J.* **53**, 1356–1361 (1989).

44. W. D. Reynolds and D. E. Elrick, Constant Head Well Permeameter (Vadose Zone), in J. H. Dane and G. C. Topp (Eds.), *Methods of Soil Analysis, Part 4. Physical Methods*, SSSA Book Series No. 5, Soil Science Society of America, Madison, WI, 2002, pp. 844–8580.

45. A. Amoozegar, Comparison of the Glover Solution with the Simultaneous-Equations Approach for Measuring Hydraulic Conductivity, *Soil Sci. Soc. Am. J.* **53**, 1362–1367 (1989).

46. D. E. Elrick, W. D. Reynolds, and K. A. Tan, Hydraulic Conductivity Measurements in the Unsaturated Zone Using Improved Well Analyses, *Ground Water Monit. Rev.* **9**, 184–193 (1989).

47. J. R. Philip, Approximate Analysis of the Borehole Permeameter in Unsaturated Soil, *Water Resour. Res.* **21**, 1025–1033 (1985).

48. W. D. Reynolds, D. E. Elrick, and B. E. Clothier, The Constant Head Well Permeameter: Effect of Unsaturated Flow, *Soil Sci.* **139**, 172–180 (1985).

49. D. B. Stephens and S. P. Neuman, Vadose Zone Permeability Tests: Steady State Results, *J. Hydraul. Div. ASCE* **108**, 640–659 (1982).

50. W. R. Gardner, Some Steady-State Solutions of the Unsaturated Moisture Flow Equation with Application to Evaporation from a Water Table, *Soil Sci.* **85**, 228–234 (1958).

51. W. D. Reynolds and D. E. Elrick, In Situ Measurement of Field-Saturated Hydraulic Conductivity, Sorptivity, and the α-Parameter Using the Guleph Permeameter, *Soil Sci.* **140**, 292–302 (1985).

52. A. Amoozegar, Comments on "Methods for Analyzing Constant-Head Well Permeameter Data," *Soil Sci. Soc. Am. J.* **57**, 559–560 (1993).

53. G. V. Wilson, J. M. Alfonsi, and P. M. Jardin, Spatial Variability of Saturated Hydraulic Conductivity of the Subsoil of Two Forested Watersheds, *Soil Sci. Soc. Am. J.* **53**, 679–685 (1989).

54. C. M. Campbel and D. D. Fritton, Factors Affecting Field-Saturated Hydraulic Conductivity Measurement by the Borehole Permeameter Technique, *Soil Sci. Soc. Am. J.* **58**, 1354–1357 (1994).

55. H. Bouwer, Intake Rate: Cylinder Infiltrometer, in A. Klute (Ed.), *Methods of Soil Analysis, Part 1. Physical and Mineralogical Methods*, 2nd ed., Agron. Monogr. 9, American Society of Agronomy, CSSA, and Soil Science Society of America, Madison, WI, 1986, pp. 825–844.

56. W. D. Reynolds, D. E. Elrick, and E. G. Youngs, Single-Ring and Double- or concentric-Ring Infiltrometers, in J. H. Dane and G. C. Topp (Eds.), *Methods of Soil Analysis, Part 4. Physical Methods*, SSSA Book Series No. 5, Soil Science Society of America, Madison, WI, 2002, pp. 821–843.

57. D. E. Daniel, In Situ Hydraulic Conductivity Tests for Compacted Clay, *J. Geotechnic. Eng.* **115**, 1205–1226 (1989).

58. S. Matula and C. Dirksen, Automated Regulating and Recording System for Cylinder Infiltrometers, *Soil Sci. Soc. Am. J.* **53**, 299–302 (1989).

59. J. R. Philip, The Theory of Infiltration: 2. The Profile of Infinity. *Soil Sci.* **83**, 435–448 (1957).

60. W. D. Reynolds and D. E. Elrick, Ponded Infiltration from a Single Ring: I. Analysis of Steady Flow. *Soil Sci. Soc. Am. J.* **64**, 478–484.

61. K. M. Perroux and I. White, Designs for Disc Permeameters, *Soil Sci. Soc. Am. J.* **52**, 1205–1215 (1988).

62. M. D. Ankeny, T. C. Kasper, and R. Horton, Design for an Automated Tension Infiltrometer, *Soil Sci. Soc. Am. J.* **52**, 893–896 (1988).

63. W. D. Reynolds and D. E. Elrick, Determination of Hydraulic Conductivity Using a Tension Infiltrometer, *Soil Sci. Soc. Am. J.* **55**, 633–639 (1991).

64. K. W. Watson and R. J. Luxmoore, Estimating Macroporosity in a Forest Watershed by Use of a Tension Infiltrometer, *Soil Sci. Soc. Am. J.* **50**, 578–582 (1986).

65. G. V. Wilson and R. J. Luxmoore, Infiltration, Macroporosity, and Mesoporosity Distribution on Two Forested Watersheds, *Soil Sci. Soc. Am. J.* **52**, 329–335 (1988).

66. M. D. Ankeny, Methods and Theory for Unconfined Infiltration Measurements, in G. C. Topp et al. (Eds.), *Advances in Measurement of Soil Physical Properties: Bringing Theory into Practice*, Soil Sci. Soc. Am. Spec. Publication No. 30., Soil Science Society of America, Madison, WI, 1992, pp. 123–141.

67. B. Clothier and D. Scotter, Unsaturated Water Transmission Parameters Obtained from Infiltration, in J. H. Dane and G. C. Topp (Eds.), *Methods of Soil Analysis, Part 4. Physical Methods*, SSSA Book Series No. 5, Soil Science Society of America, Madison, WI, 2002, pp. 879–898.

68. M. D. Ankeny, M. Ahmed, T. C. Kasper, and R. Horton, Simple Field Method for Determining Unsaturated Hydraulic Conductivity, *Soil Sci. Soc. Am. J.* **55**, 467–470 (1991).

69. I. White, M. J. Sully, and K. M. Perroux, Measurement of Surface-Soil Hydraulic Properties: Disk Permeameters, Tension Infiltrometers, and Other Techniques, in G. C. Topp et al. (Eds.), *Advances in Measurement of Soil Physical Properties: Bringing Theory into Practice*, Soil Sci. Soc. Am. Spec. Publication No. 30, Soil Science Society of America, Madison, WI, 1992, pp. 60–103.

70. K. R. J. Smettem and B. E. Clothier, Measuring Unsaturated Sorptivity and Hydraulic Conductivity Using Multiple Disc Permeameters, *J. Soil Sci.* **40**, 563–568.

71. F. J. Cook and A. Broeren, Six Methods for Determining Sorptivity and Hydraulic Conductivity with Disc Permeameters, *Soil Sci.* **157**, 2–11 (1994).

72. P. A. (TY) Ferré and G. C. Topp, Time Domain Reflectometry, in J. H. Dane and G. C. Topp (Eds.), *Methods of Soil Analysis, Part 4. Physical Methods*, SSSA Book Series No. 5, Soil Science Society of America, Madison, WI, pp. 434–444.

73. G. C. Topp and P. A. (TY) Ferré, Thermogravimetric Using Convective Oven-Drying, in J. H. Dane and G. C. Topp (Eds.), *Methods of Soil Analysis, Part 4. Physical Methods*, SSSA Book Series No. 5, Soil Science Society of America, Madison, WI, 2002, pp. 422–424.

74. G. C. Topp, J. L. Davis, and A. P. Annan, Electromagnetic Determination of Soil Water Content Using TDR: I. Application to Wetting Fronts and Steep Gradients, *Soil Sci. Soc. Am. J.* **46**, 672–678 (1982).

75. M. H. Young and J. B. Sisson, Tensiometry, in J. H. Dane and G. C. Topp (Eds.), *Methods of Soil Analysis, Part 4. Physical Methods*, SSSA Book Series No. 5. Soil Science Society of America, Madison, WI, 2002, pp. 575–608.

76. R. A. Wooding, Steady Infiltration from a Shallow Circular Pond, *Water Resour. Res.* **4**, 1259–1273 (1968).

77. C. W. Boast and D. Kirkham, Auger Hole Seepage Theory, *Soil Sci. Soc. Am. Proc.* **35**, 365–373 (1971).

78. C. W. Boast and R. G. Langebartel, Shape Factors for Seepage into Pits, *Soil Sci. Soc. Am. J.* **48**, 10–15 (1984).

79. H. P. Johnson, R. K. Frevert, and D. D. Evans, Simplified Procedure for the Measurement and Computation of Soil Permeability below the Water Table, *Agric. Eng.* **33**, 283–286 (1952).

80. D. Kirkham, Theory of Seepage into an Auger Hole above an Impermeable Layer, *Soil Sci. Soc. Am. Proc.* **22**, 204–208 (1958).

81. D. O. Lomen, A. W. Warrick, and R. Zhang, Determination of Hydraulic Conductivity from Auger Hole and Pits—An Approximation, *J. Hydrol.* **90**, 219–229 (1987).

82. C. H. M. van Bavel and D. Kirkham, Field Measurement of Soil Permeability Using Auger Holes, *Soil Sci. Soc. Am. Proc.* **13**, 90–96 (1948).

83. W. F. J. van Beers, *The Auger-Hole Method*, Int. Inst. Land Reclam. and Improve. Bull No. 1, H. Veenman & Zonen, Wageningen, The Netherlands, 1970.

84. H. Bouwer and R. D. Jackson, Determining Soil Properties, in J. Van Schilfgaarde (Ed.), *Drainage for Agriculture*, Agron. Monogr. 17, Am. Soc. Agron., Madison, WI, 1974, pp. 611–672.

85. L. L. Sanders, *A Manual of Field Hydrogeology*, Prentice-Hall, Upper Saddle River, NJ, 1998.

86. D. Downey, W. D. Graham, and G. A. Clark, An Inexpensive System to Measure Water Level Rise during the Bouwer and Rice Slug Test, *Appl. Eng. Agric. Am. Soc. Agric. Eng.* **10**, 247–253 (1994).

87. M. Maasland, Measurement of Hydraulic Conductivity by the Auger-Hole Method in Anisotropic Soil, *Soil Sci.* **81**, 379–388 (1955).

88. M. Maasland and H. C. Haskew, The Auger Hole Method for Measuring the Hydraulic Conductivity of Soil and Its Application to Tile Drainage Problems, in *Third Congress on*

Irrig. and Drain., Trans., Vol. III, International Commission on Irrigation Drain, New Delhi, India, 1957, pp. 8.69–8.114.

89. D. Kirkham, Proposed Method for Field Measurement of Permeability of Soil below the Water Table, *Soil Sci. Soc. Am. Proc.* **10**, 58–68 (1945).

90. R. K. Frevert and D. Kirkham, A Field Method for Measuring the Permeability of Soil Below the Water Table, *Highw. Res. Board Proc.* **28**, 433–442 (1948).

91. J. N. Luthin and D. Kirkham, A Piezometer Method for Measuring Permeability of Soil in Situ below a Water Table, *Soil Sci.* **68**, 349–358 (1949).

92. A. Amoozegar and G. V. Wilson, Methods for Measuring Hydraulic Conductivity and Drainable Porosity, in R. W. Skaggs and J. van Schilfgaarde (Eds.), *Agricultural Drainage*, ASA-SSSA Monogr., Soil Science of America, Madison, WI, 1999, pp. 1149–1205.

93. A. Amoozegar, Auger-Hole Method (Saturated Zone), in J. H. Dane and G. C. Topp (Eds.), *Methods of Soil Analysis, Part 4. Physical Methods*, SSSA Book Series No. 5, Soil Science Society of America, Madison, WI, 2002, pp. 859–869.

94. H. Bouwer and R. C. Rice, A Slug Test for Determining Hydraulic Conductivity of Unconfined Aquifers with Completely or Partially Penetrating Wells, *Water Resour. Res.* **12**, 423–428 (1976).

95. H. Bouwer, The Bouwer and Rice Slug Test—An Update, *Ground Water* **27**, 304–309 (1989).

96. A. Klute, Water Retention: Laboratory Methods, in A. Klute (Ed.), *Methods of Soil Analysis, Part 1. Physical and Mineralogical Methods*, 2nd ed., Agron. Monogr. 9, American Society of Agronomy and SSSA, Madison, WI, 1986, pp. 635–662.

97. R. E. Uhland, Physical Properties of Soils as Modified by Crops and Management, *Soil Sci. Soc. Am. Proc.* **14**, 361–366 (1949).

98. M. J. Vepraskas, M. T. Hoover, J. L. Beeson, M. S. Carpenter, and J. B. Richards, Sampling Device to Extract Inclined, Undisturbed Soil Cores, *Soil Sci. Soc. Am. J.* **54**, 1192–1195 (1990).

99. J. S. Rogers and C. E. Carter, Soil Core Sampling for Hydraulic Conductivity and Bulk Density, *Soil Sci. Soc. Am. J.* **51**, 1393–1394 (1987).

100. A. Amoozegar, Preparing Soil Cores Collected by a Sampling Probe for Laboratory Analysis of Soil Hydraulic Properties, *Soil Sci. Soc. Am. J.* **52**, 1814–1816 (1988).

101. J. Bouma, Measuring the Hydraulic Conductivity of Soil Horizons with Continuous Macropores, *Soil Sci. Soc. Am. J.* **46**, 438–441 (1982).

102. E. C. Childs and A. Poulovassilis, An Oscillating Permeameter, *Soil Sci.* **90**, 326–328 (1960).

103. R. L. Hill and L. D. King, A Permeameter Which Eliminates Boundary Flow Errors in Saturated Hydraulic Conductivity Measurements, *Soil Sci. Soc. Am. J.* **46**, 877–880 (1982).

104. A. W. Warrick and D. R. Nielsen, Spatial Variability of Soil Physical Properties in the Field, in D. Hillel (Ed.), *Applications of Soil Physics*, Academic, New York, 1980, pp. 319–344.

105. T. M. Zobeck, N. R. Fausey, and N. S. Al-Hamdan, Effect of Sample Cross-Sectional Area on Saturated Hydraulic Conductivity in Two Structured Clay Soils, *Trans. Am. Soc. Agric. Eng.* **28**, 791–794 (1985).

106. K. Beven and P. Germann, Water Flow in Soil Macropores. II. A Combined Flow Model, *J. Soil Sci.* **32**, 15–29 (1981).

107. J. Bouma, Use of Soil Survey Data to Select Measurement Techniques for Hydraulic Conductivity, *Agric. Water Manag.* **6**, 177–190 (1983).

108. B. C. Ball and P. Schjønning, Air Permeability, in J. H. Dane and G. C. Topp (Eds.), *Methods of Soil Analysis, Part 4. Physical Methods*, SSSA Book Series No. 5, Soil Science Society of America, Madison, WI, 2002, pp. 1141–1158.

109. B. L. Grover, Simplified Air Permeameters for Soil in Place, *Soil Sci. Soc. Am. Proc.* **19**, 414–418.

110. P. Liang, C. G. Bowers, Jr., and H. D. Bowen, Finite Element Model to Determine the Shape Factor for Soil Air Permeability Measurements, *Trans. Am. Soc. Agric. Eng.* **38**, 997–1003 (1995).

111. R. C. Reeve, A Method for Determining the Stability of Soil Structure Based upon Air and Water Permeability Measurements, *Soil Sci. Soc. Am. Proc.* **17**, 324–329 (1953).

112. J. R. Nimmo and K. S. Perkins, Aggregate Stability and Size Distribution, in J. H. Dane and G. C. Topp (Eds.), *Methods of Soil Analysis, Part 4. Physical Methods*, SSSA Book Series No. 5, Soil Science Society of America, Madison, WI, 2002, pp. 317–328.

30

PASSIVE SAMPLING

Lee Trotta

Triad Engineering, Inc.
Milwaukee, Wisconsin

Passive sampling of groundwater does not alter analyte concentrations or agitate the sample. The diffusion multilayer sampler (DMLS) offers multiple discrete zone sampling from existing wells. Operations for truly passive sampling are modeled on the use of the DMLS. Solutes enter collection cells by diffusion. Due to separation of cells by seals, which fit the inner diameter of the well, the DMLS allows contaminant versus depth profiling and detailed plume delineation.

The DMLS is used for discrete hydrochemical sampling as well as for obtaining groundwater velocity profiles. Discrete sampling allows depth profiling of chemical concentrations and detailed plume delineation. Velocity profiles allow fracture tracing and estimation of aquifer transmissivity.

Both uses are based on dialysis membrane technology and the principle of diffusion. Discrete zone sampling is accomplished in the DMLS with Viton seals, which loosely fit the inner diameter of the open interval. Two seals bracket each dialysis

Environmental Instrumentation and Analysis Handbook, by Randy D. Down and Jay H. Lehr
ISBN 0-471-46354-X   Copyright © 2005 John Wiley & Sons, Inc.

679

cell, allowing sampling of any potential stratification (vertical sequence down to 3- or 4-in. layers) that may exist in the aquifer. For discrete sampling, "a dialysis cell vial is filled with distilled water and covered by permeable membranes at both ends. When a dialysis cell is exposed to groundwater having concentrations of solutes (dissolved chemical constituents) different from inside the cell, a natural process of diffusion of solutes from higher concentrations to lower concentrations occurs. This occurs through the membrane until a dynamic equilibrium is reached. At this point the contents of the cell will be representative of the water surrounding the cell" (Ref. 1, p. 3). For velocity profiles, a cell is instead filled with a known concentration of a tracer, and the rate at which the tracer is either dissipated or picked up in another cell is measured.

Membrane technology used for sampling is sometimes referred to as *passive sampling*. Other sampling technologies involving pumps and bladders are referred to as *active sampling*. Active sampling has the tendency to incorporate air or loose soil particles (turbidity) into the sample. Incorporated air can make accurate analysis for certain volatile chemicals difficult. Incorporated sand particles can also skew an analysis and (over time) damage expensive pumping equipment. Passive sampling versus active sampling offers several advantages, as outlined below by Golder Associates (Ref. 1, p. 1):

- Purging of the well prior to sampling is not required. This results in significant cost savings and, in situations where the purged water is polluted, eliminates the need to properly dispose [of] hazardous waste.
- Since purging and/or pumping is not required, the DMLS is an excellent tool to use in wells with low yield.
- Turbid wells do not impact the sample integrity since the DMLS membrane filters the samples.

30.1 EQUIPMENT INVENTORY

The first stage in the operating procedure is to assure one has the right equipment for the well to be sampled (Fig. 30.1). The straightness and inside diameter of the existing well should be measured as accurately as possible before selecting the sampler. Samplers can be provided to fit wells from 2 to 4 in. PS in diameter [minimum 2 in. inside diameter (i.d.) required]. Samplers should not be stored on their seals. A cable or rope assembly must be available (a "hook" or eyelet hole is provided on top of the sampler) for lowering the clean sampler into the well. The use of a hoist is appropriate for well depths over 100 ft. Depending on the depth of the well and the speed with which the desired sampling operation is to be accomplished, weights may need to be attached. If more than 20 ft of connectable sampler rods is to be used in the same well, the engineered product should be used rather than the standard product.

The choice of membrane end (usually 0.2 or 10 μm) must be made relative to the solute to be sampled. If sampling organics (e.g., bacteria, hydrocarbons), the 10-μU

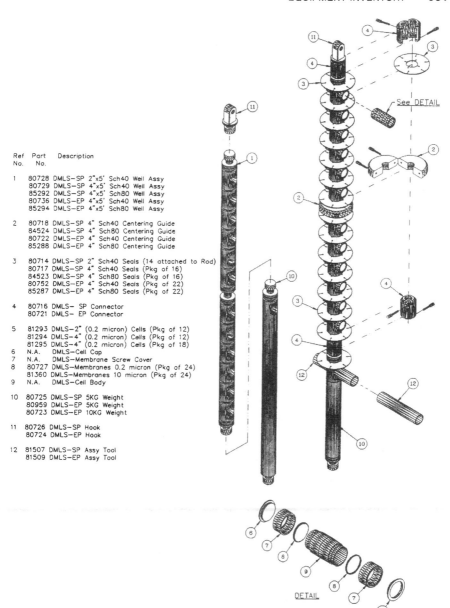

Ref   Part   Description
No.   No.

1   80728 DMLS−SP 2″x5′ Sch40 Well Assy
      80729 DMLS−SP 4″x5′ Sch40 Well Assy
      85292 DMLS−SP 4″x5′ Sch80 Well Assy
      80736 DMLS−EP 4″x5′ Sch40 Well Assy
      85294 DMLS−EP 4″x5′ Sch80 Well Assy

2   80718 DMLS−SP 4″ Sch40 Centering Guide
      84524 DMLS−SP 4″ Sch80 Centering Guide
      80722 DMLS−EP 4″ Sch40 Centering Guide
      85288 DMLS−EP 4″ Sch80 Centering Guide

3   80714 DMLS−SP 2″ Sch40 Seals (14 attached to Rod)
      80717 DMLS−SP 4″ Sch40 Seals (Pkg of 16)
      84523 DMLS−SP 4″ Sch80 Seals (Pkg of 16)
      80752 DMLS−EP 4″ Sch40 Seals (Pkg of 22)
      85287 DMLS−EP 4″ Sch80 Seals (Pkg of 22)

4   80716 DMLS− SP Connector
      80721 DMLS− EP Connector

5   81293 DMLS−2″ (0.2 micron) Cells (Pkg of 12)
      81294 DMLS−4″ (0.2 micron) Cells (Pkg of 12)
      81295 DMLS−4″ (0.2 micron) Cells (Pkg of 18)
6   N.A.   DMLS−Cell Cap
7   N.A.   DMLS−Membrane Screw Cover
8   80727 DMLS−Membranes 0.2 micron (Pkg of 24)
      81360 DMLS−Membranes 10 micron (Pkg of 24)
9   N.A.   DMLS−Cell Body

10   80725 DMLS−SP 5KG Weight
      80959 DMLS−EP 5KG Weight
      80723 DMLS−EP 10KG Weight

11   80726 DMLS−SP Hook
      80724 DMLS−EP Hook

12   81507 DMLS−SP Assy Tool
      81509 DMLS−EP Assy Tool

See DETAIL

DETAIL

Figure 30.1   DMLS components.

membrane is preferred. Either the 10 or the 0.2-µU membrane is fine for inorganics. Different membranes will result in different equilibration time for the same analytes, between 24 and 45 h (Ref. 1, pp. 10–11).

30.2 SITE CONDITIONS

Certain initial procedures may be accomplished before arriving at the field site. Others are based on site conditions. Dialysis cells are filled with distilled (or deionized) water, and membranes are screwed to the cell (preferably under water to avoid air bubbles). Cells can be either capped or left in the water until used to avoid evaporation or formation of air bubbles. Calculations are made on length of down-hole contact time for well recovery after sampler insertion based on Ficks's second law of diffusion.[2] These calculations involve the borehole diameter, the desired pass-through distance, and the transmissivity of the aquifer. Fick's second law of diffusion is given as $C = (C°2) \, \text{erfc} \, (x/2\sqrt{Dt})$. The results of this formula give the minimum duration of contact for equilibration to occur. In most cases, there is no maximum limit to the length of time the DMLS may stay in the water. Normally, a period of 7–10 days is enough (Ref. 3, p. 7) to provide a significant safety factor for the whole range of analytes. However, this period may need to be longer in low-flow wells. It is best if evaluation of well conditions (such as estimated down-hole contact time) is done prior to loading for the sampling run. This is especially true if multiple wells are to be sampled.

The well condition should be evaluated before the DMLS is installed (including total depth). Lowering a weight along with a centering guide into the well can determine potential obstacles that would prevent use of the DMLS. The cable and connection should be strong enough to not break inside the well. The water level in the well to be sampled should be measured (the DMLS must be immersed in order to function). In theory, the multilayer seals would be nonfunctional (i.e., no water quality distinctions could be made) in a thick homogeneous aquifer, but truly homogeneous aquifers are very rare. There is no restriction on the substances to be sampled, so placement in contaminated zones is encouraged.

30.3 OPERATING PROCEDURE

At the field site, the procedure is simple and efficient. A quick lock connects 5-ft-long DMLS sections. Cables and weights are attached to the sampler as needed. Dialysis cells are placed. The sampler, with cells in place, is lowered into the well to the desired sampling depth. The sampler is then secured and left in place for the calculated waiting time (a cable lock or well cap tie-off may be used for this purpose). Figure 30.2 illustrates a DMLS lowered into a monitoring zone with Viton seals sealing the well bore at critical stratigraphic intervals.

There is no need to shut in the monitoring zone with an additional borehole packer because the DMLS seals accomplish this at 3- or 4-inch intervals (depending

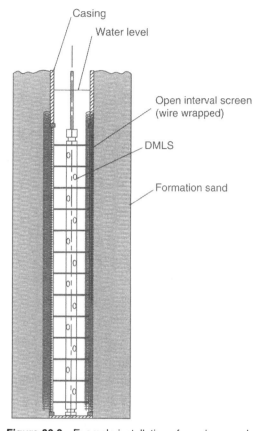

Figure 30.2   Example installation of passive sampler.

on whether the engineered or standard model is utilized, respectively). There is no need to purge the well as the contact time allows water from the formation to flow to the well. Once equilibrium is reached, the water in the sampled zone is representative of the water in the formation. There are no electronics to figure out and no pressure transducers to go bad.

After the waiting time has elapsed, the sampler is retrieved. When at the surface, dialysis cells are removed from the sampler and quickly capped at both ends. Labels identifying the well, cell number, and cell position should be placed on each dialysis cell as it is removed from the sampler rod. Labeled cells may be either tested in the field or mailed to the laboratory for analysis. Any number of available cells may be chosen for analysis from a single borehole. The remaining labeled cells may be saved for possible analysis after "hits" have been determined. Although certain organics may exhibit minor sorption to the membrane or cell, representative samples are still collected. The sample volume protocol for specific constituents may vary between laboratories.

Table 30.1 DMLS Monitoring and Research Activities

Year	Site Conditions	Region	Analyzed Parameters
1984	Sewage irrigated land	Israel	EC, pH, HCO, SO_4, Cl, NO_3
1985	Sewage irrigated land	Israel	^{18}O
1986	Sewage irrigated land	Israel	Cl, SO_4, NO_3
	Laboratory	Israel	pH, HCO_3, dissolved oxygen (DO), dissolved organic carbon (DOC) Tetrachloroethylene 1,1,1-trichloroethane
1987	Forest region	The Netherlands	EC, pH, DO, DOC, HCO_3, Cl, NO_3, SO_4, Na, K, Ca_2, Mg_2, Al
	Sewage irrigated land	Israel	Cl
		Israel	N_2O, pH, EC, HCO_3, NO_3, Cl
1988	Feedlots and pasture land	The Netherlands	pH, EC, DO, DOC, HCO_3, Cl, NO_3, SO_4, Na, K, Ca_2, Mg_2
	Leaky sewer test site	Garching, Germany	pH, DO, HCO_3, DOC, Ca_2, Mg_2, Na, K, Cl, SO_4, NO_3
	Sewage irrigated land	Israel	Al, Cd, Cu, Fe, Mn, Pb
	Hazardous waste repository site	Israel	Fe, Mn, Pb, Zn, Ag, Al, Cd, Cu
1989	Sewage irrigated land	Israel	Fe, Mn, Zn, Cu, Ag
	Groundwater research field	Mobile, Al., U.S.	^{18}O
	Oil spill site, laboratory	Long Island, NY, U.S.	Benzene, toluene, xylene, Mn, Fe, Cl, pH, DO
	Discharge area of groundwater flow system	Franklin Lake Playa, CA, U.S.	pH, EC, DO, Na, K, Cl, SO_4, ^{18}O
	Sewage irrigated land	Israel	Ba_2, Sr_2, Ca_2, Cl, SO_4, pH
1990	Sewage irrigated land	Israel	^{18}O, Cl
1991	Sewage irrigated land	Israel	Fe, Mn, Cu, Zn, Ag, carbonate, organic and oxide fraction of particulate matter Particle quantities, composition and size, DO, pH, EC
1992	Hazardous waste disposal site	Israel	Cl, Br, SO_4, NO_3, Cu, ^{18}O, D, T
	Saline freshwater interface. Dead Sea	Israel	NO_3, Cl, Br, SO_4, NO_3, Cu, ^{18}O
	Recharge areas of chalk and sandstone aquifers	Ogbourne and Boughton, England	pH, EC, DO, NO_3, Na, K, Ca_2, Mg_2, HCO_3, SO_4, Cl, DOC, Li
1994	Sewage irrigated land	Israel	Cu, Cd
1995	Acidic Cr waste infiltrated land	North Carolina, U.S.	Cr, Fe, Al, Cl
	Cotton field	Israel	^{18}O, DO, Cl, NO_3, SO_4, DOC, atrazine, bacteria
	Rocky flats	Denver, CO, U.S.	^{18}O as a tracer
1996	Pease AFB	New Hampshire, U.S.	Trichloroethylene (TCE)
	Sewage irrigated land	Israel	Turbidity, particle diameter, DOC, NO_3, Cl, specific discharge
	Acidic Cr waste infiltrated land	North Carolina, U.S.	Cr and other metals, anions—Cl, SO_4

A partial list of substances successfully sampled through 1996 was compiled by Golder Associates[1] and is summarized in Table 30.1. The table illustrates that the same operating procedure may be used under a variety of sampling goals and site conditions. Geothermal zones and extended exposure to sunlight, where temperatures exceed 110°F, should be avoided to prevent damage to the polyvinyl chloride (PVC).

For many parameters a choice of analysis methods is available. Methods like ion chromatography, which today are best suited for the laboratory, may tomorrow be considered "field" methods. If necessary, syringe transfer of samples to a field testing apparatus avoids aeration. Such choices have less to do with the DMLS tool than with economy Vs accuracy of chemical analysis and customer preference. Suffice to say it is easy to cap the cells and send them to the laboratory.

Table 30.2   Cleaning Process

Detergents

- The detergent used to clean all components is either DECON 90 concentrate or Liquinox. These phosphate-free surface-active agents are generally used to clean laboratory equipment and surgical instruments and in the pharmaceutical industry.
- Working concentration is 2–5% in tap water.
- These detergents are alkali, biodegradable, totally rinseable, and nontoxic.

Rods

1. Immerse each rod in a tub full with 5% DECON 90 in tap water and brush each rod couple of times. Leave the rod inside the detergent for at least 15 min.
2. Wash each rod with tap water until the DECON 90 leaves no traces.
3. Wash each rod with distilled or deionized water. Let it dry.
4. Replace all cleaning materials, including detergent and water, after cleaning 10–15 rods.

All Other Components Excluding Sampling Cells

Immerse all components in a tub full with 5% DECON 90 in tap water for at least 10 min.

1. Wash all the components in a tub with tap water until the DECON 90 leaves no traces.
2. Wash all the components with distilled or deionized water and dry.
3. When washing connectors and/or guide, release the screws.

Cells

A. Cells are molded under medical-grade conditions and require no initial cleaning.
B. Parts of the cells that can be reused are:
   - Cell body
   - Screw covers
   - Caps

1. Immerse all cell parts in a tub full with 5% DECON 90 in tap water for at least 15 min.
2. Wash all the parts in a tub with tap water until the DECON 90 leaves no traces.
3. Wash all the components with distilled or deionized water.

Note: Membranes cannot be washed or reused.

30.4 CLEANING

If the DMLS and the cable assembly are to be moved to another well for sampling, they must be thoroughly cleaned before reuse (Table 30.2). This cleaning procedure may be accomplished at the wellhead or at an appropriate cleaning station. Unless local regulations demand it, the cleaning procedure is unnecessary if the DMLS is to be reinserted in the same well.

New membranes should be used for every sampling event, and new dialysis cells are also recommended. Thorough washing would likely damage the membrane so repeated use of the membranes could allow cross contamination or may block the pores of the membrane and decrease the diffusion area. Avoid touching the membranes with greasy or oily hands by wearing latex surgical gloves.

30.5 CONCLUSION

The depth-discrete groundwater samples provided by passive sampling without purging or drilling additional boreholes offer cost savings to groundwater studies throughout the world.[3–5] A simple operating procedure, as described above, is part of the reason for the success of this technology. The same operating procedure may be used under a variety of sampling goals and site conditions.

REFERENCES

1. Golder Associates, *Draft DMLS User's Guide*, Golder Associates, Lakewood, CO, 1998.
2. J. Crank, *The Mathematics of Diffusion*, 2nd ed., Clarendon, Oxford, 1975.
3. G. Hanson, *DMLS User's Manual*, U.S. Filter/Johnson Screens, St. Paul, MN, 1996.
4. R. W. Puls and C. J. Paul, *Multi-Layer Sampling in Conventional Monitoring Wells for Improved Estimation of Vertical Contaminant Distributions and Mass*, USEPA National Risk Management Research Laboratory, Ada, OK, 1996.
5. E. Kaplan, S. Banerjee, D. Rosen, M. Magaritz, et al., Multilayer Sampling in the Water-Table Region of a Study Aquifer, *Groundwater*, **29**(2) 191–197 (1991).

31

INSTRUMENTATION IN GROUNDWATER MONITORING

David L. Russell, P.E.
Global Environmental Operations, Inc.
Lilburn, Georgia

Environmental Instrumentation and Analysis Handbook, by Randy D. Down and Jay H. Lehr
ISBN 0-471-46354-X   Copyright © 2005 John Wiley & Sons, Inc.

31.1 INTRODUCTION

In groundwater monitoring much is made or inferred from little data, the problem being the accuracy of the data and the cost of obtaining additional data. In any groundwater monitoring exercise, there is almost always never enough data. One would want to see what is happening between two wells to better understand the chemistry and reactions, but the cost is often prohibitive. In this chapter, we will discuss the instruments used in groundwater monitoring as well as their relative accuracy and error.

31.2 GROUNDWATER THEORY

Water moves through the ground very slowly. The ground is a porous medium. The porosity can vary locally depending upon soil type, rock, and mineral composition. The porosity of the soil often varies between 0.3 and 0.4, but it can be much lower or higher if the conditions are right. Hard rock, not fractured, may have a porosity close to zero. The porosity of fractured rock such as limestone with wide fissures may only be 0.2, but because of the fissures, water will move in the rock faster than in many soils.

Groundwater moves very slowly, often measured in meters (or feet) per year. Because the temperature of groundwater is almost always within a few degrees of 11.11°C (52°F), chemical reactions take place much more slowly than they do at the surface of the ground, where the temperature can be significantly higher. For surface water, the reference temperature is 20°C (68°F). Biochemical reactions double with each 10°C change in temperature, which means that in the soil biochemical reactions take place at about half the rate that they do on the surface.

The realm beneath the surface of the ground is not uniform in any direction. Depending upon the type of soil, one can go from miles of uniform beach sand to rock or clay in a matter of centimeters. In some soils, particularly those which have been influenced by glacial effects, soils with have shallow layers of different clays can be interspersed with sand, rocks, and cobbles, and discontinuities lenses are the rule rather than the exception.

The chemical and physical properties of groundwater can also be difficult to measure, as they are often dependent upon the soils. Clays and silts have absorptive and adsorptive properties; one measurement of the difficulty in remediation of organic compounds from a specific site is the octanol–water partition coefficient. This is an indirect measure of the adsorptive capacity of the soil and represents a measure of the difficulty of removing organic materials from the soils. Many soils, such as clintphiloite, have ion exchange capacities that will cause certain cations to be removed from the soils.

Groundwater serves most small communities in the United States and elsewhere in the world as a source of drinking water. Groundwater is characterized by natural minerals in moderate to low concentrations.

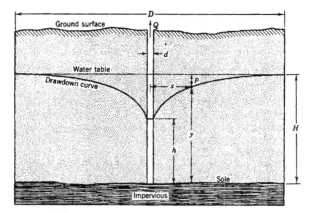

Figure 31.1   Water table well in groundwater reservoir.

Flow in groundwater is linear, and flow through porous media is analogous to heat transfer through solid a solid medium. The Darcy equations are used for laminar flow.

The groundwater flow equation, in SI units, is

$$Q(\text{flow}) = \frac{K(H^2 - h^2)}{\log_e[(D/2)/(d/2)]}$$

where the parameters apply to the illustration in Figure 31.1 and the constant K is the permeability coefficient (in gal/day ft^2, or m^3/day m^2). The model used below is the simplest of an extremely complex set of possible combinations because the ground is not a homogeneous medium. Once contaminated, groundwater is difficult, if not impossible, to decontaminate. Groundwater protection must be a plant-wide priority!

For most groundwater problems, flow and chemical concentration equations are second-order equations which are virtually identical to heat flow equations when appropriate substitutions are made. Flow can be steady or unsteady, and the set of equations can be somewhat complex. The differential equation for unconfined unsteady flow in an aquifer is given as[*]

$$K\left(\frac{\delta^2 h}{\delta x^2} + \frac{\delta^2 h}{\delta y^2} + \frac{\delta^2 h}{\delta z^2}\right) = \theta\left(\beta + \frac{\alpha}{\theta}\right)\frac{\partial P}{\partial t}$$

where b is the formation thickness; Kb is the formation transmissivity coefficient T; $S = \theta\gamma b\left(\beta + \frac{\alpha}{\theta}\right)$, the storage coefficient for the formation; and the h's are differential

---

[*]The equation is from *Engineering Hydraulics*, by Hunter Rouse, J. Wiley New York, 1950.

in water surface elevation (against a known reference) in a particular direction. The equation can also be expressed as $\frac{\partial^2 h}{\partial x^2} + \frac{\partial^2 h}{\partial y^2} = \frac{S}{T}\frac{\partial h}{\partial t}$. This equation is essentially the same as the heat flow equation for three-directional heat flow in a slab.

The problems then are to measure S and T, the storage and transmissivity coefficients. A related problem is to identify the nonhomogeneity in the formation. In most situations, flow is considered to be two dimensional, and that simplifies both data collection and modeling for the aquifer. These are all defined for a specific aquifer by pumping tests where the aquifer is pumped out for periods between 24 and 96 h and the drawdown in the main pumping well and other monitoring wells in the area is measured.

To modify the groundwater, one can try to remove or treat the chemicals or contaminants which may be in it. Air and methane are the gases most commonly pumped into the groundwater to stimulate the natural bacterial systems present in the ground to use the contaminants as food, rendering them harmless by oxidation or reduction chemistry. Methane (natural gas) is often pumped into contamination areas where there are chlorinated organics in an attempt to stimulate anaerobic dechlorination. Air is often pumped into the groundwater and withdrawn from the formation above the groundwater in an attempt to remove volatile organic chemicals trapped within the soil particles or to stimulate aerobic organic decomposition of chemicals in the groundwater. Occasionally, the groundwater is injected with other fluids to stabilize the soils or to treat specific deficiencies in the chemical composition for a treatment regimen.

Other groundwater systems include the injection of substances into the ground or groundwater, which can include air, other liquids and gels, propants,[†] and other substances used in connection with attempts to clean up groundwater contamination.

Groundwater, its contamination, and its movement cannot be separated from the geology—for example, the sand, clay, rock, and silt that hold the groundwater and the contaminants. In any groundwater cleanup situation there are three very important parts: getting the geology correct, getting the chemistry correct, and getting the engineering solution for transport correct. One of the many problems often encountered by remediation companies is that they work against the natural systems already in place. For example, the solution to a contamination problem can involve the application of bacteria and bacterial enzymes or powders which will enhance natural degradation of the contaminants. Those proposing the solution often fail to realize that sand, silts, and clays (a) are nature's perfect filter which will strain out what they are trying to add to the ground; (b) will react with the materials they are attempting to put into the ground, changing the properties of the additive; and (c) will have chemistry or physical properties which will precipitate, absorb, or adsorb the materials being added to treat the problem. This is also the case when one neglects to look at the basic biochemistry of the natural bacteria in the soil. One has to look at the system as a whole and use his or her instrumentation to come up

---

[†]Propants are used to hold open a formation and increase permeability after hydraulic fracturing has been performed on the soil matrix. The hydrofracturing often leaves a void which needs to be filled with sand or something else to hold the void open and maintain the local permeability of the worked area.

with a solution to solve the entire problem, including the pathways and the product reactants, and not just look at one aspect of the problem.

31.3 MEASUREMENTS

The principal tool used to gather information about the subsurface is the drill rig. Core sampling is one of the most reliable methods of gathering subsurface information. Unfortunately it is expensive. Therefore, a number of other methods are commonly used above ground to infer what may be happening below ground, including electrical resistivity, electrical conductance, sonic waves, and ground-penetrating radar (GPR). These methods are highly dependent upon the type of soil and rocks that lie beneath the ground as well as soil chemistry.

31.3.1 Ground-Penetrating Radar

Ground-penetrating radar sends a short-wave radar pulse into the ground and measures the reflectivity of the ground. Given the right type of soils, predominantly sands and some clays, radar can be used many meters below ground. (A colleague from the National Aeronautics and Space Agency regularly surveys the depth of groundwater below the Sahara Desert sands at a depth of up of 50 m.) In other places, highly mineralized clays and salt water can cause radars to be blinded and have a net penetration of a few centimeters. Buried objects, including anything with an electromagnetic signature, will show up very well subject to the limitations mentioned above. Radar is also useful in determining the depth of the water table below ground. One can use radar to survey a relatively large area by dragging the radar transmitter/receiver arrays across the ground, but the interpretation often needs an experienced operator to determine what is being registered and to adjust the equipment so that the right frequency is being used in the determination.

The GPR uses a variety of antennas and frequencies. Most commonly, a two-box system is employed, with the broadcast and receiver contained in separate boxes, and they are dragged over the ground at a known rate. The operator most commonly has a display which indicates the reflectivity pattern underground, and when an object is identified, the operator flags the area of interest.

A typical GPR trace is shown in Figure 31.2, whereas in an urban environment, the radar trace would resemble the one in Figure 31.3.

31.3.2 Resistivity and Conductivity

Resistivity surveys impress an electric current through a source and measure the electrical resistivity of the soils from an array. Resistivity surveys are conducted by placing steel poles in the ground and connecting them with electrical cables. By measurement of a number of points, a subsurface profile can be generated, and that profile can be related to the physical properties of the soils.

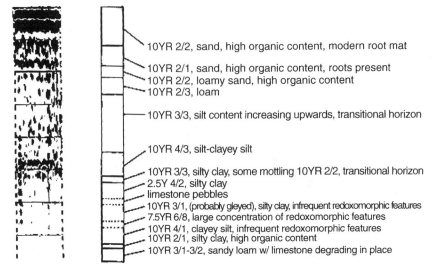

10YR 2/2, sand, high organic content, modern root mat

10YR 2/1, sand, high organic content, roots present
10YR 2/2, loamy sand, high organic content
10YR 2/3, loam

10YR 3/3, silt content increasing upwards, transitional horizon

10YR 4/3, silt-clayey silt

10YR 3/3, silty clay, some mottling 10YR 2/2, transitional horizon
2.5Y 4/2, silty clay
limestone pebbles
10YR 3/1, (probably gleyed), silty clay, infrequent redoxomorphic features
7.5YR 6/8, large concentration of redoxomorphic features
10YR 4/1, clayey silt, infrequent redoxomorphic features
10YR 2/1, silty clay, high organic content
10YR 3/1-3/2, sandy loam w/ limestone degrading in place

Figure 31.2 GPR readout at 500 MHz, 40 ns (left) of Core 3, 1.13 m total length (right) at 8Je603. *Source*: http://www.flmnh.ufl.edu/natsci/vertpaleo/aucilla11_1/gpr.JPG, article on archeological investigation.

The following is a summary of electrical resistivity surveys (from http://www.geomodel.com:

Surface Resistivity Method

Application of the surface resistivity method requires that an electrical current be injected into the ground by surface electrodes. The resulting potential field (voltage) is measured at the surface by a voltmeter between electrodes.

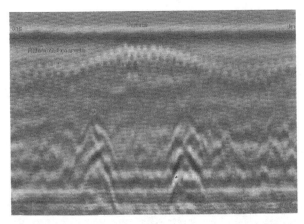

Figure 31.3 Typical radar trace in urban environment. *Source*: http:www.geophysics.co.uk/images/env9.gif. Copyright 1995–1999 Geo-Services International (UK) Ltd.

The resistivity of the subsurface materials can be calculated by knowing the electrode spacing, geometry of the electrode positions, applied current, and measured voltage. Surface resistivity measurements are reported in the units of ohm-meters or ohm-feet.

The depth of the resistivity measurement is related to the spacings of the electrodes and may vary depending on the subsurface conditions. The resistivity unit has a self-contained transmitter, capable of obtaining data to about 50 to 100 meters (160 to 300 feet), using self-contained, rechargeable batteries.

Subsurface conductivity is measured by a number of instruments which sense the difference in electrical conductivity of the soil from mobile or transportable equipment. The advantage is that the equipment can be carried around, and it is quite accurate for a variety of applications and depths. Because it measures electrical fields in the ground, it can be influenced by things like overhead power lines, fences, and other objects on the surface and buried which have a strong electromagentic profile or signature. The equipment comes in a variety of configurations, from a one-person transportable unit to one which is on a wheeled card. The differences in electrical conductivity are then related to the properties of the materials beneath the surface of the ground.

Electromagnetic methods energize the ground inductively by means of an alternating current flowing in a transmitter coil. The resulting signal, containing ground response characteristics, is detected inductively by a receiver coil. Both coils may be mounted in a traveling frame or placed on the ground or in aircraft.

Survey by electrical resistivity methods is relatively slow and expensive, particularly when used on the ground in detailed surveys, comparable to conductivity methods, but where the soil conditions are correct, it can be used to establish the boundaries of a below-ground contaminant plume with great accuracy.

If conductivity is remotely measured by using a walking survey unit, shown in Figure 31.4 data in Table 31.1 may be of some assistance. Many other sources of information are available, but the services of a competent geophysicist should be employed in the interpretation of the results.

Figure 31.4 Geophysical survey using electromagnetic resistivity for subsurface equipment. Picture shows a Geonics model EM-31 unit.

Table 31.1 Nominal Depth of Exploration for Geonics EM Instruments

| Instrument | Coil Separation (m) | Orientation | Nominal Depth of Explorationm | |
			m	ft
EM31	3.7	HD	2.5	8
		VD	5.2	17
EM34-3	10	HD	6.8	22
		VD	14.3	47
EM34-3	20	HD	13.5	44
		VD	28.6	94
EM34-3	40	HD	27	89
		VD	57	188

Source: Northwest Geophysical Associates website.

Note: VD and HD indicate relative position of dipole sensors (vertical or horizontal).

The table shows the approximate separation required for the depth of exploration based upon the equipment.

31.3.3 Sonic Waves

In this technique the difference in the refractive indices of the soils and the thicknesses of layers are measured using sound waves generated at a known point (Fig. 31.5). Sensitive microphones are set in an array on the ground. The

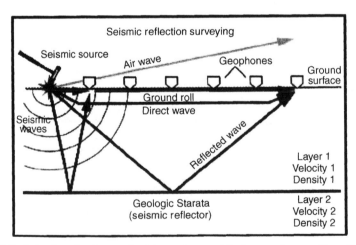

Figure 31.5 Seismic reflection methodology. *Source*: http//www.rr-inc.com/Frame%20Pages/Tech%20Pages/Seismic/seismic.htm.

microphones are then set to record the amplitude and timing of a known sound which is intermittently or regularly generated. In oil exploration, where exploration is relatively deep, the sonic surveys use small quantities of dynamite to initiate the sound pulse and the resulting refraction of the signal measured in the microphone array is used to determine the depth and thickness of rocks several kilometers below ground. A related technique using a hammer striking a steel pole has been used to measure the depth and characteristics of shallow soils. The technique is subject to interpretation and correlation with known ground elements.

The newest advances in seismic surveys combined with computer enhancement and interpretation of geophone data permit the development of three-dimensional arrays for sonic surveys. It needs to be remembered that these surveys are principally for finding large objects based upon the differences in refraction of the objects encountered. For large objects and/or deeper or layers in the subsurface, the technique is very good; however, for more shallow formations and objects with an effective diameter less than half of the wavelength of the sound wave, the resolution may be very poor.

31.4 DOWNHOLE TECHNIQUES

The movement of water in the ground is related to the hydraulic driving force, expressed as head. The head is equivalent to the elevation of the water surface above a known reference. With the height of the water head measured, the movement of the water can be determined and modeled directly.

Water velocity is seldom if ever measured in a well bore. The reason for this is twofold. First the well bore represents a disturbance in the pattern of the soils, and it is conceivable that the water flowing through the well bore might select a preferential direction which may be in response to what is in another formation rather than what is in the same formation. This is known as piping.

If, for example, a site has two layers of sand separated by an aquaclude and the water in the sand is at different pressures, the opening of a hole through the aquaclude would permit the water in the aquifer to travel to the aquifer under lower pressure, despite the overall movement of the water through either or both formations. This can happen on a macroscale as well as on a microscale. To the extent that the hole in the formation represents a large void surrounded by disturbed soil, there is the high probability that the hole will disturb the formation and affect the water flow. In the construction of wells, the practice is encouraged by screening the well and packing the annular space between the well screen and the borehole with sand to help filter the solids and maintain the permeability of the well screen.

The second reason for not measuring water velocity downhole is the magnitude of the velocity in the formation. With the right formation, water may travel feet per day down to feet per year, and the measurement equipment must be extremely sensitive. The technology for these types of low-flow measurements barely exists for these low-velocity ranges, and many professionals are not convinced that the measurement of velocity in the borehole or well will accomplish anything or that the velocity can be measured accurately.

Downhole measurement techniques are done by hand. The elements most often measured downhole are depth to surface and thickness of contaminant layer (most frequently where there is free product oil floating on the surface). In all but a few cases, these measurements are made by hand or by using a specialized conductivity probe which is lowered on a wire and which can detect the difference between water and oil. On this simple 9-V battery-powered device, the wire is calibrated in 0.01 ft or 0.001 m intervals.

Another technique developed at Hanford Nuclear Laboratory uses a low-powered Doppler laser to shine down a well and reflect from a small-diameter float residing in the well. This technique was reported to be accurate within 0.001 ft or more, and it was much faster than the taping procedure normally used. It may have been commercialized, but as of yet, it has not been put into wide use. The advantage of the laser is the rapid calibration and the accuracy. The disadvantage is the expense and inconvenience of having a float in the well which may need to be retrieved if the well is to be sampled.

Ultrasonic techniques are too inaccurate for depth measurement in a well without a permanent probe suspended in the well above the water level. Water table measurements are usually to the closest 0.1 ft (3 mm), which defines the limit of the instrumentation accuracy requirements. Most sonic probes do not have the accuracy to detect the minute fluctuations in water surface over a long distance, and they must be hung within about 1 m of the surface to work with any accuracy. This often makes them subject to seasonal high water table flooding.

Downhole measurements are not performed frequently. Most groundwater contaminant measurements are made either monthly or quarterly. The infrequency of the measurements and the slow changes in groundwater elevations often make automatic instrumentation impractical. Considering what we have said about the frequency of the measurements, this gives us practical criteria for downhole instrumentation:

- *Inexpensive.* The instrument must be inexpensive. It must be cheaper than the cost of sending one person to make measurements at whatever frequency the well is to be monitored. The criterion might be on the order of 1/2 to 2 person-hours minimum per well to get the measurements.
- *Stability.* The instrument must be highly stable. Many monitors used in chemical testing have either some drift or a measurement rate problem. For example, a dissolved oxygen probe consumes a small amount of dissolved oxygen while it is running. This is not sufficient to affect a brief exposure to a sample in the laboratory, but it will deplete the dissolved oxygen levels in a well over a period of several months.
- *Accurate.* The instrument must be able to compete with laboratory analysis for accuracy.
- *Environmently Resistant.* The instrument must be able to withstand immersion in water, electrical shock, and/or an occasional grounding from a lightning strike in the area.

- *Corrosion Resistant.* Wells have bacteria which can grow in the environment. Thus, the instrumentation should be encased in materials which are resistant to biological attack.
- *Chemcial and Slime Resistant.* For some applications, the instrumentation might be used to detect the presence of gasoline and oil or chlorinated organics in the well. Both materials can attack the instrument coatings or coat the instruments themselves.
- *Easy to Remove and Repair.* The reasons are self evident.

31.5 OTHER MEASUREMENT INSTRUMENTATION

Other groundwater measurement instrumentation is principally used in the development of a groundwater recovery situation or a groundwater or subsurface contamination remediation project. These instruments include pressure transmitters, flow measurement devices, and level measurement devices. Sometimes they have multiple uses and multiple applications with regard to the development of the process; at other times they are used for primary measurement for contamination removal. We will consider two cases of the types of processes where these devices are employed: (1) the removal of groundwater and (2) vacuum extraction for contamination remediation.

The purpose is not to provide a complete treatment train for removal or contaminated groundwater or a definitive measurement technique for the most common applications.

31.5.1 Flowmeters

Flowmeters come in many shapes and sizes, depending upon the fluid being measured. Different types of meters are used for water than for air. The most common water meter is the positive-displacement meter. This meter employs a rotating disk (Fig. 31.6), paddle wheel, turbine, or other device which measures the displacement of water through the meter. The meter is most commonly used where the water is of relatively high purity and does not contain substantial quantities of silt or solids. The most common example of this type of meter is the household water meter. The device is accurate to within 1% or less in most services and has high durability. For most applications, the openings at either end of the meter are smaller than the diameter of the pipe it is used with. The head losses are relatively low and generally amount to a few pounds per square inch pressure drop across the meter. The internal disk device is generally used in applications where the pipe sizes are less than 6 in. In very low flow applications, the meter often understates the flow because of the inherent friction in the rotating parts and the low velocity and low head losses through the device.

For larger flows the flowmeter design is often quite different. For clean-water applications, the use of propeller meters is common, although magmeters and Doppler velocity meters are also used. These flowmeters are full-pipe devices—the flow in the pipe must fill the pipe to obtain an accurate flow of the liquid. If the pipe is only partially filled, errors will result and the flowmeter will overreport the flow or may not work at all.

Figure 31.6 Disk water meter in operation. *Source*: http://www.bangorwater.org/images/singlejet.jpg.

31.5.2 Air Flow

Air is either blown into the groundwater and soil or vacuumed out of it to help accomplish remediation. Vacuum extraction is a very common remediation technique widely used for removal of volatile materials which are trapped above the groundwater in the vadose zone. The U.S. Army Corps (ACE) of Engineers has an excellent design manual on the subject which is available at http://www.usace. army.mil/inet/usace-docs/eng-manuals. If in situ stripping of the groundwater is to be used as a remediation technique, the ACE also has a manual on this technique at the same address. The reader is encouraged to consider these manuals because they are quite complete and will enable one to use these remediation options quite successfully.

Air is a compressible medium. The most often used methods of measuring air flow are by use of a venturi meter, an orifice meter, a caloric meter, or a variable-area meter. The venturi meter is a head loss device which measures the pressure drop across a known cross-sectional area. A venturi meter is shown in Figure 31.7.

31.6 OTHER TYPES OF GEOPHYSICAL MEASUREMENTS

Geologists use a wide variety of instruments to measure the properties of the formation in which they are interested. Much of this is done by downhole or well logging. The predominant use for well logging is in the petroleum exploration field, although geotechnical logging and environmental logging are coming into use more widely as their cost falls. The techniques are specialized so that they can be used even in holes which have casings. The purpose is to gain additional information about the formation and its characteristics. The most frequently used well logging technique is resistivity where electric potential is measured against depth. The method is inexpensive but cannot be used in a cased hole. Other techniques using gamma and neutron logging are substantially more expensive but can be used in a cased hole.

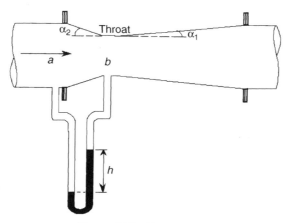

Figure 31.7   Venturi meter.

In Figure 31.7, the exit cone angle α_1 is between 2° and 8°, and the entrance cone angle α_2 may be between 10° and 12°. The head loss is a function of fluid density and flow, so the density of the fluid must be compensated for in the calculation of discharge.

More popular if slightly less accurate is the variable-area meter, or rotameter. This meter can be used for air or for clean-water flows (provided the device is manufactured for the particular service), and it relies on a carefully manufactured conical section tube and an equally carefully manufactured indicator float or even a ball of known diameter and density. A typical device is shown in Figure 31.8. Depending upon the design of the flowmeter, its use in conjunction with an orifice plate is not uncommon.

As shown in the figure, the fluid flowing around the ball creates a pressure on the underside of the ball, causing it to raise in the tube. The tube is calibrated with

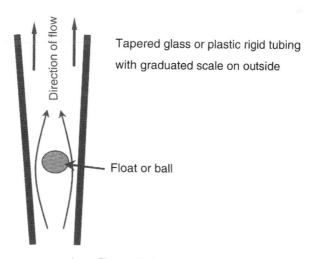

Figure 31.8   Rotameter.

markings which are generally read at the centerline of the ball in the tube. Increasing the flow through the tube increases the height of the ball in the tube.

This flowmeter is nonintegrating and can be used successfully for a fairly wide variety of flow measurements, provided accuracy is not paramount. Where the flow is pulsating or fluctuating, the device is not recommended for use.

31.7   LEVEL MEASUREMENT

Earlier in this chapter we indicated that level measurement is not generally used in well conditions. The exception to this is where there is production from the well. Static well measurements are seldom accomplished by instrumentation. Dynamic well measurements are frequently accomplished, but in specific situations, such as in product recovery wells involving pump-and-treat-product removal systems.

In a pump-and-treat product removal system, one generally has a free product layer (petroleum) on the surface of the water table. If the layer is relatively thin, one way to thicken it is to draw down the well and increase the apparent thickness of the material. This requires a two-pump system with careful control of the fluid level. It is shown in Figure 31.9 without the level sensors to simplify the drawing.

This application requires close monitoring of the levels in the groundwater if one is to avoid burning out the product recovery pump. Seasonal adjustment of the levels will also be required.

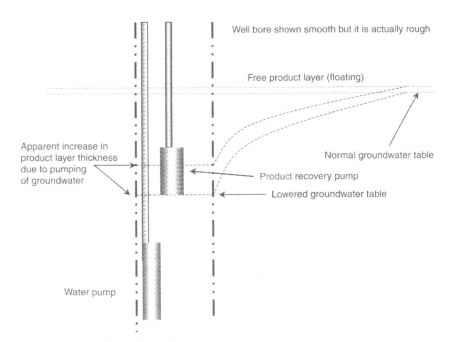

Figure 31.9   Pump-and-treat product removal system.

MICROBIOLOGICAL FIELD SAMPLING AND INSTRUMENTATION IN ASSESSMENT OF SOIL AND GROUNDWATER POLLUTION

Ann Azadpour-Keeley

U.S. Environmental Protection Agency
Ada, Oklahoma

Environmental Instrumentation and Analysis Handbook, by Randy D. Down and Jay H. Lehr
ISBN 0-471-46354-X Copyright © 2005 John Wiley & Sons, Inc.

32.1 INTRODUCTION

As groundwater serves as the source of drinking water for roughly half of the domestic needs of the country, 80% of irrigation demands, and the perennial flow of many streams and springs, it is also the recipient of natural and anthropogenic contaminants. Almost three decades ago the U.S. Environmental Protection Agency (EPA) initiated a program to prioritize environmental contaminants in the states according to their potential hazard to human heath and the environment. The resulting list included over 30 anthropogenic chemicals along with ionizing radiation. Although the biological contamination of drinking water[1] was mentioned categorically, no specific microorganisms were included.

Today, groundwater investigations are predominantly associated with chlorinated solvents and other industrial chemicals, agricultural products including pesticides and fertilizers, petrochemical products such as BTEX (benzene, toluene, ethylbenzene, xylene) and MTBE, and various metals. Although unquestionably important, these causes of risk to human health[2,3] pale in comparison to the 740 million cases of disease, 200,000 deaths, and costs of $17 billion in direct health care that occurs each year in the United States as a result of pathogenic diseases.[4]

The purposes for conducting field investigations concerning microorganisms are as varied as they are numerous. Generally these activities focus on the identification and enumeration of microorganisms either directly or indirectly, delineating conditions that encourage their proliferation or demise, determine their ability to degrade soil and groundwater contaminants or otherwise affect their transport and fate, contribute to the speciation and transport of metals, and design activities for the remediation of soil and groundwater.

Although water-transmitted human pathogens include various bacteria, protozoa, helminths, and viruses,[5] agents of major threat to human health are pathogenic protozoa (*Cryptosporidium* and *Giardia*) and enteroviruses.[6-8] In addition to having an interest in microorganisms with regard to health,[9,10] soils and groundwater investigations also include the use of indigenous or injected organisms for the purpose of contaminant degradation.

There are a number of avenues available for the introduction of microorganisms to the subsurface, among which are the land disposal of untreated and treated wastewater, land spreading of sludge, septic tanks and sewer lines, landfill leachates, irrigation with treated wastewater, and farm animals.[11,12]

The purpose of this chapter is to discuss the more prevalent reasons for microbiologically oriented soil and groundwater field-scale investigations, enumerate sample activities required to accomplish project objectives, focus on those samples that can successfully be accomplished in the field, discuss those procedures, provide information relative to their positive as well as negative attributes, and interpret sampling results.

32.2 SUBSURFACE MICROBIAL COMPOSITION AND DIVERSITY

Microbes are ubiquitous in the lithosphere, hydrosphere, and, to a greater or lesser extent, the atmosphere. Microorganisms inhabiting the subsurface environment[13] exhibit a remarkable array of metabolic capabilities enabling them to use organic or inorganic matter as energy sources and propagate under aerobic or anaerobic conditions. A large part of a microorganism's metabolism is devoted to the generation of adenosine triphosphate (ATP), an "energy currency" that is used by the microorganism for cell synthesis. Microorganisms derive energy through the oxidation of organic compounds or chemically reduced inorganic ions and compounds. The electrons or hydrogen atoms resulting from oxidation are transferred in most microorganisms by an electron transport chain to an electron acceptor, which in the case of aerobic respiration is oxygen. Some microorganisms are capable of anaerobic respiration whereby the electron acceptor is not O_2 but chemically reducible inorganic compounds such as NO_3^-, SO_4^{2-}, CO_2, Fe^{3+}, ferric oxide, or manganese oxide. A third type of metabolism, called *fermentation*, involves intramolecular oxidation/reduction without an externally supplied terminal electron acceptor. Fermentation reactions are not considered here because they are not significantly involved in subsurface transformation processes.

The metabolic diversity manifested by microorganisms is key to their being agents of geochemical change in the subsurface.[14] Subsurface biota and their metabolic characteristics are usually categorized in two major groups:

- *Heterotrophic Microorganisms.* Heterotrophs include bacteria (single- or multi celled organisms lacking internal membrane structures), fungi (mycelial or single-celled organisms possessing a cell wall but no photosynthetic capability), protozoa (unicellular, microscopic animals such as amoebae), and archaebacteria (a new group of organisms possessing a unique cell envelope, membrane structure and ribosomal RNA, distinguishing them from all other microorganisms). Heterotrophs derive energy from the oxidation of organic compounds and obtain most of their carbon from organics. Some heterotrophs respire aerobically while others respire anaerobically.

- *Chemolithotrophic Microorganisms.* Some genera of bacteria and archaebacteria are chemolithotrophs. These microorganisms derive energy from the oxidation of inorganic compounds (e.g., Fe^{2+}, S^{2-}, and H_2) and obtain their carbon as CO_3^{2-}, HCO_3^- or CO_2. Some chemolithotrophs are aerobic using O_2 as an electron acceptor while others respire anaerobically using chemically reducible inorganic matter as an electron acceptor. An example of this type of metabolism is carried out by methanogens that oxidize H_2 using CO_2 as an electron acceptor to form CH_4. Reactions carried out by chemolithotrophic microorganisms must be thermodynamically possible and must derive sufficient energy from oxidation–reduction reactions to fix carbon (i.e., reduce CO_3^{2-}, HCO_3^-, or CO_2 to organic carbon).

Photosynthetic microorganisms do not inhabit the subsurface, except for the topmost region of the soil layer, because these organisms derive their energy from sunlight.

Microbes functioning in inorganic conversions in the subsurface, that is, heterotrophs using inorganic material as an electron acceptor under anaerobic conditions and chemolithotrophic microorganisms, may be viewed as geological agents responsible for the concentration, dispersion, or fractionation of matter.[15–18]

32.2.1   Microbiological and Molecular Microbial Techniques

Analyses focused on the composition and diversity of bacterial community structures cannot rely on traditional microbiological procedures alone. This is especially critical in subsurface ecological systems because the vast majority of the microbial communities (>90%) that reside in that environment have not been cultivated using culture-dependent methods.[19] Although the importance of emerging molecular approaches in subsurface microbiology will be stressed later in this chapter, a successful program should be conducted using a combination of culture-independent methods as well as traditional cultivation strategies. Often, molecular tools are applied to pure culture isolates harvested from defined media. In other words, traditional selective and enrichment techniques are used to develop specific microbial communities to be further characterized with molecular monitoring [i.e., genes coding for 16S recombinant RNAs (rRNAs) (16S rDNA) to identify potential gene expressions].

Since standard microbiological techniques are readily available in most environmental laboratories, only a brief discussion of them is provided here:

1. *The most-probable-number (MPN)* technique (enumeration, culture) is a standard methodology used to estimate the number of specific physiological types of bacteria. Usually, a modified basal medium is amended with a carbon substrate and electron acceptors for total heterotrophic aerobes or anaerobes, denitrifying, iron-reducing, sulfate-reducing, and methanogenic bacteria.[20–23] A three-tube dilution series is often used to provide a 10-fold dilution. The same regimen is applied to semisolid media to provide plate counts.

2. *The acridine orange direct count (AODC)* (enumeration, morphological) is the most versatile technique for yielding a count of the total intact cells without differentiation for viability.[24] Differential staining for live versus dead organisms can be used via the Bac-light.[25]

3. *The phospholipid ester-linked fatty acid (PLFA)* technique (enumeration, biochemical) is a popular assay used as "biomarker" to provide a quantitative insight

into three important attributes of microbial communities: viable biomass, community structure, and metabolic activity.[26] An estimation of nonviable populations can be accompanied through the measurement of diglyceride fatty acids (DGFAs). Both assays are independent of the bias inherent in classical culturing techniques providing a more accurate estimation of in situ microbial populations. The lipid biomarker analysis is, however, incapable of identifying every microbial species in an environmental sample because many species contain overlapping PFLAs.

4. *Recombinant RNA* (enumeration) can be used for the determination of microbial biomass.[25]

5. *Microcosms* (physiological) are routinely prepared using subsurface cores and groundwater samples for the characterization of microbial communities. Recently, there has been an increased interest in the use of microcosms to perform molecular community fingerprinting such as denaturing gradient gel electrophoresis (DGGE) due to the generation of a sufficient cell mass.

6. *A community-level physiological profile (CLPP)* (physiological) can be carried out using Biolog-GN plates (Biolog, Hayward, CA). This "phenotypic fingerprinting" assay is useful for screening bacterial isolates and consortia to establish correlations between their activity and composition.

7. *Molecular microbial characterization* includes a variety of culture-independent genetic analyses that have been developed within the last decade to complement traditional culture-dependent methods (enrichment and isolation). Many of these molecular biological methods rely on 16S rDNA sequences, including in situ hybridization,[19,27] direct amplification of 16S rDNA, and additional analysis using community "DNA fingerprinting" such as temperature gradient gel electrophoresis (TGGE),[28] DGGE,[29] restriction fragment length polymorphism (RFLP),[30] single-strand conformation polymorphism (SSCP),[31] terminal RFLP,[32] or 16S rDNA cloning sequencing.[33] To date, most of the results obtained by using molecular techniques have been provided by cloning sequencing of 16S rRNA fragments. Although 16S rRNA cloning sequencing is successful on the reconnaissance of microbial diversity by detecting infrequent sequences from various habitats, thereby avoiding limitations of traditional cultivation techniques, it is time consuming and problematic for multiple sample analysis. Since the cloning approach cannot provide an immediate overview of the community structure, many environmental laboratories apply DGGE to detect population shifts. DGGE can be used for simultaneous analysis of multiple samples obtained at various time intervals to detect microbial community changes—an advantageous feature in studying microbial ecology and monitored natural attenuation (MNA).

Although the majority of the techniques used for the isolation and characterization of microorganisms such as those described above are laboratory based, the most crucial aspect of the microbiological investigation is the collection of the samples and those geochemical parameters that are auxiliary components of biotic investigations.

32.3 APPLICATIONS AND ISSUES IN SUBSURFACE ASSESSMENTS

Until recent years groundwater investigations were predominately focused on groundwater supplies with small attention given to groundwater quality.[34] As the result of an ever-increasing reliance of Americans on groundwater as a source of drinking water, agriculture, and industry, the field of subsurface microbiology was almost entirely confined to groundwater quality determination and restoration. At one time there was a general consensus that 95% or more of the bacteria and viral etiology that contaminated water could be absorbed or retained on soil through filtration. This notion has been refuted by continuing outbreaks of waterborne diseases.

Groundwater field sampling and analysis have gained considerable momentum because of recent legislation mandating microbiological sampling. The development of the Ground Water Disinfection Rule (GWDR) to meet Safe Drinking Water Act requirements began in 1987 and led to a published discussion piece (draft GWDR[35]). The deadline for the GWDR proposal was dependent upon completion of studies of the status of public health with respect to the microbial contamination of groundwater, based on studies[36–39] to generate a more descriptive nationwide picture of the problem. As there are significant differences between groundwater and surface water in terms of the type and degree of treatment, the GWDR regulatory work group realized that the assessment of vulnerability as a function of site-specific conditions (i.e., hydrogeology, land use pattern) was a key element to be addressed[48] Subsequently, on May 10, 2000, the "U.S. EPA proposed to require a targeted risk-based regulatory strategy for all ground-water systems addressing risks through a multiple barrier approach that relies on five major components: periodic sanitary surveys of ground-water systems requiring the evaluation of eight elements and the identification of significant deficiencies; hydrogeological assessments to identify wells sensitive to fecal contamination; source water monitoring for systems drawing from sensitive wells without treatment or with other indications of risk; a requirement for correction of significant deficiencies and fecal contamination, and compliance monitoring to insure disinfection treatment is reliably operated where it is used" (Ref. 41).

Therefore, current rule-making considerations for the upcoming Ground Water Rule entail a sampling program requirement. Some of the discrepancies in the reported results of these investigations may also be attributed to physical heterogeneity[42] and the earlier methods used for the detection and concentration of enteric viruses, which were usually less than 50% efficient.[43] The use of the reliable current methodologies (i.e., molecular techniques) can, however, minimize the variance between reported and actual numbers. Practices designed to ensure compliance with drinking water standards might more appropriately rely on a cadre of multi-disciplinary approaches, including geological settings that result in viral retention and microbial sampling and analysis.

An area of interest is the fate of pathogens in subsurface systems, particularly the transport of pathogenic bacteria, protozoa, and viruses through soils.[44] Once pathogens reach aquifer material, their movement and the pattern of survival[45] require characterization. The ability to determine travel distances[46,47] and survival times of viruses[48] in the subsurface is crucial for regulatory agencies attempting to maintain sources of contamination at sufficient distances from sources of drinking water.[11,49,50] Therefore, the selection of suitable indicators[51] relating to the transport of microbes in subsurface environments requires investigation. In this regard, microbial source tracking in the determination of groundwater contamination with fecal wastes is a rapidly emerging area of research. Analysis involving virus transport[52,53] diverges into two routes: selection of antibiotic-resistant populations and molecular biological techniques (e.g., ribotyping).

Since the analysis of subsurface communities has become an important area in microbial ecology, especially with respect to their mutual interactions as well as that with their environment, it is necessary to utilize novel sampling techniques followed by analyses as to culture, microscopy, biochemistry, and molecular biology. The characterization of subsurface microbial populations is an exciting area of research with respect to natural and engineered bioremedial systems. Many transformations of organic compounds and metals are carried out by subsurface microorganisms for the purpose of obtaining energy for growth and reproduction.

The use of microbiological processes in degradation of organic contaminates are environmentally positive since they often lead to the reduction of contaminant mass rather than their transfer from one medium to another, are less intrusive to the environment, and are cost effective. The characterization and monitoring of indigenous bacterial populations for biodegradation of organic compounds (e.g., petroleum hydrocarbons and chlorinated solvents) through MNA have become conventional remedial options nationwide. The use of biotic processes for the immobilization and precipitation of heavy metals is another new area of research. Two basic questions relating to the natural attenuation of inorganic contaminants in the subsurface are the level of activity of microorganisms and how rapidly the processes occur. Natural bioattenuation rates of organic contaminants in the vadose zone are relatively well defined; however, those for inorganic ions and compounds are an emerging science. Based on sampling and analysis studies, it is presumed that a diverse and dense microbial population capable of carrying out a variety of inorganic transformations exists in the vadose zone. With an availability of oxygen the rates of reaction are probably rapid with soluble metals and metalloids: Many will precipitate or otherwise be immobilized by mechanisms described earlier.

Engineered approaches to bioremediation of various organic contaminants through cometabolism and/or bioaugmentation have become important in the restoration of contaminated subsurface environments. However, not all microbial transformations are carried out to obtain energy for growth and reproduction. In addition to the generation of energy, a second crucial task for microorganisms is to protect themselves against toxic substances. In doing so, they acquire complex detoxification strategies. Field sampling activities characterizing life-sustaining bacterial resistance and detoxification processes in highly contaminated environments have demonstrated phenomena of considerable ecological significance.

Examples are bacterial resistance to mercury and organomercurials isolated from Chesapeake Bay and high concentrations of mercury in Japan.

Another rapidly emerging field of subsurface microbiology deals with microbial kinetic studies. Most of what is known about reaction kinetics in the subsurface today is based on simulating subsurface environments in laboratory bioreactors where reactions that take place are assessed and reaction rates measured. However, even in these simulated environments it is difficult to evaluate microbial composition and activity without system disruption.[54,55] System modeling is now playing an increasingly important role in a more reliable understanding of microbial kinetics in the subsurface environment.[56,57]

It is not yet possible to study a natural subsurface system in place since characterizing microbial activity is severely hampered by a lack of technology. Current methodologies rely on conventional drilling and sampling. Not only are such methods costly, but they also introduce perturbations into the subsurface environment by introducing air, water, foreign material, and extraneous microorganisms, yielding inaccurate and confusing data. Such drilling and sampling techniques can also alter permeability and porosity characteristics in the immediate drilling and sampling area. Samples that are collected and analyzed do not, therefore, represent real-time processes. New, cost effective, in situ techniques, such as miniature, nonevasive, high-resolution devices to measure real-time processes, are desperately needed.[55,58]

32.4 SOIL SAMPLING

Fertile soil is inhabited by root systems of higher plants, many animal forms, and large numbers of diverse microbiota, including bacteria, algae, fungi, protozoa, and actinomycetes that are in close association with soil particles, especially clay–organic matter complexes.[59] The bacterial composition of soil exceeds the population of all other groups in both number and variety. Heterotrophic bacteria are normally present in all soil systems in large numbers and can be used as a screening tool for assessing the microbial abundance of soil environments. Plate counts of less than 1000 colony-forming units (CFU)/gram, for example, are an indicator of toxic concentrations of inorganic or organic contaminants or depletion of TEA or other essential nutrients.

One of the most important steps in the design of a soil-sampling program is to realize that soil systems are extremely heterogeneous in their physical, chemical, and microbial content at both microscale (micrometer to millimeter) and macroscale (meter) levels. Although soil is a complex environment by the soil solid phase and varying dramatically with location and climate, the samples collected are expected to render information that is representative of the entire field being sampled. Therefore, in dealing with this heterogeneity and consequent variability, a sampling strategy should identify the precision required of the data to meet the objectives of the study. In this respect, a statistical evaluation of the variability is paramount[60]; therefore, geostatistical tools should be applied in determining how to sample in order to describe population distributions.[61] Representativeness and/or randomness of sampling are essential in making statistical comparisons and characterizing population fluctuations.[62]

Site characterization[63,64] data should be considered in the design of the field sampling and should include topography, soil type, climate and seasonal fluctuations, vegetation variability, slope, and the presence of creeks and streams. The site history should also be established since historical land use and management practices can significantly affect the soil microbial population. In the United States, invaluable site characterization data may be acquired form various governmental entities, including the Soil Science Society of America, U.S. Geological Survey topographic maps, and U.S. Department of Agriculture soil survey maps. The standards for soil sampling have been developed by the International Organization for Standardization (ISO).[65–69]

Other soil parameters that should be collected in conjunction with the microbial samples include intrinsic permeability, soil structure and stratification, soil pH, moisture content, soil temperature, and nutrient concentrations. In the event that a site is under evaluation for the application of in situ remedial technologies, sampling parameters also include the concentration and toxicity of contaminants. In general, bulk soil samples are obtained from the field site for microbial analyses. However, depending on soil heterogeneity, discrete sampling is performed when a depth profile is required. In this case, soil cores or smaller homogenized samples are collected. Although the majority of the microbiological assays require small- (up to 100-g) to medium- (100-g to several kilogram) size sample, depending on statistical considerations, data quality objective, and microbial assay requirements, large samples (over several kilograms) can be acquired. Large samples are obtained when undisturbed soil cores are required for certain industrial applications. While small- to medium-size samples can be obtained for each soil horizon by using presterilized tools (i.e., hand auger, sample cores, spade, shovel, or trowel), sampling of larger volume requires drilling and boring equipment. Soil samples are commonly processed as quickly as possible onsite, placed in thin-walled plastic bags (Whirl-Pak), shielded from direct sunlight, and transported at 4°C) to a laboratory.

32.5 SUBSURFACE SAMPLING

Knowledge of microbiological processes in the subsurface environment is especially important since microbial activity may be a major factor in controlling the fate and transport of pollutants. Although the major soil microbial activity in the processes of biogeochemical cycling of various elements (i.e., carbon mineralization, nitrogen fixation, nitrification and denitrification, and sulfur oxidation) is well established, less is known about the microbiology of groundwater. Therefore, the impact of human activities in the assessment of groundwater quality will be the focus of this chapter.

A comprehensive microbiological field sampling program should also incorporate hydrological investigations, water quality measurements, and chemical sampling. Therefore, the services of a hydrogeologist in the design of a field sampling investigation is invaluable.

32.5.1 Hydrogeology

Adequate spatial and temporal monitoring is one of the most important facets in proving that intrinsic or enhanced bioremediation is occurring in the subsurface. Recent technological advances have provided more efficient field testing methodologies and equipment that are reliable, rapid, and cost-effective for application in various hydrogeological settings. These include field screening techniques, geophysical, portable analytical, and computerized hydrologic data acquisition systems, lighter drilling and boring tools, multilevel samplers, and hydraulic "push" technologies. These advances have clearly demonstrated that conventional monitoring wells should not be used as the primary or sole source of hydrogeochemical and microbiological samples because it has been determined, through years of research, that a significant portion of both organic contaminants and microbial particles are predominantly associated with the aquifer matrix.

A major problem with monitoring wells is that a sample from an uncontaminated portion of the aquifer may be a composite of contaminated water from a contaminated plume that is drawn into the cone of depression along with clean water from the aquifer. This will result in an apparent increase in contamination. The reverse is also often possible if the monitoring well is pumped for an extended period prior to sampling; the amount of clean water coming into the well in relation to contaminated water can result in an apparent decrease in the concentration of contaminants of interest. These problems are often exacerbated by using well screens that are too long (greater than 2 m), inconsistent screened intervals, and inappropriate sampling methods.[70]

Pumping wells other than those designed for monitoring (such as those in an interdiction field or water supply wells for municipalities or irrigation) may influence the movement of a plume as well as flow lines within an aquifer. Depending on when these wells were designed and constructed, how they are pumped, and how they affect plume movement, apparent decreases in concentration may be observed in monitoring wells.[71–73] Monitoring programs should be designed such that these concerns are taken into account.[74] A good monitoring program will require sampling monitoring wells that are completed in the plume as well as monitoring wells outside of the contaminated zone in order to establish background conditions. The number and location of monitoring wells are determined not only by plume geometry and groundwater flow,[75] but also by the degree of confidence required to statistically demonstrate that natural attenuation is taking place, to estimate the rate of attenuation processes, and to predict the time required to meet established remediation goals.

32.5.1.1 Design and Installation of Monitoring Well Network. The overall objectives of a sampling program determine how the sampling points are located and installed and the choice of procedures used for sampling. The installation of

monitoring wells[76] to adequately identify contaminant concentrations is of paramount importance in determining the overall contribution of microbial processes to a reduction in either the concentration or mass of contamination. Monitoring wells that are to be used to determine the contribution of natural bioremediation to site cleanup cannot be located until sufficient knowledge of the aquifer system is obtained.[74] This includes depth to water table, hydraulic conductivity[77] and gradient, direction of groundwater flow, storage coefficient or specific yield, vertical and horizontal conductivity distribution, direction of plume movement, and the effects of any man-made or natural influences (i.e., lagoons or seeps) on the aquifer system. It is also important to determine if the hydraulic gradient is affected seasonally. Sentinel well screen depth and length are also important. Often it is advantageous to use short screens to minimize averaging of vertical water quality differences.[71]

The location, number,[78,79] and other pertinent data regarding monitoring wells[80–82] for the evaluation of MNA should be determined on a site-specific basis. The design of the monitoring network will be determined by the size of the plume, site complexity, source strength, groundwater–subsurface water interactions, distance to receptors, and the confidence limits each involved party wishes to place in the data obtained. The denser the monitoring network, the greater the degree of confidence one may place in the data. The wells should be capable of monitoring singular flow paths within a plume's course and subsequent movement of contaminants along these flow paths. One way to determine natural attenuation is to determine the concentration of appropriate parameters at one location and sample the same volume of water for the same parameters at some distance down gradient. It is generally impractical to monitor flow paths within a plume with the exception of the plume axis, which is the only flow line that can be located with any reasonable level of certainty.

Problems arise because monitoring wells even a few feet apart can differ significantly in observed concentrations of contaminants. It has been suggested that internal tracers such as a biologically recalcitrant compound can increase the confidence one might place in this method. One simply uses the difference between conservative tracer concentrations at the up-gradient and down-gradient points to measure the observed loss that can be attributed to other factors such as dispersion. To place even greater confidence on data, the time frame for analysis must be sufficient to allow for differences in the subsurface mobility of various constituents of a chemical waste as they will not arrive at the down-gradient monitoring point at the same time.

Often hazardous waste sites are monitored using wells that have previously been installed, especially if plume equilibrium is assumed. This in itself presents a number of problems, especially if the wells were installed without adequate knowledge of the subsurface and plume movement. If it can be demonstrated that contamination exists between these wells, by using a tracer study, for example, they can be used as appropriate monitoring points. Reinhard and Goodman[83] used chloride as a reference to investigate the behavior of trace organic compounds in leachate plumes from two sanitary landfills. Kampbell et al.[84] described the use of trimethylbenzene to normalize changes in BTEX concentrations due to abiotic processes of dispersion, dilution, sorption, and volatilization. It is also necessary to determine what contribution physical processes between sequential monitoring wells have on an apparent reduction of chemicals. It may also be necessary to construct new monitoring wells that are offset to present wells or to perform borings near existing wells.

32.5.2 Microbiology

Hydrogeologic conditions are of overriding significance in designing and conducting subsurface microbiological sampling programs because the selection of equipment and the location of sampling points are necessarily subject to site-specific conditions. Additionally, methods of drilling, construction, and completing wells should be examined prior to initiation of field work when sampling preexisting wells since adequate measures may not have been used. It should be cautioned that well construction activities are by nature nonaseptic. For example, although mud rotary drilling rigs are advantageous for drilling water supply wells and petroleum reservoir wells, their use should be avoided for microbiological sampling. Because drilling fluids contain a variety of bacteria as well as various additives that can serve as a source of nutrients for the exogenous bacterial contaminates that are introduced in a well,[85] the use of mud rotary drilling rigs, if necessary, should be in conjunction with the incorporation of fluorescence dyes in the drilling mud for visual inspection when paring the core samples upon extraction.

Well casings used for microbiological sampling should be an inert material that have the least potential for affecting the sample. While steel and polyvinyl chloride (PVC) are used for production wells, ceramic tile, stainless steel, polystyrene, and Teflon have been used for monitoring wells. A thorough cleaning of well casing and screen materials with clean water followed by rinsing with distilled water or steam should be instituted prior to installation.

32.5.2.1 Well Purging and Sampling. Ideally, groundwater samples for microbiological purposes are collected from a well that has been carefully located, adequately constructed, and completed. After installation, the wells are developed by purging and mechanical surging following American Society for Testing and Materials (ASTM) D 5521, *Standard Guide for Development of Ground-Water Monitoring Wells in Granular Aquifers.* The primary limitations to the collection of representative ground water samples include disturbance of the water column above the screened interval, resuspension of settled solids at the base of the casing (e.g., high pumping rates), disturbance at the well screen during purging and sampling (e.g., high pumping rates), and introduction of atmospheric gases or degassing from the water (e.g., sample handling, transfer, vacuum from sampling device). Based on these considerations, low-flow sampling protocols should be employed for sampling groundwater for geochemical and microbiological investigations.

A considerable challenge for microbiologists is to overcome well effects. Therefore, purging monitoring wells for the purpose of obtaining representative samples is necessary since groundwater chemistry can be altered through contact with the atmosphere, well casing materials, screen, gravel pack, and surface seal. It should also be noted that wells provide various habitats for microbial colonization. For example, public supply, industrial, or irrigation wells that are constantly being pumped can provide a representative sample of water in the aquifer. For monitoring purposes at hazardous waste sites, purging for water parameter stability is required.

In some instances, however, due to flow limitations, sample volume restrictions, and the desire to obtain samples with as little disturbance as possible, it may not be possible to use an in-line water quality indicator to establish well stabilization. The purge for each well should ensure the removal of several well volumes while low flow rates should be used during both purging and sampling. Under the best of circumstances direct-push technology should be used to install monitoring points when flow limitations are anticipated to achieve minimum disturbance to the formation. In addition, no filter pack should be used so that the screen will be in direct contact with formation water. Also, during sampling, due to a limitation in the amount of water that can be collected, the sampling bottles must be filled in the order that water quality parameters such as ammonia, alkalinity, and dissolved gases are collected prior to the collection of volatile organic compounds (VOCs) and microbial samples.

32.5.2.2 *Sample-Type Selection.* With regard to the type of sample (core or water), core samples provide more information in defining the horizontal and vertical distribution of microbes even though they are intrinsically disruptive and prohibit repeated sampling at the same location. Groundwater samples, on the other hand, can be obtained from the same well repeatedly but may not quantitatively or qualitatively reflect conditions in the aquifer. Since there are uncertainties that mandate caution in the extrapolation of information obtained from either type of sample, a thorough characterization may require both water and core samples.

In studies addressing the origin and nature of subsurface microbes, an important consideration is the extent to which organisms cultured or manipulated in the laboratory represent the intrinsic microbial community. Therefore, microbiologists are challenged when collecting not only representative but also microbially uncontaminated samples. To minimize contamination, after a core is obtained using strict aseptic methods, care should be taken not to disturb or contaminate the sample. Processing should be performed as quickly as possible under anaerobic and aseptic conditions while in the field. Surface layers of the core should be scraped away using a sterile sampling device and discarded so that only the center of the core is packaged in doubled sterile sample bags. The portion used for DNA/PLFA analyses must be rapidly frozen with liquid nitrogen and stored at $-70°C$. Another portion should be flushed with inert gases (N_2 or argon), sealed in canning jars and placed inside cans containing oxygen-scavenging catalyst packets (Gas Pak; BBL, Franklin Lakes, NJ), and stored at $4°C$ for microbial counts and cultural techniques for analysis within 24 h. Frozen ($-70°C$) genomic DNA can be extracted from core sample using an UltraClean soil DNA kit (MoBio, Solana, CA). Analysis involving cultural techniques should incorporate the actual crushed core materials into the microbiological media since in aquifers soluble inorganic constituents tend to concentrate on mineral particles, particularly on clays, via chemical and physical adsorption mechanisms. The mineral particles with its adsorbed concentration of inorganic ions are prime sites for biofilm development. Biofilm development and microbial activity will be enhanced if the aquifer also has oxidizable organic matter present. Even in the absence of oxygen in the saturated zone, bioattenuation of organic compounds and metals should occur at a reasonable rate as long as a sufficient concentration of electron donors and electron acceptors is present.

Filtration of ground or surface water samples enables the concentration of microorganisms from dilute samples. The use of membrane filters (MFs) is the simplest and most widely used technology for in situ indicators of fecal pollution from freshwater and drinking water samples. The technique is based on the entrapment of microbial cells by a membrane filter (0.45 μm pore size) in the field followed by the placement of the MFs on an appropriate medium followed by an appropriate incubation period in the laboratory. Based on the same principle, U.S. Environmental Protection Agency (EPA) method 1623, an immunomagnetic assay, allows for the filtration of a large volume of sample (10 L) through a capsule for the entrapment of *Cryptosporidium* (Fig. 32.1). Although the detection of this enteric protozoa is commonly associated with surface water samples, recent detections of *Cryptosporidium* oocysts from groundwater samples[86] may be combined with environmental sampling for microbial source tracking.

Groundwater samples will be unfiltered for microbial counts and cultural techniques and filtered for DNA/PLFA analysis. Community analysis involving *DNA fingerprinting* requires ultrafiltration of groundwater (50–100 L) using inorganic (Anodisc, Whatman) filters (0.2 μm pore size). The filter is placed in a sterile bag and rapidly freeze dried with liquid nitrogen. Since PCR bias can be introduced, particularly during cell lysis and PCR amplification, a physical method

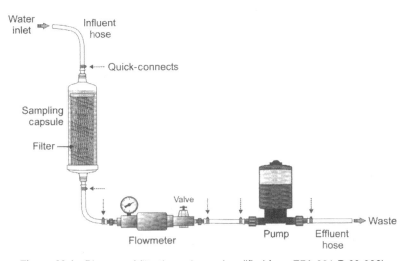

Figure 32.1   Diagram of filtration scheme (modified from EPA-821-R-99-006).

such as bead mill homogenization should be used to effectively lyse all cells types, including those that are most recalcitrant to physical and enzymatic treatments.[87]

32.5.3 Sampling Parameters

The diversity and numbers of microorganisms diminish with depth from the surface of the soil to rock strata below the aquifer. This finding is not altogether unexpected because with depth the environment becomes increasing hostile for life. At depth oxygen is rapidly depleted, temperatures and salinity may increase in certain locations, and the rock mass becomes increasingly impermeable. Therefore, it comes as no surprise that the environment is selective for those microorganisms that can oxidize chemically reduced inorganic substrates, H_2, or perhaps recalcitrant hydrocarbon compounds for energy and utilize oxidized inorganic ions such as SO_4^{2-}, NO_3^-, CO_2, Fe^{3+} and even oxidized mineral solids such as hematite (αFe_2O_3) and goethite ($\alpha FOOH$),[54,58] as electron acceptors. Therefore, sampling involving environmental microbiology should also include geochemical parameters, contaminant measurements, and hydrogeology in addition to microbial samples. The importance and relationship of these parameters are discussed in detail.

Oxidation–reduction (redox) reactions, which can be carried out by both heterotrophic and chemolithotrophic microorganisms, are either energy generating or enzymatically catalyzed but not coupled to energy generation. In theory, in any environment in which microbial activity occurs, there is a progression from aerobic to anaerobic (methanogenic) conditions. There is a definite sequence of electron acceptors used in this progression through distinctly different redox states. The rate, type of active microbial population, and level of activity under each of these environments are controlled by several factors. These include the concentration of the electron acceptors, substrates that can be utilized by the bacteria, and specific microbial populations leading to the progression of an aquifer from aerobic to methanogenic conditions.[89] This results in a loss of organic carbon and various electron acceptors from the system as well as a progression in the types and physiological activity of the indigenous bacteria.

If microbial activity is high, the aquifer environment would be expected to progress rapidly through these conditions. The following scenario outlines a general sequence of events in which aerobic metabolism of preferential carbon sources would occur first. The carbon source may be contaminants of interest or other more readily degradable carbon that has entered the system previously or simultaneously with the contamination event:

> *Oxygen-Reducing to Nitrate-Reducing Conditions.* Once available oxygen is consumed, active aerobic populations begin to shift to nitrate respiration. Denitrification will continue until available nitrate is depleted or usable carbon sources become limiting.
>
> *Nitrate-Reducing to Manganese-Reducing Conditions.* Once nitrate is depleted, populations that reduce manganese may become active. Bacterial

metabolism of substrates utilized by manganese-reducing populations will continue until the concentration of manganese oxide becomes limiting.

Manganese-Reducing to Iron-Reducing Conditions. When manganese oxide becomes limiting, iron reduction becomes the predominant reaction mechanism. Available evidence suggests that iron reduction does not occur until all Mn(IV) oxides are depleted. In addition, bacterial Mn(IV) respiration appears to be restricted to areas where sulfate is nearly or completely absent.

Iron-Reducing to Sulfate-Reducing Conditions. Iron reduction continues until substrate or carbon limitations allow sulfate-reducing bacteria to become active. Sulfate-reducing bacteria then dominate until usable carbon or sulfate limitations impede their activity.

Sulfate-Reducing to Methanogenic Conditions. Once usable carbon or sulfate limitations occur, methanogenic bacteria are able to dominate.

The ambient redox condition of the aquifer is important when determining the contribution of microbial degradation to MNA and engineered biotreatment mechanisms. In the case of petroleum hydrocarbons, because of their highly reduced condition, the preferred TEA for microbial processes would be oxygen.[90,91] From a thermodynamic standpoint, this is the most favorable reaction mechanism. When the soluble portion of petroleum hydrocarbons (BTEX) are the contaminants of concern, an inverse relationship between BTEX and dissolved oxygen concentrations within a plume is indicative of the microbial metabolism of these compounds as well as other hydrocarbons in the mixture.[92,93] Data available from various sites indicate that the natural attenuation of BTEX proceeds at higher rates under oxic conditions than normally achieved in anoxic environments, with rate constants ranging from 0.3 to 1.3% per day when modeled as a first-order process.[94–97] Although anaerobic biodegradation of toluene and xylene under nitrate-reducing,[98,99] iron-reducing,[100,101] sulfate-reducing,[102–105] and methanogenic[106–108] conditions have been extensively reported, until recently, unequivocal biodegradation of benzene under strict anaerobiosis was not demonstrated.[104,109] According to Borden et al.,[110] even though accurate description of anaerobic biodegradation of individual BTEX constituents may not follow a simple first-order decay function, biodegradation of total BTEX seems to more closely approximate a first-order decay function.

Biodegradation of chlorinated solvents, depending on the degree of halogenation, is fundamentally different from that of petroleum hydrocarbons and other oxidized chemicals.[111] The preferred redox conditions for the effective degradation of these chemicals is anaerobic (exception is vinyl chloride, VC). Effective degradation of these compounds may occur only when redox conditions are below nitrate reducing. Although under aerobic conditions cometabolism of TCE by autotrophic bacterial populations obtained from soil and groundwater has been demonstrated, it is generally accepted that this route of removal is limited only to low concentrations of the contaminant. The cometabolism of TCE proceeds in the presence of methane,[112–114] ammonia,[115] or toluene[116–118] as a cosubstrate.

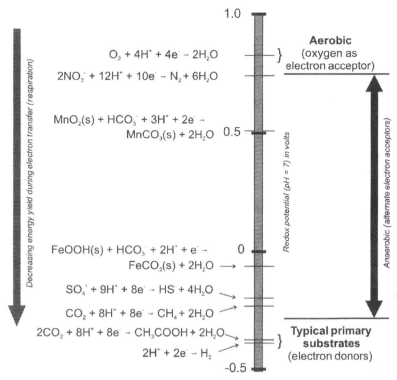

Figure 32.2 Important electron donors and acceptors in biotransformation processes. (Redox potentials data obtained from Ref. 144.)

The redox potential is a relatively simple and inexpensive indicator of the redox state of an aquifer. If the redox is positive, one can assume that dissolved oxygen is present and the system has not been stressed by biological activity.[119] If the redox potential is significantly negative, it can be assumed that processes favored under aerobic conditions (Fig. 32.2), such as BTEX degradation, are occurring at a substantially reduced rate. It should be stressed that one normally does not attempt to determine the actual redox conditions for comparison between different sites, rather the differences between points within a plume. Furthermore, due to the lack of internal equilibrium[120] and the mixed Eh potentials of natural aqueous systems, the use of any measured master Eh as an input in an equilibrium hydrogeochemical model for predicting the equilibrium chemistry of redox reactions is misleading. Instead, measuring certain sensitive species such as oxygen, Fe(II), hydrogen sulfide, or methane as qualitative guides to the redox status of the water may generate better results.[121]

In anoxic waters, where low pH and Eh exist, the reduced form of manganese, Mn(II), is favored.[122] Reduction of Fe(III) and Mn(IV), due to chemical processes or microbial metabolic reactions that couple the oxidation of organic matters to the reduction of these chemical species, has a major influence on the distribution of

Table 32.1 Range of Hydrogen Concentrations for Given Terminal Electron-Accepting Process

Terminal Electron-Accepting Process	Hydrogen (H_2) Concentration (nm/L)
Denitrification	<0.1
Iron (III) reduction	0.2–0.8
Sulfate reduction	1–4
Methanogenesis	5–20

Source: Ref. 128.

Fe(II) and Mn(IV) in aquatic sediments, submerged soil, and groundwaters.[21,123–126] Thus, measurable Fe(II) or Mn(II) may indicate suboxic conditions in the absence of detectable oxygen concentrations.

In addition to establishing background conditions away from the plume, dissolved oxygen, nitrate, manganese, iron, sulfate, and sulfide should be measured along the axis of the plume, as well as transverse to it, in order to characterize biological activity with respect to the redox state at those locations. This information will allow an estimation of the current redox state at various parts of the contaminated plume, thereby defining the types of reactions that may take place.

Another parameter that may be used to indicate the terminal electron acceptor process (TEAP) predominant in the areas of contamination is hydrogen (H_2) concentration.[127] Hydrogen concentrations for various terminal electron acceptors are given in Table 32.1.[128]

Parameters that investigators should analyze for petroleum hydrocarbons include dissolved oxygen, nitrate, Fe(II), sulfate, redox potential, pH, Mn(IV), dissolved methane, total petroleum hydrocarbons,[129] and nutrients (nitrogen, phosphorus, and other nutrients so as to not limit microbial growth). The majority of these parameters can be determined using field measurements, Hach kits, and/or CHEMetrics test kits. Since methane is produced after other TEAs (nitrate, iron, sulfate) are depleted, dissolved methane data are superior to contaminant data.[130] The parameters for chlorinated solvents may include temperature, redox potential, dissolved oxygen, sulfide, Fe(II), methane, ethane/ethene, alkalinity, pH, sulfate, nitrate, chloride, dissolved organic carbon, and hydrogen. Since Fe(III) may be dissolved from aquifer matrix, Fe(II) is measured as a proxy for biodegradation due to iron reduction. Although there may be a correlation between sediment redness and the hematite content of soil, when a soil sample represents a mixture of several iron species, the color is not a useful indicator.[131]

The microbial activities of a site are thus determined by the dissolved organic carbon, presence of macro- and micronutrients, and the TEA.[132] The presence and concentration of each will determine not only the activity but also the predominant population.

Different levels of quality assurance/quality control (QA/QC) may be required for those analyses determined in the field versus those performed under laboratory conditions.[133–136] For example, dissolved iron and oxygen, redox, and temperature must be determined on-site[137,138] using field test kits because of the deterioration that would normally occur between the time of sample collection and that of arrival at the laboratory. On the other hand, parameters such as metals, organics,[139,140] and bacteria can be properly preserved by cooling, capping, or chemical fixation and thereby subjected to a higher level of QA/QC.

32.6   FIELD INSTRUMENTS

TEMPERATURE. In order to classify microorganisms based on their temperature requirements, this parameter should be measured. For example, mesophiles are active within a range of 15–45°C. Temperature measurements can be recorded while purging with the flow-through cell; however, appurtenant tubing should be made as short as possible and the apparatus should be kept out of direct sunlight to avoid changes in temperature.

DISSOLVED OXYGEN (DO). A widely used DO meter for field investigations uses a membrane-covered electrode, although fouling is a common problem.[141] Some of the more popular instruments include the Orion 081010 electrode with the 810 meter, Orion 0083010 DO probe with model 1230 meter, and YSI 95 DO electrode and meter. It is important to note that several contaminants in groundwater have the potential to interfere with DO measurements, including hydrogen sulfide and certain organic compounds. Before use, the probes must be calibrated according to the manufacturer's recommendations and rechecked periodically.

pH. In field sampling, pH is often used to determine when pumped water has reached equilibrium with formation water because the parameter stabilizes quickly. Commonly selected meters for field use include the Orion 9107 BN low-maintenance triode or Orion 9107 WP pH electrode. It is necessary to perform a two-point calibration before sampling using, for example, a pH 4.00, pH 7.00, or pH 10.00 buffer depending on the expected range of the aquifer under investigation. It is also necessary to check the calibration periodically. Electrodes must be cleaned at the end of each sampling period and stored according to the manufacturer's recommendations.

OXIDATION–REDUCTION POTENTIAL (ORP). The ORP, measured in millivolts, is used to determine the oxidation or reduction state of the groundwater being sampled. A popular instrument is the Orion 9678 BN ORP. It is necessary to check the instrument before sampling and periodically thereafter using a standard, for example Orion 967961. Measured ORP values must be corrected according to temperature. Like DO and pH meters, redox meters have a large variance in cost, accuracy, and durability.

SPECIFIC CONDUCTANCE. Specific conductance can be used to determine when formation water is being pumped by its rapid stabilization. The meters must be calibrated before sampling and periodically thereafter using, for example, a Oakton standard 1413 μS at 25°C. The Orion 013010 conductivity cell with a model 1230 meter is a popular instrument.

TURBIDITY. A popular instrument for measuring turbidity is the Hach DR2100P portable turbidimeter, which measures in nephelometric turbidity units (NTU). The instrument must be calibrated before sampling and periodically during the sampling event.

SOIL MOISTURE. Although there are a number of direct and indirect methods for determining soil moisture content, one of the most widely used is the neutron probe. It is based on the property of hydrogen nuclei in water to alter neutrons emitted from a radioactive source. After calibration, the instrument can be used to directly determine the vertical distribution in an augured hole by converting meter readings to volumetric moisture content. The use of the equipment requires a radioisotope license.

HYDROGEN GAS ANALYSIS. Hydrogen gas measurements at monitoring wells are made after wells are pumped with a peristaltic pump at a rate of 250–300 mL/min using low-flow sampling procedures until geochemical parameters are stabilized.[143] Groundwater is then pumped through a 250-mL gas-sampling bulb leaving a gas pocket of approximately 20 mL volume. A 2-mL gas sample is collected from the gas pocket with a gas-tight syringe and injected into a RGA3 reduction gas analyzer (TRACE Analytical, Menlo Park, CA). Samples are collected and analyzed approximately every 5 min until equilibrium is established, and hydrogen concentrations are calculated from the equilibrium values.

32.7 FIELD/LABORATORY ANALYTICAL DESCRIPTION OF SECECTED PARAMETERS

CARBON DIOXIDE—STANDARD METHOD 4500-CO₂-C. In this titrimetric method, free carbon dioxide reacts with sodium carbonate or sodium hydroxide to form sodium bicarbonate. Completion of the reaction is indicated potentiometrically or by the development of the pink color characteristic of phenolphthalein indicator at the equivalence pH of 8.3.

ETHANE, ETHENE, AND METHANE—SW846 METHOD 8015B. This gas chromatography/flame ionization detection (GC/FID) method is used to determine the concentration of nonhalogenated volatile organic constituents. Samples will be introduced into the gas chromatograph by direct injection.

PHOSPHOLIPD FATTY ACID. Analysis of PLFA is based on the extraction of signature lipid biomarkers from the cell membranes and walls of microorganisms. Once extracted, lipids are concentrated and analyzed by GC/mass spectrometry (MS) for identification of each individual constituent. A profile of the fatty acids and other lipids is then used to determine the characteristics of the microbial community.

INORGANIC ANIONS—EPA METHOD 300.1 ION CHROMATOGRAPHY. The sample is introduced into an ion chromatograph, where the ions of interest (fluoride, bromide, chloride, nitrate, nitrite, orthophosphate, and sulfate) are separated and measured using a system comprised of a guard column, analytical column, suppressor device, and conductivity detector.

DISSOLVED ORGANIC CARBON—EPA METHOD 415.1. Following acidification, the sample is purged with nitrogen to remove inorganic carbon. Persulfate is injected to oxidize organic carbon to carbon dioxide that is detected by infrared (IR) spectroscopy.

SULFIDE—EPA METHOD 376.1. Excess iodine is added to a sample that may or may not have been treated with zinc acetate to produce zinc sulfide. The iodine oxidizes the sulfide to sulfur under acidic conditions. The excess iodine is back titrated with sodium thiosulfate or phenylarisne oxide.

IRON, MANGANESE, AND CATIONS—EPA METHOD 200.7. In this inductively coupled plasma–atomic emission spectroscopy (ICP-AES) method, prepared samples are nebulized and the resulting aerosol is transported to the plasma torch. Element-specific atomic-line emission spectra are produced by a radio-frequency inductively coupled plasma. The spectra are dispersed by a grating spectrometer, and the intensities of the lines are monitored.

METALS (AL, SB, BA, BE, CA CD, CR, CO, CU, FE, MG, MN, MO, NI, K, P, AG, NA, ZN, HG—EPA METHOD SW 846 3010/6010—AND MERCURY—CVAA 7470. These methods utilize ICP-AES. Groundwater samples (100 mL) for metal analyses are filtered at the field using a 0.2-μm membrane filter and preserved with HNO_3 using plastic containers.

32.8 SUMMARY

This chapter emphasizes the importance of microbiological sampling of soil and groundwater with respect to human heath risks, laws and regulations dealing with safe drinking water, and more prevalent subsurface monitoring activities associated with chlorinated organic compounds, petroleum hydrocarbons, agricultural chemicals, and metals. In highlighting the importance and difficulties of subsurface biotic sampling, the types of organisms of interest are discussed along with the processes that sustain their activation as well as the effects these processes exert on the speciation, transport, and fate of geochemical parameters and anthropogenic contaminants. In addition, an overview is provided of microbiological and molecular microbial techniques of characterizing the composition and diversity of community structures.

Subsurface sampling is addressed in considerable detail encompassing the role of hydrogeology in the design and installation of a soil and groundwater sampling program along with well purging and sampling techniques that are unique to microbiology as well as site-specific diversity. Field sampling is discussed in terms of general and specific parameters in support of microbiological investigations, and the field instrumentation commonly used in the pursuit of project goals is outlined. Finally, the interrelation of field and laboratory analyses is delineated.

REFERENCES

1. T. R. Deetz, E. M. Smith, S. M. Goyal, C. P. Gerba, J. J. Vollet, L. Tsai, H. L. Dupont, and B. H. Keswick, Occurrence of Rota- and Enteroviruses in Drinking and Environmental Water in a Developing Nation, *Water Res.* **18**, 567–571 (1984).

2. G. F. Craun, A Summary of Waterborne Illness Transmitted through Contaminated Groundwater, *J. Environ. Health* **48**, 122–127 (1985).

3. C. P. Gerba and J. B. Rose, Estimating Viral Disease Risk from Drinking Water, in C. R. Gerba (Ed.), *Comparative Environmental Risk Assessment*, Lewis Publishers, Boca Raton, FL, 1993, pp. 117–134.

4. Bureau of National Affairs, Industry Survey Reveals 79 Drugs in Development for Infectious Diseases, *BNA Health Care Policy Report* **2**(28), 1247 (1994).

5. Bull et al., 1990.

6. J. F. Schijven and S. M. Hassanizadeh, Removal of Viruses by Soil Passage: Overview of Modeling, Processes, and Parameters, *CRC Crit. Rev. Environ. Sci. Technol.* **30**(1), 49–127 (2000).

7. C. M. Hancock, J. B. Rose, and M. Callahan, *Crypto* and *Giardia* in US Groundwater, *J. Am. Water Works Assoc.* **90**(3), 58–61 (1998).

8. C. F. Brush, W. C. Giorse, L. J. Anguish, J. Y. Parlange, and H. G. Grimes, Transport of *Cryptsporium parvum* Oocyts through Saturated Columns, *J. Environ. Qual.* **28**, 809–815 (1999).

9. A. H. Havelaar, M. van Olphen, and Y. C. Drost, F-Specific RNA Bacteriophages Are Adequate Model Organisms for Enteric Viruses in Fresh Water, *Appl. Environ. Microbiol.* **59**(9), 2956 (1993).

10. T. Hirata, K. Kawamura, S. Sonoki, K. Hirata, M. Kaneko, and K. Taguchi, *Clostridium perfringens* as an Indicator Microorganism for the Evaluation of the Effect of Wastewater and Sludge Treatment Systems, *Water Sci. Technol.* **24**(2), 367 (1991).

11. B. H. Keswick, C. P. Gerba, S. L. Seco, R. and I. Cech, Survival of Enteric Viruses and Indicator Bacteria in Ground Water, *J. Environ. Sci. Health* **17**(6), 903–912 (1982).

12. C. P. Gerba and G. Bitton, Microbial Pollutants: Their Survival and Transport Pattern to Groundwater, in G. Bitton and C. P. Gerba (Eds.), *Groundwater Pollution Microbiology*, Wiley, New York, 1984, pp. 65–88.

13. R. A. Kerr, Deep Life in the Slow, Slow Lane, *Science* **296**, 1056 (2002). B. H. Keswick and C. P. Gerba, Viruses in Groundwater, *Environ. Sci. Technol.* **14**(11), 1290–1297.

14. H. L. Ehrlich, in H. L. Ehrlich (Ed.), *Geomicrobiology*, 3rd ed., Marcel Dekker, New York, 1995.

15. R. M. Atlas, *Microbiology. Fundamentals and Applications*, 2nd ed., Macmillan, New York, 1988.

16. H. L. Ehrlich, Bacterial Oxidation of Arsenopyrite and Enargite, *Econ. Geol.* **59**, 1306 (1964).

17. H. L. Ehrlich, in H. L. Ehrlich (Ed.), *Geomicrobiology*, Marcel Dekker, New York, 1990.

18. T. Hattori, *Microbial Life in the Soil*, Marcel Dekker, New York, 1973.

19. R. I. Amann, W. Ludwig, and K. H. Schleifer, Phylogenic Identification and *In Situ* Detection of Individual Microbial Cells Without Cultivation, *Microbiol. Rev.* **59**, 143–169 (1995).

20. P. M. Fedorak, K. M. Semple, and D. W. S. Westlake, A Statistical Comparison of Two Culturing Methods from Enumerating Sulfate-Reducing Bacteria, *J. Microbiol. Methods* **7**, 19–27 (1987).

21. D. R. Lovley, Dissimilatory Fe(III) and Mn(IV) Reduction, *Microbiol. Rev.* **55**, 259–287 (1991).

22. I. Mahne and J. M. Tiedji, Criteria and Methodology for Identifying Respiratory Denitrifies, *Appl. Environ. Microbiol.* **61**, 1110–1115 (1995).

23. R. S. Tanner, Monitoring Sulfate-Reducing Bacteria: Comparison of Enumeration Media. *J. Microbiol. Methods* **10**, 83–90 (1989).

24. W. C. Ghiorse and D. L. Balkwill, Enumeration and Morphological Characterization of Bacteria Indigenous to Subsurface Environments, *Dev. Ind. Microbiol.* **24**, 213–244 (1983).

25. D. Loyd and J. H. Anthony, Vigour, Vitality and Viability of Microorganisms, *FEMS Microbiol. Lett.* **133**, 1–7 (1995).

26. R. M. Lehman, F. S. Colwell, D. B. Ringelberg, and D. C. White, Combined Microbial Community-Level Analyses for Quality Assurance of Terrestrial Subsurface Cores, *Microbiol. Methods* **22**, 263–281 (1995).

27. R. I. Amann, L. Krumholz, and D. A. Stahl, Fluorescent-Oligonucleotide Probing of Whole Cells for Determinative, Phylogenetic, and Environmental Studies in Micro-biology, *J. Bacteriol.* **172**, 762–770 (1990).

28. A. Felske, A. D. Akkermans, and W. M. DeVos, Quantification of 16S rRNAs in Complex Bacterial Community by Multiple Competitive Reverse Transcription-PCR in Tem-perature Gradient Gel Electrophoresis Fingerprints, *Appl. Environ. Microbiol.* **64**, 4581–4587 (1998).

29. G. Muyzer, E. C. DeWaal, and A. G. Uitterlinden, Profiling of Complex Microbial Populations by Denaturing Gradient Gel Electrophoresis Analysis of Polymerase Chain Reaction-Amplified Genes Coding for 16S rRNA, *Appl. Environ. Microbiol.* **59**, 695–700 (1993).

30. Martinez-Murcis et al., 1995.

31. D. H. Lee, Y.-G. Zo, and S.-J. Kim, Nonradioactive Method to Study Genetic Profiles of Natural Bacterial Communities by PCR-Single Strand-Conformation Polymorphism, *Appl. Environ. Microbiol.* **62**, 3112–3120 (1996).

32. B. G. Clement, L. E. Kehl, K. L. DeBord, and C. L. Kitts, Terminal Restriction Fragment Patterns (TRFPs), a Rapid, PCR-Based Method for the Comparison of Complex Bacterial Communities, *J. Microbiol. Methods* **31**, 135–142 (1998).

33. M. G. Wise, J. V. McArthor, and L. J. Shimkets, Bacterial Diversity of a Carolina Bay as Determined by 16S rRNA Gene Analysis: Confirmation of Novel Taxa. *Appl. Environ.* (1997)

34. R. A. Freeze and J. A. Cherry, *Ground Water*, Prentice-Hall, Englewood Cliffs, NJ, 1979.

35. U.S. Environmental Protection Agency (EPA), *Draft Ground-Water Disinfection Rule*, U.S. EPA Office of Ground Water and Drinking Water, Washington, DC, 1992.

36. M. Abbaszadegan, P. W. Stewart, and M. W. LeChevallier, A Strategy for Detection of Viruses in Groundwater by PCR, *Appl. Environ. Microbiol.* **65**(2), 444–449 (1999).

37. M. Abbaszadegan, P. W. Stewart, M. W. LeChevallier, C. Rosen, S. Jeffery, and C. P. Gerba, Occurrence of Viruses in Ground Water in the United States, *AWWARF* (1999).

38. R. J. Lieberman, L. C. Shadix, B. S. Newport, M. W. N. Frebis, S. E. Moyer, R. S. Saferman, D. Lye, G. S. Fout, and D. Dahling, unpublished report, in preparation.

39. R. J. Lieberman, L. C. Shadix, B. S. Newport, S. R. Crout, S. E. Buescher, R. S. Saffrman, R. E. Stetler, D. Lye, G. S. Fout, and D. Dahling, Source Water Microbial Quality of Some Venerable Public Ground Water Supplies, *Proceedings of the Water Quality Technology Conference,* American Water Works Association, San Francisco, CA, 1994.

40. B. A. Macler, Developing the Ground Water Disinfection Rule. *J. Am. Water Works Assoc.* **3**, 47–55 (1996).

41. U.S. Environmental Protection Agency (EPA), 2000.

42. R. W. Harvey, N. E. Kinner, D. MacDonald, D. W. Metge, and A. Bunn, Role of Physical Heterogeneity in the Interpretation of Small-Scale Laboratory and Field Observations of Bacteria, Microbial-Sized Microsphere, and Bromide Transport through Aquifer Sediments, *Water Resour. Res.* **29**(8), 2713–2721 (1993).

43. C. P. Gerba, Microbial Contamination of the Subsurface, in C. H. Ward, W. Giger, and P. L. McCarty (Eds.), *Ground Water Quality*, Wiley & New York, 1985, Chapter 5.

44. K. Herbold-Paschke, U. Straub, T. Hahn, G. Teutsch, and K. Botzenhart, Behavior of Pathogenic Bacteria, Phages and Viruses in Groundwater during Transport and Adsorption. *Water Sci.Technol.* **24**(2), 301 (1991).

45. S. E. Dowd and S. D. Pillai, Survival and Transport of Selected Bacterial Pathogens and Indicator Viruses under Sandy Aquifer Conditions, *J. Environ. Sci. Health* **32**(8), 2245–2258 (1997).

46. C. J. Hurst, Modeling the Fate of Microorganisms in Water, Wastewater, and Soil, in C. J. hurst, C. R. Knudsen, M. J. McInerney, L. D. Stetzenbach, and M. V. Walter (Eds.), *Manual of Environmental Microbiology*, ASM Press, Washington, DC, 1997, pp. 213–221.

47. Y. Jin, M. V. Yates, S. S. Thompson, and W. A. Jury, Sorption of Virus during Flow through Saturated Sand Columns, *Environ. Sci. Technol.* **31**(2), 548–555 (1997).

48. Y. Chu, Y. Jin, M. Flury, and M. V. Yates, Mechanisms of Virus Removal during Transport in Unsaturated Porous Media, *Water Resour. Res.* **37**(2), 253–263 (2001).

49. B. H. Keswick, D. S. Wang, and C. P. Gerba, The Use of Microorganisms as Ground-Water Tracers: A Review, *Ground Water* **20**, 142–149 (1982).

50. M. Yates and S. Yates, A Comparison of Geostatistical Methods for Estimating Virus Inactivation Rates in Ground Water, *Wat. Res.* **21**(9), 1119–1125 (1987).

51. R. Cornax and M. A. Morinigo, Significance of Several Bacteriophage Groups as Indicator of Sewage Pollution in Marine Waters, *Water Res.* **25**(6), 673–678 (1991).

52. M. Y. Corapcioglu and A. Haridas, Transport and Fate of Microorganisms in Porous Media: A Theoretical Investigation, *J. Hydrol.* **72**, 149–169 (1984).

53. M. Y. Corapcioglu and A. L. Baehr, A Compositional Multiphase Model for Groundwater Contamination by Petroleum Products, 1. Theoretical Considerations, *Water Resour. Res.* **23**(1), 191–200 (1987).

54. D. K. Newman, How Bacteria Respire Minerals, *Science* **292**, 1312 (2001).

55. D. K. Newman and J. F. Banfield, Geomicrobiology: How Molecular-Scale Interactions Underpin Biogeochemical Systems, *Science*, **296**, 1071 (2002).

56. Q. Jin and C. M. Bethke, A New Rate Law Describing Microbial Respiration, *Appl. Environ. Microbiol.*, submitted for publication.

57. Q. Jin and C. M. Bethke, A New Rate Law Describing Microbial Respiration, *Appl. Environ. Microbiol.*, **69**(4), 2340 (2003).

58. W. C. Ghiorse, Subterranean Life, *Science*, **275**, 789 (1997).

59. R. C. Foster, Microenvironments of Soil Microorganisms, *Biol. Fertil. Soil* **6**, 189–203 (1988).

60. R. O. Gilbert, *Statistical Methods for Environmental Pollution Monitoring*, Van Nostrand Reinhold, New York, 1987.

61. E. H. Isaacs and R. M. Srivastava, *Applied Geostatistics*, Oxford University Press, Oxford, 1989.

62. Van Elsas and Smalla, 1997.

63. J. R. Boulding, *Subsurface Characterization and Monitoring Techniques, A Desk Reference Guide*, Vol. I: *Solids and Ground Water*, EPA/625/R-93/003a, U.S. Environmental Protection Agency, Washington, DC, 1993.

64. J. R. Boulding, *Subsurface Characterization and Monitoring Techniques, A Desk Reference Guide*, Vol. II: *The Vadose Zone, Field Screening and Analytical Methods*, EPA/ 625/R-93/003b, U.S. Environmental Protection Agency, Washington, DC, 1993.

65. International Organization for Standardization (ISO), *Soil-Quality—Sampling, Part 1: Guidance on the Design of Sampling Programmes*, ISO/CD 10381-1-1992, ISO, Geneva, 1992.

66. International Organization for Standardization (ISO), *Soil-Quality—Sampling, Part 2: Guidance on the Design of Sampling Programmes*, ISO/CD 10381-1-1992, ISO, Geneva, 1992.

67. International Organization for Standardization (ISO), *Soil-Quality — Sampling, Part 3: Guidance on the Design of Sampling Programmes*, ISO/CD 10381-1-1992, ISO, Geneva, 1992.

68. International Organization for Standardization (ISO), *Soil-Quality—Sampling, Part 4: Guidance on the Design of Sampling Programmes*, ISO/CD 10381-1-1992, ISO, Geneva, 1992.

69. International Organization for Standardization (ISO), *Soil-Quality—Sampling, Part 6: Guidance on the Design of Sampling Programmes*, ISO/CD 10381-1-1992, ISO, Geneva, 1992.

70. P. E. Church and G. E. Granato, Bias in Ground-Water Data Caused by Well-Bore Flow in Long-Screen Wells, *Ground Water* **34**(2), 262–273 (1996).

71. J. M. Martin-Hayden and G. A. Robbins, Plume Distortion and Apparent Attenuation Due to Concentration Averaging in Monitoring Wells, *Ground Water* **35**(2), 339–346 (1997).

72. J. M. Martin-Hayden, G. A. Robbins, and R. D. Bristol, Mass Balance Evaluation of Monitoring Well Purging: Part II. Field Tests at a Gasoline Contamination Site, *J. Contam. Hydrol.* **8**, 225–241 (1991).

73. G. A. Robbins and J. M. Martin-Hayden, Mass Balance Evaluation of Monitoring Well Purging: Part I. Theoretical Models and Implications for Representative Sampling, *J. Contam. Hydrol.* **8**, 203–224 (1991).

74. P. J. Zeeb, N. D. Durant, R. M. Cohen, and J. W. Mercer, *Site Investigation Approaches to Support Assessment of Natural Attenuation of Chlorinated Solvents in groundwater: Part 1, Sandy Aquifer*, Draft of Site Characterization Manual, U.S. Environmental Protection Agency, ORD, Washington, DC, 1999.

75. American Society for Testing and Materials (ASTM) *Standard Guide for Selection of Aquifer Test Method in Determination of Hydraulic Properties by Well Techniques*, Standard D-4043-91, ASTM, Philadelphia, PA, 1991.

76. L. Aller, T. W. Bennett, G. Hackett, R. J. Petty, J. H. Lehr, H. Sedoris, D. M. Nielsen, and J. E. Denne, *Handbook of Suggested Practices for the Design and Installation of Groundwater Monitoring Wells*, EPA/600/4-89/034, Environmental Monitoring Systems Laboratory, Las Vegas, NV, 1991.

77. F. J. Molz, G. K. Woman, S. C. Young, and W. R. Waldrop, Borehole Flowmeters: Field Application and Data Analysis, *J. Hydrol.* **163**, 347–371 (1994).

78. American Society for Testing and Materials (ASTM) *Standard Test Method for Measurement of Hydraulic Conductivity of Porous Material Using a Rigid-Wall Compaction-Mold Permeameter*, Standard D-5856-95, ASTM, Philadelphia, PA, 1995.

79. American Society for Testing and Materials (ASTM) *Standard Practice for Design and Installation of Ground Water Monitoring Wells in Aquifers*. Standard D-5092-90, ASTM, Philadelphia, PA, 1991.

80. U.S. Environmental Protection Agency (EPA), *Field Sampling and Analysis Technologies Matrix and Reference Guide*, EPA/542/B-98/002, EPA, Washington, DC, 1998.

81. American Society for Testing and Materials (ASTM), 1992.

82. American Society for Testing and Materials (ASTM), 1994.

83. M. Reinhard and N. L. Goodman, Occurrence and Distribution of Organic Chemicals in Two Landfill Leachate Plumes, *Environ. Sci. Technol.* **18**, 953–961 (1984).

84. D. H. Kampbell, T. H. Wiedemeier, and J. E. Hansen, Intrinsic Bioremediation of Fuel Contamination in Ground Water at a Field Site, *J. Hazard. Mater.* **49**, 197–204 (1995).

85. J. B. Davis, *Petroleum Microbiology*, Elsevier-North Holland, New York, 1967.

86. J. B. Rose, C. P. Gerba, and W. Jakubowski, Survey of Potable Water Supplies for *Cryptosporidium*, and *Giardia*, *Environ. Sci. Technol.* **25**(8), 1393–1400 (1991).

87. M. I. More, J. B. Herrick, M. C. Silva, W. C. Ghiorse, and E. L. Madsen, Quantitative Cell Lysis of Indigenous Microorganisms and Rapid Extraction of Microbial DNA from Sediment, *Appl. Environ. Microbiol.* **60**, 1572–1580 (1994).

88. S. K. Lower, M. F. Hochella, Jr., and T. J. Beveridge, Bacterial Recognition of Mineral Surfaces: Nanoscale Interactions between *Shewanella* and α-FeOOH, *Science*, **292**, 1360 (2001).

89. J. P. Salanitro, The Role of Bioattenuation in the Management of Aromatic Hydrocarbon Plumes in Aquifers, *Ground Water Monit. Rem.* **13**(4), 150–161 (1993).

90. R. A. Brown, R. E. Hinchee, R. D. Norris, and J. T. Wilson, Bioremediation of Petroleum Hydrocarbons: A Flexible, Variable Speed Technology, *Biorem. J.* **6**(4), 95–108 (1996).

91. D. Clark, D. Ours, and L. Hineline, Passive Remediation for Groundwater Impacted With Hydrocarbons, *Rem. Manage.* **1**, 50–45 (1997).

92. S. G. Donaldson, C. M. Miller, and W. W. Miller, Remediation of Gasoline-Contaminated Soil by Passive Volatilization, *J. Environ. Qual.* **21**, 94–102 (1992).

93. M. H. Huesemann and M. J. Truex, The Role of Oxygen Diffusion in Passive Bioremediation of Petroleum Contaminated Soils, *J. Hazard. Mater.* **51**, 93–113 (1996).

94. C. Y. Chiang, J. P. Salanitro, E. Y. Chai, J. D. Colthart, and C. L. Klein, Aerobic Biodegradation of Benzene, Toluene, and Xylene in a Sandy Aquifer: Data Analysis and Computer Modeling, *Ground Water* **27**(6), 823–834 (1989).

95. M. W. Kemblowski, J. P. Salanitro, G. M. Deeley, and C. C. Stanley, Fate and Transport of Dissolved Hydrocarbons in Groundwater—A Case Study, in Proceedings of the Petroleum Hydrocarbons and Organic Chemicals in Ground Water: *Prevention, Detection, and Restoration Conference*, Houston, TX, 17–19 November, NGWA, Dublin, OH, 1987, pp. 207–231.

96. J. P. Salanitro, *Criteria for Evaluating the Bioremediation of Aromatic Hydrocarbons in Aquifers*, Presented at the National Research Council (Water Science Technology Board) Committee on In-Situ Bioremediation: How Do We Know When It Works? Washington, DC, 26–29 October 1992.

97. P. M. McAllister and C. Y. Chiang, A Practical Approach to Evaluating Natural Attenuation of Contaminants in Ground Water, *Ground Water Monit. Rem.* **14**(2), 161–173 (1994).

98. J. R. Barbaro, J. F. Barker, L. A. Lemon, and C. I. Mayfield, Biotransformation of BTEX Under Anaerobic Denitrifying Conditions: Field and Laboratory Observations, *J. Contam. Hydrol.* **11**, 245–277 (1992).

99. R. J. Schocher, B. Seyfried, F. Vazquez, and J. Zeyer, Anaerobic Degradation of Toluene by Pure Cultures of Denitrifying Bacteria, *Arch. Microbiol.* **157**, 7–12 (1991).

100. D. R. Lovley, M. J. Baedecker, D. J. Lonergan, I. M. Cozzarelli, E. J. P. Phillips, and D. I. Siegel, Oxidation of Aromatic Contaminants Coupled to Microbial Iron Reduction, *Nature (London)* **339**, 297–299 (1989).

101. D. R. Lovley and D. J. Lonergan, Anaerobic Oxidation of Toluene, Phenol, and *p*-Creosol by the Dissimilatory Iron-Reducing Organism GS-15, *Appl. Environ. Microbiol.* **56**, 1858–1864 (1990).

102. H. R. Beller, D. Grbi-Gali, and M. Reinhard, Microbial Degradation of Toluene Under Sulfate-Reducing Conditions and the Influence of Iron on the Process, *Appl. Environ. Microbiol.* **58**(3), 786–793 (1992a).

103. H. R. Beller, M. Reinhard, and D. Grbi-Gali, Metabolic Byproducts of Anaerobic Toluene Degradation by Sulfate-Reducing Enrichment Cultures, *Appl. Environ. Microbiol.* **58**(9), 3192–3195 (1992b).

104. E. A. Edwards, and D. Grbi-Gali, Complete Mineralization of Benzene by Aquifer Microorganisms Under Strictly Anaerobic Conditions, *Appl. Environ. Microbiol.* **58**, 2663–2666 (1992).

105. R. Rabus, R. Nordhaus, W. Ludwig, and F. Widdel, Complete Oxidation of Toluene Under Strictly Anoxic Conditions by a new Sulfate-Reducing Bacterium, *Appl. Environ. Microbiol.* **59**, 1444–1451 (1993).

106. T. M. Vogel and D. Grbi-Gali, Incorporation of Oxygen From Water Into Toluene and Benzene During Anaerobic Fermentative Transformation, *Appl. Environ. Microbiol.* **52**, 200–202 (1986).

107. E. J. Bouwer and P. L. McCarty, Transformation of 1- and 2-Carbon Halogenated Aliphatic Organic Compounds Under Methanogenic Conditions, *Appl. Environ. Microbiol.* **45**(4), 1286–1294 (1983).

108. E. A. Edwards, and D. Grbi-Gali, Anaerobic Degradation of Toluene and *o*-Xylene by a Methanogenic Consortium, *Appl. Environ. Microbiol.* **60**, 313–322 (1994).

109. D. R. Lovley, J. C. Woodward, and F. H. Chapelle, Stimulated Anoxic Biodegradation of Aromatic Hydrocarbons Using Fe(III) Ligands, *Nature (London)* **370**, 128–131 (1994).

110. R. C. Borden, J. H. Melody, M. B. Shafer, and M. A. Barlaz, *Anaerobic Biodegradation of BTEX in Aquifer Material, EPA/600/S-97/003*, 1997.

111. T. H. Wiedemeier, J. T. Wilson, D. H. Kampbell, R. N. Miller, and J. E. Hansen, Technical Protocol for Implementing Intrinsic Remediation with Long-Term Monitoring for Natural Attenuation of Fuel Contamination Dissolved in Groundwater. United States Air Force Center for Environmental Excellence, Technology Transfer Division, Brooks Air Force Base, San Antonio, TX 1995.

112. M. M. Fogel, A. R. Taddeo, and S. Fogel, Biodegradation of Chlorinated Ethenes by a Methane-Utilizing Mixed Culture, *Appl. Environ. Microbiol.* **51**, 720–724 (1986).

113. R. S. Hanson and T. E. Hanson, Methanotrophic Bacteria, *Microbiol. Rev.* **60**(2), 439–471 (1996).

114. J. T. Wilson and B. H. Wilson, Biotransformation of Trichloroethylene in Soil, *Appl. Environ. Microbiol.* **49**, 242–243 (1985).

115. D. Arciero, T. Vannelli, M. Logan, and A. B. Hooper, Degradation of Trichloroethylene by the Ammonia-Oxidizing Bacterium Nitrosomonas europaea, *Biochem. Biophys. Res. Commun.* **159**, 640–643 (1989).

116. B. D. Ensley, Biochemical Diversity of Trichloroethylene Metabolism, *Ann. Rev. Microbiol.* **45**, 283–299 (1991).

117. D. Y. Mu and K. M. Scow, Effect of Trichloroethylene (TCE) and Toluene Concentrations on TCE and Toluene Biodegradation and the Population Density of TCE and Toluene Degraders in Soil, *Appl. Environ. Microbiol.* **60**(7), 2661–2665 (1994).

118. P. L. McCarty and L. Semprini, Ground-Water Treatment for Chlorinated Solvents, in R. D. Norris, R. E. Hinchee, R. Brown, P. L. McCarty, L. Semprini, J. T. Wilson, D. H. Kampbell, M. Reinhard, E. J. Bouwer, R. C. Borden, T. M. Vogel, J. M. Thomas, and C. H. Ward (Eds.), *Handbook of Bioremediation*, Lewis Publishers, Boca Raton, FL, 1994, p. 87.

119. R. C. Borden, C. A. Gomez, and M. T. Becker, Geochemical Indicators of Intrinsic Bioremediation, *Ground Water* **33**(2), 180–189 (1995).

120. J. C. Morris and W. Stumm, in W. Stumm (Ed.), *Equilibrium Concept in Natural Waters Systems*, Am. Chem. Soc., Washington, DC, 1967, p. 270.

121. R. D. Lindberg and D. D. Runnells, Ground Water Redox Reactions: An Analysis of Equilibrium State Applied to Eh Measurements and Geochemical Modeling, *Science* **225**, 925–927 (1984).

122. R. T. Wilkin, M. S. McNeil, C. J. Adair, and J. T. Wilson, Field Measurement of Dissolved Oxygen: A Comparison of Methods. *Ground Water Monitoring and Remediation*, **21**(4), 124–132 (2001).

123. A. T. Stone and J. J. Morgan, Reduction and Dissolution of Manganese (III) and Manganese (IV) Oxides by Organics. 1. Reaction with Hydroquinone, *Environ. Sci. Technol.* **18**(6), 450–456 (1984).

124. D. J. Burdige and K. H. Nealson, Chemical and Microbiological Studies of Sulfide-Mediated Manganese Reduction, *Geomicrobiol. J.* **4**(4), 361–387 (1986).

125. H. L. Ehrlich, Manganese Oxide Reduction as a Form of Anaerobic Respiration, *Geomicrobiol. J.* **5**(3), 423–431 (1987).

126. J. Di-Ruggiero, and A. M. Gounot, Microbial Manganese Reduction Mediated by Bacterial Strains Isolated From Aquifer Sediments, *Microb. Ecol.* **20**, 53–63 (1990).

127. D. R. Lovley, and S. Goodwin, Hydrogen Concentrations as an Indicator of the Predominant Terminal Electron-Accepting Reaction in Aquatic Sediments, *Geochim. Cosmochim. Acta.* **52**, 2993–3003 (1988).

128. F. H. Chapelle, P. B. McMahon, N. M. Dubrovsky, R. F. Fujii, E. T. Oaksford, and D. A. Vroblesky, Deducing the Distribution of Terminal Electron-Accepting Processes in Hydrologically Diverse Groundwater Systems, *Water Resour. Res.* **31**(2), 359–371 (1995).

129. American Society for Testing and Materials (ASTM), *Standard Guide for Remediation of Groundwater by Natural Attenuation at Petroleum Release Sites*, ASTM Designation E 1943-98, ASTM, West Conshohocken, PA, 1998.

130. Underground Storage Tank Technology Update, *Four Critical Considerations in Assessing Contaminated Plumes. Implications for Site-Specific MNA/RBCA Assessments at Petroleum-Contaminated Plumes*, EPA/600/J-98/228, 1998.

131. G. Heron, C. Crouzet, Alain C. M. Bourg, and T. H. Christensen, Speciation of Fe(II) and Fe(III) in Contaminated Aquifer Sediments Using Chemical Extraction Techniques, *Environ. Sci. Technol.* **28**, 1698–1705 (1994).

132. L. Semprini, P. K. Kitanidis, D. H. Kampbell, and J. T. Wilson, Anaerobic Transformation of Chlorinated Aliphatic Hydrocarbons in a Sand Aquifer Based on Spatial Chemical Distributions, *Water Resour. Res.* **13**(40),1051–1062 (1995).

133. U.S. EPA, *Draft Field Methods Compendium (FMC)*, OERP #9285.2-11, Office of Emergency and Remedial Response, Washington, DC 1996.

134. R. W. Klusman, *Sampling Designs for Biochemical Baseline Studies in the Colorado Oil Shale Region: A Manual for Practical Application*, Report DOE/EV/10298-2, U.S. Department of Energy, 1996, 180 pp.

135. W. J. Shampine, L. M. Pope, and M. T. Koterba, *Integrating Quality Assurance in Project Work Plans of the U.S. Geological Survey*, Open-File Report 92-162, U.S. Geological Survey, 1992, 12 pp.

136. M. Koterba, F. Wilde, and W. Lapham, *Ground-Water Data-Collection Protocols and Procedures for the National Water-Quality Assessment Program*, Open-File Report 95-399, U.S. Geological Survey, 1994, 113 pp.

137. L. R. Shelton, *Field Guide for Collection and Processing Stream-Water Samples for the National Water-Quality Assessment Program*, Open-File Report 94-455, U.S. Geological Survey, Sacramento, CA, 1994, 44 pp.

138. W. W. Wood, *Guidelines for Collection and Field Analysis of Ground-Water Samples for Selected Unstable Constituents*, U.S. Geological Survey Techniques of Water-Resources Investigations, Book 1, Chapter D2, 1981, 24 pp.

139. L. R. Shelton and P. D. Capel, *Guidelines for Collecting and Processing Samples of Stream Bed Sediment for Analysis of Trace Elements and Organic Contaminants for the National Water-Quality Assessment Program*, Open-File Report 94-458, U.S. Geological Survey, 1994, 20 pp.

140. M. J. Fishman and L. C. Friedman, Methods for Determination of Inorganic Substances in Water and Fluvial Sediments, in M. J. Fishman and L. C. Friedman (Eds.), *Techniques of Water-Resources Investigations*, U.S. Geological Survey, Book 5, Chapter A1, 1985, p. 545.

141. R. T. Wilkin, M. S. McNeil, C. J. Adair, and J. T. Wilson, Field Measurement of Dissolved Oxygen: A Comparison of Methods. *Ground Water Monitoring and Remediation*, **21**(4), 124–132 (2001).

FURTHER READING

1. R. W. Puls and M. J. Barcelona, *Low-Flow (minimal drawdown) Sampling Procedures*, EPA/540/S-95/504, U. S. Environmental Protection Agency, Washington, DC.

2. W. Stumm and J. J. Morgan, 1981. *Aquatic Chemistry.*, 2nd ed., John Wiley & Sons, Inc., New York, NY.

3. G. F. Craun, Causes of Waterborne Outbreaks in the United States, *Water Sci. Technol.* **24**, 17–20 (1989).

4. M. Y. Deng and D. O. Cliver. Persistence of Inoculated Hepatitis A Virus in Mixed Human and Animal Wastes, *Appl. Environ. Microbiol.* **61**(1), 87–91 (1995).

5. H. Dizer and U. Hagendorf, Microbial Contamination as an Indicator of Sewer Leakage, *Water Res.* **25**(7), 791.

6. E. H. Isaacs and R. M. Srivastava, Methods for Sampling Soil Microbes, in C. J. Hurst, C. R. Knudsen, M. J. McInerney, L. D. Stetzenbach, and M. V. Walter (Eds.), *Manual of Environmental Microbiology*, ASM Press, Washington, DC, 1997.

7. U.S. Environmental Protection Agency (EPA), *Method 1623: Cryptosporidium and Giardia in Water by Filtration/IMS/FA*, EPA-821-R-99-006, Office of Water, Washington, DC, 1999.

8. M. V. Yates and C. P. Gerba, Factors Controlling the Survival of Viruses in Groundwater, *Water Sci. Techol.* **17**, 681–687 (1984).

IV

WASTEWATER MONITORING

33

USE OF INSTRUMENTATION FOR pH CONTROL

Mark Lang, P.E.

The Sear-Brown Group
Rochester, New York

33.1 INTRODUCTION

Proper pH control is essential in practically every phase of water and wastewater treatment. Acid–base neutralization, water softening, precipitation, coagulation, disinfection, and corrosion control are all pH dependent. The pH is also used in determining acidity, alkalinity, and carbon dioxide concentrations.

Environmental Instrumentation and Analysis Handbook, by Randy D. Down and Jay H. Lehr
ISBN 0-471-46354-X   Copyright © 2005 John Wiley & Sons, Inc.

pH is defined as: the reciprocal of hydrogen ion concentration in gram atoms per liter, and measured on a scale from 0–14. A pH of 7 is considered to be "neutral" (neither acidic or alkaline).

Regulatory agencies often require the measurement of wastewater treatment facility influent and effluent pH to monitor overall facility conditions. It may also be necessary to monitor industrial discharges effluent pH to forecast possible toxic conditions.

Most biological wastewater treatment processes can tolerate a range of pH between pH 5 and pH 9. Some systems, such as anaerobic digestion are more sensitive to changes in pH. The anaerobic digestion process requires pH in the range of 6.6–7.6. The process is prone to failure if the pH falls below 6.2.

pH monitoring can provide feedback for control of other processes requiring pH adjustment. pH adjustment can be used to neutralize low-pH industrial wastes, enhance phosphorus removal by alum addition, or provide optimum ranges for nitrification and denitrification processes.

33.2 PRINCIPLE OF OPERATION

As shown in Figure 33.1 a pH sensor consists of a glass membrane electrode that develops an electrical potential varying with the pH of the process fluid. The difference between this potential and a reference electrode is measured and amplified by an electronic signal condition. The complete circuit includes the glass electrode wire, glass membrane, process fluid, reference electrode fill solution, and the reference electrode wire. The equivalent electric circuit for a pH sensor is shown in Figure 33.2.

A reference electrode is designed so that its potential E is constant with pH. The asymmetric potential, measured in milivolts, K_1 varies from sensor to sensor. The

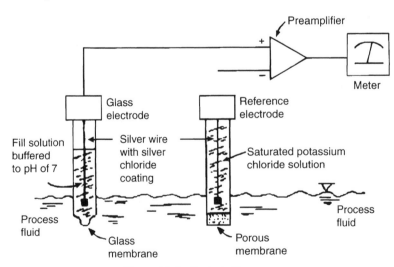

Figure 33.1 Typical pH sensor.

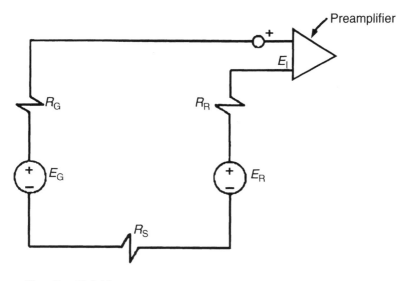

$E = K_1 + K_2 \, (\text{pH})$

$E_R = \text{Const}$

$R_G = \text{Resistance of glass electrode}$

$R_R = \text{Resistance of reference electrode}$

$R_S = \text{Resistance of process fluid solution}$

Figure 33.2   Circuit for sensor in Figure 33.1.

asymmetric potential also changes with the age of the sensor. For this reason pH sensors must be calibrated periodically using a buffer solution of a known pH.

The electrode gain K_2 is a function of temperature. Most pH sensors include temperature compensation. Without temperature compensation, an additional error of 0.002 pH/°C difference from calibration temperature should be anticipated.

33.3   PRECISION AND ACCURACY

The accuracy of pH meters typically ranges ±0.2 pH units. The precision of the pH sensor varies from 0.02 to 0.04 pH units. The stability of a pH sensor can range between 0.002 and 0.02 pH units per week. The stability indicates how often the sensor should be recalibrated.

A pH sensor with the following performance standards is suitable for use in wastewater treatment facilities: accuracy ±0.1 pH units, precision ±0.03 pH units, and stability ±0.02 pH units per week.

33.4   INSTALLATION

pH probes can be installed using two methods. Flow-through installation and in-tank or open-channel installation. Each method will be described.

33.4.1 Flow-Through Installation

Sensors are mounted in flow-through configuration when pH control is not the objective of the measurement. This can be seen in Figure 33.3. In this configuration a sampling stream is taken from the main flow. This sampling stream flows across the electrodes of the pH sensor. A sampling port is often installed as part of a sampling line to allow for sample collection for conformance testing.

33.4.2 In-Tank or Open-Channel Installation

For in-tank and open-channel installation pH sensors with integral preamplifiers are normally used as shown in Figure 33.4. The sensor assembly is normally attached to a polyvinyl chloride (PVC) or stainless steel conduit. The sensor is supported below the water's surface elevation. The conduit is typically mounted to a handrail adjacent to the tank or channel. The bracket supporting the conduit should be designed so that the conduit and the sensor can be easily removed for maintenance. It is important to note that adequate cable must be provided to allow the pH sensor and conduit assembly to be removed from the tank. A difficulty often reported with pH sensors placed in wastewater flow is that electrodes become coated with grease. Mechanical wipers have been used successfully in wastewater applications to remove grease. Ultrasonic cleaners have had limited success in removing soft coatings such as grease and oil.

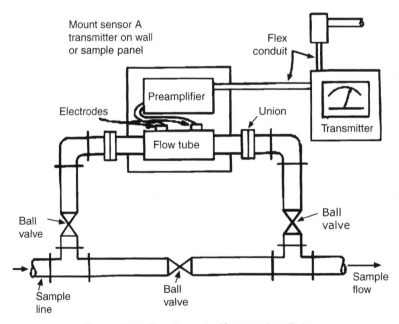

Figure 33.3 Flow-through pH sensor installation.

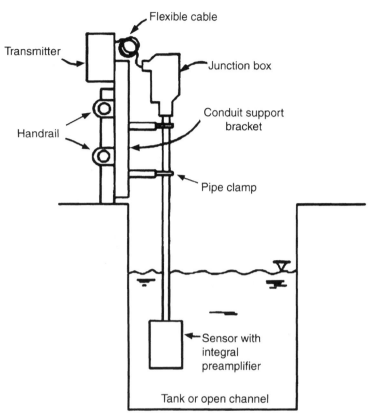

Figure 33.4   Submersion pH sensor installation.

33.5   PROCESS CONTROL USING pH SENSORS

A pH sensor will send a signal to a controller, which can send an output to produce a change in one or more systems using some form of an actuator. These controllers implement a control strategy that, while being designed in mathematical terms, is realized by mechanical or electronic means.

Control strategies may be classified as:

- Feedback
- Feedforward
- Combination of the two

33.5.1   Feedback Control

Feedback control involves the measurement of a process output to make adjustments to an input variable. The input variable then becomes the manipulated variable. The use of alum dose to control pH is an example of feedback control. Historically, operators have provided feedback control by making manual adjustments based on periodic observations of the process performance. When a feedback controller is added, these periodic observations are continuous and the system becomes a closed loop.

33.5.2 Feedforward Control

Feedforward control is a technique where variations in the process stream are converted into changes in the manipulated variable. The change in the manipulated variable reduces the effect on the process.

In theory, feedforward control is capable of very precise pH control since it is not dependant on changes of the controlled variable. In practice, however, it is virtually impossible to position the manipulated variable with respect to the exact load on the process at any given time. The response, however, can be virtually instantaneous, relative to the delay in a feedback loop. This allows feedforward control to be a useful adjunct to feedback even when limited accuracy precludes its use alone.

In a feedforward control system, the manipulated flow of reagent is calculated to balance the flow entering the process. Feedforward control, though responsive, is never accurate enough to satisfy even broad limits of effluent pH. Carbonates and other buffers may moderate these sensitivities by a factor of 10 or, even more, the accuracy of this type of system is approximately 5%. They may be more repeatable, but in a wide-ranging system overall accuracy is the critical factor since reagent demand is continuously changing.

The greatest source of error is the inability to accurately model the process. If only strong acids and bases are present, or only a *single* weak acid or base, the relationship between pH and valve position is single valued and linear. Mixtures of strong agents with weak agents or their salts impact the ability to control the process.

The feedforward settings need to be adjusted to minimize the effect of the majority of the load changes and rely on feedback to correct the balance.

33.5.3 Batch Processes

Special consideration must be given to those processes that are operated in a batch mode and those that, though continuous, are interrupted daily or weekly. Batch processing has advantages, the most important of which is an ability to retain an effluent until its quality meets specifications before release.

Economic incentives lean toward batch processes either when the plant production units are operated batchwise or when flow rates are low, typically less than 100 gal/min. A single vessel can be used, although in many cases, duplicate vessels have to be used, with one being filled while treatment is proceeding in the other.

In a continuous process, the controlled variable is to be maintained at the set point by balancing the manipulated variable against the load. In a batch process, however, the only time the measurement is at the set point is at the end of the run. At this point reagent addition may be in excess of that required. If a controller with reset action is used on a batch process, it will drive the valve fully open and keep it there until the controlled variable crosses the set point.

"Overshooting" a set point is possible since a delay always exists between the motion of the valve and the resulting response in the controlled variable. To reduce the potential of overshooting a set point, the valve must be closed before the controlled variable reaches the set point.

REFERENCES

1. F. G. Shinskey, *Process-Control Systems*, McGraw-Hill, New York, 1967.

2. L. K. Spink, *Principles and Practices of Flowmeter Engineering*, 9th ed., Foxboro Company, Foxboro, MA, 1967.

3. V. L. Trevathan, pH Control in Waste Streams, Paper No. 72-725, Instrument Society of America, Research Triangle Park, NC.

4. F. G. Shinskey, A Self-Adjusting System for Effluent pH Control, paper presented at the 1973 Instrument Society of America Joint Spring Conference, St. Louis, MO, April 24–26, 1973.

5. F. G. Shinskey, *pH and pION Controls in Process and Waste Streams*, Wiley, New York, 1973.

6. *Instrumentation in Wastewater Treatment Facilities, Manual of Practice (21)*, Water Environment Federation, Alexandria, VA, 1993.

7. R. Skrentner, *Instrumentation Handbook for Water and Wastewater Treatment Plants*, Lewis Publishers, Chelsea, MI, 1989.

34

AUTOMATIC WASTEWATER SAMPLING SYSTEMS

Bob Davis and Jim McCrone

American Sigma
Loveland, Colorado

Environmental Instrumentation and Analysis Handbook, by Randy D. Down and Jay H. Lehr
ISBN 0-471-46354-X   Copyright © 2005 John Wiley & Sons, Inc.

34.1 INTRODUCTION

As its name implies, a water *sampler* is an instrument that collects a sample or samples of water or wastewater. The samples are collected from municipal and industrial sources, as well as urban and agricultural runoff.

A wide variety of water samplers are manufactured, both manual and automatic, ranging from basic models to refrigerated units with advanced monitoring capabilities and from battery-operated portable samplers to permanently installed equipment connected by modem to plant management systems. Sophisticated automatic sampling systems are capable of performing a variety of functions for diverse applications.

34.2 USES OF SAMPLING SYSTEMS

Samplers—along with flow meters and other water quality monitoring equipment—are used extensively in publicly owned treatment works (POTWs) and various industries and by consulting engineers, laboratories, universities, and federal and state governments.

Monitoring water and wastewater quality for environmental analysis helps set environmental policies, develop pollution prevention and remediation programs, and assess their effectiveness. Samplers allow users to comply or enforce compliance with environmental regulations and permits, conduct research, and obtain information about the character of a discharge or receiving body of water. This water may contain pollutants that adversely impact the environment. Industrial discharges, for example, may deplete oxygen in the receiving waters or cause toxic fish kills. Thermal discharge from a power plant might negatively impact the biological population, and nutrient discharges can lead to eutrophication (void of oxygen) of lakes, preventing aquatic life. A thorough sampling program helps determine the current conditions of—and spot changes and trends in—water resources, identify sources of pollutants, and evaluate the impact of these discharges on the environment.

Sampling also helps measure the performance of municipal and industrial treatment plants. Sampling in wastewater treatment plants is performed to validate National Pollution Discharge Elimination System (NPDES) permit compliance and to monitor and improve process control. Automatic samplers are particularly useful for characterizing influent and effluent flows.

Sampling is also performed for safety and health considerations. (Likewise, when sampling and analyzing wastewater samples, personnel must adhere to strict guidelines and follow specific procedures to ensure their own safety.)

34.3   MANUAL SAMPLERS

Although there are still conditions where manual sampling is performed, it has in general been proven to be less effective, less reliable, and more labor intensive than sampling with an automatic system. It is especially unreliable when sampling frequency, sample location repeatability, and compositing of samples are considered. When wishing to collect samples that are representative of the source—where both the makeup of the flow and the sample itself can vary—it is safe to say that automatic samplers should be the instruments of choice. Modern automatic sampling systems are more capable instruments, able to perform out-of-limits sampling based on such parameters as flow, pH, temperature, conductivity, and dissolved oxygen or based on such events as rainfall. For these and other reasons, automatic sampling systems decrease costs and increase accuracy and versatility when compared with manual sampling.

34.4   AUTOMATIC SAMPLING SYSTEMS

Automatic samplers range from basic to sophisticated units and may be portable or refrigerated. Portable samplers are generally employed in tight locations such as manholes and at remote sites without alternating current (AC) power. These are usually operated with a direct current (DC) battery pack. Refrigerated samplers are permanently installed where AC power is available. These locations may be indoor or outdoor sites.

A basic automatic sampling system is comprised of a controller, pump, intake line, and a bottle (Fig. 34.1) Basic systems are appropriate when a repeatable sample volume at a timed interval is needed or after a predetermined volume of flow.

More advanced systems, explained below, greatly expand the capabilities of the sampler for a wide range of monitoring applications, including set-point sampling, sampling based on characteristics of the waste stream, monitoring industrial discharges, collecting out-of-range samples, and sampling for parameters that normally require grab samples due to rapid degradation, such as pH (Fig. 34.2).

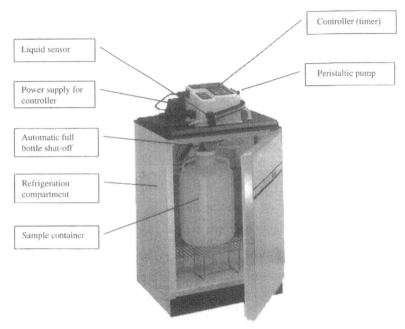

Controller (timer)

Liquid sensor

Peristaltic pump

Power supply for controller

Automatic full bottle shut-off

Refrigeration compartment

Sample container

Figure 34.1 Components of basic automatic samplers.

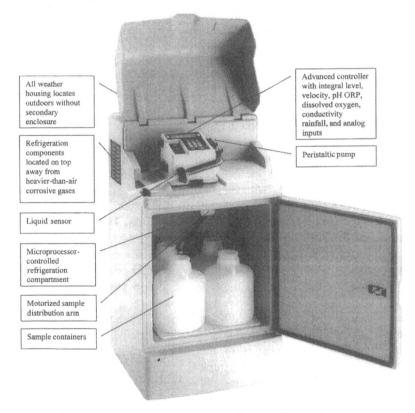

All weather housing locates outdoors without secondary enclosure

Advanced controller with integral level, velocity, pH ORP, dissolved oxygen, conductivity rainfall, and analog inputs

Refrigeration components located on top away from heavier-than-air corrosive gases

Peristaltic pump

Liquid sensor

Microprocessor-controlled refrigeration compartment

Motorized sample distribution arm

Sample containers

Figure 34.2 Components of advanced automatic sampling systems.

Most commercially available automatic samplers function with the following five interrelated subsystem components:

- Sample intake subsystem
- Sample transport subsystem
- Sample delivery subsystem
- Sample storage subsystem
- Controls and power subsystems

34.4.1   Sample Intake and Transport Subsystems

The sample intake is essentially a strainer, immersed in the water, where the liquid enters the sampler. The intake is connected to the end of suction tubing (the sample transport line), which transports the sample from the intake to the storage subsystem (collection bottle or bottles).

The strainer and tubing should be chemically compatible with the liquid in the stream and be constructed of a material that will not alter the sample by leaching, absorption, or desorption. They should also resist or be protected from potential damage from flowing debris. The strainer may be constructed of stainless steel for sampling priority pollutants or polypropylene for general use. The most commonly used intake tubing material is nonleaching tygon (although Teflon tubing must be used when sampling for priority pollutants).

The inside diameter should be at least $\frac{1}{4}$ in.; however, $\frac{3}{8}$ in. is more commonly used—large enough to allow water to enter freely without clogging or plugging, yet small enough to maintain sufficient flow velocity to prevent solids settling. [The U.S. Environmental Protection Agency (EPA) *Manual NPDES Compliance Sampling Inspection Manual*, MCD-51, states that sample train velocity should exceed 2 ft/s. The U.S. EPA *Manual Sampling of Water and Wastewater* states the sampler flow–velocity should be at least 2 ft/s.] The tubing should also resist physical damage from large objects in the flow stream. The end of the tubing should be fixed to maintain its location throughout the sampling period.

The line must be kept free of kinks, twists, and sharp bends to avoid plugging and low-flow velocities.

To prevent contamination of subsequent samples, the intake line should be thoroughly purged before and after every sample collection, a task ideally suited to automatic samplers. Some samplers also offer the option to select a line rinse where the intake is preconditioned with the source liquid prior to collecting the sample. Another desirable feature is the ability to detect a failed attempt to collect a sample because of a plugged intake and immediately repeat the cycle (starting with a high-pressure purge) so that no samples will be missed.

34.4.2   Sample Delivery Subsystem

The sample delivery system transfers the sample from the source to the bottle. Automatic samplers offer mechanical, forced-flow and suction lift methodologies as detailed in Figure 34.3 and Table 34.1. In early samplers, sample volume was

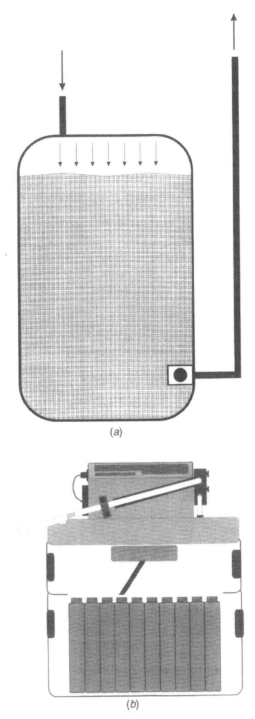

(a)

(b)

Figure 34.3 Mechanical sample delivery systems.

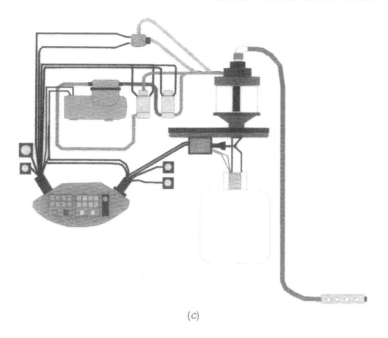

(c)

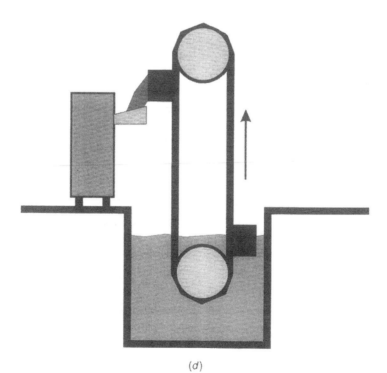

(d)

Figure 34.3 (*Continued*)

Table 34.1   Mechanical Sample Delivery Systems

Mechanical gathering subsystems	Often built in place. Include such devices as cups or calibrated scoops on cables and paddle wheels that are lowered mechanically into the water to collect a sample. Such devices may obstruct the stream flow. Capable of high sampling lifts—as much as 200 ft (60 m) or more—and wide or extremely deep channel flows. Their many moving parts and exposed mechanisms tend to easily foul and require periodic inspection and maintenance to prevent failure and inaccurate monitoring results.
Forced-flow gathering subsystems	Often built in place as permanent sampling facilities. A positive force—usually supplied by pneumatic pressure or a submersible pump—transfers the liquid sample from the source, through the transport line, into the sample containers. Can extract samples at very low depths and lift them great distances. Because they use air pressure to transport samples, they may be an appropriate choice for sample collection in potentially explosive environments. They are susceptible to periodic failures, which can affect the quality of the sampling program. May obstruct the stream flow and require periodic inspection and maintenance.
Suction lift	The most widely used method because of its versatility and minimal effect on flow patterns. Applied vacuum transfers the liquid to the storage container through positive displacement with a peristaltic pump or a vacuum pump with a metering chamber. Peristaltic pumps are extremely versatile because they have a minimal effect on the flow stream and can purge transport lines by reversing pumping direction. Lift capability, however, is limited to around 25–29 vertical feet (9 m) for peristaltic and vacuum pumps because of internal friction losses and atmospheric pressure. When the pressure or a liquid that contains dissolved gases is reduced, the dissolved gases may pass out of the solution, leaving suspended solids at the liquid's surface layer. To minimize the concentration effect, collect at least 100 mL per sampling unit.

determined based on time calibration; today's more advanced samplers more accurately measure volume by using a liquid sensor.

34.4.3 Sample Storage Subsystem

Sample storage is essentially a bottle or bottles (or bags) that store samples for future laboratory analysis. Samples may be stored in a single large collection bottle or a number of smaller collection bottles for discrete collection, with a rotating distributor to fill the appropriate bottle at the time of sampling. Each bottle may hold from a few milliliters to several liters of the sample liquid, with a total sample volume storage capacity of at least 2 gal (7.6 L).

The size and quantity of bottles depend on several factors:

- Type of samples needed
- Number of samples needed
- Laboratory analyses to be performed
- Requirements of the laboratory that will perform the analyses
- Parameters to be monitored
- Regulations or permit requirements

Sample containers must be manufactured from chemically resistant material that will be unaffected by the concentrations of the pollutants measured. Collection bottles are generally made of glass, polypropylene, polyethylene, or Teflon. Choose the appropriate material to maintain the integrity of the sample; considerations include chemical compatibility, leaching, absorption, desorption, and legal requirements. Most wastewater samples for chemical analyses are collected in plastic (polyethylene or polypropylene) containers. Oil and grease samples, pesticides, phenols, polychlorinated biphenyls (PCBs), and other organic pollutant samples, however, are collected in glass jars or bottles and sealed. Bacteriological samples must be collected in properly sterilized plastic or glass containers, and samples that contain constituents that oxidize when exposed to sunlight should be collected in dark containers. Sample containers must also have a closure to protect the sample from contamination.

Unless refrigerated samplers are used, storage subsystems should be large enough to provide space for ice to chill the samples for preservation after collection.

Sample containers should be properly cleaned and free of contamination. (Some analytical procedures specify particular cleaning procedures.) Precleaned and sterilized disposable containers are also available.

The EPA has set strict guidelines for the use of sample containers. Required sample containers, sample preservation, and sample holding time are described in the Chart of Selected Parameters and Required Containers, U.S. EPA 40 Code of Federal Regulations (CFR) Part 136, July 1, 1990, or EPA 300B 94-014, Table 5-3.

34.4.3.1 Bottle Sequencing and Configuration. Flexibility of bottle configurations is important for differing routines, based on sampling requirements for any individual application; differing applications require different bottle sizes and sequences.

When sampling stormwater, for example, the first flush must be separated from the rest of the event. An industry may share a sample with a municipality, using two-bottle split sampling, or a laboratory may require a particularly large discrete sample for analysis. Trouble bottle capability permits a normal sampling routine to proceed, while segregating one or more out-of-limit discrete samples during an upset, to help pinpoint and analyze particular violations.

It is therefore helpful to choose equipment appropriate to the required sampling application or equipment with flexibility in sequencing when it will be used for a variety of applications. The more sophisticated samplers permit the greatest versatility for multiple usage over time. The equipment may be capable of multiple-bottle sequential as well as multiple-bottle composite sampling (with two, four, or eight bottles) and sampling on a time or flow proportional basis.

34.4.4 Controls and Power Subsystems

The electronic controller is the "command center" of the sampling operation. The control units usually allow operators to select either time- or flow-compositing methods or continuous sampling methods. In advanced systems, the controller can be programmed to run a variety of sampling programs that control:

- Frequency of sampling
- Volume of samples
- Allocation of samples to storage bottles

The controller can also serve as a data logger, storing:

- The sampling program
- The sampling data
- Date of sampling
- Time of sampling
- Conditions of sampling
- Flow, pH, temperature, conductivity, dissolved oxygen (DO), and other parameters

The versatility of automatic sampling systems is further enhanced when the controller interfaces with a variety of other water monitoring instruments. Through modems, for example, one can download or upload sampling programs to personal computers and printers. This is especially helpful for advanced analysis and reporting as part of total system management. Built-in software allows modern automatic sampling systems, for instance, to take discrete samples during the first half hour of a storm event, followed by composite samples (see Section 34.5.1).

The most popular automatic samplers use environmentally sealed solid-state controls to minimize the effects of such highly unfavorable field environments as humidity and corrosiveness. These units are also sealed to protect the controls if the sampler is accidentally submerged. When field conditions permit, samplers operating from a power supply usually require less maintenance than battery-operated models.

34.4.4.1 *Power Sources.* A variety of power sources and accessories are available for portable and permanent samplers, including:

- AC power converter
- Lead–acid batteries
- Nickel–cadmium batteries
- Battery chargers
- Multistation battery chargers
- Solar panels

34.4.4.2 *Battery Power and Longevity.* Portable samplers must conserve power so they can operate on batteries for a sufficient time to complete the investigations for which they are employed. An important consideration is a long battery life, which is a result of a low power draw from the equipment itself. An optimum standby state is 5 mA. When available, integral trickle chargers maintain batteries at full charge for continued operation.

34.4.4.3 *Sample Delivery Subsystem (Pumps).* The most common sample delivery systems (pumps) for automatic sampling systems are vacuum, bladder, and, most often, peristaltic pumps. Modern high-speed peristaltic pumps produce the recommended EPA criteria for representative intake velocity and have become the preferred method for most applications. (See Fig. 34.4 and Table 34.2.)

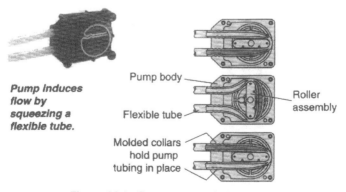

Figure 34.4 Pump types and advantages.

Table 34.2 Pump Types and Advantages and Disadvantages

Pump Type	Operation	Advantages	Disadvantages
Peristaltic	A rotating roller squeezes tubing to create a vacuum that draws the sample directly from the source into the bottle. (See Fig. 00.)	Recommended for most applications. The sample touches only the sample tubing, which can be automatically purged or rinsed, decontaminated, or replaced. (There are no metering chambers that can cause cross-contamination between samples.) They produce a sufficient intake line velocity for representative sampling (2–3 ft/s), reducing the potential for scouring high levels of solids at the sample intake point. Their extremely simple design reduces the need for maintenance and almost eliminates the potential for failure.	Peristaltic pumps have maximum sample lifts of approximately 26 ft (8 m), and the vacuum created during operation may pull out some of the volatile organic compounds (VOC) in the sample.
Bladder pumps	The pump is submerged in water and compressed gas—usually air—is cycled on and off, squeezing and expanding the bladder. The water is drawn through the line and into the bottle. (See Fig. 00 for the operation of bladder pumps.)	Air does not touch the sample, serving its integrity and preventing it from deterioration. Since they are extremely gentle on the sample, they can be an excellent choice for VOC sampling. They have a high sample lift capability of more than 240 ft (76 m).	Bladder pumps are susceptible to damage in waters with high levels of solids. This limits their application unless fitted with a screen.

| Vacuum pumps | Vacuum applied to a metering chamber forces the liquid sample to be drawn through the sample intake into the chamber. A float or conductive sensor in the metering chamber senses the liquid level and cuts off the vacuum, stopping delivery to the chamber. The liquid in the chamber is then drained to the sampling bottle. | Popular in Europe; no longer popular in North America. | The high intake velocity of vacuum samplers tends to scour solids at the sampling point and overstate solids in samples, and the vacuum they create tends to pull out dissolved gases and VOCs in samples. Their intricate system of moving parts and sensors renders them unreliable in rough environments, and their sample lines and metering chambers cannot be properly rinsed, creating the potential for cross-contamination and nonrepresentative samples. Vacuum samplers are generally out of favor for North American enforcement agencies and industrial dischargers. |

34.4.4.4   Casing. The casing, or housing, of automatic sampling equipment should be rugged and durable and capable of withstanding the harsh and corrosive environments under which they frequently operate. The controller should be tightly sealed, ideally conforming to NEMA 4x, 6 standards (capable of withstanding hose spray and submersion under 6 ft of water for a minimum of 30 min), and all electronic components should be isolated from hostile environments.

34.4.4.5   Advanced System Features and Capabilities. In addition to the basic elements already discussed, the advanced automatic water sampling systems available today include additional capabilities and enhanced features that make them more functional, more versatile, and easier to use. These features can include one or more of the following elements.

34.4.4.6   Full Graphic Display. Large, easy-to-read graphical displays allow users to quickly set up and program samplers with step-by-step menu prompts. The displays also permit users to instantly review program settings in the field and remain up to date with the current program status, including:

- Program start time and date
- Bottle number
- Time to next sample
- Samples collected
- Samples remaining

34.4.4.7   Parameter Monitoring and Multitasking Capability. With additional data-logging channels and multitasking capabilities, the more advanced sampling systems available today can—in addition to sampling—also monitor rainfall, level, velocity, flow, pH, temperature, oxygen reduction potential, dissolved oxygen, conductivity, and other parameters on a combination of data-logging channels.

These capabilities can, for example, improve the control of pH, temperature, ORP, DO, and conductivity. Conventional pH, temperature, ORP, DO, and conductivity meters may provide readings and indicate unacceptable levels, but it is difficult to determine the cause of the upset without continuous monitoring of these key parameters. A sophisticated automatic sampler, however, can monitor them with the following options:

- Sample normally and log chosen parameters.
- Sample normally and deliver any out-of-limit samples to one or more designated "trouble bottles."
- Sample only when a parameter goes outside preset limits.

34.4.4.8   Set-Point Sampling and Alarm Capabilities. Advanced sampling systems can alert operators to selected conditions or potential problems with two types of alarm relays:

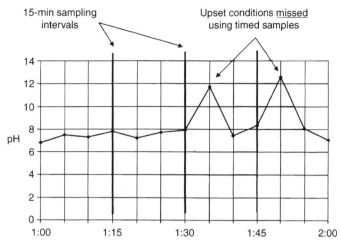

Figure 34.5 Upset conditions missed during timed samples.

34.4.4.8.1 Set-Point Alarms. Initiate an action when user-definable trip points (high, low, or both) are reached. Set-point sampling can also act as a screening tool. For example, pH swings in a waste stream may indicate other problems. By triggering a sampler when pH falls outside the normal range, the problem can be quantified. In addition, since samples are collected only when there is a cause, sampling and analytical costs are reduced.

A particularly useful application for set-point sampling is pinpointing violations. Changes in pH, temperature, and conductivity can indicate process changes resulting in permit violations or other problems. This is one reason why industrial discharges, which are seldom routine or predictable, are difficult to identify; they can even be missed entirely when samplers attempt to identify an upset condition based on time or flow (Figs. 34.5 and 34.6).

If a parameter goes out of limits for brief periods—as happens with batch discharges—a frequent sampling interval is necessary to detect the problem. This increases container and battery consumption and increases demands on sampling personnel. Even with a short sampling interval, a dumped tank can remain undetected. One effective solution is to set high and low trip points. Should the parameter exceed the preset limits, a sample is immediately collected, and sampling can be continued during the duration of the program or until the discharge returns to normal limits. One or more trouble bottles for out-of-limit samples may be segregated from the rest and, with the time, date, and source of the trouble sample logged, determining the problem source is simplified. Set-point sampling is arguably the most cost-effective tool currently available for industrial discharge monitoring.

34.4.4.8.2 Trouble Alarms. Initiate a specific action when a trouble condition occurs, such as a missed sample or a purge failure. These alarm conditions can be assigned to set a relay or generate a report.

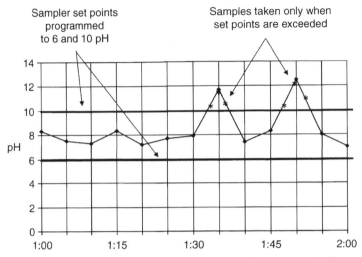

Figure 34.6 Samples taken only when set points are exceeded.

34.4.4.9 Modem. Computer interfaces eliminate handwritten logs by letting users download sample data for remote retrieval. Certain sophisticated samplers can also dial out under user-definable alarm conditions. (See Section 34.4.4.8).

34.4.4.10 Software and Reporting. Advanced software and a communications interface (RS 232, modem, cell, radio) that allow operators to download data can eliminate the need for handwritten logs, saving time and manpower and avoiding human error. Such software not only controls the operation of automatic sampling systems, it also simplifies the analysis and conversion of raw data into usable information and greatly aids in the preparation of standard as well as customized reports.

34.4.4.11 Programming Versatility. Other desirable programming features that may be available on certain automatic samplers as of the time of this writing include:

- Real-time clock
- Program delay
- Intake line rinse
- Intake line fault
- Ability to present multiple program starts and stops
- Manual sample taken independently of the program in progress
- Ability to preset a number of variable intervals between samples
- Automatic positioning of the distributor arm to the appropriate bottle position
- Automatic shut off

- Displays capable of displaying current data, including program feedback
- Internal lithium battery, backup battery, and low-battery indicator
- Timed cycle/flow proportional operation
- Program lock to prevent tampering
- Multiple program storage
- Cascade (using multiple samplers in a cycle)
- Flash memory for software updates and upgrades
- User diagnostics

The most advanced software is particularly useful as part of proactive total system management programs.

34.4.4.12 Sampler with Integral Flow Meter. Whereas samplers and flow meters are often used in tandem, it is possible to obtain a sampler with an integral flow meter as a single unit. This can be a convenient, versatile, and space-saving solution for a variety of applications, among them storm water, combined sewer overflow, and industrial monitoring.

34.4.4.13 Totalizers. When equipped with this integral flow option, liquid-crystal display (LCD) and mechanical totalizers (resettable and nonresettable) track total flow.

34.4.4.14 Online Analyzers. Whether separate devices or integral to the automatic sampling system, continuous parameter analyzers and control systems are particularly effective and efficient devices. When used for a wastewater treatment plant, for example, they can provide early warnings of industrial slug discharges and automatically monitor and control DO and chlorine concentrations.

34.5 TYPES OF SAMPLING

34.5.1 Composite Versus Discrete Samples

Once collected, samples can be retained as individual samples, or combined to form a single sample. Single samples are *grab* or *discrete* samples. Combined samples are *composite* samples. The U.S. EPA also has specific definitions for composite and grab samples and sample preservation techniques. In general, a composite (single-bottle) sample is appropriate when the flow makeup is consistent, and discrete (multiple-bottle) samples are appropriate when the flow makeup is variable (See Fig. 34.7 and Table 34.3.)

If the sample type for a particular analysis is not specified in 40 CFR Part 136, the NPDES permit writer generally determines and specifies the appropriate sample type.

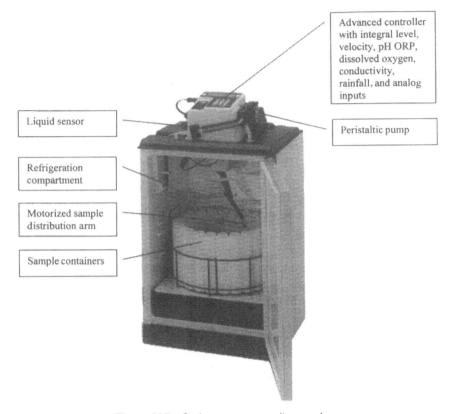

Advanced controller
with integral level,
velocity, pH ORP,
dissolved oxygen,
conductivity,
rainfall, and analog
inputs

Liquid sensor

Peristaltic pump

Refrigeration
compartment

Motorized sample
distribution arm

Sample containers

Figure 34.7   Grab versus composite samples.

Composite samples can be collected either manually or with automatic samplers. A variety of compositing methods are discussed briefly below and in more detail in EPA 300B 94-014, Table 5-2, Compositing Methods. The method available for compositing sample type may be specified by the permit. Carefully consider variability in wastestream flow rate and parameter concentrations when choosing compositing methods, sampling equipment (tubing and containers), and quality assurance procedures.

34.5.2   Flow-Weighted Versus Time-Weighted Samples

Generally speaking, time-weighted samples are appropriate when the flow amount is consistent, and flow-weighted samples are appropriate when the flow amount varies.

34.5.3   Time Composite Sample Method

The time composite sample method combines discrete fixed-volume samples by collecting them in one container at constant time intervals. A task exceptionally

Table 34.3 Grab Versus Composite Samples

	Characteristics	Applications
Grab samples	Grab samples are discrete (individual) samples, collected over 15 min or less, that are representative of conditions at the time of collection. They are usually hand-collected, although some automatic samplers allow for discrete samples to be taken and isolated in specified individual bottles, without interrupting an ongoing sampling program. Sample volume depends on the type and number of analyses to be performed.	Show characteristics of process stream or unstable parameters soon after collection, or variations of waste stream over time. Verify whether composite sample masks short-term variations in wastewater characteristics. Sample intermittent flows. Analyze unstable or difficult-to-preserve sample constituents. Sample an effluent discharge that is not on a continuous basis. Provide information about instantaneous concentrations of pollutants at a specific time. Collect variable sample volume. Corroborate composite samples. Monitor parameters not amenable to compositing.
Composite samples	Composite samples represent the average characteristics of the wastestream during the compositing period. They are formed by compositing (combining) grab samples collected durig a specified time period, either by continuous sampling or by mixing the discrete samples. They may be time composite or flow proportional and are easily prepared with automatic samplers.	When stipulated in a permit and when: Average pollutant concentration during the compositing period is determined. Mass per unit time loadings is calculated. Wastewater characteristics are highly variable. Provide information on average characteristic of the sample over a specified period. Average data needed to monitor and adjust plant processes.

well suited to automatic sampling systems, time composite sampling is appropriate when the flow of the sample stream is constant or when flow monitoring equipment is not available. The total volume of the composite sample depends on the required analyses.

34.5.4 Flow-Proportional Sample Method

To produce a flow-proportioned composite sample, the volumes among the grab samples or the sampling frequency must be varied. In other words, a constant sample volume is collected at varying time intervals (constant volume/variable time) proportional to stream flow, or a sample is collected by increasing the volume of each aliquot as the flow increases, while maintaining a constant time interval between the aliquots (constant time/variable volume). This is done in order to weight the final sample in proportion to the various flow rates that occur during the sampling period. This type of sample is more representative of a varying process stream than is a fixed-volume composite sample. As accuracy is essential, automatic samplers are preferred over manual sampling whenever possible.

34.5.5   Sequential Composite Sample Method

The sequential composite method, although labor intensive when performed manually, is commonly used in certain automatic samplers. It requires discrete samples to be collected in individual containers at constant time intervals or discharge increments, which can then be flow proportioned to form the composite sample.

34.5.6   Continuous Composite Sample Method

The continuous composite sample method—not widely used—collects continuous samples from the wastestream. The sample may be constant volume, or the volume may vary in proportion to the flow rate of the wastestream.

34.5.7   Sample Volume

The volume of samples collected depends on the type and number of analyses needed for the measured parameters. Along with sufficient volume for all required analyses, an additional amount should be collected for split samples or repeat analyses. Consult the laboratory for specific volumes required. Specific recommended minimum sample volumes for different pollutant parameters can be found in EPA's *Methods for Chemical Analysis of Water and Wastes* (EPA, 1979b) and *Handbook for Sampling and Sample Preservation of Water and Wastewater* (EPA, 1982) and the current EPA-approved edition of *Standard Methods for the Examination of Water and Wastewater* [American Public Health Association (APHA), American Water Works Association (AWWA), and Water Environment Federation (WEF)]. For a guide to sample volumes required for determining the constituents in wastewater, see EPA 300-B 94-014, Table 5-1.

34.5.8   Preserving Sample Stability

Samples may be preserved through pH adjustment, chemical fixation, and refrigeration. The latter is most frequently used since it has no detrimental effect on the sample composition and does not interfere with any analytical methods. For additional information, see *NPDES Compliance Monitoring Inspector Training: Sampling EN-338 21W-407*, August 1980 or later editions.

Wastewater samples usually contain one or more usable pollutants that require immediate preservation and/or analysis to prevent deterioration, bacterial degradation, or volatilization of constituents or reactions that can form different chemical species. Proper preservation and holding times are essential to ensure sample integrity; preservation procedures include pH adjustment, chemical treatment, or fixation and, most frequently, cooling. (See EPA 300-B 94-014 Table 5-3 and refer to 40 CFR Part 136.)

When composite samples may be stored for as long as 24 hrs, sample preservation must be provided during compositing, usually by refrigeration to 4°C (or icing). With this technique, the sample is quickly chilled to suppress biological activity and volatilization of dissolving gasses and organic substances. Refrigeration begins at the start of sample collection in the field and continues through sample shipment until the sample is received in the laboratory for analysis. Sample temperature should be verified and recorded, especially if the analytical results are to be used in an enforcement action.

As ice used in automatic samplers must be replaced frequently to maintain low temperatures, it is helpful to choose a model with a double-walled insulated base large enough to hold a sufficient quantity of ice. A light color that reflects sunlight also helps the ice last longer. An attractive alternative, especially suited to permanent installations, is a refrigerated sampler. Self-contained "all-weather" models are available that provide precise temperature control, while protecting the sampler from harsh environments and other conditions. These should be provided with environmentally safe non-chlorofluorocarbon (CFC) refrigerant. Enclosures are also available.

In addition to preservation techniques, 40 CFR Part 136 indicates maximum allowable holding times between sample collection and analysis for samples to be considered valid. Not paying attention to holding times can throw off the accuracy of measurements. Certain parameters begin to change or degrade the moment they are collected, including temperature, pH, and dissolved oxygen, as well as total residual chlorine and sulfite; the EPA requires these parameters to be analyzed within 15 min of collection time.

This is another area in which automatic sampling systems are especially helpful. When monitoring the pH of a waste stream, for example, a portable handheld meter can provide misleading results since pH can be highly variable and must be measured in the field, preferably in situ. A continuous monitoring program can, however, be performed using an automatic sampler with pH monitoring capabilities. This procedure allows for reliable accuracy, as users can see fluctuations throughout the day and easily determine the number of minutes out of compliance. (40 CFR 401.17 stipulates no more than 60 min/day or 7 hrs/month of out-of-compliance levels.)

34.6 SAMPLING PROCEDURES AND TECHNIQUES

Sample collection is an integral element of the compliance monitoring program under the NPDES Compliance Monitoring Program. Even with the most precise and analytical measurements, proper sample collection, quality assurance and quality control procedures are essential to ensure useful and valid results. These are discussed in detail in NPDES *Compliance Inspection Manual*, EPA 300-B 94-014 SEPT 94 or later editions.

34.6.1 Quality Assurance and Quality Control Procedures for Sampling

Quality assurance (QA) and quality control (QC) maintain a specified level of quality when measuring, documenting, and interpreting data. QC is a set of preferences that provides precise and accurate analytical results. QA ensures that these results are adequate for their intended purposes.

The first step is to plan and follow appropriate calibration and preventive maintenance procedures and document them in a QC logbook, including equipment specifications, calibration date and expiration date, and maintenance due date. Calibrate automatic samplers for sample quality, line purge, and timing factor. Other factors to consider include collecting representative samples, maintaining their integrity through proper handling and preservation, and adhering to adequate chain-of-custody and sample identification procedures.

34.6.2 EPA Sample Identification Methods

Samples must be completely and accurately identified with waterproof pens on moisture-resistant labels or tags able to withstand field conditions. Each sample should include:

- Facility name and location
- Sample site location
- Sample number
- Name of sample collector
- Date and time of collection grab or composite sample with appropriate time and volume information
- Parameter to be analyzed
- Preservative used

34.6.3 Custody Transfer and Shipment of Samples

To ensure the validity of permit compliance sampling data in court, written records must accurately trace the custody (possession and handling) of each sample through all phases of the monitoring program, from collection through analysis and introduction as evidence. (See EPA 300 B 94-014, Section 5-8.)

34.6.4 Commonsense Points and Precautions

34.6.4.1 Handling and Maintenance. As most equipment failures are caused by careless handling and poor maintenance, sampling equipment should always be handled carefully and maintained in accordance with the manufacturer's instructions. Prepare and follow a detailed checklist, including such periodic procedures as changing the pump tube and checking the line for kinks, cracks, and blockages, and follow all necessary safety procedures as a matter of course.

34.6.4.2 Equipment Security. Equipment security is always a consideration, and equipment should be protected from vandalism, theft, and other tampering. Manhole locations where battery-operated equipment may be installed and the cover replaced will aid in maintaining security. Provide any sampling equipment left unattended with a lock or seal that, if broken or disturbed, would indicate tampering.

34.7 TROUBLESHOOTING

Although specific troubleshooting procedures will vary from model to model, a few general guidelines can be given. For more details, consult the automatic sampling system manufacturer.

Unit Will Not Power Up. Check appropriate fuses, circuit breakers, and AC outlets. Verify that batteries are fresh, fully charged, and correctly inserted.

Short Battery Life. Verify that batteries are fresh and fully charged and that the sampler is set to the proper mode. A sampler set to the "awake" mode all the time will cause a faster drain on the battery than one programmed to respond to set points. Also check the condition of the mechanical elements of the sampler to be certain it is operating efficiently. Be certain that the correct equipment is chosen for the specific application and that it is properly set up. (See appropriate sections in this chapter, for optimum locations and general operating tips.)

Sampler Will Not Create Sufficient Lift. Make sure the anchor strainer—if used—is sufficiently submerged and that the intake tubing is free of cuts or holes. Check for worn pump tubing and roller assemblies, if appropriate. Also be certain the sampler is correctly situated in relation to the sampling source.

Inaccurate Sample Volumes. Verify that the most appropriate equipment and techniques have been chosen for the application in question, that the equipment is properly programmed and well maintained and that correct procedures are consistently followed. Be sure the sampler is correctly situated in relation to the sampling source. Ascertain that the volume has been correctly calibrated.

34.8 CHOOSING A SYSTEM

34.8.1 General Considerations

This first step in choosing an automatic water sampling system is to consider what must be monitored, taking into account:

- Regulatory requirements
- Site requirements
- Reporting requirements

Next, it helps to consider what else could be monitored during the same application. As monitoring is seldom a one-time operation, it is also wise to consider potential future needs; it may prove prudent to purchase a sampler with more capabilities than are needed for the short term but that will also be helpful or essential for later applications. Other broad categories to consider include:

- Type of sites monitored (For example, if there will be multiple sites and a variety of field conditions, consider a lightweight, easy-to-set-up system.)
- Ease of use, including training of staff
- Speed of setup, especially in the field
- Portability
- Frequency of use
- Battery life
- Durability, including the ability to withstand highly corrosive environments

34.8.2 Sample Integrity

As the primary task of a water sampler is to collect water representative of the actual conditions of the source, with the ultimate goal of obtaining accurate information, the sample integrity should be as high as possible. A number of factors contribute to high sample integrity, from the capabilities of the operators and the capabilities of the equipment to precise and complete follow-up procedures. Some of these factors are:

- Using a reliable pump, such as a positive displacement peristaltic pump
- Maintaining a minimum 2 ft/s intake velocity
- Properly preparing sample bottles
- Sampling in well-mixed flow
- Preserving samples
- Following chain-of-custody procedures

34.8.3 Accuracy

While all of these factors are important, the accuracy of the sampler itself is perhaps one of the most vital for most applications. When sampling in a wastestream, for example, volume accuracy and repeatability should be dependable regardless of changes in head or composition.

34.8.4   NPDES Criteria

The NPDES *Compliance Monitoring Inspector Training: Sampling* (J-1) includes the following Criteria for Selection of Automatic Sampling Equipment:

- Capability for AC/DC operation with adequate dry battery energy storage for 120-hrs operation at 1-hr sampling intervals
- Suitability for suspension in a standard manhole while accessible for inspection and sample removal
- Total weight, including batteries, under 18 kg (40 lb)
- Sample collection interval adjustable from 10 min to 4 hrs
- Capability for flow-proportional and time-composite samples
- Capability for collecting a single 9.5-L (2.5-gal) sample and/or collecting 400-mL (0.11-gal) discrete samples in a minimum of 24 containers
- Capability for multiplexing repeated aliquots into discrete bottles
- One intake hose with a minimum inner diameter of 0.64 cm (0.25 in.)
- Intake hose liquid velocity adjustable from 0.61 to 3 m/s (2.0- to 10-ft/s) with dial setting
- Minimum lift of 6.1 m (20 f)
- Explosion proof (for hazardous rated locations)
- Watertight exterior case to protect components in the event of rain or submersion
- Exterior case capable of being locked, including lugs for attaching steel cable to prevent tampering and to provide security
- No metal parts in contact with waste source or samples
- An integral sample container compartment capable of maintaining samples at 4–6°C for a period of 24 hrs at ambient temperatures ranging from −30 to 50°C
- With the exception of the intake hose, capability of operating in a temperature range from −30 to 50°C
- Purge cycle before and after each collection interval and sensing mechanism to purge in the event of plugging during sample collection and then to collect the complete sample
- Field repeatability
- Interchangeability between glass and plastic bottles, particularly in discrete samplers, desirable
- Sampler exterior surface painted a light color to reflect sunlight

Please note that these are minimum criteria and do not necessarily reflect the needs of any individual user, advancements made in automatic sampling equipment technology since the manual was published, and the capabilities of more advanced systems now available. Other criteria for selecting equipment include its current and future uses, specific permit requirements, and other factors, including convenience of installation and maintenance, equipment security, the compatibility of the sampler with other water quality monitoring equipment, cold weather operation, and total life-cycle costs.

34.8.5 Total Cost of Ownership

When choosing an automatic sampling system, more than the initial purchase price must be taken into account. It is also important to consider the total life cycle or total ownership costs over time, with the following factors in mind:

- Initial purchase price
- Longevity, durability, and reliability of the equipment
- Capabilities of the system, including versatility and accuracy
- Potential future sampling applications
- Ease or difficulty of in-field installation, programming, and usage
- Costs of training workers
- Costs, availability, and ease of replacing major components
- Costs, availability, and ease of replacing minor components
- Responsiveness and reliability of technical and customer support
- Ability to acquire and use timely, accurate, and versatile information
- Operating and maintenance costs
- Ability to enhance system performance (upgrades)
- Costs of not having the information
- Costs of reactive versus proactive management

When budgeting, overall equipment costs can be lowered by considering a broad sampler line with one-source standardization. In addition, multitasking instruments allow expenditures to be spread across many departmental budgets and amortized over multiple projects.

An alternative to purchasing automatic sampling systems and other water monitoring equipment outright is to rent or lease them. Programs are available that convert a significant part of the lease or rental charges to the purchase of the equipment.

34.8.6 Additional Purchase Considerations

Any equipment is subject to occasional problems; these can cause downtime, which can lead to considerable expenses, as well as missed opportunities for sampling during critical events. To minimize or avoid such problems, it is important to choose equipment of reliable quality from reputable manufacturers or suppliers and to consider the availability of after-sale support. A responsible vendor should offer such policies as:

- A toll-free 24-h help line, staffed by knowledgeable personnel.
- Parts kept in stock and shipped without delay. (Also consider whether parts are interchangeable and easily replaceable in the field.)
- Return policies to ensure satisfaction.
- No fault warranties for repair or replacement of defective equipment.
- Customer satisfaction programs.
- Upgrade programs as sampling needs or technology changes.

34.8.7   Preparing for Future Technologies

Advances in automatic sampler technology are continually taking place. Of the particular equipment features discussed in this chapter, some are currently available only on the most advanced models but may become standard or may be replaced by future technology. It is advisable to consult with reputable manufacturers and other reliable sources of current information and to check up-to-date product specifications before purchasing or using water quality monitoring equipment. Manufacturer's programs that permit upgrading as emerging technologies become available, and products and software designed to expand with growing needs, help users maintain state-of-the-art monitoring equipment and procedures.

34.8.8   Consulting Engineers

Engineers are an important part of designing treatment plants with water monitoring equipment and should be consulted when planning an equipment purchase. They may even design a sampling location or locations into the plant. In addition, the advice of a knowledgeable consulting engineer is advised when modifications to pipes are necessary or prudent (such as in closed pipe sampling).

34.8.9   Selection of Representative Sampling Sites

Samples should be collected at the location specified in the permit. If said location is not adequate to collect a representative sample, determine the most representative sampling point available. Collect samples at both locations, documenting the reason for the conflict for later resolution by the permitting authority.

35

OPTIMUM WASTEWATER SAMPLING LOCATIONS

Bob Davis and James McCrone

American Sigma
Loveland, Colorado

Environmental Instrumentation and Analysis Handbook, by Randy D. Down and Jay H. Lehr
ISBN 0-471-46354-X   Copyright © 2005 John Wiley & Sons, Inc.

35.1 INTRODUCTION

Although the title of this chapter refers to "optimum" wastewater sampling locations, the best site in terms of collecting representative samples is not always the most convenient, accessible, or safe location. This chapter will therefore address general considerations for choosing sites conducive for collecting representative samples as well as specific considerations for collecting samples at specific sites for particular applications.

35.1.1 Permit-Specified Locations

In many cases, sample locations are specified in a National Pollution Discharge Elimination System (NPDES) permit and must be followed. If the specified location is not adequate to collect a representative sample, choose and negotiate the most representative sampling point available. When determining such a site, the considerations discussed herein should be taken into account as well as knowledge of the site itself. At a plant, for instance, knowledge of the production process and the outfalls may be considered. Samples should be collected at both the specified and the alternate locations, and the reason for the conflict must be documented for later resolution by the permitting authority. (See Chapter 33.)

35.1.2 General Note on Sampling Equipment

Representative samples accurately reflect the true conditions and characteristics of the sampling source. For sampling at most locations, consider that automatic sampling is virtually always more representative of the source. Automatically collected samples can standardize the collection method and collection point and assure repeatable and comparable data from sample to sample.

(For more information about automatic water sampling systems, see Chapter 33.)

35.2 SITE SELECTION: GENERAL GUIDELINES

Careful site selection is critical to acquiring representative samples that can provide accurate data. Here are general guidelines for choosing locations, followed by site-specific recommendations:

1. *Accessibility.* This is especially important when frequent trips are required to collect data. In sampling sewer flow, for example, avoid manholes located in busy traffic areas or with a history of surcharging. Manholes with ladders in good condition are helpful. Also consider physical site restrictions such as 18-in. manholes; in these cases a compact sampler is necessary. Sophisticated equipment is available that utilizes a single microprocessor to combine a portable flow meter, pH/temperature monitor, oxygen reduction potential, dissolved oxygen or conductivity meter with a sampler into a single compact unit designed to sample within manholes and other confined spaces. When appropriate—as in situations where crews must frequently carry equipment over distances—choose equipment that is lightweight and easy to carry and set-up. Some units feature flip-up handles for more convenient carrying. Remember, however, that a representative sample is more important than a convenient sample.

2. *Flow.* The flow at the location should be known or measurable, turbulent, and well-mixed.

3. *Turbulent Flow.* Suspended solids and floating oil and grease create problems when gathering samples. Some materials tend to stratify and can—depending on the position of the sample intake—bias samples. To minimize this problem, choose a well-mixed location; sites immediately downstream from a drop or hydraulic jump (such as a flume) generally provide the mixing necessary for representative sample collection.

4. *Well-Mixed Flow.* Avoid stagnant areas (especially if the wastewater contains suspended solids or immiscible liquids) in favor of areas where there is hydraulic turbulence, such as in aerated channels. To reduce the possibility of contamination, collect samples near the center of the flow. To prevent solids settling, avoid intake locations where solids can collect in a disproportionate amount, or where there are large pieces of floating debris. In a channel, for example, place sample lines in a straight turbulent section. In a wastestream, the sampling depth may range from a few inches below the surface to 60% of the total flow. Sample intakes should offer little or no flow restriction. Strainers can help avoid clogging intake lines.

5. *Placement Near Source.* Place the sampler as close to the source as possible to increase pump life, optimize overall sampler performance, and collect the most representative samples. Locate the sampler higher than the source, with the intake tubing sloping downward to the sampling point, to facilitate complete drainage of the intake line and to help prevent cross-contamination between samples. Keep the vertical and horizontal distance from the sampling source within the capabilities of the sampler. For example, the basic rated capability of a high-speed peristaltic pump with proper pump tubing to achieve consistent lift is a maximum of 27

vertical feet before it loses its vacuum. It should be noted that operation at or near maximum lift puts more wear and tear on the unit if done on a regular, continued basis.

A rule of thumb for maximum vertical lift is to choose the minimum horizontal distance. For the maximum horizontal distance, minimize the vertical lift. If sampling must be performed far from the source either vertically or horizontally, a remote or "intermediate" pump (placed between the source and the sampler) permits sampling at greater distance. (see Fig. 35.1).

6. *Other Considerations*

- Choose a sampling site far enough downstream from tributary inflow to ensure full mixing of the tributary with the main stream. Collect influent samples upstream of recycle stream discharges.

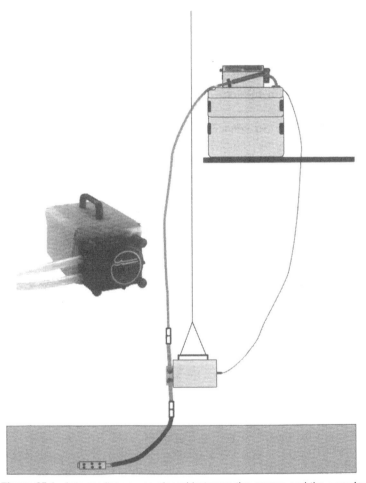

Figure 35.1 Intermediate pump placed between the source and the sampler.

- Avoid areas where nonrepresentative deposits or solids accumulate on channel walls.
- When sampling from flowing pipes, short sample lines help ensure adequate flushing between samples.
- For wide channels or flow paths, dye testing may help determine the most representative sampling site. If the testing is inconclusive, consider cross-sectional sampling.
- The minimum intake velocity should be 2 ft/s.
- Mark selected sample sites for consistency, and whenever feasible keep the same sampling equipment and containers for each site to minimize the risks of cross-contamination of samples.

35.2.1 When to Choose Multiple Versus Single Sites

Multiple monitoring sites may be called for when the size of the area to be monitored is large. Whereas a small treatment plant may be sampled at the influent and effluent, a larger facility could be sampled at multiple internal points for process control (Fig. 35.2).

For industrial storm water monitoring, the industry may be required to monitor several point source outlets (such as their parking lot) to measure the runoff at each one. Occasionally, the state or the Environmental Protection Agency (EPA) allows sampling from only one of similar sites, such as a roof where the runoff characteristics will probably be the same from the front as from the back.

35.2.1.1 Frequency. The frequency with which samples will be taken should also be considered when choosing equipment. If sampling is needed quarterly for short periods, a portable sampler may be sufficient. For more frequent sampling, permanent installations make more sense. Some parameters may require more frequent sampling than others; this is usually specified by permit.

35.2.2 Temporary Versus Permanent Sites and Related Equipment

When deciding between portable temporary sampling equipment at a location or installing permanent equipment, considerations include the physical characteristics of the site [such as the availability of alternating-current (AC) power], frequency of sampling, parameters being tested, and permit requirements. When samplers must be used at multiple sites for short periods, the portability and versatility of the equipment are important considerations.

35.3 SAMPLING AT MUNICIPAL LOCATIONS

35.3.1 Municipal Treatment Plants

Federal and state regulations dictate that the influent (incoming waters) and effluent (outgoing waters) of wastewater treatment plants (WWTPs) of publicly owned treatment works (POTWs) be sampled daily to assure compliance with discharge permits. Well-operated plants also sample a number of intermediate points within the facility to determine the effectiveness of certain points in the process. Sampling provides data so that operators can make operational adjustments, ensuring that the final effluent meets federal and state statutes at the least possible cost. Permanently installed refrigerated samplers are typically utilized in waste treatment plants because of the availability of AC power and the 4°C sample temperature requirement.

Wastewater treatment plants are generally monitored to sample influent and effluent for compliance with NPDES permits and to sample intermediate points within WWTPs to evaluate the treatment process, identify unit inefficiencies, and correct operational problems. The NPDES permit often specifies that the internal wastestreams be sampled. (See Table 17.2 in Ref. 1, Suggested Sampling Locations, Typical Analyses Performed, Sampling Frequencies, and Type of Sample required For Wastewater Treatment Unit Processes.)

In proactively minded municipalities, monitoring is also used for planning.

35.3.1.1 Optimum Locations

For Influent Sampling

- At permit-specified sites if adequate to collect representative samples.
- At points of turbulent flow (to ensure good mixing—such as an influent line upflow distribution box from the plant wet well, or a flume throat—upstream of sludge or supernatant recirculation. When samples are taken from a closed conduit (with a vale or sample tap), or from a well equipped with a hand or mechanical pump, sufficient flushing time is necessary to ensure a representative sample, considering the flow velocity, diameter, and length of pipe.
- Above plant return lines.

For Raw Wastewater

- Flowing from the final process in manufacturing operations
- Pump wet wells (if turbulent)
- Upstream collection lines, tank, or distribution box following pumping from wet wells or sumps
- Flume throats
- Aerated grit chambers (if available; otherwise upstream siphons following the comminutors)

For Effluent Sampling

- At permit-specified sites and/or at the most representative site downstream from all entering wastestreams before they enter the receiving waters. To prevent liquid from pooling when sampling in an effluent stream, the sampler should be positioned above the stream so that the tubing runs in a taut, straight line.

- Collected after chlorination. (If the permit specifies sampling prior to chlorination, all parameters can be monitored at the upstream location except fecal coliforms, pH, and total residual chlorine. Collect wastewater for use in bioassays at the location specified in the facility's NPDES permit.)
- Collected facing upstream to avoid contamination.
- Placed to avoid collecting large nonhomogeneous particles and objects.

35.3.1.2 *Specifications of Optimum Sampling Equipment.* When choosing sampling equipment for WWTPs, look for their ability to operate in highly corrosive atmospheres. Multiparameter capabilities put more functionality in one place and maximize equipment investment, which can be particularly helpful in smaller plants. For example, samplers that monitor pH and flow with sampling make it easier to bring data into a consistent format. Also, operators need learn only one piece of equipment instead of three.

For process control, sampling is usually performed along with such other parameters as pH and flow values. Advanced samplers that measure flow and water quality parameters with alarm capabilities help operators react quickly to changes in flow characteristics. Automatic reactions are also possible with set-point technology. As an example, an output signal can be sent to control chemical releases to maintain proper levels of key parameters. This permits the optimal usage of chemicals and prevents operators erring on the high side and using more chemicals than they need.

35.3.1.3 *Tips.* Consolidate instrumentation into one unit (see above). Since some plants are highly corrosive, there are benefits to the superior corrosion resistance of newer technology, such as an "all-weather" device with a top-mounted controller, all fiberglass construction, and other features.

Be certain to choose a reliable refrigerated sampler (preferably microprocessor-controlled) that consistently maintains 4°C. Temperature is an easy specification for inspectors to check; if it's off, the permitee is out of compliance, subject to fines, and may have erroneous data and invalid samples. Some advanced samplers show the current sample temperature through the controller and can provide documentation that the sampler has maintained 4°C over a specified period of time.

35.3.2 Pretreatment

General pretreatment regulations have been implemented to make sure that industrial discharges to POTWs are within their permitted parameters and in compliance with local sewer use ordinances. The purpose of the regulations is to protect POTWs and the environment from the damage that may result from discharges of pollutants to sanitary sewer systems.

The three specific objectives cited in 40 CFR 403.2 of the General Pretreatment Regulations[2] are to:

- Prevent the introduction of pollutants that would cause interference with the POTW or limit the use and disposal of its sludge.
- Prevent the introduction of pollutants that would pass through the treatment works or be otherwise incompatible.
- Improve the opportunities to recycle or reclaim municipal and industrial wastewater and sludges.

Local sewer use ordinances are designed to (1) assure that no incompatible pollutants enter the system and cause damage to the sewer or treatment plant, (2) assure that an industrial discharge does not cause the POTW to violate its NPDES permit, and (3) to prevent the contamination of municipal sludges that might limit sludge disposal options.

Sampling is also used in pretreatment to:

- Determine and document compliance with all applicable federal requirements.
- Identify and locate all possible industrial users that may be subject to pretreatment programs.
- Identify the character and volume of pollutants contributed to POTWs.
- Inspect and sample industrial users.
- Investigate instances of noncompliance and perform sampling and analysis to produce evidence admissible in enforcement proceedings or judicial actions.
- Improve POTW worker health and safety.
- Reduce influent loading to sewage treatment plants.
- Assess industries' "fare share" of the cost of treatment as a revenue source for the POTW.

Pretreatment monitoring may be performed by the industry itself and verified by the municipality or by the municipality for the industry. The industry may be required by the municipality to install a sampler, from which the municipality may routinely collect the samples. Municipalities also generally use portable, battery-operated samplers for industrial discharge verification.

35.3.2.1 *Optimum Locations.*
The optimum placement of sampling equipment for pretreatment monitoring is at the final discharge of an industrial facility into a municipal sewer system. This may be at a manhole in the street or an access point in the facility or on the property. The final point gives the municipality the total discharge. Certain industries today are becoming more proactive about controlling and pretreating their process before it gets to the final discharge point. Plant size and complexity should also be considered; at a large site, it is also prudent to sample at a number of locations. Some industries may be required to have a certain level of pretreatment equipment.

Although some smaller municipalities may monitor at the influent of the treatment plant, this placement does not give sufficient time to react to a problem discharge. Municipalities want to be certain that the makeup of the flow can be treated; industrial slugs could thwart the biological treatment process.

35.3.2.2 *Specifications of Optimum Sampling Equipment.* Perhaps the most important consideration in selecting equipment is that the samples are both representative and defendable. Since the data must potentially hold up in court, documentable accuracy is crucial. Choose equipment that is easy to use, to avoid operator error, with simple, automatic documentation. (For example, sample times and dates, bottle numbers, sample temperature, and other information can be captured with data logging and reporting features.) Portable equipment must conserve power so that it can operate on batteries for sufficient time to complete the investigation. Long battery life is best achieved through a lower power draw from the equipment itself, with an optimum standby state of 5 mA.

35.3.2.3 *Tips.* Use set-point sampling to reduce the number of samples taken and to have a higher rate of success at identifying upsets (Fig. 35.2). A municipality, for example, can sample an industry four times an hour all day long, take the samples to a laboratory, analyze them, and can still miss an upset based on time or flow. Using an indicator like pH, however, grab samples may be collected only when parameter limits are exceeded—such as when a discharge goes out of compliance. One community, for example, put samplers upstream and downstream of a particular plant suspected of illegal discharges. When the facility discharged in the middle of the night, samples were taken and the sampler set off an alarm.

Note that when samplers use set-point technology, a very low power drain allows the equipment to operate for a long time.

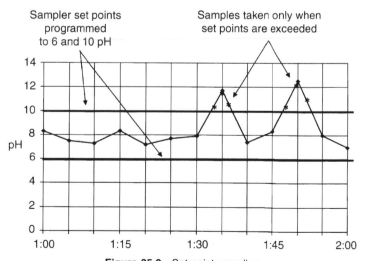

Figure 35.2 Set-point sampling.

Modern automatic samplers are also capable of collecting a composite sample in one bottle, segregating a discrete upset sample in a trouble bottle, then continuing with the regular sampling routine using the first bottle.

It should be remarked that because monitoring is used in part to protect a very expensive treatment process at the main treatment plant, the investment in superior technology can be easily justified. In addition, the same technological tool that helps municipalities determine industries that are out of compliance helps those that want to stay in compliance do so, by functioning as a proactive diagnostic tool to help pinpoint and solve problems.

35.4   COMBINED SEWER OVERFLOW

Before discussing combined sewer overflows (CSOs, Fig. 35.3) and sanitary sewer overflows (SSOs) it will be helpful to consider the following key definitions.

Sewers are pipes that carry wastewater and storm water runoff from the source to a treatment plant or receiving stream.

Sanitary sewers carry domestic and industrial wastewater, without storm water runoff.

Collection sewers are common lateral sewers primarily installed to convey wastewater from individual structures to trunks or intercepting sewers.

Combined sewers—remnants of America's early infrastructure—collect both wastewater and storm water.

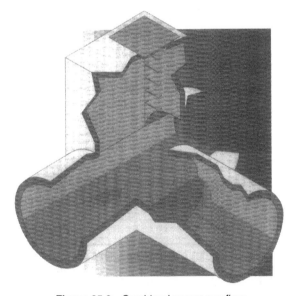

Figure 35.3   Combined sewer overflow.

Storm sewers are systems that collect and carry rain or snowfall runoff to a point where runoff can soak back into the groundwater or flow into surface waters.

Wastewater is the spent part of a community. From the standpoint of source, it may be a combination of liquid and water-carried wastes from residences, commercial buildings, industrial plants, and institutions together with any groundwater, surface water, and storm water that may be present.

Combined sewer overflows (CSOs). Dry weather CSOs indicate maintenance problems or other serious deficiencies in the wastewater treatment system; they are prohibited and must be eliminated. In wet weather, these systems may become overloaded, causing untreated water to overflow into the nearest body of water, posing significant health and environmental risks.

Combined sewer overflows—as well as sanitary sewer overflows (SSOs)—can spill raw sewage into public and private areas, causing substantial health and environmental risks. In addition, property and environmental damage can be extensive and expensive to rectify. Overflows can happen occasionally in almost every sewer system. Frequent occurrences, however, can be caused by:

- Too much rainfall or snowmelt infiltrating into leaky sanitary sewers
- Excess water inflowing into sewers
- Sewers and pumps too small to carry sewage from newly developed sub divisions or commercial areas
- Blocked, broken, or cracked pipes and other equipment or power failures
- Deteriorating sewer systems due to age or improperly installed or maintained sewers

On April 19, 1994, the EPA published the National CSO Control Policy in the *Federal Register* (40CFR, Part 122), containing the agency's objectives, control plans, and alternative approaches for addressing CSOs. In order to meet the provisions of the Clean Water Act, the *Federal Register*, Part VII, Combined Sewer Overflow Control Policy states that permittees with CSOs should accurately characterize their sewer systems and demonstrate implementation of nine minimum CSO controls.

The ninth control is: "Monitor to effectively characterize CSO impacts and the efficacy of CSO controls."[3]

35.4.1 Sampling Requirements

Communities with combined sewer systems are also expected to develop long-term CSO control plans, of which sampling can be a vital element. The stated minimum elements of the long-term CSO control plan includes: Characterization, monitoring and modeling of the Combined Sewer System."[3] The comprehensive, representative monitoring of the CSO discharges should include:

- Frequency
- Duration
- Flow rate
- Volume
- Pollutant concentration

Characterizing combined sewer overflows requires a unique monitoring protocol that involves sampling along with measuring rainfall, pH, temperature, and flow. The automatic sampling equipment should collect a timed first-flush grab sample within the first 30 min of the overflow (representing the early stages of the event, when pollutant levels are expected to be higher), while concurrently depositing a flow-proportional composite in one or more containers (representing the duration of the event).

The impact of the CSOs on the receiving waters should also be assessed. The monitoring program should include necessary CSO effluent and ambient in-stream monitoring and, where appropriate, other monitoring protocols such as biological assessment, toxicity testing, and sediment testing. Automatic samplers allow sampling at the initiation of an event and improve personnel safety during storm events. Since multiple parameters may be monitored, advanced automatic water sampling equipment should be considered for this task (see Chapter 33).

35.4.2 Optimum Locations

Samplers must be located as near to the overflow point as possible to achieve samples representative of the discharge. If placed too far upstream, the flow may be mixed with other elements. Since these locations can be dangerous or difficult to access, an outdoor enclosure is sometimes used above a nearby manhole. If the sampler is placed in or above a nearby manhole, locate the intake at the overflow point.

35.4.3 Specifications of Optimum Sampling Equipment

In order to characterize the pollutant loading over the duration of the discharge, the permit may require frequent sampling (e.g., every 5 min)—a task ideally suited to automatic systems. Since the samplers are often left out for long durations, it is important that the sampler maintain sufficient power to complete the event. A low power draw (in a standby state with set-point activation based on level, flow, or other parameters) allows samplers to run for a longer time on batteries. Another option is using a larger battery, such as a marine or automobile battery or similar 12-V power source. Test the sampler before leaving the site to be sure it is set up properly and will not miss an event.

Other capabilities to consider are:

- A sample pump that provides high lift capabilities to maintain EPA-required transport capabilities for representative sampling
- Software that ties together flow and samples to more easily document pollutant loading with storm intensity and duration
- The ability to combine sampling and flow and rain logging in one unit so that it is easier to set up and secure the equipment
- Remote communication for event notification

35.5 SANITARY SEWER OVERFLOW

Much of the information discussed above under Combined Sewer Overflow is also relevant to sanitary sewer overflows (SSOs). SSOs are typically created by buildups or blockages in the system. SSOs can cause untreated wastewater to overflow into basements or out of manholes onto city streets and into the receiving stream prior to treatment.

Municipalities should be aware of their obligations regarding SSOs, especially during wet-weather conditions. SSOs of untreated or partially treated wastewater from collection systems that may reach U.S. waters are violations of Section 301 of the Clean Water Act (CWA) and the provisions of NPDES permits, and are therefore subject to enforcement actions. In addition, federal regulations 40 CFR Part 122.41(1)(6) require that all discharges that may endanger health or the environment must be reported to the EPA. SSO monitoring is discussed in more detail in Chapter 42.

35.6 STORM WATER

In recent years, POTWs have established Storm Water Utilities and Flood Control Districts to provide urban storm water management. As part of the management plan, the utility monitors rainfall along with flow and water quality to assess the impact of wet-weather events on receiving water. Monitoring also helps verify treatment effectiveness. A typical storm water monitoring system consists of a rain gauge, flowmeter, sampler, and pH and temperature monitors.

The EPA defines a storm water sewer system as a drain and collection system designed and operated for the sole purpose of collecting storm water that is segregated from the process wastewater collection system. A *segregated* storm water sewer system is defined as a drain and collection system designed and operated for the sole purpose of collecting rainfall runoff at a facility and that is segregated from all other individual drain systems. Sampling helps determine, during or after a rain event, where the runoff goes and how it is characterized. This can come from a single or multiple locations.

When sampling for a storm event, a discrete sample of the first flush taken during the first half-hour can identify pollutants that run off immediately following the rainfall. After the first flush, a flow-weighted composite sample can be used to estimate pollutant loading into water resources.

35.6.1 Sampling Requirements

Monitoring for storm events should be conducted when there is sufficient rainfall to create a discharge of pollutants that have been allowed to collect during the preceding dry period. Allowable events may be permit specified and may, for example, include a storm that results in at least 0.1 in. of rainfall, preceded by at least 72 h without rainfall greater than 0.1 in., where the rainfall is between ±50% of the local average rainfall depth and duration. The EPA requires a flow-weighted composite sample for the storm duration or the first 3 h, whichever is less. (This, when combined with the volume discharged, provides the pollutant loading to receiving waters.) Municipal dischargers must provide data for 3 storm events, meeting EPA's representative criteria, at 5–10 outfalls. Industry must provide data for one event. (The number of monitored storm events may vary between states; it is important to check with the relevant state regulatory group.)

35.6.2 Optimum Locations

Many industries must monitor discharges that contain "stormwater discharges associated with industrial activity." (For a complete definition, consult the glossary in Appendix B of Ref. 4.) These samples must be collected from the storm water discharges leaving the property, at or near the property boundary, at a point that is representative of the total site discharge. Other accessible locations on or off the property may be selected—perhaps in a channel or ditch that collects runoff from a property, but the storm water collected should include all the industrial activity area under storm water regulations.

One or multiple sampling sites may be permit specified at points believed to be representative of the runoff, based on drainage, topology, location of storm sewers, land use, and other factors.

According to the *NPDES Storm Water Sampling Guidance Document*[5] (EPA 833 B-92-001, section 2.8), storm water samples should be taken at a storm water point source—defined as "any discernible, confined and discrete conveyance, including (but not limited to) any pipe, ditch, channel, tunnel, conduit, well, discrete fissure, container, rolling stock, concentrated animal feeding operation, landfill leachate collection system, vessel, or other floating craft from which pollutants are or may be discharged (as per 40 CFR 122.2)." The term also includes "storm water from an industrial facility that enters, and is discharged through, a municipal separate storm sewer."

Industrial applicants submitting individual applications should consult the *NPDES Storm Water Sampling Guidance Document*[5] (EPA 83-B-92-001, July 1992, section 2.8.1) for detailed information regarding specific sampling requirements. Municipalities should consult section 2.8.2 of the above-mentioned source.

Ideal sampling locations include:

- The lowest point in the drainage area where a conveyance discharges storm water in to U.S. waters or to a municipal separate storm sewer system
- Locations easily accessible on foot
- Nonhazardous locations
- Sites on the applicant's property or within the municipality's easement
- The discharge at the end of a pipe, channel, or ditch

35.6.2.1   *Tips.* One tip to follow when locating a sampler for storm sampling is to look for a high-water mark to help ensure operator safety and to prevent the sampler from being washed away.

Should logistical problems arise with sample locations, refer to the *NPDES Storm Water Sampling Guidance Document*,[5] Exhibit 2-12, "Solutions to Sample Location Problems."

35.6.3   Specifications of Optimum Sampling Equipment

At a minimum, automatic water sampling equipment should adhere to EPA's criteria for representative storm water monitoring. In addition, a sampler with an integral flowmeter simplifies transporting, setting up and retrieving the equipment. It is helpful to employ a system with the flexibility to initiate sampling activity based on depth only, rainfall only, depth or rainfall, or depth and rainfall. A reliable liquid sensing system can eliminate sample volume calibration and assure consistent volumes even with changes in head at the sample intake. A subcompact size—capable of fitting in 18-in. manholes—makes handling easier, and a sealed NEMA 4X,6 design can withstand submersion, hose spray, and the tough corrosive environments often encountered when monitoring storm water. Also helpful are data-logging and reporting capabilities of storm-water-friendly software.

As flow is also measured, it is useful to have an automatic sampling system that includes an integral flowmeter; this minimizes the amount of equipment that must be transported and set up and can also simplify data collection and reporting (Fig. 35.4).

The EPA storm water regulations demand a unique monitoring protocol involving, at a minimum, sampling as well as the measurement of rainfall, pH, temperature, and flow. The automatic sampling equipment should collect a timed first-flush grab sample within the first 30 min of rain (representing the early stages of the storm, when pollutant levels are expected to be higher), while concurrently depositing a flow-proportional composite in one or more containers (representing the duration of the event). The sampling may be initiated by rainfall and/or depth in the channel. Since storm events can occur unexpectedly, automatic equipment can eliminate the need for personnel to instantly respond and then to "baby sit" the site throughout the sampling period. It can also eliminate the need for personnel to be exposed to lightening or other hazardous conditions during storms and heavy

Figure 35.4 Storm water setup.

rainfalls. Automatic monitoring equipment can be placed at the outfall, programmed to a standby/ready state in advance of the storm, in the safety of dry weather during daylight.

Electronic data logging—with the data stored in electronic memory—avoids having to create written records in wet conditions.

35.6.3.1 Tips. Many of the tips mentioned above for CSO sampling concerning such factors as battery life also apply to storm water sampling. Also consider:

- Locate the sampler far enough away from the storm flow so it will not become submerged or washed away, to avoid losing the samples and the event.
- Make sure the sampler is secure and vandal-proof.
- When choosing a location, consider accessibility for retrieving samples.
- Since the sampler should be able to operate during a rain event and withstand temporary submersion, consider such features as a sealed controller.
- Since there are specific monitoring protocols for storm water and storm event monitoring, software designed for this use can be particularly helpful.
- Be certain the sampler will actually meet the complete EPA requirements for a representative storm event. Reliable equipment is a sensible investment; since one of the highest costs of sampling is missed events (taking into account manpower, laboratory analysis, and other expenses of repeating the sampling) because of unreliable equipment.

For more details on storm water sampling and monitoring, refer to the November 16, 1990 *Federal Register*, as well as the Phase II update.

35.7   NON–POINT SOURCE RUNOFF

Controlling non–point source (NPS) pollution requires different strategies—and can be more difficult—than controlling pollution from point sources. Whereas pollutants from point sources enter the environment at well-defined locations and in relatively even, continuous discharges, NPS pollution can take place over thousands of acres, resulting from a variety of human activities (see Ref. 6). NPS pollutants usually enter surface and groundwaters in sudden surges, often in large quantities, and often associated with rain and storm events or snowmelt.

35.7.1   Optimum Location

Locating sampling equipment at the optimum location to collect truly representative samples can be challenging. An effective strategy is to start at a problem area and continue working upstream to locate the discharge point. Examining the topography and looking for low points where the runoff is likely will provide the greatest sampling of the overall area. Rural farm runoff is frequently the source of NPS pollution and is receiving increasing attention at federal and state levels for improved best management practices.

35.7.2   Specifications of Optimum Sampling Equipment

Specifications of optimum sampling equipment and tips for effective NPS sampling are similar to those recommended for storm water above.

35.8   SAMPLING AT INDUSTRIAL LOCATIONS

35.8.1   Industrial Discharge

Like municipal waste treatment plants, industries that use water in their production processes are limited by permits as to the contaminants that may be discharged. These discharges must be sampled regularly to assure the regulatory bodies that effluent limits are adhered to. Industries must also frequently monitor the volume of their discharge, as permits typically require that samples be collected in proportion to flow. In these cases, a flowmeter controls the actuation of a sampler based on the passage of a preset flow volume. Alternately, a sampler may be chosen that features an integral flowmeter.

35.8.2   Sampling Requirements

Industrial dischargers to POTWs must comply with:

- Prohibited discharge standards (40 CFR 403.5).
- Appropriate pretreatment standards (40 CFR Part 405-471), along with state requirements or locally developed discharge limitations per 40 CFR 403.5.

- Reporting requirements, as specified in 40 CFR 403.12 and/or by the POTW. For more details, see the *NPDES Compliance Inspection Manual*[5] (EPA 300-B 94-014, September 1994, 9-6).

Industries that discharge directly to surface waters and indirectly to POTW sewer systems monitor their discharges in order to:

- Satisfy NPDES or pretreatment monitoring requirements.
- Satisfy federal and state storm water regulations.
- Trace the source of problem discharges to a specific production process.
- Determine the cost of waste treatment to be attributed to each production unit.

35.8.3 Optimum Location

Most of the information conveyed and considerations discussed above in reference to pretreatment are equally applicable to industrial discharge monitoring. A key factor is this: It is the industry's responsibility to ensure that discharges are in compliance with regulations. Municipalities generally monitor industries to verify the industry's compliance. The more proactive industries, and those who act responsibly as corporate citizens, will also employ monitoring procedures to improve their plant processes and take action to minimize any potential for out-of-compliance discharges.

For specifications of optimum sampling equipment and effective operating tips, please see the section on pretreatment above.

35.8.4 Commercial Laboratories

Commercial laboratories, as well as engineering firms performing field monitoring services, also monitor water for a variety of reasons, including NPDES and pretreatment compliance sampling for industrial clients. General information conveyed and considerations explored for specific applications in this work are also applicable for these uses.

35.8.5 Difficult or Challenging Conditions

As sampling locations are often in reality less than optimum, it is wise to be prepared for the following conditions.

35.8.5.1 Closed Pipe Sampling. Whereas an open pipe, such as with a manhole or sewer line, is usually a gravity flow condition, a closed pipe is usually pressurized (pumped flow). Setting up equipment for and collecting representative samples from a pressurized line is difficult and offers limited equipment choices. When it is called for, the applications are more typically industrial than municipal. Examples are when a sample is needed near a wet well, or from a pressurized line in an industrial plant, to sample at a variety of process points.

35.8.5.2 Specifications of Optimum Sampling Equipment. Automatic samplers with peristaltic pumps can sample from a pressurized line if the pressure is less than 12 psi. If the pressure is greater than 12 psi, an option is to divert flow to a discharge point with a pressure-reducing valve and sample from that point (Fig. 35.5). Make certain that the sample volume and the location within the pipe are representative of the flow. Keep in mind that the laboratory determines the sample volume required for analysis and the sampler should be capable of drawing that volume.

For monitoring in closed pipes, a standard sampler can be modified with a solenoid valve that can be activated by the pump control to open and close in order to collect timed samples. It is advisable to involve an engineer to design the setup specific to the application. In the event of a break in the sampling line, a readily accessible shut-off valve is prudent to prevent leaking or spraying of pressurized wastewater.

35.8.5.3 Low flow. A low-flow strainer can be helpful when sampling under such conditions.

35.8.5.4 Distance. When the sampler is located far from the source, a remote pump can permit higher lift and longer distance.

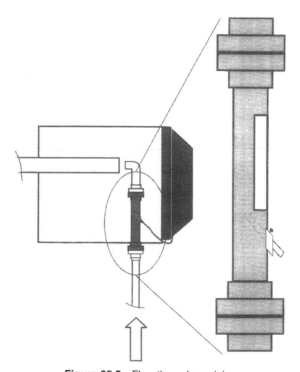

Figure 35.5 Flow through module.

35.8.5.5 Small Manholes. Some manholes are as small as 18 in. in diameter. In such cases, compact sampling systems can be employed. When flow must also be monitored, samplers are available with integral flow measurement, necessitating only one piece of equipment instead of several. Samplers can also be placed in an enclosure above the manhole, with the intake line dropped into the manhole.

35.8.5.6 Inclement Weather. Using most automatic samplers during cold weather can present such problems as sampler malfunctions and frozen intake lines. These problems may be handled with heat tape or by placing the sampler inside a weatherproof, thermostatically controlled, electrically heated enclosure. Walk-in enclosures also allow sufficient space for the operator and may be equipped with insulation, a heater, an exhaust fan, and light. Unless special equipment is available, freezing may be prevented by installing the sampler in a manhole or wet well or by wrapping it with insulation and wind protection. The superior solution, however, where AC power is available, is a self-contained "all-weather" model that provides precise temperature control, while protecting the sampler from harsh environments in temperatures from -40 to $120°F$.

35.8.5.7 Highly Corrosive Environments. When sampling must be performed where there are poor ambient temperatures and harsh environmental conditions, choose equipment specifically designed for such applications. The features and specifications available on advanced self-contained refrigerated sampling systems include a top-mounted compressor, chemically resistant coating on refrigeration lines, a NEMA 4X,6 electronics enclosure rating, stainless steel hardware, controller compartment, heater, sealed electrical connections, weather-resistant cables, insulated sample compartment, temperature-resistant liquid crystal diodes (LCDs), and precise temperature controls that guarantee a constant $4°C$. Enclosures may also be considered to protect a sampler from harsh environments.

35.9 LOCATIONS TO AVOID

35.9.1 General Characteristics

In general, following the general and site-specific recommendations above will go a long way toward effectively collecting representative samples under most conditions.

Safety, however, should always be a consideration. Some sites may be favorable for sampling but hazardous, such as the likelihood of lightening strikes, high-flow conditions, erosion, or storm events. In such cases carefully consider the balance or trade-off between an optimum site and optimal safety.

Note that, most samplers are *not* certified for operation in hazardous locations in the United States.

Wherever possible, the following site characteristics should be avoided:

- Locations where the point of sampling is above the sampler
- Unsafe locations

- Sites that are difficult to get to for installing equipment as well as collecting the samples
- Sites lacking well-mixed flow
- Locations where the sampler can be easily vandalized or tampered with
- Locations where the equipment is subject to being submerged or washed away

When the above conditions cannot be avoid, practice care and diligence to ensure representative sampling and the safety of all personnel installing, operating, or inspecting the equipment or retrieving data and samples. If possible, locate the sampler away from the dangerous area. Be certain the pump capabilities are sufficient for the length of the intake line.

35.9.2 Hazardous Locations

Class I, Div I. (Class one, Division One) of Hazardous (Classified) Locations in accordance with Article 500, National Electrical Code-1990 is defined as locations with flammable gasses or vapors in which hazardous conditions occur at all times.

Confined Spaces. According to the U.S. Department of Labor Occupational Safety and Health Administration (OSHA) in OSHA 3138, 1993, a confined space has "limited or restricted means of entry or exit, is large enough for an employee to enter and perform assigned work, and is not designed for continuous occupancy by the employee." A permit-required confined space contains or has the potential to contain serious safety or health hazards. Special training in preentry testing, ventilation, entry procedures, evacuation/rescue procedures, and safety work practices is necessary to ensure against the loss of life in confined spaces.

Sewers. Instead of entering the manhole (which is a hazardous location) the intake line can be dropped into the manhole.

Headworks. The front end, at the influent of a treatment plant, is typically defined by OSHA as a hazardous location, largely due to a preponderance of H_2S gas. When sampling influent at the headworks is necessary, it can sometimes be performed from an adjacent room or just outside the building. Because this requires a longer intake line, a high-speed peristaltic pump is recommended.

Outdoor. Example: where lightning may strike.

35.9.3 Commonsense Points

Some general considerations for sampling at practically any location are:

- Make certain the line is not kinked.
- Be certain the battery is fully charged for portable units.

- Test the sampler before leaving the site to be sure it is working properly and that events will not be missed.
- Secure and protect the sampler to avoid vandalism and tampering.
- Locate the sampler away from possible high-flow events, which could submerge and wash away the equipment and other dangerous and hazardous conditions
- Locate samplers to avoid vandalism and tampering. Consider lockable equipment with password protection. In enforcement applications, consider installing samplers at night. In storm water applications, install in protective, locked enclosures.

REFERENCES

1. *Operation of Municipal Wastewater Treatment Plants*, 5th ed., Vol. 2, Water Environment Federation.
2. 40 Code of Federal Regulations (CFR) Part 403, General Pretreatment Regulations, Federal Register.
3. *Meet EPA Requirements by Preventing Overflows*, SIG 4884, American Sigma, 1998.
4. U.S. Environmental Protection Agency (EPA), *SWP3 Manual*, EPA, Washington, DC.
5. U.S. Environmental Protection Agency (EPA), *NPDES Storm Water Sampling Guidance Document*, EPA, Washington, DC.
6. U.S. Environmental Protection Agency (EPA), *National Water Quality Inventory*, EPA, Washington, DC, 1991.
7. *Storm Water Monitoring Application Handbook*, American Sigma, November 16, 1990, Federal Register.
8. *Compliance Monitoring Manual*, 21W-4007.
9. *NPDES Compliance Inspector Manual*, EPA 300-B 94-014, September 1994.

36

WASTEWATER LEVEL MEASUREMENT TECHNIQUES

Ernest Higginson

Greyline Instruments, Inc.
Massena, New York

Environmental Instrumentation and Analysis Handbook, by Randy D. Down and Jay H. Lehr
ISBN 0-471-46354-X   Copyright © 2005 John Wiley & Sons, Inc.

Level monitoring and control is a fundamental requirement in any wastewater treatment process. You will find level instrumentation installed at even the simplest treatment plants for pump control, chemical storage tanks, and process level controls or alarms. Operators have only limited or emergency control over treatment plant influent, so level controls and instrumentation play an important role in managing the wastewater treatment process.

By careful planning of process elevations, engineers can design treatment systems where gravity manages wastewater levels—and therefore flow—through most of the treatment process. The type of plant and topography of the site are major factors, but the requirement for pumps, controls, instruments, and operator intervention can be minimized by good treatment plant design.

Wastewater treatment plant operators overwhelmingly endorse simple plant designs and instrumentation. Susan Salvetti, Instrument Technician at the Albany County Sewer District, states, "We try to keep things simple. More complicated equals more expensive" (Fig. 36.1). Corrosive atmospheres in wastewater treatment plants attack everything from instrumentation to the mechanical structures and fixtures at a treatment plant. To be reliable, instruments must be designed to withstand the harsh operating environment.

Figure 36.1 Albany County North Sewage Treatment Plant, Albany NY.

36.1   LEVEL CONTROL AND ENVIRONMENTAL REGULATIONS

Environmental standards for wastewater treatment plants are becoming increasingly stringent. Although enforcement varies from region to region, both public and private treatment plant operators may now be held legally responsible for proper operation of their wastewater treatment facilities. New regulations are constantly being introduced with higher and higher standards of safety and effectiveness of the treatment process. Tank overfill protection and combined sewer overflow monitoring are current examples where regulators are imposing new requirements for level controls and instrumentation.

36.2   COMMON TERMS USED IN WASTEWATER LEVEL INSTRUMENTATION

Sensor. Any device that directly measures a physical condition. A float is an example of a sensor.

Transducer. A device that contains a sensor and converts its signal into an electrical signal for communication with other devices. A hydrostatic pressure sensor is an example of a transducer.

Instrument. The combination of sensors or transducers into a communication device that may transmit a signal, display a variable, and provide control functions. An ultrasonic level indicating transmitter is an example of an instrument.

Transmitter. An instrument that transmits a continuous process signal to a remote device. A 4–20-mA analog signal is the most common where each transmitter uses a pair of wires ("twisted pair") to deliver an analog current proportional to the measured variable. Modern serial or digital transmitters are rapidly entering the market where a number of "multidrop" transmitters can share the same wire by taking turns or by transmitting data only when "polled" by a remote controller.

Accuracy. How closely an instrument measures the actual value of a material being sensed.

Resolution. The smallest increment of change that the instrument can measure.

36.3   CONSIDERATIONS IN SELECTION OF LEVEL INSTRUMENTS

1. Price is always important, but operational life and maintenance requirements of an instrument should be the first factors in the selection process.
2. Point level or continuous level may be required.
3. Compatibility of exposed sensor materials, temperature and pressure specifications should be determined.

4. Whether the instrument or sensor locations are rated hazardous should be checked.

5. Any restrictions in sensor mounting location should be determined.

6. The application should be considered and the appropriate technology selected.

7. Instrument functions should be selected 4–20 mA, display, control relays, serial outputs, etc.

36.4 LEVEL INSTRUMENT TECHNOLOGIES USED IN WASTEWATER TREATMENT

36.4.1 Ultrasonic Level

Common Level Applications

- Pump control in wet wells and sumps
- Chemical storage tank inventory
- Aeration basin level
- Aerobic digester level
- Chlorine contact tank level
- Sludge tank level
- Bar screen differential level

The operating principle of ultrasonic level instruments is quite simple (Fig. 36.2). They use a transducer with combined transmit and receive capability. The instrument measures the time it takes for a sound pulse to travel from the transducer to a target and then for the echo to return. Because we know the speed of sound in air (1086 ft/s or 331 m/s), the distance to the target can be accurately calculated.

Ultrasonic level instruments have become one of the most popular technologies used in wastewater treatment applications. The obvious advantage is that they use a noncontacting acoustic sensor. That means minimal fouling of sensors and consequently little or no maintenance. Ultrasonics also offer relatively high accuracy (±0.25% of full scale is typical) at low cost.

Adoption of ultrasonic instruments in wastewater treatment did not come about easily. In the early 1980s most operators considered them an exotic solution to wastewater level application problems and applied them where a noncontacting sensor seemed to be the only solution. "If nothing else works, try an ultrasonic" was a typical attitude. Early converts to ultrasonic instrumentation were pioneers, often developing calibration, installation, and maintenance skills that were passed back to manufacturers.

With the advent of microprocessors, ultrasonic technology moved into the level instrumentation mainstream, and today it is one of the most common techniques in use at wastewater treatment plants. Many of the characteristics unique to ultrasonics can be managed automatically by signal-processing algorithms programmed into

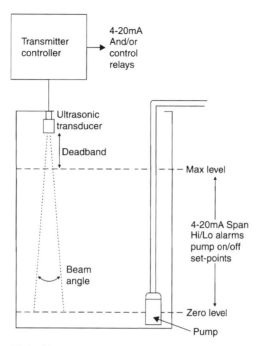

Figure 36.2 Noncontacting ultrasonic level controller transmitter.

each instrument. Today's operators can take successful performance of an ultrasonic instrument for granted assuming the instrument is properly applied and installed.

The term *ultrasonic* is used because the frequencies of most instruments are above the audible range of the human ear, although one can also usually hear a lower frequency clicking from ultrasonic transducers. Operating frequencies range from about 12 to 90 kHz. Lower frequencies are used by sensors measuring the greatest ranges, while higher frequencies are best for short measurement ranges.

Ultrasonic level instruments all have a "blind" space immediately in front of the sensor. Manufacturers often describe it as a sensor "deadband" or "blanking zone." As a rule of thumb, the lower the frequency, the further this blind space will extend. Ultrasonic transducers generate sound when the ceramic crystal is energized by a pulse of electricity. The crystal vibrates and emits an acoustic pressure wave—much like a stereo speaker. But following the law of momentum, once set in motion, the crystal continues vibrating for a brief period of time (milliseconds). Echoes received from targets too close to the transducer will return very quickly and arrive before the crystal stops vibrating from its transmit pulse. Hence the blind space, or deadband.

Ultrasonic sensors also have a *beam width*, or *beam angle*, which is listed in manufacturer's specifications. This factor in ultrasonics is often misunderstood but is crucial for successful operation of ultrasonic instruments.

By convention, manufactures specify a transducer's beamwidth angle at −3 dB (the point where sound intensity is half the maximum). But sound energy diffuses very easily. It fills the nearby space and is reflected off all hard surfaces that it contacts. A good everyday example is that you can face the wall and speak in a closed room and anyone else in the room will have no difficulty hearing you.

Thus ultrasonic manufacturers design their transducers to focus as much energy as possible into a narrow cone emanating from the sensor. The width of this cone at half power is the beam angle.

The instrument's signal-processing software plays an equally important role in minimizing echoes from adjacent targets. Ricochet echoes are filtered out and various processing algorithms are applied to reject false echoes and lock onto the correct target.

Ultrasonic technology is also used for point level control with "gap"-type sensors, where the presence of a liquid in the gap between transmit and receive crystals can be detected. Applications in wastewater treatment are for high-level alarms in chemical storage tanks. Similar gap-type ultrasonic sensors can also be used for sludge-blanket level measurement.

36.4.2   Differential Pressure Sensors

Common Level Applications

- Wet wells and sumps
- Chemical tank inventory

With access to the bottom of a tank, liquid levels are often measured with pressure sensors. They operate by measuring the weight or pressure exerted on the sensor diaphragm by the liquid above the sensor. Three types of pressure sensor are available:

Gauge. One side of the pressure diaphragm is open to the atmosphere.

Differential. The open side of the diaphragm is connected to a pressure other than the atmosphere.

Absolute. The open side is sealed off.

Gauge-type pressure sensors are the most common in wastewater treatment applications for unpressurized tanks. The higher the liquid level, the greater the force exerted on the pressure sensor diaphragm. Typical accuracy is $\pm0.25\%$ of full-scale rating of the sensor.

Pressurized tanks can also be measured by connecting the vent side of the differential pressure sensor to the vapor space at the top of the tank (Fig. 36.4).

Lighter weight fluids exert less pressure on the pressure sensor than heavy fluids, so specific gravity of the liquid must be known in order to calculate level properly. Liquids also expand and contract relative to temperature. Pressure sensors cannot detect this change because the weight or pressure exerted on the sensor will not change.

Pressure sensors are typically two-wire devices transmitting a 4–20 mA current output with 24 V direct current (DC) excitation (Fig. 36.3). Some models include

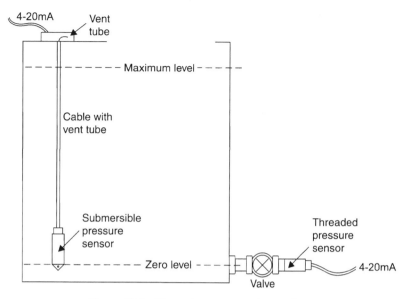

Figure 36.3 Gauge-type pressure transmitters.

adjustment potentiometers allowing operators to trim output according to their application. Gauge-type pressure sensors can be either threaded into a port at the base of a tank or suspended on a cable from the top of the tank ("submersible"). The cable is reinforced so that there is no stretching and also contains a small air tube so that the sensor can reference the liquid pressure to atmospheric pressure.

Figure 36.4 Differential pressure transmitter for thermophilic aerobic sludge digesters at the Long Sault Sewage Treatment Plant, Long Sault, Ont.

Pressure rating at 20 mA output is configured by the manufacturer, so maximum pressure in an application must be defined before purchase. In most pressure sensor designs the face or sensor diaphragm will be directly exposed to the liquid being measured, so chemical compatibility must also be considered. For hazardous rated locations, intrinsically safe and explosion-proof models are available from many manufacturers.

Solids buildup on or around the face of a pressure sensor will reduce its sensitivity and cause measurement errors. Thus pressure sensors are generally selected for relatively clean fluids. Some manufactures have developed designs with *flush-mount* and larger diameter diaphragms or installation techniques that protect the sensor diaphragm from solids buildup. With this feature pressure sensors can be used in high-solids applications like sludge level.

Because pressure sensors rely on the physical deflection or strain of a material supporting the strain gauge, there may be some drift in readings over time. Most manufacturers include reference to this effect in their product specifications. Typical "stability" for pressure sensors is ±0.5% of span over 6 months. Where accuracy is important, instrument technicians set up a maintenance schedule so that pressure sensors are calibrated periodically. Most operators report overall installed accuracy of ±1% of full scale.

36.4.3 Bubblers

Common Level Applications

- Wet wells and sumps
- Bar screen differential level

Bubblers measure the depth of a liquid by forcing air down a tube mounted in a tank or wet well. The air discharges or bubbles out of the tube at its opening near the bottom of the vessel. The pressure required to force air down the tube is proportional to the liquid level.

A typical bubbler system consists of an air compressor or pressure source, a pressure sensor and control and calibration electronics, and a bubble tube made of rubber, plastic, or stainless steel (Fig. 36.5). A source of clean, dry gas is required. Accuracy is generally 0.5–1% of full scale.

Bubblers are often used in wastewater treatment for basin and wet well level applications. They are ideal for deep wet wells, narrow vessels, or tanks with many obstructions. The bubble tube can be lowered and mechanically fastened to the sidewall. Most liquids, including sewage with high solids content, will form crystallized deposits at the discharge point of the bubble tube. A buildup of deposits will cause erratic readings and eventually block the bubble tube. Bubbler vendors can offer valuable advice on application of their systems in wastewater treatment.

Some bubbler systems are designed so that the compressor runs in repeated cycles. Most models offer an adjustment for the time between cycles—anywhere

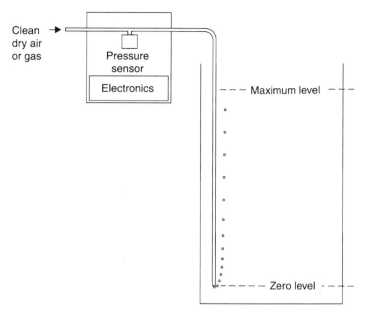

Figure 36.5 Bubbler level instruments.

from a few seconds to an hour. Between cycles static pressure in the bubble tube is measured. Other bubbler designs use a regulator to adjust the compressor so that just enough pressure is applied to force a bubble from the tube. Minimizing the amount of air discharged from the bubble tube reduces the formation of deposits in the bubble tube. Typical accuracy is ±0.5% of full scale.

36.4.4 Capacitance

Common Level Applications

- Chemical tank inventory

Capacitive sensors are available for point level detection or for continuous level measurement with a long probe immersed in a tank or silo. Capacitance sensors typically consist of insulated electrical conductors and use a radio-frequency (RF) signal on the sensor. To calculate level, they measure a change in electrical capacitance between the sensor and ground—usually the tank wall. The change is relative to level or immersion depth of the sensor.

Within design ratings, capacitance sensors are not affected by pressure or temperature as they have no moving parts. Capacitance sensors can be used for liquids or solids level measurement, but they must be selected and calibrated according to the dielectric constant of the material being measured.

Material buildup on capacitance sensors may be detected as false levels. So viscous liquids or materials such as wastewater scum or grease that coat the sensor may cause measurement errors. Manufacturers have various techniques to help compensate for the effects of buildup.

Modern RF capacitance level instruments can be bench calibrated by inputting calculated levels of capacitance or after installation by physical measurements at two levels—normally a high and low point. Typical installed accuracy is $\pm 1\%$ of full scale.

36.4.5 Sight Glasses or Gauges

Common Level Applications

- Chemical tank inventory

Tank levels that are not automated or transmitted to the treatment plant control system can be fitted with a simple, low-cost sight gauge (Fig. 36.6). They are generally glass or transparent plastic pipes $\frac{1}{2}$ or $\frac{3}{4}$ in. diameter mounted vertically on the outside wall of a tank with entry points near the top and bottom of the tank. Translucent, plastic tanks can have a staff gauge attached to or scribed directly on the

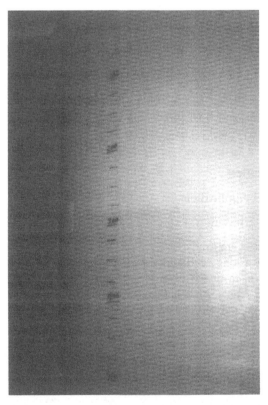

Figure 36.6 Sight gage scribed on polyethylene chlorine contact tank, Cornwall Sewage Treatment Plant, Cornwall, Ont.

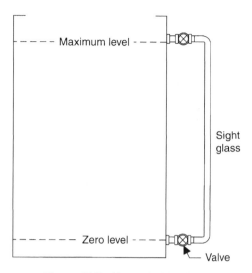

Figure 36.7 External sight glass.

tank wall. Operators can visually check the level by looking at the sight gauge instead of climbing the tank.

External sight gauges are not recommended for hazardous liquids where a broken or ruptured gauge could result in a spill or for tanks without spill containment. They are not suitable for highly viscous or solids-bearing fluids where coating or blockage of the gauge would result in false readings. In such cases, the sight gauge pipe is normally installed between valves so that it can be easily removed for cleaning or replacement (Fig. 36.7).

For high visibility, some manufactures offer sight gauges with magnetically activated, colored flags. A floating magnet inside the sight gauge rises and falls with the fluid level. Flags below the magnet are rotated to show one color, while flags above the magnet are another color.

36.4.6 Infrared

Common Level Applications

- Sludge blanket level

Portable, battery-powered sludge blanket detectors often use infrared probes. The sensor contains an infrared light emitter or light-emitting diode (LED) and a photocell receiver (Fig. 36.8). Light is transmitted across a gap to the receiver. As the sensor is lowered into the clarifier sludge blanket, the instrument will indicate a gradual decrease in opacity. Some models just detect the point where infrared light is not received by the photocell and provide indication to the operator. The cable suspending the infrared probe is normally marked off in feet or centimeters.

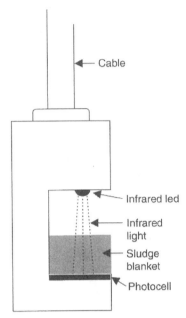

Figure 36.8 Infrared sludge blanket level detector.

36.4.7 Microwave Radar

Common Level Applications

- Digester level

Radar tank gauges have been in use since the 1960s primarily for petroleum products (Fig. 36.9). The radar microwave signal is not affected by vapor or changes in the media being measured. Recent advances in design and lower costs are resulting in more widespread use of radar in other applications. Although still very new in wastewater treatment, as prices continue to fall, use of radar may increase.

Modern radar instruments consist of an antenna inside the tank and remote mounted electronics. They work by transmitting high-frequency microwave signals that reflect off of the liquid surface and are returned to a receiver, much like ultrasonics use sound echoes.

Microwave radar signals travel at the speed of light, so the reflected signal is returned in just nanoseconds from relatively nearby targets in a tank level application. The instrument's timing circuitry has to have very high resolution to be able to measure level accurately.

Radar signals can also be transmitted through a probe or wire suspended in the tank. This technique permits measurement in confined spaces where noncontacting antenna-based systems might have their signals reflected off of obstructions such as ladders, agitators, or mixers.

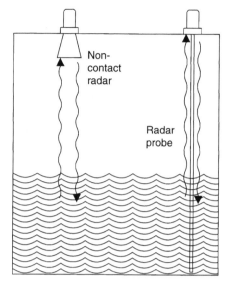

Figure 36.9 Microwave radar.

36.4.8 Weight/Load Cells

Common Level Applications

- Chlorine cylinder level
- Chemical mixing or batching

Weight-based level measurement offers high accuracy and high resolution (Fig. 36.10). Weigh meters include one or more load cells that deliver an electrical output proportionate to the force or weight exerted on them. Weigh meters or scales are common in wastewater treatment plant laboratories but are also used for chemical level measurement in the treatment process. The weight of the empty vessel must be deducted or "tared" from the final weight reading to determine the weight of the tank contents.

Installation of load cells requires raising the vessel, and so this method is not practical for large tanks or wastewater process basins. Load cells and weigh meter systems are also relatively expensive and so are normally limited to specific applications in wastewater treatment (e.g., chlorine cylinder level).

36.4.9 Point Level Floats

Common Level Applications

- Point level or pump control in wet wells and sumps
- High-level alarms in all wastewater process basins
- Level controls in chemical storage tanks

Figure 36.10   Chlorine gas cylinder on load cells, Cornwall Sewage Treatment Plant.

Floats or "displacers" are the point level control of choice in most wastewater collection systems and treatment plants. From pump stations to process tanks, floats are widely used because of the relative low cost and simplicity. Even tanks or wet wells with transmitting instruments often include a float as a backup high- or low-level alarm.

There are many different configurations, but all point level floats rely on the buoyancy of a noncorroding cylinder—usually plastic or porcelain—attached to a control relay. Designs vary, but action of the float will act on a mercury switch or a spring connected to a control relay.

Floats can also be used for continuous level measurement where the float is attached to a cable and pulley (Fig. 36.11). As the float rises and falls with the liquid level, the pulley rotates proportionally and electronics convert the rotation into an analog signal.

The obvious disadvantage of floats is the potential for buildup of solids or grease. Since the float must rise and fall freely to operate, deposits on the float or its linkage can cause a malfunction. Most plants develop a maintenance program where floats are periodically washed with high-pressure spray. Installations with a history of float failures are often retrofitted with noncontacting sensor like ultrasonics to reduce maintenance.

36.4.10   Conductance

Common Level Applications

- Pump control in wet wells and sumps
- High-level alarms in process basins

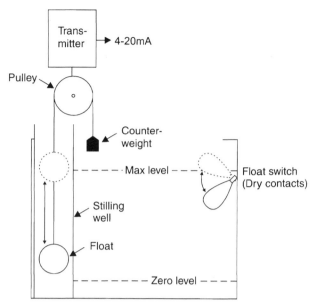

Figure 36.11 Float level controls.

Conductance sensors work by using water—or any conductive fluid—to transmit a low-voltage electrical current from one probe to another or to the steel wall of a tank. Most conductance sensors consist of two or more small-diameter steel rods inserted into a tank at desired set points (Fig. 36.12). When water or any conductive liquid contacts a transmitting probe, it completes the circuit by conducting the electrical current through the water back to a receiving probe or to the steel wall of the tank.

Conductance sensors are simple and inexpensive and are frequently included as level controls with equipment packages such as ultraviolet (UV) disinfection systems or sump pump packages (Fig. 36.13).

36.5 LEVEL APPLICATIONS AND CHOICE OF INSTRUMENT TYPE

36.5.1 Wet Well and Sump Level Control

Ultrasonic instruments and floats dominate this application. Ultrasonics are chosen for their noncontacting feature and when a 4–20-mA proportional signal is required for PLCs or variable-speed pumps. Most manufacturers offer intrinsically safe or explosion-proof transducers for hazardous rated locations.

Ultrasonic transducers must be mounted in a position to receive an unobstructed echo from the wastewater level (Fig. 36.14). Locations where false echoes may be received from ladders, pipes, or falling influent should be avoided. Ultrasonic

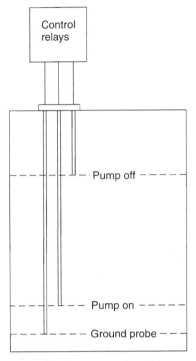

Figure 36.12 Conductance probes for pump control.

Figure 36.13 Conductance point level control in UV disinfection channel, Long Sault Sewage Treatment Plant, Long Sault, Ont.

Figure 36.14 Ultrasonic sensor mounting in a sewage wet well, Iroquois Sewage Treatment Plant, Iroquois, Ont.

transducers are designed to withstand accidental submersion, so they should be mounted as low in the wet well as possible.

Bubblers or submersible pressure transducers (Fig. 36.15) are a good alternative to ultrasonics for narrow wet wells where obstructions limit the range of ultrasonics and for wet wells where foam is constantly present.

Floats are selected for simple on–off pump control applications and high- or low-level alarms. It is not unusual to find an ultrasonic or pressure instrument and a high-level float installed in pump stations. The float provides added security in case the primary instrument fails.

36.5.2 Plant Bypass Alarm/Control

Conductance, capacitance, or floats sensors are often installed at the plant headworks to activate an alarm at the high-water or plant bypass level. The alarm may simply alert an operator or activate bypass flowmeters and other instrumentation.

36.5.3 Chemical Storage Tank Inventory

Ultrasonics are a common choice for liquid chemical storage and mixing tanks. Models from most manufacturers offer local display for operators, 4–20-mA output

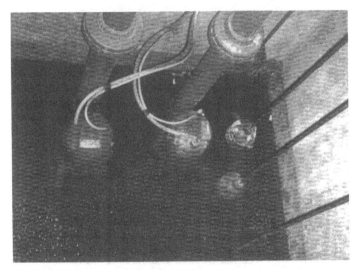

Figure 36.15 Float level controls for seal water sump, Massena Wastewater Treatment Plant, Massena, NY.

to the plant PLC or SCADA system, and control relays for pump or valve control and level alarms. Ultrasonics are ideal for viscous chemicals such as polymer and alum, and sensor materials can be selected for compatibility with chemicals such as potassium permanganate, ferric chloride, and sodium hypochlorite.

With access to the bottom of the tank, gauge-type pressure transmitters with 4–20-mA output and optional displays are also popular (Fig. 36.16).

36.5.4 Dry-Solids Level Monitoring

Lime powder level at wastewater treatment plants is often monitored with point level controls, including:

Figure 36.16 Gage type level indicating pressure transmitter.

- Rotating paddle switches where contact with powder is sensed as a torque change
- Tuning forks where contact with the powder dampens the probe's vibration
- Diaphragm sensors where pressure of the solids activates a control switch
- Capacitance probes that are designed for nonconductive powders

Level or powder or dry solids can also be measured with continuous level instruments, including:

- Mechanical cable sensors that raise and lower a probe continuously to track powder level
- Capacitance probes designed for nonconductive powders
- Radar transmitters with either contacting probes or noncontacting antennae
- Ultrasonics—selecting only models designed for dry-solids measurement with dust-shedding sensors

Level instrumentation may also be installed in process equipment, for example, gravity belt thickeners to monitor and control sludge level.

36.5.5 Bar Screen Differential Control

The inlet side of a bar screen will normally have a high-level alarm using a conductance or capacitance sensor. Inlet water level rises as debris builds up on the bar screen. The level switch activates an alarm to alert plant operators.

Automatic systems may include a differential level instrument where level is measured both up and downstream of the bar screen (Fig. 36.17). The instrument monitors and calculates both levels and activates the screen trash rake at preset differential levels. Ultrasonic, bubbler, and capacitance systems are generally selected.

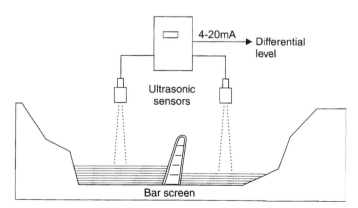

Figure 36.17 Bar screen differential level with ultrasonic transducers.

Figure 36.18 Ultrasonic transducer monitoring aeration tank level in the Cardinal sewage treatment plant's sequential batch reactor, Cardinal, Ont.

36.5.6   Process Basins Including Aeration, Chlorine Contact, Skimmer Tanks, Sedimentation, and Flotation Thickeners

For continuous level measurement, ultrasonic and pressure transmitters are common throughout the wastewater treatment process (Fig. 36.18). Floats or conductance probes are also used for simple high-level alarms and point level control (Fig. 36.19).

36.5.7   Sludge Blanket Level

Wastewater treatment plant operators must maintain the proper sludge concentration and level in clarifiers to optimize the activated-sludge process. The interface between sludge and water can be gauged by hand-held tubes. A sample of the sludge layer is captured and raised for operator viewing. However, this technique depends on the skill and judgment of the operator, and plant operators now have new choices in instrumentation to simplify and improve the accuracy of sludge blanket level measurement.

The simplest detectors use a sensor lowered on a cable by the operator. Sludge detectors can use ultrasonic sensors, where change in sound transmission from a transmitter crystal to a receiver is measured. Or they can use an infrared light source and photocell receiver (Fig. 36.8). The operator lowers the sensor into the sludge blanket, and the instrument indicates when the sensor falls to a preset sludge density. Normally there is an adjustment for sensitivity. This method can be quite accurate, but the sludge blanket can be easily disturbed or the sensors fouled.

Continuous sludge blanket level instruments are gaining acceptance in modern, automated wastewater treatment plants. They range from mechanically actuated

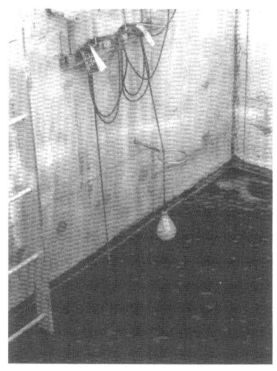

Figure 36.19   Float level control for the effluent contact chamber, Ingleside Sewage Treatment Plant, Ingleside Ont.

systems that raise and lower an infrared or ultrasonic sensor to continuous sonar-based ultrasonic systems that can profile the sludge blanket and provide operators with a detailed view of sludge density.

36.5.8   Sludge Holding Tanks

Ultrasonic instruments and flush-mount pressure sensors are common choices in sludge holding tanks (Fig. 36.20). Ultrasonics are ideal because the sensor mounts above the sludge and measures without contact. However, splashing in tanks where sludge fills from the top can coat ultrasonic transducers, resulting in signal loss. Operators can choose flush-mount pressure sensors for sludge tanks where there is access to the bottom of the tank. The flush-mount sensor design helps reduce solids buildup or coating of the diaphragm.

36.5.9   Anaerobic Digester Level

Scum, methane, steam, and solids buildup make anaerobic digester level measurement one of the most difficult applications in wastewater treatment. Differences in digester design mean that instruments that succeed in one plant may fail in another.

Figure 36.20   Sludge holding tank with ultrasonic level transducer, Ingleside Sewage Treatment Plant.

Many digesters displace supernatant from the top of the tank as sludge is pumped in from the clarifier. Often a simple high-level alarm using a conductance or capacitance probe is sufficient.

For continuous level measurement in digesters the choice of instruments ranges from simple, mechanical level gauges for floating roofs (Fig. 36.21) to pressure transmitters with an intrinsically safe or explosion-proof sensor.

Figure 36.21   Digester roof position indicator, Massena Sewage Treatment Plant, Massena, NY.

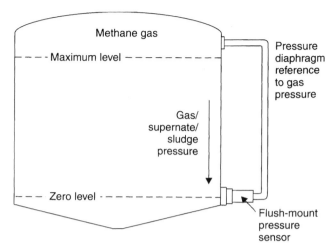

Figure 36.22   Differential pressure level monitoring in an anaerobic digester.

Flush-mount differential pressure sensor designs (Fig. 36.22) are normally preferred to minimize sludge buildup at the sensor. Because of methane pressure, the atmospheric side of the sensor's diaphragm must be referenced to the vapor layer at the top of the tank. Therefore, a tube or pipe is connected from the top of the digester down to the pressure sensor.

Floating digester roofs can be fitted with a mechanical roof position indicator or a noncontacting ultrasonic instrument. Ultrasonic sensors require a flat reflective surface, so domed digester roofs should be fitted with a flat "target" to reflect the ultrasonic signal.

Ultrasonics are rarely selected for mounting inside a digester. The speed of sound in methane varies according to the gas concentration and pressure, so ultrasonics are not repeatable. Where noncontacting level measurement is necessary, some digesters have been fitted with stilling wells for ultrasonic sensor mounting (Fig. 36.23). The well is ventilated, so that the space above the supernatant level is filled with outside air rather than digester gas. Regular cleaning of the stilling well is necessary to prevent false readings from solids buildup or solidified scum.

Noncontacting radar level instruments offer a new choice in instrumentation for digester level measurement. The radar signal is unaffected by methane gas concentration. As the price of radar instruments continues to fall, this technique may become more common.

36.5.10 Lagoons and Settling Ponds

Ultrasonic transmitters are the first choice for many treatment plant operators. Most ultrasonic instruments provide a local digital display, 4–20-mA analog signal, and control relays. They are selected for their noncontacting sensor and simple installation (Fig. 36.24).

Figure 36.23 Anaerobic digester with ventilated stilling well and ultrasonic transducer, Cornwall Sewage Treatment Plant.

Submersible pressure sensors are also popular for sewage lagoons. Pressure sensors normally transmit a 4–20-mA analog signal to a PLC or SCADA system but can also be equipped with displays and control relays.

Figure 36.24 Ultrasonic level indicating transmitter on aeration lagoon at the EMOS Wastewater Treatment Plant, Santiago Chile.

V

AIR MONITORING

37

DATA ACQUISITION SYSTEMS FOR AMBIENT AIR MONITORING

Matthew Eisentraut, M.Sc., P.Eng.

RSLS Engineering
Calgary, Alberta, Canada

Martin Hansen, B.Sc.

Jacques Whitford Environment
Calgary, Alberta, Canada

Environmental Instrumentation and Analysis Handbook, by Randy D. Down and Jay H. Lehr
ISBN 0-471-46354-X   Copyright © 2005 John Wiley & Sons, Inc.

37.1 INTRODUCTION

37.1.1 Note About the Authors

The majority of the authors' experience concerns ambient air quality monitoring. Both authors have been involved in the processing of data from a variety of systems and have participated in the development of custom systems. By choosing the specific topic of data acquisition systems (DASs) for ambient air monitoring, the authors are narrowing their focus to what they know from many years of experience. Nonetheless, much of the material presented here will have relevance to many other environmental contexts and should be of interest to a broad readership.

37.1.2 Goals

A search on the Internet for "dataloggers" will yield dozens of results. Selecting a well-known company, one will find a dozen or more different models of dataloggers. In addition to hardware, a wide variety of software is usually offered. Software from one company is often compatible with components from another company, leading to another multiplication of choices. A wide range of well-developed technology makes it possible to bypass standard commercial solutions and create entirely custom solutions. A less ambitious program will create a unique blend by combining some components from standard commercial technology with custom components. With a range of choices available for each system component, the range of possible configurations for a complete DAS is enormous and overwhelming.

The goals of this chapter are as follows:

1. To define and discuss the concept of a DAS and acquaint the reader with some of the many possible functions such a system may perform in the real world of air quality monitoring.
2. To define and discuss other important related concepts and provide some historical context for the development of DASs.
3. To present and discuss the wide spectrum of choices that are available for the creation of a DAS.
4. To raise questions and issues that help define the requirements needed from a system and influence decisions about its design.
5. To provide information that will aid in the evaluation of the suitability of various choices to meet the scope of requirements.

37.1.3 Datalogging and Reporting versus Data Acquisition System

A typical traditional datalogger is a rugged box that sits in a remote, harsh environment and reliably collects and stores data being produced from some nearby instrumentation. Periodically, by some means, the stored data are transferred out of the datalogger and eventually end up on some computer that processes the data and creates reports. Even today many of these reports are paper and are stored in large rooms with many shelves.

What has developed during the last few years is the concept of a DAS that includes all the hardware and software needed to collect the data, process the data in an increasingly sophisticated manner, and then, besides printing a summary report, distribute data to a wide variety of interested parties through convenient electronic access via various types of disks, the Internet, computer telephony, and so on. In the traditional system data collected today is not readily accessible for perhaps several weeks. Today's DASs provide clients access to data that are not more than a few minutes old.

The core components of datalogging and reporting are the same as those in a DAS. What distinguishes the two is the level of integration and automation that can be achieved as a result of current computer and communication technology. The traditional system does not automatically phone a technician first thing in the morning and verbally tell him or her that a reading from an instrument is either too high or too low. Today's DASs do this and much more.

The question of where a DAS begins and ends is becoming increasingly unclear. Many modern analyzers include as part of their operating system datalogging, self-diagnostic, and control capabilities. The analyzers themselves may be considered part of the DAS. The other end of the data flow is also unclear. Is the limit of the system at that point where data pass to some modeling program or is that modeling program itself part of the DAS? There is a trend that increasingly sophisticated software tools for processing and presenting data are being included as a standard part of the DAS package.

37.1.4 Data

Data are collections of quantified information about phenomena. The creation of data is part of a process that begins with a need, defined by an issue or a purpose. Investigation of the need identifies key factors and phenomena relating to the issue. Understanding the measurable attributes of the phenomena and appropriate measurement methods leads to implementation of a data collection program. With sufficient data collected, evaluation and analysis yield conclusions which are disseminated as information about the specific issue. This information, considered with respect to disciplines related to the issue and the world in general (e.g., physical, social, economic), is validated and integrated as general knowledge. Information and knowledge are applied to address the original need and resolve the issue.

The process may be summarized as follows: issue or purpose, phenomenon, measurement, data, information, knowledge, and application. Within the context of this process, data can be seen to be an intermediate representation of a phenomenon necessary to generate a solution to a problem. Some important limitations of data's ability to represent phenomena are apparent by considering the relationship between phenomena, the measurement process, and data acquisition. In the diagram below each of these elements is represented by a circle and arrows, representing their dynamic and cyclical nature.

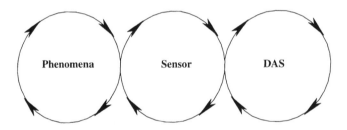

A sensor is a reactionary device where contact with the phenomenon induces a response analogous to the state of the phenomenon. The accuracy of the analogy in the sensor with the phenomenon is limited by the sensor capabilities, such as sampling frequency, sensitivity, range of capability, and response time. The DAS reacts to the sensor by creating and storing a number that represents the state of the sensor. The small regions of contact between the circles in the diagram represent the periodic and discontinuous interactions between the elements. The illusion of continuity can be obtained by sampling at high rates. Typically, the DAS summarizes the values of its samples in the form of averages and other statistical information and writes just these numbers to its storage device.

Data are series of snapshots of a measurement system responding to limited contact with dynamic phenomena. The challenge in designing a successful system is to make selections that ensure the data sufficiently represent the phenomena of interest.

37.1.5   QA/QC

QA/QC (or just QAQC) is an acronym for the term *quality assurance/quality control.*

Quality control (QC) generally refers to operational methods and procedures of the measurement program designed to ensure quality. Quality control in continuous air quality monitoring is accomplished operationally by following standard criteria related to siting and station design and operation, including monthly multipoint and daily zero/span calibrations. In monitoring air quality by integrated and passive sampling methods, quality control is accomplished largely by following established protocols in sample preparation, handling, exposure, and analysis.

Quality assurance (QA) refers to reviewing quality control information and undertaking actions necessary to provide confidence that requirements for quality of data are being met. Feedback of information into operations for improvement is essential for effective quality assurance. Quality assurance also includes validation of collected data. Validation is the process of reviewing and judging final data to be representative of ambient conditions based on constraints of the measurement system and qualified knowledge of measurement system operation and the physical phenomena being measured.

More generally, QAQC is applied to almost any activity when being related to the creation of a final data set that is complete, accurate, and scientifically defensible.

37.1.6   Metadata

Metadata are secondary data describing attributes or the state of data values from primary measurement. Quality control information may be considered metadata relative to measurement data. Quality assurance information may be considered metadata relative to quality control data. Project documentation information may be considered metadata relative to all of these.

Most analyzers operate best under conditions of constant temperature and a steady power supply. Hence, two common metadata parameters are room temperature and line voltage. Other important sources of metadata are logs that document human observations and activities. Unusually high readings of a pollutant may be due to a local construction activity involving the regular presence of large diesel trucks rather than emissions from an industrial process. Such information is an important part of a complete set of usable data.

Information regarding security, such as when a door is opened, may be included as metadata. While such information may not directly affect the interpretation or adjustment of primary data, it is important for the practical operation of the network.

37.1.7   Other Basic Concepts and Definitions

An *airshed* is the region of space in which air quality information is wanted. Typically, an airshed will cover several thousand square kilometers. However, it could also be a single building. Airsheds can be delineated based on considerations that are cultural, physical, geographical, socioeconomic, and political in nature. Airsheds may encompass areas where emissions occur and where concerns exist about their effects.

A *station* is a building containing instrumentation for the purpose of measuring air quality. A typical station might include several air pollutant analyzers; one or more meteorological sensors, including temperature, wind speed and direction, relative humidity, solar radiation, leaf wetness, and precipitation; and other devices measuring station or state parameters.

An *air-monitoring network* is a collection of stations that are located within the airshed. To gain adequate information about a large airshed, several dozen stations may be required. Decisions about the location of these stations are not within the scope of discussion of this chapter. However, these decisions have an impact on any implementation of a DAS. To gather relevant information, stations may be located where there is neither readily available power nor standard telephone service.

A *host* is a location where data from a set of stations are collected. Typically, in an airshed there would be one host that collects all the data from all the stations. However, there may be operational and economic advantages to having several hosts in large airsheds. Since it is usually desirable to eventually have all the data in a single database, airsheds operating with several hosts need to consider the mechanisms by which the data will be combined.

37.1.8 Purposes of Collecting and Distributing Data

In a compliance network air pollutant levels are measured to ensure conformance with regulatory guidelines. If a guideline is exceeded, then, within some set time frame, various government and/or industry contacts must be advised. If the time frame is on the order of 1 or 2 h, then either the system must be manually inspected 24 h per day or an automatic alarm and call-out system must be implemented. Usually a part of compliance networks is to eventually assign responsibility to the industry causing the exceedance. Therefore, supporting information such as accurate wind direction data is essential. In this chapter, when reference is made to a compliance network, it will be assumed that immediate response (less than 2 h) to violations is required.

A common task that usually needs to be accomplished is the regular generation of standard government reports. For example, the government agency may require a monthly summary of hourly averages, details of monthly calibrations, discussion of any abnormal events, and various other statistics. To the extent that the DAS automatically generates these reports, substantial savings in time can be realized.

Scientists such as those interested in the environment or human, animal, or vegetation health will want access to data. They will best be served by data that are in some electronic form that can be easily imported into various computer programs for further analysis.

Another special group of data users consists of those who need real-time data. A radio station may want current temperature data as part of its reporting. An operator of a large gas turbine may want access to the barometric pressure data at a nearby air-monitoring station. Plant operators may want access to real-time data to ensure their operations do not cause pollutant exceedances. The authors have encountered a surprisingly wide variety of people who have good reasons to access real-time data.

The general public is a group who often wants access to data. Many air-monitoring networks make their real-time data available to the general public on a website. Modern DASs should give standard Web servers easy access to data.

If the goal of a network is to cost effectively collect scientifically defensible data with a minimum of downtime, then the most important group in terms of convenient access to detailed real-time data is comprised of the technicians who maintain and operate the network. In any modern DAS, technicians should be able to sit down at their home computers and access detailed information about all aspects of the network in graphical and tabular form. Besides access to the most current and historical data, summaries of alarms, results of zero/spans, and various system status indicators should be available. Technicians should be able to call stations directly and force various events to occur (cycle power to an instrument) or adjust control parameters (change the time of a zero/span cycle).

As previously mentioned, there is a class of data known as metadata that are collected for reasons of security, maintenance and operation, troubleshooting, and/or QAQC. Correct interpretation and adjustment of the primary data often depend on metadata. Technicians rely on metadata for troubleshooting problems and generally maintaining the network.

Data are collected for purposes relating to calibration and/or zero/span cycles. In many jurisdictions pollutant analyzers must undergo complete four- or five-point calibrations every month. Technicians spend a large proportion of their time performing these calibrations and reporting the results. A DAS should be designed to make these calibrations and their reports as convenient as possible. In addition to multipoint monthly calibrations, many jurisdictions require daily zero/span cycles to be performed. Baseline correction is done based on the zero point, and the span point provides confirmation that the analyzer is working properly. If calibration gases are used for the spans, then slope correction can be done as well. In a well-designed DAS, the process of baseline correction and span evaluation should be a balanced mix of automatic processes and human supervision.

The design and operation of a data collection program and its DAS are related to program purpose. For example, in a program whose goal is regulatory, compliance monitoring, the purpose is to determine if any and how often regulatory concentration levels are exceeded. Calibrations and operations may be optimized to ensure accuracy at the higher concentrations, and concern for the accuracy of low concentrations around the sensor detection limit may not be great. However, in a program, to study effects of concentrations on vegetation or health, it may be recognized that high concentrations near the regulatory levels occur infrequently and receptors may be sensitive to long exposures of low concentrations. Hence the accuracy of low level-concentrations and the need to determine when concentrations are detectable and nondetectable may be of prime importance. The operational requirements of these two programs are very different, and the resulting data from one may not be applicable to the other unless objective standard procedures are adopted in both.

37.1.9 Brief History

Air quality monitoring in the 1970s was driven largely by regulatory needs. The general environmental awareness of society was undeveloped and the information age had not yet arrived. The data processing system started with recording the sensor signals on a strip chart recorder. Periodic inspection was required by operators to minimize the occurrence of jammed or expired charts or dry chart pens. Once a month (a frequency determined by the regulatory reporting requirement), charts were collected and physically transported to the "host," an office where clerks scrutinized the charts with templates, visually averaging traces over various distances corresponding to averaging periods, and recording numerical values on keypunch forms. These forms went to a data center and the data keypunched onto computer cards. The data were now in the form of a summary text listing and a card deck with each 4×8-in. card containing the data for one observational period. Square holes in the card were read optically to digitize the data, and a text string at the top of the card allowed reading of the data by the human eye. The total depth of the deck could be several feet for many parameters and monitoring sites. The clerk then verified the keypunch numbers in the output listing with the original forms and submitted changes for correction. When this correction cycle finally yielded an accurate deck of cards, the deck was submitted back to the data center and processed by a mainframe computer application. Hard-copy monthly report tables were produced and returned. The clerks manually organized the tables by site and parameter into reports for submission to the various clients and regulatory agencies.

The charts, keypunch forms, and cards were stored in boxes and on shelves. Reports were stored in a file cabinet. Together these paper records formed the database. As the need for access to materials in the database decreased with their age, the physical location of the database migrated to facilities with decreasing accessibility. Eventually they arrived at a remote location where they could be forgotten unless a rigorous indexing system was implemented. Often the indexing system resided in someone's memory.

Retrieval of data for subsequent use was usually accomplished using data listings in reports. Data were analyzed by tabulating manually, placing tick marks in cells on a hand-lined sheet where horizontal and vertical cells corresponded to various categories of the parameters being tabulated. Analysis consisted of counting tick marks and summarizing results in a table. Statistics were generated with a calculator and graphics were created manually.

Data recovery is a ratio of the time for which data were present to the time for which data should be present. Data recovery is an important statistic in evaluating any data collection program. Data recoveries of 60% were not uncommon in the 1970s, especially for remote sites. In the worst case, the strip chart consisted of torn paper fragments due to jamming in its transport mechanism and destruction by the action of the pen over and over at the same location on the immobile chart.

Electronic data capture devices became available in the early 1980s. These desktop-sized devices were specifically designed for environmental monitoring data capture. The earliest units were stand alone but were followed by devices that provided an interface between a personal computer and the instrumentation. Features of all these systems included signal processing, data storage, reporting functions, and remote control of daily zero/span cycles. Operational staff needed to become computer literate and acquire a sense for electronic data management. Issues associated with chart recorder reliability virtually disappeared, and typical data recovery increased from 60 to 70% to more than 90%.

Electronic data capture improved the quality of data by recording secondary parameters, giving remote access to data, and using smaller averaging periods. New QAQC activities emerged from recording secondary parameters such as power fluctuation and maximum and minimum instantaneous values. As memory became less expensive, the amount of data collected increased. Remote access to data meant that technicians could connect to a station via a telephone and retrieve summaries of hourly averages and results of zero/span cycles. It became possible to detect problems within a day or less without having to visit the station. Smaller averaging periods made the data useful for the study of emerging issues of environmental and health effects where receptors are often influenced by the magnitude and frequency of short-term exposures.

One author was fortunate enough to be involved in an acid deposition study in the mid-1980s where all these advances were applied operationally for the first time. A two-year monitoring program recorded 2-min averages as well as the instantaneous minimum and maximum values during these 2-min periods. These values were recorded for each air pollutant concentration as well as analyzer flows, internal shelter temperature, line voltage, and other miscellaneous parameters. The monitoring program recorded about 100 parameters at each of three stations. Detailed scrutiny of the data allowed the detection of many nuances specific to individual sensors and to their interaction within the system. Recording of calibration parameters allowed systematic baseline correction and, in many cases, correction of operational errors.

The creation of reports became much less labor intensive with electronic data capture. Many of the manual tasks from the 1970s were eliminated. Data-editing and report generation programs were written. Crude spreadsheet programs and the eventual availability of mainframe programs in personal computer (PC) versions made analysis and application of data much easier.

The artifacts of data collection changed from boxes of paper to boxes of diskettes. Proper management of the data continued to be a challenge. Several experiences of manually scanning the contents of dozens of diskettes to find needed data were usually enough to provide incentive to develop good filing and labeling systems.

The trends through the 1990s until now are mostly a continuation of those started in the 1980s. Storage capacities have become immense and processor speeds are orders of magnitude faster. Computer telephony technology and the Internet are ubiquitous and inexpensive. Operating systems, higher level programming languages, and readily available software components have drastically cut the time to develop applications with sophisticated graphical interfaces. The authors are involved with airsheds that have had data recovery rates of 98% for more than five years. Much of this success is due to a data acquisition system that gives the technicians convenient access to detailed real-time data.

37.1.10   Robustness and Reliability

In the discussions of hardware and software that follow, the issues of reliability and robustness appear frequently.

The need for reliability in a system and which parts of a system need to be especially reliable depend upon many factors. For example, in compliance networks requiring exceedance notification in 2 h or less, part of the DAS must include a reliable means of alerting the contacts. Another example is the case in which a station is in a remote area and visited only once every few weeks. If such a station is not connected to a communications network, then the status of the station is unknown until it is visited. Having some component fail shortly after a station visit means that weeks of data can be lost.

The concept of robustness is closely related to reliability. Robustness is the ability of a system to recover from unusual circumstances. If one component fails, can the rest of the system continue to function? How well does a system recover from frequent power losses? How well does the system survive or function if the heating system fails and the temperature drops to $-20°C$ or the air conditioning fails and the temperature rises to $50°C$? A system may be very reliable sitting on a counter in a laboratory being supervised and used by a knowledgeable technician. If the system is also robust, then it will function reliably in real conditions while being used by persons who make occasional errors.

In large systems consisting of many programs running on several computers it is useful for a supervisory program to monitor the status of the other programs. If a program locks up or someone accidentally turns it off, then alarms can be used to alert staff of the malfunction or error.

A compliance network can only be as reliable as the callout system. A robust system will have an independent backup callout system to alert staff if the primary system fails. The backup system does not need to have the same sophistication as the primary system. Any persistent alert will do.

37.1.11   Recovery from Error

Suppose a technician performs some maintenance that requires the physical disconnection of several analyzers from the DAS. Suppose that upon reconnection the wires are mixed up so signals from various analyzers are going to the wrong channels. Suppose the technician is in a hurry, under pressure, or distracted so that the mixup is not noticed. An important consideration is how much effort and time are required to correct the records.

Anyone familiar with running a network knows that errors will happen. At least a couple of times per year, errors will be made that were never contemplated, anticipated, or imagined possible. Many of these errors will be a result of human action; some will come from sources never expected or identified, such as a computer that suddenly and inexplicably changes its clock time by 1 or 2 hours.

The capacity a system gives its users to recover from errors is an important consideration in choosing or designing a DAS. Providing users with the ability to make alterations to correct errors increases the flexibility of the system but also increases the potential to create even more errors. For this reason it is advisable that access to some editing features be restricted.

37.1.12 Custom, Generic, and Specialized Components and Products

At every point in the selection of hardware or software the question arises as to whether the device or code should be a custom creation or obtained as a generic or specialized product. A generic product is one that is widely available for use in a variety of applications. Copper wire, resistors and electrical relays, various computer languages, and operating systems are, by this definition, generic products. One important feature of generic products is that they are usually available. So, if a system uses such product and that product fails after 15 years of use, then it should be easy to find an acceptable replacement. Because they are mass produced, generic products are comparatively inexpensive. Specialized products are designed for a much narrower range of application and are often manufactured by a single manufacturer. They tend to be more expensive and may at anytime cease to exist in the marketplace. If such a product plays a key role in a system, then the failure and unavailability of the product may mean a substantial and expensive redesign of the system.

Some basic questions should be considered in evaluating the purchase of any product or creating a custom product. Is an expensive and specialized product required or can a cheaper generic product be adapted to do an adequate job? Is the product going to be available for replacement in several years? Can a complex device or system be custom built from a combination of generic parts or should the entire device or system be obtained as a single specialized product? Does a specialized product have the flexibility to interact easily with other products?

Suppose a custom software application is to be created. This does not usually mean creating a new operating system or programming language to create the application. However, sometimes it does mean exactly this if there is no operating system or programming language that will do the job. Assume that appropriate operating systems and languages are available and the nontrivial decision of selecting them has been made. Most systems and languages will come with dozens of prewritten components that can be simply inserted into the new program. Other components will be available for sale by external developers. If the needed component is not available, then a custom component may have to be created. The point is that at every stage of creating a large custom device the question of how to obtain the components is again a decision about whether to create custom components or otherwise obtain them as generic or specialized products.

Creating custom systems usually takes more time but gives greater control. There are exceptions. Trying to adapt a generic or specialized product or component to meet some special need may take longer than creating a custom solution. In the end the adaptation may fail anyway due to limitations of the component with respect to system requirements and the multiplicity of limitations when several components are combined. The cost of finding a component and then purchasing it and having it delivered may far exceed the cost of designing and building the component from simple subcomponents in the first place.

The discussion of custom, generic, and specialized devices is important because most modern DASs are to some degree custom systems because the requirements and operating constraints for these systems are so varied. New requirements for existing systems usually appear as time goes on. Any given system that meets only today's requirements will have to be upgraded or customized in some way to meet tomorrow's requirements. The ease with which such enhancements can be made is an important factor in designing or evaluating a system.

37.2.   AIR-MONITORING STATIONS

37.2.1   Location, Communication, and Power

The location of a station affects many aspects of the design of a DAS. Consider the time and expense to actually visit the station with (or without) a full set of tools. Consider the effect of the seasons on this accessibility. For cold climates, winter brings solid roads but the possibility of ice and heavy snow. Spring brings mud. The cost of repairing trucks that are regularly buried to their axles in mud is significant. For stations located in swamps helicopter access may be the only summertime access. A station may be remote and require many hours of driving to access. The greater the time and cost for access, the greater the reward in investing in a highly reliable and robust DAS that can be customized to do every conceivable task without a station visit. The question of redundant systems appears here. A secondary system may monitor the primary system and do something as simple as cycling power to the primary system in the event of some kind of lockup. A more ambitious secondary system will perform many or all the functions of the primary system.

The authors recently encountered a situation in which an important device was locking up for unknown reasons. To return the device to normal operation simply required cycling power to it. Driving to the station and back took about 2 h. The best solution to this problem would be to fix the device or circumstances so that it no longer malfunctioned. However, the reason for the malfunction was not obvious. Trips to the station to cycle the power were beginning to cause considerable inconvenience and expense. Fortunately the DAS could be quickly and cheaply modified to cycle the power by phoning the DAS from anywhere and giving it a simple signal. The important point here is that the initial scope of requirements for the DAS for this station did not include the need to cycle power to anything. This need arose later. By having a DAS with more capability and flexibility than originally specified, the new requirement was accommodated with minimal expense.

Location will influence the nature of communication with the station. The process of regularly moving data from the station to the host is part of the DAS. At the time of this writing telephone systems have unparalleled reliability and availability. High-speed Internet is a great way to move data; however its availability is comparatively limited and it is not yet as reliable. Having a station connected to a standard local telephone line is close to ideal. In such cases it is possible, if desired, to maintain continuous contact with the station. Flat-rate long-distance plans may also allow such close supervision. Otherwise, some sort of periodic connection will be implemented.

For stations in more remote areas communication via radio, cell phone, or satellite phone is an option. Assuming appropriate geography, a pair of broadband radios will move data about 20 km without the need for licenses. Most air-monitoring stations have a tower of at least 10 m in height for mounting wind speed and direction instruments. This provides a place to mount an antenna. Such radios typically provide a link to a location where there is a telephone line that can take the data the rest of the way to the host. Whether the radio link is by broadband or by licensed frequency, baud rates of 9600 or more are possible and there are no delays in transmission. Once the initial cost for the radio setup is paid, there are no ongoing fees and, when it is working properly, the radio link is transparent to the system. Various factors, such as weather conditions, may cause problems for periods of time and occasional cycling of power may be needed. If the station is sufficiently remote, then a DAS should be selected that is intelligent enough to recognize the lack of communication and cycle power or take other action.

Data transmission over cell phones is another option. Slower modem transmission protocols and baud rates of 4800 or less are usually required, and there are significant delays in transmission. In a typical polling situation the host will request a specific record, the remote unit will respond with the record, and the host will process the record into the database and then request the next record or set of records. Because of the transmission delays inherent in cell phone transmission, requests for larger sets of data at one time will speed up the process. However, if transmission errors are common, trying to transmit too much data in a single request–response cycle will make matters worse. Unless the quality of the cell phone connection rivals that of a land line, it is important to have a good system of retries in place.

Where cell phones are used in combination with long-distance calls, monthly telephone bills can be significant. One method of reducing these costs is to divide the data into two parts, that which is necessary for real-time operations and that which is necessary only for reports due next month. Transmit only the data needed now over the cell connection and download the remaining data onto a disk or mobile laptop hard drive during a routine visit to the station.

Do not assume that hanging up one side of a cell phone connection will necessarily hang up the other side. Although telephone companies will deny this possibility, the authors have seen it happen more than once, and such malfunctions can cause significant bills. It is usually at the host end that calls are initiated and terminated. It is a wise practice to design the station system with the ability to send commands to the modem to hang up at the station end independent of the host.

Satellite phones are an option when stations are in locations with no cell coverage. Satellite connections can suffer from many of the same complications as cell phone connections and are usually much more expensive. The many options offered by satellite communication are beyond the scope of this discussion. Besides, whatever is written here will be out of date in a matter of a few months.

Regardless of the means of communication, it is good practice to always set up a thorough bench test before any commitment is made. Expect such tests to reveal unanticipated problems. Also, expect that in a robust and well-designed DAS being operated and maintained by competent technicians, problems arising from communication breakdown will be the most common problems. A shovel driven into the ground and accidentally severing a telephone line is one of many possible events that are beyond your control and that will stop communication. A well-designed DAS will minimize the inconvenience and expense of such events.

One other means of collecting data, alluded to above, is to transport the data by going to the station, copying the data to a disk, and carrying the disk back to the host. If the station is remote and the need for real-time data does not justify the expense of setting up a real-time communication system, then such a solution can be satisfactory. If the period between visits is a long time and the data are regarded as very important, then redundant collection systems should be considered.

Another issue that arises, especially in the case of remote locations, is who has access to the station. When stations are remote, they often serve other functions that have nothing to do with air monitoring. The fact that persons not concerned with air monitoring may have regular access to a station creates security issues not limited to the DAS.

Location will be a determining factor regarding sources of electrical power. For small stations in remote areas there are dataloggers available that run on a minimal amount of low-voltage DC power. In most cases, however, an abundant supply of standard AC power must be available because of the requirements of the pollutant analyzers and the additional requirement to heat or cool the building. It is recommended that the DAS have its own uninterrupted power source (UPS) that is connected into any other UPS systems that may be in place to provide power to the instrumentation itself. This way, in the event of a prolonged power failure, the DAS is the last system to fail. Furthermore, it is recommended that line voltage to the building be monitored. If a UPS is used to supply the analyzers, then the line voltage output from this UPS should also be monitored. Monitoring line voltage need not involve anything more sophisticated than monitoring the output from a simple and inexpensive AC-to-DC transformer.

The DAS should record maximum and minimum instantaneous values of line voltage. Averages of 1 or 5 min are of little value without the maximum and minimum values. A voltage that is varying between 100 and 140 V has a very different effect on instrumentation compared with one that is varying between 119 and 121 V. If available, most technicians will look at such voltage statistics as the first step in troubleshooting analyzer problems. Especially for stations having their power provided by portable sources or grids that are affected by large industry, anomalies in power supply are often the most common cause of anomalies in the data.

37.2.2 Data-Averaging Options and Flags

The outputs from devices should be sampled at least once per second. From these samples various averages can be created. In the case of devices used to measure precipitation, totals rather than averages need to be computed. Common averaging times are 1 min, 5 min, and 1 h. Other choices might include 6-, 10-, or 15-min averaging times. Whatever the selection, it is recommended that the averages be calculated independently of one another. If this is done, then showing that a 1-h average is indeed equal to the average of the twelve 5-min averages can be part of a simple QAQC procedure. The calculation of this comparison is neither simple nor trivial in the case of wind direction standard deviation. Besides averages, it is useful for the system to record the highest and lowest instantaneous sample values during either the 1-min or 5-min periods. These values will give useful information about how much variation is in the signal. As mentioned above, this can be vital information concerning line voltage. If needed, standard deviations can be calculated and recorded as well. In the case of wind speed and wind direction, calculation of standard deviations is almost always required. Averages of wind speed and direction can be calculated as independent scalar quantities and/or treated as vectors. Calculation involving wind direction requires the availability of trigonometric functions. If a custom DAS is being assembled, these functions must be available.

One-hour averages are most commonly wanted by regulatory agencies, some researchers, and the public. If averages over periods of 8 or 24 h are needed, these are derived from the 1-h data. Five-minute averages provide more useful resolution to technicians, especially if accompanied by minimum and maximum values. Five-minute averages of temperature are also ideal for reporting to local radio stations, if they are providing current temperatures to their listeners. During zero/span cycles and calibration, the resolution of 1-min averages is needed. During calibration it is useful to also present and record sample values (not averages) taken every 10 s or less. This gives the technician almost immediate feedback as adjustments are made and also gives a useful picture of signal noise when graphed.

At some point a slope and intercept must be applied to measured voltages to convert them to engineering units. In some systems the application of this transformation is done immediately and only the result is recorded. In other systems the unaltered voltage measurement is recorded and the linear transformation is done later. The original voltage data are never altered. Immediate application of slopes and intercepts is simpler but can create a mess when errors are inevitably made. Keeping an unaltered record of the original raw output will give the data greater credibility.

During zero/span cycles, multipoint calibration, and special maintenance, it is convenient to be able to set flags in the station system so that this flagging information becomes part of the data records. These flags may also affect the way the averages are calculated. Consider one possible way the flags might be used. Suppose the output from an analyzer is sampled once per second and 80% of these 1-s samples must be valid to form a valid average. Suppose that during the first 11 min of an hour maintenance is being performed and the output from the analyzer is invalid. During the first 11 min a maintenance flag is set. The calculation of 1- and 5-min data proceeds as normal but a flag is set that indicates the data are invalid. Set the invalid flag for the average even if just one sample during the averaging period is flagged as invalid. Eleven 1-min and three 5-min averages will therefore be marked invalid. The 5-min average from minute 11 to minute 15 is invalid because samples during minute 11 are invalid. For the calculation of the 1-h average, use only samples that are valid. So, the hour average will be calculated on the basis of the samples from minute 12 to minute 60. Since 49 min of valid data out of a possible 60 is more than 80%, the hour average will be valid and can be marked as such. Such a system means that the 1- and 5-min averages always show the technician what is being output by the analyzer and the hour averages are formed only from valid data. Such a system is one way to maximize the number of valid hourly averages created.

37.2.3   Collecting Data and Controlling Analyzers

In the simplest case a station will contain instruments that output some easily measured signal, such as a voltage, that corresponds to the value of the measurement. A simple DAS would periodically record the value of the signal. The system could consist of a person sitting in the station with a voltmeter. The person would record the values measured with the voltmeter on a pad of paper with a pen. If data need to be recorded for a longer period of time, then the person could be replaced with a strip chart recorder. In either case the data are written on a piece of paper. This paper could be transported to an office (the host) where someone manually transfers the data to some database.

Why would such a system be discussed in the twenty-first century?—because elements of a DAS may still include what is described above. Recently one of the authors was developing a system that needed to include daily information about the number of individuals in a building. The daily count is done by a person and recorded on a blackboard. Every few days a picture of the blackboard is taken with a digital camera. The electronic image is sent by email to the host where someone types the information from the photo into a spreadsheet. The data on the spreadsheet are then imported into the database. The majority of data in this database come from a fully automatic system. However, there are small parts of it that must be done manually and the DAS must accommodate these special aspects.

Returning to the station, a higher level of complexity is caused by the fact that the interaction between the DAS and the instrumentation is usually more involved than just measuring an output signal. Typically, for pollutant analyzers, a zero/span cycle will be performed on a daily basis to check its calibration and to provide data for baseline/slope correction. The control of these cycles is a normal part of the function of the DAS. Additionally, the DAS often provides power for various devices, especially those that require low-voltage DC.

Besides zero/span cycles, the DAS may be required to control other more unusual events. For example, relight sequences for some THC (total hydrocarbon) analyzers may require control. Control of the heating and cooling of the shelter may be best done by the DAS. The remote ability to cycle power to a device may be extremely convenient, as mentioned in an earlier example.

37.2.4 Personal Computers

Personal computers are readily available and inexpensive. Hard-drive capacity is enormous in comparison to any potential data storage needs for typical air quality monitoring. Personal computers will run a variety of operating systems. Programming can be written in any of dozens of languages. The speed of PC processors far exceeds any imaginable demands within the context of gathering air quality data. However, of the options considered here, PCs are the least reliable and robust. Also, the standard inputs and outputs of PCs do not include anything that will connect to most instrumentation. This lack of any interface to instrumentation will likely change in the next few years as more analyzers have options for ethernet connections or interrogation via RS232. However, today the most common outputs from analyzers are voltage outputs, and control of zero/span is done with relays. Personal computers have no readily available means of reading voltages or controlling relays. Companies offering PC-based packages for data collection usually manufacture special conversion boards to create the interface between the PC and instrumentation. The cost of such boards can easily exceed the cost of the PC. Such a board, after being plugged into a PC, is useless until appropriate software drivers are written and installed. This is another complex task usually done by the company offering the package.

Once the problems concerning the hardware interface are solved by either custom development or purchasing a package, PCs offer great advantages with respect to programming. Some problems that have required considerable ingenuity in the past can be entirely bypassed. For example, consider the calculation of wind direction standard deviation over a period of an hour assuming the wind direction data are sampled once per second. If each of the 3600 samples is stored in high-speed memory, an exact value for the wind direction standard deviation can be easily calculated with two passes through the data. However, storing all these data in older datalogging systems was an unacceptably large use of their limited random-access memory (RAM). So, in order to avoid storing all the values, sophisticated single-pass algorithms were developed that yield good approximate results.[1] Today, a standard PC has at least 256 MB of RAM. Storing even 100,000 values for a two-pass calculation puts no strain on resources. The question of implementing complex algorithms to achieve approximate results becomes irrelevant. (Ironically, in systems that have the resources to easily use a two-pass algorithm, the approximate single-pass algorithm is still often used.)

Having a PC at the station usually means having a monitor there as well. With data stored on a hard disk, viewing current and historical data should be convenient and fast. This contrasts with other hardware options which may not support a monitor and which may store historical data on a media with much slower access time. Note that the need to view historical data more than a couple of days old at the station using the local database is of debatable importance. If the data are readily accessible via the Internet or a dial-in connection, then it is unnecessary to view historical data with the station hardware. Most technicians prefer to review all the data in the network from the comfort of their home or office before venturing out into the field.

37.2.5 Programmable Logic Controllers

Programmable Logic Controllers (PLCs) are readily available and vary widely in terms of price and quality. If one wishes to gain real benefits from this option, then expect to pay the price that usually comes with quality. Like all technologies based on computer components, the price of PLCs has dropped dramatically during the last decade. It is this price drop that makes the PLC especially worthy of consideration today as a generic solution to data acquisition. Programmable logic controllers are designed to control machines and processes in industrial settings. Compared to PCs, PLCs are at the other end of the spectrum regarding reliability and robustness. There are dozens of manufacturers of PLCs. The manufacturer supplies an operating system and programming software. The ease of programming and the features of the operating system vary greatly between manufacturers and need to be thoroughly investigated before any choice is made.

A standard PLC is a solid-state device with no moving parts—one of the reasons they are so reliable. This means there is no mass-storage hard disk and memory for local data storage becomes an important consideration. Recently, some PLCs come with a built-in interface to flash ram (the memory cards used in digital cameras). Since large-capacity flash ram storage has become inexpensive and available at every store that sells cameras, issues concerning long-term data storage do not apply to these PLCs.

The reason PLCs exist is to interact with instrumentation. Therefore there are many standard devices for connecting analog voltage signals and controlling relays. A common way of constructing PLCs is to plug together various modules or cards. High-speed counter cards may be used to directly read the pulses from anemometers. Special cards for reading resistive temperature devices, thermistors, and very low voltages such as those from solar radiation transducers are available from most PLC manufacturers. The quality, specifications, and range of configuration for such specialty modules vary greatly between PLC manufacturers.

The operating system of a typical PLC makes the process of accessing data from modules almost automatic. The software drivers that interact with the modules are part of the operating system. The act of plugging in an analog module and connecting the output from an analyzer is enough to make the voltage appear in the central processing unit (CPU) memory ready for use. Traditionally the programming language for a PLC is based on some sort of ladder logic. Writing routines to form arithmetic averages and control outputs is an easy task. Complex wind speed and wind direction calculations are a bit more challenging. Certainly, the programming environment of PLCs is less powerful than those available in PCs. However, this disadvantage is greatly offset by the convenient interface for input and output.

Programmable logic controllers usually do not support a standard monitor. If a visual display is required, then expensive panel displays can be purchased. An inexpensive alternative is to connect to the PLC with a laptop and view the internal addresses of the PLC with the laptop. It is with such a connection and appropriate software that the PLC is programmed in the first place.

The reader should be aware that some PLC manufacturers give away their programming software while others charge a hefty price for it. Also, the communication protocols for communicating with the PLC are in some cases in the public domain and in other cases proprietary.

Major PLC manufactures have been in business for a long time. Support for their products is usually guaranteed for 15 years or more. In the world of PCs do not expect support and parts for more than a few years. Just imagine trying to find parts for a PC manufactured in 1990.

37.2.6 Commercial Dataloggers

Whereas a PLC is a general-purpose device designed for control of industrial processes and PCs are general-purpose devices for home, office, and laboratory, commercial dataloggers are devices specifically designed for collecting data. In many cases companies have designed dataloggers with specific attention to the needs of air monitoring. A typical datalogger has much in common with PLCs; they are usually a solid-state box with no monitor and they can be as robust and reliable as PLCs. Programming the datalogger usually requires learning a particular language developed by the manufacturer. Like ladder logic, such languages are simple and pose no real barriers to using the hardware. Like PLCs, dataloggers can be quite expensive. However, you can also expect them to be supported and functioning for 15 years or more.

37.2.7 Station Hardware in Context of Entire System

The device that collects data in the station should be robust and reliable. Data lost at this point can never be recovered. The accuracy and precision at which data are measured is set at this point. Station data collection and control hardware must connect and interact with the instrumentation. Programmable logic controllers and commercial dataloggers are designed to do this out of the box; PCs require special hardware and software to accomplish this. Data are not processed at the station in any really complex way. The software capabilities of PLCs and commercial dataloggers are adequate to meet these data processing needs. The software capabilities of PCs far exceed the requirements at the stations.

Both commercial dataloggers and PLCs have characteristics that match the requirements for data collection and control at the station. The characteristics of PCs do not match these requirements. In industry hundreds of thousands of PLCs operate machines and processes often in severe environments for years without interruption or failure. No sane engineer would try to replace them with PCs. Likewise, there are hundreds of thousands of commercial dataloggers collecting data in environments and circumstances that are entirely inappropriate for PCs.

Suggesting that PCs are a poorer choice for collecting data in the station compared to PLCs or dataloggers does not entail that they are necessarily a bad choice. In a station that is geographically close at hand and that has an officelike environment, PCs-based systems can perform adequately and offer some advantages.

37.2.8 Note on Multipoint Calibrations

Monthly multipoint calibrations and detailed reports of these calibrations are a requirement for many air-monitoring networks operating in the context of regulatory agencies. In such cases, doing these calibrations constitutes the largest single commitment of time and expense in operating the network. Having a sophisticated and dedicated calibration component as part of the DAS can reduce the time required to complete calibrations by 1 or 2 h per analyzer per calibration and reduce to almost zero the possibility of making calculation errors. In a network with 50 or more analyzers the initial cost of buying or creating such a calibration program is quickly recovered. Without a calibration program technicians typically use word processing and spreadsheet programs to create graphs of analyzer response, least-squares graphs, and summaries of results. Besides being time consuming, such tools make the process of creating reports mind-numbingly tedious.

Dedicated calibration programs should collect 1-min averages as well as instantaneous samples taken at intervals of 10 s or less. Real-time graphs and tables of these values should be readily available. Ranges of data to be used for points on the least-squares regression should be easily selectable. Relevant information from previous calibrations should be importable to the current calibration. All standard calculations of input concentrations should be automated. If a network contains NO_X analyzers and the calibration of the NO_2 channel is done with the usual gas-phase titration method, then the calibration program should be able to handle these complexities.

In some DASs, the calibration program is the single largest software component. This is often the case where a custom system has been developed by those who are actually operating a large network requiring frequent calibrations and extensive reports. In other systems a dedicated calibration program may not exist because its creation is not justified by operational and reporting needs.

37.3 CENTRAL SYSTEM ACQUISITION AND DISTRIBUTION

In most cases the data recorded from the analyzers in the stations need to be collected from the stations and brought to a central location. Without any further processing, all of the data should be available to technicians in charge of operations. Partial or complete sets of these real-time raw data may be made available to the public and special functions may make specific subsets of raw data available in specific formats for particular clients. Raw data are transformed into final data by passing through the processes of correction and flagging.

37.3.1 Connecting to Stations and Obtaining Data

One simple connection scheme is to have a dedicated line to each station that remains continuously connected. Most commercial quality modems have a redialing option that will cause the modem itself to make several attempts to redial and connect if the connection is broken. In this case neither the central system nor the station system need to be able to dial. Sending commands to modems to cause them to dial is a function easily configured in PLCs, standard dataloggers, and PCs. Thus, even if the modems have a built-in redialing mechanism, this can be backed up with programmed redialing from either the central system or the station.

A more common connection scheme is to have the central system phone the station, collect the data, hang up, and then move on to the next station. Occasionally, the arrangement may have the station system phoning into the central system at specific times or at times when the station detects some alarm condition. The authors recently tested a satellite phone system for its suitability for communication with remote stations. In this particular case the station end of the system was incapable of answering calls but could initiate calls. This satellite system would require the station to make the connections to the central system. Usually, there is no impediment to having the central system making the connections, and having it perform this function is preferred.

Exactly when connections are made is subject to many variables. Since cost is usually an issue, connection costs become an important consideration. To illustrate some of the complexities, assume connections for a compliance network are long distance and made with a cell phone to an area where reception is sometimes poor even with a high-powered phone and a Yagi antenna. Assume that connection fees are dependent upon the duration of the call and that significantly lower rates apply if the call is initiated in the evening or on a weekend. Assume the station creates 1-min, 5-min, and hourly averages and the zero/span adjustment and evaluation depends on the 1-min data. Because compliance depends usually on hourly averages, these must be collected every hour. It is best to collect them a few minutes past the top of the hour to maximize the time available for responding to exceedances. The system should be aware of both the time and whether the day is a weekday or weekend and be able to use this awareness to suppress collection of 5- and 1-min data during times of high connection rates. The collection of 1-min data takes by far the most time and consideration should be given to daily collection of only 1-min data that occurs during the zero/span cycle. The remainder of the 1-min database can be filled in later by performing quick manual downloads directly from the station system during visits to the station. The authors have implemented such a system and realized savings of several hundred dollars per month per station in connection fees. In the case of pollutant exceedance events the user should have a convenient option to force a connection and have the 5-min data collected. Although the compliance regulation may refer only to hourly data, having the resolution given by 5-min data adds a lot to understanding the nature of the event. The initial assumptions noted that reception is sometimes poor. This means that after initiating the phone call the modems do not always connect because of the poor quality of the connection. The system should recognize this, wait a specified amount of time, and then try again. Endless numbers of retries can be expensive so the number of retries should be limited.

Suppose communication between the central system and station is accomplished through a reliable high-speed cable Internet connection that is always available and costs some trivial amount. This represents the opposite situation to that described in the preceding paragraph. No complications arise and the central database can poll the stations every few minutes. There is no need whatsoever to develop the kind of system described above.

37.3.2 Corrections to Data and Flagging

Many pollutant analyzers, by their physical nature, are subject to baseline drift. For this reason many jurisdictions require a daily introduction of zero gas to determine the value of this baseline drift. A baseline correction, calculated by interpolation between the daily zero values, is then applied to the data. In some cases span gas is of sufficient accuracy to apply slope corrections as well. Regardless of what type of correction is done, the DAS should present the user with result from each daily zero/span and allow the user to accept, reject, or edit any data corrections based on these.

The DAS may also include a general means to edit the values of data, although circumstances are rare when such editing is justified. Regardless of the type of modification to the data, the modification should be kept separate from the original data. In viewing data, it should be possible to simultaneously view original unaltered data and baseline or otherwise corrected data.

Individual data observations should be represented by two entities, a value and a flag that indicates the status of the value. Data that are invalid need to be flagged as such. The practice of removing invalid data from the database or changing their value should be avoided. Flags generated by activity at the station will be part of the records transferred from the station. The user should have the ability to view these flags and then accept, reject, or edit them. There may be other periods that the user may want to flag as invalid.

Usually flags not only distinguish between valid and invalid data but also give several reasons why the data are invalid. It can be useful to know that data are invalid because of a power outage as opposed to a routine calibration. Since power outages usually affect every instrument in the station, time can be saved if the flag editor allows the global application of flags to the data from a station for a period of time.

37.3.3 Distribution of Data

The most reliable and secure means of providing unattended real-time access to data is over telephone lines. One way to do this is with computer telephony technology. The human client dials the host and the host answers. By making selections with the keypad, the clients requests the needed data and an automated voice supplies the data. Such systems are everywhere and require only moderate effort to customize. If the client is another computer, then there is no need for voice generation and just a pair of modems connected to serial ports is required for hardware. Command strings are sent to the host, which responds with strings containing the data. If several such telephone clients need to be serviced at once, hardware is available to convert single-serial or USB ports to multiple serial ports. Connecting modems and telephone lines to each serial port yields the needed capacity.

Using the Internet instead of direct communication through telephone lines provides less reliability and less security but is otherwise advantageous. The Internet is usually faster and less expensive, provides more options for implementation, and provides for multiple simultaneous connections with only one physical connection. Obtaining data through the Internet can be accomplished by many different means. Although a discussion of the various options is beyond the scope of this chapter, there are two basic approaches to Internet communication. The first is to run a standard Web server at the host and have the client access data with a standard Web browser. The second method is to develop a custom Internet server application and a custom client application communicating with each other through a normally unassigned Internet port. Each of these methods has its advantages and disadvantages. Both methods can be simultaneously implemented on the same machine. The Web server/browser combination is best suited to simple presentations of data. Development of such presentations in this environment is quick and easy. Problems with firewalls are minimal because most firewalls are configured to allow such communication. The custom server/application pair makes possible extremely complex presentations of data that would be very difficult to create in the Web server/browser context. This option gives the developer full control over every aspect of the behavior of the server and the application. Mild complications often arise when clients are behind firewalls. Explicit permission for the connection must be granted by the administrator. In a system developed by the authors, the public and clients with uncomplicated needs, such as the local radio station, have Internet access to graphs and tables of data using a standard browser. The technicians operating the system and others who require a more powerful interface use the custom applications.

Moving data from the host to another application or database is a common request. Independent researchers will want subsets of data to import into their own programs for further analysis. Data acquisitions systems need to accommodate these requests. At a minimum the system should be able to export data along with flags in comma-separated text format. Be aware of whether exports to any particular format can be done on an automatic and regular basis, can be easily initiated by other computer applications, or must be done manually. Having to provide a person to manually export data every hour of the day so that it can be read by some real-time modeling program is not desirable. Another option for providing access to data is to provide clients with a DLL (dynamic link library) that will give clients direct access to the database in the language of their choice.

Data may also be distributed in the form of paper reports. While this method is the least desirable in terms of using the data, it can create a secure and long-term set of unalterable records. Properly stored, ink on paper will last for centuries in a form that can be read by anyone. The same cannot be said for various types of electronic storage.

37.3.4 Alarming

Regular searches through the database for alarm conditions serve two purposes. The first is to provide alerts for exceedances in compliance networks. The second is to alert technicians to anomalies that may require maintenance or other action. Assume for this discussion that both purposes are to be met and that immediate responses are required for the compliance aspect. A well-designed alarm program can be configured to search for high and low values of all types of averages and instantaneous sample values. So, if someone wants to be alerted to gusts of wind so dust control activities can be initiated, then the system can search for high values in 1-min averages or maximum sample values or both. Natural background THC levels are about 1.8 ppm. If much lower values are being recorded, then technicians should be alerted. On/off conditions such as the status of a door (opened or closed) may be represented in a more compact form in the database records than analog signals. The alarm system needs to be able to detect these conditions as well. Ice on a wind direction device or a voltage spike causing an analyzer's software to malfunction may cause an unusually steady signal. Detection of such signals is a useful option. Some compliance standards are based on 24-h averages or 8-h running averages. Since these averages may not be directly contained in the database, the alarm program must be able to compute them from hourly averages.

An alert for a high level of H_2S may need to be sent out immediately regardless of the time of day. However, few technicians want to know that a THC reading is a little low at 3:00 AM. Alarms of lower severity can likely wait for the beginning of regular business hours. In addition to being able to control when alerts are issued, it is useful to be able to identify a specific person or group to whom the alert is directed. The high wind conditions mentioned above are likely of no importance to the technicians operating the air-monitoring network but important to staff in a nearby open-pit mine.

Finding alarms in the database and the action of alerting relevant people to these alarms are two separate processes often done by two separate programs. The alarm program detects the alarm and sends all relevant information at the appropriate time to a callout program. The callout program then begins phoning people on the appropriate callout list and, depending on the severity of the alarm, does not stop until the callout is acknowledged or stops after a specific sequence of numbers has been attempted. For severe alarms is it useful to have the callout system begin by phoning a small group of people several times and then, if none of them are available, broaden the search to ensure someone is eventually contacted within the desired time constraints. The program should write a log describing the alarm and identifying who acknowledges the alarm. Given the state of computer telephony technology, computer-generated voice messages should include considerable detail of the alarm and perhaps even allow the user to obtain more detail by selecting options by pressing the keypad of a touch-tone phone.

The alarm program may also include features that evaluate the status of the central system and have a backup option for alerting staff if the primary callout system fails. A simple backup system will just send a dial command to a modem and let the phone ring a few times before hanging up. Since most people have caller ID, they will recognize the call to be from the backup system and know the primary system has failed and needs attention.

A properly designed alarm system can play an important role in the operation of a network whether or not the network must meet compliance regulations.

37.3.5 Hardware and Software

Whereas there is debate about what hardware to use in the station, there is no debate about what to use for the central system or host. Personal computers are the only choice. If available, any mainframe computers could be used as well. These are not included as a choice because typical air quality data acquisition systems do not require the computing power of a mainframe. Today PCs have many of the capabilities of mainframes just a few years old. Personal computers are well suited to the tasks required by the central system and have ample memory and speed to handle networks with dozens of stations.

For complex systems attending to many varied needs, the central system will likely be a network of computers rather than just one. One computer may be dedicated to collecting data and identifying alarm conditions. Another computer may have computer telephony hardware installed along with multiple serial ports and modems to handle all aspects of telephone communication. Another computer may be attached to the Internet and have a Web server and various custom server applications running. Still other computers may be attached to this network and serve as workstations for staff. The cost of PCs is small compared with the problems associated with having one PC overloaded with hardware and software.

An evaluation of operating systems and programming languages is beyond both the scope of this chapter and the expertise of its authors. Nonetheless, a few comments are in order. If a system is being put together from commercial packages, then these will stipulate the options for the operating systems. If the system is being custom developed, then greater choice exists. Custom programming for data acquisition can certainly be complex but does not require features that are not available in virtually every operating system and major programming language. Also, given the memory capacity and speed of current computers, there is usually little need for code that is optimized for speed or memory efficiency. Most operations will be almost instantaneous to the human user. During data collection the limiting process will be the transmission of data rather than the rate at which the computer can process the data. So, in choosing a computer language, it is recommended that the choice be made on the basis of ease and speed of programming rather than the presence of features that optimize the speed of the running code or give the programmer unnecessary control over low-level operations.

37.3.6 Security from Hackers and Viruses

For any system that is intended to distribute data over the Internet, security becomes a primary consideration. There are two basic approaches. The first is to use firewall and antivirus software to protect the network. The second is to physically isolate the key parts of the network from the Internet. The first method is common and can give a high degree of security; the second gives total security from external threat.

Any network connected to the Internet will be subjected to a constant bombardment from external threats. Keeping these at bay requires continuous updating of firewall, antivirus, and operating system software. Even then, expect occasional attacks to get through and cause some inconvenience. Be sure you have backup disks and the knowledge to wipe your hard disks clean and reinstall and configure all of your system from scratch. This is good advice in light of not only external threats but also routine physical failure of components.

The second method is practical because there is usually no pressing need to have the computers doing data gathering and alarm functions to be connected to the Internet. Buy a separate computer for Internet communication and maintain an up-to-date database on this computer by moving records from the host system to the Internet computer over a serial connection using software designed only for that purpose. This means the host computers are not part of any external network and cannot be subjected to any external threat. The only threat of contamination is from inserting contaminated disks into the host system.

37.3.7 Database Considerations

Books are written on database design and implementation. The goal here is to present some basic information and highlight considerations especially relevant to air-monitoring DASs.

A box containing strip chart recordings is a database. Retrieval of information from such a box of paper is manual and slow. Writing numbers electronically in text files is a common and simple method of storing data. Such files can be viewed and edited with any word processor. Usually such text files are comma or tab delimited and can be imported into spreadsheet applications or more sophisticated database applications. Text files are bulky in terms of memory usage, and retrieval of information from them directly is not much faster than shuffling through a box of paper. Spreadsheet applications have many useful features and can be used to manipulate and present small quantities of data. However, for the quantities of data created by a moderately sized air-monitoring network, spreadsheets are an unsuitable database solution.

Developers of commercial DASs will either incorporate database systems from other software vendors or custom create their own database system. This same choice exists for those who are custom creating their DAS. The level of effort and training required to use existing commercial database applications can be close to that required to design and create a simple custom database within a general programming environment.

Commercial databases are all relational databases. The rules and mathematics of relational databases were established several decades ago in response to general database problems. Air-monitoring data can certainly be stored in a relational database. However, since air-monitoring data are static and historical, the sophistication and power of a relational database is not necessary. A simple nonrelational hierarchical database structure is well suited to storing air-monitoring data and give extremely fast access to the data. The most common database query in air monitoring is to find the concentration of a particular pollutant at a particular time at a particular place. This single query is all that is necessary to create the tables and graphs most users need to see.

If there is some special requirement or issue that requires the power and complexity of a relational database management system, then make that choice. Otherwise, consider as a reasonable option the development of a custom database structure that suits the simple nature of air-monitoring data.

37.3.8 Reporting

Data that are in the form of a simple comma-separated file can be imported into a spreadsheet application and then formatted as desired. This is an acceptable method for producing small reports or special reports. However, in many cases, standard reports of several hundred pages are issued every month. In this case the user should be able to select a parameter (or range of parameters) and a time range, press a button, and watch the report roll off the printer. The expense of creating such programs for automating the production of standard reports is quickly returned.

Some circumstances and types of reports warrant a two-step approach. In the first step the required data and statistics are generated and placed in an easily editable text format. Manual adjustments can then be made before initiating the second step of producing a fully formatted output.

Many clients now prefer to receive reports electronically as a PDF file. Software can be purchased that will enable any program that can write to the printer to write to PDF. Understanding the details of creating and combining PDF files will help design report programming that eases the production of reports in PDF.

A common time interval for reports is one month. In most cases creating a report that spans from the first day of the month to the last day of the month is what is needed. However, mobile stations may be at a site from, for example, the eighth day of one month to the eleventh day of the next month. In this case it is handy to be able to set a specific date range for the report rather than just the month.

A subtle problem that arises when writing software for reports concerns rounding and its effect on calculation of averages. The issue is best presented with a simple example. Suppose four concentrations are stored in the database as 3.6, 3.6, 3.6, and 2.6. The average of these four numbers is 3.35, which becomes 3 when rounded to the nearest whole number. If the four values in the database are rounded to the nearest whole number, the result is 4, 4, 4, and 3. The average of these numbers is 3.75, which becomes 4 when rounded to the nearest whole number. This issue arises in regulatory contexts in which reports of hourly averages and daily averages are presented with specific rules for rounding results. An average of 4 instead of 3 may imply the violation of a standard. It is important to be clear how the regulatory agency wants calculation of averages to be done and ensure the software doing the calculations complies with the method.

Data can be presented in tables or graphical form. Besides the usual rectangular graph, wind and pollutant roses are common graphical displays. Following is a list of some of the reports the authors have had to produce in their experience: table of hourly averages for an entire month with diurnal and daily averages; graph of hourly averages for a month; table and graph of maximum 1-min averages for each hour for a month; percentile, frequency, average, maximum, and minimum value statistics; graph of daily zero response and daily span response for a month; count of violations; calculation of uptime; and annual summaries of monthly statistics.

37.4   CONCLUSION

The concept of a data acquisition system (DAS) for environmental monitoring has been explored in the context of ambient air quality monitoring networks. Continued developments of relevant technologies and increasing expectations have broadened the concept of a DAS. The boundary of what a DAS encompasses is unclear but certainly expanding.

A DAS is not a single entity but a collection of devices, software, and operational procedures. In the station a DAS fulfills the traditional roles with respect to data, including sampling of and translation of sensor signals into data values, processing, and storage. The DAS interacts with station components, controls automated station functions such as daily zero/span operations, and facilitates sensor calibration and other maintenance-related tasks. At the network level the DAS encompasses communications with sites for data transfer, site and event surveillance, and alarm condition detection and response, including notification of appropriate parties. It also provides or facilitates data functions, including QAQC operations, validation, reporting, and dissemination. Other network level considerations include system security and the archiving of primary and documentation data.

The creation or selection of a DAS involves the consideration of a broad range of issues. Fundamental questions about the reasons for creating and collecting the data should be raised. Practical issues arising from such factors as station location and availability of communication services need to be addressed. There is always the importance of cost effectively maintaining and operating a network with a minimum of interruption. Many networks have the additional burden of meeting strict regulatory requirements. The effects of human fallibility and the failure of components due to normal operating conditions or extraordinary events need to be taken into account. Most monitoring programs have a life of many years during which circumstances and needs will change and advances in technology will be made. Therefore consideration should be given to the capacity of a DAS to adapt and accept upgrades.

After giving such issues due consideration, most networks are discovered to be sufficiently complex and unique that it is unlikely to find a match with any easily configurable, out-of-the-box system. The designer and/or purchaser is faced with a wide variety of options. At every level, system components may be created or purchased and purchased components may be generic or specialized/custom products. The challenge is finding a balance that creates a flexible system at a reasonable cost to meet the needs of the network over its life span.

ACKNOWLEDGMENTS

The authors would like to thank their employers and colleagues for their support and suggestions regarding the writing of this document.

REFERENCES

1. R. J. Yamartino, A Comparison of Several 'Single-Pass' Estimators of the Standard Deviation of Wind Direction, *Journal of Climate and Applied Meteorology* **23**, 1362–1366 (1984).

38

AIR POLLUTION CONTROL SYSTEMS

Randy D. Down, P.E.
Forensic Analysis & Engineering Corp.
Raleigh, North Carolina

38.1 INTRODUCTION

Applying the proper instrumentation to a pollution control system requires a basic understanding of the process conditions in which it will be used. This is true,

Environmental Instrumentation and Analysis Handbook, by Randy D. Down and Jay H. Lehr
ISBN 0-471-46354-X   Copyright © 2005 John Wiley & Sons, Inc.

regardless of the type of process control application. Since environmental instrumentation is applied to pollution control systems to control the rate of treatment and monitor removal efficiency, it is appropriate to include a chapter that covers the general nature of these systems.

38.2   THEIR PURPOSE

Airborne pollutants, such as the emission of solvent vapors and other forms of VOCs (volatile organic compounds) into the atmosphere, have been identified as a major factor contributing to the depletion of the ozone layer. The ozone layer filters harmful ultraviolet radiation from the sun that can be damaging to plant and animal life.

Pollution control systems are designed to significantly reduce the concentration of pollution emission exhausted from an industrial plant or utility plant to a level deemed acceptable by local, regional, and national regulatory agencies. These systems are typically rated in terms of their removal efficiencies (percentage of pollutant removed) and generally must be effective and reliable in removing 90–99% of VOCs in the exhaust stream. Some systems are designed to remove particulate from the air stream, which can also be harmful to the environment.

Note: When involved with the selection or design of these systems, it is important that you review all applicable federal, state, and local codes and regulations that apply to the use of these systems or consult someone familiar with them.

38.3   GENERAL SYSTEM TYPES

Systems capable of this kind of removal efficiency for gaseous vapors include some form of oxidizer (to destroy VOCs and other contaminants using high temperature) wet scrubber systems and carbon adsorption systems.

Systems designed specifically to remove particulate include filtration systems such as baghouses, cartridge filter systems, electrostatic precipitators (ESPs), and cyclone separators.

Each system has certain advantages and disadvantages when applied to a pollution abatement application. Proper measurement of the process conditions and control of the process are critical if the system is to function at its maximum removal efficiency.

38.4   THERMAL OXIDIZERS

Thermal oxidizers (TOs) function in the manner that their name implies. They use thermal energy to oxidize (burn) the pollutant in an air stream as it passes through a heated chamber. A thermal oxidizer is essentially a sophisticated incinerator and the simplest form of oxidation system.

There are multiple oxidizer technologies that achieve the oxidation of organics in alternate ways:

- Thermal oxidizers
- Recuperative thermal oxidizers
- Catalytic oxidizers
- Regenerative thermal oxidizers (RTOs)
- Rotor concentrators
- Flares

Oxidation causes compounds (in this case, contaminated air pollutants) to be broken down and re-formed into new, harmless compounds. Adding the right amount of heat and oxygen to hydrocarbons results in oxidation. Chemically, the process is

$$C_nH_{2m} + \left(n + \frac{1}{2}m\right)O_2 \phi n\, CO_2 + H_2O + \text{heat}$$

In a thermal oxidation process, contaminated air is heated, breaking apart the contaminated compounds. The elements bond into harmless compounds of carbon dioxide, water vapor, and released energy. During catalytic oxidation, contaminants in the air react with a catalyst material (platinum, palladium, or rhodium), which breaks down the contaminant bonds more easily. This applies to all oxidation technologies: the breaking down of harmful compounds into harmless by-products and energy.

Thermal oxidation requires high temperatures to break apart the pollutant compounds. The large amount of auxiliary fuel needed to maintain high temperatures can be expensive. Utilizing a different oxidation technology can help reduce the long-term operational costs of the equipment. Catalysts, for example, react and oxidize the VOCs at a lower temperature than TOs and RTOs, resulting in less fuel consumption and therefore lower operating costs.

An oxidizer has a burner system and combustion controls that regulate the chamber temperature at a temperature of 1400–1600°F, which is sufficient to destroy VOCs by means of oxidation. Depending upon the fuel content of the VOCs, the oxidizer may require little or no auxiliary fuel to maintain a destructive temperature.

A heated chamber (or chambers) within the oxidizer is filled with ceramic chips in a shape that allows a high degree of surface contact with the pollutant vapor (VOCs) as the exhaust air is forced or drawn through it. When the VOC comes into contact with the hot surface of the ceramic media, it is destroyed. The time the vapor stream spends traveling through the ceramic bed where oxidation occurs is referred to as the "residence time."

Thermal oxidizers must be designed to provide adequate residence time and temperatures to achieve the destruction of the organic contaminants. Thermal oxidizers are typically two to four times larger in size than catalytic oxidizers because of their higher residence time requirements.

Older TOs were constructed with a carbon steel outer shell and refractory or brick as the interior, thermal liner. New TOs use a ceramic fiber insulation on the inside, which is lighter and more resilient.

38.4.1 Oxidation Process

The preheated air enters the combustion zone. The air temperature must be raised to the required oxidation temperature, usually between 1400 and 1600°F. The additional heat that is required to reach the oxidation temperature is provided by a burner using a supplemental fuel such as natural gas, propane, or fuel oil. Preheating the contaminated air in the preheater greatly reduces the amount of supplemental fuel required.

The primary fuel source for heating a TO is a combination of an auxiliary fuel (such as natural gas) and the pollutant it is burning. Pollution control systems are typically designed to burn natural gas as a supplementary fuel to compensate for variations in pollutant concentrations. The variations are caused by process changes that are occurring at the source of the VOCs (such as a process variation, shift change, or maintenance downtime). As the pollutant concentration entering the air stream drops, more secondary fuel (natural gas) must be burned to maintain the same destructive temperature level (typically over 1400°F) required within the chamber to reliably destroy the pollutant.

Because concentrations of volatile pollutants in the contaminated air stream can approach the lower explosive limit (LEL) of the vapor, it presents a potentially hazardous condition within the exhaust ductwork and pollution control system. Therefore, depending on the flash rate of the flammable VOC, a flame arrestor or a detonation arrestor may be required in the air duct.

The conversion rate of VOC to $CO_2 + H_2O$ depends on three key factors: the temperature, time at temperature (resident or dwell time), and turbulence. Together, the "three T's" determine the rate of destruction for a contaminated stream. In general, with proper turbulence, the minimum dwell time is 0.5 s and the minimum temperature is 1400°F.

The chamber is designed such that pressure drop (restriction of exhaust air flow) through the chamber is minimized. However, the ceramic bed induces a high pressure drop (up to 25 in. w.c.), and sufficient energy is needed to push or draw the contaminated exhaust air through the ceramic media.

As the now "clean" air exits the heat exchanger, it passes through an exhaust stack before being discharged to the atmosphere.

If multiple process lines exist, a means to combine the air streams and control the total volumetric flow to the oxidizer must be utilized. For example, several process exhausts may feed into a common collection plenum, pass through a filter, and then pass through a system fan. The pressure drop across the filter is monitored by means of a differential pressure switch (equipped with remote alarm annunciation) to determine when the filter must be cleaned or replaced. The amount of flow being moved by the system fan is controlled by means of a vortex damper or variable-speed drive. Because the amount of air from multiple process lines into the common plenum can vary, static pressure in the plenum is continuously sensed by a pressure transmitter whose output is sensed by a pressure controller. The pressure controller varies the speed of the fan (through the variable-speed drive or by positioning the vortex damper). This pressure control loop regulates the volumetric flow rate of exhaust air drawn from the plenum based on static pressure. The fan slows down or speeds up until the amount of air drawn from the plenum is the same as the amount of air entering. If the amount of air being drawn from the plenum is occurring at the same rate as the exhaust air entering from the exhaust fans, then the net static pressure in the plenum will be zero inches of water. If the fan draws less from the plenum than the air being exhausted into the plenum, then the plenum's static pressure will exhibit a slightly positive pressure. If more air is being drawn by the fan than the amount of air being exhausted into the plenum, then the plenum's static pressure will exhibit a slightly negative pressure. This indirect method of tracking the flow rate (monitoring static pressure in the plenum) works well if the rate of change in the combined exhaust rates is relatively slow (taking 10 s or more for a 10% change in flow rate). If rapid, drastic changes in flow rate are anticipated (this is a very critical question to be asked during the system's design), then a feed-forward signal may be needed to prevent exposure of the pollution control system to large swings in the rate of air flow.

An example of a feed-forward signal in an application of this type would be to install some type of instrument to monitor damper position or sash position (in the case of a laboratory) or perhaps to install individual flow-monitoring stations at the sources.

This type of pressure control loop is somewhat standard on all systems that have multiple process lines and variable-flow conditions. It is not required when a constant-flow, single process line exists but is beneficial to minimize air volume variations drawn from the process. When a fixed speed fan is used, variation in organic loading causes differing pressure drops across the oxidizer. The differing pressure drops result in varying flow from the process unless a pressure control loop or bypass ductwork and diverting dampers are provided to direct the excess air flow around the pollution abatement system.

Systems with variable exhaust rates clearly add complexity to the pollution abatement system. Bypass dampers, variable-speed drives, or some other means of accommodating variable flow rates must be provided. Oxidizers are designed for a constant rate of air flow through them. A flow rate lower than the design flow rate can potentially damage the pollution abatement system by causing it to overheat. Safety limits are provided as part of the system package to shut the abatement system down at low flow rates or excessively high temperatures.

An unplanned shutdown of the system can also result in problems, such as a shutdown of the manufacturing process where the VOCs originated. This may need to occur to avoid emitting the untreated VOCs to the atmosphere. Generally, provisions in the clean-air regulations will allow this to occur for a brief period of time without penalty (while the system problem is being resolved). Regulatory agencies have acknowledged that there may be an occasional situation due to a system failure in which contaminants will be emitted to the atmosphere.

Other types of oxidation systems have been developed to reduce the total energy required (i.e., attain a better overall energy efficiency).

Excess heat can cause damage to the preheater and unscheduled oxidizer shutdowns. To prevent these, a hot-side bypass is provided to direct the clean, hot oxidizer air around the heat exchanger, rather than allowing it to go through the preheater. This redirection of hot air creates an unbalanced flow in the heat exchanger (more dirty cold air than clean hot air). The result is a lowering of the preheat temperature and return to stable operating conditions.

In some cases, there is a need to recover the remaining waste heat energy, building heat or steam. In these cases, a second heat exchanger, often referred to as an economizer, is used to recover even more of the heat from the oxidized exhaust.

38.5 REGENERATIVE THERMAL OXIDIZERS

Regenerative thermal oxidizers are a newer, hybrid form of TO. By providing two oxidizing chambers with ceramic beds, rather than one, the lead bed can be used to pass through and destroy pollutant vapors as the second chamber is regenerating. This is accomplished by isolating the air stream (through diverting dampers) and allowing it to heat up to a temperature that burns any residual substances from the air stream off of the surface of the ceramic chips prior to the chamber's reuse.

After the ceramic heat transfer beds have reached an operating temperature of 1500°F, the unit is ready for the process air stream. As the process air stream enters the ceramic heat transfer beds, the heated ceramic media preheat the process air stream to its oxidation temperature. Oxidation of the air stream occurs when the autoignition of the hydrocarbon is reached. At this point the heat released from oxidation of the process hydrocarbons is partially absorbed by the inlet ceramic heat transfer bed. The heated air passes through the retention chamber and the heat is absorbed by the outlet ceramic heat transfer bed. If the oxidizer is self-sustaining, the net increase in temperature (inlet to outlet) is 100°F.

During the normal mode of operation of the system the process air enters the RTO system fan and passes through the inlet diverter valve where the process air is forced into the bottom of the left ceramic heat transfer bed. As the process air rises through the ceramic heat transfer bed, the temperature of the process stream will rise. The top of the beds are controlled to a temperature of 1500°F. The bottom of the beds will vary depending upon the temperature of the air that is coming in. If it is assumed that the process air is at ambient conditions, or 70°F, then as the air enters the bottom of the bed, the bottom of the bed will approach the inlet air temperature of 70°F. The entering air is heated and the media are cooled. As the air exists the ceramic media, it will approach 1500°F. The process air then enters the second bed at 1500°F and now the ceramic media recover the heat from the air and increase in temperature. At a fixed time interval of 4–5 min or based on thermocouple control, the diverting valves switch and the process air is directed to enter the bed on the right and exit the bed on the left. Prior to valve switching the air heated the right bed and now this bed is being cooled. The cooling starts at the bottom and continues upward because the media are hot and the energy is transferred. The process air then goes through the purification chamber and exits through the second bed.

An advantage of this type of system over a conventional oxidizer is its improved energy efficiency through reduced natural gas consumption. It tends to have a higher initial capital cost but a lower life cycle cost due to its energy efficiency and good reliability.

Instrumentation and controls play an important part in the proper operation of oxidizer systems. Combustible gas detectors are installed in the exhaust plenum supplying VOC-laden exhaust air from the process to the oxidizer to monitor the percent LEL. A typical control system (PLC, DCS, or stand-alone controller) will begin to open an automated inlet damper to allow "dilution" (outdoor) air into the duct if the %LEL exceeds 10%. If the inlet (dilution) damper modulates fully open and the %LEL continues to climb to 25%, at 25% LEL the thermal oxidizer shuts down and the exhaust air is diverted to a second exhaust blower that discharges the air to the atmosphere. An added precaution involves the addition of a flame or detonation arrester in the exhaust duct supplying the thermal oxidizer. These devices act as a sort of check valve to prevent a flashback occurrence (the vapor igniting and burning back along the vapor path to the source of the emissions). Such a flashback occurrence could have catastrophic results, such as a major plant fire or explosion. The differences between a flame arrester and detonation arrester are reaction time (detonation arresters react much faster and are required where highly volatile vapors with a fast flame spread rate are used) and cost (detonation arresters typically cost 10 times that of a flame arrester).

Recuperative heat exchangers can also be added to thermal and catalytic oxidizers to recover between 50 and 75% of the heat released during oxidation. Another system advance is the Regenerative oxidizers use multiple ceramic chambers to recover as much as 90–95% of the heat released from oxidation.

38.6 CATALYTIC OXIDIZERS

Catalytic oxidizers are another hybrid form of TO (Fig. 38.1). A catalytic converter, much like those used to reduce emissions from automobile exhaust, is used to

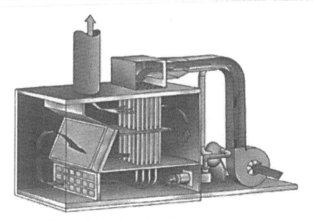

Figure 38.1 Recuperative catalytic oxidizer.

destroy VOCs. The advantage is that the destruction can occur at a lower temperature due to the chemical reaction of a catalyst (such as platinum) with the pollutant. The vapors are heated as they pass through a ceramic honeycomb and contact and react with the platinum. The advantage of this is reduced energy costs (which translates into reduced operating costs) due to less natural gas being required to achieve and maintain the lower oxidizing temperature.

Catalytic treatment of VOCs and other air pollutants works by reacting the harmful air pollutants over a specially designed precious metal catalyst where VOCs are converted to CO_2, water vapor (H_2O), and usable heat. These harmless by-products are released to atmosphere or use an energy recovery technique to further lower the operational costs.

A disadvantage of this type of system is that it cannot be used on applications where there is a significant concentration of chlorinated compounds in the exhaust stream. Chlorinated compounds will attack the platinum catalyst and quickly render it useless.

A catalyst can lose its activity by loss of surface area due to sintering (exposure to very high temperatures) or poisoning/masking by compounds such as silicone and phosphorus. To ensure that a catalytic oxidizer is meeting VOC conversion requirements, it is important that catalyst deactivation is measured and procedures are put in place to continually monitor the performance of the catalyst. There have been some issues and concerns about the best possible method of field catalyst sampling so that a representative sample could be obtained.

Catalyst samples are shown in Figure 38.2. Some catalyst is deposited on a ceramic substrate. These ceramics are extruded in a malleable state and then fired in ovens. The process consists of starting with a ceramic and depositing an aluminum oxide coating. The aluminum oxide makes the ceramic, which is fairly smooth, have a number of bumps. On those bumps a noble metal catalyst such as platinum,

Figure 38.2   Catalyst samples.

palladium, or rubidium is deposited. The active site, wherever the noble metal is deposited, is where the conversion will actually take place.

An alternate to the ceramic substrate is a metallic substrate. In this process, the aluminum oxide is deposited on the metallic substrate to give it a wavy contour surface. The precious metal is then deposited onto the aluminum oxide to form what are called "monoliths."

An alternate form of catalyst is pellets. The pellets are available in various diameters or extruded forms. The pellets can have an aluminum oxide coating with a noble metal deposited as the catalyst. The beads are placed in a tray or bed and have a depth of anywhere from 6 to 10 inch. The larger the bead size, the less the pressure drop through the catalyst bed. However, the larger the bead, the less surface area present in the same volume, which translates to less destruction efficiency. Higher pressure drop translates into higher horsepower required for the oxidation system. The noble metal monoliths have a relatively low pressure drop and are typically more expensive than pellets used for the same application.

As the air passes through the heat exchanger, it is heated and will then exit into a section called the "reactor." As it enters the reactor, the process stream will be further heated by a burner controlled by a thermocouple measuring the temperature of the air and a temperature controller regulating the burner firing to bring the process stream up to the catalyzing temperature of 300–700°F. The catalyzing temperature depends on the organic, the requirement for the destruction of the organic, and the type and volume of catalyst. At the catalyzing temperature, the process stream will pass through a series of beds that have catalyst in them. As the air containing organics comes across the catalyst, the organic is converted to CO_2 and water vapor and an exothermic reaction occurs. This reaction will raise the temperature of the stream exiting the catalyst bed. Hence the catalyst outlet temperature will be higher than the temperature going into the catalyst bed. The process stream is then directed though the shell side of the heat exchanger where it preheats the incoming air and is then exhausted to the atmosphere.

Much like the RTOs, the catalytic oxidizers have a higher capital cost but a lower life-cycle cost due to their energy-efficient operation.

It is necessary to determine if there are compounds in the exhaust stream which can be detrimental to the catalyst, thereby precluding the use of this technology. As an example, when automotive catalytic mufflers and converters were introduced a number of years ago, the automobile industry required the petrochemical industry to eliminate lead from gasoline, since lead degraded and reduced the effectiveness of the catalyst and the destruction of the gasoline. Industrial compounds that can harm catalysts are halogens, a family of compounds which includes chlorine, bromine, iodine, and fluorine. Bromine, while not prevalent in industry, is present in some chemical plants. Freons are fluorine compounds. Silicone is another compound which is detrimental to a catalyst. It is used as a slip agent or a lubricant in many industrial processes. Phosphorus, heavy metals (zinc, lead), sulfur compounds, and particulates can shorten the life of the catalyst. It is necessary to estimate the volume or amount of each of these contaminants to assess the viability of catalytic technologies for the application:

- Catalytic recuperative oxidizer designs assume a 65% efficient heat exchanger.
- Thermal recuperative oxidizer designs assume a 65% efficient heat exchanger.
- Regenerative thermal oxidizer designs assume a 95% efficient heat exchanger.
- Regenerative catalytic oxidizer designs assume a 95% efficient heat exchanger.

38.7   ROTOR CONCENTRATORS

Rotor concentrators consist of a slowly rotating concentrator wheel that uses a zeolite or carbon media to adsorb organics from the contaminated exhaust air stream as it passes through the wheel. A portion of the concentrator wheel is partitioned off from the rest of the rotor, and hot, clean air is passed through this section to desorb (remove) the organics. The clean, heated air is approximately 10–15% of the total air stream volume. The desorption air drives off the organics in that particular sector of the concentrator wheel. As the wheel continues to rotate, the entire wheel is gradually desorbed. As the stripped, concentrated organic exits the wheel, it is processed through a thermal recuperative oxidizer. This approach greatly reduces the total volume of air that a TO must heat, ultimately resulting in lower supplementary fuel costs.

A concentrator wheel is shown in Figure 38.3 with exhaust passing through it. A 10–15% of the original process stream is drawn through the rotor on the cooling pass. A secondary fan forces this air through the secondary heat exchanger, where the reduced air volume temperature is raised to the required desorption temperature. The preheated air is then used to desorb the air in the pie-shaped sector of the wheel. As the air exits the desorption section, the organic concentration is approximately

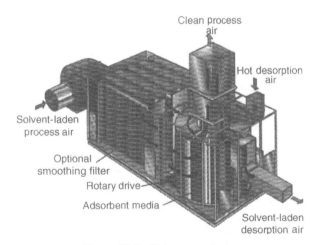

Figure 38.3   Rotor concentrator.

10 times the concentration of the original process stream. This low-volume, higher concentration stream then enters the induced draft section of a catalytic or thermal recuperative oxidizer, where the organics are destroyed through oxidation.

The total hydrocarbon reduction efficiency for the rotor/concentrator is the adsorption efficiency of the rotor/concentrator times the destruction efficiency of the oxidizer. Hence, if the rotor can adsorb 98% of the organics in the air stream and the oxidizer can destroy 99% of the organics that pass through it, then the overall system organic reduction is given as

$$\text{Destruction efficiency} = 0.98 \times 0.99 = 0.97 = 97\%$$

38.8 FLARES

Flares are an oxidation technology that continues to be used both domestically and internationally. They are used in the petroleum, petrochemical, and other industries that require the disposal of waste gases of high concentration in both a continuous and intermittent basis. As for other thermal oxidation technologies, the three T's are necessary to achieve adequate emission control.

Flares ideally burn waste gas completely and therefore without smoke. Two types of flares are normally employed: *open flares* and *enclosed flares*. The major components of a flare consist of the burner, stack, water seal, controls, and ignition system. Flares required to process variable air volumes and concentrations are equipped with automatic pilot ignition systems, temperature sensors, and air and combustion controls.

Open flares have a flare tip with no restriction to flow and are effectively a burner in a metal tube (Fig. 38.4). Combustion and the mixing of air and gases take place above the flare, with the flame being fully combusted above the stack.

Enclosed flares are composed of multiple gas burner heads placed at ground level in a stacklike enclosure that is lined with refractory material. Many flares are equipped with combustion control dampers that regulate the supply of combustion air to maintain the temperature monitored upstream but within the stack. This class of flare is becoming an industry standard due to its ability to more effectively control pollutant emissions.

Requirements on emissions include carbon monoxide limits and minimal residence time and temperature. Exhaust gas temperatures may vary from 1000 to 2000°F.

38.9 SCRUBBER SYSTEMS

There are a variety of scrubber system designs on the market. Most fall into one of two categories: "packed-tower" or "spray" scrubbers.

A packed-tower scrubber uses a principle similar to the ceramic chips in an oxidation chamber (Fig. 38.5). The chips, often consisting of plastic geometric shapes,

Figure 38.4 Open flare.

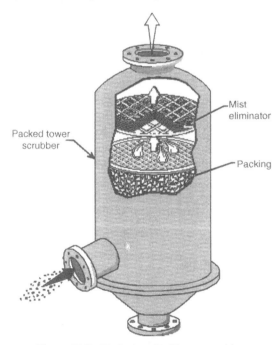

Figure 38.5 Typical packed-tower scrubber.

are designed to maximize contact of the vapor with the surface of the media with minimal air flow restriction. The media tends to "scrub" the air as it passes through, depositing the pollutant on the surface of the media. The media must then be washed of the pollutant. The collected pollutant, now contained in the water sump at the base of the scrubber, is pumped to a wastewater holding tank for treatment or disposal.

A spray scrubber uses a fine spray of water to knock pollutant molecules out of the air stream as the exhaust air passes through the spray. Often the spray contains water and either a caustic or an acid to neutralize caustic or acidic vapor in the exhaust air. A pH sensor in the basin detects the pH level of the wash water and adds the neutralizing agent (caustic or acid) to neutralize the vapor. The pollutants and reagent left in the sump are then removed as wastewater from the basin of the tower.

A disadvantage of scrubber systems is the wastewater produced from the cleaning (scrubbing) process that must then be disposed of as hazardous waste. This disposal of wastewater must be considered when determining the operating cost of such a system. Scrubber efficiencies also tend to be lower than those of oxidizers. But, in general, their operating costs are lower.

38.10   RACT AND MACT ANALYSIS

A reasonable available control technology (RACT) or maximum available control technology (MACT) analysis must be conducted to determine which feasible control method yields a more attractive dollar per ton of pollutant removed. The U.S. Environmental Protection Agency (EPA) will accept credible engineering calculations (the so-called engineering assessment) to demonstrate MACT compliance. By having an engineering assessment, which can be provided by either the equipment vendor or themselves, operators can avoid unnecessary performance testing.

Maximum available control technologies typically require an abatement system with a HAP (hazardous air pollutant) reduction of 95.0% or greater.

38.11   INSTRUMENTATION

A wide range of instrumentation is applied to pollution control systems for measurement and control. To properly control a pollution control system and monitor the system's removal efficiency (overall performance) as well as prove compliance with regulatory requirements, instrumentation becomes a necessity. A list of these instruments includes:

- Thermocouples
- Temperature transmitters
- Thermostats

- Level transmitters
- Level switches
- Flow transmitters
- Flow switches
- Combustible gas monitors
- Gas analyzers

Temperature measurement is used in a variety of environmental applications. Temperature measurement of exhaust emissions allows the environmental technician to use a temperature correction factor when determining the total volume of air being exhausted from the pollution control system.

Temperature measurement is also used to determine the entering air temperature and temperature within the oxidation chamber of an oxidizer. These measurements are used by the oxidizer's combustion control system to regulate the burner and maintain the optimum chamber temperature (high enough to destroy VOCs but no higher—in order to conserve secondary fuel costs).

Flow transmitters, typically a differential or draft range pressure transmitter attached to a flow station or Pitot tube, are used to continuously monitor the rate of volumetric air flow through the pollution control system during operation. This information is used to control the operation of the pollution control system as well as provide data for compliance with environmental permitting.

Level transmitters are used to continuously measure liquid level in the basin of a scrubber system. The transmitter output serves as a remote indication of level and acts as the input to a level controller, allowing the controller to modulate a makeup water valve and maintain a constant level in the tower.

Combustible gas monitors are used to monitor the percentage of LEL of a flammable vapor concentration in a section of ductwork or enclosed area. These are typically areas where concentrations of combustible gas can potentially build up to a hazardous level under normal or abnormal circumstances. These instruments are equipped with two sets of alarm contacts with independent settings. The first, lower setpoint alarm is typically set at 10% of the LEL and used to trigger a "warning" alarm (such as a yellow flashing beacon) to indicate to the operator that a potential problem exists. The second, higher alarm is set at 25% of the LEL. This set of contacts activates a red, "critical" alarm and is often interlocked to shut down the process equipment in a safe mode or activate emergency exhaust fans to remove the high vapor concentrations from the building. Combustible gas monitors, sometimes referred to as LEL monitors, are housed in explosion-proof enclosures with the exception of the electrochemical sensor. The sensor operates at such a low electrical current level that it cannot be the source of ignition.

Gas analyzers come in a variety of types and applications. Whole books have been developed discussing their design and application. You will find them as well as the other instruments discussed in this section in other chapters of this handbook. They basically consist of two types: electrochemical sensing and optical systems. Electrochemical systems detect a physical change in the sensor media (caused by a chemical reaction of the sensor material with the detected substance) and convert it to a scalable electrical signal. Optical systems typically project a filtered light at the measured substance and measure the amount of light at a certain frequency that is absorbed (not reflected back to an optical sensor). The light source is typically infrared or ultraviolet, depending on the application. Both are capable of analyzing a wide range of gases.

Note: When selecting any instrument, you must take into consideration the area in which it will be installed. Determine the hazard classification of the area and work with a reputable instrument supplier or engineering consultant to specify and obtain one that will be consistent with the National Electrical Code (NEC), National Fire Prevention Association (NFPA) International Society of Measurement and Controls (ISA), and other relevant regulatory and standards agencies.

38.12 GLOSSARY

Catalyst	Substance that increases the rate of a chemical reaction without itself being consumed in the reaction process.
Active Catalytic Material	A chemical that provides the desired catalytic reaction. The most active materials for the oxidation of VOCs are noble metals (Pt, Pd) and certain metal oxides
HAP	Hazardous air pollutants—known cancer-causing compounds in the air.
Hydrocarbon	Compound found in all organic compounds. It is the bond that is broken during oxidation.
Incineration	Also known as oxidation.
LEL	Lower explosive limit—the lowest organic concentration in a stream that would yield a combustible mixture in the presence of an ignition source.
Monolith Catalysts	Ceramic or metal honeycombs that provide support for the active catalytic material. Exhaust gases flow through the individual channels of the monolith.
Packed Beds	Loose, bulk materials packed into a container and retained by perforated plates. The exhaust gases flow through the bed, providing mixing and contact between gas and solid.
Recuperative	Oxidation technology (thermal or catalytic) that uses a plate, shell, and tube (or other conventional type of heat exchanger) to heat incoming air with air from the oxidation process. Recuperative systems are capable of recovering up to 75% of the heat released by oxidation.
Regenerative	Oxidation technology that uses two or more ceramic heat transfer beds that act as heat exchangers and a retention chamber where the organics are oxidized. It can recover up to 95% of the heat released by oxidation.

SCFM   Standard cubic feet per minute—flow conditions at standard
                     conditions (usually defined at 70°F, sea level, and 1 atm).
VOCs   Volatile organic compounds—organic chemicals that exist as
                     vapor in air and react in the atmosphere with nitrogen oxides
                     in the presence of sunlight to form ozone (O_3).

39

MEASUREMENT OF AMBIENT AIR QUALITY

Gerald McGowan

Teledyne Monitor Labs, Inc.
Englewood, Colorado

Environmental Instrumentation and Analysis Handbook, by Randy D. Down and Jay H. Lehr
ISBN 0-471-46354-X   Copyright © 2005 John Wiley & Sons, Inc.

The quality of the ambient air that we breathe has been a concern since the beginning of the industrial age. Some of the earliest air quality episodes which captured the public interest resulted from the extensive use of coal with substantial sulfur content for heating in both homes and industrial facilities. When weather conditions were right, sulfur dioxide (SO_2) concentrations became extremely high and many related health problems and deaths were noted. In more modern times, the widespread use of combustion processes to provide energy for both industrial processes and vehicle power sources has degraded our ambient air quality. With the establishment of the U.S. Environmental Protection Agency (EPA) and the passage of the 1970 Clean Air Act, ambient air quality became an important part of U.S. regulatory activity. In the United States, all of the easiest targets of pollutant emission reductions have now been controlled and regulated, and yet there are many areas of the country which are not in compliance with each component of the NAAQS (National Ambient Air Quality Standards). Thus, further monitoring and modeling are required in order to understand the source of the problems and identify the optimum pollutant control strategies. Thousands of air-monitoring stations have been implemented to determine the extent and magnitude of the pollution problems which must be resolved. The control of the ambient air quality inherently pits the economic interests of inexpensive energy, vehicles, and industrial processes against the health-related interests of the populace. Thus the resolution of ambient air quality problems requires many difficult compromises.

This chapter will address the more common measurement techniques used in the field today to monitor ambient air quality. Emphasis is on the criteria pollutants, sulfur dioxide, nitrogen oxides, carbon monoxide, ozone, and particulate and the regulatory requirements associated with monitoring such pollutants.

39.1 REGULATORY BACKGROUND (EPA)

39.1.1 Primary and Secondary Ambient Air Quality Standards

The EPA has established NAAQS for the criteria pollutants of interest to protect our natural environment. Similar air quality standards have been adopted in most other countries where regulatory agencies have been established to protect their national environmental air quality. The EPA establishes primary air quality standards at levels designed to protect the public health and secondary standards to protect the public welfare. The U.S. standards are specified in EPA 40 CFR 50 and are listed in Table 39.1. They are also available at the EPA website (http://www.epa.gov/ttn/amtic/).

Ozone is generally measured only during the ozone season, which varies from June to October for states in the northern, colder climes of the United States to year around for southern, warmer states. The OTC (Ozone Transport Commission) region, consisting of 13 northeastern states, has adopted May through September as the ozone season for purposes of NO_X source monitoring and allowance trading. Cold weather generally reduces the formation of ozone so that concentrations measured under such conditions are relatively low compared to the standards.

Table 39.1   EPA Primary and Secondary Ambient Air Quality Standards

Pollutant	Primary Standard	Secondary Standard
Sulfur dioxide (SO_2)	Annual standard of 0.030 ppm not to be exceeded in a calendar year (CY). The 24-h standard of 0.14 ppm not to be exceeded more than once per CY.	The 3-h standard of 0.5 ppm not to be exceeded more than once per CY.
Particulate matter, PM_{10} (mean aerodynamic diameter ≤ 10 μm)	The 24-h standard of 150 μg/m^3, with an allowed exceedance of only one day per CY. Annual standard is 50 μg/m^3.	Same as primary standard
Particulate matter, $PM_{2.5}$ (mean aerodynamic diameter ≤ 2.5 μm)	The 24-h standard of 65 μg/m^3 (98th percentile); annual average standard is 15.0 μg/m^3.	Same as primary standard
Carbon monoxide (CO)	For an 8-h average concentration 9 ppm not to be exceeded more than once per year; for a 1-h average concentration 35 ppm not to be exceeded more than once per year.	Same as primary standard
Ozone (O_3)	The 1-h standard of 0.12 ppm, with only one day per year allowed to have higher hourly averages. The 8-h standard is 0.08 ppm, and the average of the annual fourth highest daily maximum 8-h average must be less than or equal to 0.08 ppm	Same as primary standard
Nitrogen dioxide (NO_2)	Standard annual arithmetic mean of 0.053 ppm	Same as primary standard
Lead (Pb)	1.5 μg/m^3, maximum arithmetic mean averaged over a calendar quarter	Same as primary standard

Note: To convert from ppm to μg/m^3 at 25°C and 760 mm Hg, multiply ppm by $M/0.02447$, where M is the molecular weight of the gas ($SO_2 = 64.07$, $NO_X = 46.01$, $O_3 = 48.00$, $CO = 28.01$).

39.1.2   Reference and Equivalent Method Determination

Reference methods have also been specified by the EPA for the measurement of the above pollutants. Reference methods are the "golden standard" for measurement of specific pollutants against which all other measurement methods are compared. Analyzers designated as reference methods or equivalent methods are required to show compliance or noncompliance with the air quality standards. Instruments certified for such regulatory compliance use under EPA standards must exhibit a front-panel label indicating the EPA designation number. The reference methods

Table 39.2 U.S. Reference Method Analytical Techniques and Typical Automated Methods

Pollutant	Reference Method	Typical Automated Method Equivalent Method
SO_2	Wet chemistry, pararosaniline method	UV fluorescence
TSP (total suspended particulate)	High-volume sampler	Same as reference method
CO	Nondispersive infrared (NDIR) photometry, with gas filter correlation	Same as reference method
NO_2	Gas-phase chemiluminescence with ozone and NO_2-to-NO converter	Same as reference method
Pb	Collected on filter, extracted with nitric acid, and analyzed by AAS (atomic absorption spectroscopy)	Inductively coupled argon plasma–optical emission spectrometry
O_3	Chemiluminescent reaction with ethylene, calibrated with ultraviolet (UV) photometer	UV photometer

are listed in Table 39.2 in addition to the most common EPA-certified equivalent automated methods.

The EPA 40 CFR 53 contains the specified test procedures which must be used to qualify a candidate instrument for reference method or equivalent method status. A list of all such analyzers approved by the EPA for ambient air monitoring can be obtained from http://www.epa.gov/ttn/amtic/files/ambient/criteria/. The manufacturer typically tests a candidate analyzer according to the specified procedures and submits the required data to the EPA, and the EPA reviews the test procedures and results and makes one of the following determinations: more testing and/or data are required to evaluate and/or confirm the test results, the test data do not justify the requested designation, or the test data do justify the requested designation. In the latter case the EPA designates the specific manufacturer and model as an equivalent or reference method and specifies the conditions under which it must be operated. Such test procedures evaluate the instruments range, noise, lower detectable limit, interference equivalent for each interferent and all interferents in total, zero drift and span drift (20 and 80% of range), lag time, rise and fall times, and precision (20 and 80% of range). Zero drift, span drift, lag time, rise and fall times, and precision are all measured under test conditions which involve various line voltages from 105 to 125 V AC and ambient temperatures from 20 to 30°C, at a minimum. Some analyzers are qualified for operation over 15–35°C or even 0–40°C, which simplifies HVAC (heating, ventilating, and air conditioning) requirements for the instrument enclosure. Further, the operator's manual is also reviewed to ensure the adequacy of calibration and operation instructions/procedures. The results of such testing must fall within the EPA required specifications for that analyzer (pollutant) in order to be designated as an equivalent or reference method. Minimum-performance specifications for automated methods (analyzers) are given in Table 39.3. It should be noted that most modern ambient air analyzers far exceed (a factor of 10 or better in some cases) the minimum requirements specified in these regulations. Competitive pressure among vendors and user requirements have driven the performance of such analyzers to higher and higher standards as new models have evolved.

At a minimum, the EPA requires a measurement range of 0–0.5 ppm for SO_2, NO_X, and O_3, as shown in Table 39.3. Other ranges are often supplied and tested/approved as part of the EPA analyzer certification process. In some cases the analyzer may have higher or lower ranges than those which were tested and approved by the EPA. Maximum measurement ranges for such analyzers generally range from 1 to 20 ppm, depending on the analyzer. The lowest measurement range available on such analyzers generally is down to 0.1 or 0.05 ppm; however, it should be

Table 39.3   EPA Performance Specifications for Automated Methods

Parameter	SO_2 and O_3	NO_X	CO
Range (ppm)	0–0.5	0–0.5	0–50
Noise (ppb)	5	5	500
Lower detectable limit (ppb)	10	10	1000
Interference equivalent, each (ppb)	20	20	1000
Interference equivalent, total (ppb)	60	40	1500
Zero drift, 12 and 24 h (ppb)	20	20	1000
Span drift 24 h, 20% of upper range (%)	20	20	10
Span drift 24 h, 80% of upper range (%)	5	5	2.5
Lag time (min)	20	20	10
Rise time, 95% (min)	15	15	5
Fall time, 95%(min)	15	15	5
Precision, 20% upper range level (ppb)	10	20	500
Precision, 80% upper range level (ppb)	15[a]	30	500

Notes: To convert from ppm to g/m^3 at 25°C and 760 mm Hg, multiply ppm by $M/0.02447$, where M is the molecular weight of the gas ($SO_2 = 64.07$, $NO_X = 46.01$, $O_3 = 48.00$, CO $= 28.01$).

[a]Value is 10 ppb for O_3 automated methods.

noted that lowering the measurement range, or span, of a given analyzer without changing the basic gain of the measurement reaction and/or optical detection scheme does not generally improve the measurement performance—it just increases the analog output resolution. If measurements are transmitted to external recording equipment in digital form (RS232), the span adjustment may have no perceptible effect on the analyzer since microprocessor-based analyzers output the measured value from zero to the maximum range of the analyzer. Drift, noise, and other measurement errors are often the same, when expressed as ppm, regardless of the selected measurement range unless the physical parameters of the measurement system are altered to provide higher or lower measurement ranges. For CO measurement, the typical CO levels and standards are higher than for SO_2, NO_X, and O_3, so that the EPA requires a range of 0–50 ppm CO at a minimum. Such CO analyzers may have additional measurement capabilities, including a maximum measurement range up to several hundred ppm or more and minimum measurement ranges of 1–10 ppm, or less.

39.1.3 Air-Monitoring Networks

Most environmentally sensitive countries or regions where ambient air quality has been compromised because of industrialization and mobilization have installed a number of ambient air–monitoring sites to evaluate the quality of the air. Measurements from these sites are typically connected to a network so that monitoring information is fed to a central computer where data are collected and summarized on a regional basis. Such networks generally provide information for the public and regulatory agencies regarding air quality in order to facilitate identification and prediction of air quality indices upon which certain public activities, which may further compromise air quality, may be prohibited or restricted, confirm whether ambient air quality standards are being met, allow health effects studies to be correlated with air quality, and provide data to evaluate the efficacy of specific pollutant control strategies.

In the United States, there are several types of monitoring sites as designated by the EPA (see EPA 40 CFR 58, Subparts C, D and E). State and local air-monitoring stations (SLAMs) typically monitor the criteria pollutants described in Table 39.1. The number of SLAMS sites in a given metropolitan statistical area are determined according to population. For example, in a city of just over a million, three different SLAM monitoring sites are required at a minimum. If the population is over 8 million, then at least 10 sites are required. In addition, there are NAMSs (national air-monitoring stations), which are a subset of SLAMSs. The SLAMSs provide information required for a state or region to control its ambient air quality and the NAMSs provide information at the federal level as necessary for the EPA to ensure adequate air quality. Because of the abundance of ozone nonattainment areas (areas which do not meet the ambient ozone air quality standards) and the difficulty in reducing ozone, the EPA has also implemented PAMSs (photochemical air-monitoring stations), which are used for the purpose of enhanced monitoring in O_3 nonattainment areas listed as serious, severe, or extreme. Such classifications

Table 39.4   Ozone Precursors Measured by PAMSs

Ethene	Acetylene	Ethane
Propene	Propane	Isobutane
1-Butene	*n*-Butene	*trans*-2-Butene
cis-2-Butene	3-Methyl-1-butene	2-Methylbutane
1-Pentene	*n*-Pentane	Isoprene
trans-2-Pentene	*cis*-2-Pentene	2-Methyl-2-butene
2,2-Dimethylbutane	4-Methyl-1-pentene	Cyclopentane
2,3-Dimethylbutane	Cyclopentene	2-Methylpentane
3-Methylpentane	2-Methyl-1-pentene	Hexane
trans-2-Hexene	*cis*-2-Hexene	2,4-Dimethylpentane
Methylcyclopentane	Benzene	2-Methylhexane
Cyclohexane	2,3-Dimethylpentane	3-Methylhexane
2,2,4-Trimethylpentane	Heptane	Methylcyclohexane
2,3,4-Trimethylpentane	2-Methylheptane	Toluene
3-Methylheptane	Octane	Ethylbenzene
m-Xylene	*p*-Xylene	Styrene
Nonane	*o*-Xylene	Isopropylbenzene
α-Pinene	Propylbenzene	1,3,5-Trimethylbenzene
Decane	1,2,4-Trimethylbenzene	β-Pinene

are determined by how high the ozone levels are compared to the standard. The PAMSs are designed to monitor VOCs (volatile organic compounds) which contribute to the formation of ozone. The ozone precursor compounds measured by PAMSs are shown in Table 39.4 and include C_2–C_{10} hydrocarbon compounds. Several hundred of these stations have been implemented throughout the major ozone nonattainment areas in the United States to determine the role played by VOCs in the formation of ozone. This provides information for selecting the optimum control strategy, which can be either the reduction of NO_X or the reduction of VOCs. Selecting the wrong control strategy can actually exacerbate the ozone problem.

In addition to the above monitoring sites, there are also PSD (prevention of significant deterioration) stations, with their own criteria for siting. These sites are provided in response to PSD regulations which are designed to protect the air quality in critical areas, such as national parks, scenic vistas, and so on.

39.1.4   Quality Assurance/Quality Control Requirements

Quality assurance/quality control (QA/QC) requirements are generally imposed to ensure that data obtained from network stations are representative of ambient concentrations in designated urban areas; that measurements are accurate, precise, and traceable to international standards; that data obtained at each station within a network are reproducible and comparable to similar data obtained from other networks; and that data are consistent and available for near 100% of the time frame of interest. This requires that data should be of known quality and based on documented quality programs so that the limits of the accuracy of such data can be obtained. In the United States, each SLAMS gas analyzer must be tested at least every two weeks at a single upscale value (see EPA 40 CFR 58, Appendix A). At least every year, and preferably each quarter, each analyzer must be tested at three or four upscale values, depending on the analyzer. Calibration gases must be of known accuracy, preferably NIST (National Institute of Standards and Technology), CRMs (certified reference materials), or GMIS (Gas Manufacturers Internal Standard), the latter only one reference level away from NIST-calibrated gases. The gas standards for these tests are of sufficient accuracy that the results of these calibrations may be used to back correct measurement data for errors in span and/or zero calibration of the analyzer. In some cases the results of the biweekly calibration check are used to ramp in span corrections starting with the prior calibration check and ending with the current calibration check. If the span drift of the analyzer changes linearly with time, this type of correction provides the most accurate method of correction.

Most sites also incorporate a daily check of zero and span for each analyzer. This is done to ensure the day-to-day performance of the analyzer and to indicate the need for maintenance, which otherwise may go unnoticed for up to two weeks if only the required biweekly QA/QC checks are provided. The span gas standards used for such daily checks need not be of the quality of the biweekly QA/QC checks. Similarly, the results of the daily calibration checks are not generally used to correct the calibration of the analyzer, although zero gas can be of such integrity that the analyzer may be automatically rezeroed based on the daily zero gas check.

39.1.5 Station Siting and Sample Inlet Considerations

Siting requirements are specified according to the type of site (SLAMS, NAMS, PAMS, PSD) and the objective of the monitoring station. In the United States, the EPA has provided these guidelines in 40 CFR 58, Appendix E.

A sample "cane" is often used to extract the sample from the atmosphere, which is shaped as a walking cane with downward-curved sample inlet to minimize the entry of wind-blown material such as leaves, large dirt particles, rain, and snow. Particulate monitors generally have their own sample inlet which is optimized for particulate transport, which is not the case for gas analyzer inlets. Sample inlets for gas analyzers are generally required to be 3–15 m above ground level, although inlets for CO monitoring must be 3 m above the ground. Sample inlet materials must be inert, and the EPA has recommended either borosilicate glass or FEP Teflon for monitoring the normal pollutants. Volatile organic compound monitoring is more demanding with only borosilicate glass or stainless steel allowed. Sample residence time in the inlet tubing/manifold is limited to 20 s to preclude further chemical reactions prior to measurement. In some cases the residence time is even further limited. In the United Kingdom, for example, the residence time is limited to 5 s.

39.2   ANALYZER REQUIREMENTS, FEATURES, AND OPTIONS

Analyzers used for measuring the quality of ambient air with respect to regulatory requirements generally must have the approval or certification of an applicable agency with such jurisdiction. In the United States, the EPA has the authority to designate a given analyzer as a reference or equivalent method, as previously discussed. In other countries the Eignungsgepruft approval of the TUV/UBA German test agencies or Mcert designation of the United Kingdom may be required or other suitable designations. All analyzers sold in the EC (European Community) must be CE marked, which requires testing against a variety of electrical specifications, including electromagnetic radiation susceptibility and electromagnetic emission characteristics. These approvals are analyzer manufacturer/model specific and such specifications are only tested on the analyzers involved in the certification testing process.

Such analyzers are generally supplied in bench-mount configurations but can be mounted in 19-in. rack assemblies using optional rack-mount hardware. Rack-mounting slides are preferred because of the easy slide-out/slide-in access to the analyzer and components. A typical ambient air analyzer is shown in Figure 39.1. Ambient gas analyzers using optical techniques generally require a particulate filter at their inlet to ensure the cleanliness of the sample prior to being analyzed. This is done to minimize the rate of buildup of contamination on exposed optical surfaces. Typically, this is done with a replaceable/disposable filter cartridge of about 5 μm pore size. A 47-mm-diameter filter provides sufficient filter area that it can be used for weeks to months without causing a significant flow restriction in moderate-flow-rate applications. Such particulate filters may be either internal to the analyzer or external user-supplied equipment.

The sample flow rate and calibration gas flow rates of analyzers are somewhat variable among different vendors. Sample flow rates generally cover the range from about 0.5 to 3–5 L/min, and it should be noted that the life of particulate filters and calibration gas bottles both decrease as flow rates increase. In addition, each

Figure 39.1   Typical ambient air analyzer. (Courtesy of Monitor Labs, Inc.)

analyzer generally must have a means for checking zero and span on a daily basis as part of the QA/QC requirements. Thus, valving is often built in to the analyzer to allow use of external gas sources, especially for CO. Additionally, standard or optional equipment is often available to provide an internal source of zero air and to incorporate permeation tube–based span gas generators for NO_X (NO_2) and SO_2 analyzers. An internal ozone generator may be available on ozone analyzers to provide the span check capability.

Sample pumps are also sometimes available as built-in standard equipment or as add-on external equipment which can be supplied by the system integrator. For multigas analysis systems, it is generally less expensive to use one large sample pump to pull sample through a variety of analyzers than to use individual pumps for each analyzer. Care must be exercised to ensure that the specific analyzer flow controls are suitable for such operation.

With the advent of microprocessor-controlled analyzers, a variety of features have been developed which can be added to such analyzers at nominal cost. Such features can considerably simplify the complications involved in system integration and allow the user to handle a variety of application requirements with only setup changes—not hardware changes. Further, they may reduce the amount of hands-on time required by field service personnel to ensure normal operation of the instrument. Features which may be found on the more advanced instruments include one or more of the following:

(a) Single, dual, or multiple ranges with automatic overranging—analyzer automatically changes the analog output range to a higher value when the range is exceeded and similarly reduces the range when applicable.

(b) Automatic ranging—analyzer continuously adjusts internal measurement parameters to optimize accuracy for the range of measurements encountered from zero to the maximum concentration for which the analyzer is scaled.

(c) Predictive diagnostics—analyzer predicts when maintenance may be required based on current trends and/or past performance.

(d) Operation with 12 V DC—allows the analyzer to be used in portable or remote applications with battery and/or solar battery charger.

(e) Internal floppy disk drive or equivalent solid-state memory for continuous storing of measurement data without the need for an external datalogger.

(f) Remote access, serial RS485/RS232 port for PC or modem/telephone line access which provides remote control of analyzer diagnostic, configuration, and operational information. The RS232 ports can also be provided with a multidrop capability so that one modem and telephone line is used to access a family of ambient air analyzers.

(g) Software which can be used on a remote PC to obtain a virtual analyzer front-panel interface capability with full control and monitoring capability at a remote location.

(h) Specific network interfaces—allow the operator to select a network interface configuration suitable for the system specific requirements. Ethernet has gained extensive market share in the past few years and is offered on many analyzers. It is easily coupled to the Internet via the TCP/IP (Transmission Control Protocol/Internet Protocol) which provide wide access to measurement and control information.

(i) Event log—provides an automatically recorded log of significant maintenance events or setup changes which are written to a file which can be accessed later as needed to facilitate troubleshooting.

(j) Graphics display—provides graphical trend charts for easy interpretation of trends of critical measurements and internal diagnostic parameters.

(k) Automated diagnostics—provides detailed information about conditions internal to the analyzer which could result in degraded performance if left unattended.

(l) Temperature and pressure compensation—includes continuous measurement and compensation for sample gas and/or measurement cell temperature and pressure for most accurate measurements.

(m) Sample flow measurement and control—provides automatic programmable control of sample flow, which may include ramping up flow rates during the beginning of calibration cycles to speed up the purging process and reduce off-line time.

(n) Voltage/current analog outputs—allows the field selection of 0–1 V DC or 0–20 or 4–20 mA DC outputs without additional hardware.

(o) Digital contact inputs and outputs to indicate and/or control measurement or calibration status; indicate maintenance alarms, warnings, or faults; and activate external equipment.

(p) Printer output—provides serial or parallel printer output for recording measurement averages and diagnostic information in a printed record.

(q) Adaptive or standard measurement filtering/averaging—allows a choice of low-pass signal filtering which includes traditional fixed-time-constant or adaptive filters, where Kalman filters are a specific type of adaptive filter. Kalman filters are digital filters which can be configured to adaptively modify the digital filter parameters as appropriate to speed up responses to step changes and minimize longer term zero noise/drift. Fixed/conventional filters can provide very low zero noise/drift at the expense of very long response times, or vice versa; adaptive filters can achieve a better compromise of speed of response and low noise.

(r) Adjustments completely controlled by the microprocessor through the use of digital potentiometers, as opposed to analog potentiometers, which require manual operator adjustment.

(s) Password protection of critical analyzer setup and calibration data to prevent unauthorized access to information which should only be changed by someone with proper training and authorization.

(t) Measurement averaging which can extend from seconds to minutes to hours to days.

(u) Automatic zero and/or span compensation, based on the response of the instrument to programmed selection of zero and span gases when such gases are of sufficient integrity to justify resetting the instrument zero and gain adjustments.

(v) Indication of periodic maintenance items based on operating time schedules.

(w) Serial data port for downloading firmware upgrades from a remote computer, typically requires a flash read-only memory (ROM).

(x) Internal battery backed-up clock and calendar for time-stamping measurement data and establishing calibration time periods.

(y) Automatic diagnostic testing of a variety of essential measurement and control functions.

Major supliers of classical bench-type ambient analyzers for ozone, sulfur dioxide, nitrogen oxides, and carbon monoxide can be seen in the EPA web page showing the certification or designation numbers for such analyzers (http://www.epa.gov/ttn/amtic/).

39.3 NO$_X$ MONITORING

Nitrogen oxides are reported as NO$_2$ with regard to molecular weight (46.01), which is used for concentration values expressed in mg/m^3 or lb/μBtu for combustion applications. Nitrogen oxide is actually measured as the sum of NO and NO$_2$. This is appropriate because NO$_X$ is produced primarily in the form of NO from combustion sources due to the high temperatures involved in combustion. After being emitted into the atmosphere and subject to atmospheric chemistry, it oxidizes to form NO$_2$. Thus, the NO$_X$ measurement is a measure of NO$_2$ (nitrogen dioxide) and its precursor, NO (nitric oxide). Other pollutants in the atmosphere may alter this simplified analysis.

The chemiluminescence NO$_X$ analyzer is the most complex of all the analyzers used to measure the criteria pollutants. When properly implemented, however, the resulting performance is outstanding with noise and lower detectable limits available at well under 1 ppb. Special versions of such analyzers have been developed with ppt capability for background level surveys. The chemiluminescent technique is based on the reaction of O$_3$ (ozone) and NO, which produces an excited NO$_2$ molecule. This excited molecule emits a photon of energy when it reverts to the relaxed state. Such luminescent emissions occur in the band of 500–3000 nm with a maximum near 11–1200 nm. This reaction can be expressed as

$$NO + O_3 = NO_2^* + O_2 \quad \text{and} \quad NO_2^* \rightarrow NO_2 + h\nu$$

where NO$_2^*$ is in the excited state and $h\nu$ is the energy associated with the emitted photon, where h is Planck's constant and ν is the frequency of emitted radiation.

Typically, a photomultiplier tube (PMT) or solid-state detector is used to detect the light emitted during the reaction, which is proportional to concentration of NO. Thus, the analyzer must contain a built-in ozone generator. Generally, a high-voltage-driven corona discharge device is used to obtain the relatively high levels of ozone (a few thousand ppm) required. As clean air is passed through the corona discharge, a certain fraction of the O_2 is converted to O_3 and that supply is used as the reactant. The ozone then must be mixed very thoroughly with the sample gas so that the chemiluminescent reaction can be observed, or detected, by the PMT. The chemiluminescent reaction can be quenched by gases which absorb energy in the band of radiation emitted by the excited NO_2 molecule and by collision with other molecules. Because of the quenching action of several gases which are present in atmospheric gases, most notably O_2, H_2O, and CO_2, it is imperative that the environment in which the reaction takes place is a relatively high vacuum (near 100–200 torr, for example) if sub-ppb sensitivity is to be obtained. Since the efficiency of the chemiluminescent reaction increases with decreasing absolute pressure, higher vacuums enhance the signal-to-noise ratio. Uncontrolled changes in the vacuum level will cause span drift in the analyzer unless measurement and correction for reaction cell pressure are provided within the analyzer. For higher levels of NO$_x$ measurement, some reaction cells have been operated at near-atmospheric pressure. With the reaction cell operated at relatively high vacuum, the controlled blending of the ozone and sample gas in the reaction cell is easily accomplished with critical orifices. By using orifices under critical operating conditions, the flow through the orifices is sonic and the resulting flow is primarily determined by the size of the orifice, thereby allowing a very precise blending of ozone and sample flow. The temperature of the orifices and gas streams is also temperature controlled to ensure precise mixing. The analytical technique is flow dependent in that it is sensitive to the number of ozone-reacted NO molecules which can be presented in the field of view of the PMT per unit time. Thus, ozone and sample flow rates and reaction cell pressure must be well controlled. The reaction cell is designed to control the mixing of the ozone and sample gas in a small volume as close to the PMT as possible. The resulting emission is a diffuse source which occurs in the mixing zone. Because of the high sensitivity desired of such monitors, the PMT is typically thermoelectrically cooled to near 0°C. The reaction cell is often heated to a constant temperature of about 55°C to stabilize the reaction and prevent contamination of the reaction cell.

To measure NO$_x$, it is necessary to measure not only NO but also the NO_2 content of the sample gas. This requires a heated catalytic converter which converts NO_2 to NO. Such converters generally use a molybdenum catalyst within an oven operating at near 315°C and provide 98 or 99% conversion efficiency—the EPA requires at least 96% efficiency. Other converters using different catalysts and/or temperatures may also be used, but care must be used to establish what gases are converted to NO for the particular design. Thus, the analyzer measures NO$_x$ (NO + NO_2) when measuring the gas stream which has been reduced by the heated catalytic converter. When it measures the raw (unconverted) sample stream, it measures only NO. The difference between the two measurements is attributed to NO_2.

However, this difference value only represents the true value of NO_2 if the NO_X and NO measurements are both with respect to a common volume of gas. Care must also be exercised to ensure that the heated catalytic converter is not run at excessive temperatures which may convert other nitrogen compounds to NO, which would cause the NO_X and NO_2 measurements to be in error. Further, the catalyst can become poisoned or degraded with time and will require periodic replacement.

As a result, the accurate and precise measurement of NO_X requires a stable source of ozone, controlled flows of ozone and sample, thorough mixing of ozone and the sample gas in front of the PMT, a stable vacuum source and a constant-temperature reaction cell, a cooled constant-temperature PMT, and a stable heated catalytic converter. In addition, there are many supporting measurements and voltage/current sources which must be implemented with similar precision and accuracy.

There are generally no significant interferences when using the chemilumines-cence technique for measuring NO_X in ambient applications. The NH_3, SO_2, and CH_4 may be detectable, but not at significant levels in typical ambient applications. Some chemiluminescence NO_X analyzers have been modified with a high-temperature stainless steel converter which converts NH_3 in the sample stream to NO to obtain an ambient NH_3 measurement. Care must be exercised to ensure that quenching effects are similar under both calibration and measurement conditions to obtain the most accurate NO_X measurements.

Two types of analyzers evolved using the chemiluminescence analytical techni-que. The first was a dual-reaction-cell analyzer where one reaction cell was used to measure the converted (NO_X) sample stream and the other to measure the uncon-verted (NO) sample stream. By balancing the flow streams into each reaction cell, the measurements of NO_X and NO were made in parallel, and the difference, NO_2, was a valid measurement under virtually all conditions. To reduce cost and improve accuracy, other analyzers were designed to incorporate a single reaction cell and stream switching so that NO_X and NO could be measured sequentially with the same hardware. This configuration can provide erratic measurements of NO_2 unless the flow streams are balanced so that the same gas volume is measured during each pair of NO_X and NO measurements. To further enhance accuracy, either a zero con-centration gas sample or just the ozone reactant can be flowed into the reaction cell on a periodic basis to allow automatic zero correction. This zero or background cor-rection cycle is periodically added to the normal NO_X and NO measurement cycles. This allows the NO_X analyzer to automatically correct for zero drift and provide an extremely stable zero measurement. Such measurement cycles may be as short as a few seconds to facilitate keeping up with changing dynamics of NO and NO_2.

Daily zero and span calibration checks of such analyzers are often done with an optional internal zero/span (IZS) module. Zero air, free of NO_X, can be generated by first converting/oxidizing NO to NO_2 and then scrubbing all the NO_2 in the given gas stream. Oxidizing NO is accomplished by either chemical reaction or ozona-tion. Chemical oxidation is accomplished using agents such CrO_3 on alumina or purafil. Purafil is also effective in scrubbing the oxidized gas stream of NO_2. Ozo-nation involves reacting the gas with an internal source of ozone. Analyzers can also obtain a zero backgound measurement by venting the sample and just allowing the ozone to enter the reaction cell. With such a technique any interaction between the reaction cell walls and the ozone will also be compensated for. Span gas can be obtained from bottled calibration gas standards or from an NO_2 permeation-tube-based span gas generator. The permeation tube contains a liquid which permeates through the tube walls at a constant rate for a specific temperature. Thus, by containing the permeation tube in a constant-temperature environment and purging the volume surrounding the tube with a known constant flow of zero gas, the resulting concentration of the pollutant-laden gas stream can be determined. The NO_2 permeation tubes are readily available, but care must also be exercised to ensure that the purge air associated with the permeation tube is dry. Such a calibration gas is suitable for daily check of analyzer operation but is generally not sufficiently accurate to be used as a standard unless it is first calibrated with respect to a known standard.

Gas-phase titration (GPT) is often used for complete NO_X analyzer calibration and is based on the controlled mixing of O_3 and NO, which results in converting a portion of the NO to NO_2. Thus, if you start with a bottled gas concentration of 2 ppm of NO and mix it with an equal volume of ozone-controlled dilution air, the initial measurement with no ozone will indicate 1 ppm of NO and essentially no NO_2. As you gradually increase the ozone level in the dilution gas stream, the concentration of NO will decrease and the concentration of NO_2 will increase proportionally. Thus, the NO_X measurement should stay constant as the GPT is adjusted from no ozone to the maximum ozone level which converts all the NO to NO_2. Thorough mixing and a suitable equilibration time are required to obtain the desired chemical reaction.

Another term used in the measurement of nitrogen oxides, NO_y, is sometimes mentioned as a measure of the odd nitrogen compounds, including such things as PANs (peroxyacylnitrates), HNO_3, and HONO. Whereas NO_X is measured near the back end of the sample train (where the analyzers are typically located) and uses the heated catalytic converter as an internal part of the analyzer, NO_y typically requires the converter to be placed at the front end of the sample train so that the nitrogen compounds which may be otherwise lost in the sample train are converted to NO prior to being transported to the analyzer. Some designs use a different converter design to optimize the desired NO_y conversion.

Other types of converters, such as high-temperature stainless steel and photolytic devices, have been used for measurements of nitrogen oxides which include other nitrogen compounds besides those mentioned above.

39.4 SO₂ MONITORING

Sulfur dioxide is primarily emitted from large stationary sources which burn sulfur-containing fuels such as oil and coal. Through fuel switching to lower sulfur fuels and the use of FGD (flue gas desulfurization) systems, SO_2 source emissions have been substantially reduced in the United States.

Ultraviolet fluorescence has proven to be the method of choice for the continuous measurement of ambient air levels of SO_2. This technique uses the well-known fluorescence properties of SO_2. When an SO_2 molecule is exposed to radiation at the wavelengths where it absorbs energy, the SO_2 molecule becomes excited (SO_2^*), and it then emits a photon of energy when it reverts to the relaxed state. This can be expressed as

$$SO_2 + h\nu_a \rightarrow SO_2^* \quad \text{and} \quad SO_2^* \rightarrow SO_2 + h\nu_e$$

where h = Planck's constant

ν_a = absorption/excitation frequencies

ν_e = emission frequencies

The absorption and emission spectra of a given molecule are generally mirror images of each other with a shift in emission wavelengths toward the longer wavelengths (Stokes's law). Thus fluorescence always results in the emission of energy at longer wavelengths than the wavelength used for excitation. Several molecules of interest in the atmosphere exhibit fluorescent properties; however, the interference response of the reaction can be minimized by properly selecting both the excitation wavelength and the resulting wavelengths used for measuring the fluorescence emissions. The SO_2 molecule has strong absorption characteristics in the vicinity of 200–240 nm, and the emission is generally measured in the 300–450-nm area. A zinc arc lamp with a major emission line at 214 nm is generally the preferred source for the SO_2 excitation. The fluorescent properties of hydrocarbons have similarly been observed, and used as a geochemical analytical technique to detect trace amounts of oil in soil, rock, sand, etc.

In practice, ambient SO_2 is measured by illuminating the sample gas with UV radiation in a reaction cell which is closely coupled to a PMT. Typically, the illumination is done in one axis and the PMT observation of fluorescence is done in an orthogonal axis to minimize cross coupling of excitation energy into the field of view of the detector. The excitation energy is many orders of magnitude higher than the resulting fluorescence emissions, dictating a high degree of isolation between the two beams. In some analyzers a shutter is used to block the light entering the PMT and provide for correction of PMT zero drift. Since the fluorescent emissions are proportional to the excitation energy as well as the concentration of SO_2 gas molecules in the reaction cell, the intensity of the excitation radiation must also be measured so that the fluorescence emissions can be normalized for the source intensity. Again, because of the high sensitivity desired of an ambient air analyzer, the PMT detector is often cooled via a thermoelectric cooler to optimize its sensitivity. Such instruments generally exhibit noise and lower detectable concentration of less than 1 ppb. This analytical technique is largely independent of sample flow rate since the detected energy is only dependent on the number of excited SO_2 molecules seen by the PMT.

Because of potential aromatic hydrocarbons in the atmosphere and their tendency to fluoresce when exposed to UV light, the sample gas is scrubbed of these compounds using a membrane (Silastic) filter prior to measurement. The hydrocarbon filter is composed of a permeable membrane tube within a tube. The outer chamber of the coaxial tube filter is purged with zero gas containing none of the hydrocarbons of interest, and the hydrocarbons in the sample gas flowing through the inner chamber/tube are preferentially drawn out into the purged volume. Thus the sample is cleaned up to the extent that the measurement of SO_2 fluorescence is not substantially affected by aromatic hydrocarbons which may be present in the incoming sample stream. This membrane separator or scrubber is often referred to as a "kicker." This filter is not expended or consumed, as it is continuously refreshed by the purge air supply.

Other potentially interfering gases include NH_3 and NO, but both responses are generally negligible. If the coexisting NO concentration is very high, some analyzers are available with an optional spectral filter to minimize the response to that gas. Quenching the fluorescence reaction is not generally a problem; however, O_2 has been shown to modify the reaction. This effect is not significant in ambient air applications where the O_2 is 20.9% (dry basis) but may affect other applications where O_2 is at or near zero.

Daily calibration checks are easily performed using a locally or internally generated zero gas and span gas. Zero gas for an SO_2 analyzer is easily generated through the use of an activated charcoal filter. Span gas can also be provided using a permeation-tube-based span gas generator. Permeation tubes are available for SO_2 with a variety of specified permeation rates. By controlling the permeation rate through the use of a constant-temperature oven and providing a controlled/known purge flow rate, the final span gas concentration is easily established. Thus IZS is often an allowable option for SO_2 analyzers.

39.5   O₃ MONITORING

Ozone is a photochemical oxidant which is formed in the atmosphere and not emitted per se from combustion sources. It has proven to be the most difficult of the criteria pollutants to maintain within the NAAQS. It is linked with hydrocarbon and NO_X emissions, but the chemistry involved in the formation of ozone is very complex. The typical equivalent method for measuring ozone uses the UV absorption of ozone to quantify the concentration of ozone in the sample gas stream. The relationship between absorption and concentration is established by Beer's law, which can be stated as

$$\frac{I}{I_0} = e^{-acl}$$

where I is the intensity of the light detected at the output of a measurement cell (or path) of length l, with a concentration c of the gas of interest, and with that gas exhibiting an absorption coefficient a. The intensity of light detected at the output of the same measurement cell with none of the absorbing gas(es) present in the measurement cell is I_0.

It should be noted that Beer's law is based on the use of a monochromatic light (single-wavelength) source which is never practically realized unless a laser is used as the light source. Further, the absorption coefficient a is a unique characteristic of each specific gas as well as its temperature and pressure and is further dependent on the specific wavelength(s) of light used for the measurement. To keep the exponential terms as independent as possible, the concentration c should be expressed in units of density, which are representative of the number of molecules of the gas of interest per unit volume. For most spectroscopic measurement systems, the absorption coefficient is empirically established from experimental data. If other gases absorb at the same wavelength, the product of the absorption coefficient and concentration (ac) shown above becomes the sum of all the individual ac products of the gases which absorb at that wavelength.

In practice, the measurement of absorption via Beer's law to determine very small concentrations of a given gas has been challenging because of its sensitivity to small changes in I_0, or the clear path energy. Thus most modern ozone analyzers use a stream-switching technique to update the I_0 measurement on a very frequent basis. In such analyzers, the unknown gas sample is flowed through the measurement cell and an I measurement is established. Then, a few seconds later, the stream is switched to a zero concentration gas for establishing I_0. In this way, the value of I_0 is kept updated so that the small changes which occur in I_0 and I as a function of the lamp, detector, alignment, cell contamination, and so on, are canceled out in the ratiometric (I/I_0) determination of ozone. The zero gas is obtained by scrubbing either the ambient gas or sample gas of ozone using an ozone scrubber. Thus, the measurement of ozone by UV absorption can be made very accurate and precise. Similar results can be obtained with a dual-channel, dual-cell measurement configuration. In such systems, the reference cell is filled with zero gas and the measurement cell contains the unknown sample. By making parallel dual-cell measurements of I and I_0, many of the same advantages of a single-cell, dual-channel measurement can be achieved.

The ozone scrubbers are generally fabricated by coating a copper substrate with manganese dioxide (MnO_2). This ozone scrubber only scrubs the ozone and passes the other coexisting gases without significant modification. This is important because there may be coexisting gases which also absorb at the 254-nm measurement wavelength. If the sample gas is only scrubbed of ozone to obtain the ozone-free reference gas, these other gases will also be present in the zero gas. As a result, their interference effects are essentially canceled out of the ozone measurement since they will affect I and I_0 nearly equally. This interference cancellation results from the ratiometric, dual-measurement technique. Such ozone scrubbers have a finite lifetime proportional to the product of time, flow rate, and ppm of ozone. The scrubbers must either be checked or replaced periodically to ensure zero-based ozone measurements. Sulfur dioxide and some hydrocarbons may be detectable as potential interferents, but their effect is generally negligible.

The daily check of zero and span concentrations of ozone can be accommodated by use of an ozone scrubber for generating zero air and a UV-emitting lamp to generate a span gas. Activated charcoal is very effective at removing ozone as well as a variety of other gases. By exposing air to UV radiation, a certain small percentage of the oxygen is converted to ozone. Thus, by flowing a controlled flow of zero air by a UV lamp operating at a fixed lamp current, a repeatable concentration of ozone can be established. Such gas can be fed to the analyzer for a check of its daily calibration. Such standards are not of sufficient integrity that they should be used for actual calibration; rather they should only be used to check that the analyzer calibration is within the normal range of operation.

Some analyzers are also configured to include a variable-concentration ozone gas generator for use as a calibration standard. A calibration photometer must use the same source of zero air for the analyzer clear-path reference measurement as used to supply air to the ozone generator. Such photometers can be used to generate calibrated concentrations of ozone for calibrating other analyzers. Thus, they become a transfer standard. This is important since ozone cannot be stored in a gas bottle and used, as such, for the calibration of ozone analyzers. Ozone is a difficult gas to handle and care must be exercised to ensure that only inert materials such as glass or kynar are used in contact with the gas stream.

The EPA maintains several ozone primary standards at its various regional offices. Users of ozone analyzers should establish a secondary or transfer standard which is calibrated against a primary standard prior to being used to calibrate individual field-installed analyzers. Ozone analyzers or photometers can be used as calibration standards when properly configured and operated; however, they should also be compared against a primary standard maintained by a standards organization or regulatory agency in order to be assured of quality calibration standards. The specific requirements to establish a calibration photometer are described in Appendix D of EPA 40 CFR 50. The specified approach uses the 254-nm line to measure the ozone absorption. This line is easy to define since it is the major emission line generated by mercury arc lamps. The specific relationship which defines the ozone absorption versus concentration for a calibration photometer is prescribed by the EPA in accordance with the Beer–Lambert law as

$$C \text{ (ppm)} = \frac{-10^6}{al} \left(\ln \frac{I}{I_0} \right)$$

where $\ln$ = natural log, base e

a = absorption coefficient at 254 nm, or 308 ± 4 atm/cm at 0°C, 760 torr

C = concentration, ppm

l = path length, cm

I = detected intensity of UV light with an unknown zone gas mixture

I_0 = detected clear-path intensity with no ozone in the measurement cavity

The EPA further prescribes the correction for temperature and pressure of the gas as measured in the optical measurement cavity and allows a small correction factor for ozone losses within the instrument ($<5\%$).

39.6   CO MONITORING

Carbon Monoxide is normally caused by incomplete combustion either in large stationary sources or in mobile sources (vehicles or automobiles). When carbon is completely combusted, it converts to CO_2, which is of no concern as a pollutant, except as a "greenhouse gas." Since automobiles have become very sophisticated with regard to fuel–air mixture control, their emissions of CO have decreased dramatically and the ambient air has benefited from these advances. The reference method for measuring CO uses a combined NDIR and GFC (gas filter correlation) analytical technique. Similarly, most automated methods also use the same technique. *Nondispersive* means that the optical train does not use a spectrally dispersive element such as a diffraction grating or prism, but instead uses an optical filter to select the CO measurement bandwidth of interest.

In such an analyzer, an electrically heated resistive element is used as the source of IR energy, and the detector is usually a thermoelectrically cooled PbSe detector, which has excellent sensitivity at the 4.7-μm band where CO is measured. Carbon monoxide exhibits a relatively low absorption coefficient in this band, so that a relatively long optical path (5 m or more) must be used to obtain sufficient absorption at the concentrations of interest. A white cell is typically used for the measurement cavity since it can have a physical length of less than 0.3 m, but multipass techniques (e.g., 20 passes) provide an optical path of several meters with a low enclosed volume. An optical selection filter is used to restrict the measurement spectral bandwidth to the region of interest, and two gas cells are used to provide the measurements necessary for a GFC measurement of CO. One gas cell is filled with N_2 and the other with CO, and they are alternately placed in the optical train. The column density (concentration times path length) of the CO-filled cell should be sufficient that it is at least somewhat comparable to the maximum column density of the measurement cell. Such measurement systems are based on Beer's law for the measurement of absorption under the two separate conditions, that is, with the N_2 cell in place or with the CO cell in place. Beer's law does not very closely approximate the absorption-versus-concentration relationship in GFC analyzers since the measurement bandwidth is relatively wide and certainly outside of the monochromatic light which is the basis for Beer's law. If multiple CO absorption lines are included in the gas model, the concentration versus absorption can be more accurately calculated. Empirical data are usually used to characterize a given analyzer design.

Gas filter correlation analyzers are based on a ratiometric measurement of $I(N_2)/I(CO)$, or the ratio of the intensity detected with the N_2 cell in the optical train and the intensity detected with the CO cell in the optical train. Typically, a rotating wheel is used to move the two gas-filled cells in and out of the optical train.

This ratiometric technique provides automatic continuous compensation for variations in the IR source, the sensitivity of the detector, optical contamination of the measurement cell, misalignment, and so on, since they affect the two terms in the ratio equally. Thus, it provides a robust measurement of CO without being affected by the normal component variations which result from time and temperature.

The ratio $I(N_2)/I(CO)$ under clear-path conditions (free of CO) is used as the basis for determining absorption due to an unknown gas and the resulting concentration. For example, if $I(N_2)$ under clear-path conditions is normalized to 1.00 and $I(CO)$ is 0.8 of $I(N_2)$, then we have a ratio $I(N_2)/I(CO)$ of 1.25. If we designate this 1.25 value as K, then, for example, we can establish a relationship as

$$CO(abs) = \frac{1}{(K-1)} * \left(K - \frac{I(N_2)}{I(CO)} \right)$$

where $K = 1.25$ in this example. Thus, at zero CO gas concentration CO(abs) is zero, and as the CO gas concentration increases, the $I(N_2)$ level decreases much more than the $I(CO)$, approaching an asymptotic value of $I(N_2)/I(CO)$ equal to 1.00. When the ratio is 1.00, the resulting CO(abs) is 1.00. This is similar to $1 - I/I_0$ as expressed in terms of Beer's law. After computing CO(abs) for a variety of known gas concentrations, that correlation of CO(abs) versus CO concentration is the basis for establishing the concentration of the unknown gas. Other relationships can be established which can be correlated with CO concentration as well.

There are two significant advantages to such a measurement technique. First, the absorption measurement is compensated for short-term variations in detected light level due to source and detector changes, measurement cell contamination, alignment, and so on. Second is the inherent interference rejection which results from this measurement technique. If an interfering gas is present which partially overlays both the CO absorption spectrum and the residual spectra which results from subtracting out the CO spectrum from the spectral band seen by the detector, then the measurement of $I(N_2)$ with the nitrogen-filled cell in the optical path will be affected about the same as the measurement of $I(CO)$ with the CO-filled cell in the optical path. Under such conditions, the ratio $I(N_2)/I(CO)$ remains the same with or without the interferent present. Thus the effect of that interferent is virtually eliminated. On the other hand, if an interfering species largely overlays the CO absorption lines, $I(N_2)$ will be affected (reduced) more than $I(CO)$, and with the presence of such a gas there will be a positive interference on the measured value of CO. In this case, the CO measurement will read higher than the actual value due to the positive interference. Alternately, if an interfering species largely overlays the residual spectra (the spectral measurement band minus the CO spectral lines), $I(CO)$ will be affected (reduced) proportionately more than $I(N_2)$ and there will be a negative interference from the effect of such a gas. Thus GFC has the potential to reduce interferences due to gases which overlay the spectrum of the particular gas of interest, depending on the specific spectral features involved. In the case of diatomic molecules, such as CO, the GFC technique is particularly effective because of the well-resolved fine structure involved in the IR CO absorption spectrum. Further, in the IR measurement spectrum, H_2O and CO_2 are broad-band absorbers which absorb energy over much of the available bandwidth. In the case of the CO measurement, the potential interfering effects of H_2O, CO_2, and N_2O are all greatly reduced by the use of GFC techniques, and they are of no real significance in well-designed instruments used for ambient monitoring applications. Ambient applications provide a background gas matrix, including a few percent H_2O and typically about 350 ppm CO_2, which is much less interference prone than stack applications where both gases can be significantly higher.

This analytical technique is largely independent of sample flow rate since the detected energy is only dependent upon the number of CO molecules in the sample cell which interact with the impinging IR light source. The sample cell is typically temperature stabilized at an elevated temperature to minimize drift and contamination.

Zero gas for such an analyzer is easily obtained by flowing the ambient or sample gas through a heated catalytic oxidizer which converts CO into CO_2. Palladium operated at moderate temperatures (near 100°C) has been shown to be effective in providing this conversion, and Hopcalite is often used for scrubbing CO, although it must be replaced from time to time. Thus an internal source of CO free zero gas can be easily incorporated into the analyzer. The span gas for daily calibration checks must be obtained from a bottled gas supply. Permeation tubes are not available for CO, and it is otherwise not easily generated in a controlled manner. Thus, valving is often built in to such analyzers to accommodate an external span gas bottle.

39.7 PARTICULATE MONITORING

Particulate monitoring is particularly difficult due to the variable size of the dust particles as well as the vagaries involved in defining specifically what constitutes a particulate. Particulate matter is normally thought of as the solid material suspended in air, and TSP (total suspended particulate) is a measure of the concentration or density of such material. However, depending on the measurement conditions, other chemical compounds which are normally in the vapor phase can condense or precipitate out of the sample gas stream, resulting in higher concentrations of observed particulate.

The first particulate size fraction to become of interest was that particulate with a mean aerodynamic diameter of 10 µm or less (PM_{10}). This was thought to be of major significance with regard to respiratory diseases. As a result of legislation in this area, a generation of size-selective sample inlets was designed and installed to allow collection/filtering of just the particles less than or equal to 10 µm in size. However, in the last few years there has been significant evidence that the smaller size fractions are much more important from a health effects standpoint. Consequently, there is now a standard for $PM_{2.5}$ (particulate matter equal to or smaller than 2.5 µm mean aerodynamic diameter).

The most basic and least cost approach for an instrument used to measure ambient particulate is the so-called high volume sampler. This is fundamentally a high-volume blower coupled with some form of flow control and a timer along with a platform where a filter can be placed to collect particulate. Thus, by weighing the filter prior to insertion in the high-volume sampler and weighting the filter again after a known number of cubic meters of air has been pulled through the filter, the density of particulate material (mg/m^3) can be determined. Temperature and pressure measurements can be used to refine the concentration measurements. Such instruments are obviously limited by the time resolution they can provide and require periodic manual maintenance to replace filter material before flow rates are affected by the dirt accumulation. Further sophistication of such instruments has provided automatically changeable filters and size selective inlets for PM_{10} or $PM_{2.5}$ particle size fractions.

Two other techniques have gained considerable market acceptance for monitoring the ambient particulate size fractions of interest. These are the beta gauge instruments and the TEOM (tapered element oscillating microbalance) instruments manufactured by Rupprecht and Patashnik (www.rpco.com). Both the beta gauge and TEOM particulate measurement techniques have achieved wide market acceptance and certification/approval by major testing agencies in the United States and Europe. The TEOM has become the preferred measurement technique for many European nations.

Optical techniques can also be used to measure particulate, a good example of which is the opacity monitor used to monitor smoke stack visible emissions. In general, however, such optical absorption or transmission measurements are not sufficiently accurate to measure the low levels of particulate found in ambient level applications without special long-path-length configurations. Particulate matter also scatters incident light in back, side, and forward directions, and such instruments have been used for measuring lower levels of particulate density. The relative sensitivity of such scattering measurements are dependent on the optical characteristics of the particulate, including chemical composition, index of refraction, density, measurement wavelengths, shape and size, and color. The optical techniques are all indirect measurements of particulate matter and as a result are not as accurate as direct measurements. Nephelometers, special optical instruments for measuring scattered light, have been rather widely used for particle size and visual range measurement applications. They use an extractive measurement technique where the gas of interest is drawn into the instrument for measurement. Transmissometers, which measure the transmission of visible light, have also been developed for visibility (visual range) measurements in airport, tunnel, and highway applications.

39.7.1 Beta Radiometric Measurement

Beta gauges are based on the radiometric measurement of particulate collected on a filter paper which is normally a fiberglass-based material. Typically, the filter material is contained on a reel so that it can be automatically fed to the analyzer particulate collection system as required for quasicontinuous measurements of particulate. The reels allow unattended operation for weeks to months depending on measurement conditions. The instrument also contains a pump and flow controls so that a prescribed flow rate can be established for a programmable time interval. In operation, an empty filter location (spot) is first physically positioned in the filter weighing station where it is weighed according to the absorption of beta radiation which it produces. This is accomplished by placing the filter spot in between the beta radiation source and detector, which is typically a Geiger–Mueller tube. Then the clean filter spot is incremented until it is in the measurement position where clamps seal it off so that the full sample flow of the analyzer is drawn through the filter material. This sampling station is usually heated so as to preclude condensation of moisture and/or other compounds on the filter. After a prescribed time interval, the filter material is unclamped and repositioned in the weighing station. Again, it is radiometrically weighed, and the difference between the loaded and empty weight is divided by the product of time and flow rate to determine the particulate concentration or density. Then the above process is repeated with another clean filter spot. The time resolution of the particulate measurements can be on the order of minutes to hours. The key to the operation of the beta gauge is the fact that beta radiation is quite uniformly attenuated by particulate matter of a variety of typical compositions. This results from the basic measurement relationship, which can be approximated as

$$\ln\left(\frac{N_0}{N}\right) = \frac{h}{p}d_{avg}$$

where N_0 = rate at which beta rays pass through filter when it is clean

N = rate at which beta rays pass through filter with a sample of particulate

h = linear attenuation coefficient, cm^{-1}

p = density of absorbing matter, cm^3

d = surface density, mg/cm^2

The mass absorption coefficient h/p of the beta rays depends on the density of the electrons in the absorbing material and is therefore proportional to the relation Z/A, where A is the atomic number of the element and Z is the mass number. The Z/A factors for common elements of interest are listed below:

Al	C	Ca	Fe	K	Mg	O	S	Si
0.482	0.500	0.499	0.499	0.486	0.494	0.500	0.499	0.498

An exception to this uniform response is hydrogen, for which $Z/A = 1$. Fortunately, hydrogen forms only a small part of common particulate matter and is generally insignificant in the overall analysis of typical ambient aerosols. Since h/p, or Z/A, is nearly constant for most particulates, it is possible to use the above expression to find the mass of the particulate deposited on the filter. Also, since the coefficient h/p increases with decreasing beta maximum energy of the emitter, mass determination by way of beta absorption becomes more sensitive with weaker beta emitters. Because C-14 is a weak emitter (maximum energy of 0.156 MeV) and has a long half life (5730 years), it is particularly suitable for such measurements. The radiation source is often sized so that its strength is sufficiently low (<100 μC) so that it avoids most, if not all, of the regulations regarding controlled radioactive substances and does not warrant any special handling other than the obvious safety precautions.

Zero and span calibration checks are easily implemented in the beta gauge. If the same clean filter spot is measured successively (without drawing a sample), this condition should produce a zero measurement check. If during the second measurement of the clean sample described above an absorber of known attenuation is placed in front of the filter paper, it should produce a repeatable upscale calibration check. Thus, it is very straightforward to provide a real measurement of a zero and upscale check condition.

The radiometric absorption can also be accurately correlated with gravimetric weight, and in fact that is one method of checking the radiometric analysis. This can be done by manually placing individual filters in the collection and weighing positions and comparing the gravimetric weight with the radiometric weight. The filter spots on the reel of tape can also be cut out and measured using more sophisticated laboratory analysis to determine the chemical composition of the resultant particulate.

The inlet to the beta gauge is usually separate from the gas analyzer inlet so that it can be optimized for transport of particulate. It can incorporate size-selective inlets so that the units can be configured to measure TSP or particulate size fractions corresponding to PM_{10} or $PM_{2.5}$.

39.7.2 TEOM Measurement Technique

The TEOM-based particulate monitor incorporates a patented technique wherein the resonant frequency of an oscillating tapered quartz tube is used to quantify the mass load on one end of the tube. The tapered element is basically a hollow cantilever beam with an associated spring rate and mass. Such a system exhibits the relationship

$$f = \left(\frac{K}{M}\right)^{0.5}$$

where f = natural vibrating frequency of beam

M = mass loading on beam

K = calibration constant

The filter and collected particulate are attached to one end of the oscillating tapered quartz tube, and as their weight increases on the oscillating element, the resonant frequency decreases proportional to the square root of the mass. Since the tube is hollow, the sample flow stream can be drawn through the tube, thereby collecting the particulate on the filter resting on the open end of the tube. Since the natural frequency of the loaded quartz tube is continuously measured, the difference in frequency measured over a short time interval provides a measure of the change in particulate weight observed over that time interval. The time resolution of particulate density is from minutes to hours. As the particulate accumulates on the filter paper, the sensitivity of the analyzer decreases, thereby requiring periodic replacement to maintain proper operation. Such analyzers track the accumulated weight and provide operator notification when filter maintenance is required. Other versions of this measurement technique are available which incorporate an automatic filter-changing mechanism. Up to 16 filter cartridges can be stored in a cassette for continuous sequential use in the unattended mode of operation.

A very key element of such a particulate monitor is the size-selective inlet. Particles are aerodynamically sized with different options to collect TSP, PM_{10}, $PM_{2.5}$, and even $PM_{1.0}$, (particulate smaller than 1 μm). One configuration designated as a dichotomous sampler has two parallel particle size fraction separators so that PM_{10} can be collected on one filter and $PM_{2.5}$ on another simultaneously. Flow rates are appropriately controlled to maintain required accuracy under changing temperature and pressure conditions. The filter collection platform is heated and temperature is controlled so that further chemical changes in the nature of the particulate can be minimized.

39.8   TOTAL HYDROCARBON MONITORING

The FID (flame ionization detector) is normally used for the measurement of THCs (total hydrocarbons) at ambient levels. It uses a hydrogen-fueled flame, a polarizing voltage source, and an ion collector and amplifier. It is the analytical method of choice for most THC measurements because it has very high sensitivity and wide dynamic range and is insensitive to water, carbon dioxide, and other inorganic compounds. It is specifically sensitive to C–H-containing organic molecules. Further, it is a carbon mass-sensitive detector so that its output represents the total number of carbon atoms per unit time in the sample stream. It is often used with gas chromatography (GC) columns to allow measurement of specific organic compounds. In some cases it can be equipped with a methane scrubber or oxidizer as needed to measure TNMHCs (total nonmethane hydrocarbon compounds), which is a more useful measurement than THCs in some circumstances. Methane is a biogenic gas which occurs from natural sources in many areas. The FID has a relatively uniform response to most of the typical hydrocarbon compounds of interest and can be calibrated specifically for any one of interest. Care needs to be exercised in the calibration of the analyzer since the analyzer may provide a nontypical response to some compounds. Acetylene, for example, has an anomalous high response with the FID. In general, most such analyzers are calibrated in ppm of propane or methane.

39.9 VOC MONITORING

The major thrust to implement ambient VOCs monitoring on a broad scale was the large-scale nature of the ozone nonattainment problem and the lack of detailed knowledge as to the specifics of the formation of ozone under a variety of regional conditions. Volatile organic compounds generally have a relatively high vapor pressure. Two different control strategies have been proposed for the control of ozone: One is to reduce the hydrocarbons in the atmosphere and the other is to reduce NO_X. Both of these approaches can reduce ozone under the right conditions, but the reduction of NO_X under some conditions may actually cause ozone to increase. Thus, additional data are needed to ensure that the proper control strategies are used.

Sorbent tubes and passivated stainless steel canisters have been routinely used for sample collection and off-line analysis of VOCs. They are very effective and easy to use; however, they do not provide a continuous record of such concentrations. As a result, an analyzer was specifically developed under a contractual relationship with the EPA to provide this capability. Typically, the incoming air sample is preconcentrated on a filter (cooled in some cases); then the compounds are periodically thermally desorbed and finally measured by GC analysis. The analysis provides an indication of the 55 ozone precursor compounds for which affected states must provide measurement data. A list of the ozone precursors measured with this technique is given in Table 39.4. Using the preconcentration approach, sensitivity is excellent. Perkin-Elmer worked with the EPA under a cooperative development program to develop PAMS/VOC monitoring hardware suitable for unattended operation in the field. These systems, as used for the PAMSs, were also equipped with a data system so that data could be collected on the listed individual VOCs plus other pollutant and meteorological data and transmitted to a central station on a periodic basis. Several hundred of these stations have been installed in the more difficult ozone nonattainment areas.

39.10 OPEN-PATH MONITORING

Ambient air has traditionally been monitored with fixed-single-point located analyzers. Such analyzers gave a good analysis of the ambient conditions at that point, and the places where they were located were carefully controlled to ensure conditions representative of the desired environment. An open-path monitoring technique was pioneered by Opsis, a Swedish company. This technique has been widely used in the EC but has only recently (1995) been tested and certified in the United States by the EPA. The objective of such an approach is to obtain long open-path measurements which are more representative of an area than a point. If the long open path of the analyzer contains a homogenous gas, then the analyzer provides a measure of that concentration based on the Beer's law relationship. However, if the long open path contains a nonhomogeneous gas, it then provides a line average of the concentration. Using Beer's law and assuming the variable gas distribution can be represented as a series of segments of concentrations c_n existing over a distance b_n, the exponential term becomes $a(b_1c_1 + b_2c_2 + \cdots + b_nc_n)$. But, since we only know the total absorption over the total path length b_t, the analyzer output corresponds to a concentration $c = (b_1c_1 + b_2c_2 + \cdots + b_nc_n)/b_t$. So, even though a point measurement may not be located so as to see a localized maximum concentration area, a line average measurement going through that localized area may or may not provide much information about that localized maximum either, since that part of the line is averaged with the length-weighted concentration existing over the rest of the long open path. The EPA has provided additional guidelines for the use of open-path techniques to ensure representative measurements. Obviously, the inherent limitations of the open path necessitate certain restrictions, such as making certain that the light path does not become obscured (tree limbs, cranes, people, birds, etc.) or misaligned, and understanding that data may become invalid if sufficient light is not available for measurement which limits the amount of, for example, fog, rain, snow, hail, and aerosols (which scatter, reflect, and absorb light) which can be tolerated.

There are a variety of analytical techniques which can be used for long open-path measurements. Both FTIR (Fourier transform infrared) and IR derivative techniques have been applied to such measurements; however, the UV/VIS dispersive technique used by Opsis has gained the most market acceptance for ambient air quality–monitoring applications. This technique uses a remote source located several hundred meters away from the receiver. The source is pulsed so that the analyzer can distinguish the difference between absorption caused by ambient light and the measurement light source. The receiver focuses the light onto a fiber-optic line which feeds the energy to a UV/VIS spectrometer for analysis. With this design approach, it is possible to time share the spectrometer/analyzer with several different pairs of sources/receivers using a fiber-optic switch/multiplexer. These systems have also been configured to utilize a common receiver on a steerable, or rotatable, computer-controlled platform with elevation adjustments so that it can be precisely pointed to one of several remote sources located around the receiver. In this way a single receiver/spectrometer/analyzer can be shared with multiple sources, or measurement lines. Shorter path measurements have been configured with the source and receiver colocated on one end of the path and a retroreflector located at the remote end of the measurement path.

The spectrometer and analytical engine use a differential optical absorption spectroscopy (DOAS) measurement technique. Such a technique uses a pair of measurements for each gas where one measurement is of a reference wavelength and the other is of a major feature of the spectrum of the gas of interest. Such wavelengths are very closely located, in wavelength, so as to ensure that the effect of the atmospheric variables (dust, mist, rain, etc.) and analyzer variables (alignment, light level, optical contamination, detector gain, fiber-optic losses, etc.) are the same on both measurements. If the absorption coefficient at the measurement wavelength is high for the gas of interest and the absorption coefficient at the reference wavelength is low, then the difference between the two measurements can be correlated with the concentration of the gas of interest. In the United States, Opsis has been certified by the EPA for the measurement of SO_2, NO_2, and O_3 in ambient air. The analyzer is also capable of measuring a wide variety of hydrocarbons including BTX (benzene, toluene, and xylene), NO, NH_3, HF, Hg, and H_2O.

Calibration of such open-path analyzers can be done with a measurement cell inserted either into the normal open path or into a light path illuminated with a calibration light source. Such a measurement cell has transparent windows on each end and gas ports so that it can be filled with zero gas or a given calibration gas. Since the sensitivity of such analyzers is basically in units of ppm-m (ppm multiplied by meters of path length), for 0.1 ppm concentration over 1000 m measurement path length it is 100 ppm-m. Such optical absorption can be simulated by using a 1-m cell and 100 ppm gas concentration. The gas cell temperature should be the same as the path length gas for optimum accuracy. If such a cell is interposed on the normal measurement path length, it should cause an increase in the analyzer measurement of the stated ppm-m of the gas cell. However, since the background concentration of the open path is not static but continually changing, there may be considerable uncertainty in the use of such a technique in a changing environment. The use of measurements before and after the cell insertion may reduce such uncertainties. To circumvent such uncertainty, the cell can also be used with its own calibration light source, which just illuminates the shorter measurement cell path length. Thus there is no significant background concentration. In such conditions, the light source must be spectrally the same as the measurement light source but will have to be attenuated to simulate the light level seen in the actual application. Under such conditions, the calibration is much more stable, albeit somewhat dependent on the exact spectral match of the two light sources.

39.11 METEOROLOGICAL MONITORING

Meteorological information is needed in conjunction with pollutant measurements in order to provide some visibility of the nature and location of the source or cause of unusual pollutant conditions. Typical meteorological parameters measured and reported at each monitoring site include wind speed and direction, barometric pressure, temperature, relative humidity, and solar radiation. Solar radiation spans the electromagnetic spectrum from the UV to the IR, so that care must be used in selecting the proper spectral measurement band. Wind speed and direction are typically mounted on a mast which extends upward from the roof of the monitoring station.

39.12 DATALOGGERS AND NETWORK INTERFACES

Ambient monitoring stations are typically linked together by telephone line and the data retrieved and processed on a central computer system. The interface between the analyzer and the telephone data transmission set can be done in different ways. Traditionally, dataloggers have been used as the intermediary device which accepts a variety of analog and digital input/output from the analyzers and sensors and converts the data into digital format for processing and storage. Further the datalogger retrieves data from the analyzers on a selectable time interval, calculates specific time-based averages, implements and identifies the daily zero and span calibration checks, retrieves analyzer and system status information, and buffers the data until requested by the central or polling computer. The datalogger then interconnects to a modem and responds when interrogated via a dial-up telephone line. With this buffering of data, the data transmission lines or central computer can be down for extended periods of time without losing measurement data. When the polling computer sends out a request for data, the datalogger then sends the data collected since the last polling request. Alternately, some analyzers can buffer their own data and send data to the polling computer when interrogated via the RS232 serial data port available in some analyzers. Alternately, a local PC at the monitoring site can communicate with the analyzers via the RS232 serial interface and buffer data for transmission to the central computer upon request. The datalogger provides more flexibility in accommodating different analyzers and their often unique input/output options and formats, meteorological measurements, and other sensors and can store data in intermediate memory for short-term intervals and on removable media for longer term storage. Further it provides more flexibility in establishing different sampling rates and setting up the control sequence for daily zero/span calibration checks or linearity checks. The datalogger typically has the capability to communicate with a local PC at the analyzer site so that operators can review measurement data, status information, calibration sequences, and other information necessary to facilitate station maintenance. In some cases where telephone lines are not available, the datalogger can provide sufficient storage on removable storage media so that the service technician can pick up the storage media on a biweekly or monthly basis and manually enter it into a suitable computer system without loss of data. For maximum reliability, many dataloggers and analyzers are battery backed up so that loss of facility AC power does not result in a loss of air quality data. Unfortunately, most dataloggers have their own serial data transmission protocol, and drivers must be provided at the central collection site to communicate with whatever types of dataloggers are in use. This has led some network users to standardize on specific brands of dataloggers which are accommodated by their specific network data collection and processing software. In the United States, some specific network data collection protocols have been established to simplify the problem of collecting data from a wide variety of ambient monitoring sites. SAROAD (Storage and Retrieval of Aerometric Data) was one of the early network protocols, and it has been replaced by AIRS (Aerometric Information Retrieval System) in the EPA computerized system for storing and reporting of information relating to ambient air quality data. Other countries and agencies have established their own network software and data collection protocols. Consult your regional authority to determine what data collection protocols are available and the types of air quality summary data which can be made available.

Other data output options have been implemented by some monitoring site operators. One involves collecting and processing the ambient monitoring data and presenting them in a format suitable for public access over the Internet via a web page. Another is to provide a driver for a large, outdoor flat-panel display which can locally indicate ambient air quality indices and specific measurement data. Timely public access to such air quality data is becoming more important in many environmentally sensitive areas.

39.13 PORTABLE VERSUS STATIONARY AIR-MONITORING STATIONS

Most ambient monitoring stations have been installed as permanent sites because of the need for long-term data. In some situations, there has been a need to have portable monitoring stations so that they can be moved from site to site in a given region with the goal of providing more monitoring information with respect to major emission sources and locations of high automobile congestion. For these cases, ambient monitoring stations have been installed in trailers which are moved as needed to further clarify pollutant characteristics in a given locale. When installing ambient monitoring analyzers and meteorological equipment in/on a trailer, there are a variety of concerns which become more significant than in permanent locations. One of the major concerns is for security, not only for the trailer, but also for the contents of the trailer, the meteorological gear installed on the roof, and the personnel. Trailers can be obtained with wheels which retract into the wheel wells, thereby making their removal (stealing) difficult and further making the unauthorized removal of the trailer even more difficult. The walls of the trailer can be made bullet proof for small-caliber rifles and all doors can be locked with high-security locks/deadbolts. Windows are generally minimized since they are difficult to secure, thus requiring internal illumination via lights. Further, it may be desirable to provide movement or intrusion alarms which activate local audible alarms and telephone emergency security numbers. Access to AC power and a telephone line are, of course, a necessity. Alternatively, some installations use portable generators and cellular telephones. Portable generators may be problematic because of the emissions of the typical internal combustion powered generator. If engine emissions become part of the sample inlet stream, measurement integrity will be compromised. Safety is another concern which impacts how the trailer is laid out, particularly with respect to where the calibration gases are stored and vented. Gas leaks from the bottles must not compromise personnel safety, and gas sensors may be used internally to ensure a safe working environment. Care must also be used to discharge gases used during calibration so that they do not interfere with the measurement of ambient air quality. Heating, ventilating, and air conditioning are other important considerations since most trailers are totally enclosed without substantial outside air circulation. Care must be exercised to ensure that temperature rise within the equipment rack is tolerable. A stable ambient environment within the trailer is necessary for optimum analyzer performance and operator comfort. This can impose substantial AC power requirements, particularly under air-conditioning compressor startup conditions. Road shock and vibration during transport are other concerns which should be addressed if system reliability and longevity are of concern. Shock-mounted racks are often required if the effects of road shock and vibration are to be minimized. All movable components must be securely tied down or they are at risk of being irreparably damaged during transport. Roof construction and associated strength and durability are also of concern since meteorological equipment is generally roof mounted and requires personnel access to mount and dismount. Security of roof-mounted equipment must be considered as well. With a well-constructed and equipped trailer and appropriately laid-out interior, trailers can be very well adapted to this portable use. Their use thereby allows one set of monitoring equipment to be shared among sites, thereby lowering the cost of obtaining survey information. In some cases, this approach has been carried a step further with pole-mounted enclosures which can be installed close to roadways to monitor vehicle-related pollution.

VI

FLOW MONITORING

40

AIR FLOW MEASUREMENT

Randy D. Down, P.E.
Forensic Analysis & Engineering Corp.
Raleigh, North Carolina

40.1 OVERVIEW

Air flow measurement is a somewhat common, yet very important application in the field of environmental measurement and control. Accurate and reliable air flow measurement is critical in quantifying and controlling the rate of pollutant emissions.

Environmental Instrumentation and Analysis Handbook, by Randy D. Down and Jay H. Lehr
ISBN 0-471-46354-X   Copyright © 2005 John Wiley & Sons, Inc.

A number of methods have been established to measure air flow rates. This chapter will provide examples of standard environmental air flow measurement applications and the proper instruments and procedures to use to obtain accurate and reliable readings. There is no universal air flowmeter that is appropriate for any type of measurement application. This chapter will help guide you through the selection process, and what is involved in obtaining a useful measurement.

Air flow measurement instruments to be discussed in this chapter include:

- Pitot tubes and manometers
- vane-type (turbine) flowmeters
- hot-wire anemometers
- differential pressure elements (such as Venturis and orifices)
- stationary arrays (flow stations)

40.2   FLOWMETER OPTIONS

The air flow measurement instruments covered in this chapter are the instruments most often used in environmental applications. The first three types are available as either stationary or portable instruments. The latter (differential pressure instruments and arrays) are generally designed to be permanently installed in a section of ductwork.

The average duct velocity and duct cross-sectional area must be known to determine a rate of air flow. Flowmeters directly or indirectly measure air velocity, which can then be multiplied by the cross-sectional area of the duct or plenum through which the air is traveling to determine the volumetric flow rate in cubic feet per minute (cfm) or cubic meters per minute (cmm).

40.3   PITOT TUBES

Named after its inventor, Henri Pitot, a Pitot tube measures the difference between the static and the dynamic pressure of the air flow as it traverses a duct or pipe cross section. This measured pressure differential, coupled to the true gas density at the ensuing pressure and temperature conditions, yields the velocity of the air within the duct.

A standard Pitot tube consists of a stainless steel tube within a tube with very small openings (or perforations) along one side of the outer tube wall and a 90° radius bend near the end that tapers to a point at the tip of the inner tube (Fig. 40.1). The tubes are generally available in lengths ranging from 12 to 60 in. Selection of tube length is dependent upon the diameter of the ducts or pipes to be traversed.

The instrument is used to sense velocity pressure and static pressure, which is in turn converted to an air velocity measurement by the use of a connected manometer.

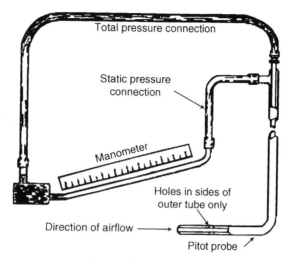

Figure 40.1 Pitot tube.

The manometer measures the pressure differential between the static and velocity pressure.

Because the outside holes are perpendicular to the direction of travel, the outer tube is pressurized by the local random component of the air velocity. The pressure in these tubes is the static pressure (p_s), as discussed in Bernoulli's equation. The tip of the center tube is pointed in the direction from which the air is flowing and is pressurized by both the random and the ordered air velocity. The pressure in this center tube is the total pressure (p_t) discussed in Bernoulli's equation. The pressure transducer, or manometer, measures the difference between the total and static pressure:

$$\text{Velocity pressure measurement} = p_t - p_s$$

A Pitot tube measures air velocity in a very small area within a duct. The velocity of air near the interior walls of a duct (where friction of the wall surface is a factor) is lower than that of air velocity near the center of the duct (where there is no friction). It is necessary to take multiple "traverse" readings horizontally and vertically (across the width and height of the duct) to get a true representation of the average air velocity through the duct. Once velocity is measured (typically in feet per minute), the volumetric flow rate in cubic feet per minute can be determined by determining the cross-sectional area of the duct and multiplying it times the velocity. Recommended measurement locations for traverse readings are shown in Figure 40.2.

To reduce the effort in determining average duct velocity, an averaging tube may be used. An averaging tube operates in the same manner as a Pitot tube to determine average duct velocity pressure. However, instead of measuring one point within the duct, the averaging tube measures multiple points to determine a true average

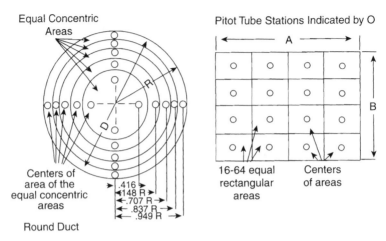

Equal Concentric Areas

Pitot Tube Stations Indicated by O

Centers of area of the equal concentric areas

.416
.148 R
.707 R
.837 R
.949 R

Round Duct

16-64 equal rectangular areas

Centers of areas

Figure 40.2 Pitot tube duct traverse.

velocity. Some averaging tubes are designed so that the differential pressure measured is 2–3 times the actual velocity pressure. This amplification is accomplished by the shape and design of the tube, resulting in greater resolution and the ability to measure lower air velocities.

Always install air flow measurement devices at least $7\frac{1}{2}$ duct diameters downstream and $1\frac{1}{2}$ diameters upstream of anything in the duct that could cause turbulence.

40.4 MANOMETERS

A liquid-filled (inclined or U-tube type) or a digital manometer is used to measure the difference between the static and total pressures sensed by a Pitot tube, averaging tube, or stationary array.

An inclined manometer is a liquid-filled instrument that has graduation marks at various points along an inclined tube to indicate the velocity that corresponds to the measured velocity pressure in relation to atmospheric pressure (Fig. 40.3). These instruments are rather cumbersome to use because they can leak colored fluid

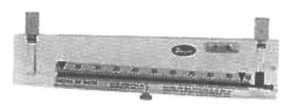

Figure 40.3 Inclined manometer.

Figure 40.4   Digital manometer.

unless carried upright and are somewhat susceptible to vibration. They are gradually being replaced with Magnehelic and Photohelic gauges as well as digital manometers.

Digital manometers (Fig. 40.4) offer a higher degree of resolution and accuracy than inclined or other styles of liquid-filled manometers. However, they can be more difficult to read if turbulence in the air flow stream causes the digital display indication to fluctuate.

The measured pressure difference can be very small at lower velocities, and an inclined manometer or digital manometer with a low range may be required to obtain accurate measurements.

Obtaining a highly accurate velocity measurement can be a challenge with a digital instrument because there is almost always some turbulence in the system. This causes the digitally displayed readings to fluctuate. The instrument must have damping abilities if it is to be of any real use. Various companies manufacture digital manometers. The cost of these instruments is generally higher than liquid-filled manometers. Generally speaking, under realtively stable flow conditions the digital manometer will provide more precise measurement than a liquid-filled manometer.

40.5   VELOMETERS

Most velometers are basically digital micromanometers that are calibrated to a specific Pitot tube. Their packaging and displays vary and often are integrated into a convenient hand-held box. They have been used almost exclusively in the HVAC (heating, ventilation, and air conditioning) industry for balancing air conditioning ducts.

Figure 40.5 Vane anemometer.

40.6 VANE ANEMOMETERS

Vane or turbine meters employ a free-spinning propeller (turbine blades) that spins at a rate proportional to the velocity of air passing through it (Fig. 40.5). The higher the air velocity, the faster the propeller's number of rotations per minute (rpm). A magnet mounted on the tip of one of the propeller blades comes in very close proximity to a magnetically coupled (*Hall effect*) semiconductor device mounted in the flowmeter once per revolution. As the magnet on the propeller tip passes by the Hall effect device, the device produces an electronic pulse that is detected by the meter's electronic circuitry. The frequency of the pulses generated by the magnet and Hall effect semiconductor is proportional to the rotational speed of the propeller, which is also representative of the velocity of the air passing through the propeller. As with the Pitot tube and manometer, by multiplying the duct air velocity by the duct cross-sectional area, the volumetric flow rate in cubic feet per minute or cubic centimeters per minute (ccm) can be calculated.

The useful velocity range of a vane anemometer is between 200 and 2000 feet per minute (fpm), and the accuracy of the instrument will depend on the precision of use, the type of application, and its calibration. Calibration should be checked frequently against a manometer. Every 6 months the unit should be recalibrated by the manufacturer to National Institute of Standards and Technology (NIST) traceable standards.

40.7 HOT-WIRE ANEMOMETERS

A hot-wire anemometer senses air velocity by indirectly measuring the amount of heat dissipated from a heated temperature element (Fig. 40.6). The higher the air velocity across the heat source (element), the more heat that is dissipated. The

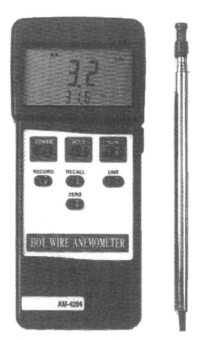

Figure 40.6 Hot-wire anemometer.

cooling effect of faster flowing air (due to higher heat dissipation) changes the resistance of the wire, which in turn is indicative of a higher air velocity. As the air velocity decreases, the measured temperature increases proportionally. A thermistor located near the hot wire measures the temperature of the air being sensed and is used to electronically provide temperature correction since the temperature of the flowing air can vary significantly. The hot wire and thermistor are mounted on the end of a stainless steel, telescoping probe that is inserted into the air stream through a small-diameter (typically $\frac{1}{2}$ in.) hole in the duct wall.

The hot-wire anemometer measures a fluid velocity by noting the heat convected away by the fluid. The core of the anemometer is an exposed hot platinum or tungsten wire that is either heated up by a constant current or maintained at a constant temperature. In either case, the heat that is lost to fluid convection is a function of the fluid's velocity.

Hot-wire anemometers are limited to applications where the air stream they are measuring is no greater than 140° F and there are low levels of particulate. Above that temperature limit, the amount of heat dissipation cannot be measured accurately by the thermistor.

With proper calibration, it is possible to measure air velocities with an accuracy of 0.05% or greater, depending upon the measurement range and quality of the instrument and its calibration. The response time between measurement and instrument output is very short when compared with other methods of air flow measurement and can reach a minimum of $\frac{1}{2}$ µs.

40.8 DIFFERENTIAL PRESSURE ELEMENTS

Differential pressure elements, such as a Venturi or orifice plate, when used with a draft range differential pressure transmitter or transducer, are a highly reliable and low-maintenance method of air flow measurement due to the fact that they essentially have no moving parts and are less affected by particulate or other contaminants in the air stream.

As air flows across the element, there is a pressure drop (differential pressure) created between the front and back of the element that is a square-root function of the air velocity (Fig. 40.7). The greater the velocity of air across the element, the greater the pressure drop that occurs. Another advantage of differential pressure flow elements is their ability to measure temperature in a high-temperature air stream (such as a boiler flue duct, where the temperature can reach 800° F) without being damaged.

A significant disadvantage is the pressure drop a differential pressure flow element creates within the duct. This pressure loss must be considered when a blower or fan is selected during system design in order to provide adequate static pressure and still attain the design flow rate. Since the fan must work harder (to produce more static), this in turn may require a fan with a higher fan curve and a motor with a higher horsepower rating, resulting in more capital cost and higher energy consumption. The indirect cost of using a differential pressure element may outweigh the cost of going to a more sophisticated air flow measurement instrument that induces a much lower static pressure loss.

40.8.1 Orifice Plates and Venturis

Orifice plates and Venturis have been used extensively for many years in the measurement of steam and water flow. Particularly in large chilled-water, hot-water, and steam plants. They consist of a thin metal plate with a hole in the center having a substantially smaller diameter than the duct in which they are installed. By constricting the fluid flow, they produce a pressure drop greater than that of a Venturi or other shape in the flow stream but work on a similar principle. The greater the air flow, the higher the pressure drop across the orifice plate. Pressure drop across the

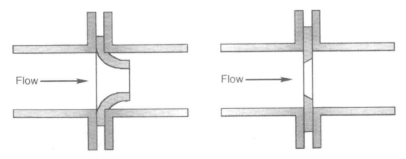

Figure 40.7 Differential pressure flow elements.

plate is a square-root function of the fluid velocity. They have found much less use in the measurement of air flow largely because there are other methods that are less energy consuming and similar or lower in cost.

Among the disadvantages of orifice plates and Venturis are (1) the relatively high pressure loss they create by constricting flow to create a pressure differential (usually resulting in higher energy costs to overcome the pressure loss); (2) the relatively low turndown (typically a 3-to-1 ratio) that they can provide with any degree of accuracy using a single differential pressure transmitter and plate (in other words, they are accurate down to one third of the maximum flow they are designed to measure); and (3) the plate cannot be removed without fluid (in this case air or other gases) leaking from the duct or conduit.

Sensing is accomplished by piping the impulse lines of a differential pressure transmitter across the plate to sense the pressure drop (differential) created by the orifice plate restriction. The high-pressure impulse line is connected upstream of the plate and the low-pressure impulse line is piped downstream. The sensed pressure drop is transmitted back to a controller or digital control system that performs a square-root extraction of the signal (which is proportional to the pressure drop, e.g., 4–20-mA DC signal over a 0–5-in. pressure drop).

If a greater turndown ratio is required using an orifice plate, multiple differential pressure transmitters may be stacked (piped in parallel) and set for different pressure measurement ranges. The turndown ratio is the rangeability of the instrument—in other words, the full measurable range of the instrument you can achieve without recalibration. At low flow rates, a more sensitive transmitter range is desired (typically referred to as a "draft range" transmitter), with differential pressures measured in terms of inches of water column.

40.9 STATIONARY ARRAYS

An array is a cross-sectional matrix of sensors (Pitot tubes or hot-wire anemometers) designed to fit within a specifically sized round or square duct and sense air velocity at multiple points across the duct (Fig. 40.8). This allows a more

Figure 40.8 Flow station (stationary array).

accurate measurement of average flow rate throughout the duct than a single-point measurement can provide. It is essentially the same as taking an average of multiple traverse measurements across the duct.

Too often one quick "spot reading" is recorded, either with a Pitot static tube and manometer in the duct or with a vane anemometer at the outlet of the system. A single measurement taken in the middle of the air stream (often the highest flow) will result in a misleading reading and inaccuracies in all subsequent calculations and adjustments.

40.10 VANE ANEMOMETERS VERSUS THERMAL ANEMOMETERS

As is often the case with a successful technology, development has progressed in different directions such that now both high-performance and low-cost, poorer performing instruments exist within the marketplace. Each has its place and field of application. The problem that arises, however, is that often all modern anemometers are considered to be similar and the precision end of the range is erroneously miscounted. A parallel can be drawn with ultrasonic measuring techniques and early Doppler flow metering mistakes. Here we look at the distinct advantages of modern vane anemometers in terms of their fundamental digital technology, wide operating range, and averaging capabilities. The principal advantage of independence from flowing media density is stressed together with the certainty of true velocity realization. All this is compared with the specific application advantages of thermal anemometers.

At low velocities, typically less than 300 fpm (1.5 m/s), anemometer blade design and bearing quality become increasingly significant. Extensive work has led to the adaptation of 4 in. (100 mm) as the optimum vane diameter for standards acceptance. A well-designed 4-in. (100-mm) vane head can be yawed and pitched through angles of $\pm10°$ and still maintain an accuracy to within 2%.

The low-velocity performance can be further enhanced by the use of software correction of the electronically sensed speed of rotation. By these methods today's 4-in. (100-mm) vane anemometers can provide accurate, repeatable results down to as low as 40 fpm (0.2 m/s), but this achievement is dependent upon good blade and bearing design.

Air has a mass, and therefore density, and at any given velocity of this mass there is an impact on the rotating vane blades which induces them to turn until blade and mass velocity are in synchronization. This allows the electronics to convert the rotation into a velocity reading. As the blade rotation is induced by such forces impacting upon them until velocity synchronization is achieved, the rotating vane is virtually independent of temperature and pressure.

The big advantage of thermal anemometers is the ability to provide the sensing head in a comparatively small size, which can then be inserted through small access holes in duct walls and within confined spaces and for precise point measurement.

Unlike the vane anemometer, which has a direct linear relationship with velocity, thermal anemometers have a relationship that approximately represents a fourth-power equation. Therefore, a sophisticated microprocessor is required to calculate an adequately accurate reading across a significant turndown range. Thermal devices also suffer from fundamental dependence on temperature, density, and composition changes in the flowing media.

Contaminants in the measured air stream typically have a minimal effect with the shrouded bearing design of a vane anemometer. Contaminants can have a much larger effect on thermal (hot-wire) anemometers. Contamination of the sensing element with a surface coating is generally not prevented by thermal cleaning because the normal operating temperatures are not high enough to vaporize commonly encountered substances. This coating can have a significant effect on calibration factors and long-term repeatability due to the resultant insulating layer. This shortcoming can be overcome by regularly returning the thermal anemometers to a reputable service and calibration firm. These devices should be calibrated to NIST-traceable standards.

40.11   APPLICATIONS

As stated at the beginning of this chapter, numerous environmental measurement and control applications require the use of an accurate and reliable instrument for monitoring air flow. Examples of these applications include the following:

1. *Measurement of Exhaust Air Flow Rates.* Exhaust air flow is measured in order to establish the rate of pollutant emissions necessary for:

- Determination of the rate at which pollutants are being emitted to the atmosphere (by measuring the exhaust air flow rate as well as concentration of contaminants in the air stream).
- Sizing and selecting a pollution abatement system (such as a scrubber, oxidizer, or carbon recovery system). The larger the exhaust air rate, the larger (and more costly the pollution control system.
- Regulation of the air flow rate through the pollution abatement system using a feedback control loop or using the measured air flow rate as a feedforward signal to regulate the combustion rate of an oxidizer or rate of flow through a carbon adsorption bed.
- Complying with the Clean Air Act.

Stationary and portable air flow measurement instruments can both be used for these applications.

2. *Test Ports.* In the case of portable instruments, test ports must be installed in the wall of the ductwork at each measurement location. The test port consists of a hole in the duct wall that is large enough to insert a Pitot tube, hot-wire anemometer, or other flow measurement device. When not in use, the port is either capped or plugged to prevent air from escaping or infiltrating the duct.

3. *Single Versus Multiple Air Flow Stations.* If there are multiple sources of exhaust air at a facility and a stationary flowmeter is prefered, it is often more economical to install a single, larger flow station in a main branch or main plenum than multiple smaller flow stations in multiple branch ducts. Using multiple flow stations to determine the total air emissions requires the use of a computer or other means to add the measurements of the individual flow stations together. A word of caution regarding this approach—the measurement error will also be cumulative, often resulting in a total measurement that is less accurate than that of a larger single-station measurement.

4. *Accessibility.* Often air flow stations are mounted in exhaust air ducts that would be very difficult to access for measurements using a portable instrument. When selecting a method of air flow measurement, health and safety issues should be given primary consideration. It often makes more sense to install a stationary instrument in difficult-to-access areas to avoid the risk of injury when a ladder, human lift, or other equipment must be used to reach the measurement location.

5. *Contamination of Stationary Pitot Arrays.* The measurement capability of an air flow station that uses a Pitot array can be compromised by small particulate suspended in the air stream. There is a tendency for the small openings in the outer tube to become blocked by a buildup of particulate on the tube's surface. To overcome this problem, compressed air (instrument air) can be piped to the stations and equipped with a solenoid air valve and a timer that periodically sends a pulse of compressed air through the outer tube to clear the perforations. The frequency of the pulsed air is determined by the density of the particulate in the measured air stream and is established through a process of trial and error.

6. *Measuring Variable-Flow Conditions.* Flow measurement is particularly important when dealing with flow rates that have a relatively high rate of variability. If a fixed volumetric flow rate of pollutant air is being treated, the system can be designed for constant-flow conditions, and flow measurement is needed for record-keeping purposes. When a variable-air-flow condition occurs, which is fairly common in many facilities, the flow measurement devices that are used must have adequate turndown capabilities to measure the lowest design flow condition.

7. *Use of Vane Anemometers.* Vane-type anemometers are useful in measuring the discharge velocity from an air source but are somewhat impractical for the measurement of flow rates in a duct due to their size (the optimum impeller size, as stated earlier, being 4 in. in diameter). Using vane anemometers in a duct requires a larger access port than a hot-wire anemometer. They are also susceptible to yaw and pitch. They should be positioned within 10% of perpendicular to the air flow pattern in order to get an accurate measurement.

40.12 SUMMARY

Whether selecting a stationary or portable air flow measurement instrument, many issues need to be considered, including:

- Process conditions (heat, humidity, vibration, contaminants, etc.)
- Area harzard classification
- Accessibility
- Accuracy
- Reliability
- Cost
- Output signal compatability (4–20-mA dc, 1–10 V, etc.)
- Measurment requirements

When in doubt about the proper instrument to use, consult an engineer or instrument supplier who does not represent a single product line or have any other reason to give a biased recommendation regarding the instrument selection.

40.13 GLOSSARY

Absolute Pressure	The total of the indicated gauge pressure plus the atmospheric pressure. Abbreviated psia for pounds per square inch absolute.
Air Differential Pressure Gauge	Gauges that contain no liquid but instead work on a diaphragm and pointer system which move within a certain pressure differential range.
Anemometer	An instrument used to measure air velocies.
Atmospheric Pressure	The pressure exerted on Earth's surface by the air due to the gravitational attraction of Earth. Standard atmosphere pressure at sea level is 14.7 pounds per square inch (psi). Measured with a barometer.
Barometer	An instrument for measuring atmospheric pressure.
Cubic Feet Per Minute (CFM)	The volumetric rate of air flow.
Differential Pressure Gauge	An instrument that reads the difference between two pressures.
Feet per Minute (fpm)	The velocity of air movement.
Gauge	An instrument for measuring pressure.
Hot-Wire Anemometer	An instrument that measures instantaneous air velocity in feet per minute using an electrically heated wire.
Inches of Water Gauge or Water Column (in. WG or in. WC)	A unit of air pressure measurement equal to the pressure exerted by a column of water 1 in. high.
Manometer	An instrument for measuring pressures. Can serve as a differential pressure gauge.

Micromanometers	Instruments used to measure very low pressures accurately down to plus or minus one thousandth (0.001) of an inch of water.
Pitot Tube	A sensing device used to measure total pressures in a fluid stream. It was invented by the French physicist Henri Pitot.
Sensitivity	A measure of the smallest incremental change to which an instrument can respond.
Standard Cubic Feet per Minute (scfm)	The volumetric rate of air flow at standard air conditions.
Vane Anemometer	Gives instantaneous, direct readings in feet per minute based on the amount of force that the flowing air exerts on the rotating vanes. The higher the air velocity, the more force exerted and the faster the rotation of the vanes, which translates into air velocity.
Velocity Pressure	The subtraction of static pressure from total pressure, and because total pressure is always greater than or equal to static pressure, velocity pressure will always be a positive value.

41

GAS FLOW MEASUREMENT

Ashok Kumar and Jampana Siva Sailaja

University of Toledo
Toledo, Ohio

Harish G. Rao

LFR Levine Fricke
Elgin, Illinois

41.1 INTRODUCTION

Measurement of gas flow is important for various types of applications, including process controls, environmental monitoring, and compliance, and is a critical operational parameter in a number of industries—aerospace, nuclear, pharmaceutical, industrial and chemical process—as well as in residential and commercial heating, ventilation, and air conditioning systems. In environmental applications, accurate

Environmental Instrumentation and Analysis Handbook, by Randy D. Down and Jay H. Lehr
ISBN 0-471-46354-X   Copyright © 2005 John Wiley & Sons, Inc.

measurement of gas flow rate is a necessary element in the determination of air pollutant concentrations in a gas stream or in the ambient air. A variety of instruments are available to measure the gas flow rate. The objective of this chapter is to describe and discuss these different devices. The information has been drawn from the literature[1-6] and websites listed at the end of the chapter.

41.2 TYPES OF AIR-MEASURING DEVICES

Air-measuring devices can be categorized into the following groups:

1. *Volume Meters.* These devices measure the total volume V of gas passed through the meter over some specific duration of time. Flow rate can be calculated by the equation

$$Q = \frac{V}{t}$$

  where Q = flow rate $(L^3\,T^{-1})$

  V = total volume of gas passed over some specific measured period (L^3)

  t = time period of passage of gas that is measured with a timing device (T)

2. *Rate Meters.* These devices are used to calculate the time rate of flow through them and the computation is based on using some property of the gas.

3. *Velocity Meters.* These devices measure the linear velocity of a gas and the cross-sectional area of the duct through which the gas is flowing to calculate the volumetric flow rate Q using the formula

$$Q = A\bar{u}$$

  where A = cross-sectional area of conduit through which gas is flowing (L^2)

  $\bar{u}$ = linear velocity of gas through the conduit (LT^{-1})

4. *Other Devices.* These other devices measure flow using a variety of novel and innovative techniques to overcome some of the technical or economic constraints of regularly used devices.

Commonly used devices in each category are described in the following sections.

41.2.1 Volume Meters

There are seven common volume meters commonly used in air sampling and analysis.

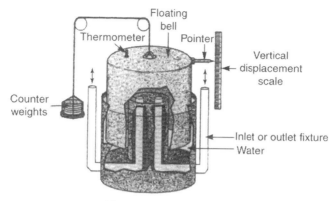

Figure 41.1   Spirometer.[6]

41.2.1.1 *Spirometer (Bell Prover): Primary Standard.* This consists of a cylinder of known volume with one end closed and the open end submerged in a circular tank of fluid. The cylinder can be opened or closed to the atmosphere by a valve. The water displaces the air to be measured as the cylinder is lowered into the water and causes it to be discharged from the cylinder; the rate of discharge can be regulated. The displaced volume of air is proportional to the vertical displacement of the cylinder measured by a pointer and scale (see Fig. 41.1). The volume of the cylinder is calculated from its dimensions. A counterweight and cycloid counterpoise allow pressure differentials across the spirometer as low as 0.02 in. of water, as shown in Figure 41.2.

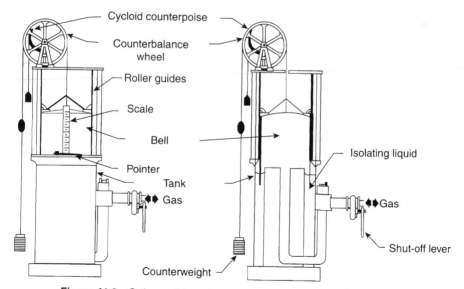

Figure 41.2   Orthographic and cross-sectional views of 5-ft^3 spirometer.

The volume of air passed through a spirometer is calculated by the formula

$$V = \frac{\Pi d^2 h}{4} \tag{41.1}$$

where V = volume of air passed through spirometer (L^3)

Π = a constant = 3.14

d = diameter of bell (L)

h = vertical displacement of bell (L)

Spirometers use water or, in some cases, oil as the fluid of choice. The fluid in the spirometer should be maintained at the same temperature as the room. This is to ensure that the fluid and the air will be in thermal equilibrium and hence minimize spirometer fluid evaporating into the air. The pressure inside the bell is also brought into equilibrium with room conditions. The exact importance of thermal equilibrium is that the air displaced from the bell must be at the same temperature as the room for volume calculations, which minimizes the need for volume corrections, as the temperature is constant during the calibration procedure.

A conversion to standard conditions must be made to find the true volume of air that has passed through the spirometer once the volume of air is determined using room conditions. This conversion to standard conditions is made using the formula

$$V_2 = V_1 \frac{P_1}{760 \, \text{mm Hg}} \frac{298 \, \text{K}}{T_1} \tag{41.2}$$

where V_2 = volume of gas at standard conditions of pressure (P_2) and temperature (T_2)

V_1 = volume of gas at measured room conditions of pressure (P_1) and temperature (T_1)

T_1 = measured room temperature of gas, K

T_2 = standard temperature of gas, K (generally 298 K)

P_1 = measured room pressure of gas, mm Hg

P_2 = standard pressure of gas, mm Hg (generally 760 mm Hg)

The spirometer is generally calibrated by the manufacturer against a National Bureau of Standards (NBS) "cubic foot" bottle. An owner who doubts that the spirometer is in error can check the calibration with an NBS-certified cubic-foot bottle. Alternatively, a "strapping" procedure may be used, which involves the measurement of the dimensions of the bell with a steel tape and subsequent calculation of the volume. Experienced personnel routinely obtain accuracies of ±0.2% when calibrating a spirometer by the "strapping" procedure.

Flow rates can be calculated by timing the volume of air passing to or from the spirometer and determining the rate of flow. The spirometer is simple, inexpensive, and dependable and is used almost solely as a primary standard for calibration of other types of flow- and volume-measuring devices. As the spirometer can be produced in large sizes, it has typically been used to calibrate Roots meters, which are positive-displacement meters.

41.2.1.2 Displacement Bottle Technique: Primary Standard.

In this technique, the displacement bottle consists of a bottle filled with water (or some other liquid) and a tube through which air can enter the bottle from the top and a drain pipe from which the displaced liquid can be collected and measured (see Fig. 41.3). Air that is drawn into the bottle takes the place of the volume of liquid lost, as the liquid in the bottle is drained or siphoned out. The volume of gas sampled is equal to the volume of liquid displaced. The fluid in the displacement bottle must be in thermal equilibrium with the room temperature. This equilibrium will ensure no liquid evaporation from the bottle water to the air and simplify volume corrections for T and P. The volume of displaced liquid can be measured with a graduated cylinder or Class A volumetric flask, depending on the accuracy with which the volume needs to be measured. Accuracy can range from 1 to 5%, depending on the measuring device used.

Again, once the volume of air has been determined at room conditions, it should be converted to the volume at standard conditions.

41.2.1.3 Frictionless Pistons: Primary Standards.

Two frictionless pistons, the soap-bubble meter and the mercury-sealed piston, are discussed in this section. Accurate and convenient measurement of lower flow rates between

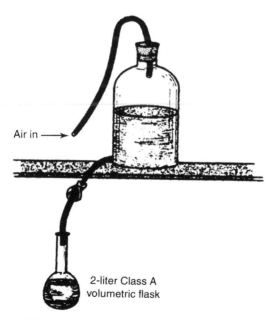

Figure 41.3 Displacement bottle technique.[6]

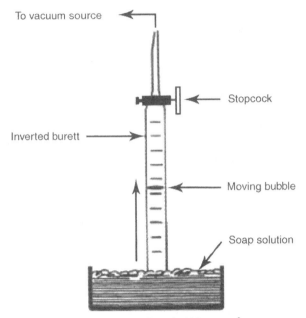

To vacuum source

Stopcock

Inverted burett

Moving bubble

Soap solution

Figure 41.4   Soap-bubble meter.[6]

1 and 1000 mL/min can be made with a soap-bubble meter. Mercury-sealed pistons that are available can accurately measure higher flow rates, from 100 to 24,000 cm^3/min.

41.2.1.3.1 Soap-Bubble Meter. A bubble meter consists of a cylindrical glass tube with graduated markings, usually in milliliters. Inverted buretts (see Fig. 41.4) are frequently used as soap-bubble meters; however, buretts cannot be used with anything other than a vacuum source. Simple bubble meters can be purchased commercially, though the basic design can be created conveniently by a competent glass blower. Figure 41.5 shows two different types of soap-bubble meters. The soap-bubble meter stands as one of the simplest of primary standards. The inside walls of the tube are wetted with a soap solution. A bubble is formed by touching the tip of the burett to the soap solution or by squeezing the rubber bulb until the soap solution is raised above the gas inlet.

   Either a vacuum at the top or slight positive pressure at the bottom of the tube moves the bubble (a frictionless piston) up the tube. Volumetric flow rate can be calculated by timing this movement and noting the volume traversed by the bubble over a measured time span.

   The volume measured by a soap-bubble meter must be corrected for two conditions. First, if the room temperature and pressure are different from standard atmospheric conditions, the volume must be corrected by using Equation (41.2). Second, the measured volume can be slightly larger than the actual volume because water

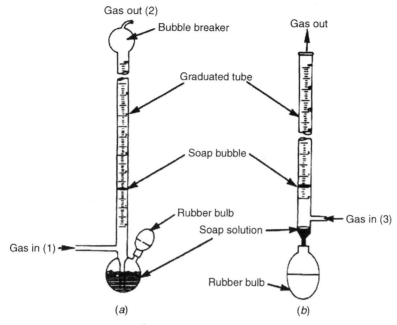

Figure 41.5 Soap-bubble meters:[6] (a) with bubble beaker and capable of handling vacuum at 2 or pressure at 1; (b) capable of handling only pressure at 3.

from the bubble evaporates into the gas behind the bubble. If the gas behind the bubble has a relative humidity greater than 50%, the error is small. If the gas is dry, the error can be large and must be corrected by the formula

$$V_c = V_{meas} \frac{P_b - P_w}{P_b} \tag{41.3}$$

where V_c = corrected volume

V_{meas} = measured volume

P_b = atmospheric pressure (mm Hg)

P_w = vapor pressure of water at room temperature (mm Hg).

Note that P_b and P_w must have the same units.

Vapor pressure tables can be found in any chemistry handbook. Soap-bubble meters can be calibrated by measuring the dimensions of the tube, but the poor control on glass dimensions in manufacturing makes this inaccurate. The bubble meter is usually calibrated by filling the tube with a liquid (e.g., water or mercury), then draining away the liquid from the top graduation to the bottom graduation. The volume or weight of the collected liquid can be measured. With proper corrections for temperature this calibration is accurate. The soap-bubble meter should only be used to measure volumes between graduations that have been calibrated.

The bubble meter is used almost exclusively in laboratory situations for calibration of other air-measuring instruments. In average laboratory conditions the soap-bubble meter is accurate to about ±1%, depending on how accurately it is calibrated. Accuracy decreases for flows below 1 mL/min and above 1 L/min, mainly because of gas permeation through the bubble. Increased accuracies have been reported for bubble meters fitted with automatic sensing devices that start and stop a timer.[6]

41.2.1.3.2 Mercury-Sealed Piston. Where a bubble meter is unsuitable, an electronically actuated mercury-sealed piston may meet the need. Although the mercury-sealed piston is expensive, its accuracy (±0.2% for time intervals greater than 30 s) and simple operation make it an extremely useful tool. Figure 41.6 shows a sketch of the mercury-sealed piston volume meter.

The mercury-sealed piston consists of a precision-bored, borosilicate glass cylinder with a close-fitting polyvinyl chloride (PVC) piston. The space between the piston and cylinder wall is sealed with a ring of mercury placed in a groove around the piston that stays in place because of its high viscosity and the closeness of the fit between the cylinder and piston. Gas entering the solenoid valve is allowed to vent until the measurement cycle is actuated. When the measurement cycle is started, the solenoid valve closes the vent, allowing gas to enter the cylinder. A timer is started and stopped as the mercury seal passes the lower (stationary) and upper (movable) proximity coils (metal detectors). The volume displaced can be set by adjusting the

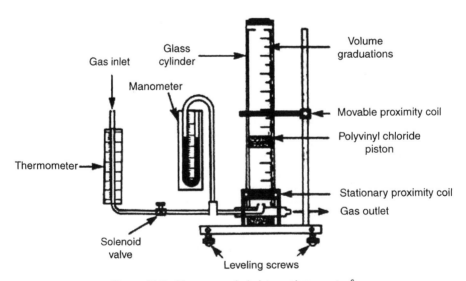

Figure 41.6 Mercury-sealed piston volume meter.[6]

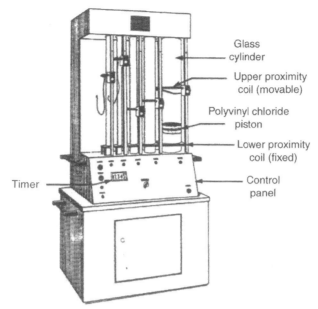

Figure 41.7   Calibrator console, front view (Brooks model 1051). (Courtesy Brooks Instruments Division, Emerson Electric Co.)[6]

upper proximity coil. The volume is corrected to standard conditions using the pressure drop across the piston (usually <3 in. of water). The measured time and the corrected volume can be used to calculate volumetric flow rate. The system has a reported accuracy of ±0.2%.

The manufacturer usually performs calibration of the mercury-sealed piston volume meter. The borosilicate glass cylinder is bored to a precise diameter. The inside diameter is air gauged at least every inch to check for consistency. Before the instrument is sent out, it is compared to a standard mercury-sealed piston volume meter that is traceable to NBS. Cylinders must be aligned if a multicylinder unit is purchased. One cylinder is chosen to be correct and all others are aligned with the set screw located on top of the piston, which changes the displaced volume slightly. Figure 41.7 shows a Brooks Instruments Division calibrator console with multiple cylinders, control panel, and timer.

Figure 41.8 shows the configuration of a typical mercury-sealed PVC piston. Erratic movement of the instrument can break the mercury seal. For this reason the mercury-sealed piston instrument is used as a primary standard in laboratory settings. Mercury-sealed piston volume meters are available for accurate flow measurement over a wide range of flow rates (100–24,000 cm^3/min).

41.2.1.4   *Wet Test Meter: Intermediate Standard.*   The wet test meter consists of a series of inverted buckets or traps mounted radially around a shaft and partially immersed in water. The location of the entry and exit gas ports is such

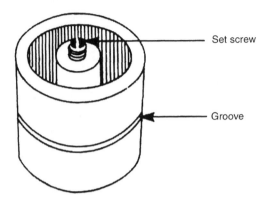

Figure 41.8 PVC piston for mercury-sealed piston volume meter.[6]

that the entering gas fills a bucket, displacing the water and causing the shaft to rotate due to the lifting action of the bucket full of air. The entrapped air is released at the upper portion of the rotation and the bucket again fills with water. In turning, the drum moves index pointers that register the volume of gas passed through the meter.

Once the meter is leveled, the proper water level is achieved by using the filling funnel, fill cock, and drain cock to bring the meniscus of the water in touch with the tip of the calibration index point. The calibration gas must be passed through the meter for 1 h to saturate the water with the gas. The water in the meter should be at the same temperature as the surrounding atmosphere. If any water is added, sufficient time must be allowed for complete equilibrium to be reached. Figure 41.9 shows the main components of a wet test meter.

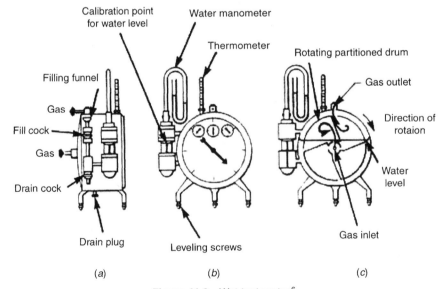

Figure 41.9 Wet test meter.[6]

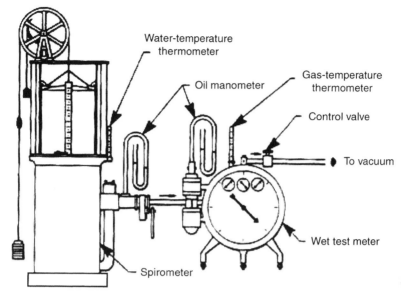

Figure 41.10 Setup for calibrating wet test meter against spirometer.[6]

Once the water level is set and the meter is equilibrated, the wet test meter is ready for calibration (or use if it is already calibrated). An accurate calibration of a wet test meter can be done with a spirometer, as shown in Figure 41.10. The wet test meter can also be calibrated against a mercury-sealed piston, as shown in Figure 41.11.

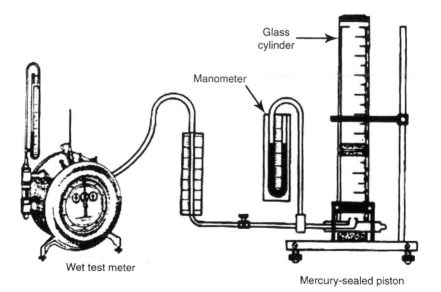

Figure 41.11 Calibration of wet test meter (WTM) against mercury-sealed piston.[6]

Enough gas is drawn through either system to turn the wet test meter at least three revolutions and to significantly move the spirometer drum or piston. This measurement is made several times. Atmospheric pressure and temperature and the temperature and pressure differential for both the wet test meter and calibrating device are needed to correct the volume to standard conditions (taking pressure differential into consideration). There is no need to correct for water vapor effects, since both the calibration device and wet test meter are measuring gas saturated with water vapor.

If a wet test meter is used to measure a dry gas stream, a significant error is introduced if the measured volume is not corrected to dry conditions. This correction is made using Equation (41.3).

A simple calibration check can also be performed using a displacement bottle, as shown in Figure 41.12. After all the water is in thermal equilibrium with the surroundings, the wet test meter is properly set up and connected to the displacement bottle. As the air flows into the bottle, the drain tube of the displacement bottle is filled, and the pinch clamp is opened, allowing 2 L of water to drain into a 2-L Class A volumetric flask. The corresponding wet test meter readings are taken. This is repeated several times (usually four).

The following optional wet test meter calibration procedures can be used:

Option 1: Draw a multipoint calibration curve for flow for each of the following setups:

   1. Setup for calibrating a wet test meter against a spirometer

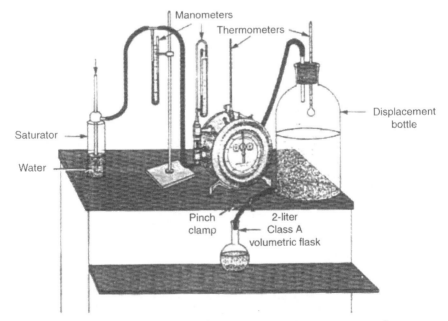

Figure 41.12   Calibration of wet test meter with displacement bottle.[6]

2. Setup for calibrating a wet test meter against a mercury-sealed piston
Option 2: Adjust the calibration index point so that the meter volume is correct.
Option 3: Calculate a correction factor for the wet test meter.

Wet test meters should check within ±0.5% if option 2 is used.

All volumes measured by a wet test meter should be corrected to standard conditions using Equation (41.2).

Wet test meters are used as transfer standards because of their high accuracy (less than ±1%). Because of their bulk, weight, and equilibration requirements, they are seldom used outside a laboratory setting. Wet test meters are useful for laboratories that need an accurate standard yet do not have the funds or space for a spirometer or mercury-sealed piston. Calibrated wet test meters can be used to measure flow rates upto 3 rev/min, at which point the meter begins to act as a limiting orifice and obstructs the flow. Typical ranges of wet test meters are 1, 3, and 10 per revolution.

41.2.1.5 *Roots Meter: Intermediate Standard.* The Roots meter is a positive-displacement, rotary-type meter for measuring volume flow. It is suitable for handling most types of clean, common gases. It is not suitable for handling liquids, and excessive particulate matter carried in the gas stream can impede its operation. Roots meters consist basically of two adjacent "figure 8" contour or two-lobe impellers rotating in opposite directions within a rigid casing (Fig. 41.13). The casing is arranged with inlet and outlet gas connections on opposite sides. Impeller contours are mathematically developed and accurately produced and are of such form that a continuous seal without contact can be maintained between the impellers at all positions during rotation. To accomplish this, the correct relative impeller positions are established and maintained by precision-grade timing gears. Similar seals exist between the tips of the impeller lobes and two semicircular parts of the meter casing. As a result of this design, the gas inlet side of the meter is always effectively isolated from the gas at the outlet side of the impellers. Consequently, the impellers rotate even with a very small pressure drop across the meter.

The direction of rotation and relative position of the impellers are as indicated in Figure 41.13. As each impeller reaches a vertical position (twice in each resolution), it traps a known specific volume of gas between itself and the adjacent semicircular portion of the meter casing at A and B (Fig. 41.13, position 2). Thus, in one complete revolution the meter will measure and pass four similar gas volumes, and this total volume is the displacement of the meter per revolution.

The displacement volume of the Roots meter is precisely determined by the manufacturer, both by calculation and by testing it using a known volume of air or other gas. Roots meters are usually calibrated against large spirometers prior to shipment. Users do not usually have a way to calibrate Roots meters and must depend on the supplied calibration data. Volumetric accuracy of the Roots meter is permanent and nonadjustable (except for linkage adjustment) because its measuring characteristics are established by the dimensions and machined contours of

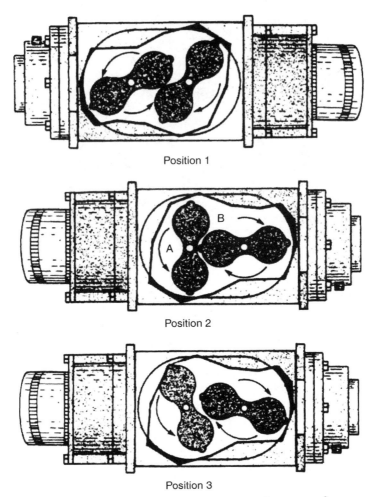

Position 1

Position 2

Position 3

Figure 41.13   Principle of gas flow through Roots meter.[6]

nonwearing fixed and rotating parts. The revolutions of the impellers are indexed with the meter reading and calibrated to a specific volume unit (i.e., cubic feet or cubic meters). Units are sold that have temperature compensation devices, but corrections to standard temperature and pressure conditions are easily made using Equation (41.2).

The measurement pressure, P_1, in this instance, is the atmospheric room pressure (P_1 in millimeters of mercury) minus the pressure drop across the Roots meter, Δp, in millimeters of mercury.

$$P_1 = P_b - \Delta p \qquad (41.4)$$

The metering unit is magnetically coupled to the impellers. The entire counting unit is enclosed in a plastic cover. The cover also holds oil that lubricates the metering device. The proper oil level is set by the inscribed oil level lines on the ends of the plastic covers. The user of a Roots meter must be careful not to severely tilt the meter when oil is in the plastic cover, as this can force oil into the impeller casing. If the oil gets into the impeller casing, flushing with a solvent such as kerosene will need to be used to remove the oil.

Although Roots meters are widely used in a number of industrial applications, they have been used almost exclusively as the standard for measuring high-volume sampler flow rate in atmospheric sampling applications.

41.2.1.6 *Dry Test Meter: Intermediate Standard.* The dry test meter (an intermediate standard) works on the same principle as the dry gas meter (a secondary standard), but a different indexing method (read-out) makes the dry test meter more accurate (usually 1–2% when new). Dry test meters are an improvement over the more common dry gas meters that are most commonly used in residential and industrial settings to measure gas flow (e.g., natural gas). The dry test meter in Figure 41.14 shows the new read-out mechanism. Figure 41.15 is a cut-out view showing the inside components of a dry test meter.

The interior of the dry meter consists of two or more movable partitions, or diaphragms, attached to the case by a flexible material so that each partition may have a reciprocating motion (Fig. 41.16). The gas flow alternately inflates and deflates each bellows chamber, simultaneously actuating a set of slide valves that shunt the incoming flow at the end of each stroke. The inflation of the successive

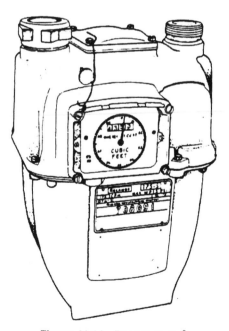

Figure 41.14 Dry gas meter.[6]

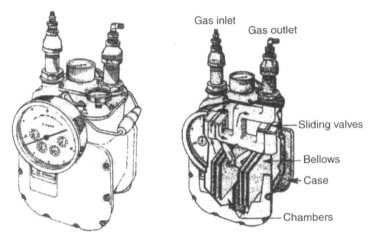

Figure 41.15 Dry test meter. (Courtesy of Western Precipitation Division, Joy Manufacturing Company.)[6]

chambers also actuates, through a crank, a set of dials that register the volume of gas that has passed through the meter.

The dry test meter is calibrated against a spirometer, mercury-sealed piston, or displacement bottle similar to the wet test meter. One big advantage of the dry test meter over the wet test meter is that calibration errors larger 2% can be corrected by

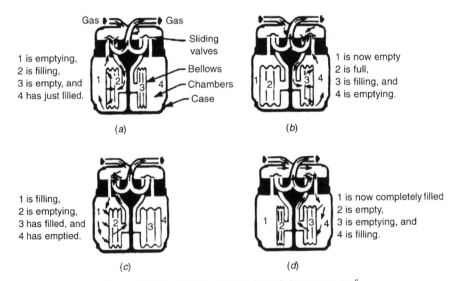

Figure 41.16 Working mechanism of dry test meter.[6]

adjustment of the meter linkage. If linkage adjustment cannot correct the problem, then the dry test meter must be returned to the manufacturer for repairs.

Dry test meters are used in the field as well as in laboratory calibrations. Since the dry test meter does not contain water, it is lighter and easier to use than the wet test meter. Also, the dry test meter is more rugged than the wet test meter. Accuracy of the dry test meter does, however, worsen with age.

41.2.1.7 *Hot-Wire and Vane Anemometer.*

An anemometer is a device used to measure air speed. Two types of anemometers are common, depending on the type of air flow being measured: hot-wire anemometers and vane anemometers.

The hot-wire anemometer (HWA) works by having air flow over a hot-wire. The instrument maintains the wire at a fixed temperature so that, as it is cooled by the air flow, the current increases to maintain the temperature of the wire. The current drawn through the hot-wire is measured as a result of the cooling effect of the air flow, which extracts heat from the wire. The current is thus a function of the gas velocity. The HWA is used for measurements in the range of approximately 20–100 ft/min (1–5 m/s) to determine the air flow rate in ducts or pipes. Figure 41.17 shows a HWA.

A vane anemometer (VA) is a more rugged instrument that is better suited to measuring the wind speed across larger areas such as over the face of a fan, large

Figure 41.17   Hot-wire anemometer.

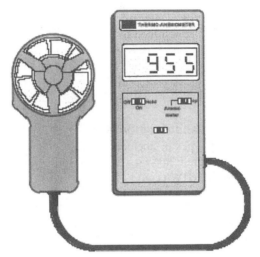

Figure 41.18   Vane anemometer.

duct, or sidewall opening (e.g., wind energy sites, livestock applications, large intake hoods). The velocity of the air mass on the vanes induces them to turn until the mass and blade velocity are in synchronization. The rotation of the vanes has a direct linear relationship to wind speed and is virtually independent of temperature and pressure. The instrument can be used to average true flow rate over a sample area. A VA is not ruined by dust and small airborne debris since it can be carefully cleaned. As the mass of the vane requires a fair amount of air movement, it does not measure low air speeds. These instruments are not considered accurate below 50–70 ft/min, although the meter provides readout at these lower air speeds. They should be used in the air streams that are at least as wide as the vane diameter. Accurate measurements are not provided in the case of narrow-inlet air jets that are smaller than the vane anemometer propeller. In this device velocity can be displayed as a running average value over time. This will help in scanning a fluctuating air stream. A modern VA detects the vane speed of rotation.

A wind tunnel is used for the calibration of anemometers. A scientific calibration procedure, in which the interaction between wind tunnel and anemometer is known, can result in a calibration accuracy of up to 0.1 m/s (if carried out in a very accurate wind tunnel). To ensure accurate measurements, it is required that the complete calibration curve be developed over the range of wind speeds that will be measured. Figure 41.18 shows a VA.

41.2.1.8   Mass Flow and Tracer Techniques

41.2.1.8.1   Thermal Meters. Thermal meters are used for measuring mass flow rate with negligible pressure loss. These instruments are based on the principle that the rate of heat absorbed by a flow stream is directly proportional to its mass flow. As the moving air/gas comes into contact with the heat source, it absorbs heat and cools the source. At increased flow rates, more of the gas comes into contact with the heat source, thereby absorbing even more heat. Thus the amount of heat dissipated from the heat source is a measure of the flow characteristics of the gas being measured. The meter consists of a unit containing a heating element that is placed in a duct section between two points at which the temperature of the air or gas stream is measured. The temperature difference between the two points and the heat input are used to measure the mass flow rate, from which the volumetric flow rate can be derived.

41.2.1.8.2 Mixture Metering. The principle of mixture metering is similar to that of thermal metering. Instead of adding heat and measuring temperature difference, a contaminant is added to an air/gas stream and the increase in concentration is measured or clean air is added and the reduction in concentration is measured. This method is useful for metering corrosive gas streams. The measuring device may react to some physical property such as thermal conductivity or vapor pressure, which is used to measure mass flow rate.

41.2.1.8.3 Ion-Flowmeters. These meters generate ions from a central disc that flow radially toward the collector surface. Air flow through the cylinder causes an axial displacement of the ion stream in direct proportion to the mass flow.

41.2.2 Rate Meters

The most popular devices for measuring flow rate are the rate meters. Rate meters measure, indirectly, the time rate of the fluid flow through them. Their response depends on some property of the fluid related to the time rate of the flow.

41.2.2.1 Variable Pressure Meters: Head Meters. Head meters are those in which the stream of fluid creates a significant pressure difference that can be measured and correlated with the time rate of flow. The pressure difference is produced by a constriction in the flow conduit causing a local increase in velocity.

41.2.2.2 Orifice Meters: Noncritical, Secondary Standard:. An orifice meter can consist of a thin plate having one circular hole coaxial with the pipe into which it is inserted (Fig. 41.19). Two pressure taps, one upstream and

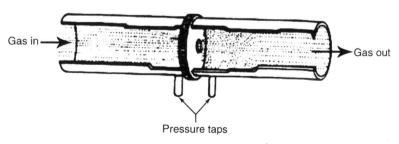

Figure 41.19 Orifice meter.[6]

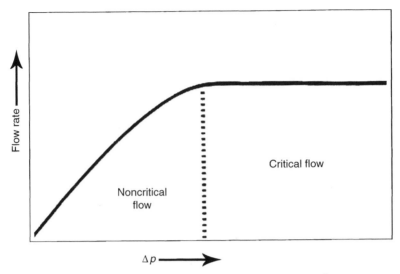

Figure 41.20 Typical orifice meter calibration curve.[6]

one downstream of the orifice, serve as a means of measuring the pressure drop, which can be correlated to the time rate of flow. Watch jewels, small-bore tubing, and specially manufactured plates or tubes with small holes have been used as orifice meters. The pressure drop across the orifice can be measured with a manometer, magnehelic, or pressure gauge. Flow rates for an orifice meter can be calculated using Poiseuille's law; however, this is not done for practical use. Instead, the orifice meter is usually calibrated with either a wet or dry test meter or a soap-bubble meter. A typical calibration curve is shown in Figure 41.20. Calibration curves for orifice meters are nonlinear in the upper and lower flow rate regions and are usually linear in the middle flow rate region.

Laboratories with a minimum of equipment can make orifice meters. They are used in many sampling trains to control the flow. Care must be exercised to avoid plugging the orifice with particles. A filter placed upstream of the orifice can eliminate this problem. Orifice meters have long been used to measure and control flows from a few milliliters per minute to 50 L/min.

41.2.2.3 Orifice Meter: Secondary Standard. If the pressure drop across the orifice is increased until the downstream pressure is equal to approximately 0.53 times the upstream pressure (for air and some other gases), the velocity of the gas in the constriction will become acoustic, or sonic. Orifices used in this manner are called critical orifices. The constant 0.53 is purely a theoretical value and may vary. Any further decrease in the downstream pressure or increase in the upstream pressure will not affect the rate of flow. As long as the 0.53 relationship exists, the flow rate remains constant for a given upstream pressure and temperature, regardless of the value of the pressure drop. The probable error of an orifice meter is in the neighborhood of 2%.

Only one calibration point is needed for a critical orifice. The critical flow is usually measured with a soap-bubble meter or a wet or dry test meter. Corrections for temperature and pressure differences in calibration and use are made using the formula

$$Q_2 = Q_1 \left(\frac{P_1 T_2}{P_2 T_1} \right)^{1/2}$$ (41.5)

where Q = flow
    P = pressure
    T = temperature

and the subscripts 1 and 2 refer to initial and final condition, respectively. This formula can be used to correct measured orifice meter flows to standard conditions by substituting $P_2 = 760$ mm Hg and $T_2 = 298$ K. Note the square-root function of T and P. *Any time that rate meters are corrected for T and P, this square-root function is needed.* Critical orifices are used in the same types of situations as noncritical orifices. Care must also be taken not to plug the orifice.

41.2.2.4 *Venturi Meter: Secondary Standard.*

The Venturi meter consists of a short cylindrical inlet, a converging entrance cone, a short cylindrical throat, and a diffuser cone (Fig. 41.21). Two pressure taps, one in the cylindrical inlet and one in the throat, serve to measure the pressure drop. There is no abrupt change of cross section as with an orifice; thus the flow is guided both upstream and downstream, eliminating turbulence and reducing energy losses. The advantage of this meter is in its high energy recovery, allowing it to be used where only a small pressure head is available. Venturi meters are, of course, more difficult and expensive to fabricate. The probable error of a Venturi meter is 1%.

The Venturi meter is calibrated in the same manner as the orifice meter. The calibration curve is a plot of pressure drop across the venturi versus flow rates determined by the standard meter.

41.2.2.5 *Laminar Flowmeters.*

Laminar flowmeters are used only for very low flow rates. In these devices the flow goes through the device in thin parallel streams, dominated by wall friction instead of turbulence. Within the laminar

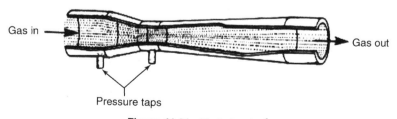

Figure 41.21   Venturi meter.[6]

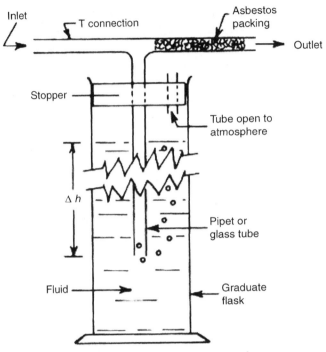

Figure 41.22   Laminar flowmeter.[1]

flow region, the pressure drop is directly proportional to the flow, instead of being proportional to the square root of the pressure as it is across an orifice or a cylinder head. One of the main disadvantages of this device is that the viscosity of the air changes as it flows through the device. A slight change in the temperature also drastically changes the flow readings. In addition, calibration is much more difficult and surface build-up can change the calibration. Hence, this device is not commonly used.

Figure 41.22 depicts a "home-made" laminar flowmeter. This consists of a T connection, pipet or glass tubing, cylinder to hold fluid, and packing material. The gas to be measured enters via the inlet arm of the T. The outlet arm of the T is covered with material such as asbestos. The leg of the T is attached to a tube or pipet projecting down into a stoppered cylinder filled with water or oil. The flow rate is directly proportional to the change in the liquid level (Δh) in the pipet or tube.

41.2.2.6   Pressure Transducers.   A transducer is a device for converting energy from one form to another for the purpose of measuring a physical quantity. Flow rate determination based on differential-head measuring devices requires a pressure sensor or transducer, sometimes referred to as the secondary element. Any type of pressure sensor can be used, with the three most common types being manometers, mechanical gauges, and electric transducers.

Properly aligned liquid-filled manometer tubes (liquid whose density is properly known) provide the most accurate measurement of differential pressure. In practice, it is not feasible to use liquid-filled manometers in the field. So, the pressure differentials are measured with mechanical gauges with scale ranges in centimeters or inches of water. The most commonly used gauge for low-pressure differentials is the Magnehelic. These gauges are accurate up to ±2% of full scale and are reliable when their connecting hoses do not leak, and their calibration is periodically rechecked.

41.2.2.7 *Bypass Flow Indicators.* The bypass flowmeter operates on the differential pressure method (Fig. 41.23). A ring with an orifice plate is installed in the pipeline. The orifice plate constricts the air flow, resulting in a pressure drop which is proportional to the square of the flow rate. A bypass pipe connecting the pressure taps upstream and downstream of the orifice will carry partial flow. This partial flow is proportional to the flow in the main pipe. Two ball valves in the bypass line can be used to turn the partial flow on or off.

Flow rate is highly dependent on the flow resistance in most high-volume samplers. Flowmeters which can operate with low flow resistance are usually expensive. For such samplers a bypass flowmeter or rotameter is the commonly used metering

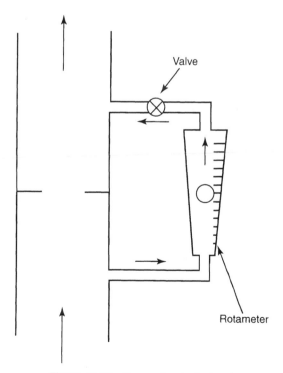

Figure 41.23 Bypass flow indicators.[1]

element. A bypass rotameter actually meters only a small fraction of the total flow. A bypass flowmeter consists of both a variable-head element and a variable-area element. The pressure drop across the fixed orifice or flow restrictor creates a proportionate flow through the parallel path containing the small rotameter. The scale on the rotameter usually reads directly in cubic feet per meter or liters per meter of total flow. Bypass flow indicators used in portable high-volume samplers have an adjustable bleed valve at the top. This valve should be set initially and periodically readjusted in laboratory calibrations so that the scale markings can indicate overall flow. If the rotameter tube accumulates dirt or the bleed valve adjustment drifts, the scale readings can depart greatly from the actual flows.

41.2.2.8   Variable-Area Meters. The variable-area meter differs from the fixed-orifice meter in that the pressure drop across it remains constant while the cross-sectional area of the constriction (annulus) changes with the rate of flow. Examples of this type of flow rate meters are the rotameter, orifice-and-plug meter, and cylinder-and-piston meter.

41.2.2.8.1   Rotameter: Secondary Standard.   The best known of the variable-area flow rate meter family is the rotameter. The rotameter consists of a vertically graduated glass tube, slightly tapered-in bore with the diameter decreasing from top to bottom, containing a float of the appropriate material and shape (Fig. 41.24). The

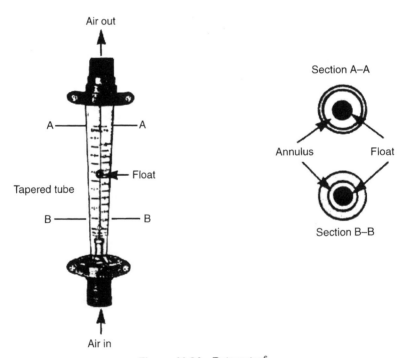

Figure 41.24   Rotameter.[6]

fluid to be measured passes upward through the conical tube, which is inserted in the flow circuit. A specially shaped float with a diameter slightly greater than the minimum bore of the conical tube is carried upward by the passage of the fluid until it reaches a position in the tube where its weight is balanced by the upward forces due to the fluid flowing past it. A variable ring or annulus is created between the outer diameter of the float and the inner wall of the tube. As the float moves upward in the tube, the area of the annulus increases. The float will continue to move upward until a pressure drop across the float, which is unique for each rotameter, is reached. This pressure drop across the float is constant regardless of the flow rate. A measure of the flow rate is noted by the float position on a vertical scale compared with a calibration chart.

The flow rate through a rotameter can be calculated from the tube diameters, float dimensions, float composition, and gas characteristics; this is not commonly done for calibration purposes. Manufacturers generally provide accurate calibration curves for rotameters. However, it is advisable to calibrate a rotameter under its operating conditions.

Most rotameters are used and calibrated at room temperature with the downstream side at atmospheric pressure. Corrections for pressure and temperature variations can be made using Equation (41.5).

If a gas being measured has a different density than the calibration gas, the flow rate can be corrected using the formula

$$Q_1 = Q_2 \left(\frac{\rho_2}{\rho_1}\right)^{1/2} \qquad (41.6)$$

where Q_1 = corrected flow rate of gas $(L^3 T^{-1})$
Q_2 = measured flow rate of gas $(L^3 T^{-1})$
ρ_1 = density of calibration gas (ML^{-3})
ρ_2 = density of gas being measured (ML^{-3})

Because corrections of this type are cumbersome and add inaccuracies, rotameters are usually calibrated under normal operating conditions against a primary or intermediate standard.

Rotameters are the most widely used laboratory and field method for measuring gas or liquid flow. Their ease of use makes them excellent for spot flow checks. Many atmospheric sampling instruments use rotameters to indicate the sample flow rate. With proper calibration, the rotameter's probable error is 2–5%.

41.2.2.8.2 Orifice-and-Plug Meters. Most "shielded" or "armored" rotameters operate under the orifice-and-tapered-plug principle. A tapered plug generally of a V shape, held in place by a vertical, center-mounted follower moves up and down in a sharp-edged annular orifice as flow changes. Flow is always in an upward direction, and the height of the plug in relation to the orifice is proportional to the flow rate. The follower is connected (usually by magnetic coupling) to a local indicator, which may be either mechanical or electronic or even a transmitter.

"Armored rotameters" can be used for steam and gases but most often are used for liquids (including corrosive liquids) under high pressure, which requires that they be made from specialty alloys, such as Hastelloy. These orifice-and-plug meters can be made from a variety of opaque plastics as well.

41.2.2.8.3   Cylinder-and-Piston Meters. The cylinder-and-piston meter is a commonly used variable-area meter. These are used for many in-plant fluid flow applications, where a flow rate indicator is the only acceptable solution. Piston-type variable-area meters can be operated on high-viscosity liquids, and even gases and compressed air at relatively high pressures and temperatures, provided the materials of construction are compatible.

The piston-type meter has an open centered piston with a tapered metering cone. This cone rides in a sharp-edged annular orifice and the piston is spring opposed. Velocity causes a pressure differential across the orifice, which moves the piston and the metering cone. An externally mounted magnetic-follower indicator tracks the movement of the metering cone. The use of the retention spring to hold the piston in the "zero" flow position when there is no velocity allows this type of meter to be installed inline in any orientation.

41.2.3   Velocity Meters

Velocity meters measure the linear velocity or some property that is proportional to the velocity of a gas. Several instruments exist for measuring the velocity of a gas. In this chapter only the Pitot tube and mass flowmeter are discussed.

Volumetric flow rate information can be obtained from velocity data, if the cross-sectional area of the duct through which flow takes place is known, using the formula

$$Q = A\bar{v} \tag{41.7}$$

where Q = volumetric flow rate (L^3/T)
   A = cross-sectional area (L^2)
   $\bar{v}$ = average velocity (L/T)

41.2.3.1   Pitot Tube. The Pitot tube is a simple pressure-sensing device used to measure the velocity of a fluid flowing in any closed conduit or an open channel. The complexity of the underlying fluid flow principles involved in Pitot tube gas velocity measurements is not apparent in the simple operation of this device. The Pitot tube should, however, be considered and treated as a sophisticated instrument. Unlike an orifice, which measures the full flow stream, the Pitot tube detects the flow velocity at only one point in the flow stream. This single point of measurement has the advantage of allowing a slender Pitot tube to be inserted into existing and pressurized pipelines without requiring a shutdown.

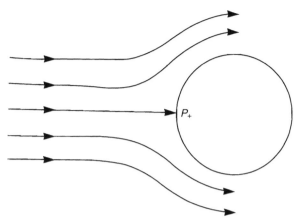

Figure 41.25 Gas stagnation against an object.[6]

The Pitot tube is used in a wide range of flow-measuring applications. While the accuracy is relatively low, the Pitot tube is easy and inexpensive to install, reliable, and suitable for use in a range of environmental conditions (wide pressure range and high temperatures). A Pitot tube measures two pressures: the impact or stagnation pressure (directly facing the oncoming flow stream) and the static or operating pressure (at right angles to the flow direction, preferable in a low-turbulence section of the conduit).

The Pitot tube opening facing the flow actually measures the impact pressure (sum of the static and kinetic pressures) of a gas stream. Gas streamlines approaching a round object placed in a duct flow around the object except at a point directly perpendicular to the flow where the gas stagnates in front of this object, and this stagnation pressure, P_+, can be measured as shown in Figures 41.25 and 41.26a.

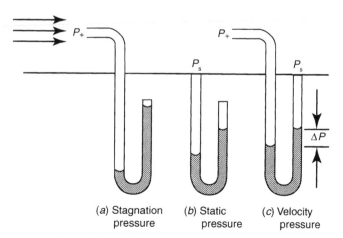

(a) Stagnation pressure   (b) Static pressure   (c) Velocity pressure

Figure 41.26 Pitot tube pressure components.[6]

The Pitot tube opening that is perpendicular to the flow direction measures the static pressure in a gas stream, which is defined as the pressure that would be indicated by a pressure gauge if it were moving along with the stream so as to be at rest, or relatively "static" with respect to the fluid. The static pressure can be measured as shown in Figure 41.26b. The difference between the stagnation pressure (P_+) and the static pressure (P_s) is termed the velocity pressure or differential pressure (Δp), which can be measured as shown in Figure 41.26c.

The differential pressure measured in a Pitot tube is proportional to the square of the velocity. Bernoulli's theorem relates Pitot tube velocity pressure (Δp) to gas velocity in the following equation:

$$v = K_p C_p \left(\frac{T \Delta p}{PM} \right)^{1/2} \tag{41.8}$$

where v = velocity of gas stream, ft/s

T = absolute temperature, R (°F+460)

P = absolute pressure, in. Hg

M = molecular weight of gas, lb/lb-mol

Δp = velocity pressure, in. H_2O

$K_p = 85.49 \text{ ft/s} \sqrt{\dfrac{\text{lb/lb-mol})(\text{in. Hg})}{(\text{in. } H_2O)(R)}}$

C_p = Pitot tube coefficient, dimensionless

As the flow velocity increases, the velocity profile in the pipe changes its shape from an elongated parabola (laminar flow) to one that is flat (turbulent flow). In order to obtain the flow rate using the Pitot tube gas velocity and the area of the conduit, it is critical that the insertion point in flow regimes that exhibits a varying velocity profile be placed at a point where the average velocity can be measured. Generally Pitot tubes are recommended for highly turbulent flows (Reynolds number Re > 20,000), where the velocity profile is flat enough that the location of the insertion point is not critical to flow rate determination.

For testing stacks using standard U.S. Environmental Protection Agency methods, traverses across the stack are needed to obtain the velocity profile. To overcome the problem of finding the average velocity point, a multiple-opening averaging Pitot tube can be used, which has multiple impact and static pressure points designed to extend across the diameter of the conduit. Such an arrangement, known as an annubar, allows all the readings to be combined to obtain an average flow velocity. Another option is to use area-averaging Pitot stations for measuring large, low-pressure air flows from boilers, dryers, and ventilation systems. Such stations come with port locations to fit standard sizes of circular or rectangular piping or ductwork.

The two most commonly used Pitot tubes are the standard and S-type Pitot tubes, which are described below.

41.2.3.2 Standard Pitot Tube: Primary Velocity Standard.

The standard Pitot tube consists of two concentric tubes. The center tube measures the stagnation or impact pressure, and the static pressure is measured by the holes located on the side of the outer tube. The Pitot tube must be placed in the flowing air stream so that it is parallel with the streamlines. The velocity pressure differential (Δp) can be measured with a U-tube manometer, an inclined manometer, or any suitable pressure-sensing device. Only velocities greater than 2500 ft/min can be measured with a U-tube manometer. Pitot tube velocity pressures are typically 0.14 in. H_2O at 1500 ft/min and 0.56 in. H_2O at 3000 ft/min.

The standard Pitot tube was first calibrated against an orifice meter using Bernoulli's theorem. Repeated calibration proved that different standard Pitot tubes have the same characteristic flow calibration. If the static pressure holes are six outer tube diameters from the hemispherical tip and eight outer tube diameters from the bend (Fig. 41.27), then the C_p value is computed using Equation (41.8).

Standard Pitot tubes can be used to measure linear velocity in almost any situation except in particle-laden gas streams. The particulate matter will foul the carefully machined tip and orifices. The velocity of gas streams with high particulate matter concentrations can be measured better with an S-type Pitot tube.

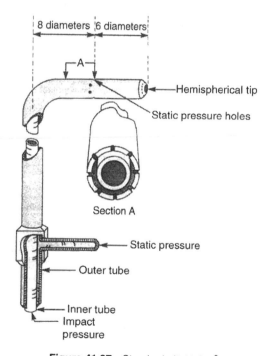

Figure 41.27 Standard pitot tube.[6]

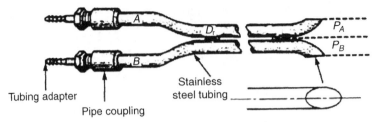

Figure 41.28 S-type pitot tube.[6]

41.2.3.3 S-Type Pitot Tube.

The S-type Pitot tube consists of two identical tubes mounted back to back. The sampling ends of the tubes are oval with the openings parallel to each other. In use, one oval opening should point directly upstream, the other directly downstream. The tubes should be made of stainless steel or quartz if they are used in high-temperature gas streams. The alignments of the tubes (Fig. 41.28) need to be checked before use or calibration, as misalignments may result in variations in the Pitot tube calibration coefficient (C_p).

Calibration of the S-type Pitot tube is performed by comparing it to a standard Pitot tube. Both the standard and S-type Pitot tubes are placed alternately into a constant air flow. Pressure readings are taken for the standard Pitot tube and for leg A of the S-type tube, facing the direction of flow, and leg B, facing the direction of flow. The Pitot coefficient (C_p) is calculated using the equation

$$C_{p(S)} = C_{p(STD)} \left(\frac{\Delta p_{STD}}{\Delta p_S} \right)^{1/2} \tag{41.9}$$

where $C_{p(S)}$ = S-type Pitot tube coefficient
$C_{p(STD)}$ = standard Pitot tube coefficient
Δp_S = S-type Pitot tube velocity pressure
Δp_{STD} = standard Pitot tube velocity pressure

The average $C_{p(S)}$ is calculated from several readings and should have a value of approximately 0.84. The C_p for leg A and then leg B should differ by less than 0.01. Equation (41.8) can then be used to calculate velocity.

The S-type Pitot tube maintains calibration in abusive environments. The large sensing orifices minimize the chance of plugging with particulates. The S-type Pitot tube also gives a high manometer reading for a given gas velocity pressure, which can be helpful at low gas velocities. These features make the S-type Pitot tube the most frequently used instrument to measure stack gas velocity.

41.2.3.4 *Mass Flowmeter: Secondary Velocity Standard.* Mass flow-meters work on the principle that when a gas passes over a heated surface, heat is transferred from this surface to the gas. This type of meter, similar in concept to the meter described earlier in the section on volumetric flowmeters, contains a heating element placed in a duct section. The amount of current required to keep the surface at a constant temperature is a measure of the velocity of the gas. The amount of heat transferred is proportional to the mass and velocity of the gas passing over the heating element. These meters can be used to measure both mass flow rate and velocity.

Atmospheric sampling applications of the mass flowmeter are usually limited to the measurement of volumetric flow rate. Since these devices measure mass flow rate directly, they should be calibrated against a primary, intermediate, or secondary volumetric standard. The standard flowmeter is corrected to standard conditions and compared to the mass flow rate measured. No corrections for temperature and pressure need to be made to the mass flowmeter reading. Calibration must be done with the same gas as will be measured in use, because different gases have different thermal properties.

Mass flowmeters are most often used for flow measurement or as calibration transfer devices in the field and laboratory. Their insensitivity to temperature and pressure makes them a useful tool for standard conditions measurement.

41.2.4 Other Devices

41.2.4.1 *Ultrasonic Flowmeters.* Ultrasonic flowmeters use ultrasonic waves ($>20,000$ cycles per second) at frequencies that are beyond what humans can hear to measure flow rate in closed pipes and open channels. Ultrasonic flowmeters contain transducers that send and receive these ultrasonic waves. The transducers are located outside the pipe or conduit and do not come in contact with the air/gas flow being measured. Ultrasonic waves travel more quickly when they travel with the flow stream than when they travel against it. Ultrasonic waves are sent across the flow stream, and the time it takes for the wave to cross from one side of the pipe to the other is measured. Using this information, the flowmeter can calculate flow rate.

Ultrasonic flowmeters are used to measure both liquid and gas flow. Ultrasonic meters can be connected to a centralized or mobile computer by telephone lines or by a radio link. Time saving and increased accuracy are obtained by replacing the mechanical gas meters with ultrasonic meters as these devices support the automated collection of information. The clamp-on designs are less expensive and are adaptable to any situation in which gas flow needs to be measured and/or analyzed, including domestic, industrial, and medical applications. Pipe fouling can degrade accuracy. With careful selection, taking into account the application and meter constraints, an ultrasonic meter can be more reliable than the standard mechanical equivalent and consume less power.

41.2.4.2 *Turbine Flowmeters.* In a turbine flowmeter, gas passing through it causes a rotor to spin. The rotational speed of the rotor is proportional to the

velocity of the gas. The volumetric flow rate is obtained by multiplying the velocity with the cross-sectional area of the turbine through which the gas flows. Turbine flowmeters provide excellent measurement accuracy for most clean liquids and gases. Turbine flowmeters are similar to positive-displacement meters as they have moving parts that are subject to wear and the pressure loss in the device is nonrecoverable.

41.2.4.3 *Vortex-Shedding Flowmeters.* This type of flow-measuring device is based on the principle that a bluff body (obstruction) placed in a flow stream creates vortices on the downstream side of the obstruction. The frequency of vortices shed from the obstruction is proportional to the velocity of the fluid. Volumetric flow rate is obtained as the product of velocity and cross-sectional area of flow. Temperature or pressure sensors are used for measuring the vortices created. Vortex-shedding meters require flow rates that are high enough to generate vortices. Vortex flowmeters can be used for fairly accurate measurement of flow rate with liquids, gases, or steam. They have no moving parts and tolerate fouling but are sensitive to pipeline noise.

41.2.4.4 *Coriolis Flowmeters.* The Coriolis flowmeter is based on the drift or deflection of flowing gases (to the right of the motion in the Northern Hemisphere and vice versa in the Southern Hemisphere) caused by the rotation of the earth. Coriolis force can be used to measure the mass flow rate of liquids, slurries, gases, or vapors. Fluid flowing through a vibrating flow tube causes a deflection of the flow tube proportional to mass flow rate. These meters provide excellent measurement accuracy. However, flow tube material needs to be carefully selected to minimize the effects of wall erosion and corrosion.

41.2.4.5 *Semiconductor Flowmeters.* In this type of flowmeter, a small silicon chip–based gas flow velocity sensor is used. The sensor consists of a very thin (30-μm) silicon chip placed at the end of an equally thin silicon beam. A plastic material, such as polyimide, serves as a mechanical support as well as a thermal insulation between the chip and the beam. The chip is electrically connected to the supporting beam by thin metal conductors on the beam and across the polyimide joint. The chip is heated with an integrated resistor and the temperature is measured with an integrated diode. The temperature measured by the chip is kept constant, at a fixed level above that of the gas temperature. The heating power required to maintain the temperature is a measure of the gas flow velocity. The small mass of the chip gives the sensor a fast thermal response.

This type of sensor, in a hand-held version, has found applications such as the measurement of air flow conditions in ventilation ducts. One model (Swema Flow 233) is a fast and accurate flow-measuring instrument for exhaust flows in ventilation systems from 2 to 65 L/s (7–233 m^3/h) and uses a net of hot wires. The net allows a wide opening, which minimizes the throttle effect of the flow. The net of hot wires covers the whole opening and gives an accurate mean value of the flow.

41.2.4.6 LFE (Laminar Flow Elements) Calibration Chain. A pressure drop is created as gas or water enters a laminar flow meter's inlet and is channeled into laminar flow elements (LFEs). These elements force the flow into thin parallel streams that are dominated by wall friction. Within this laminar flow region, the pressure drop is directly proportional to the flow velocity. The LFE is ideal as a transfer standard to obtain traceability from gravimetric determinations. The LFEs measure flow continuously so the gravimetric test does not have to be interrupted for resetting or recycling and they are very stable. However, calibration of many LFEs directly by gravimetric methods in a laboratory can be burdensome.

A calibration chain of LFEs has been created to be used as a direct transfer from gravimetric determinations to other transfer standards used to calibrate process instruments. The objective of the calibration chain is to verify the coherence of a large number of independent gravimetric calibrations on different-range LFEs and different gases. The LFE calibration chain concept is the accumulation of metrology research work. An LFE calibration chain that is maintained provides a unique resource that establishes direct traceability to mass and time with a high degree of confidence by using techniques that exploit the LFE's repeatability and stability.

41.3 CONCLUSION

This chapter has presented an overview of gas flow-measuring devices. These devices are classified into four groups (volume meters, rate meters, velocity meters, and other devices) based on the techniques used to measure the gas flow. Care should be taken in choosing a particular device for any application based on sample volume, sample mass, or sample concentration and desired accuracy.

REFERENCES

1. American Conference of Governmental Industrial Hygienists, *Air Sampling Instruments for Evaluation of Atmospheric Contaminants*, 6th ed., ACGIH, Cincinnati, OH, July 1983.

2. M. Bair, The Dissemination of Gravimetric Gas Flow Measurements through an LFE Calibration Chain, paper presented at the 1999 NCSL Workshop and Symposium, Charlotte NC, July 13, 1999.

3. R. L. Chapman and D. C. Sheesley, *Calibration in Air Monitoring*, ASTM Pub. 598, American Society for Testing and Materials, Philadelphia, PA, 1976.

4. H. N. Cotabish, P. W. McConnaughey, and H. C. Messer, Making Known Concentrations for Instrument Calibration, *Am. Ind. Hyg. Assoc. J.* **22**, 392–402 (1961).

5. L. Silverman, in P. L. Magill, F. R. Holden, and C. Ackley (Eds.), Experimental Test Methods, *Environmental Instrumentation and Analysis Handbook*, McGraw-Hill, New York, 1956, pp. 12:1–12:48.

6. U. S. Environmental Protection Agency (EPA), *Air Pollution Instrumentation*, Office of Air, Noise and Radiation, Office of Air Quality Planning and Standards, EPA, Research Triangle Park, NC, June 1983.

ELECTRONIC REFERENCES

http://www.age.psu.edu/extension/factsheets/g/G81.pdf

http://www.superflow.com/support/support-flowbench-laminar-flowmeters.htm

http://dougall3.tip.csiro.au/IMP/SmartMeasure/FlowMeters/FlowMeters1.htm

http://www.flowmeterdirectory.com/flowmeter_artc/flowmeter_artc_02032401.html

http://www.manufacturingtalk.com/news/ifl/ifl100.html

http://www.flowmeterdirectory.com/flowmeter_orifice_plate.html

http://www.swema.com/Instrument/Luftflode/SWFlow233_eng.htm

http://www.dhinstruments.com/prod1/pdfs/papers/mf/tpmf9907.pdf

http://www.test-and-measurement-instrumentation-hire.com/flow_measuring_equipment.htm

http://www.test-and-measurement-instrumentation-hire.com/air_velocity_indicators.htm

42

NON-OPEN-CHANNEL
FLOW MEASUREMENT

Randy D. Down, P.E.

Forensic Analysis & Engineering Corp.
Raleigh, North Carolina

42.1 INTRODUCTION

Measurement of liquid flow through a pipe is commonly found on numerous environmentalapplications. This chapter will examine the methods used to obtain an accurate and reliable flow measurement for a variety of environmental applications.

Environmental Instrumentation and Analysis Handbook, by Randy D. Down and Jay H. Lehr
ISBN 0-471-46354-X   Copyright © 2005 John Wiley & Sons, Inc.

42.2 DIFFERENTIAL PRESSURE FLOWMETERS

Differential pressure flowmeters work on the premise that a pressure drop across the meter is proportional to the square of the flow rate. The flow rate is obtained by measuring the pressure differential and then extracting the square root. Differential pressure flowmeters have a primary element that causes a change in kinetic energy (differential pressure in the pipe) and a secondary element that measures the differential pressure and provides an output signal that is converted to a flow value.

Velocity depends on the pressure differential that is forcing the liquid through a pipe or conduit. Since the pipe cross-sectional area is a known constant, by determining the average velocity, we have an indication of the flow rate. The basic relationship for determining the liquid's flow rate in such cases is

$$Q = V \times A$$

where Q = liquid flow through pipe
   V = average velocity of flow
   A = cross-sectional area of pipe

Therefore, to convert the measured velocity of the liquid to a volumetric flow rate (in gallons per minute),

$$GPM = 2.45 \times (\text{pipe inside diameter in inches})^2 \times \text{velocity in feet per second}$$

Additional factors that affect liquid flow rate include liquid viscosity and density and the friction of the liquid in contact with the inner wall of the pipe. Direct measurements of liquid flows can be made using positive-displacement flowmeters.

The Reynolds number (Re) is a dimensionless ratio of inertial forces to viscous forces in a liquid. If the Reynolds number is high (>2000), inertial forces dominate and turbulent flow exists. If it is low (<2000), then viscous forces prevail and a laminar flow condition is observed.

Flow rate and specific gravity are considered *inertial forces*, while pipe diameter and viscosity are considered *drag forces*. A pipe's diameter and the specific gravity of the liquid generally remain constant for most applications. At very low velocities, Re is low and the liquid flows in smooth layers, with the highest velocity at the center of the pipe and lower velocities at the pipe wall, where the viscous forces restrain it. This type of flow is called a laminar flow condition. Reynolds number values at laminar flow are below approximately 2000. A characteristic of laminar flow is the parabolic shape of its velocity profile.

Most environmental flow applications involve turbulent flow, with Re values that are above 2000. Turbulent flow occurs at high velocities. Under a turbulent flow condition, flow breaks up into turbulent eddies that flow through the pipe with the same average velocity. Fluid velocity is less significant, and the velocity profile is more uniform in shape. A transition zone exists between turbulent and laminar flows. Depending on the piping configuration and other installation conditions, the flow may be in either a turbulent or laminar flow condition within this zone.

The use of differential pressure as an indirect measurement of a liquid's rate of flow is very common. Differential pressure flowmeters are, by far, the most common type of liquid flowmeter in environmental applications.

Differential flowmeters typically used in environmental applications include orifice plates, Venturis, Pitot tubes, and target meters.

42.3 ORIFICE PLATES

A differential pressure element, the orifice plate consists of a flat plate with an opening inserted into a pipe so that it is perpendicular to flow. As the flowing fluid passes through the orifice plate, the restricted cross-sectional area causes an increase in velocity and a decrease in pressure. The pressure difference before and after the orifice plate is used to calculate the flow velocity.

Bernoulli's equation allows us to correlate an increase in velocity head with a decrease in pressure head:

$$v_b = \frac{1}{\sqrt{1 - \beta^4}} \sqrt{\frac{2(p_a - p_b)}{\rho}} \quad \text{where } \beta = \frac{D_b}{D_a} = \left(\frac{A_b}{A_a}\right)^{0.5}$$

An orifice plate is the simplest flow path restriction used in flow detection as well as the most economical. Orifice plates are flat plates $\frac{1}{16} - \frac{1}{4}$ in. thick that are normally mounted between a pair of flanges and installed in a straight run of smooth pipe to avoid disturbance of flow patterns from fittings and valves (Fig. 42.1).

Flow through a sharp-edged orifice plate is characterized by a change in velocity. As the fluid passes through the orifice, the fluid converges and the velocity of the fluid increases to a maximum value. At this point, the pressure is at a minimum value. As the fluid diverges to fill the entire pipe area, the velocity decreases back to the original value. The pressure increases to about 60–80% of the original

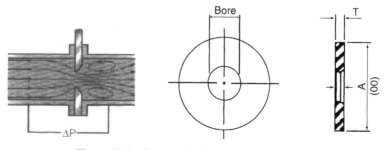

Figure 42.1 Sharp-edged concentric orifice plate.

Concentric orifice   Eccentric orifice   Segmental orifice

Figure 42.2 Standard orifice plate configurations.

input value. The pressure loss is irrecoverable; therefore, the output pressure will always be less than the input pressure. The pressures on both sides of the orifice are measured, resulting in a differential pressure which is proportional to the flow rate.

Three kinds of orifice plates are used: concentric, eccentric, and segmental (as shown in Fig. 42.2). The concentric orifice plate is the most common. As shown, the concentric orifice is equidistant (concentric) with the inside wall of the pipe.

42.4 VENTURIS

A section of tube forms a relatively long passage with a smooth entry and exit. A Venturi tube is connected to the existing pipe, first narrowing in diameter then opening back up to the original pipe diameter. The changes in cross-sectional area cause changes in the velocity and pressure of the flow.

A classical Venturi tube (Fig. 42.3) consists of (1) a straight inlet section of the same diameter as the pipe and in which the high-pressure tap is located; (2) a converging conical inlet section in which the cross section of the stream decreases and the velocity increases in conjunction with an increase of velocity head and decrease of pressure head; (3) a cylindrical throat section containing the low-pressure tap in which the flow velocity is neither increasing nor decreasing; and (4) a diverging

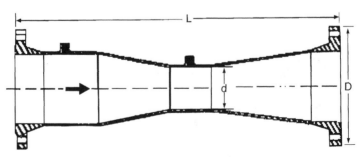

Figure 42.3 Venturi tube.

recovery cone in which velocity decreases and the decreased velocity head is recovered as pressure head. A calibrated accuracy of $\pm 0.25\%$ can be obtained using a Venturi.

42.5 PITOT TUBES

The outer impact tube has a number of pressure-sensing holes facing upstream that are positioned at equal annular points in accordance with a log-linear distribution.

The "total pressures" developed at each upstream hole by the impact of the flowing medium is averaged within the outer impact tube and then more accurately averaged within the internal averaging tube. This averaged pressure is represented at the head as the high-pressure component of the differential pressure.

42.6 VORTEX-SHEDDING FLOWMETERS

A vortex-shedding flowmeter (Fig. 42.4) operates on the following principle: When a blunt object is placed in the middle of a liquid flow stream, the flow stream must separate to move past the stationary object. In doing so, it rolls up into a vortex (or swirl) that forms around the side of the object, which is referred to as a bluff body or shedder bar. As the volume of the vortex builds, it eventually will reach a point where it is too large to remain there. The vortex then sheds and continues

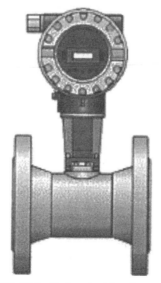

Figure 42.4 Vortex-shedding flowmeter.

downstream. At the point of shedding, a momentary low-pressure/high-velocity condition occurs on one side of the shedder bar, with the reverse taking place on the other side of the bar. The next swirl forms on the opposite side of the shedder bar, repeating this process. Each vortex swirl contains the same volume of material. The frequency of the vortex-shedding process is proportional to the velocity of the flow through the meter.

Although many other sensing techniques have been used over the years, in most designs the sensor portion of the flowmeter detects a differential pressure across the shedder bar, rather than measuring the frequency of the vortices. The flowmeter uses a piezoelectric crystal sensor to detect the pressure exerted by the vortices. The piezoelectric crystal converts the vortex-shedding frequency into an electrical signal. The vortex-shedding flowmeter's output signal may be analog (4–20 mA DC), a pulse, or digital.

42.7 TURBINE FLOWMETERS

Invented by Reinhard Woltman in the eighteenth century, the turbine flowmeter is an accurate and reliable flowmeter for both liquids and gases. It consists of a multi-bladed rotor mounted at right angles to the flow and suspended in the fluid stream on a free-running bearing. The diameter of the rotor is very slightly less than the inside diameter of the metering chamber, and its speed of rotation is proportional to the volumetric flow rate. Turbine rotation can be detected by solid-state devices (reluctance, inductance, capacitive, and Hall effect pickups) or by mechanical sensors (gear or magnetic drives).

Turbine meters should be selected so that the expected average flow is between 60 and 75% of the meter's maximum-flow capacity. Flow velocities under 1 ft/s can be insufficient, while velocities in excess of 10 ft/s can result in excessive wear on the turbine. Most turbine meters are designed for a maximum velocity of 30 ft/s.

Turbine flowmeters should also be selected so that 3–5 psid pressure drop at a maximum-flow condition can be obtained. Because pressure drop increases with the square of the flow rate, reducing the meter to the next smaller size will increase the pressure drop.

Density changes do not significantly affect turbine meters. On low-density fluids (specific gravity <0.7), the minimum flow rate is higher due to a reduced torque.

Turbine meters are sensitive to upstream piping objects that can cause vortices and turbulence. Specifications typically call for 10–15 pipe diameters of straight pipe run upstream and 5 pipe diameters of straight run downstream of the meter.

To reduce the straight-run requirement, flow straighters can often be installed. Flow straighters consist of tube bundles or radial vane elements within the pipe that are installed at least 5 diameters upstream of the meter.

Under certain conditions, the pressure drop across the turbine can cause flashing or cavitation. Flashing causes the meter to read high, and cavitation can result in turbine damage. To avoid this, the downstream pressure must be held at a value equaling 1.25 times the vapor pressure plus twice the pressure drop across the meter. Small amounts of air entrainment will tend to make the meter read slightly high, while large quantities of air entrainment can destroy the turbine.

Solids entrained in the fluid can also damage turbine meters. If the amount of suspended solids exceeds 100 mg/L of +75 μm size, a flushing y-strainer or a motorized cartridge filter must be installed at least 20 diameters of straight run upstream of the flowmeter.

When a flowmeter is operating too near its vapor pressure, bubbles can form and collapse. This physical phenomenon is called cavitation, and it can cause serious damage to piping and equipment, including turbine meter rotor blades.

When liquid passes the turbine's rotor blades, its velocity increases and creates a low-pressure area on the backside of the rotor blades. When this low-pressure area is less than the liquid's vapor point, bubbles form. As the bubbles move downstream, pressure recovery allows the bubbles to collapse (return to a liquid state). Collapsing of the vapor bubbles releases energy and produces a noise similar to what would be expected if gravel were flowing through the pipe. When the energy is released in close proximity to solid surfaces, small amounts of material can be torn away, leaving a rough, cinderlike surface.

Besides the wear and tear, cavitation usually causes the rotor to spin faster than it should for the liquid-flowing conditions, thus producing an inaccurate flow measurement signal.

Specific gravity is another factor affecting differential pressure across rotor blades. As the specific gravity of the liquid decreases, the pressure differential also decreases. Fluids with very low specific gravity and a low flow rate produce very low differential pressure across the rotor blades, leaving very little energy to turn the rotor.

Regardless of a fluid's ability to lubricate turbine flowmeter bearings, manufacturers use varying techniques to minimize bearing wear. For example, ball-bearing units contain nonmetallic bearing retainers, and the sheer hardness of tungsten carbide and ceramic bearings tends to extend bearing life.

Bearing life is approximately inversely proportional to the square of bearing speed. Therefore, to prolong turbine flowmeter bearing life, it is best to operate the flowmeter at lower flow rates. For example, if a flowmeter is operated at 33% of its maximum flow rate, bearing life will be extended by a factor of 10.

For turbine flowmeter ball bearings, it is recommended the bearings be inspected every 6 months.

Like most flowmeters, the accuracy of turbine flowmeters is highly dependent on the ability of the installation to ensure nonswirling conditions.

Even at constant flow rates, swirl can change the angle of attack between the fluid and the rotor blades, causing varying rotor speeds and thus varying flow rate indications.

Effects of swirl can be reduced or eliminated by ensuring sufficient lengths of straight pipe, a combination of straight pipe and straightening vanes, or specialized devices such as a Vortab's flow conditioner are installed upstream and downstream of the turbine flowmeter. (See "Installation rules of thumb" table.) Continued on p.43

Turbine flowmeters for liquid applications perform equally well in horizontal and vertical orientations, while gas applications require horizontal flowmeter orientation to achieve accurate performance.

When installing turbine flowmeters in intermittent liquid applications, it is recommended that the flowmeter be mounted at a low point in the piping.

Turbine flowmeters are designed for use in clean-fluid applications. Where solids may be present, installation of a strainer/filter is recommended. Also, because the strainer/filter can introduce swirl, it needs to be located beyond the recommended upstream straight pipe lengths.

42.8 MAGNETIC FLOWMETERS (MAGMETERS)

Magnetic flowmeters (Fig. 42.5) operate using Faraday's law. Shaped coils are used to set up a magnetic field at right angles to the direction of flow. The passage of conductive fluid through the magnetic field induces a voltage on electrodes set perpendicular to the field.

Most magnetic flowmeters use a switched DC (direct-current) magnetic field that approximates a square wave. It requires power to generate this field, sometimes as much as 20 W. Typically, AC (alternating-current) line voltage is brought to the "converter" or transmitter and the converter produces the switched DC field to the coils. It then receives and amplifies the small voltage induced on the electrodes. The converter typically produces either a pulse or current (4–20 mA DC) output or both. In some cases, the converter is also a Fieldbus transmitter.

Magnetic flowmeters are widely used because of their inherent accuracy and reliability. They are inherently accurate because the output of the flowmeter closely

Figure 42.5 Magnetic flowmeter (magmeter).

approximates the average velocity in the entire pipeline and the flowmeter is resistant to many flow perturbations. While other flowmeters often require a minimum of 10 diameters of straight run upstream and 5 downstream, many magnetic flowmeters can operate accurately with less than 5 diameters upstream of the electrodes and 3 downstream. Since the electrodes are near the device centerline, this can result in virtually no straight-run requirements in many applications involving pipe sizes. Magnetic flowmeters have replaced many orifice plates and turbine meters in conductive fluid applications because they are nonintrusive and have very low wear even in abrasive and corrosive services.

A major problem with magnetic flowmeters is that a significant amount of electrical energy is required to excite the coils while two-wire transmitters are limited to the power than can be obtained from an instrument loop that has limited current (4-20 mA dC) and voltage (typically 24 V DC). In some installations, higher voltages can be used, but the current is still limited to between 4 and 20 mA DC. Yet, to perform accurately, magnetic flowmeters use approximately the same amount of power regardless of flow rate. This is because the power used to operate the magnetic field is approximately the same at zero flow and at maximum flow.

42.9 FLOW STRAIGHTENERS

Flow straighteners (Fig. 42.6) and conditioners are used in conjunction with flowmeters to smooth turbulent and transitional flows and help meters measure flow more accurately.

42.10 NONINTRUSIVE FLOWMETERS

A unique flow-monitoring device utilizes the frequency shift (Doppler effect) of an ultrasonic signal to measure the flow velocities of liquids (Fig. 42.7). A generating crystal in the transducer transmits a high-frequency signal through the pipe wall into the liquid. This signal is reflected by suspended particles or gas bubbles in the moving liquid and is then detected by a receiving crystal located in the sensor.

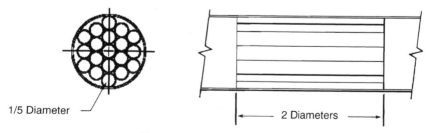

1/5 Diameter

2 Diameters

Figure 42.6 Flow straighteners.

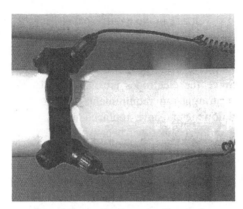

Figure 42.7   Nonintrusive flowmeter.

The difference between transmitted and detected frequencies is directly proportional to the liquid's flow velocity.

A Doppler flowmeter can be used to measure the flow rate of virtually any type of fluid, regardless of its chemical or physical properties. Sensors do not come in contact with the fluid, making this flowmeter ideal for measuring flow velocities of sewage waste or sludge that will clog in-line sensors.

These flowmeters can be used to measure the flow within pipe sizes ranging from 1 to 120 in. in diameter. The sensors are strapped onto the pipe, then flow velocity is measured in feet per second. By entering the pipes' inside diameter, the flow rate can be calculated in gallons per minute or MGD (million gallons per day) and totaled in gallons. In clean-liquid application, the sensors must be attached to the pipe 1–3 pipe diameters downstream from a 90° elbow to reflect the Doppler signal off of entrained bubbles induced by turbulence at the pipe elbow.

42.11   FLOWMETER SELECTION

To aid in the selection process, Table 42.1 provides a cursory, comparative description of the various types of flowmeters commonly encountered in environmental applications. This table is intended not to be used alone but rather as an aid in narrowing the list of applicable flow measurement instruments that should be considered in the selection process.

Other considerations that need to be considered when selecting an appropriate flowmeter for your application include:

- Viscosity of the media being measured (if something other than water)
- Ambient conditions where the flowmeter will be installed:
  (a) Temperature
  (b) Humidity

Table 42.1 Flowmeter Selection

Flowmeter Type	Recommended Service	Rangeability	Pressure Loss	Accuracy (%)	Required Upstream Pipe Diameters	Relative Cost
Orifice	Clean, dirty liquids and slurries	3 : 1	Medium	±2 to ±4 of full scale	10–30	Low
Venturi	Clean, dirty liquids and slurries	4 : 1	Low	±1 of full scale	5–20	Medium
Pitot tube	Clean liquids	3 : 1	Very low	±3 to ±5 of full scale	20–30	Low
Target meter	Clean, dirty liquids and slurries	10 : 1	Medium	±1 to ±5 of full scale	10–30	Medium
Turbine	Clean, viscous liquids	20 : 1	High	±0.25 of rate	5–10	High
Vortex	Clean or dirty liquids	15 : 1	Medium	±1 of rate	10–20	High
Magmeter	Clean or dirty liquids and slurries	40 : 1	None	±0.5 of rate	5	High
Nonintrusive	Dirty liquids and slurries	12 : 1	None	±5 of full scale	5–30	High

(c) Corrosivity

(d) Vibration

(e) Pipe Stress

- Position/location in which flowmeter must be installed
- Output signal requirements (4–20 mA DC, 1–10 V DC, pulse output)
- Power source (Is it loop powered or does it require its own power supply?)
- Temperature of measured liquid (Does it require a remote head to isolate the electronics from extreme temperatures in excess of 140° Fahrenheit?)
- Whether a flow straightener is required
- Whether it has rangeability to accurately measure flow rate throughout the entire design range of the application
- Type of service, maintenance, and calibration requirements required for proper performance

42.12 APPLICATIONS

As stated at the beginning of this chapter, there are numerous applications that require the use of an accurate and reliable flowmeter. The following are examples of commonly encountered environmental flowmeter applications:

Wastewater Flow. Flowmeters are used to monitor and record (using a data logger or chart recorder) to meet regulatory requirements (record keeping).

Makeup Water. To monitor water consumption by measuring the amount of makeup water supplied to spray towers in a scrubber system.

Bypass Water. To measure the rate of water passing through a side-stream filter system.

Fill Control. To measure the rate of wastewater supplied to a tank for temporary storage or pretreatment.

Chemical Treatment. To monitor the water or wastewater flow rate as a feed-forward signal to a controller that injects an acid, base, or biocide for chemical treatment or neutralization downstream.

Totalization. Multiple flowmeters are installed in wastewater pipelines or drains to monitor their individual flow rates. The output of these multiple meters are then added together (totaled) using a processor or signal conditioner to obtain a net wastewater discharge to storage tanks or some other collection system.

42.13 GLOSSARY

Accuracy Range The flow range over which a meter can operate and still maintain its accuracy within an acceptable tolerance.

Density	The mass per unit volume of a fluid.
Differential Pressure	The difference between a reference pressure and a variable pressure.
Digital Signal	A series of electronic pulses being sent out through the wires attached to the flowmeter. The pulses correspond to the velocity of the liquid passing through the meter.
Flowmeter	A device installed in a pipe or conduit to measure the fluid flow rate. Flowmeters can be used to measure the flow rates of liquids or gases and are available in various configurations and with differing operating principles.
GPM	Gallons per minute.
Pitot Tube	A sensing device used to measure total pressures in a fluid stream.
Rangeability	The ratio of the flowmeter maximum operating capacity to its minimum operating capacity within the meter's specified tolerance.
Repeatability	The ability of a meter to deliver a consistent reading or output at any given flow rate within close tolerances.
Reynolds Number	A relative number that represents a combination of factors that influence the flow behavior in a pipe, including the pipe diameter, and density, velocity and viscosity of the liquid.
Traceability	the "chain" of instruments used to calibrate other items, provided through certification. Calibration certificates from NIST facilities are said to have direct traceability to national standards.
Velocity Profile	As a fluid moves through a pipe, friction slows down the fluid near the inner wall. That means the flow near the wall is moving slower than the flow nearer the center of the pipe.
Viscosity	A relative measure of the extent that a fluid resists flowing.

43

OPEN-CHANNEL WASTEWATER FLOW MEASUREMENT TECHNIQUES

Bob Davis and Jim McCrone

American Sigma
Loveland, Colorado

Environmental Instrumentation and Analysis Handbook, by Randy D. Down and Jay H. Lehr
ISBN 0-471-46354-X   Copyright © 2005 John Wiley & Sons, Inc.

43.1   INTRODUCTION

Flow measurements are a critical element of water quality monitoring for diverse applications. This chapter focuses on flow monitoring in open channels and provides specific recommendations for achieving accurate measurements in general as well as for specific situations.

43.1.1   Open Channel

The two types of flow that are monitored are open channel (gravity flow) and closed pipe (pressurized flow in completely close conduits). For the most accurate measuring device performance, both types require approach conditions free of obstructions and abrupt changes in size and direction; these can produce velocity profile distortions that can lead to inaccuracies.

Open-channel flow, in which the liquid flows by gravity, includes

- Rivers
- Irrigation ditches
- Canals
- Flumes
- Other uncovered conduits
- Partially closed channels (such as sewers and tunnels) when flowing partially full and not under pressure

Open channels may be frequently found in sewage treatment plants, most storm and sanitary sewer systems, many industrial waste applications, some water treatment plants, and most irrigation water.

43.1.2   Wastewater

Wastewater is the spent water of a community. From the standpoint of source, it may be a combination of liquid and water-carried wastes from residences, commercial buildings, industrial plants, and institutions together with any groundwater, surface water, and storm water that may be present.

43.1.3   Flow Measurement

A variety of techniques are available for measuring flow. This chapter focuses on the most prevalent methods, with particular emphasis on techniques employing modern flow monitoring equipment, since these generally provide the greatest ease of use, overall cost effectiveness, and accuracy. In essence, flow is typically measured by using a combination of primary devices (weirs, flumes, nozzles) and secondary devices (flowmeters).

43.2 PRIMARY DEVICES

A primary measuring device is a man-made, calibrated, hydraulic structure with an engineered restriction, placed in an open channel, with a known depth-to-flow (or level-to-flow) relationship. The structure creates a geometric relationship between the depth and the rate of the flow. The flow rate over or through the restriction can be related to the liquid level so that the flow rate through the open channel can be derived from a single measurement of that level.

The two basic types of primary measuring devices are weirs and flumes.

43.2.1 Weirs

Weirs are damlike structures over which the liquid flows. They are easier and less expensive to install than flumes. Weirs are normally classified according to the shape of the measurement point, the most common types being:

- V-notch (triangular)
- Rectangular
- Trapezoidal (Cipolletti)

The rate of flow over a weir is directly related to the hydraulic head, the vertical distance from the crest of the weir to the water surface (see Fig. 43.1). To measure the hydraulic head, a measuring device is placed upstream of the weir at a distance of at least four times the head. Accuracy of a weir can be affected by varying approach velocities.

43.2.2 Flumes

Flumes are artificial structures that force water through a narrow channel. These specially shaped open-channel flow structures increase the velocity through the

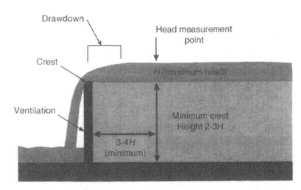

Figure 43.1 Side view of weir.

converging upstream section and change the level of the liquid flowing through the throat before decreasing the level in the diverging section to create a known level-to-flow relationship. The wastestream's flow is proportional to the depth of water in the flume and is calculated by measuring the head upstream from the throat. The increased velocity of flow through a flume is an advantage as it creates a self-cleaning effect.

The most common types of flumes are the following:

Parshall. Typically used in permanent installations, commonly in concrete-lined channels, these flumes are accurate for a relatively wide range of flow rates. The size of the flume is determined by the throat width. The recommended depth measurement point is upstream from the throat at two-thirds the distance of the converging section in free-flow conditions (Fig. 43.2). Parshall flumes are typically more expensive to install than weirs or Palmer-Bowlus flumes.

Palmer-Bowlus. For installation in existing circular channels, these flumes measure flow rates accurately only over a narrow flow range. The flume is sized to the existing pipe diameter, so an 8-in. Palmer-Bowlus flume is used in an 8-in. pipe. The recommended depth measurement point is one-half the pipe diameter upstream from the entrance of the flume (Fig. 43.3). These flumes are easier and less costly to install than Parshall flumes. Palmer-Bowlus flumes can vary by manufacturer, so users should always be familiar with the individual manufacturer's rating for the flume.

Other types of flumes include Leopold-Lagco and H-flumes.

For a concise examination of the various types of primary devices and their uses, consult Ref. 1, pages 42–52.

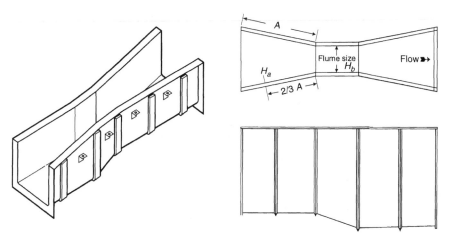

Figure 43.2   Parshall flume.

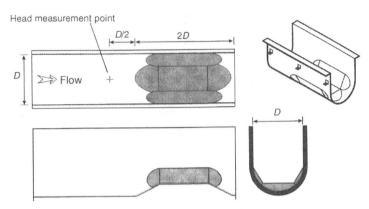

Figure 43.3 Palmer-bowlus flume.

43.3 SECONDARY DEVICES

A secondary device is a flowmeter that measures the liquid level of a primary device and converts it to an accurate flow rate (level-to-flow conversion) (Fig. 43.4). To perform this function, modern flowmeters use computer-generated tables or online devices that perform instantaneous flow rate calculations based on established mathematical relationships.

Figure 43.4 Open-channel flowmeter.

The secondary device transmits its output to recorders and/or totalizers to provide flow data (historical and instantaneous). Output may also be transmitted to sampling systems to facilitate flow proportioning.

Flowmeters frequently activate other instruments, such as samplers, in proportion to the flow rate.

Flow data may be collected on an instantaneous or continuous basis. The latter requires automatic flow measurement technology.

Basic automatic flow measurement equipment consists of:

- A level sensor
- A control and conversion device
- A recorder
- A totalizer

More advanced equipment may also include other capabilities and specifications—such as parameter measurements—that extend their ability, enhance their accuracy, and increase their versatility for use under a variety of applications. For example, a system that is compatible with laptop computers simplifies downloading of stored data from remote field sites. Some of these features are discussed below in Section 43.13 on choosing equipment, with specific capabilities recommended when appropriate for measuring flow at particular sites.

43.4 LEVEL-SENSING TECHNOLOGIES

A variety of devices are used by open-channel flowmeters to measure the liquid levels at appropriate points in or near primary devices. The most popular are:

- Ultrasonic transducers
- Submerged pressure transducers
- Bubblers

See Table 43.1 for a summary of these technologies. Other devices are:

- Floats
- Conductive sensors

43.4.1 Ultrasonic Transducers

PRINCIPLE OF OPERATION. The ultrasonic sensor is positioned above the channel. A series of acoustic pulses are transmitted to the liquid surface, then reflected back to a receiver. The liquid level is measured by determining the time required for

Table 43.1   Variety of Devices Used by Open-Channel Flowmeters

Level Sensing Technology	Advantages	Disadvantages	Recommended Applications
Ultrasonic	• Noncontact • Installable in manhole without entry • Sensor located remotely • Measures to bottom of channel • Level readings not affected by velocity	• Not preferable in windy or turbulent applications • Requires special treatment in narrow, deep channels • Expensive sensor subject to vandalism	Use in medium and larger pipes or in presence of silt, suspended solids, or suspended grease and where chemical compatibility is a concern.
Submerged Sensor	• Available in various pressure ranges and levels of accuracy	• Sensor may collect debris • Readings may be affected by velocity • Potential sensor damage if frozen in water • Remote sensor subject to vandalism • Must be under more than $\frac{1}{2}$ in. level to begin reading	Use where floating oil, grease, foam, steam, silt, solids, or turbulence is present.
Bubbler	• Able to measure down to $\frac{1}{4}$ in. of pipe invert • Inexpensive vinyl tubing • Less prone to vandalism	• Tube in contact with stream may collect debris • May be affected by velocity • Frozen water affects reading	Use in low or intermittent flow or in presence of floating oil, grease, foam, steam, surface turbulence, and/or excessive wind.

the pulse to travel from a transmitter to the liquid surface and back to a receiver; the time of travel for each pulse is directly proportional to the distance, from which the liquid level is calculated. The flowmeter converts the level reading to a flow rate based on the level-to-flow relationship of the primary device or channel configuration. Temperature sensors are important to compensate for variations in the speed of the acoustic pulses.

SUGGESTED APPLICATIONS. Ultrasonic, the most popular level-sensing technology, is ideal where heavy solids are present in the wastestream; for use in medium and large pipes or in the presence of silt, suspended solids, or suspended grease; for permanent applications; for process control within waste treatment plants; and where chemical compatibility is a concern.

ADVANTAGES. The noncontacting sensor provides accurate, open-channel flow measurements without contacting the liquid as well as being low maintenance, easy to mount, not affected by chemicals, not affected by liquid velocity, and able to measure low flows to the bottom of the pipe invert. The low power consumption is desirable in portable, battery-powered applications.

CONCERNS. Wind can cause loss of echo reception. Flow-stream turbulence can cause inconsistent echoes; steam or foam above the flow stream can give false readings. Air temperature gradients can give inaccurate measurement.

NOTE ON ACCURACY. There are four parameters critical to ultrasonic performance: pulse strength, pulse width, detector sensitivity, and frequency. In modern flowmeters, these key operating parameters are microprocessor controlled to optimize operation; advanced technology, for example, can automatically compensate should the echo signal strength deteriorate.

43.4.2 Submerged Pressure Transducers

PRINCIPLE OF OPERATION. The submerged sensor uses a differential pressure transducer positioned directly in the flow stream to measure liquid level. The pressure over the transducer is proportional to the liquid level and changes when the liquid level changes. The flowmeter converts the level reading to a flow rate based on the level-to-flow relationship of the primary device or channel configuration.

SUGGESTED APPLICATIONS. It is excellent for foaming or turbulent liquid surfaces and for use where floating oil, grease, foam, steam, silt, sand, solids, or turbulence is present. It is good choice for natural channels where silt and suspended solids are present.

ADVANTAGES. It is quickly installed. Available low-profile sensors present only a minimal obstruction to the flow and are accurate even when covered with silt and sand. It is not affected by wind and air temperature gradients. Low power consumption makes it suitable for portable use.

CONCERNS. Flow-stream chemicals may be a concern, and periodic cleaning may be required. Accuracy can be affected by changes in flow-stream temperature.

43.4.3 Bubblers

PRINCIPLE OF OPERATION. Tubing is positioned in the flow at a known depth. The integral air pump pushes a small volume of pressurized air or other gas through the tubing to the measurement point. The pressure in the tubing changes in proportion to liquid level. (As level rises, the pressure needed to maintain air-bubble flow increases.) The flowmeter converts the level reading to flow based on the level-to-flow relationship of the primary device or channel configuration.

SUGGESTED APPLICATIONS. Use in low or intermittent flows or in the presence of floating oil, grease, foam, steam, surface turbulence, and/or excessive wind. It is not affected by air temperature gradients; therefore it is useful in dry applications such as storm sewers.

ADVANTAGES. Due to easy setup and installation, it is ideal for temporary flow applications such as infiltration and inflow (I&I) studies or publicly owned treatment works (POTW) monitoring of industry for permit compliance. Only inexpensive tubing is exposed, making the bubbler well suited for unsecured areas where vandalism is a concern. The bubble tubing can be installed in the flow and then routed to a point under a manhole cover for quick connection to a flowmeter without entering the manhole.

CONCERNS. Tube may plug (usually alleviated by automatic purges). Bubblers exhibit higher power consumption. Periodic maintenance is required to regenerate desiccant used to prevent moisture from being drawn into the flowmeter.

NOTE ON ACCURACY. Because it can automatically compensate for transducer drift, a bubbler is the most accurate measurement technology. However, a bubbler flowmeter is accurate only to the extent that it maintains a constant bubble rate with changing water level. A microprocessor-controlled unit that maintains a constant differential pressure between the internal air reservoir and the bubble line can assure accuracy.

43.4.4 Floats

A float measures the liquid level with a graduated tape or beaded cable, a counterweight, and a pulley or a pivoting arm and converts it into an angular shaft position that is proportional to liquid level. As the water fluctuates, the float moves the tape or cable up or down and rotates the pulley. Float-operated devices are sometimes used for measuring tides and water level fluctuations in canals, rivers, flumes, and other uncovered conduits. The readings are generally recorded on paper charts.

43.4.5 Conductive Sensors

Conductive level measurement systems rely on a changing level to cause a change in an electrical circuit and thereby indicate the liquid level.

The most appropriate technology should of course be selected for any particular application. When flow may be measured at a variety of sites for differing applications, a flowmeter that combines bubbler, ultrasonic, and submerged sensor technologies in a single instrument should be considered.

43.5 MANNING EQUATION

Before affordable area–velocity measurement was introduced, most short-term studies in sewer pipes used the Manning equation to roughly estimate flow in a round pipe without a primary device since it is not generally feasible to dig up sewers and install primary devices for anything other than permanent monitoring.

The Manning formula is

$$\text{Flow} = \frac{1.486}{n} \frac{A5/3}{P2/3\sqrt{\sin \alpha}}$$

where A = cross-sectional area (ft^2)

P = perimeter,

n = roughness coefficient

and flow is in cubic feet per second (cfs).

Rather than manually compute the equation, discharge tables were developed based on the diameter of the pipe, the level, the slope, and the roughness coefficient. Many flowmeters include the formula in their software and automatically calculate the estimated flow measurement.

The equation is not a precise measurement, however, with inaccuracy of 10–30%. This may be acceptable when an approximation of flow is permitted, such as for storm water measurement. Modern area–velocity flowmeters provide greater accuracy at a comparable investment to level-only flowmeters.

43.6 AREA–VELOCITY FLOW MEASUREMENT

Primary devices are not always available or appropriate and cannot be used to measure:

- Full-pipe conditions
- Surcharged conditions
- Reverse flows

The area–velocity (AV) method, however, can accurately determine the flow rate in these and other conditions—including free-flow conditions—without a weir or a flume. This technique measures the cross-sectional area of the flow stream at a certain point as well as the average velocity of the flow in that cross section. To determine the flow rate, AV flowmeters measure flow by multiplying the area of the flow by its average velocity—a calculation known as the continuity equation:

$$Q = AV$$

where Q = flow rate

A = cross-sectional area of flow

V = average velocity of flow stream

This calculation is automatically performed by modern flowmeters, which offer reliable accuracy in a variety of applications.

There are two techniques to measure velocity—reference velocity and average velocity:

> *Reference Velocity.* Reference velocity techniques calculate an average velocity based on a single reference point. "Point velocity" methods measure one point in the flow while "peak-velocity" methods seek the highest velocity (Fig. 43.5). These reference velocity measurements are modified through profiling and flow equations to establish an average velocity.

> *Direct Average Velocity.* This technology measures velocity across a cross section of the flow and establishes an average velocity.

The velocity of liquid is not constant from point to point in a flow stream. In general, water flows more quickly the farther it is from the conduit wall; the change in velocity is greater near the conduit wall than it is toward the center and decreases near the surface due to surface effects. Since velocity varies in the flow cross section, point velocity and peak methodologies require lengthy, costly, and error-prone site calibrations or profiling to determine or estimate average

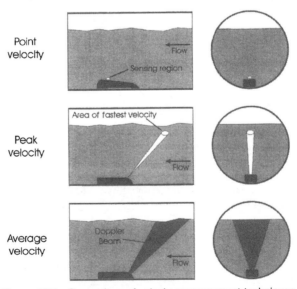

Figure 43.5   Comparison of velociy measurement techniques.

velocity by calculating an average or mean of all the varying velocities from measurements taken at various positions in the flow. Each step required to eventually establish the average velocity introduces its own percentage of error and adds to setup time. These methods vary depending on the site selected and the type of flow (low, rapidly changing, asymmetrical, etc.).

Direct average velocity technology that spans the flow profile to determine an average velocity avoids calibration and profiling procedures.

Two primary methods that measure velocity in the flow stream use Doppler meters and electromagnetic probes.

43.6.1 Doppler Meters

Doppler meters measure frequency shifts caused by liquid velocity. A signal of known frequency is sent into the liquid to be measured. Solids, bubbles, or any discontinuity in the liquid causes the sound's energy to be reflected back to the receiving crystal. Because the reflectors are moving, the frequency of the returned sound energy is shifted. The frequency shift is proportional to the liquid's velocity. For example,

$$\text{Signal sent} = 1\,\text{MHz}$$
$$\text{Signal received} = 1.01\,\text{MHz}$$
$$\text{Doppler shift} = 0.01\,\text{MHz}$$

Then 0.01 is converted to velocity.

The Doppler beam is similar to that of a flashlight in that it spreads out as it travels a greater distance. In a large pipe, the beam continues to spread until it reaches the surface. (Surface reflections can be filtered out with appropriate software.)

Because Doppler measurement relies on signal strength, higher frequency Doppler signals and amplified return signals significantly enhance signal reliability, which is particularly important in low-flow and low-velocity situations. A high signal-to-noise ratio reduces filtering and speeds signal processing to accurately record rapid changes in flow such as occurs during a storm event.

43.6.2 Electromagnetic Probes

Based on Faraday's law of induction, a moving conductor (water) in a magnetic field (created by the probe) will generate a voltage in the conductor proportional to the velocity of the moving conductor.

Due to exposed electrodes, electromagnetic probes may become coated with oil and grease in typical wastewater applications and may need to be cleaned more frequently. Profiling to establish average velocity increases setup time and the potential for human error.

43.7   SPOT-FLOW MEASUREMENT VERSUS CONTINUOUS-FLOW MEASUREMENT

Spot-flow measurements are appropriate when quick manual checks are called for to verify flow on a primary or secondary device or to check flow in a particular area. For continuous or more advanced monitoring, more advanced devices should be used.

Two methods for spot-flow measurement are dilution and point velocity.

43.7.1   Dilution Method

This method may be used to resolve a challenge or dispute; when one municipality bills another for water usage or treatment, for example, a third party can perform a dye dilution test. One reading is taken to establish a number and verify the performance of the flowmeter.

The dilution method measures the flow rate by determining how much the flowing water dilutes with an added brine, radioactive, or fluorescent dye tracer solution. This indicates flow rate directly by simple theoretical formulas:

The constant-rate injection method injects the tracer solution into the flow stream at a constant flow rate for a given period of time.

The total-recovery method places a known quantity of the tracer solution in the flow stream with a continuous sample removed at a uniform rate.

43.7.2   Point Velocity Method

Portable point velocity meters are used to check and calibrate primary devices and flowmeters and for spot measurements in sewers, streams, and irrigation channels. Using Doppler or electromagnetic technology, these hand-held meters can measure fluid point velocity or time-averaged velocity.

43.8   REMOTE VERSUS ON-SITE DATA COLLECTION

A variety of communications options are available to retrieve data from flowmetering equipment. The simplest communication format is an RS-232C port for programming and downloading data directly to a PC or laptop. Modems are also available for remote retrieval of data, as are low-power cellular communications on some meters for sites at which telephones cannot be installed or where they are not desired. Telephone line modems should include interfaces that provide surge protection and high-frequency filtering as well as telephone line fuses that open during a fault condition and self-reset after the fault has been removed from the line to avoid having to send personnel into the field to replace blown fuses. When the communications hardware and software permit programming of the meter and data downloads, such meters can call out during user-specified events either by paging selected individuals or by calling a PC connected to a phone line and informing the recipient of the event.

43.9 REAL-TIME SUPERVISORY CONTROL AND DATA ACQUISITION (SCADA): CONTINUOUS VERSUS INTERMITTENT INCREMENTAL POLLING

The central station for the waste and water industry must have the capability to monitor pumping stations and load requirements throughout its system to serve the needs of its customer base. SCADA systems—a communications loop continuously polling data for control of a system—can perform a variety of functions, including flow and volume calculations, electronic billing, and operational switching. Users of SCADA-capable flow monitoring equipment can connect multiple units at many remote sites and query those units in real time. This allows the user to display numerous, real-time sensor readings from multiple units on a single PC monitor in a central station. End users have real-time, read-only access to all available data channels in the flowmeter or meters and the engineering units associated with the logged data channels.

Implementation of the industry standard Modbus ASCII protocol (a messaging structure widely used to establish master–slave communication between intelligent devices) allows a PC, SCADA, or distributed control system (DCS) to directly communicate with flowmeters to interface with a host computer, programmable controller, or processor-based equipment using the same protocol without the requirement or additional cost of a PLC. This implementation allows users to easily incorporate level, flow, velocity, dissolved oxygen, pH, ORP, temperature, rainfall, and other parameters into a real-time monitoring program.

43.10 SINGLE VERSUS MULTIPLE SITES

Remote monitoring is usually appropriate for multiple sites, such as within a large plant, or for collection system monitoring where measurements from multiple sites help establish flow patterns within a system. Single-site monitoring is often appropriate as part of industrial pretreatment monitoring where one site is monitored at a time, as when inspecting several industries at individual sites.

43.11 PORTABLE TEMPORARY METERS VERSUS PERMANENT INSTALLATIONS

All methods of measuring flow can be performed using either portable or permanent flowmeters. An example of permanent monitoring for level only is using an ultrasonic flowmeter over a flume in a treatment plant to measure the influent, or at a weir measuring the effluent. Examples of applications are as follows:

Portable Temporary	Continuous Permanent
Industrial pretreatment enforcement	Flow in and out of a treatment plant
Short-term collection system studies [I&I, combined sewer overflow (CSO), sewer system evaluation study (SSES)]	Billing sites Long-term collection system capacity analysis

43.12   FREQUENCY OF FLOW MONITORING

In determining the frequency of flow monitoring, acquiring permits and other regulatory requirements are considerations. Another factor is the decision to be proactive or reactive in response to applications. The particular needs of the situation should also be considered; a small I&I problem, for example, may be identified through a short-term study. Budgets should also be taken into account.

43.13   CHOOSING A FLOW MONITORING SYSTEM

As a general note, the quality of a decision requiring flow information depends largely on the quality of the data. To state the obvious, better data come from better technology.

43.13.1   Accuracy

Users must determine the degree of accuracy necessary to match the objectives they wish to accomplish. When measuring storm water, for example, an approximation of flow may be sufficient. The choice of flow technology and the type of flowmeter, will either assure or compromise accuracy. It is important to choose the best technology for the particular application.

43.13.2   Elements of a Flow Monitoring System

Flow monitoring equipment ranges from simple, basic devices to sophisticated and versatile equipment for a variety of applications:

Basic Systems

- Meter
- Sensor
- Power source
- Simple software

Advanced Systems

- Full graphic display and keypad
- Alarm capabilities
- Modem
- Advanced Software package
- Reporting
- Multiple sensors
- Advanced programmability

Advanced programmability provides the capability to choose between the following options:

- Monitor only
- Monitor and record (requires internal datalogging)
- Measure level only
- Measure level and velocity

It is important to note that one meter may not fit all applications. Sometimes versatility in a meter is called for, sometimes a variety of meters are called for. When selecting equipment, follow these guidelines:

1. Decide what you must monitor, considering regulatory requirements, site requirements, and reporting requirements.
2. Decide what else you could monitor.
3. Consider future needs.
4. Consider site types.
5. Consider site conditions.
6. Consider ease of use and other factors, including speed of setup, portability, battery life, durability, and the ability to withstand highly corrosive environments.
7. Factor in total cost of ownership, or life cycle costs, over time (see Chapter 33 of this work).

Also consider whether flow monitoring equipment should be rented, leased, or purchased. The former options may be appropriate for short-term studies (less than 6 months), occasional studies, and situations in which the meters are not expected to be used for other applications. Consultants may not wish to carry an inventory or want specialized equipment for a particular application, preferring to rent or lease the most appropriate equipment for any given project. Or, a certain inventory of meters may be kept on hand, with additional equipment rented for specific projects.

When evaluating equipment, pertinent engineering journals, consultants, manufacturers, and operators of flow monitoring equipment under similar conditions as the anticipated usages should be consulted.

For guidelines on evaluating flow measurement devices (including design, installation and operation), see Ref. 2 pages 48–49.

43.14 APPLICATIONS OF FLOW MEASUREMENT

Flow measurement is a major element of water monitoring for community planning, water treatment, pollution control, billing, and other purposes, including:

- Compliance
- Gathering information
- Reporting
- Monitoring
- Controlling
- Planning
- Billing

43.14.1 Publicly Owned Treatment Works

At POTWs, flow monitoring is performed for a number of purposes:

1. *Collection Systems.* Flowmeters help municipal managers better understand the hydraulic loadings in their sewer systems and better manage the systems. Effective monitoring can help determine:

- How rainfall affects flow capacity in a sewer
- Why backups and overflows occur
- At what volume of flow overflows occur
- Which parts of the sewer system suffer from the worst (I&I).
- How much flow is entering the lift stations
- What parts of the sewer subsystem are in greatest need of rehabilitation

2. *Waste Treatment Plants.* Virtually every process in the plant is controlled based on flow. A higher flow entering the plant, for example, means increased chemical feed, aeration, and chlorination. Flowmeters can control these devices in response to changing flows.

3. *Pretreatment Groups.* Pretreatment groups regularly monitor industrial discharges to assure compliance with NPDES permits. Portable, battery-operated flowmeters are often used for this activity.

4. *Storm Water Utilities.* POTWs have established storm water utilities and flood control districts to manage urban storm water. As part of the management

plan, the utility monitors rainfall as well as flow and water quality to assess the impact of wet-weather events on receiving water and to verify the effectiveness of treatment. A typical storm water monitoring system consists of a rain gauge, flowmeter, sampler, and pH and temperature monitor.

In addition to the above, flow measurement and water quality monitoring are regularly performed for other applications:

Industries that utilize water in their production processes are limited by permit as to the contaminants that may be discharged. In addition to sampling the discharge to verify compliance, industries frequently use flow measurements to monitor their discharge volume. Permits typically require that samples be collected in proportion to flow, which can be easily achieved by setting flowmeters to control the actuation of the samplers based on the passage of preset flow volumes. Industries also measure flow because they are billed based on the volume of their discharge to the sewer.

Consulting engineers and laboratories perform a variety of monitoring studies for public and private clients, such as determining the extent and location of I&I to a municipal sewer, developing a watershed management plan, or assessing the impact of a commercial development on sewer capacity.

Universities conduct a wide variety of water quality research, including the impact of storm water runoff from agricultural sites.

43.14.2 Sewer System Evaluation Study

The SSESs use flow monitoring as one of the inputs to build a computer model of the sewer system. An engineer then builds scenarios of what will happen to the sewer system under different event conditions.

These studies are designed to analyze I&I problems in sanitary sewer systems so that repairs, maintenance, and other corrections can be prioritized and performed. They help determine:

- The amount of extraneous water present in the system
- Where the water enters the system
- What portion of the water can be eliminated

Objectives of the SSESs are to:

- Determine the extent of I&I problems through continuous flow monitoring programs
- Locate and identify I&I sources
- Categorize I&I sources and quantify I&I attributable to each source
- Reconcile results with I&I quantities measured during continuous flow monitoring programs

- Determine and prioritize the most economical rehabilitation methods for each located I&I source
- Help manage the entire system

Where I&I can be a short-term study to identify, prioritize, and fix problems, an SSES is a more detailed study of the entire system. An example of the wisdom and efficacy of this approach is the western New York town of Amherst, which was able through an ongoing flow monitoring program to demonstrate its system's ability to handle increased capacity and achieve a rerate of the existing plant without investing millions of dollars in additional treatment capacity.

43.14.3 Total System Management

Flow monitoring can also play a major role in total sewer system management, especially when flow, rain, and water quality instrumentation, communications, and software are integrated and coordinated for planning and implementing proactive management. The benefits of this approach include:

- Proactively preventing collection system overflows rather than reacting to them
- Addressing capacity concerns before they become issues.
- Budgeting, reducing overall expenses, and making and justifying cost-effective purchases
- Managing I&I
- Meeting regulatory requirements
- Providing a basis for informed decisions.
- Acting as early warning systems by alerting management early to potential collection system problems, such as sewer backups, irregularities in the flow pattern, dry-weather discharges from CSOs, and toxic industrial slugs

The most effective and proactive total system management programs involve continual monitoring at permanent sites. (See the Amherst, New York, example above.)

43.15 FLOW MONITORING FOR MUNICIPAL LOCATIONS

43.15.1 Treatment

When attempting to control the wastewater treatment process, one key measurement is the incoming flow to the facility. WEF MOP 11[13] states, "Flow is one of the most important wastewater parameters to be measured." This measurement, when combined with the amount of food contained in the flow, tells the operator the exact amount of biochemical oxygen demand (BOD) arriving at the treatment plant. This critical measurement determines how the rest of the treatment process is controlled.

The BOD measurement is typically made through laboratory analysis of a sample taken from the plant influent, while the influent flow rate is typically measured in an open channel using a primary device and a secondary flow-metering device.

In addition to monitoring the incoming flow to a treatment facility, federal and state regulatory agencies require that each treatment facility obtain and comply with an NPDES permit. Part of this compliance requires the measurement of plant effluent. Plant effluent flow rates are typically measured at the discharge to the chlorine contact chamber, converting liquid level to flow over a weir.

According to the *NPDES Compliance Inspection Manual*,[4] "Permittees must obtain accurate wastewater flow data in order to calculate mass loading (quantity) from measured concentrations of pollutants discharged as required by many NPDES permits." (For further details and steps necessary to evaluate the accuracy of primary and secondary flow measurement devices for wastewater treatment plants along with other details of compliance, refer to Chapter 6 in Ref. 4)

RECOMMENDED EQUIPMENT SPECIFICATIONS AND FLOW MONITORING SUGGESTIONS. At the influent, flow is often measured with an ultrasonic meter over a flume. In addition to measuring flow rate, sampling is also performed to assess other parameters like pH, and newer technology can be put to use that integrates flowmeters with other water quality parameters. Since the rest of the treatment plant process will be based on the influent flow, an output (e.g., 4-20 output) that can be used for alarms is helpful. The operator can then set the amount of chemicals used, the amount of holding time in the tanks, and other processes to operate based on the volume of flow. Temperature compensation is a desirable feature for equipment used in these applications. When possible, a unit that features electronics sealed from the environment should be chosen to avoid the need for repairs. Direct sunlight should be avoided when feasible when installing the meter, particularly in hotter climates, to protect the digital display.

43.15.2 Pretreatment

Open-channel flow (where the flow occurs in conduits that are not full of liquid) is the most prevalent type of flow at NPDES-regulated discharge points. According to the U.S. Environmental Protection Agency (EPA), "pollutant limits in a NPDES permit are usually specified as a mass loading. To determine a mass loading and thereby evaluate compliance with permit limits, it is necessary for the inspector to obtain accurate flow data."

In addition to verifying compliance with permit limits, flow measurement serves to:

- Provide operating and performance data on the wastewater treatment plant
- Compute treatment costs, based on wastewater volume
- Obtain data for long-term planning of plant capacity

In pretreatment applications, the amount of discharge is generally measured at an industrial facility as part of the NPDES permit at or as close as possible to the point of discharge. An industry may be required to install a flowmeter, the municipality may do a spot check, or both conditions may apply.

RECOMMENDED EQUIPMENT SPECIFICATIONS AND FLOW MONITORING SUGGESTIONS. Because sampling is such an important element of pretreatment monitoring, a good choice for this type of application is a sampler with integral flow capabilities, rather than a dedicated flowmeter. This is helpful for several reasons, including ease of use. (In a manhole, e.g., only one piece of equipment need be installed and set up instead of several.) With integral AV in a sampler, more accurate open-channel flow measurement may be performed without a primary device, as is most often the case when performing a spot check by a manhole outside an industrial plant. If sampling is not performed, additional water quality parameters (such as pH) can be added to many modern flowmeters.

Power conservation is another important consideration since equipment may be kept in place over a long period of time to capture the entire event.

Flowmeters at industrial sites should be easy to use for operators that do not typically perform these operations. They should provide a reliable, continuous, non-interrupted recording of flow for the NPDES permit compliance. Good reporting capabilities, including site identification, help provide defendable data for court should it become necessary.

43.15.3 Combined Sewer Overflow

Across America, sewer systems serve millions of people at work and at home every day. Increasingly, however, wet-weather sewer system overflows are a national problem. EPA statistics show that the 18,500 municipal separate sanitary collection systems serving 135,000,000 Americans can, under certain circumstances, overflow. In addition, 1100 communities serving 42 million people with combined sewer systems have as many as 20,000 designed discharge points. One difficulty with sanitary sewer overflows (SSOs) and CSOs is that the problem is often unseen. Sewer systems are never completely watertight. Settling pipes, cracking manholes, crumbling brick and mortar, and leaky joints can contribute to I&I. In fact, I&I can add as much as 100 gal per person per day to the normal wastewater flow. Since more capacity is required at the plant, treating this extra water can be expensive. Efforts to correct I&I can be expensive and less than effective. Normal plant operations are often disrupted, lowering treatment efficiency.

SSOs and CSOs can also spill raw sewage into public and private areas, causing substantial health and environmental risks. In addition, property and environmental damage can be extensive and expensive to rectify.

The problem is more preventable than may be generally realized. The Water Environment Federation estimates that approximately 80% of the problem can be corrected by repairing 20% of the leaks. Flow monitoring programs can help identify the most critical leaks and prioritize repairs. They can also help pinpoint potential problems before they grow into major problems. In essence, finding leaks and performing preventive maintenance is much more cost effective than fixing them and their consequences.

In 1994, the EPA published the National CSO Control Policy in the *Federal Register* (40 CFR 12), containing the Agency's objectives, control plans, and alternative approaches for addressing overflows from CSOs. Water monitoring and flow measurements are critical elements of these plans.

Monitoring wet-weather events to establish design level flow and pollutant loadings requires the following basic elements:

- Historical rainfall data (obtainable from the National Oceanic and Atmospheric Administration)
- Continuous rainfall measurement (over several months to obtain enough differently sized rain events to properly correlate variables)
- Continuous flow measurement (conducted over a range of rainfall intensities and durations, synchronized from the beginning of an event and representative of total wet-weather amounts)
- Flush-weighted discrete samples (Split stream sampling—sampling one of multiple streams representing the watershed flow—can be applied if all flows are known.)

Part of the CSO regulations is that municipalities are typically required to measure the amount and duration of discharges, at the discharge point, since the goal is to control, reduce, or eliminate out-of-compliance discharges.

RECOMMENDED EQUIPMENT SPECIFICATIONS AND FLOW MONITORING SUGGESTIONS. Useful features for a flowmeter include the added ability to monitor rainfall. Modern equipment can combine sampling, flow measurement, and rainfall monitoring. AV flow measurement is an advantage since there may not be a primary device. An AV flowmeter should be capable of measuring and recording rapid changes in level and velocity and preferably be able to measure and process the data in seconds. Since it is not practical to enter a pipe during a wet-weather event to obtain average velocity, a meter that will measure direct average velocity should be chosen. Ideally, flowmeters should operate under free-flowing, reverse-flow, and surcharged conditions. Low-profile probes and cables help avoid solids. (Logs and other large objects have been known to come through a CSO discharge and high-profile obstructions can be easily damaged. In the event of such damage, detachable, interchangeable probes that are quickly and easily changed are a decided benefit.) Datalogging and software capabilities should produce detailed reports, ideally combining rainfall and flow so that a correlation can be made between the two.

43.15.4 Sanitary Sewer Overflow

Separate sanitary sewers have, in most cases, been designed not to surcharge under peak-flow conditions. However, the combination of growth and I&I often results in surcharging after intense storm events.

To determine the extent of SSOs within the collection system, permittees need a thorough understanding of:

- The capacity of the sanitary sewer system
- The response of the system to various rain events
- The characteristic of the overflow events
- The extent of inadequate capacity in the system for handling wet-weather flows and other details

The permittee must also develop adequate flow monitoring data through installation of continuous flow monitoring stations throughout the system. The flow monitoring stations should be located in such a manner that the data collected will indicate where and to what extent problems exist and may be temporary, permanent, or a combination of both. Without adequate flow monitoring of various storm events, it is impossible to determine the peak inflow rates. Also, the existing system capacity must be compared with projected peak flows from various storm events to evaluate the need for relief sewers and other modifications necessary to ensure that the system will not overflow.

Continuous flow monitoring of the collection system along with rainfall monitoring enables permittees to identify subbasins with I&I problems and capacity deficiencies and to prioritize subbasins for further field study. Computer modeling can also be an effective tool in evaluating the capacity and the response of a sewer system to storm events of varying frequency and duration. The modeling of the sewer system may include hydrologic/hydraulic models utilizing both steady-state and dynamic computer simulations of flow through the sewer network and computing hydraulic grade lines for storm events of varying frequency and duration. This may be accomplished by computer analysis of the flow monitoring data gathered during storm events and extrapolating flow for the range of anticipated storm events.

In short, flow monitoring provides the information needed to establish the plan of action and is far less expensive than increased capacity that does not address or correct the problem.

RECOMMENDED EQUIPMENT SPECIFICATIONS AND FLOW MONITORING SUGGESTIONS. In SSOs, any overflow will be a problem. Whereas CSOs have designed overflows, sanitary sewers will overflow into basements and residential streets where there is a higher risk of human contact. Flow is therefore measured as a primary criterion to establish system capacity. Municipalities will decide whether to perform short-term studies or continuous monitoring; the latter is helpful for proactive programs to prevent overflows and predictive modeling. Permanent monitoring can provide a picture of a complete collection system to correctly identify the areas that need the most work and prioritize repairs and maintenance. Rainfall monitoring capabilities are helpful here, as in CSO monitoring. Low-profile probes are advantageous, particularly in the smaller lateral lines (the majority of U.S. sewer lines are under 18 in. diameter). Most flow studies start in the larger pipes, but as studies move down into the smaller pipes to obtain more detailed information, equipment should be capable of handling a range of conditions from low-flow to high-velocity, full-pipe flow. "One size fits all" is not always the best answer in the sewer system, and accuracy may be compromised if the equipment is not the best for a particular site. Technology that can be tailored to the site conditions should be considered, such as multiple sensors for multiple pipes in one flowmeter and different options for level measurement technology. The equipment should be able to be subjected to prolonged surcharges, which can be a relatively common occurrence.

43.15.5 Storm Water

Regulations for storm water monitoring require collecting data that characterize the flow rate (quantity discharged from an outfall per unit of time) and flow volume (the total volume of storm water runoff discharged during a rain event) for each storm water discharge that is sampled. Regulations further mandate that flow-weighted composite samples be collected for the storm duration.

The EPA's *Storm Water Sampling Guidance Document*[2] outlines the four most common methods for estimating storm water flow rate:

The runoff coefficient method uses estimated factors such as size of the runoff area, soil conditions, and rainfall history. This method is the least accurate and is used only when other measurements and estimation techniques are not appropriate.

The float method (a velocity method) is appropriate where flow is open and easily accessible. An average velocity is determined by measuring the time it takes a float to travel between two points. The area is determined by measuring the depth and width of the flow.

The slope-and-depth method, based on Manning's equation, is appropriate for flow from a pipe or ditch where the conveyance slope is known, the flow does not totally fill the ditch or pipe, and the channels are of uniform slope and characteristics. The depth of the flow in the middle of the channel is recorded, as is the pipe inside diameter and slope of the pipe or the ditch width.

The volumetric method ("bucket and stopwatch"), appropriate when free-flowing flow from a pipe or ditch is small enough to be captured in a bucket, is rarely suitable for storm water flow measurement. Flow rate is calculated by dividing the collected volume of flow in the container by the time required to collect it.

Part of an inspector's procedure for examining illicit connections (point source discharges of pollutants to separate storm sewer systems that are not composed entirely of storm water and that are not authorized by an NPDES permit) includes evaluating the flow pattern of and characterizing the discharge.

RECOMMENDED EQUIPMENT SPECIFICATIONS AND FLOW MONITORING SUGGESTIONS. Due to the nature of storm water monitoring, the combination of sampling and flow measuring in one unit is an advantage, as is the capability to record rainfall and flow together and to report flow in relation to rainfall. Having less equipment to transport and set up is especially practical for remote sites and confined manhole spaces. Equipment reliability and the ability to withstand moisture and possible submergence are important criteria. A low-profile depth sensor is less likely to interfere with the flow and be affected by solids, silt, or floating materials.

43.15.6   Non–Point Source Runoff

The USGS (U.S. Geological Survey) is the primary group for monitoring rivers and streams in the United States. State agencies often have non–point source monitoring groups, looking for pollutant loading due to agricultural activity, for example, in order to improve on best farm management practices. They may set up continuous monitoring stations to assess pollutant loading during appropriate seasons and events.

Flow monitoring is probably not as common as sampling for this application. It is, in fact, typically harder to perform because with channels and streams, there really is not particular point (it is *nonpoint*) for flow measurement. There is more of a tendency to perform stream gauging, which is more level measurement; for this purpose, the level recording feature of a flowmeter with a programmable datalogger will provide continuous level measurement over time.

RECOMMENDED EQUIPMENT SPECIFICATIONS AND FLOW MONITORING SUGGESTIONS. As with storm water monitoring, level sensing capabilities are required that can cover the expected range of events, including fairly great depths; a stream may flow at 2 ft on average but at 10–15 ft during a storm event. Consider remote power capabilities, with a low power draw (a marine battery or solar power may be added to trickle charge the battery). Capability for datalogging over longer periods of time is important, with sufficient memory for the duration of time that data will be recorded. At some remote sites, the information may only be collected, for example, at monthly intervals. The ability to combine rainfall and level or flow measurements is helpful.

43.15.7   Interagency Billing

Not all communities have their own wastewater treatment plant; flow may be sent to a nearby or neighboring facility. Interagency billing—the cost of treating another township, city, or municipality's flow—requires the ability to measure that flow accurately. Whereas 5–10% accuracy for modeling sewer flow provides a good idea of system conditions, inaccuracy in this application creates erroneous billing and disputes between communities. In some cases, communities have found flow measurement inaccurate by as much as 30%, which translates in to substantial over- or underpayment.

Flow should typically be measured at the last or closest point before crossing community borders, which could be a pumping station or a flume. Flumes, however, depending on how well they are installed, can affect the accuracy. If they have been in place for years, they can be undersized for current conditions; they can also be submerged and miss peak flow. (As an example, one western New York community was about to put in a supplemental sewer line to handle more flow, until it determined the problems were actually coming from the neighboring community whose flow it was treating. Instead of spending more than $1 million for the supplemental line, the neighboring community was required to fix its I&I problem.)

Accurate flow is also important because additional flows from a neighboring community become part of the total system flow and may tax the system performing the treatment, increasing the cost of running the plant and risking violations should the plant get more flow than it can handle. Also significant is the ability to prove the accuracy of the data collected. Documentation (software and reports) should be comprehensive and conclusive to avoid costly, aggravating, lengthy delays and disputes.

RECOMMENDED EQUIPMENT SPECIFICATIONS AND FLOW MONITORING SUGGESTIONS. Probably the most commonly used equipment over the years is an ultrasonic flowmeter over a flume. The flowmeter should feature temperature compensation. The trend today is toward redundant measurement for a higher degree of accuracy and data defendability. Advanced meters can provide an additional channel for a backup AV measurement upstream; this allows comparing readings and having another measurement if the ultrasonic measurement fails or the flume becomes submerged.

In older existing flumes, where there may be submerged conditions, the installation of a more advanced flowmeter with redundant measurement allows accurate measurement capabilities without having to dig up the flume. Where open-channel pipes exist without a primary device, area–velocity technology can allow accurate measurement without the expense of installing a primary device.

Open-channel transit time flowmeters include bidirectional (reverse-flow) measurement capability and can be configured for multiple acoustic paths. This method compares the time it takes an acoustic pulse traveling across a pipe downstream with the time it takes to travel upstream and calculates an average flow velocity. Transit time is a permanent installation, with significantly higher installation and service costs than area–velocity technology. It is sometimes considered in large-diameter pipes, such as interceptor lines; the technique can be very accurate for higher flows but less so in the lower third of the pipe.

In short, for accuracy over a range of conditions from low flow to peak flows, redundancy is recommended to reduce opportunities for dispute. Consider alternating current (AC) with a power back-up or permanent battery-powered units with a long life and the capability to use telephone, radio, or cell-phone modems. Alarm capabilities permit early notification of irregular or heavy flows in the collection system, so operators can store flow in pump stations longer or take other actions to use more of the system capacity.

Software should be capable of handling a significant amount of data and be able to easily turn the data into reportable information. Look for report templates and automatic polling and software in a commonly used, easy-to-understand format (as opposed to proprietary software that raises issues of training, upgrading, compatibility, and customizability).

43.16   FLOW MONITORING FOR INDUSTRIAL AND OTHER APPLICATIONS

43.16.1   Industrial Discharge

Much of the content discussed above for municipal treatment applies to industrial discharge monitoring. Most of the differences have to do with the fact that municipalities are generally much more familiar with flow measurement equipment and techniques than are industrial users. For this reason, equipment should be easy to set up, program, and use—with minimal if any maintenance.

RECOMMENDED EQUIPMENT SPECIFICATIONS AND FLOW MONITORING SUGGESTIONS. Accurate equipment capable of producing defendable data is of paramount importance. Accuracy can depend on the installation of the primary device in combination with the accuracy of the flowmeter. Consider an ultrasonic flowmeter over a flume or other primary device. One benefit of an overhead ultrasonic meter in some industrial discharges is that some submerged probes can have trouble with such factors as temperature and the nature of the discharges.

Proper use of AV flowmeters allows for accurate measurement without a primary device, making it possible to perform spot checks, assess the accuracy of the industry's installed equipment, perform short-term studies, and help solve disputes between actual and reported discharge amounts.

Also important is continuous-flow totalization, since industries must report an accurate total flow. Where possible, backup AC power and backup totalizers ensure against gaps in reporting total flow. Software totalizers are available that cannot be tampered with.

A flowmeter should be installed at locations that provide convenient access for both the industry and the municipality that must verify the flow. For plants with SCADA systems, flowmeters that will interface with that system should be used for simplified communication and reporting.

43.16.2   Storm Water

In addition to the information conveyed above for municipal storm water monitoring, other conditions exist for industrial applications. The issue for storm water is not to continually monitor; rather, it is to monitor, identify, and solve problems. Solutions can be as simple as installing a roof over a storage area that creates oily runoff, putting in a containment system that controls the amount of runoff, or initiating some type of treatment.

Whereas industrial wastewater monitoring is ongoing, storm water monitoring may often be discontinued or reduced by providing proof-of-management practices that alleviate storm water problems.

For storm water monitoring (essentially rain runoff), the amount of rain is correlated to the resulting amount of flow. Therefore, both rainfall and flow over the duration of the storm must be measured. The resulting information indicates that, for example, $\frac{1}{2}$ in. of rainfall produces x amount of flow, or 2 in. of rainfall produces *y* flow. Sampling must also be tied in to determine the level of pollutants associated with the event.

In choosing locations to measure flow so that runoff can be captured, a site survey determines a final discharge point or points (based on topography and other factors) where the greatest amount of flow is anticipated. For industrial applications, flow will typically be measured in:

- A type of ditch or culvert
- A stream
- A storm drain
- A roof downspout or parking lot drain runoff into manholes

Open-channel measurement techniques are valid at such locations. Permanent primary devices may be installed, or AV methods, or the Manning equation used. The latter has been accepted in some cases because storm water is an *estimate* of flow, so there may be some leniency on accuracy. Working with the enforcement agency to get prior approval of a system and monitoring points is particularly helpful. For example, an agency may say that, although the original request was to monitor five sites, two of them are similar, and it may allow monitoring in only one of them; this may be negotiable depending on the agency and the circumstances.

RECOMMENDED EQUIPMENT SPECIFICATIONS AND FLOW MONITORING SUGGESTIONS. Equipment should be rugged, able to hold up in the environment (including temporary submergence), easy to use, and able to log rainfall or at least tie in rainfall with the report. Permittees may sometimes be allowed to use local rainfall reports as opposed to monitoring rainfall themselves, depending on such factors as proximity to airports.

Considering where rainfall and flow are being monitored, the rain gauge can often be put with the flowmeter, either as a stand-alone rain gauge or with a cable that connects to the flowmeter and is logged into the flowmeter. Also, since there can be rainfall differences and variations even within very short distances, monitoring rainfall at the actual site provides the greatest accuracy and allows automatic initiation of sampling. Software that compares flow to rainfall is helpful, as it allows verification of the amount and intensity of a rain and the subsequent amount of surface discharge.

A final consideration is acquiring accurate, defendable data that will not have to be redone. The more complete, accurate, and professional the reporting is, the less risk there is of being challenged.

43.17 HAZARDOUS LOCATION MONITORING

The potential for hazards in municipal sewer systems is widespread. In fact, a sewer can become an explosive underground tunnel. Explosive and flammable substances can be dumped down manholes or generated in sewers. The sludge-handling areas of many wastewater treatment plants produce the possibility of methane and hydrogen sulfite gases, which could also be produced within a municipal collection system, although these are normally vented.

A high risk can result from an illicit discharge from an industry or an accidental spill. At industrial sites as well as residential areas, for example, gasoline can leak into the sewer system. Consumer wastes can also be hazardous.

Other potentially hazardous locations include the headworks (the front end) at treatment plants, where there is often a high concentration of H_2S gas, and outdoor locations where lightning may strike. For industrial applications, hazardous locations can be created by certain chemicals manufactured at a plant.

RECOMMENDED EQUIPMENT SPECIFICATIONS AND FLOW MONITORING SUGGESTIONS. Intrinsically safe equipment and wiring are available that will not release sufficient electrical or thermal energy to cause these flammable or combustible mixtures to ignite. Intrinsically safe equipment does not have sufficient capacity to arc or spark to create an explosion due to a low-power use and/or sealed design.

Intrinsically safe portable and long-term area–velocity flowmeters are commercially available in the United States that are CSA-NRTL/C certified for operation in Class I, Division 1, Groups C and D hazardous locations and CENELC approved. Such a rating can be on the meter or a sensor. Until fairly recently, such meters were quite expensive or were placed in explosion-proof containers that were exceptionally large, heavy, and expensive.

Certain sites necessitate intrinsically safe meters and sensors because all electrical equipment on the property must be so rated. At other times, a nonrated flowmeter could be mounted outside a manhole, with only a sensor rated for hazardous areas actually put into the manhole.

Choosing a properly rated flowmeter that is reasonably priced is sensible and desirable from a safety and liability standpoint. If intrinsically safe equipment is available but not used, a municipality may face a greater liability in the event of an accident. Intrinsically safe flowmeters should be rated by an independent agency through detailed testing.

Modems are also available for intrinsically safe applications that feature an integral fiber-optic link to the flowmeter, so that it can be safely connected to a telephone line without the potential danger of a hazardous surge voltage from the telephone line entering the explosive zone.

43.18 FLOW MONITORING FOR DIFFICULT AND CHALLENGING CONDITIONS

43.18.1 Low Flow

In collection systems in particular, low nighttime flows can be difficult to monitor. A low-profile probe has a much better chance of measuring these. Ultrasonic meters are also effective measuring devices for low flows. When there is no primary device, the combination of a low-profile velocity probe and an ultrasonic sensor provides the most accurate measurement.

When measuring velocity with a Doppler probe in low-flow conditions or clean water (which causes a reduced area of reflectors), a system that uses a higher frequency with signal-amplifying electronics in the probe will perform better in lower flows. For a short-term study in the collection system, an AV flowmeter designed to accurately measure high and low flows should be chosen.

43.18.2 Multiple Pipes

A collection system often has more than one incoming pipe at the same manhole. This presents a challenge because the flow at outgoing pipe is the total flow, not the contribution from each area, which is often required or useful information. Rather than install separate meters in the manhole, flowmeters with multiple-sensor capabilities allow accurate measurement in these situations (Fig. 43.6).

43.18.3 Dry to Wet (No Flow to Peak Flow)

The most important equipment specification to consider in these conditions is a rapid reaction time. For example, CSOs are typically dry pipes during nonevents; the meter is essentially just sitting there waiting. When the flow suddenly rushes through, measurement must begin as soon as possible, preferably within seconds. If the meter responds slowly (due to temperature changes, a slow signal filtering system, etc.), all or part of the event may be missed; time-accurate flow measurements or complete data representative of the entire event may not be acquired. Another example is storm water, where a sensor may be in 80°F air temperature

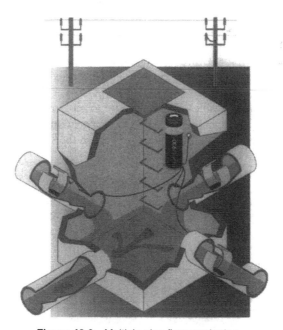

Figure 43.6 Multiple-pipe-flow monitoring.

when storm water suddenly comes in at 50°F. For applications such as these, fast, reliable temperature stabilization is necessary for accurate measurement.

43.18.4   Full Pipe and Surcharge

In collection system monitoring (as well as CSO, storm water, and wet-weather monitoring) the potential exists for the pipe being measured to become full. In some cases, there is a surcharge where there is so much pressure in a sewer during a storm event that it forces water up out of the manhole opening. Meters being used in these conditions can become submerged for hours; they must withstand this or operators may frequently replace meters, repeat studies, or acquire incomplete data. One may also want to be able to measure flow in a full-pipe condition; surcharge, full-pipe, and reverse-flow conditions are applications for which level-only measurement will be inaccurate. AV meters rated for prolonged submersion are good choices for surcharge conditions.

43.18.5   High Velocity

It becomes more difficult for a submerged sensor to measure accurately in high flows due to the Bernoulli principle, which states that as the velocity of a fluid increases, its pressure decreases (Fig. 43.7). The increased velocity across the surface of the probe creates a vacuum near the level measurement points, causing inaccurate level measurement. Level measurement using a submerged depth sensor is similarly affected by increases in fluid velocity. For example, in a 12-in. pipe at 4 fps, an error in level from 3.5 in. to 3 in. equals a difference of 66 gpm or 19% of the flow rate.

Drawdown correction capabilities compensate for the effects of velocity in level measurement accuracy. Drawdown may be corrected somewhat through the hydrodynamic shape of a probe and to a greater degree through sophisticated software using specific algorithms.

Low-profile probes and cables also contribute to more accurate measurement in higher velocities. Sensors should be secured to proper mounts. Ultrasonic measurement is also appropriate for higher velocities because the sensor is out of the water and unaffected by velocity.

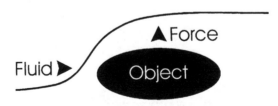

Figure 43.7   Bernoulli principle.

43.18.6 Distance

If the meter must be distant from the measuring point (due to safety or accessibility), there may be extremely long cable lengths on the sensor. Ultrasonic sensors may have maximum cable lengths separating the sensor from the meter to continue to provide accurate measurements. Signal strength and clarity are critical considerations for such applications.

43.18.7 Silting, Solids in Flow, and Ragging

Silting, ragging, and solids in the flow are very common issues in collection systems. Low-profile probe designs avoid having matter catch on them and are not easily fouled by oils and greases. Easily replaceable probes are also helpful in these situations.

43.18.8 Backup Conditions

For applications where backup conditions may occur (such as coastal areas during high tide or pipe blockages), meters that can recognize forward as well as reverse flows should be considered. AV meters have an advantage over level-only flowmeters in these cases, as level measurements will not tell whether the flow is moving forward or backward. If there are blockages in a pipe or tidal influences that could cause reverse-flow conditions, most AV flowmeters can identify flow in both directions and also log and identify it as such. Not only is this important for accurate flow measurement, it can also help identify a problem and trigger an alarm.

43.18.9 Outdoor, Permanent Mount

When flowmeters may be permanently outdoors, such as measuring influent flow at a treatment plant that is not close to a building (Fig. 43.8), equipment specifications

Figure 43.8 Influent flow monitor.

to consider include UV resistance, temperature compensation, remote communications, and a NEMA 4×6 controller.

43.19   COMMONSENSE CONSIDERATIONS OF LOCATION

In conclusion, a few brief points can be made concerning the choice of locations for flow monitoring:

- Seek a straight run of pipe without obstructions that is at least three times the pipe diameter upstream from the sensor.
- Monitoring in the invert of a manhole can cause problems due to surcharges and turbulence.
- Seek a free-flow condition, with no backups.
- Avoid pipe joints, curves in the pipe, and lateral connections.
- If the location specified by the NPDES permit is not the best location for the most representative accurate flow monitoring, consider choosing and negotiating an alternate site.

REFERENCES

1. Manual of Practice, No. OM-1, Water Environment Federation, 1996.
2. *NPDES Storm Water Sampling Guidance Document*, EPA 833-B-92-001, July 1992, Chapter 3.
3. WEF MOP 11.
4. Flow Measurement, in *NPDES Compliance Inspection Manual*, EPA 300-B 94-014, September 1994, Chapter 6.
5. *NPDES Compliance Monitoring Inspector Training: Sampling*, 21W-4007, August 1990.
6. *Wastewater Sampling for Process and Quality Control*, Manual of Practice No. OM-1, Water Environment Federation, 1996.
7. *NPDES Compliance Flow Measurement Manual*.
8. *Wet Weather* **2**(1), (Spring 1997).
9. *Sigma 950 Series Open Channel Flow Meters*, GIV TP 7 M 4337-A 12/94, pp. 4–6.
10. *Meet EPA Requirements by Preventing Overflows*, American Sigma, SIG 4884, 1998.
11. *Clean Water for Today: What Is Wastewater treatment?* Water Environment Federation, Public Education Program.
12. *Sanitary Sewer Overflows*, USEPA Compliance Assurance and Enforcement Division, Water Enforcement Branch.
13. *EPA Region 6 Water Management Division's Strategy for Wet Weather Sanitary Sewer Overflows*, EPA Region 6.
14. *Increase Plant Efficiency Through More Effective Process Control*, American Sigma, SIG 5037, 1998.

44

COMPLIANCE FLOW MONITORING IN LARGE STACKS AND DUCTS

Richard Myers, Ph.D.

Teledyne Monitor Labs
Englewood, Colorado

Environmental Instrumentation and Analysis Handbook, by Randy D. Down and Jay H. Lehr
ISBN 0-471-46354-X   Copyright © 2005 John Wiley & Sons, Inc.

A number of flow monitoring technologies have been used successfully in a wide variety of applications. Similarly, they have also often been misapplied, resulting in poor performance coupled with maintenance and reliability problems. It is the intent of this chapter to provide sufficient information that potential users can better understand the capabilities and limitations of the most commonly used technologies and thereby avoid problematic installations. To be useful in today's environmental and process monitoring environment, flow measurement systems must be augmented with automated calibration and diagnostic systems that enhance their accuracy and reliability. Calibration systems must provide a complete check of the normal measurement system to ensure the integrity and accuracy of resulting measurements. The situation is further complicated by the fact that the reference methods by which flow monitoring systems are evaluated are themselves limited in absolute accuracy and are dependent on the specific application. In other words, there does not exist an absolute reference material analogous to certified [U.S. Environmental Protection Agency (EPA) protocol verified] gases such as are used for gas monitoring systems. The typical applications of interest are those associated with EPA continuous emission monitoring system (CEMS) requirements, and such regulatory compliance is assumed as a basis of comparison for these systems of interest. The EPA requires that the performance of each flow monitoring system be individually certified in the actual installation. The performance certification requirements are described in the Code of Federal Regulations, 40 CFR 75, Appendix A.

Note: In this chapter, we deal with general technologies and avoid identifying specific suppliers of flow monitoring systems. For a complete list of suppliers, you may search the Internet via a search engine or consult web pages with industry information, such as www.awma.org, www.isadirectory.org, www.thomasregister.com, www.manufacturing.net, www.industry.net, and www.pollutiononline.com.

The EPA emission measurement and regulatory information is available at www.epa.gov/ttn/emc, www.epa.gov/acidrain, www.epa.gov/acidrain/otc/otcmain. html, www.access.gpo.gov, and http://ttnwww.rtpnc.epa.gov/html/emticwww/index.htm.

Some consultants experienced with the issues discussed in this chapter are:

- Emission Monitoring, 8901 Glenwood Avenue, Raleigh, NC 27617.
- Source Technology Associates, P.O. Box 12609, Research Triangle, NC 27709.
- RMB Consulting & Research, 5104 Bur Oak Circle, Raleigh, NC 27612.

44.1 REASONS FOR MEASURING STACK FLOW

Gas flow, or velocity, measurement in smoke stacks became one of the earliest concerns of the EPA in monitoring and controlling emissions of the primary criteria pollutants, sulfur dioxide (SO_2), total oxides of nitrogen (NO_X), and particulate. A method for measurement of gas flow was defined in the New Source Performance

Standards, EPA 40 CFR 60, Appendix A, Reference Methods (RM) 1 and 2. Such methods involved traversing the stack, point by point, in order to obtain a true average of the velocity across the stack cross-sectional area. These methods were necessary to support the particulate reference method (RM 5) in that it is necessary to know the velocity of the stack gas in order to sample it isokinetically. Isokinetic sampling requires that the velocity of the gas entering a sampling probe is the same as the velocity of the gas going past the sampling probe. This ensures that the same distribution of particle sizes is maintained in the sample stream, which is filtered to obtain the particulate component, as in the main stack gas stream. Particulate measurement and reduction in smoke stack effluent were early targets of the EPA Clean Air Act. These flow measurements were derived from a type-S Pitot tube (to be described in a later Section), which is the reference method for flow measurement. These measurements were normally performed annually and not on a continuous basis.

With the advent of the Acid Rain Program, EPA 40 CFR 75, it became necessary to continuously monitor either the heat input rate (mmBtu/h) or the stack gas flow rate (scf/h). With the appropriate calculations (40 CFR 75, Appendix F), one can then determine the SO_2 and/or CO_2 mass flow rate (lb/h) from the gas concentration measurements. The Acid Rain Program established a nationwide SO_2 allowance trading program (tons/yr of SO_2) for fossil fuel–fired electric utility boilers as defined in 40 CFR 73. This requirement drove demand for obtaining improved gas flow monitors that could be used continuously, with little maintenance, and with excellent relative accuracy. Accuracy of flow monitors is checked by comparison with the Pitot tube reference method, which is called a RATA (Relative Accuracy Test Audit). RATA accuracy requirements were initially 15% but phased down to 10% in later years, with incentives to achieve 7.5% (40 CFR 75, Appendix A) as the flow monitoring technologies improved. Daily calibration errors were limited to 3% during the 7-day instrument certification test period and 6% following that certification. A flow-to-load test is also required under 40 CFR 75, Appendix B. The flow to load ratio is computed on a quarterly basis to ensure that the basic accuracy of the flow monitor is maintained with respect to the normal load supported by the boiler.

Following the Acid Rain Program, the NO_X Budget Program was established by the OTC (Ozone Transport Commission) to set up a NO_X allowance and trading program (tons/yr of NO_X), similar to the Acid Rain Program SO_2 allowance trading system but affecting a wider variety of sources. This program primarily affected states east of the Mississippi River whose emissions are carried by the prevailing winds to the northeastern United States. This region consequently experienced significant ambient ozone problems (ozone nonattainment designation) due to the transport of ozone precursors (NO_X and hydrocarbons). The subsequent reduction of NO_X allowances and the trading value of such allowances are improving the ambient air quality in the affected region. In addition, the operating permits for large emission sources in many parts of the country often also incorporate mass emission measurements (lb/h) for CO, HCl, and other pollutants. This allows the regulatory agency to put a cap (but not yet a trade program) on the emissions in terms of tons per year of a given pollutant. Thus, the regulation-driven need for gas flow monitoring in smoke stacks is widespread.

The gas flow measurement technologies initially used to satisfy the Acid Rain Program requirements were primarily of the Pitot tube (differential pressure based), ultrasonic time-of-flight, and thermal heat transfer instruments. Such systems had to incorporate a means for performing, in addition to the normal calibration zero and span checks, a daily "interference check" (40 CFR 75, Appendix A, paragraph 2.2.2.2) that provides some indication that the flow monitoring systems are operating properly. Pitot tube systems included a single Pitot tube and a variety of multiple, or averaging, pitot tubes. Thermal heat transfer velocity probes were also used singly and in multiple, or averaging, configurations. Laser velocimeters entered the market but never gained a significant market share. Ultrasonic techniques eventually became the prevailing method for such measurements. They provide line average measurement across the stack and are time-based measurements that can yield a high degree of measurement accuracy. Stratification and nonaxial flow (sometimes referred to as cyclonic flow) are potential problems in many applications that must be accommodated by the selected measurement technology. The available reference methods for flow monitoring have been expanded from the initial RM 2 and 2A through 2H and include additional capabilities to handle such problems as well as wall effects. All of the above continuous flow measurement techniques will be described in more detail below.

Gas flow measurements have also been used in the ambient air monitoring networks dictated by the EPA; however, this monitoring is generally much less complex since the ambient air is a much better medium in which to measure than stack gases, which are very hot, dirty, and corrosive. Further, it is only necessary to measure the total velocity of the wind and direction. Anemometers of the rotating cup variety have traditionally been used in such applications, but recently, ultrasonic transducers have also made an impact in this market. These measurements will not be addressed in this discussion.

44.2 APPLICATION ISSUES THAT AFFECT SOME FLOW MEASUREMENT TECHNOLOGIES

44.2.1 Nonaxial Flow Patterns

In a real smokestack, the flow pattern is almost never axial (flow going straight up the stack or straight through a section of ductwork). There is usually some degree of swirling (cyclonic flow tangential to the wall, or "yaw"), sideways movement toward or away from the wall ("pitch"), or nonuniform stratified flow ("profile problem") (Fig. 44.1). This is an important consideration because what we want to measure is indeed the axial component of the flow. One way to think about this is to imagine a stack that is capped at the top with a cyclone trapped inside: There is plenty of flow going on inside the stack, but we would want the flow monitor to indicate zero flow because nothing is actually exiting the stack. Similarly, if there is

Two types of flow:

– Cyclonic flow is the whirlwind pattern

– Pitch flow is the fishtail pattern

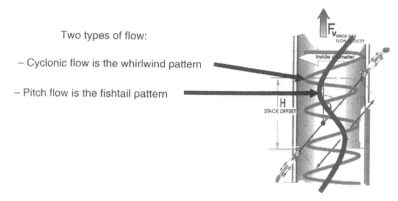

Figure 44.1   Typical nonaxial flow patterns.

"pitch" in the flow (e.g., imagine flow at a 45° angle at the location of the flow monitor), we only want the vector component of the flow that is parallel to the axis of the stack or duct to be reported. Finally, if the distribution of axial flow across the stack or duct is not uniform, the possibility exists that we will get a non-representative measurement.

As we will see, different technologies behave differently with respect to different types of nonaxial or stratified flow. This in itself is not a reason to rule out a technology if (*and only if*) the nonaxial or asymmetrical flow pattern is very repeatable in the sense that it is a single-valued function of the net axial flow out of the stack or through the duct. This can sometimes occur if there is only one source of flow entering the stack.

Because the calibration of a flow monitor in most pollution monitoring applications is based on correlation against a reference method (EPA methods 2, 2F, 2G, 2H), a repeatable nonaxial (or asymmetric profile) flow situation can be in effect "calibrated out." Of course, dealing with the situation in this manner will always result in residual errors that can be avoided or minimized by choosing a technology that is relatively immune to these effects. EPA RM 2H provides a method using a three-dimensional probe to minimize such errors. Any source of inconsistency in the reading of a flow monitor can (in 40 CFR 75 applications) work to the distinct disadvantage of the user because, when a test is performed, positive bias is allowed to stand (a penalty for the user who is selling emission credits based on the result of the flow monitor) but negative bias is dealt with by applying a positive "bias adjustment factor" that penalizes the user until the next time that the reference method is used. The use of BAFs (bias adjustment factors) for flow measurement is more fully described in 40 CFR 75, Appendix A, paragraph 7.6.

44.2.2 Effects of Particulate, Gas Composition, and Contaminants

The composition of flue gas varies greatly. It will be shown in subsequent sections that the accuracy of some technologies is dependent on the molecular weight and heat capacity of the gas and that technologies that involve physical contact with the flue gas can be compromised by accumulation of contamination on the contacting surfaces.

44.2.3 Wet- Versus Dry-Basis Measurements: Correction to Standard Conditions

All technologies for measuring flow give a "wet"-basis reading only. A wet-basis measurement is one that includes the volume of the water vapor component of the effluent in the flow measurement. A dry-basis measurement is one that removes the water vapor and makes the measurement on the remaining components of the effluent. Where regulations require it [e.g., when emissions are stated in terms of dry standard cubic feet (DSCF)], the stack water vapor content must be estimated or measured and the appropriate calculation made.

Once the linear flow velocity V has been determined, it is multiplied by the cross-sectional area of the stack or duct to determine the actual volumetric flow. The units may be English or metric, per hour, minute, or second. Wall effects (the decrease in velocity caused by friction as one approaches the wall containing the flow stream) often cause such simple calculations to be biased high. This effect has been estimated at 1–2% of the total flow. EPA 40 CFR 60, Appendix A, RM 2H provides a method to account for wall effects in determining total flow.

Effluent pressure and temperature can vary dramatically from one process to another. For the sake of uniform reporting, most regulatory applications require the volume and mass flow to be reported in standard pressure and temperature units. To do this, the monitor's measurement must be corrected according to the ideal gas law for the agency specified pressure and temperature. The EPA standard conditions are 68°F and 29.92 in. Hg (English units) or 20°C and 101.32 kiloPascal (metric units). Accordingly, flow monitoring systems need the ability to take inputs from a pressure transducer and a temperature measurement. Some monitors incorporate an internal temperature measurement that may also be used for correction.

To convert actual flow volume (acf/hr or al/min) to standard flow volume (scfh or slpm),

$$\text{Flow}_{\text{Standard}} = \text{Flow}_{\text{Actual}} \times \left[\frac{P_{\text{Actual}}}{P_{\text{Ref}}} \times \frac{T_{\text{Ref}}}{T_{\text{Actual}}}\right] \quad (44.1)$$

(*Note*: The temperature and pressure must be in absolute terms: for example, in absolute units $T_{\text{Actual}} = 459.69 + °\text{F}$).

44.2.4 Volumetric Measurements Versus Point or Line Average Measurements

The objective of using flow measurements is to determine the total flow rate in scfh or pounds per hour of gas exiting the stack. Stack gas flow is rarely constant across the stack profile in either of the two dimensions and results in a velocity profile that must be taken into account. Thus, we would ideally segment a circular stack into several coaxial circles of equal volume and measure the velocity in each volumetric component. This is the approach used in the reference methods. But, velocity varies with time and such an approach depends on the velocity profile staying constant unless all the points are measured at the same time, which is not practical. A single point measurement is only accurate if the point is representative of the velocity across the stack. A line average measurement is certainly much better than a point measurement but is still a line average measurement as opposed to an area-averaged measurement. The imperfections of all such devices are usually taken into account with an empirically derived profile curve that corrects flow measurements by a factor determined from the flow rate.

44.3 SPECIFIC STACK FLOW MEASUREMENT TECHNIQUES

44.3.1 Differential-Pressure Systems

44.3.1.1 Basic Principles. When a gas stream of mass m moving with velocity V impacts upon an obstacle, the change in momentum of the impacting stream of gas creates a force on the surface:

$$\text{Force} = \frac{\Delta(mV)}{\Delta t} \tag{44.2}$$

If the gas stream were completely stopped, the force would be proportional to the mass of gas per unit time that underwent the change in velocity. This would be

$$\text{Force} = \frac{V\,\Delta(m)}{\Delta t} = V A V \rho \tag{44.3}$$

where ρ is the gas density and A is the cross section of the surface being impacted. But the force per unit area is just the pressure ΔP relative to the static pressure in the vicinity, so, in this idealized case,

$$\text{Pressure} = \Delta P = \rho V^2 \tag{44.4}$$

or

$$V = \left(\frac{\Delta P}{\rho}\right)^{0.5} \tag{44.5}$$

However, because the aerodynamic shape of the surface will govern how much the gas stream actually slows down, a "fudge factor" K is used:

$$V = K\left(\frac{\Delta P}{\rho}\right)^{0.5} \tag{44.6}$$

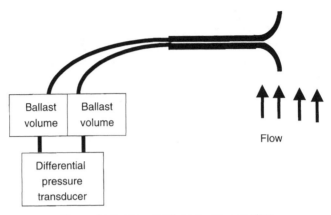

Figure 44.2 Type S Pitot tube flow monitor.

This is the basic equation that governs the way a Pitot tube works. A Pitot tube is a piece of open-ended tubing that is pointed into the flow, with the pressure buildup measured relative to the static pressure in the vicinity. In a practical instrument, a second tube is employed facing downstream, so that the venturi vacuum created on the tube facing downstream added to the pressure from the upstream side creates a larger delta pressure and the factor K is adjusted accordingly. This variation of a delta pressure flowmeter is called a *Type S Pitot tube* (Fig. 44.2).

44.3.1.2 Single-Point and Array-Based Pitot Tubes. Because the flow profile across a real-life stack or duct can vary quite a bit, the use of a single-point Pitot tube can be problematic, and much work and testing might be involved in finding the location most representative of the total flow. For better accuracy, one would install multiple Pitot tubes, each with its own delta pressure transducer, and average the velocities calculated from the individual systems.

44.3.1.3 Averaging Pitot Tubes. One approach to dealing with flow stratification in a Pitot tube–based system is to install an array of Pitot tubes and connect the delta pressure plumbing in parallel to a single delta pressure transducer (Fig. 44.3). Alternatively, some designs will employ a pair of pipes with a number of holes spaced so as to cover the diameter of the stack or duct. Either of these approaches, however, can be problematic due to internal flow circulation into the higher pressure (corresponding to higher velocities) holes and out of the holes where there is lower pressure. This not only complicates the analysis of the averaging process but can also result in accumulation of particulate matter and condensed material inside the plumbing. Further, because of the square-root relationship between pressure and flow, averaging the separate pressures (via use of a single transducer) before conversion to velocity does not produce the same

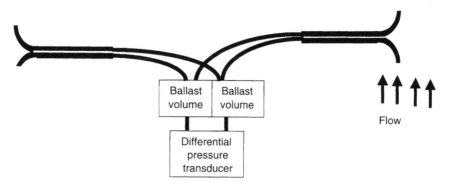

Figure 44.3 Multiprobe Pitot tube–based flow monitor.

result as averaging the velocities corresponding to the separate Pitot tubes. This error increases at low-flow velocity.

44.3.1.4 *Other Limitations of Pitot-Based Flow Monitors.* It has been

well established that cyclonic (off-axis) flow causes a significant positive bias in the reading from a Pitot tube: of the order of 1% (of reading) error per each 2° of off-axis flow. When flow measurements are being used for calculating mass emissions for trading credits, the financial penalty of unnecessary positive bias is substantial. Because this source of error is also present in EPA RM 2, it was not well understood until a series of papers were published in the mid-1990s. Ultimately, several improved procedures (RM 2F, 2G, and 2H) were adopted. If (*and only if!*) the presence of cyclonic or nonaxial flow is consistent (as a function of total flow) from test to test, this effect in a static Pitot tube can be calibrated out. Otherwise, it will always work to the disadvantage of the user (see Section 44.2.1).

Another source of bias in Pitot-based systems occurs when the differential pressure is measured downstream of a large "ballast volume" (Fig. 44.2) used to smooth out the substantial fluctuations (that occur due to turbulence) in the measured differential pressure. Recall from Equation (44.4) that ΔP is proportional to the square of the velocity; therefore, if the device averages the differential pressure, it is in fact averaging the square of the velocity instead of the velocity itself. This always results in positive bias in the reading. A better approach is to measure ΔP as close to the Pitot device as possible and with the shortest time constant possible and then electronically calculate the average of the square root of ΔP. This will reduce this source of positive bias.

Any leaks in the plumbing between the Pitot openings and the pressure transducer can create measurement errors, particularly if the process is at a substantial positive or negative pressure. If the system plumbing also includes a large buffer volume, a leak large enough to cause a meaningful error in the flow velocity calculation can be difficult to detect unless the buffer volume is isolated from the system and the plumbing needed to accomplish this adds other potential sources of leaks. Note that the Pitot tube differential pressure is low, often as low as 0.10 in. of water, necessitating the use of very low level pressure transducers that are not as accurate or robust as higher level measuring devices.

Delta pressure systems, like any technologies that contact the flow, are subject to the possibility of an orifice plugging with particulates, corrosion/erosion by stack acids, and the like. In many applications, this has been successfully dealt with via periodic back-purging using instrument air.

For any monitor used for EPA compliance, there exists a requirement that there be a mechanism whereby the essential components of the monitor be tested on at least a daily basis. In the case of a Pitot tube–based system, the options are somewhat limited. Typically, the pressure transducer is isolated and subjected to one or more known pressure differentials, and the electronics and software used to calculate the flow and correct it to standard conditions are tested. There is, however, not much that can be done to check the existence of leaks in the plumbing, partial blockage of the Pitot probe by particulate, or erosion/corrosion of the probe. These are typically addressed via quality control (QC) procedures (checking for leaks and probe deterioration) and by a daily (or more frequent) "interference check," which is essentially a periodic blowback of the probe using instrument air, assumed (but not guaranteed) to clear out deposits.

44.3.2 Heat Transfer Systems (Thermal Dispersion "Mass" Flowmeters)

44.3.2.1 Basic Principles.
These devices are still sometimes referred to as *hot-wire anemometers*, having evolved from earlier technologies that used thin (typically <0.001-in.) heated platinum (or other material) sensor wires, usually heated by a constant-current electrical source, the temperature of which would be reduced by the cooling effect of the moving air stream that is flowing past the wire. As applied to EPA compliance applications, where the gas stream contains a number of corrosive and abrasive gases and particulates, the sensors take the form of ruggedized probes (typically metal-clad sensors with internal heaters and accompanying Resistive Temperature Devices (RTDs) to measure temperature) with sophisticated electronics packages (Fig. 44.4). Thermal dispersion flowmeters are

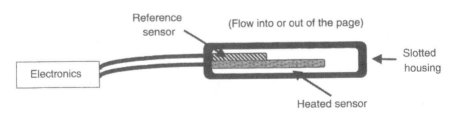

Figure 44.4 Thermal dispersion flow monitor.

often sold as mass flowmeters because the fundamental equation governing the loss of thermal energy from a heated surface is

$$Q = hA(T_s - T_a) \tag{44.7}$$

where Q = thermal energy lost

h = heat transfer coefficient

T_s = surface temperature of heated surface

T_a = ambient temperature of gas stream

A = cross-sectional area of element exposed to flow

and h, the heat transfer coefficient, is given by

$$h = \left(\frac{CK}{d}\right)\left(\frac{\rho V d}{\mu}\right)^m \tag{44.8}$$

where C = constant relevant to specific design

K = thermal conductivity of gas

d = diameter of probe (assuming it is cylindrical in shape)

ρ = density of flowing gas

μ = gas viscosity

V = velocity of moving gas stream

m = coefficient related to specific design

Note that the factors ρV are grouped together in the equation. This means that if (*and only if!*) all of the other factors such as thermal conductivity and viscosity are constant, this kind of monitor measures the product of velocity and density, which is indeed the mass flowing through the cross-sectional area of the heated surface. In typical fossil fuel applications where the oxygen, carbon dioxide, and water vapor content can vary, this claim should be evaluated with caution.

Typically, two sensors are used in each probe. One is an unheated reference sensor used to measure the temperature T_a of the flowing gas. The other is an active (heated) sensor used to determine T_s.

Equations (44.7) and (44.8) are actually simplified compared to reality. There are three additional sources of heat loss from the heated sensor. One is thermal conduction out of the sensor (it has to be attached mechanically to the probe). Another is radiative heat loss from the heated sensor: This is typically small. A third, and perhaps most significant, is that in the absence of forced convective flow (the overall gas stream is not moving), the heated sensor heats the local region around itself, creating free convection flow and an accompanying zero offset that is a function of the viscosity and heat capacity of the gas. For these reasons, each specific design of a thermal flowmeter is tested and uniquely calibrated.

44.3.2.2 Constant-Power Method (Often Called Constant Current in Literature).

By this method, a fixed amount of electrical power is applied to the heated sensor. At zero flow, the temperature difference between the heated and reference sensors will be greatest, diminishing as the flow increases. This makes for a simple design, with two primary disadvantages. First, because the temperature of the heated sensor is greatest at zero flow, the error from free convection is also greatest at zero (or very low) flow—just where one would want good accuracy. Second, because the entire heated sensor must heat up or cool down in response to a change in flow velocity and because both the heated and reference sensors must heat up or cool down in response to any change in the ambient temperature of the gas, this constant-power design cannot respond quickly to changes. Special care in the design must be taken to minimize the heat capacity of the sensors if rapid response is desired.

44.3.2.3 Constant-Temperature Method.

By this method, the electrical power to the heated sensor is continually adjusted so as to maintain a constant temperature differential between the heated and reference sensors, and the flow is calculated from the amount of power needed. This means that when the flow velocity changes, the temperature of the heated sensor does not need to change, and therefore its heat capacity (thermal mass) does not matter. The chief advantage of this technique is a faster response time to changes in flow velocity. A secondary advantage is that, by having a consistent temperature difference between the heated and reference sensors, the measurement error tends to be a consistent percentage of the actual reading. Finally, by limiting the amount of temperature elevation of the heated sensor, the error at zero or very low flow (due to free convection; See the final paragraph of Section 44.3.2.1) is less than is the case using the constant-power method.

44.3.2.4 Limitations of Thermal Dispersion Flow Monitors.

Properly designed thermal dispersion monitors can provide accurate monitoring of clean gases over a wide dynamic range of flow velocities with low maintenance. Where the flow is not uniform, arrays of thermal elements can be installed. The limitations of thermal dispersion monitors in EPA stack monitoring compliance applications fall into essentially four categories:

1. The active (heated) sensor cannot distinguish between a reduction in the flow velocity and a coating of particulate matter—both will reduce the heat loss and be interpreted by the circuitry as a reduction in flow. Some commercially available designs include a compressed-air "puffer" that periodically attempts to blow any accumulated materials off the sensors, but there is no way to know whether the desired cleaning has been accomplished. Accordingly, thermal dispersion flow monitors are inappropriate in applications in which the flow contains "sticky" particulates that cling to surfaces.

2. The active (heated) sensor cannot distinguish between an increase in the flow velocity and the effect of liquid droplets impinging and evaporating on the surface—both will increase the heat loss and be interpreted as an increase in

flow. Accordingly, thermal dispersion flow monitors are inappropriate in applications in which the flowing gas is saturated with water or acid, and entrained liquid droplets are likely to be present—a condition common in CEMS applications.

3. The flowing gas will cool the heated sensor to the same extent even if it is not flowing axially: Off-axis, cyclonic flow will also produce cooling. Indeed, one manufacturer presents this as a good feature (it "sounds right" to say that a sensor's response would be independent of orientation), but it is *not* a good feature (see Section 44.2.1). Thermal dispersion–based flow monitors will always be biased high in the presence of nonaxial flow.

4. As with the Pitot tube–based system, the options for performing a meaningful daily calibration check are quite limited. Most commercially available systems simply switch a precision resistor into the circuitry in place of the RTDs that measure the temperatures of the heated and reference sensors. This checks the electronics and computational software but not the sensor elements themselves. Apart from removing and testing the sensors, there is no way to ensure that drift in their response to flow has not occurred, other than detection of catastrophic failure (e.g., a totally unreasonable reading of one or both of the RTDs).

44.3.3 Ultrasonic Time-of-Flight Systems

44.3.3.1 *Basic Principles: Time-of-Flight Theory.* This approach measures the transit times of ultrasonic tone bursts traveling through the gas stream to determine flow velocity, temperature, and volume. Typically, two transducers, each with transmit and receive capability, are used to eliminate the need to calculate the speed of sound in the specific gas of interest. The basic operation is easy to understand: A tone burst traveling downstream is "helped"—speeded up—by the flow, while a tone burst traveling upstream will be slowed down by the flow. The difference in transit times is then used to calculate the velocity of the gas as described below.

Each effluent path monitored uses two transducers placed on opposite sides of the stack or duct pointed at each other with one transducer located diagonally upstream from the other (Fig. 44.5). Each transducer acts alternately as a transmitter and a receiver with the ultrasonic waves typically passing through the centroid of the stack or duct to the other transducer. This makes the measurements a line average of the tone burst path length. When a tone burst is sent through the gas stream from the upstream transducer to the downstream transducer, the movement of the gas stream reduces the time required to traverse the distance. When the tone burst is traveling against the gas stream from the downstream to the upstream transducer, the traverse time is increased. When there is no gas flow, the time required for the ultrasonic tone bursts to traverse the gas stream in either direction is the same. The root measurement is transit time. The difference between upstream and downstream transit times through the gas stream is directly proportional to

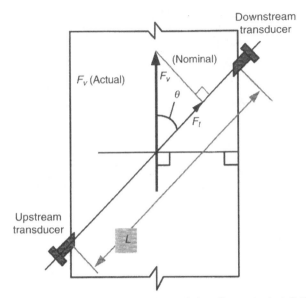

Figure 44.5 Simplified diagram of transit time flowmeter installation.

the velocity of the gas stream. The system calculates flow velocity from transit time and the following physical equations:

Velocity of sound from upstream to downstream transducer:

$$V_1 = C_s + F_v(\cos \theta) \tag{44.9}$$

Velocity of sound from downstream to upstream transducer:

$$V_2 = C_s - F_v(\cos \theta) \tag{44.10}$$

where C_s = Speed of Sound
F_v = flue gas velocity
θ = transducer angle to flow
V_1, V_2 = velocity of respective tone bursts

Subtracting Equation (44.10) from Equation (44.9) yields

$$V_1 - V_2 = 2 \cos \theta (F_v) \tag{44.11}$$

and solving for F_v,

$$F_v = \frac{V_1 - V_2}{2 \cos \theta} \tag{44.12}$$

Substituting $V_1 = L/t_1$ and $V_2 = L/t_2$,

$$F_v = \frac{L}{2\cos\theta}\left(\frac{t_2 - t_1}{t_2 t_1}\right) \tag{44.13}$$

where F_v = line average velocity
  t = transit times of sound between transducers
  L = distance between transducers

Note that the final equation for the velocity measurement contains no reference to the speed of sound or any other direct or indirect reference to the gas composition. This means that the measurement is inherently independent of the temperature, density, viscosity, and particulate concentration since these terms drop out of the equations. Also note that the final equation contains no reference to the intensity of the tone burst. This means that the measurement is inherently independent of factors that would attenuate the tone burst: So long as it gets across the stack with enough intensity that the transit time can be measured, the calculated velocity will be unaffected.

An additional advantage is that the flow is measured across an entire diameter of the stack and substantially averages out irregularities in the flow profile. Finally, because Equation (44.13) contains only data relating to the geometry of the installation (which cannot drift) and the measured times (which can be done to high precision using nondrifting digital signal processing), the calibration of a transit time flow monitor, once established, tends to be exceptionally stable.

Now consider the case of simple cyclonic flow in a cylindrical stack. This is simple "corkscrew" motion of the gas up the stack. This kind of flow often occurs when the flow from a duct does not enter the stack exactly straight on or when a flow deflector has been installed to prevent erosion of the far-side wall. Simple cyclonic flow can be factored into two vectors: one aligned axially with the stack and one aligned "sidewise" at right angles to both the axis and the axis of the stack (refer again to Fig. 44.5). In this very common situation, the axial vector will affect the transit times in exactly the way desired to measure the flow exiting the stack. However, the sidewise (yaw) vector component will not have any effect on the measurement because it is at right angles to the path traveled by the ultrasonic tone bursts and thereby neither helps or hinders the tone bursts. Recall from Section 44.2.1 that this is the ideal way that a flow monitor should behave.

By knowing the precise time required for the tone bursts to traverse the gas stream, the speed of sound can also be calculated. The speed of sound in a gas is mainly dependent on two things: the temperature and the molecular weight of the gas. If the gas composition is predictable, then the measured speed of sound can be used to calculate the gas temperature, eliminating the need to measure the temperature for the purpose of correcting the flow to standard conditions (Section 44.2.3). If, on the other hand, temperature is measured, the molecular weight of the gas can be calculated, which is useful in some process applications.

44.3.3.2 *Alternative Configuration for Nonuniform Flow Profiles.* Ideally, the flow profile across a circular stack would be the same in both the east–west and north–south directions. But, because of the layout of ductwork, fans, confluences of two or more flow streams, flow obstructions, and so on, the flow profiles may be substantially different in the two axes and flow may be skewed in the pitch and/or yaw axis. To maximize accuracy in such applications, an X pattern consisting of two sets of two transducers each can be used to obtain an enhanced measurement of total flow. Each pair of transducers produces a measure of the vertical flow in their orientation, and overall accuracy can be enhanced by combining these two measurements.

44.3.3.3 *Alternative Ways to Measure Ultrasonic Transit Time.* There are several reasons that ultrasonic sound, with frequencies typically below 100 kHz, is usually used for flow measurements. An obvious one is that ultrasonic frequencies cannot be heard by the human ear and thereby pose no environmental and safety issues. A second is that most acoustic noise typical of industrial processes occurs at frequencies of 10 kHz and lower and therefore interfere less when higher frequency ultrasound is used to make the measurement. A final reason is that, at higher frequencies, the wavelength of the sound is shorter, enabling a shorter duration tone burst to be created, making for a more precise measurement. However, a practical upper limit on the frequencies that can be used in any specific application is imposed by the fact that higher frequency sound is attenuated and scattered to a significantly greater degree than lower frequency sound, and this becomes important when measuring flow in large, hot stacks and ducts with entrained particulates and/or droplets.

Before the advent of high-speed digital processors, a common way to measure transit time was analog correlation. An ultrasonic transducer would generate a wave packet (a tone burst with typically 6–10 sinusoidal oscillations) and send that packet into the stack. The drive circuit that generated the tone burst would save the shape of the "sent" waveform and then attempt to match it up with the shape of the "received" waveform. One way that this was done was by applying a time delay to the sent waveform and multiplying it by the received signal in an analog circuit, continuously adjusting the time delay until the product of multiplication was maximized. This worked satisfactorily on small stacks where the signal-to-noise ratio in the received signal was good but had difficulties on large, hot stacks where background noise was substantial.

The family of ultrasonic instruments most successfully applied to the emission trading program in the United States uses a technique known as *boxcar integration.* Each time a tone burst traverses the stack, a time-based window of interest within the received signal is digitized by an analog-to-digital (A/D) converter at frequent (e.g., 0.5-µs) intervals. This results in a window of interest composed of a very large number of digital values or "boxcars." For each tone burst during the integration period the value of each boxcar is added to the sum of all previous boxcars having the same location in the window. For example, on the fourth tone burst of the integration period, the value of boxcar 100 will be added to the sum of the three previous boxcar 100's. This technique greatly enhances the signal-to-noise ratio. The signal processing algorithms then determine the center of the receive signal and thus the exact transit time across the stack.

Because the addition of many received pulses into a set of boxcars takes time, the earliest incarnations of this approach tended to have much slower response times—typically 90 s to 5 min—than the analog correlation method. However, the emergence of much faster data processing hardware has now made possible instruments, using the excellent noise rejection capability of boxcar integration, with response times as fast as a few seconds.

44.3.3.4 Wetted Transducers.

The most precise measurement of transit time will occur if the transit distance L (Fig. 44.5) is known precisely. This is easiest to do if the transducers are inserted directly into the flow. The disadvantages associated with this approach include:

- Corrosion of the transducer by the stack gases
- Erosion of the transducer and coating by particulates
- Deterioration of the transducer: Ultrasonic transducers based on piezoelectric materials can be used only at temperatures well below the "Curie temperature," above which an ultrasonic transducer irreversibly loses its ability to convert acoustic pressure into electricity and vice versa.
- Changes in the characteristic frequency at elevated temperatures and corresponding changes in the shape of the wave packet
- Requirement to use special, more expensive transducers

44.3.3.5 Purged Transducers.

A much broader range of transducer types and designs can be used if the transducers are protected from direct contact with the stack gas by use of an "air curtain" of purge air from a relatively simple and inexpensive blower system. The issues associated with this approach include:

- Two portions of the ultrasonic transit path are now not within the stack flow or at the stack temperature but instead are affected by the temperature and motion of the purge air. On large stacks, where the stack measurement path is large compared to the purged regions, this effect is minimal and can be accounted for in the calculation algorithm.
- Instead of the ultrasound coupling directly from the transducer to the stack gas, it must couple into the purge air and from there into the stack gas, which is at a different temperature. This will reduce signal substantially, which is why the first systems using purged transducers and analog correlation had difficulty with the signal-to-noise ratio. The desire to use purge-air-protected transducers practically necessitated the development of the more sophisticated boxcar integration techniques.

44.3.3.6 Calibration Check Capability.

Although not all designs necessarily incorporate the necessary software and controls, the transit time approach affords the potential for performing a comprehensive full system evaluation check of all active components of the system. One example of how this can be implemented follows.

During a "zero" check, only the upstream transducer transmits, but it transmits at twice the normal rate. The downstream transducer is receiving only. The electronics and software process each pair of tone bursts *as though* they were typical upstream and downstream signals. But since all of the received tone bursts go through the gas stream in the same direction, the time required to traverse the gas stream should be essentially the same. With no difference in the transit times, the flow calculation algorithm should indicate zero velocity. The system has now tested the upstream transducer as a sender and the downstream as a receiver and the overall ability of the system to measure transit times with the accuracy necessary for satisfactory normal operation.

During a "span" check, the opposite process occurs. Only the downstream transducer transmits, again at twice the normal rate. The upstream transducer is receiving only. The electronics and software process each pair of signals *as though* they were upstream and downstream signals. This would be expected to produce another zero-flow indication. However, in this mode every other received tone burst is delayed by a predetermined amount. This means that the system should output a difference in the time required for the upstream and downstream tone bursts to traverse the gas stream. The system has now tested the downstream transducer as a sender and the upstream transducer as a receiver and the ability of the system to measure small differences in transit time—and with the accuracy desired at a given upscale flow rate in the stack.

44.3.3.7 *Limitations of Ultrasonic Transit Time Flow Monitoring.* We

have seen that tangentially sideways, or "yaw," flow is handled correctly by the transit time approach. However, "pitched" flow (flow that is leaning either toward or away from the downstream transducer) does have an undesirable effect because this kind of off-axis flow has a vector component in the direction of propagation of the tone bursts. In general, pitch across the entire diameter of a stack or duct does not persist after about five diameters of straight run. The effect is most likely to occur just after a bend in a duct or just up from where a duct feeds into a stack, before the flow has had time to straighten out. When pitch flow is present or suspected, the solution is either to mount the transducers at a steeper angle (so that the ratio of the axial effect to the pitch effect is increased) or to mount two sets of transducers in an X pattern (so that one set of transducers reads high and the other low, with the average exactly canceling out).

Apart from pitch flow, the only substantial limitation of transit time flow monitoring is the obvious requirement that the tone bursts make it across the flow to the receiving transducer with enough intensity—and with their waveform shape not substantially altered—that the system can calculate the transit times accurately. Factors that attenuate the tone burst include long measurement path, high process temperature, very high flow velocity, and high particulate loading and/or lots of acid mist following an SO_2 scrubber or a catalytic reducer for NO_x control. All of these factors interact to determine whether an application is appropriate for an ultrasonic transit time monitor, so that, for instance, a high temperature may be acceptable if the measurement path is not too long. Manufacturers typically maintain nomographs for use in deciding whether an application can be handled.

44.3.4 Optical Scintillation

44.3.4.1 Basic Principles. This is a fairly new technology for monitoring flow in stacks and ducts for EPA compliance. The core technology, however, has been used in the past for certain meteorological purposes and for detecting turbulence on airport runways. The basic approach is illustrated in Figure 44.6. A beam of light is directed into a stack and is observed by a detector on the opposite side. Due to turbulence accompanied by temperature striations in the stack, the receiving detector does not see a steady signal: instead, it sees a time-varying, "scintillating" signal (this is what one sometimes sees across the rooftops of automobiles in a parking lot, or on a highway, on a sunny day). A second detector located a bit further up the stack (as little as an inch or two) from the first will also see a time-varying scintillating signal. If the pattern of the scintillation has not changed in the time required for the overall flow to move from the elevation of the first detector to the elevation of the second detector, the two detectors should see the same time-varying scintillation but with a delay directly related to the flow speed.

Another variation of this approach does not use any light source, instead using an infrared detector to detect infrared radiation emanating from the stack gas. Obviously, this technique works better at high process temperatures and was originally developed for use in the steel industry.

Compared to the technologies described earlier, a scintillation-based approach is one of the least expensive to purchase, install, and maintain.

A variation of this approach has also been implemented that uses the normal time variation of opacity in lieu of a true scintillation phenomenon. Two laser beams with a given vertical displacement can also be used to determine the time it takes for a given opacity signature in time measured at one station to appear at the other station.

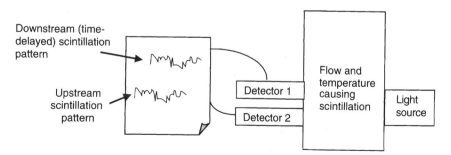

Figure 44.6 Scintillation-based flow monitor.

44.3.4.2 Limitations of Scintillation Technology. It goes without saying that, to measure the speed at which a scintillation pattern propagates up a stack, there must first exist a scintillation pattern. For instance, if the gas flowing up a stack were of a completely uniform temperature, there would not be any scintillation to measure. The implication of this is that scintillation technology will work quite well in some stacks but not in others. In many cases a suitable scintillation pattern may be created at certain boiler loads or process conditions on a specific stack but not at different loads or conditions on the same stack. The science of "tuning" the signal processing for each specific application is undoubtedly a work in progress, and it is reasonable to expect that the range of applications in which scintillation can be made to work reliably will grow over time.

One key limitation is that existing designs need to project the light into the stack, and receive it at the other side, through windows that contact the stack gas and that therefore will get dirty. The reason for this requirement is that if purge air is used between the windows and the stack gas, the interaction of the purge air with the hot stack gas creates substantial scintillation right at the stack wall: The system would be measuring the characteristics of the purge–stack interface rather than the flowing stack gas in the middle of the stack. This makes this technology most suitable for relatively clean applications.

44.4 CONCLUSIONS, RECOMMENDATIONS, AND PRECAUTIONS

We have discussed four completely different flow monitoring technologies: pneumatic (Pitot tubes), thermal dispersion, ultrasonics, and optical scintillation. Each has been used, to some degree, in coal-burning utility smokestacks, cement kilns, pulp and paper plants, and other similar applications. In terms of market share, transit time ultrasonics has been the most-used technology for emission trading applications, followed by Pitot tubes, thermal, and optical, in that order. Most of these systems are less than 10 years old, and time will tell whether they will be replaced by "like instrumentation" versus alternate technologies. When a technology is working even moderately well, it will tend to be replaced with similar instruments because of the sunk cost of platforms, mounting fixtures, sample lines, electrical cables, data-handling systems, and so on, associated with that specific technology.

Users would do well to consult with independently found references in addition to any provided by vendors. Across the entire range of installations associated with EPA compliance and large duct/stack process applications, there will always be some successful installations for any specific technology as well as some that are problematic. The information sources cited in the introduction to this chapter would be a good place to get more data and appropriate references.

One very significant but not widely understood issue associated with stack flow monitoring is that the earliest version of EPA RM 2 (manually operated type S Pitot tube with no correction for nonaxial flow) is very limited in terms of accuracy and repeatability. When a "lowest-bidder" test team performs RM 2, the result is often that the installed flow monitor is merely being readjusted every time the test is run to agree with the latest RM 2 result. This author was involved during the mid-1990s in educating agencies and users of this fact. Happily, the newer and improved RM 2F, 2G, and 2H provide better accuracy. In addition, automated devices have been developed that substantially eliminate human error from the testing process. In consulting with references, a potential user would do well to ask whether the required annual or semiannual testing of an installed flow monitor has been done using these improved methods and in addition whether an automated version of RM 2 has been used.

REFERENCES

1. http://www.epa.gov/ttn/emc/ is a source of information on performance specifications and EPA test methods.

2. http://www.gpoaccess.gov/cfr/index.html is the main page for the Code of Federal Regulations.

3. U.S. EPA emission measurement and regulatory information is available at www.epa.gov/ttn/emc, http://www.epa.gov/ttn/emc, www.epa.gov/acidrain, www.epa.gov/acidrain/otc, http://www.epa.gov/acidrain/otc/otcmain.html>/otcmain.html, www.access.gpo.gov http://ttnwww.rtpnc.epa.gov/html/emticwww/index.htm.

4. *An Operator's Guide to Eliminating Bias in CEM Systems* contains some information about application of flow monitors for EPA compliance: http://www.epa.gov/airmarkets/monitoring/bias/bias.pdf.

5. R. W. Miller, *Flow Measurement Engineering Handbook*, McGraw-Hill, New York, 1983.

6. R. L. Myers, Error-Analysis of EPA Method 2, Paper 92-66.15, *Proceedings, AWMA*, June 1992 Meeting, Kansas City.

7. J. J. Page, E. A. Potts, R. T. Shigehara, *3-D Pitot Tube Calibration Study*, EPA Contract No. 68-D1-0009, Work Assignment No. I-121, March, 1993.

8. J. E. Traina and R. L. Myers, Fully Automated Probe Performs EPA Methods 1 and 2 for Volumetric Flow, paper presented at the ASME Joint Power Generation Conference, Phoenix, AZ, 10/2/94.

9. J. E. Traina and R. L. Myers, Fully Automated Probe Performs EPA Methods 1 and 2 for Volumetric Flow: Recent Field Experiments and Technical Enhancements, paper presented at the A&WMA Conference, Tempe, AZ, January 24, 1995.

10. R. L. Myers and J. E. Traina, EPA Method 2: Technical Advances that Greatly Reduce Overall Measurement Bias and Eliminate All Negative Bias, A&WMA CEM Conference, Chicago, October 1995.

11. J. E. Traina and R. L. Myers, Determination of Stack Gas Velocity and Volumetric Flow Rate (Type S Pitot, Enhanced), document submitted to US EPA, October, 1995, available upon request to the authors.

INDEX

Environmental Instrumentation and Analysis Handbook, by Randy D. Down and Jay H. Lehr
ISBN 0-471-46354-X   Copyright © 2005 John Wiley & Sons, Inc.

Printed in the USA/Agawam, MA
February 26, 2013

573138.080